PAPERS IN HONOUR OF BERNHARD BANASCHEWSKI

Proceedings of the BB Fest 96, a Conference Held at the University of Cape Town, 15–20 July 1996, on Category Theory and its Applications to Topology, Order and Algebra

Edited by

GUILLAUME BRÜMMER AND CHRISTOPHER GILMOUR

Department of Mathematics and Applied Mathematics,
University of Cape Town, South Africa

Reprinted from Applied Categorical Structures, Vol. 8, Nos. 1–2 (2000), with additional materials

SPRINGER-SCIENCE+BUSINESS MEDIA, B.V.

A.C.I.P. Catalogue record for this book is available from the Library of Congress

ISBN 978-90-481-5540-8 ISBN 978-94-017-2529-3 (eBook)
DOI 10.1007/978-94-017-2529-3

Printed on acid-free paper

PAPERS IN HONOUR OF
BERNHARD BANASCHEWSKI

Table of Contents

Applied Categorical Structures **8:** 1, 2000.
G. Brümmer & C. Gilmour (eds), Papers in Honour of Bernhard Banaschewski.

Preface

In July 1996 a conference, named 'BB Fest 96', was held at the University of Cape Town in honour of the seventieth birthday of Professor Bernhard Banaschewski of McMaster University, Canada. The University of Cape Town hosted the conference in recognition of Banaschewski's inspiring contribution over many years to the Cape Town school of categorical topology.

The proceedings of the BB Fest 96 are taking the twofold shape of a selection of papers in *Applied Categorical Structures* as well as a larger selection in a hard-cover book published by Kluwer Academic Publishers. Both selections bear the title 'Papers in Honour of Bernhard Banaschewski'. We are grateful to Bob Lowen, Editor-in-Chief of *Applied Categorical Structures*, for inviting us to act as guest editors. Likewise we are grateful to all the authors and the many referees for their work, and to all members of the Cape Town group for their help in organising the BB Fest 96.

All papers appearing here have been subjected to the regular refereeing procedures of *Applied Categorical Structures.*

GUILLAUME BRÜMMER
CHRISTOPHER GILMOUR

Applied Categorical Structures **8:** 3–6, 2000.
G. Brümmer & C. Gilmour (eds), Papers in Honour of Bernhard Banaschewski.

Relatively Parametrizable Subalgebras

Dedicated to Bernhard Banaschewski for his 70th birthday

JOHN ISBELL
Department of Mathematics, State University of New York, Buffalo, N.Y. 14214 U.S.A.

(Received: 1 October 1996; accepted: 3 August 1997)

Abstract. m-parametrizable algebras are characterized by a projectiveness property.

Mathematics Subject Classification (2000): 08A99.

Key words: parametrizable, relatively parametrizable, projective, k-projective, projectively filtered.

Introduction

An algebra A, of any type, is called *parametrizable* (respectively *m-parametrizable*) if for all k (for $k = m$), the solution set of any system of algebraic equations in A^k is a union of parametric sets. The basic characterization theorem is [3] that A is parametrizable if and only if in the smallest variety of algebras $\mathcal{V}_A$ to which A belongs, A is projective. The central result of this paper is a localized characterization; oversimplifying drastically, a particular solution set is a union of parametric sets if and only if it is analyzable by mappings into projectives. The main application is to m-parametrizability of A, which is characterized as follows: for every m-generated subalgebra S of A, there exists a factorization of $i : S \subset A$ across a projective algebra P of $\mathcal{V}_A$, $S \to P \to A$.

The key definition: a subalgebra S of A is *relatively parametrizable* if for each point $p = (p_\alpha)$ of S^k (k any cardinal) there is a parametric mapping $\pi\colon A^j \to A^k$ representing p ($p \in \pi(A^j)$) and identically satisfying all algebraic equations satisfied by p (each $\pi(x)$ satisfies them). Precise statement of the main result: S is relatively parametrizable if and only if $i : S \subset A$ factors across a projective algebra P of $\mathcal{V}_A$. That does not quite tell us when a particular solution set is a union of parametrized sets, but (a) this is where the theorem is: the natural generalization of "parametrizable = projective in $\mathcal{V}_A$", and (b) it has at least the indicated application to m-parametrizability.

Note, relative parametrizability is very much a positional property. It is neither necessary nor sufficient that S be parametrizable. Indeed, being parametrizable and generic in $\mathcal{V}_A$ is sufficient, as that makes S projective in $\mathcal{V}_A$. Every subalgebra of a relatively parametrizable subalgebra is relatively parametrizable.

Of course k-parametrizability is related to κ-freeness in abelian groups, cf. [4]. It is more nearly like κ-projective filtration in Boolean algebras [1]. I do not know if they are equivalent (k-parametrizability to k^+-projective filtration). The filtration version seems much more restrictive; it requires projective subalgebras, and there must be a neatly arranged family of them. But a counterexample would be desirable. Anyway, the evident one-way relation gives two k-parametrizability corollaries of theorems of S. Fuchino and S. Shelah [1]. (1) For Boolean algebras, ω_2-parametrizability does not imply ω_3-parametrizability. (2) If $V = L$ and there are no weakly compact cardinals, then for each regular aleph k there are Boolean algebras which are m-parametrizable if and only if $m < k$.

In [3] I stressed that the results apply to infinitary algebras, even such badly behaved ones as complete Boolean algebras. Globally, complete Boolean algebras have a proper class of ω-ary operations, but our viewpoint is never global. At most, we work with a variety, often the smallest variety that includes a given algebra, which is no problem for complete Boolean algebras (or worse, such as complete closure algebras [2]). This applies equally to results in [4].

Results

The *algebraic envelope* of a point $u \in A^k$ is defined [4] as the set $E(u)$ of all points v which satisfy every equation satisfied by u. A *parametric analysis* of u is defined as a parametric mapping ϕ: $A^j \to A^k$ which takes u as a value and has $\phi(A^j) \subset E(u)$. So the new definition of a *relatively parametrizable* subalgebra S of A, given above, says just that every point u of S^k, considered as a point of A^k, has a parametric analysis.

We shall use Lemma 1.2 from [4]:

For every point $p \in P^k$ over a projective algebra in a variety $\mathcal{V}$ there is a k-tuple in a $\mathcal{V}$-free algebra on generators x_β ($\beta < m$), $\phi = (\phi_\alpha(x_\beta))_{\alpha<k}$, such that the parametric mapping Φ: $P^m \to P^k$ defined $\Phi((u_\beta)_{\beta<m}) = (\phi_\alpha(u_\beta))_{\alpha<k}$ takes p as a value, and every algebraic equation satisfied by p is satisfied by ϕ.

THEOREM 1 *The following properties of a subalgebra $S \subset A$ are equivalent.*

(a) *S is relatively parametrizable.*

(b) *Some point $p \in S^n$ whose coordinates generate S has a parametric analysis in A^n.*

(c) *The inclusion i: $S \subset A$ factors across a projective object of $\mathcal{V}_A$.*

Proof. (a) $\Rightarrow$ (b); if every point over S has a parametric analysis, a generating point p does.

Assume (b), with a parametric analysis π: $A^m \to A^n$ of a generating $p \in S^n$. Choose a point t of A^m for which $\pi(t) = p$; more fully, $p = (p_\alpha)_{\alpha<n}$, $\pi = (\pi_\alpha)$, $t = (t_j)_{j<m}$, and $\pi_\alpha(t) = p_\alpha$. The free algebra F of $\mathcal{V}_A$ on m generators z_j ($j < m$) can be described as a subset of the set of all functions from A^m to A, viz. those expressible as (composite) operations. We define homomorphisms Π: $S \to F$ and T: $F \to A$. At each point s of S, $\Pi(s)$ is defined by expressing s anyhow as $w((p_\alpha))$ and taking $\Pi(s) = w((\pi_\alpha(z_j)_{j<m})_{\alpha<n})$. If s can be expressed also as $v((p_\alpha))$, then $E(p)$ is contained in the algebraic set where $v = w$; $v((\pi_\alpha(x_j))) = w((\pi_\alpha(x_j)))$ for all $(x_j) \in A^m$, hence also for $x_j = z_j \in F$. So Π is well defined, and therefore homomorphic. T is just the substitution $z_j \mapsto t_j$, $T(w((z_j))) = w((t_j))$, homomorphic. We have $T\Pi(p_\alpha) = T(\pi_\alpha(z_j)_{j<m}) = \pi_\alpha(t) = p_\alpha$, so as p generates S, $T\Pi = i$. (c) is proved.

Assume (c), i: $S \subset A$ factoring as $T\Pi$ across P projective in $\mathcal{V}_A$. For any point $v = (v_\alpha) \in S^k$ we have $p = (\Pi(v_\alpha)) \in P^k$. By Lemma 1.2, recalled above, there is $(\phi_\alpha(x_\beta))_{\alpha<k}$ in a free algebra on m generators x_β, with p representable as $(\phi_\alpha(u_\beta))$ for some $u = (u_\beta) \in P^m$, and ϕ satisfying all equations that p satisfies. I claim that v has a parametric analysis Φ: $A^m \to A^k$ defined $\Phi((a_\beta)_{\beta<m}) = (\phi_\alpha(a_\beta))_{\alpha<k}$. Φ takes the value v at $t = T^m(u)$, since each $\phi_\alpha(t) = \phi_\alpha(T(u_\beta)_{\beta<m}) = T(\phi_\alpha(u_\beta)_{\beta<m}) = T(p_\alpha) = T\Pi(v_\alpha) = v_\alpha$. As for the equations, $\phi = (\phi_\alpha)$ satisfies all equations that p satisfies; $p = \Pi^k(v)$ satisfies all equations that v satisfies; and each $\Phi((a_\beta)) = (\phi_\alpha((a_\beta)))$ satisfies all equations that ϕ satisfies. Thus (a) holds, and the proof is complete. □

Theorem 1 gives us two characterizations of k-parametrizable algebras. We call an algebra A *k-projective* in a variety $\mathcal{V}$ if given any surjective morphism of $\mathcal{V}$, t: $T \to Y$, and any morphism f: $A \to Y$ and k-generated subalgebra i: $S \subset A$, there exists f^*: $S \to T$ such that

$$\begin{array}{ccc} T & \xrightarrow{t} & Y \\ {\scriptstyle f^*}\uparrow & & \uparrow{\scriptstyle f} \\ S & \xrightarrow{i} & A \end{array}$$

commutes: $tf^* = fi$. By a *singular subalgebra* of A we mean just a morphism $B \to A$. It is *finitely generated*, or *projective*, or ..., if B is so.

THEOREM 2 *The following properties of an algebra A are equivalent.*
(a) *A is k-parametrizable.*
(b) *Every k-generated subalgebra of A is a subalgebra of a $\mathcal{V}_A$-projective singular subalgebra of A.*
(c) *A is k-projective in $\mathcal{V}_A$.*

Proof. Evidently A is k-parametrizable if and only if every point of A^k has a parametric analysis. Evidently again, this is equivalent to every k-generated subalgebra being relatively parametrizable. So the equivalence of (a) and (b) follows from Theorem 1. Given (b), consider surjective t: $T \to Y$ in $\mathcal{V}_A$ and f: $A \to Y$, and any k-generated subalgebra i: $S \to A$. We have a $\mathcal{V}_A$-projective P and a factorization of i as vu, u: $S \to P$ and v: $P \to A$. Then by projectiveness, fv: $P \to Y$ lifts to c: $P \to T$, tc being fv. Put $f^* = cu$; tcu is $fvu = fi$ as required. Finally, (c) implies (b) by putting $Y = A$ and taking a surjection t: $P \to A$ from a $\mathcal{V}_A$-projective. □

There is an example in [4] of a 1-parametrizable algebra A (an M-set, for a suitable monoid M) whose cyclic subalgebras are not contained in projective *subalgebras* of A. In fact, it was a study of that example that suggested Theorem 1.

Another example in [4] shows that in Theorem 2(b), the projective cannot be taken to be k-generated. This is just the 16-element group $G = Z_4 \times_\sigma Z_4$, where $\sigma\colon Z_4 \to Aut\ Z_4$ is the surjective homomorphism. The only nontrivial algebraic set in G is the center $C = Z_2 \times Z_2$, defined by "$x^2 = e$"; it is the parametric set $\{t^2u^2\}$, but 1-parameter sets do not represent it, and one inclusion $Z_2 \subset G$ fails to factor through a cyclic $\mathcal{V}_G$-projective. To characterize the 1-parametrizable finite groups without using a *deus ex machina* like "P, projective in $\mathcal{V}_A$" is probably a challenging problem.

Finally, for Boolean algebras, (1) there is a non-projective one, A, of power ω_3, which is ω_3-projectively filtered [1, p. 192]. Such a filtration puts every ω_2 elements in a projective subalgebra; so A is ω_2-parametrizable. Like every non-singleton Boolean algebra, A is generic. Hence if it were (ω_3-) parametrizable it would be projective.

In Boolean algebras, ω_2-parametrizability does not imply ω_3-parametrizability.

(2) If $V = L$, then examples like that just described exist for every regular cardinal κ that is not weakly compact [1, p. 184]. So

If $V = L$ and there are no weakly compact cardinals, then for every regular aleph k there is a Boolean algebra of power k which is m-parametrizable just for $m < k$.

The set-theoretic assumption here runs counter to what set theorists seem to prefer; but unlike the popular large cardinals, it is known to be consistent with ZFC. The reader will note the curious position of ω_2; in this family of examples, the example for ω_3 exists in every model of ZFC, and examples for ω_1 (just non-projective algebras of power ω_1, since all smaller ones except the singleton are projective) are long known, but for ω_2 it seems that $V = L$ is needed.

Note also: S. Fuchino's result (2) ((1) is S. Shelah's) will sound familiar to specialists in near-free abelian groups, being a precise Boolean version of a theorem of Gregory. But for a universal algebraist, there is a striking difference. Finitely generated Boolean algebras are finite; roughly, Boolean theory can be held in your hand. Nobody but William Blake can hold a free abelian group in his hand.

References

1. Heindorf, L. and Shapiro, L. B.: *Nearly Projective Boolean Algebras*, Springer Lecture Notes 1596, Springer, Berlin, 1994.
2. Isbell, J.: A note on complete closure algebras, *Math. Systems Theory* **4** (1969), 310–312.
3. Isbell, J.: Parametrizable algebras, *J. London Math. Soc.* **8** (1974), 750–752.
4. Isbell, J.: k-parametrizable algebras, *Alg. Universalis* **36** (1996), 122–134.

Applied Categorical Structures **8:** 7–15, 2000.
G. Brümmer & C. Gilmour (eds), Papers in Honour of Bernhard Banaschewski.

An 'Unsitely' Result on Atomic Morphisms

Dedicated to Bernhard Banaschewski, who taught me the importance of always looking for the right *proof*

PETER JOHNSTONE
Department of Pure Mathematics, University of Cambridge, Cambridge CB2 1SB, U.K.

(Received: 8 December 1996; accepted: 28 April 1997)

Abstract. We give an 'elementary' proof, without mentioning sites, that any section of an atomic geometric morphism is open, and any section of a connected atomic morphism is an open surjection. Previously, these results were known only for bounded morphisms. As a by-product, we obtain a proof that any connected atomic morphism with a section is necessarily bounded.

Mathematics Subject Classification (2000): 18B25.

Key words: atomic morphism, topos theory.

Introduction

A geometric morphism $f: \mathcal{F} \to \mathcal{E}$ between (elementary) toposes is said to be *atomic* if its inverse image $f^*: \mathcal{E} \to \mathcal{F}$ is a logical functor (i.e. preserves the elementary topos structure). Many important results (pullback-stability, descent theorems, etc.) are known about the class of atomic morphisms, but the known proofs of most of them rely on the characterization of atomicity for bounded morphisms (that is, for those f which have the property that $\mathcal{F}$ may be represented as the topos of sheaves on an internal site in $\mathcal{E}$) in terms of 'atomic sites', as originally introduced (in the case when the base topos $\mathcal{E}$ is the topos of sets) by Barr and Diaconescu [1], and later extended to more general base toposes by Joyal and Tierney [6]. In particular, this means that the results are known only in the case of bounded morphisms.

This latter restriction is not in practice a very irksome one, since very few examples of unbounded geometric morphisms are known. Also, since many of the results involve reference to pullbacks of geometric morphisms, and since the pullback of a pair of morphisms is known to exist only in case at least one of the pair is bounded, it is at least unclear whether some of them could be formulated at all for unbounded morphisms. Nevertheless, it remains of interest in at least some cases to see whether the results can be proved by purely 'elementary' means, without making reference to sites.

In this paper, our aim is to provide a site-free proof of the result that any point of an atomic $\mathcal{E}$-topos is open, and any point of a connected atomic $\mathcal{E}$-topos is an open surjection. The only previous proof of this result (see [6], VII 4.1) makes use of atomic sites, and hence establishes the result only for bounded morphisms. (However, when $\mathcal{E}$ is the 'classical' topos **Set** of constant sets, both halves of the result are trivial: for if a topos $\mathcal{F}$ admits an atomic morphism to **Set** then it is Boolean (and so every morphism with codomain $\mathcal{F}$ is open), and if $\mathcal{F} \to$ **Set** is also connected then $\mathcal{F}$ is two-valued, so that every morphism from a non-degenerate topos to $\mathcal{F}$ is surjective.) Our result is therefore a genuine extension of what was previously known; but it has to be admitted that it is unlikely to lead to any major new applications. Rather, the interest lies in the method of proof itself: the fact that such 'elementary' site-free arguments can be made to work in a context where they have not been employed before.

The paper is organized in two sections: the first collects together some results from the literature (plus one result which, although we have not been able to find it in print, is surely well-known) which are relevant to the proof of our main result. Section 2 contains the proof of the main result itself, together with a few comments on it.

1. Preliminaries

We recall first that, for any cartesian (= finite-limit-preserving) functor $F: \mathcal{E} \to \mathcal{F}$ between toposes, we have for each $A \in \text{ob}\,\mathcal{E}$ a comparison map

$$\phi_A : F((\Omega_{\mathcal{E}})^A) \longrightarrow (\Omega_{\mathcal{F}})^{FA},$$

corresponding to the monomorphism $F(\in_A) \rightarrowtail F(\Omega^A \times A) \cong F(\Omega^A) \times FA$. We say F is *logical* if this comparison map is an isomorphism for all A; as already indicated, we call a geometric morphism *atomic* if its inverse image is a logical functor. We recall also the following result from [4]:

LEMMA 1.1. *For a geometric morphism $f: \mathcal{F} \to \mathcal{E}$, the following are equivalent:*

(i) *For every object A of $\mathcal{E}$, the comparison map $\phi_A : f^*((\Omega_{\mathcal{E}})^A) \to (\Omega_{\mathcal{F}})^{f^*A}$ is monic.*

(ii) *For every object A of $\mathcal{E}$ and every subobject $B \rightarrowtail f^*A$ in $\mathcal{F}$, the image of the composite geometric morphism*

$$\mathcal{F}/B \longrightarrow \mathcal{F}/f^*A \xrightarrow{f/A} \mathcal{E}/A$$

is an open subtopos of $\mathcal{E}/A$. (Here f/A is the geometric morphism whose inverse image is obtained by applying f^* to morphisms with codomain A.)

A geometric morphism with the equivalent properties of 1.1 is said to be *open*. The following weak version of the descent theorem for atomic morphisms (cf. [7],

1.10), though not required for the proof of our main result, turns out to be quite closely related to it.

LEMMA 1.2. *Let*

$$\begin{array}{ccc} \mathcal{G} & \xrightarrow{h} & \mathcal{F} \\ & \searrow^{g} & \downarrow{\scriptstyle f} \\ & & \mathcal{E} \end{array}$$

be a commutative triangle of geometric morphisms, such that g is atomic and h is an open surjection. Then f is atomic.

Proof. For any object A of $\mathcal{E}$, we have a commutative triangle

$$\begin{array}{ccc} g^*((\Omega_{\mathcal{E}})^A) & \xrightarrow{h^*(\phi_A)} & h^*((\Omega_{\mathcal{F}})^{f^*A}) \\ & \searrow^{\psi_A} & \downarrow{\scriptstyle \chi_{f^*A}} \\ & & (\Omega_{\mathcal{G}})^{g^*A} \end{array}$$

where ϕ, ψ, χ are the comparison maps for the functors f^*, g^*, h^* respectively. Since ψ_A is an isomorphism, it follows that χ_{f^*A} is (split) epic; but χ_{f^*A} is monic in any case since h is open, and hence it is an isomorphism. So $h^*(\phi_A)$ is an isomorphism; but h^* reflects isomorphisms since h is surjective, so ϕ_A is an isomorphism. □

A logical functor between toposes has a left adjoint if and only if it has a right adjoint (see [2]); thus a sufficient condition for a logical functor to be the inverse image of an atomic morphism is that it should have either a left or a right adjoint. The following result on left adjoints of logical functors is not new, but we have not been able to find a reference for it.

LEMMA 1.3. *Let* $f^*: \mathcal{E} \to \mathcal{F}$ *be a logical functor between toposes having a left adjoint* $f_!$. *Then* $f_!$ *preserves monomorphisms; and in fact, for every object A of* $\mathcal{F}$, *the assignment* $m \mapsto f_!m$ *is a bijection from (isomorphism classes of) subobjects of A in* $\mathcal{F}$ *to subobjects of* $f_!A$ *in* $\mathcal{E}$.

Proof. We clearly have a bijection between subobjects $A' \rightarrowtail A$ classified by morphisms $\chi : A \to \Omega_{\mathcal{F}} \cong f^*\Omega_{\mathcal{E}}$ and subobjects of $f_!A$ classified by morphisms $\overline{\chi}: f_!A \to \Omega_{\mathcal{E}}$; the problem is to show that it is induced by the action of $f_!$ on monomorphisms. For the moment, let us write $\lambda A' \rightarrowtail f_!A$ for the subobject corresponding to $m : A' \rightarrowtail A$. Since the diagram

$$\begin{array}{ccccc} f_!A' & \longrightarrow & f_!1 & \xrightarrow{\epsilon_1} & 1 \\ \downarrow{\scriptstyle f_!m} & & \downarrow{\scriptstyle f_!\top} & & \downarrow{\scriptstyle \top} \\ f_!A & \xrightarrow{f_!\chi} & f_!\Omega_{\mathcal{F}} & \xrightarrow{\epsilon_\Omega} & \Omega_{\mathcal{E}} \end{array}$$

is readily seen to commute (where ϵ is the counit of $(f_! \dashv f^*)$), we have a canonical morphism $f_!A' \to \lambda A'$; to show that this is an isomorphism, we shall demonstrate that $\lambda A'$ has the same universal property as $f_!A'$.

For any object B of $\mathcal{E}$, morphisms $A' \to f^*B$ can be regarded as partial maps $A \rightharpoonup f^*B$, and so correspond to morphisms $g\colon A \to \widetilde{f^*B}$ satisfying $\delta g = \chi$, where $\delta\colon \widetilde{f^*B} \to \Omega_{\mathcal{F}}$ classifies the canonical monomorphism $i_{f^*B}\colon f^*B \rightarrowtail \widetilde{f^*B}$. But since f^* is logical, it commutes with the construction of partial-map representers; so we have an isomorphism $\widetilde{f^*B} \cong f^*(\tilde{B})$, which identifies δ with $f^*(\delta')$ where $\delta'\colon \tilde{B} \to \Omega_{\mathcal{E}}$ classifies i_B. Thus morphisms $A' \to f^*B$ correspond to morphisms $\overline{g}\colon f_!A \to \tilde{B}$ satisfying $\delta'\overline{g} = \overline{\chi}$, and these in turn correspond to morphisms $\lambda A' \to B$, as required. □

A geometric morphism $f\colon \mathcal{F} \to \mathcal{E}$ is called a *local homeomorphism* if there is an object A of $\mathcal{E}$ and an equivalence $\mathcal{F} \simeq \mathcal{E}/A$ which identifies f^* and f_* with the pullback functor $A^*\colon \mathcal{E} \to \mathcal{E}/A$ and its right adjoint Π_A. Local homeomorphisms are atomic; conversely, an atomic morphism f is a local homeomorphism provided the left adjoint $f_!$ of f^* preserves equalizers ([3], 1.47), or equivalently provided $f_!$ is faithful. We note in passing that a local homeomorphism $\mathcal{E}/A \to \mathcal{E}$ is surjective iff $A \to 1$ is epic in $\mathcal{E}$ ([3], 4.12(iii)).

We recall that a geometric morphism $f\colon \mathcal{F} \to \mathcal{E}$ is said to be *surjective* if f^* is faithful, and *connected* if f^* is full and faithful. f is called *hyperconnected* if it is connected and the image of f^* is closed under subobjects in $\mathcal{F}$. A connected atomic morphism is hyperconnected: this follows easily from 1.3, since for any object A of $\mathcal{E}$ the functor $f_!$ induces a bijection from subobjects of f^*A to subobjects of $f_!f^*A \cong A$, whose inverse must be induced by f^*. Corresponding to 1.3, we have:

LEMMA 1.4. *For a geometric morphism $f\colon \mathcal{F} \to \mathcal{E}$, the following are equivalent:*

(i) *f is hyperconnected.*

(ii) *f_* preserves Ω; i.e. the comparison map $\phi_1\colon f_*(\Omega_{\mathcal{F}}) \to \Omega_{\mathcal{E}}$ is an isomorphism.*

(iii) *For any object A of $\mathcal{E}$, f^* induces a bijection from the subobject lattice of A in $\mathcal{E}$ to that of f^*A in $\mathcal{F}$.*

Proof. See [5], 1.5. □

We recall also from [5] that every geometric morphism factors, uniquely up to unique isomorphism, as a hyperconnected morphism followed by one which is localic. For atomic morphisms, this factorization has a particularly simple form, resulting from the fact that a localic atomic morphism is necessarily a local homeomorphism.

LEMMA 1.5. *Let $f\colon \mathcal{F} \to \mathcal{E}$ be an atomic geometric morphism. Then*

(i) *f is connected iff $f_!$ preserves* 1.

(ii) *f can be factored, uniquely up to unique isomorphism, as a connected atomic morphism followed by a local homeomorphism.*

Proof. (i) If f is connected, then the counit map $f_! f^* \to 1_{\mathcal{E}}$ is an isomorphism, so $f_! 1 \cong f_! f^* 1 \cong 1$. Conversely, from the fact that f^* preserves exponentials we deduce the 'Frobenius reciprocity' condition that the natural map $f_!(B \times f^* A) \to f_!(B) \times A$ is an isomorphism for all A and B; putting $B = 1$, we deduce that the counit $f_! f^* \to 1$ is an isomorphism, i.e. that f is connected.

(ii) Given f, we set $A = f_! 1$. Then we have a canonical factorization of f through the local homeomorphism $\mathcal{E}/A \to \mathcal{E}$, whose inverse image is given by applying f^* to objects over $f_! 1$ and then pulling back along the unit $1 \to f^* f_! 1$. It is easy to see that this factorization is atomic, and (using (i)) that it is connected. This establishes the existence of the factorization; the uniqueness follows from [5] and the fact that connected atomic morphisms are hyperconnected. □

2. Proof of the Theorem

We are now ready to state our main result.

THEOREM 2.1. *Let $f: \mathcal{F} \to \mathcal{E}$ be an atomic morphism, let A be an object of $\mathcal{E}$ and let $h: \mathcal{E}/A \to \mathcal{F}$ be a geometric morphism over $\mathcal{E}$. Then*

(i) *h is open.*
(ii) *If f is connected and $A \to 1$ is epic, h is surjective.*

This is a slight generalization of the result as usually stated for bounded f (see, for example, [6], Proposition VII 4.1), which corresponds to the special case $A = 1$. However, the general case stated here follows easily from the special one plus known stability properties of atomic morphisms under pullback along local homeomorphisms. Our reason for stating the result in the more general form is that we shall need to allow the object A to vary in the course of our proof.

As a first step in the proof, we show:

LEMMA 2.2. *Let $f: \mathcal{F} \to \mathcal{E}$ be an atomic geometric morphism, and let η denote the unit of the adjunction $(f_! \dashv f^*)$. Then*

(i) *For every monomorphism $B' \rightarrowtail B$ in $\mathcal{F}$, the naturality square*

$$\begin{array}{ccc} B' & \xrightarrow{\eta_{B'}} & f^* f_! B' \\ \downarrow & & \downarrow \\ B & \xrightarrow{\eta_B} & f^* f_! B \end{array}$$

is a pullback.
(ii) *If f is connected, η_B is epic for all B.*

Proof. (i) Let $\chi : B \to \Omega_{\mathcal{F}} \cong f^*(\Omega_{\mathcal{E}})$ be the classifying map of $B' \rightarrowtail B$. By the proof of 1.3 we have a pullback square

$$\begin{array}{ccc} f_!B' & \longrightarrow & 1 \\ \downarrow & & \downarrow \top \\ f_!B & \xrightarrow{\overline{\chi}} & \Omega_{\mathcal{E}} \end{array}$$

in $\mathcal{E}$, where $\overline{\chi}$ is the transpose of χ; but f^* preserves this pullback, and the composite $f^*(\overline{\chi}).\eta_B$ equals χ, so the result follows.

(ii) Since both f^* and f_* preserve Ω, their left adjoints $f_!$ and f^* both induce isomorphisms of subobject lattices, by 1.3 and 1.4; and by part (i) the inverse of the induced bijection $\mathrm{Sub}(B) \to \mathrm{Sub}(f^*f_!B)$ is given by pullback along η_B. But the image of η_B and the top subobject of $f^*f_!B$ both pull back to the top subobject of B; so they must be equal. □

We may now deduce the second assertion of 2.1:

Proof. For any $m : B' \rightarrowtail B$ in $\mathcal{F}$, we have a pullback square

$$\begin{array}{ccc} B' & \twoheadrightarrow & f^*f_!B' \\ \downarrow m & & \downarrow f^*f_!m \\ B & \twoheadrightarrow & f^*f_!B \end{array}$$

by 2.2(ii). Applying h^* to this (and recalling that $h^*f^* \cong A^*$), we obtain a pullback square

$$\begin{array}{ccc} h^*B' & \twoheadrightarrow & A^*f_!B' \\ \downarrow h^*m & & \downarrow A^*f_!m \\ h^*B & \twoheadrightarrow & A^*f_!B \end{array}$$

in $\mathcal{E}/A$. So

$$\begin{array}{llll} h^*m \text{ is an isomorphism} & \Rightarrow & A^*f_!m \text{ is an isomorphism} & \\ & \Rightarrow & f_!m \text{ is an isomorphism,} & \text{since } A^* \text{ is faithful} \\ & \Rightarrow & m \text{ is an isomorphism,} & \text{by 1.3.} \end{array}$$

In other words, h^* 'preserves properness of subobjects'; but since it also (unlike $f_!$) preserves equalizers, it follows that it is faithful. □

We note that the above proof still works if we replace the topos $\mathcal{E}/A$ by any topos $\mathcal{G}$ admitting a surjective geometric morphism to $\mathcal{E}$. Similarly, the proof of 2.1(i) which follows can be made to work when $\mathcal{E}/A$ is replaced by a topos $\mathcal{G}$ admitting a bounded open morphism to $\mathcal{E}$.

For the latter proof, we first observe that we may reduce to the case $A = 1$ (i.e. to the case $\mathcal{G} = \mathcal{E}$): for we have a pullback square of toposes and geometric morphisms

$$\begin{array}{ccc} \mathcal{F}/f^*A & \longrightarrow & \mathcal{F} \\ \downarrow{\scriptstyle f/A} & & \downarrow{\scriptstyle f} \\ \mathcal{E}/A & \longrightarrow & \mathcal{E} \end{array}$$

where f/A is atomic, and thus h induces a morphism $\mathcal{E}/A \to \mathcal{F}/f^*A$ over $\mathcal{E}/A$, whose openness implies that of h itself. Having reduced to the case $A = 1$, we note that we have a natural transformation $h^* \to f_!$, obtained by applying h^* to the unit $1_{\mathcal{F}} \to f^*f_!$. We now show that h satisfies condition (ii) of 1.1.

Proof. Let B be an object of $\mathcal{F}$, and suppose we are given a subobject $C \rightarrowtail h^*B$. By 1.3, there is a unique subobject $B' \rightarrowtail B$ such that $f_!B' \rightarrowtail f_!B$ is the image of the composite $C \rightarrowtail h^*B \to f_!B$. By 2.2(i) and the fact that h^* preserves pullbacks, we have a pullback

$$\begin{array}{ccc} h^*B' & \longrightarrow & f_!B' \\ \downarrow & & \downarrow \\ h^*B & \longrightarrow & f_!B \end{array} \quad ;$$

so B' is the unique smallest subobject of B such that $C \leqslant h^*B'$.

Now, if we apply the factorization of 1.5(ii) to the composite $\mathcal{F}/B' \to \mathcal{F} \to \mathcal{E}$, we obtain $\mathcal{F}/B' \to \mathcal{E}/f_!B' \to \mathcal{E}$; so the first factor of the latter is connected and atomic. Also, the composite $\mathcal{E}/C \to \mathcal{E}/h^*B' \to \mathcal{F}/B'$ is a morphism over $\mathcal{E}/f_!B'$; and since $C \to f_!B'$ is epic by the definition of B', it follows from 2.1(ii) that this latter composite is a surjection. So the image of $\mathcal{E}/C \to \mathcal{E}/h^*B \to \mathcal{F}/B$ (in the sense of the surjection–inclusion factorization) is the open inclusion $\mathcal{F}/B' \to \mathcal{F}/B$. □

Combining 2.1(i) with 1.2, we immediately obtain:

COROLLARY 2.3. *Let $f: \mathcal{F} \to \mathcal{E}$ be a geometric morphism, and suppose there exists an object A of $\mathcal{E}$ and a surjection $h: \mathcal{E}/A \to \mathcal{F}$ over $\mathcal{E}$ (that is, '$\mathcal{F}$ has a sufficient set of $\mathcal{E}$-valued points'). Then f is atomic iff h is open.*

As we remarked in the Introduction, the extension of the result of 2.1 from bounded morphisms to arbitrary ones does not seem likely to lead to any important new applications. There are two main applications of (the bounded version of) 2.1 in the existing literature: the first is the characterization of atomic morphisms as those open morphisms whose diagonals are open ([6], Theorem VII 4.1), and the second uses 2.1(ii), in conjunction with the descent theorem for open surjections, to characterize connected atomic $\mathcal{E}$-toposes with points (for a given base topos $\mathcal{E}$) as the categories of continuous actions of localic groups in $\mathcal{E}$ ([6], Theorem VIII 3.1). The first of these clearly requires the existence of a pullback of f along itself before we can even state it for a morphism f; and so, in the absence of any known method of constructing pullbacks of unbounded morphisms, we cannot hope to extend it to unbounded atomic morphisms.

However, the second application does not require the assumption of boundedness for the morphism f from which we start, even though the descent theorem is only valid for bounded open surjections. The point is that, if we are given $f: \mathcal{F} \to \mathcal{E}$ equipped with a morphism $h: \mathcal{E}/A \to \mathcal{F}$ over $\mathcal{E}$ (for some object A of $\mathcal{E}$), then h is necessarily bounded (indeed, localic), since the composite fh is localic (cf. [3], 4.44(ii)). Thus we can form the kernel-pair of h (and prove that it is localic over $\mathcal{E}$) in any case; and, provided h is an open surjection, we may then apply the descent theorem to obtain an equivalence between $\mathcal{F}$ and the topos of continuous $\mathcal{G}$-actions in $\mathcal{E}$, where $\mathcal{G}$ is a localic groupoid in $\mathcal{E}$ whose object of objects is the discrete locale A (in particular, if $A = 1$, then $\mathcal{G}$ is a localic group in $\mathcal{E}$). However, a topos of this particular form is necessarily bounded over $\mathcal{E}$ (cf. [7]). We therefore obtain the following somewhat unexpected result:

COROLLARY 2.4. *Let $f: \mathcal{F} \to \mathcal{E}$ be a connected atomic geometric morphism, and suppose there exists a morphism $h: \mathcal{E}/A \to \mathcal{F}$ over $\mathcal{E}$ where $A \to 1$ is epic in $\mathcal{E}$. Then f is bounded.*

Similarly, we may deduce from 2.1(i) that any atomic $\mathcal{E}$-topos with a sufficient set of $\mathcal{E}$-valued points is bounded over $\mathcal{E}$.

The unexpectedness of 2.4 derives from the fact that the two best-known examples of unbounded geometric morphisms $\mathcal{F} \to$ **Set** (those obtained by taking $\mathcal{F}$ to be the topos of set-actions of a 'large' group which is an inverse limit of small ones, or by taking it to be the topos of 'uniformly continuous' actions of a topological group with no smallest open subgroup – cf. [3], 4.49) are both connected and atomic. However, neither of them has a section: although, in both cases, the forgetful functor $\mathcal{F} \to$ **Set** 'looks as if it should be an inverse image functor', in that it preserves finite limits and all colimits which exist in $\mathcal{F}$, it fails to have a right adjoint in either case. (In the first case, the solution-set condition required for the Adjoint Functor Theorem is not satisfied, and in the second the category $\mathcal{F}$ is not cocomplete.) Thus, although 2.4 itself is a new result, it does not seem likely to lead us to any better understanding of unbounded geometric morphisms in general, or even of connected atomic ones in particular.

References

1. Barr, M. and Diaconescu, R.: Atomic toposes, *J. Pure Appl. Algebra* **17** (1980), 1–24.
2. Johnstone, P. T.: Adjoint lifting theorems for categories of algebras, *Bull. London Math. Soc.* **7** (1975), 294–297.

3. Johnstone, P. T.: *Topos Theory*, London Mathematical Society Monographs No. 10, Academic Press, 1977.
4. Johnstone, P. T.: Open maps of toposes, *Manuscripta Math.* **31** (1980), 217–247.
5. Johnstone, P. T.: Factorization theorems for geometric morphisms, I, *Cahiers Top. Géom. Diff.* **22** (1981), 3–17.
6. Joyal, A. and Tierney, M.: *An Extension of the Galois Theory of Grothendieck*, Mem. Amer. Math. Soc., 1984, p. 309.
7. Moerdijk, I.: The classifying topos of a continuous groupoid, I, *Trans. Amer. Math. Soc.* **310** (1988), 629–668.

Applied Categorical Structures **8:** 17–44, 2000.
G. Brümmer & C. Gilmour (eds), Papers in Honour of Bernhard Banaschewski.

Quotient Maps of Locales

TILL PLEWE*
Faculty of Science and Engineering, Ritsumekan University, Kusatsu-shi, Shiga 525-77, Japan

(Received: 4 October 1996; accepted: 26 November 1997)

Abstract. We consider regular epimorphisms in the category **Loc** of locales. Closed surjections with subfit domain are regular epimorphisms. However, there exists a closed surjection which is the composite of two regular epimorphisms without being regular epic itself. This example answers both of the following questions in the negative: Y. Li's question of whether weak quotient maps are necessarily regular epimorphisms, and P. Johnstone's related question of whether regular epimorphisms compose. It follows that not all extremal epimorphisms in **Loc** are regular. The weak quotient maps of Y. Li and the equationally closed subframes of A. Pultr and A. Tozzi are shown to be dual notions. We also give a new characterization of regular epimorphisms in **Loc**.

Mathematics Subject Classifications (2000): 18A20, 18B30, 54B15, 54C10.

Key words: locales, closed surjections, simple coverings, singly generated frame extensions, extremal epimorphisms, regular epimorphisms, quotient maps, weak quotient maps, equationally closed subframes.

Introduction

The question which gave rise to this paper is the question of whether there is a characterization of regular epimorphisms in the category **Loc** of locales, which is similar to the characterization of regular epimorphisms in the category **Top** of topological spaces as quotient maps. The problem of characterizing regular epimorphisms in **Loc** turns out to be much more complex than its topological counterpart. The additional difficulties one encounters in **Loc** can perhaps be seen most clearly in the following somewhat unusual proof of the fact that quotient maps of T_0 spaces are coequalizers of their kernel pair (the extra work needed to deal with non-T_0 spaces is purely set theoretic). This proof shows in particular the important role played by the forgetful functor $\mathbf{Top} \xrightarrow{|\,|} \mathbf{Set}$.

Recall that a continuous surjection $Y \xrightarrow{f} X$ of topological spaces is called a quotient map if it has the following property:

(QO) $\quad \forall S \subseteq X,$ if $f^{-1}(S)$ is open in $Y,$ then S is open in X.

Now let $Y \xrightarrow{f} X$ be a quotient map in **Top** and let $Y \times_X Y \underset{\pi_2}{\overset{\pi_1}{\rightrightarrows}} Y$ be its kernel pair.

* Financially supported by the British Science and Engineering Research Council under the 'Foundational Structures in Computing Science' reseach project.

(i) In order to show that $Y \times_X Y \overset{\pi_1}{\underset{\pi_2}{\rightrightarrows}} Y \xrightarrow{f} X$ is a coequalizer diagram it suffices to show that the universal property holds for maps into the Sierpinski space $\mathbb{S}$ because $\mathbb{S}$ is a cogenerator in the category of T_0 spaces. Now a map $Y \xrightarrow{g} \mathbb{S}$ is a cocone if and only if the inverse image U of the open point in $\mathbb{S}$ has isomorphic pullbacks along π_1 and π_2, i.e. $\pi_1^{-1}(U) \simeq \pi_2^{-1}(U)$, which in turn is equivalent to the statement that U is equipped with descent data relative to f.

(ii) Since the forgetful functor **Top** $\xrightarrow{|\,|}$ **Set** preserves limits and in the category **Set** of sets all surjections are effective descent morphisms, there exists a subset $S \subseteq |X|$ whose pullback along $|f|$ is equal to $|U|$.

(iii) Since surjections are stable, S is unique.

(iv) Equipping S with the subspace topology, we get that $f^{-1}(S) = U$, because embeddings are preserved by pullbacks, and subspaces are determined by their points.

(v) Since f is quotient, S is open.

(vi) If $X \xrightarrow{h} \mathbb{S}$ is the characteristic map of S, i.e. the unique map for which the inverse image of the open point of $\mathbb{S}$ is equal to S, then $hf = g$ and h is the unique map with this property.

Now (i) also works in the localic case, but there is no analogue of (ii). One of the examples of localic surjections given in this paper (4.1) shows that such an S need not exist, even if the localic version of (QO) is satisfied. (Note that for maps of sober spaces the localic version of (QO) is quite a bit stronger than the topological version because the set of sublocales of a spatial locale is generally much larger than the set of subspaces of the corresponding sober space.) (iii) doesn't hold for all regular epimorphisms of locales because regular epimorphisms are not necessarily stable surjections. If we assume that the problem of finding a sublocale S of X which pulls back to U has been solved, then we still need to construct from S a unique open sublocale V which pulls back to U. A condition which accomplishes that is described in 4.8. Step (vi) of the proof above works also for locales.

In the present paper we will only address some of the problems mentioned above. We will consider some variations on condition (QO) for localic surjections, which all deal with the question of how to show that, provided an open sublocale descends to any sublocale (respectively, complemented sublocale) at all, it also descends to an open sublocale. But before doing that we will consider two restricted classes of surjections: simple coverings and closed surjections. Restricting our attention to the first class gives a simplified picture of surjections, which nonetheless gives some first insights. The second class provides some counterexamples.

A more detailed description of this paper is as follows. In the first section we recall facts about sublocale lattices and stability of epimorphisms under pullbacks. We also describe (the externalization of) Artin-glueing of locales in a localic topos. This will mainly be used to construct a closed surjection which is not a regular epimorphism. After that we consider two special classes of surjections. In the second section we look at simple coverings; these are surjections $B \xrightarrow{f} A$ which factor through $A \times \mathbb{S} \xrightarrow{\pi_A} A$ by an embedding ($\mathbb{S}$ the Sierpinski space). They are the localic version of the singly generated frame extensions introduced by B. Banaschewski [1]. We characterize the restrictions of several classes of surjections to simple coverings. The most noteworthy result here might be that a simple covering which is a stable regular epimorphism is necessarily a retraction. We also give an example of a simple covering between Hausdorff spaces which is a regular epimorphism in **Loc** but not in **Top**. This shows that regularity of epimorphisms is not reflected by the embedding of sober (T_D) spaces into **Loc**. The third section concerns closed surjections. We show that closed surjections with subfit domain are regular epimorphisms. However, we also construct a closed surjection which is the composite of two regular epimorphisms without being regular epic itself. This answers both of the following questions in the negative: Y. Li's question of whether weak quotient maps are necessarily regular epimorphisms, and P. Johnstone's related question of whether regular epimorphisms compose. In the last section we compare four different conditions on localic surjections which are all similar to the definition of quotient maps in **Top**. Two of them, the weak quotient maps of Li and the localic equivalent of the equationally closed subframes of Pultr and Tozzi, turn out to be equivalent notions. We also introduce a property which solves half of the problem of how to give an internal characterization of regular epimorphisms in **Loc**.

1. Background

For general results in locale theory the reader may consult [4, 7, 8]. Notation will follow for the most part [6]. The frame which determines a given locale X and the topology for a topological space X will both be denoted by $T(X)$. The open sublocale corresponding to an element $x \in T(X)$ will be denoted by U_x and its closed complement by F_x. Conversely, if U is an open sublocale of X, then the corresponding element of the frame $T(X)$ will be denoted by x_U. All spaces considered in this paper are sober and will be identified with the corresponding spatial locale.

The lattice $S(X)$ of all sublocales of a given locale X is complete and satisfies the (coframe) distributive law: $S \vee \bigwedge T_i = \bigwedge(S \vee T_i)$. Each locale X has a smallest dense sublocale $D(X)$. A sublocale $S \in S(X)$ is said to be nowhere dense in X if and only if it is disjoint from $D(X)$, i.e. if $S \wedge D(X) = 0$. The closure of a nowhere dense sublocale is again nowhere dense. If all points of X are locally closed (i.e. if X is a T_D space), then X has a largest pointless sublocale $pl(X)$ [6]. It can be computed either as the meet of all point complements, or as the union of all pointless Boolean sublocales of X.

Sublocales do not necessarily have complements. But there is a natural substitute which can be described in two ways. For $S, R \in S(X)$, put $R \backslash S = \bigvee\{T \in S(X) \mid T \leq R \text{ and } T \wedge S = 0\}$ [6] and $sup(S) = \bigwedge\{R \in S(X) \mid R \vee S = X\}$

($sup(S)$ stands for the supplement of S). Since $S(X)$ is a coframe $S \vee sup(S) = X$. So $sup(S)$ is the smallest sublocale whose join with S is X. We will show below that $X\backslash S = sup(S)$.

A sublocale C is called complemented if $C \wedge sup(C) = 0$. Complemented sublocales can be partitioned into locally closed ones. If C is complemented, then there exists an ordinal-indexed chain of pairwise disjoint locally closed sublocales whose union is C and for which the union of any initial segment is open in C [6, 12]. Each sublocale is the meet of all complemented sublocales containing it; in fact, it suffices to consider sublocales of the form $U \vee F$ for open U and closed F.

The following lemma collects some further properties of $S(X)$.

LEMMA 1.1. *Let X be any locale and $S, S_i \in S(X)$ $(i \in I)$ be arbitrary sublocales, then*

(1) $S \wedge \bigvee S_i = \bigvee S \wedge S_i$ *if S is complemented;*
(2) $S \wedge \bigvee S_i = \bigvee S \wedge S_i$ *if all S_i are open;*
(3) $X\backslash S = sup(S)$;
(4) *If C and D are complemented sublocales of X, then $C \vee S = D \vee S$ if and only if $C \wedge (X\backslash S) = D \wedge (X\backslash S)$.*

Proof. A proof of (1) is already contained in [4] and for a proof of (2) see [5] or [9]. As for (3), the inclusion $\bigvee\{T \in S(X) \mid T \wedge S = 0\} \leq \bigwedge\{R \in S(X) \mid R \vee S = X\}$ follows from the fact that for any such T and R we have that $T = T \wedge X = T \wedge (R \vee S) = T \wedge R \vee T \wedge S = T \wedge R \leq R$. To show that the reverse inclusion holds let C be any complemented sublocale of X such that $C \geq X\backslash S$. We need to show that $C \geq sup(S)$, equivalently that $C \vee S = X$. Let (D, E) be a complementary pair of sublocales of X such $D \geq C \vee S$. Then $E \wedge S \leq E \wedge (C \vee S) \leq E \wedge D = 0$. Hence $E \leq X\backslash S \leq C \leq C \vee S \leq D$, which implies that $E = 0$. To show (4), let C and D be any two complemented sublocales of X such that $C \vee S = D \vee S$. If $T \wedge S = 0$, then

$$
\begin{aligned}
C \wedge T &= (C \wedge T) \vee (S \wedge T) = (C \vee S) \wedge T \\
&= (D \vee S) \wedge T = (D \wedge T) \vee (S \wedge T) = D \wedge T,
\end{aligned}
$$

hence,

$$
\begin{aligned}
C \wedge (X\backslash S) &= C \wedge \bigvee\{T \in S(X) \mid T \wedge S = 0\} \\
&= \bigvee\{C \wedge T \mid T \in S(X) \text{ and } T \wedge S = 0\} \\
&= \bigvee\{D \wedge T \mid T \in S(X) \text{ and } T \wedge S = 0\} \\
&= D \wedge \bigvee\{T \in S(X) \mid T \wedge S = 0\} \\
&= D \wedge (X\backslash S).
\end{aligned}
$$

Conversely, if $C \wedge (X\backslash S) = D \wedge (X\backslash S)$, then

$$
C \vee S = (C \vee S) \wedge (X\backslash S \vee S) = (C \wedge X\backslash S) \vee S
$$

$$= (D \wedge X\backslash S) \vee S = (D \vee S) \wedge (X\backslash S \vee S) = D \vee S. \qquad \square$$

Every map $f\colon X \to Y$ in **Loc** can be factored as a surjection followed by a sublocale inclusion $X \twoheadrightarrow f(X) \hookrightarrow Y$ (see e.g. [7]). Images with respect to this factorization are not quite as well behaved as images in **Top**. (Another consequence of the "absent" forgetful functor to **Set**.) In particular, this factorization is not stable under pullback. But we have at least the following results:

PROPOSITION 1.2. *Let $B \xrightarrow{f} A$ be any map in* **Loc**. *Then for all complemented $C \in S(A)$ and all $S \in S(B)$ we have:*

$$f(S \wedge f^{-1}(C)) = f(S) \wedge C.$$

Proof. This is [15, Prop. 1.4]. □

COROLLARY 1.3. *Surjections are stable under pullback along complemented inclusions. In particular, if C and D are complemented sublocales of A, then $f^{-1}(C) = f^{-1}(D)$ implies that $C = D$ for any surjection $B \xrightarrow{f} A$. Hence f^{-1} preserves and reflects inclusions between complemented sublocales.*

Proof. Let C be complemented in A and let g be the pullback of f along the inclusion $C \to A$. If $B \xrightarrow{f} A$ is a surjection then $g(f^{-1}(C)) = f(f^{-1}(C)) = f(B \wedge f^{-1}(C)) = f(B) \wedge C = C$. Finally, for C and D as above both $f^{-1}(C - D) = f^{-1}(C) - f^{-1}(D)$ and $f^{-1}(D - C) = f^{-1}(D) - f^{-1}(C)$ are empty. Because surjections are stable under pullback along complemented inclusions, $C - D$ and $D - C$ must also be empty, i.e. $C = D$. □

Recall that a map $B \xrightarrow{f} A$ of locales is said to be closed if the image of each closed sublocale is again closed. An equivalent condition is that f_* satisfies the Frobenius identity $f_*(b \vee f^*(a)) = f_*(b) \vee a$. If f is closed, then f_* and the map $f()$ restricted to closed sublocales can be expressed directly in terms of each other. For any open sublocale U of B and its closed complemented F we have $a_{A-f(F)} = f_*(b_U)$.

COROLLARY 1.4. *Closed maps are stable under pullback along complemented inclusions.*

Proof. Let C be complemented in A and let g be the pullback of f along the inclusion $C \to A$. If $B \xrightarrow{f} A$ is a closed map and if F is a closed sublocale of $f^{-1}(C)$, then $g(F) = f(F) = f(cl_B(F) \wedge f^{-1}(C)) = f(cl_B(F)) \wedge C$ which is closed in C because $f(cl_B(F))$ is closed in A. So g is also closed. □

LEMMA 1.5. *Regular epimorphisms are stable under pullback along open or closed inclusions.*

Proof. A surjection $B \xrightarrow{f} A$ is a regular epimorphism if and only if for all open sublocales U of B equipped with descent data, there exists a unique open sublocale V of A such that $f^{-1}(V) = U$. If $B' \xrightarrow{f'} A'$ is the pullback of $B \xrightarrow{f} A$ along an open inclusion $A' \to A$, then any open sublocale U' of B' equipped with descent data relative to f' is also open in B and the descent data also lift to descent data relative to f. So U' descends to some unique open sublocale V of A, which is necessarily contained in A'. To show stability under pullback along closed inclusions the analogous arguments apply, using the fact that regular epimorphism can also be described as surjections such that for all closed sublocales F of B equipped with descent data, there exists a unique closed sublocale G of A such that $f^{-1}(G) = F$. □

We conclude this section with a description of Artin-glueing of locales in $\mathbf{Sh}(Z)$ where Z is an arbitrary locale. A locale X in $\mathbf{Sh}(Z)$ is determined by its sheaf of frames $\mathcal{T}(X)$. Recall that for any pair $z, z' \in T(Z)$ with $z \leq z'$ the restriction map $\mathcal{T}(X)_{z'} \xrightarrow{\rho_z^{z'}} \mathcal{T}(X)_z$ is a surjective frame homomorphism which has a left adjoint $\Sigma_z^{z'}$ which satisfies the Frobenius identity $\Sigma_z^{z'}(a \wedge \rho_z^{z'}(b)) = \Sigma_z^{z'}(a) \wedge b$ [8].

A fringe map (a map which preserves all finite meets) $\mathcal{T}(X) \xrightarrow{\mu=\{\mu_z | z \in T(Z)\}} \mathcal{T}(Y)$ is completely determined by its component μ_1 because $\mu_z = \mu_z \rho_z^1 \Sigma_z^1 = \rho_z^1 \mu_1 \Sigma_z^1$. Furthermore, we have $\Sigma_z^1(1) = \Sigma_z^1 \mu_z(1) \leq \mu_1 \Sigma_z^1(1)$ because each μ_z preserves 1 and $\mu_z \leq \rho_z^1 \mu_1 \Sigma_z^1 \Leftrightarrow \Sigma_z^1 \mu_z \leq \mu_1 \Sigma_z^1$. (Since the context will always make it possible to determine the sheaf to which a given restriction map, respectively left adjoint, belongs, there is no need for the notation to specify the sheaf.)

Conversely, any finite meet preserving map $\mathcal{T}(X)_1 \xrightarrow{m} \mathcal{T}(Y)_1$ which satisfies $\Sigma_z^1(1) \leq m\Sigma_z^1(1)$ gives rise to a fringe map $\mathcal{T}(X) \xrightarrow{\mu} \mathcal{T}(Y)$ of internal posets in $\mathbf{Sh}(Z)$ by putting $\mu_z = \rho_z^1 m \Sigma_z^1$. To show that $\rho_z^{z'} \mu_{z'} = \mu_z \rho_z^{z'}$ it suffices to consider the case $z' = 1$ because all maps $\rho_{z'}^1$ are surjections. Now $\rho_z^1 \mu_1 = \rho_z^1 m$ and $\mu_z \rho_z^1 = \rho_z^1 m \Sigma_z^1 \rho_z^1$. For all $a \in \mathcal{T}(X)_z$ we have

$$\begin{aligned} \rho_z^1 m \Sigma_z^1 \rho_z^1(a) &= \rho_z^1 m\big(a \wedge \Sigma_z^1(1)\big) = \rho_z^1 m(a) \wedge \rho_z^1 m \Sigma_z^1(1) \\ &\geq \rho_z^1 m(a) \wedge \rho_z^1 m(1) = \rho_z^1 m(a). \end{aligned}$$

Since $\mathrm{id} \geq \Sigma_z^1 \rho_z^1$ we get the reverse inequality $\rho_z^1 m \geq \rho_z^1 m \Sigma_z^1 \rho_z^1$ immediately.

Externalizing these data, we obtain continuous maps $X \xrightarrow{f} Z$ and $Y \xrightarrow{g} Z$ and a finite meet preserving map $T(X) \xrightarrow{m} T(Y)$ satisfying $mf^* \geq g^*$, where $T(X) = \mathcal{T}(X)_1$, $T(Y) = \mathcal{T}(Y)_1$, $f^*(z) = \Sigma_z^1(1)$ and $g^*(z) = \Sigma_z^1(1)$ for all $z \in T(Z)$ and $m = \mu_1$. Let $X +_m Y$ be the locale obtained by glueing X onto Y along m. Its topology $T(X +_m Y) = \{(x, y) \in T(X) \times T(Y) \mid m(x) \geq y\}$. The map $X + Y \xrightarrow{\langle f,g \rangle} Z$ factors through $X +_m Y$ because $mf^*(z) \geq g^*(z)$ for all $z \in T(Z)$. So we get the following commutative diagram:

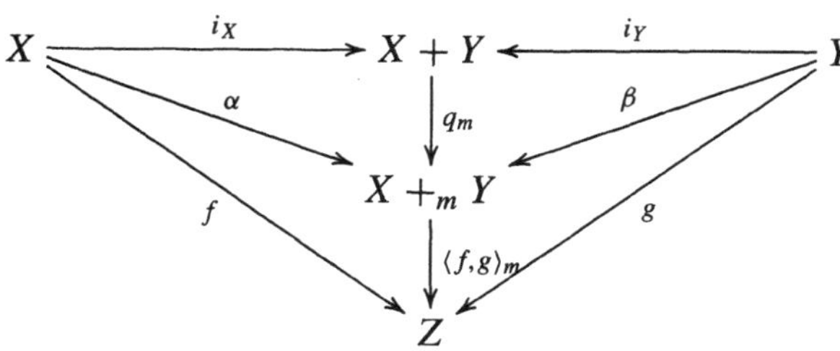

(†)

If n is another fringe map satisfying $nf^* \geq g^*$, then q_n factors through q_m if and only if $n \leq m$ in the pointwise ordering.

In this paper we are mainly interested in the case where $m = g^*f_*$. The map g^*f_* is the smallest fringe map which satisfies $mf^* \geq g^*$. That $g^*f_*f^* \geq g^*$ is clear, and if n is any map satisfying $nf^* \geq g^*$ then $n \geq nf^*f_* \geq g^*f_*$. (Of course, the largest such map is just $m = 1$ and then $X +_m Y = X + Y$.) If $m = g^*f_*$ then we will abbreviate $X +_{g^*f_*} Y \xrightarrow{\langle f,g\rangle_{g^*f_*}} Z$ to $X +_\bullet Y \xrightarrow{\langle f,g\rangle_\bullet} Z$. The factorization of a map $A + B \xrightarrow{h+k} X + Y \xrightarrow{q_\bullet} X +_\bullet Y$ through $A +_\bullet B$ will be denoted by $A +_\bullet B \xrightarrow{h+_\bullet k} X +_\bullet Y$ irrespective of the locales over which $A +_\bullet B$ and $X +_\bullet Y$ are obtained by glueing along smallest fringe maps.

The next lemma shows that this construction is stable under pullback.

LEMMA 1.6. *Let t: $T \to Z$ be any map in* **Loc**. *If the first two of the following diagrams are pullbacks then so is the third*:

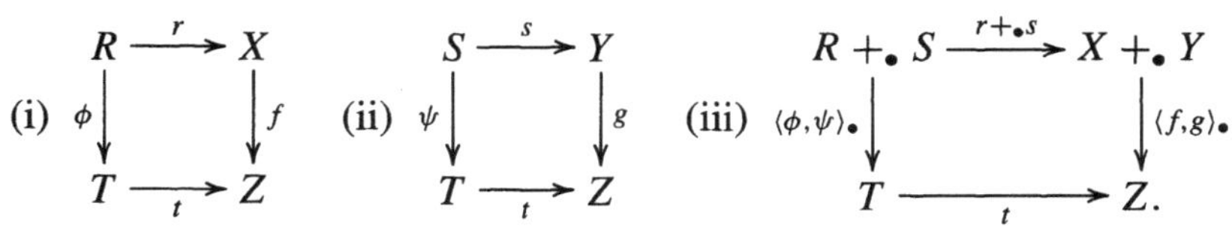

Proof. It is clear that the pullback of $X +_\bullet Y$ along t has an open sublocale homeomorphic to R with closed complement homeomorphic to S. So we have to show that R is glued onto S by $T(R) \xrightarrow{\psi^*\phi_*} T(S)$. Since $\psi^*\phi_*$ is the smallest meet preserving map n for which there is a continuous map $R +_n S \to T$ which restricts to ϕ and ψ on R, respectively S, it suffices to show that $R + S \xrightarrow{r+s} X + Y \to X +_\bullet Y$ factors through $R +_\bullet S$. In order words we have to show that whenever $y \leq g^*f_*(x)$ for $x \in T(X)$ and $y \in T(Y)$ then also $s^*(y) \leq \psi^*\phi_*r^*(x)$. Since $g^*t_* \leq s_*\psi^*$ ($t_* \leq t_*\psi_*\psi^* = g_*s_*\psi^*$), $y \leq g^*f_*(x)$ implies that $y \leq g^*f_*r_*r^*(x) = g^*t_*\phi_*r^*(x) \leq s_*\psi^*\phi_*r^*(x)$ which in turn is equivalent to $s^*(y) \leq \psi^*\phi_*r^*(x)$. □

More generally, one can show that, given diagrams (i) and (ii) above and any fringe map $T(X) \xrightarrow{m} T(Y)$, the pullback of $\langle f, g\rangle_m$ is given by $R +_n S \xrightarrow{\langle\phi,\psi\rangle_n} T$ where n is the least fringe map satisfying $n\phi^* \geq \psi^*$ and $nr^* \geq s^*m$.

By applying the preceding lemma twice we can compute the kernel pair of the map $X +_\bullet Y \xrightarrow{\langle f,g\rangle_\bullet} Z$ in terms of the kernel pairs of f and g. The following lemma gives details under the assumption that $X \times_Z Y = 0$.

LEMMA 1.7. *Let* $X \xrightarrow{f} Z$ *and* $Y \xrightarrow{g} Z$ *be continuous maps of locales. If* $X \times_Z Y = 0$, *then the kernel pair of* $\langle f, g\rangle_m$ *is given by*:

$$(X \times_Z X) +_\bullet (Y \times_Z Y) \underset{\pi_2^f +_\bullet \pi_2^g}{\overset{\pi_1^f +_\bullet \pi_1^g}{\rightrightarrows}} X +_\bullet Y \xrightarrow{\langle f,g\rangle_m} Z,$$

where

$$X \times_Z X \underset{\pi_2^f}{\overset{\pi_1^f}{\rightrightarrows}} X \xrightarrow{f} Z \quad \text{and} \quad Y \times_Z Y \underset{\pi_2^g}{\overset{\pi_1^g}{\rightrightarrows}} Y \xrightarrow{g} Z$$

are the kernel pairs of f *and* g *respectively. If*

$$X \times_Z X \underset{\pi_2^f}{\overset{\pi_1^f}{\rightrightarrows}} X \xrightarrow{\phi} A \quad \text{and} \quad Y \times_Z Y \underset{\pi_2^g}{\overset{\pi_1^g}{\rightrightarrows}} Y \xrightarrow{\psi} B$$

are the coequalizers of Ker f, *respectively* Ker g, *then*

$$(X \times_Z X) +_\bullet (Y \times_Z Y) \underset{\pi_2^f +_\bullet \pi_2^g}{\overset{\pi_1^f +_\bullet \pi_1^g}{\rightrightarrows}} X +_\bullet Y \xrightarrow{\phi +_\bullet \psi} A +_\bullet B$$

is also a coequalizer.

Proof. Applying 1.6 twice we get the following pullback diagram, where α and β are the maps defined in (†):

$$\begin{array}{ccc} X \times_Z X +_\bullet Y \times_Z Y & \xrightarrow{\pi_1^f +_\bullet \pi_1^g} & X +_\bullet Y \\ {\scriptstyle \langle \alpha\pi_2^f, \beta\pi_2^g\rangle_\bullet} \downarrow & & \downarrow {\scriptstyle \langle f,g\rangle_\bullet} \\ X +_\bullet Y & \xrightarrow[\langle f,g\rangle_\bullet]{} & Z. \end{array}$$

Here $X \times_Z X$ is glued onto $Y \times_Z Y$ along $\pi_2^{g*}\beta^*\alpha_*\pi_{2*}^f$. Since $\beta^*\alpha_*(x) = \beta^*(x, g^*f_*(x)) = g^*f_*(x)$, we get $\pi_2^{g*}\beta^*\alpha_*\pi_{2*}^f = \pi_2^{g*}g^*f_*\pi_{2*}^f = \pi_\varepsilon^{g*}g^*f_*\pi_{\varepsilon'*}^f$ for $\varepsilon, \varepsilon' \in \{1, 2\}$. So it is not really necessary to keep track of the locale over which the pullback is obtained by glueing along a least fringe map. Since $\langle \alpha\pi_2^f, \beta\pi_2^g\rangle_\bullet = \pi_2^f +_\bullet \pi_2^g$ we get the first half of the lemma. As for the second half, since $(\pi_1^f +_\bullet \pi_1^g)^*(X) = X \times_Z X = (\pi_2^f +_\bullet \pi_2^g)^*(X)$, there exists an open sublocale A of the coequalizer which gets pulled back to X. Since coequalizers are stable under pullback along open inclusions, A is necessarily the coequalizer of the kernel pair of f; the same argument shows that the complement of A is the coequalizer of the kernel pair of g. So it remains to determine the fringe map $T(A) \xrightarrow{m} T(B)$. There exist (unique) maps $A \xrightarrow{a} Z$ and $B \xrightarrow{b} Z$ such that $f = a\phi$ and $g = b\psi$. The unique factorization of $\langle f, g\rangle_{\bullet}$ through $X +_{\bullet} Y \to A +_m B$ gives a map which necessarily restricts to a on A and to b on B. Universality of the coequalizer now implies that $m = b^* a_*$. □

2. Simple Coverings

To get a first approximation of how various classes of quotient maps in **Loc** look like we restrict our attention in this section to a class of "simple" surjections: surjections whose corresponding frame monomorphism is a singly generated frame extension. These were introduced by B. Banaschewski in [1] in order to analyze the gap between Hausdorff and strongly Hausdorff spaces, respectively strongly Hausdorff and regular spaces. For lack of a better name ("localic surjections which correspond to singly generated frame extensions" just doesn't sound right) we call these surjections simple coverings. Among these maps we characterize several classes of quotient maps, for instance, proper surjections, open surjections, regular epimorphisms and stable regular epimorphisms. We also give an example of a simple covering between Hausdorff spaces which is a regular epimorphism in **Loc**, but not in **Top**. This shows that the embedding of sober T_D spaces into **Loc** does not reflect regularity of epimorphisms.

DEFINITION 2.1. A surjection $B \xrightarrow{f} A$ of locales is called a simple covering if and only if there exists an embedding $B \xrightarrow{\iota} A \times \mathbb{S}$ ($\mathbb{S}$ the Sierpinski space) such that $f = \pi_A \iota$; equivalently, $B \xrightarrow{f} A$ is a simple covering if and only if the corresponding inclusion of frames is a singly generated frame extension.

The reason why the restriction of a class $\mathcal{C}$ of quotient maps to simple coverings can serve as a first approximation of $\mathcal{C}$, lies in the fact that many such classes satisfy the following property (modulo some assumption on g):

$$\text{if } f \in \mathcal{C} \text{ and } f \text{ factors as } Z \xrightarrow{f} X = Z \xrightarrow{g} Y \xrightarrow{h} X, \text{ then } h \in \mathcal{C}. \qquad (*)$$

The simplest examples of such maps h are given by simple coverings. If g is a surjection, then (∗) holds for each of the following classes: the class **Prop** of proper surjections [15], the class **Op** of open surjections [8], the class **Cl** of closed surjections [2] and the class **RegE** of regular epimorphisms. To see that (∗) holds for **RegE** let $Z \xrightarrow{f} X$ be a regular epimorphism and U any open sublocale of Y equipped with descent data relative to h. Then $g^{-1}(U)$ is equipped with descent data relative to f which implies that there exists an open sublocale V of X such that

$f^{-1}(V) = g^{-1}(U)$. Since $g^{-1}h^{-1}(V) = g^{-1}(U)$ 1.3 implies that $h^{-1}(V) = U$. An example which shows that $(*)$ fails for **RegE** without any assumption on g is given by the factorization of the regular epimorphism $X \xrightarrow{f} \mathbb{Q}$ of 3.4 as $X \to pl(\mathbb{Q}) +_\bullet X \xrightarrow{\langle\iota,f\rangle_\bullet} \mathbb{Q}$, see also 3.7. For the class **Triq** of triquotient maps [13], the class **Ret** of retractions, the class **WQ** of weak quotient maps (see the last section for the definition) and the class **ExtE** of extremal epimorphisms (we will show later on that **RegE** is properly contained in **ExtE**) $(*)$ holds without any assumption on g. For $(*)$ to hold for the class **SRegE** of stable regular epimorphisms one has to assume that g is a stable surjection.

In order to be able to express properties of simple coverings as painlessly as possible we will introduce some further notation. First, unless the choice of factorization through $A \times \mathbb{S}$ matters, we will assume that a fixed factorization $B \to A \times \mathbb{S} \xrightarrow{\pi_A} A$ has been chosen. Pulling back the open sublocale $A \times \{0\}$ (we assume that $\mathbb{S} = \{0, 1\}$ and that $\{0\}$ is the only non-trivial open subspace) along the chosen embedding $B \to A \times \mathbb{S}$ decomposes B into an open sublocale U and its closed complement F. (Of course, U and F depend on the chosen embedding $B \to A \times \mathbb{S}$.) We denote the images of U and F under f by A_o and A_c, their meet $A_o \wedge A_c$ by A_m, and $cl(A_m)$ by $A_{\bar{m}}$. The sublocales A_o and A_c completely determine the simple covering $B \xrightarrow{f} A$ and a factorization through $A \times \mathbb{S}$. Every element of $T(B)$ can be written as $f^*(a) \vee (f^*(a') \wedge b_U)$ for $a, a' \in T(A)$, where we may also assume that $a \leq a'$. In Banaschewski's notation we have that A_o is the sublocale of A generated by the congruence relation Φ defined as follows: $a\Phi a' \Leftrightarrow f^*(a) \wedge b_U = f^*(a') \wedge b_U$. Similarly, A_c is generated by the congruence Θ defined by $a\Theta a' \Leftrightarrow f^*(a) \vee b_U = f^*(a') \vee b_U$. Joins and finite meets of open sublocales are computed pointwise, i.e. for any set I, any finite set F, and $a_i \leq a'_i$ we have:

$$\bigvee_{i\in I}(f^*(a_i) \vee (f^*(a'_i) \wedge b_U)) = \left(\bigvee_{i\in I} f^*(a_i)\right) \vee \left(\bigvee_{i\in I} f^*(a'_i) \wedge b_U\right);$$

$$\bigwedge_{i\in F}(f^*(a_i) \vee (f^*(a'_i) \wedge b_U)) = \left(\bigwedge_{i\in F} f^*(a_i)\right) \vee \left(\bigwedge_{i\in F} f^*(a'_i) \wedge b_U\right).$$

Note that choosing a generator b_U for the singly generated frame extension and choosing a factorization through $A \times \mathbb{S}$ amount to the same thing.

Artin glueing yields another description of simple coverings. A map $X +_\bullet Y \xrightarrow{\langle f,g\rangle_\bullet} Z$ is a simple covering if and only if f and g are sublocale inclusions with join Z, and each simple covering arises in this way. In this case $X \xrightarrow{f} Z$ is the inclusion of A_o into A and g the inclusion of A_c.

To conclude this first discussion we look at the kernel pair $B \times_A B \underset{\pi_2}{\overset{\pi_1}{\rightrightarrows}} B \xrightarrow{f} A$ of a simple covering $B \xrightarrow{f} A$. The locale $B \times_A B$ is the union of four pairwise disjoint locales: open $U \times_A U \simeq A_o$, locally closed $U \times_A F \simeq A_m$ and $F \times_A U \simeq A_m$, and closed $F \times_A F \simeq A_c$. If we identify these sublocales of $B \times_A B$ with the corresponding sublocales of A and if $\iota_o, \iota_c, \iota_m$ are the respective embeddings of A_o, A_c, A_m into A, then π_1^* and π_2^* are given by:

$$\pi_1^*(f^*(a) \vee (f^*(a') \wedge b_U)) = (\iota_o^*(a'), \iota_m^*(a'), \iota_m^*(a), \iota_c^*(a));$$
$$\pi_2^*(f^*(a) \vee (f^*(a') \wedge b_U)) = (\iota_o^*(a'), \iota_m^*(a), \iota_m^*(a'), \iota_c^*(a)).$$

Similarly, if S is any sublocale of B, then

$$\pi_1^{-1}(S) = (\iota_o^{-1}(f(S \wedge U)), \iota_m^{-1}(f(S \wedge U)), \iota_m^{-1}(f(S \wedge F)), \iota_c^{-1}(f(S \wedge F)));$$
$$\pi_2^{-1}(S) = (\iota_o^{-1}(f(S \wedge U)), \iota_m^{-1}(f(S \wedge F)), \iota_m^{-1}(f(S \wedge U)), \iota_c^{-1}(f(S \wedge F))).$$

So a sublocale S is equipped with descent data if and only if $A_m \wedge f(S \wedge U) = A_m \wedge f(S \wedge F)$.

PROPOSITION 2.2. *Simple coverings are pullback stable epimorphisms. The pullback $B \times_A X \xrightarrow{g} X$ of a simple covering $B \xrightarrow{f} A$ along any map $X \xrightarrow{\phi} A$ is again a simple covering. For any factorization of f through $A \times \mathbb{S}$ there exists a factorization of g through $X \times \mathbb{S}$ for which $X_o = g^{-1}(A_o)$, $X_c = g^{-1}(A_c)$ and $X_m = g^{-1}(A_m)$.*

Proof. This follows by considering the following diagram (and its counterpart obtained by replacing "0" by "1", "A_o" by "A_c" and "X_o" by "X_c") all squares of which are pullbacks, and then observing that pulling back along any map preserves finite joins and (arbitrary) meets of sublocales.

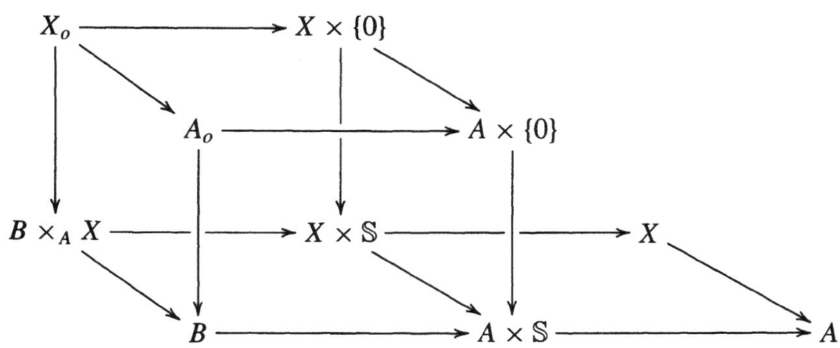

□

One could also use the description of simple coverings as maps $A_o +_\bullet A_c \xrightarrow{\langle \iota_o, \iota_c \rangle_\bullet} A$ and 1.6 to prove the preceding proposition.

Recall that a map $B \xrightarrow{f} A$ is called closed if the image $f(F)$ of each closed sublocale F of B is closed, and similarly that a map $B \xrightarrow{f} A$ is called open if the image $f(U)$ of each open sublocale U of B is open. The next propositions show that a number of properties of simple covering maps $B \xrightarrow{f} A$, such as being open or closed, can be inferred directly from the sublocales A_o, A_c and A_m (irrespective of the chosen factorization through $A \times \mathbb{S}$).

PROPOSITION 2.3. *If* $B \xrightarrow{f} A$ *is a simple covering then*

(1) f *is monic if and only if* $A_m = 0$;
(2) f *is a closed surjection if and only if* A_c *is closed in* A;
(3) f *is an open surjection if and only if* A_o *is open in* A;
(4) f *is a retraction if and only if there exists a closed sublocale* H *of* A *such that* $H \leq A_c$ *and* $A - H \leq A_o$.

Proof. (1) has been shown in [1], but I think that it is easier to consider the kernel pair of f as described above, and observe that the projections are isomorphisms (which is equivalent to f being monic) precisely if both $U \times_A F = 0$ and $F \times_A U = 0$, i.e. if and only if $A_m = 0$. (2) and (3) have dual proofs, so it suffices to do one of them. That A_o has to be an open sublocale of A for f to be open is clearly necessary. It is also sufficient because each open sublocale of B is of the form $f^{-1}(V) \vee (f^{-1}(W) \wedge U)$ for open sublocales V and W of A, and by 1.2 we have $f(f^{-1}(V) \vee (f^{-1}(W) \wedge U)) = V \vee (W \wedge A_o)$ which is open if A_o is. To see that (4) holds, consider any section $A \xrightarrow{s} B$ of f. Then $H = s^{-1}(F)$ is a closed sublocale which satisfies $H \leq A_c$ and $A - H \leq A_o$. Conversely, if there exists closed $H \leq A_c$ such that $A - H \leq A_o$, then the inclusion of $(f^{-1}(H) \wedge F) \vee (f^{-1}(A - H) \wedge U)$ into B is a section for f. □

Recall that a map $B \xrightarrow{f} A$ is proper if and only if it is compact as an internal locale in the topos $\mathbf{Sh}(A)$. Two equivalent conditions are that f is stably closed, or that f is closed and f_* preserves directed suprema [15]. Triquotient maps are a generalization of open and proper surjections. A map $B \xrightarrow{f} A$ is called a triquotient map if there exists a map $T(B) \xrightarrow{f_\#} T(A)$, called a t-assignment, which preserves directed suprema and satisfies the two Frobenius identities $f_\#(f^*(a) \star b) = a \star f_\#(b)$ for $\star \in \{\vee, \wedge\}$. (These conditions imply that $f_\# \circ f^* = \mathrm{id}$ [13].) For simple coverings we have the following equivalences:

PROPOSITION 2.4. *If* $B \xrightarrow{f} A$ *is a simple covering, then*

(5) f *is a closed surjection if and only if* f *is proper*;
(6) f *is a retraction if and only if* f *is a triquotient map.*

Proof. (5) By (2) f is a closed surjection if and only if A_c is closed in A. If $X \xrightarrow{g} A$ is any map then the pullback of f along g is by 2.2 again a simple covering and $X_c = g^{-1}(A_c)$. So X_c is closed in X and hence so is the map $X \times_A B \to X$.

(6) We show that any t-assignment is necessarily a frame homomorphism, i.e. $f_\# = s^*$ for some map $A \xrightarrow{s} B$ of locales. Then $(fs)^* = s^* f^* = f_\# f^* = \mathrm{id}$ implies that s is a section for f. Let $f^*(a) \vee (f^*(a') \wedge b_U) \in T(B)$. The Frobenius

identities imply that $f_\#(f^*(a) \vee (f^*(a') \wedge b_U)) = a \vee f_\#(f^*(a') \wedge b_U) = a \vee (a' \wedge f_\#(b_U))$. So $f_\#$ is completely determined by $f_\#(b_U)$. Now

$$\begin{aligned}
\bigvee f_\#(f^*(a_i) \vee (f^*(a_i') \wedge b_U)) &= \bigvee (a_i \vee (a_i' \wedge f_\#(b_U))) \\
&= \Big(\bigvee a_i\Big) \vee \Big(\Big(\bigvee a_i'\Big) \wedge f_\#(b_U)\Big) \\
&= f_\#\Big(f^*\Big(\bigvee a_i\Big) \vee \Big(f^*\Big(\bigvee a_i'\Big) \wedge b_U\Big)\Big) \\
&= f_\#\Big(\bigvee (f^*(a_i) \vee (f^*(a_i') \wedge b_U))\Big)
\end{aligned}$$

for any set $\{f^*(a_i) \vee (f^*(a_i') \wedge b_U) \mid i \in I\}$, so $f_\#$ preserves arbitrary suprema. A similar calculation shows that $f_\#$ preserves finite meets. □

PROPOSITION 2.5. *If $B \xrightarrow{f} A$ is a simple covering, then*

(7) *$T(B)$ is a sublattice of $S(A)$ containing $T(A)$ if and only if $A_c = sup(A_o)$;*
(8) *$T(B)$ is a subframe of $S(A)^{\mathrm{op}}$ containing $T(A)$ (as the opposite of the lattice of all closed sublocales of A) if and only if $A_o = sup(A_c)$;*

and f^* is the respective embedding of $T(A)$ into $T(B)$.

Proof. (7) Let L be the sublattice of $S(A)$ generated by all open sublocales of A and an arbitrary sublocale S. All elements of L can be written in the form $U \vee (V \wedge S)$ with U and V open sublocales of A and $U \leq V$. The lattice L has arbitrary joins and they are computed as in $S(A)$. (Writing elements of L in the form given above, and using the fact that any sublocale distributes over arbitrary joins of open sublocales, it is easy to check that $\bigvee(U_i \vee (V_i \wedge S)) = (\bigvee U_i) \vee ((\bigvee V_i) \wedge S)$.) Distributivity of meets over arbitrary joins is also straightforward. Hence L is a singly generated frame extension of $T(A)$ which is generated by S. Now A_c is the sublocale generated by the congruence $U \Theta V \Leftrightarrow U \vee S = V \vee S$, equivalently, by 1.1(4), if $U \wedge (A \backslash S) = V \wedge (A \backslash S)$, hence $A_c = A \backslash S$. Conversely, for any simple covering $B \xrightarrow{f} A$ which satisfies $A_c = X \backslash A_o$ (for some factorization of f through $A \times \mathbb{S}$), a straightforward computation shows that $T(B)$ is isomorphic to the sublattice L of $S(A)$ generated by A_o and $T(A)$. The proof of (8) is a similar; it is also given in [1, Prop. 2.4]. □

To characterize those simple coverings which are regular epimorphisms is a bit more difficult than the corresponding tasks for proper or open surjections, but see also 4.5.

LEMMA 2.6. *A simple covering $B \xrightarrow{f} A$ is a regular epimorphism if and only if it satisfies the following condition*:

(RE) $\quad \forall H \in S(A)$: *H is closed $\Rightarrow H_o - H_{\bar{m}}$ is open in H.*

Proof. Let $B \xrightarrow{f} A$ be a regular epimorphism and H a closed sublocale of A. We want to show that $H_o - H_{\bar{m}}$ is open in H. Since regular epimorphisms are stable under pullback along closed inclusions we may assume that $H = A$. Now $f^{-1}(A_o - A_{\bar{m}})$ is open and has equal pullbacks along both projections because

$$\begin{aligned} A_m \wedge f(U \wedge f^{-1}(A_o - A_{\bar{m}})) &\leq A_m \wedge f(U) \wedge f(f^{-1}(A_o - A_{\bar{m}})) \\ &\leq A_m \wedge A_o \wedge (A_o - A_{\bar{m}}) = 0, \end{aligned}$$

and

$$\begin{aligned} A_m \wedge f(F \wedge f^{-1}(A_o - A_{\bar{m}})) &\leq A_m \wedge f(F) \wedge f(f^{-1}(A_o - A_{\bar{m}})) \\ &\leq A_m \wedge A_c \wedge (A_o - A_{\bar{m}}) = 0. \end{aligned}$$

So there exists an open sublocale V of A such that $f^{-1}(V) = f^{-1}(A_o - A_{\bar{m}})$. Because $A_o - A_{\bar{m}}$ is complemented (its complement is $A_c \vee A_{\bar{m}}$) 1.3 implies that $V = A_o - A_{\bar{m}}$, and hence that $A_o - A_{\bar{m}}$ is open.

Conversely, assume that f is a simple covering satisfying (RE), and let $f^{-1}(V) \vee (f^{-1}(W) \wedge U)$ be an open sublocale of B (V and W open in A) whose pullbacks along π_1 and π_2 are equal. This means that

$$\begin{aligned} &A_m \wedge f((f^{-1}(V) \vee (f^{-1}(W) \wedge U)) \wedge U) \\ &\quad = A_m \wedge f(f^{-1}(W) \wedge U) \\ &\quad = A_m \wedge W \wedge f(U) = A_m \wedge W \end{aligned}$$

and

$$\begin{aligned} &A_m \wedge f((f^{-1}(V) \vee (f^{-1}(W) \wedge U)) \wedge F) \\ &\quad = A_m \wedge f(f^{-1}(V) \wedge F) \\ &\quad = A_m \wedge V \wedge f(F) = A_m \wedge V \end{aligned}$$

are equal. For $H = A - V$ (RE) implies that $H_o - H_{\bar{m}}$ is open in H, and hence that both $V \vee (H_o - H_{\bar{m}})$ and $V \vee (W \wedge (H_o - H_{\bar{m}}))$ are open in A. We will show that $f^{-1}(V \vee (W \wedge (H_o - H_{\bar{m}}))) = f^{-1}(V) \vee (f^{-1}(W) \wedge U)$. To show that LHS $\leq$ RHS it suffices to show that $f^{-1}(W) \wedge f^{-1}(H_o - H_{\bar{m}}) \leq f^{-1}(W) \wedge U$ which follows because $f^{-1}(H_o - H_{\bar{m}}) \wedge f^{-1}(A_c) = f^{-1}((H_o \wedge A_c) - H_{\bar{m}}) = f^{-1}(H_m - H_{\bar{m}}) = 0$ and hence $f^{-1}(H_o - H_{\bar{m}}) \leq U$.

As for the reverse inequality, we have $V \vee (W \wedge (H_o - H_{\bar{m}})) = V \vee (W \wedge H_o) = V \vee (W \wedge A_o)$. The first equality holds because $A_m \wedge W = A_m \wedge V$ implies that $H_m \wedge W = H_m \wedge V = 0$, and hence that $H_{\bar{m}} \wedge W = 0$; for the second it suffices to observe that $H_o \vee V \geq A_o$. So $f^{-1}(V \vee (W \wedge (H_o - H_{\bar{m}}))) = f^{-1}(V \vee (W \wedge A_o)) = f^{-1}(V) \vee (f^{-1}(W) \wedge f^{-1}(A_o)) \geq f^{-1}(V) \vee (f^{-1}(W) \wedge U)$. □

COROLLARY 2.7. *A simple covering $B \xrightarrow{f} A$ is a regular epimorphism stable under pullback along inclusions (semi-stable regular epimorphism for short) if and only if the following condition is satisfied:*

(SSRE) $\forall S \in S(A)$: *$S_o - S_{\bar{m}}$ is open in S.*

Proof. This follows directly from the preceding lemma and the description of pullbacks of simple coverings. □

PROPOSITION 2.8. *A simple covering $B \xrightarrow{f} A$ is a stable regular epimorphism if and only if it is a retraction.*

Proof. By (4) it suffices to show that $A - int(A_o) \leq A_c$. Pulling back along the closed sublocale $A - int_A(A_o)$ if necessary, we may assume that $int_A(A_o) = 0$. Let $X = \coprod\{S \leq A_c \mid S \wedge A_o = 0\}$ and let Y be any locally closed sublocale of A disjoint from A_c. Let $X \xrightarrow{\alpha} A$ be the map induced by the various inclusions $S \to A$ and $Y \xrightarrow{\beta} A$ be the inclusion of Y. Observe that α and β factor uniquely through B, say by γ and δ, because β and each of the inclusions $S \to A$ used to construct α have images disjoint from A_m. Consider the map $X +_\bullet Y \xrightarrow{\langle\alpha,\beta\rangle_\bullet} A$. The following diagram is a pullback because $(X +_\bullet Y)_o = (\langle\alpha, \beta\rangle_\bullet)^{-1}(A_o) = Y$ (or by 1.6).

$$\begin{array}{ccc} X + Y & \xrightarrow{g} & X +_\bullet Y \\ {\scriptstyle\langle\gamma,\delta\rangle}\downarrow & & \downarrow{\scriptstyle\langle\alpha,\beta\rangle_\bullet} \\ B & \xrightarrow[f]{} & A \end{array}$$

Since f is a pullback stable regular epimorphism the splitting $X + Y \xrightarrow{g} X +_\bullet Y$ has to be an isomorphism, hence X is a clopen sublocale of $X +_\bullet Y$. At the same time X is dense in $X +_\bullet Y$ because $\beta^*\alpha_*(0) = 0$, which follows from the fact that $\bigvee\{S \leq A_c \mid S \wedge A_o = 0\} = sup(A_o)$, as the supplement of a sublocale with empty interior, is dense in A. Hence $Y = 0$. Because $A_c = \bigwedge\{A - Y \mid Y \wedge A_c = 0 \text{ and } Y \text{ locally closed in } A\}$ this implies that $A_c = A$. □

The preceding results seem to imply that simple coverings are not useful as far as distinguishing different classes of surjections is concerned: stable regular epimorphisms are just retractions, closed maps are proper, etc. .... But the following example shows that they are, nevertheless, useful in this respect. It shows that the embedding of the category of sober T_D spaces as spatial locales into **Loc** does not reflect regular epimorphisms. (For arbitrary sober spaces not even epimorphisms are reflected.)

Pultr and Tozzi show in [14, 5.2] that the restriction of this embedding to metrizable spaces reflects regular epimorphisms. More precisely, they first introduce the property (AP) of approximation of points. A space X is said to satisfy (AP) if for all subspaces M of X and all points $x \in M\backslash int_X(M)$ there exists $C \subseteq X$ such that $C \cap M = \emptyset$ and $cl_X(C) \cap M = \{x\}$. Then they show that for any continuous map $Y \xrightarrow{f} X$ of spaces for which the codomain is a T_D space which satisfies (AP) the following are equivalent: (i) f is a quotient map in **Top**, and (ii) f^* is the inclusion of an equationally closed subframe (see Section 4 for the definition).

This implies that the inclusion above restricted to T_D spaces which satisfy (AP) does reflect regular epimorphisms. They also show that maps of the same type as the map constructed in the proposition below satisfy (ii) but not (i) [14, 4.4].

PROPOSITION 2.9. *There exists a continuous surjection $Y \to X$ of Hausdorff spaces which is not a quotient map, but which is a semi-stable regular epimorphism and a stable epimorphism in* **Loc**.

Proof. Recall that a Bernstein set B is a subspace of a completely metrizable separable space X which meets every Cantor set in X but contains none. If B is a Bernstein set then so is $X\backslash B$. Choose any complemented pair (B_1, B_2) of Bernstein sets in the unit interval $\mathbb{I}$ equipped with the usual topology. Let X be the space with the same underlying set as $\mathbb{I}$, but whose topology is generated by the topology of $\mathbb{I}$ and all sets $X\backslash A$, where A is a scattered (hence countable) subspace of B_1. (Recall that a space is scattered if each nonempty (closed) subspace contains an isolated point, and that scattered spaces are necessarily sober.) So all open sets in X are of the form $U - A$ where U is open in $\mathbb{I}$ and A is a scattered subspace of B_1. Note that X is Hausdorff (Hausdorffness is preserved under refinement of the topology) but not regular: if $x \in B_2$ and $\{x_n\}_{n\in N} = F \subseteq B_1$ is a sequence which converges (in $\mathbb{I}$) to x then F is closed in X but any pair of open neighborhoods (in X) of $\{x\}$ and F meets in a set of cardinality $\mathfrak{c}$ because any such pair of open neighborhoods in $\mathbb{I}$ does and open neighborhoods in X are open in $\mathbb{I}$ modulo a countable set. Note also that X does not satisfy (AP) because for $M = B_2, int(M) = \emptyset$ and any closed subspace of X is either disjoint from M or meets M in a set of cardinality $\mathfrak{c}$.

Any subset $S \subseteq [0, 1]$ is scattered as a subspace of X if and only if it is scattered as a subspace of $\mathbb{I}$. Assume that there were a set which is scattered considered as subspace S of X but not when considered as a subspace S' of $\mathbb{I}$. Restricting to its dense-in-itself part in $\mathbb{I}$ we may assume that S' is dense in itself. For S to be scattered there has to exist an open subspace $U - A$ (U and A as above) of X which meets S in a single point x. Since A is scattered in $\mathbb{I}$, so is $A \cup \{x\}$ and hence $U \wedge S'$ is a nonempty, open and scattered subspace of S', contradicting the fact that S' was dense in itself.

The natural map $X \to \mathbb{I}$ induces an isomorphism $pl(X) \to pl(\mathbb{I})$ between the respective pointless parts. This follows because X is the spatial part of the pullback $X' \to \mathbb{I}$ of all maps $\mathbb{I}_A \to \mathbb{I}$, where A is a scattered subspace of B_1 and where $\mathbb{I}_A$ is the simple covering given by $\mathbb{I}_c = A$ and $\mathbb{I}_o = \mathbb{I} - A$. Both B_1 and $pl(\mathbb{I})$ are constant under pullback along any of these maps $\mathbb{I}_A \to \mathbb{I}$. Since pullbacks preserve limits, we get that both B_1 and $pl(\mathbb{I})$ are constant under pullback along the map $X' \to \mathbb{I}$. Since the pointless parts are the intersection of all point complements and pullbacks preserve intersections we also have that the pullback of $pl(\mathbb{I})$ is just $pl(X')$. Hence $pl(X') = pl(\mathbb{I})$, but since $pl(X') \leq B_1 \leq X$ we also get $X = X'$.

Now let $Y \xrightarrow{f} X$ be the simple covering determined by the spaces $X_o = B_2$ and $X_c = B_1$. Then $X_m = pl(X)$. As a map of topological spaces f is not quotient because $f^{-1}(B_2)$ is open but B_2 is not. Since f is a simple covering it is a stable epimorphism. To show that f is a semi-stable regular epimorphism in **Loc** we use 2.7. So let S be any sublocale of X. Then $S_m = S \wedge pl(X) = pl(S)$. Hence $S_o - S_{\bar{m}} = S_o - cl_S(pl(S)) = S_o \wedge (S - cl_S(pl(S)))$ is a scattered subspace of S. Since S_c and S_o have no points in common and since $S - cl_S(pl(S))$ is scattered, $(S_c - cl_S(pl(S))) \wedge (S_o - cl_S(pl(S))) = 0$ and $S - cl_S(pl(S)) = (S_c - cl_S(pl(S))) \vee (S_o - cl_S(pl(S)))$. Because $S_c - cl_S(pl(S))$ is a scattered subspace of B_1 it is closed in X, hence also in S, and therefore $S_o - cl_S(pl(S))$ is open in $S - cl_S(pl(S))$, and hence in S. So (SSRE) is satisfied by all sublocales of X. □

Note that 2.8 implies that the simple covering constructed in 2.9 is not a stable regular epimorphism.

3. Closed Surjections

In this section we consider closed surjections of locales. In **Top** closed surjections are quotient maps, hence regular epimorphisms. So the question of whether closed surjections are also regular epimorphisms in **Loc** is a natural one. In this section we show that this is true at least for all closed surjections with subfit domain. However, we also give an example of a closed surjection which is the composite of two regular epimorphisms, but not itself a regular epimorphism.

Before proving the fact that binary products of closed surjections with subfit domain in **Loc**$/X$ have again global support, a fact which is needed to prove that closed surjections with subfit domain are regular epimorphisms, we give an example which shows that this is not true for arbitrary surjections.

EXAMPLE 3.1. Two surjections $Y \xrightarrow{f} X$ and $Z \xrightarrow{g} X$ of spatial (respectively, metrizable) locales for which $Y \times_X Z = 0$ (but $X \neq 0$).

Let X be the partially ordered set of finite sequences of natural numbers ordered by

$$(i_1, \ldots, i_n) \leq (j_1, \ldots, j_m) \Leftrightarrow n \geq m \text{ and } i_k = j_k \text{ for } k = 1, \ldots, m.$$

Equip X with the weak topology in this order, i.e. the topology generated by all sets $X \backslash \downarrow ((i_1, \ldots, i_n))$. Every closed subspace of X is the intersection of families consisting of finite unions of principal downward closed sets; irreducible closed sets are necessarily the intersection of singleton families, and nonempty irreducible closed sets are therefore principal downward closed sets in the partial order. So X is sober. The spaces Y and Z are discrete and consist of all finite sequences of natural numbers of even, respectively odd, length. f and g are the obvious set inclusions. Continuity is clear, as is the fact that f and g have empty product. f and g are surjections in **Loc** because Z contains no nonzero complemented sublocale disjoint from the set theoretic image of either f or g. (Any complemented sublocale of Z contains with every set $\{(i_1, \ldots, i_n, k) \mid k \in \mathbb{N}\}$ also the point $(i_1, \ldots, i_n)$.)

One can build a similar example replacing the points of X by Boolean sublocales of the rationals. Each sequence $(i_1, \ldots, i_n)$ will be replaced by a Boolean sublocale $D(F_{i_1,\ldots,i_n})$ where $\{F_{i_1,\ldots,i_n,k} \mid k \in \mathbb{N}\}$ is a sequence of closed nowhere dense sublocales of $F_{i_1,\ldots,i_n} (F = \mathbb{Q})$ whose union is dense in $F_{i_1,\ldots,i_n}$. The locale Z is the union in $\mathbb{Q}$ of all $D(F_{i_1,\ldots,i_n})$, while X and Y are the coproducts of all $D(F_{i_1,\ldots,i_n})$ for which the indexing sequence is of even, respectively odd, length. The remainder of the construction works as before.

Recall that a locale is said to be subfit [4] if each open sublocale is the join of closed sublocales.

LEMMA 3.2. *If* $Y \xrightarrow{f} X$ *and* $Z \xrightarrow{g} X$ *are closed surjections with subfit domains, then the induced map* $Y \times_X Z \to X$ *is again a surjection.*

Proof. First note that it is actually sufficient to prove that $Y \times_X Z = 0$ implies $X = 0$ because (i) closed surjections are stable under pullback along complemented sublocales (1.4), (ii) subfitness is inherited by complemented sublocales, and (iii) the image of $Y \times_X Z$ in X is the intersection of those complemented sublocales C of X for which the restrictions of f and g to $X - C$ have zero pullback.

The topology of the pullback $Y \times_X Z$ can be presented as follows:

$$T(Y) \otimes_{T(X)} T(Z) = (T(Y) \times T(Z) \mid Cov),$$

where

$$\begin{aligned} &Cov(y, z) \\ &\quad = \left\{\{(y_i, z) \mid i \in I\} \mid \bigvee y_i = y\right\} \cup \left\{\{(y, z_i) \mid i \in I\} \mid \bigvee z_i = z\right\} \\ &\quad\quad \cup \left\{\{((y \wedge f^*(x)), z)\} \mid g^*(x) \geq z\right\} \cup \left\{\{(y, (g^*(x) \wedge z))\} \mid f^*(x) \geq y\right\}. \end{aligned}$$

It is straightforward to check that these covers are stable. Now let $\mathcal{S}$ be the sieve consisting of those pairs (y, z) which satisfy the following condition:

$$\forall \text{ closed sublocales } F \leq Y, G \leq Z : F \leq U_y \text{ and } G \leq U_z \Rightarrow f(F) \wedge g(G) = 0.$$

I claim that the sieve $\mathcal{S}$ is closed. To show that it is closed under the covering relations of the first type let $\{(y_i, z) \mid i \in I\} \subseteq \mathcal{S}$ be arbitrary, and let $F \leq U_{\bigvee y_i}$ and $G \leq U_z$ be a closed sublocales of Y, respectively Z. If $U_{y_i} = \bigvee F_{i,j}$ where each $F_{i,j}$ is closed, then

$$\begin{aligned} f(F) \wedge g(G) &= f\Big(F \wedge \bigvee_{i,j} F_{i,j}\Big) \wedge g(G) = f\Big(\bigvee_{i,j} F \wedge F_{i,j}\Big) \wedge g(G) \\ &= \bigvee_{i,j} f(F \wedge F_{i,j}) \wedge g(G) \leq \Big(\bigvee_{i,j} f(F_{i,j})\Big) \wedge g(G) \\ &= \bigvee_{i,j} (f(F_{i,j}) \wedge g(G)) = 0, \end{aligned}$$

because each $F_{i,j} \leq U_{y_i}$ and $G \leq U_z$. Closure of $\mathcal{S}$ under covers of the second type follows by symmetry. For covers of the third (and by symmetry of the fourth) type let $((y \wedge f^*(x)), z) \in \mathcal{S}$ and assume that $g^*(x) \geq z$. Let $F \leq U_y$ and $G \leq U_z$ be arbitrary closed sublocales. We want to show that $f(F) \wedge g(G) = 0$. Because X is subfit (subfitness is preserved under closed surjections), there exists a set $\{H_i \mid i \in I\}$ of closed sublocales of X such that $U_x = \bigvee_{i \in I} H_i$. Now $f(F) \wedge H_i \wedge g(G) = f(F \wedge f^{-1}(H_i)) \wedge g(G) = 0$, because $F \wedge f^{-1}(H_i) \leq U_{y \wedge f^*(x)}$ for all $i \in I$. Since $g^*(x) \geq z$ means that $g^{-1}(U_x) \geq U_z$ and hence $U_x \geq g(U_z) \geq g(G)$, we finally have $f(F) \wedge g(G) = f(F) \wedge g(G) \wedge U_x = f(F) \wedge g(G) \wedge \bigvee_{i \in I} H_i = \bigvee_{i \in I}(f(F) \wedge H_i \wedge f(G)) = 0$.

Now $\mathcal{S}$ is maximal if and only if $(1, 1) \in \mathcal{S}$, i.e. if $f(Y) \wedge f(Z) = 0$. But since f and g are surjective, the latter statement is true (if and) only if $X = 0$. Therefore we conclude that $Y \times_X Z = 0$ if and only if $X = 0$. □

THEOREM 3.3. *Closed surjections $X \xrightarrow{f} Y$ with subfit domain X are regular epimorphisms.*

Proof. We want to show that f is the coequalizer of its kernel pair (π_1, π_2). So let F be any closed sublocale of X such that $\pi_1^{-1}(F) \simeq \pi_2^{-1}(F)$. Restricting f to a map $f^{-1}f(F) \to f(F)$ if necessary (subfitness is inherited by closed subspaces), we may assume that $f(F) = Y$. We want to show that F is necessarily equal to X. This will follow if we can show that $F \wedge G = 0$ implies $G = 0$ for any closed sublocale G of X, because then subfitness of X implies that $X - F = \bigvee\{G \text{ closed in } X \mid F \wedge G = 0\} = 0$. So let G be any closed sublocale of X disjoint from F. We restrict the domain of f to $F \vee G$. The resulting map $g = f \restriction_{F \vee G}$ is still a closed surjection onto Y, and the domain of g is the disjoint union of F and G. Furthermore, both F and G have isomorphic pullbacks along π_1 and π_2 (the kernel pair of g) because $F \times_Y G \leq F \times_Y (X - F) = 0$. Finally, we restrict the codomain to $Y' = g(G)$ to arrive at a closed surjection $g^{-1}g(G) = X' \xrightarrow{h = g \restriction_{g^{-1}g(G)}} Y'$. Then both $F' = F \wedge X'$ and G are mapped by h onto Y' and because $F' \times_{Y'} G = 0$ both F' and G are equalized by the pair (π_1^{-1}, π_2^{-1}) (this time the kernel pair of h). But since the restrictions of h to F' and G are also closed surjections the preceding lemma implies that $F' \times_{Y'} G$ maps onto Y', hence we conclude that $Y' = 0$ and therefore also $G = 0$. □

Next we show that the restriction to closed surjections with subfit domain is not superfluous. To do so we first construct a closed surjection from a scattered space to the rationals. Using this map we then construct a closed surjection of locales which is not a regular epimorphism by Artin glueing. The first construction is a slightly simplified version (sufficient for our needs here) of a construction of E. van Douwen [3]. Using the axiom $[\mathfrak{b} = \mathfrak{c}]$ (a weak version of the continuum hypothesis) he constructs for an arbitrary first countable regular space Y of cardinality at most

$\mathfrak{c}$, a closed surjection with countably compact fibers from a scattered regular space X onto Y. Since we do not require regularity of X we can stay within ZFC.

PROPOSITION 3.4. *There exists a closed surjection $X \xrightarrow{f} \mathbb{Q}$ from a scattered space onto the rationals. Furthermore, for any pointless sublocale $S \in S(\mathbb{Q})$ we have $X \times_{\mathbb{Q}} S = 0$.*

Proof. If we are given any surjection from a scattered space X onto the rationals, then $X \times_{\mathbb{Q}} S = 0$ for any pointless sublocale S of $\mathbb{Q}$ because $X \times_{\mathbb{Q}} S$ has to be pointless and every nonzero sublocale of a scattered space contains a point. As the underlying set for X we take the product $\mathbb{Q} \times \mathfrak{c}$, and the map $X \xrightarrow{f} Y$ will simply be the projection $\pi = \pi_{\mathbb{Q}}$ onto $\mathbb{Q}$. The topology $\mathcal{T}_{\mathfrak{c}}$ on $\mathbb{Q} \times \mathfrak{c}$ will be defined by recursion on $\alpha \in \mathfrak{c}+1$. Let ϕ be any surjection $\mathfrak{c}\backslash\{0\} \to \mathcal{A}$ with fibers of cardinality $\mathfrak{c}$, where $\mathcal{A}$ is the set of all countable subsets $A = \{a_n \mid n \in \mathbb{N}\} \subseteq \mathbb{Q} \times \mathfrak{c}$ with the property that the sequence $\{\pi(a_n)\} = \{q_n\}$ converges, and its limit $q \notin \pi(A)$. Replacing all $A_\eta = \phi(\eta)$ for which $A_\eta \not\subseteq \mathbb{Q} \times \eta$ by $\{(1/n, 0) \mid n \in \mathbb{N}\}$ we may assume that $\phi(\eta) \subseteq \mathbb{Q} \times \eta$ for all η. For limit ordinals λ, the topology $\mathcal{T}_\lambda = \langle \bigcup_{\alpha<\lambda} \mathcal{T}_\alpha \rangle$. At successor ordinals $\alpha+1$ we construct $\mathcal{T}_{\alpha+1}$ as the topology generated by the base $\mathcal{T}_\alpha \cup \{\{q, \alpha\} \mid q \neq q_\alpha = \lim(q_{\alpha,n})\} \cup \mathcal{N}((q_\alpha, \alpha))$ where the neighborhood base $\mathcal{N}((q_\alpha, \alpha))$ of the point (q_α, α) consists of sets $U_{\alpha,n} = \{(q_\alpha, \alpha)\} \cup \bigcup_{k\geq n} V_{\alpha,k}$ where $V_{\alpha,k}$ is some open neighborhood of $a_{\alpha,k}$ in $\mathbb{Q} \times \alpha$ such that $\pi(V_{\alpha,k})$ has diameter less than $1/k$. The resulting space is scattered since for each $\alpha < \mathfrak{c}$ at most one point in $\mathbb{Q} \times \{\alpha\}$ is not isolated and $\mathbb{Q} \times \eta$ is open in $\mathbb{Q} \times (\eta+1)$ for all η. The map f is continuous because $\lim_{n\to\infty}(\pi(U_{\alpha,n})) = 0$ for all α; f is closed because if $F \subseteq X$ is a closed subspace and $\{q_n\}$ is a sequence in $f(F)$ converging to $q \notin \{q_n \mid n \in \mathbb{N}\}$, then there exists a sequence A_η in F lying over $\{q_n\}$ and hence $(q_\eta, \eta) \in F$ and therefore $q = q_\eta \in f(F)$. □

Note that there is no such map from a completely metrizable space X to the rationals because completeness is preserved under images of closed surjections of metrizable spaces. (If there were such a map then the second sentence of 3.4 above would automatically hold because the pullback $X \times_{\mathbb{Q}} pl(\mathbb{Q})$ would then be a pointless O_δ sublocale of X and hence empty. This argument would also apply a bit more generally to strongly Baire spaces. Bit I suspect that strong Baireness is also preserved under images of closed surjections of metrizable spaces.)

LEMMA 3.5. *If $Y \xrightarrow{g} Z$ is a closed surjection, then so is $X +_\bullet Y \xrightarrow{\langle f,g\rangle_\bullet} Z$ for any map $X \xrightarrow{f} Z$.*

Proof. A sublocale T is closed in $X +_\bullet Y$ if and only if $T \wedge X$ and $T \wedge Y$ are closed in X, respectively Y, and if furthermore $g^{-1}(cl_Z(f(T \wedge X))) \leq T \wedge Y$. So in order to show that $\langle f, g\rangle_\bullet$ is a closed surjection it suffices to observe that $\langle f, g\rangle_\bullet(T) = g(\beta^{-1}(T))$ for any closed sublocale T of $X +_\bullet Y$. □

COROLLARY 3.6. *There exists a closed surjection in* **Loc** *which is not a regular epimorphism.*

Proof. The map $pl(\mathbb{Q}) +_{\bullet} X \xrightarrow{\langle \iota, f\rangle_{\bullet}} \mathbb{Q}$ where f is the map constructed in 3.4 is a closed surjection by 3.5, but it is not a regular epimorphism because 1.7 implies that the coequalizer of the kernel pair of $\langle \iota, f\rangle_{\bullet}$ is $pl(\mathbb{Q}) +_{\bullet} X \xrightarrow{\mathrm{id}_{pl(\mathbb{Q})} +_{\bullet} f} pl(\mathbb{Q}) +_{\bullet} \mathbb{Q}$. □

COROLLARY 3.7. *The composite of two regular epimorphisms in* **Loc** *is not necessarily a regular epimorphism.*

Proof. The closed surjection $pl(\mathbb{Q}) +_{\bullet} X \xrightarrow{\langle \iota, f\rangle_{\bullet}} \mathbb{Q}$ from the preceding corollary is not a regular epic, but factors as the regular epimorphism $pl(\mathbb{Q}) +_{\bullet} \mathbb{Q} \xrightarrow{\mathrm{id}_{pl(\mathbb{Q})} +_{\bullet} f} pl(\mathbb{Q}) +_{\bullet} \mathbb{Q}$ followed by the simple covering $pl(\mathbb{Q}) +_{\bullet} \mathbb{Q} \xrightarrow{\langle \iota, \mathrm{id}\rangle_{\bullet}} \mathbb{Q}$ which is proper by (2) and (5) of Section 2. □

COROLLARY 3.8. *The inclusion* **RegE** $\subseteq$ **ExtE** *is proper.*

Proof. Recall that an epimorphism $X \overset{f}{\to} Y$ is said to be (i) an extremal epimorphism if, whenever h is a monomorphism in a factorization $X \overset{g}{\to} Z \overset{h}{\to} Y$ of f, then h is necessarily an isomorphism, and (ii) a strong epimorphism if for every commutative square

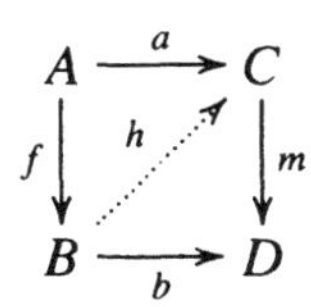

in which m is monic there exists a diagonal fill-in h (necessarily unique) making both triangles commute. Strong epimorphisms are closed under composition and hence so are extremal epimorphisms in **Loc** (because in any category with pullbacks these notions coincide). So the closed surjection in 3.6 is extremal, but not regular. □

The next question which should be answered about closed surjections is the question of whether all closed surjections are extremal epimorphisms. I believe this to be true, but have no proof.

4. Weak Quotient Maps, Extremal and Regular Epimorphisms

As already indicated in the introduction, the problem of characterizing regular epimorphisms in **Loc** consists of two almost orthogonal parts. The more difficult part, it seems, is the problem of finding properties which ensure that open sublocales equipped with descent data descend (step (ii) and (iv) of the proof given in the introduction). We will not pursue this question here.

The more accessible part of this problem is to determine properties which ensure that the set $\mathcal{S}_U$ of all sublocales of A which pull back to a given open sublocale U of B (equipped with descent data) contains an open sublocale provided that $\mathcal{S}_U$ is nonempty. This corresponds to step (v) of the proof given in the introduction taking into account that step (iii) no longer works.

In this section we compare four properties of localic surjections which are potential answers to the second half of this problem. The first notion we consider is a direct translation of the definition of quotient maps in **Top**. This notion turns out to be too restrictive: there exist regular epimorphisms in **Loc** which do not satisfy it. The next two notions we consider avoid the problems caused by the lack of uniqueness of S above by restricting their attention to complemented sublocales of A. But the corresponding classes (which turn out to be identical, see 4.4), the class of weak quotient maps [9] and the class of localic surjections whose corresponding frame monomorphisms are equationally closed frame inclusions [14], contain all closed surjections and hence non-regular epimorphisms. The last notion, which I think provides the correct answer to this problem, is a modification of the first notion which uses the fact that if $\mathcal{S}_U$ contains an open sublocale V, then V is necessarily the smallest sublocale in $\mathcal{S}_U$.

We start by considering the straightforward translation of the definition of topological quotient maps into **Loc**. We say that a surjection $B \xrightarrow{f} A$ of locales satisfies (QO) if and only if the following holds:

$$\text{(QO)} \qquad \forall S \in \acute{S}(A)\colon\ f^{-1}(S) \text{ is open } \Rightarrow S \text{ is open.}$$

The next two examples show that (QO) is not satisfied by all regular epimorphisms in **Loc**, and also that (QO) is not enough to imply that a given map is a regular epimorphism.

EXAMPLE 4.1. A surjection $Y \xrightarrow{f} \mathbb{Q}$ of locales which is not a regular epimorphism, but which satisfies (QO).

Let $X \xrightarrow{f} \mathbb{Q}$ be the map constructed in 3.4. The map $X +_\bullet pl(\mathbb{Q}) \xrightarrow{h=\langle f,\iota\rangle_\bullet} \mathbb{Q}$ satisfies (QO) but is not a regular epimorphism. h is not regular because $\pi_1^*(X) = \pi_2^*(X)$ ($X \times_{\mathbb{Q}} pl(\mathbb{Q}) = 0$ see 3.4), but since $h(X) = f(X) = \mathbb{Q}$, X is not the inverse image of any open sublocale of $\mathbb{Q}$.

To show that h satisfies (QO) let $S \in S(\mathbb{Q})$ be any sublocale for which $h^{-1}(S)$ is open. Since $f^{-1}(S)$ is open in X, it suffices to show that S is a spatial sublocale of $\mathbb{Q}$ because then we can use the fact that f is a quotient map of topological spaces to conclude that S is open in $\mathbb{Q}$. Since $h^{-1}(S)$ is open in $X +_\bullet pl(\mathbb{Q})$, $f^{-1}(S)$ is open in X and $\iota^{-1}(\mathbb{Q} - f(X - f^{-1}(S))) \geq \iota^{-1}(S) \simeq pl(S)$, hence in particular $\mathbb{Q} - f(X - f^{-1}(S)) \geq pl(S)$. Since $\mathbb{Q} - f(X - f^{-1}(S))$ is an open sublocale of $\mathbb{Q}$ it is spatial. If we can show that $S \geq \mathbb{Q} - f(X - f^{-1}(S))$ then we are done because this implies $S \geq pt(S) \geq \mathbb{Q} - f(X - f^{-1}(S)) \geq pl(S)$ and hence that S is spatial because any locale is the join of its largest pointless sublocale and its points (provided, of course, that the largest pointless sublocale exists). But this follows from the fact that $S \vee f(X - f^{-1}(S)) \geq f(f^{-1}(S)) \vee f(X - f^{-1}(S)) = f(f^{-1}(S) \vee (X - f^{-1}(S))) = f(X) = \mathbb{Q}$.

EXAMPLE 4.2. A surjection $X \xrightarrow{f} \mathbb{Q}$ of spatial metrizable locales which is a regular epimorphism, but which does not satisfy (QO).

Take any quotient map (in **Top**) from a completely metrizable space X onto the rationals $\mathbb{Q}$. A simple example of such a map is the map $X = \coprod_{q \in \mathbb{Q}} \mathbb{Q}_q \xrightarrow{f} \mathbb{Q}$ induced by the natural maps $\mathbb{Q}_q \xrightarrow{f_q} \mathbb{Q}$ where $\mathbb{Q}_q$ ($q \in \mathbb{Q}$) has the same underlying set as $\mathbb{Q}$, but all points except q are isolated and q has the same neighborhoods in $\mathbb{Q}$ and $\mathbb{Q}_q$. Since the embedding of the category **Sob** of sober spaces into **Loc** as spatial locales has a right adjoint, it preserves all regular epimorphisms, so $X \xrightarrow{f} \mathbb{Q}$ is a regular epimorphism in **Loc**. All pointless sublocales of $\mathbb{Q}$ pullback to the empty sublocale of X because (i) $pl(\mathbb{Q})$, the largest pointless sublocale of $\mathbb{Q}$ is an O_δ in $\mathbb{Q}$ (a countable intersection of open sublocales), (ii) pulling back along f preserves all meets and (iii) O_δ's in complete spaces are spatial [6], hence pointless O_δ's are empty. (In fact, for the example above this argument can be simplified because $\coprod_{q \in \mathbb{Q}}$ is scattered and scattered spaces have no nonzero pointless sublocales.) So if S is any nonzero pointless sublocale of $\mathbb{Q}$, then $f^{-1}(S) = 0$ is open in X, but S is not open in $\mathbb{Q}$, which means that f does not satisfy (QO).

The reason why (QO) is not satisfied by all regular epimorphisms is that regular epimorphisms are not necessarily semi-stably epic. In the example above all pointless sublocales of $\mathbb{Q}$ are contained in $\mathcal{S}_0$. For (QO) to be satisfied all those sublocales would have to be open in $\mathbb{Q}$. But for f to be a regular epimorphism it is only necessary that among all those sublocales there is one sublocale, in this case 0, which is open.

If we replace "open" by "closed" in (QO) then we get a further non-equivalent condition, call it (QC). These conditions are not equivalent because, unlike subspaces, sublocales do not always have complements. (QO) and (QC) are independent conditions: h in 4.1 satisfies (QO) but not (QC), while the closed surjection in 3.6 satisfies (QC) but not (QO).

The next two notions are weakenings of (QO) for which the "open" and the "closed" version are equivalent. These two notions avoid dealing with the fact that the sets $\mathcal{S}_U$ may contain more than one element by restricting (QO) to classes of sublocales on which pulling back along any surjection induces an injective map.

DEFINITION 4.3. Let $B \xrightarrow{f} A$ be a surjection of locales. We say that f satisfies condition (WQ), respectively (EQC), if

$$f^{-1}(S) \text{ is open in } B \Rightarrow S \text{ is open in } A$$

for all complemented sublocales S of A, respectively for all sublocales $S = U \vee (V - W)$ where U, V and W are open sublocales of A.

Surjections which satisfy (WQ) were introduced by Li in [9] as weak quotient maps. She introduced this notion in order to give a sufficient condition which guarantees that local connectedness descends. Every extremal (and hence every regular) epimorphism $B \xrightarrow{f} A$ in **Loc** satisfies (WQ) because if C is a complemented sublocale and $f^{-1}(C)$ is open in B, then f can be factored as $B \xrightarrow{g} A_C \xrightarrow{h} A$ where $A_C \xrightarrow{h} A$ is the simple covering given by $A_o = C$ and $A_c = A - C$. So if f is extremal, then h is necessarily an isomorphism which in turn means that C has to be open.

Surjections which satisfy (EQC) were introduced in a different guise by Pultr and Tozzi in [14]. First of all, they were working in the category of frames. For these they defined the notion of equationally closed subframe. A subframe M of a frame N is said to be equationally closed in N if

$$\forall m \in M, n \in N : m \vee n \in M \text{ and } m \wedge n \in M \text{ implies } n \in M.$$

A localic surjection satisfies (EQC) if the corresponding frame monomorphism is the inclusion of an equationally closed subframe [14]. Just as for distributive lattices one can show that a frame is epimorphically embedded in its equational closure in any superframe. So localic surjections which satisfy (EQC) are a weak version of extremal epimorphisms in **Loc**. They satisfy the condition characterizing extremal epimorphisms where the monomorphisms occurring as the second half of a factorization $B \xrightarrow{g} C \xrightarrow{h} Z$ are restricted to those monomorphisms for which h^* has "equationally dense" image in $T(Z)$.

It is clear that each surjection which satisfies (WQ) also satisfies (EQC) because any sublocale of the form $U \vee (V - W)$ (U, V and W open) is complemented. But in fact both notions are equivalent.

LEMMA 4.4. *A surjection $B \xrightarrow{f} A$ in* **Loc** *satisfies* (EQC) *if and only if satisfies* (WQ).

Proof. That (WQ) is stronger than (EQC) is clear. For the converse assume that $B \xrightarrow{f} A$ satisfies (EQC) and that C is a complemented sublocale of A for which $f^{-1}(C)$ is open. We want to show that C is already open. Let $U = int(C)$, $H = cl(C - U)$ and $P = int_H(C - U)$. Then (i) P is dense in H; (ii) P is locally closed in A; (iii) $U \vee P$ is open in C. The last two statements are clear, and the first one follows from the fact that every dense and complemented sublocale has dense interior (the complement D of $C - U$ in H is disjoint from $D(H)$, and hence so is $cl_H(D)$). Now $f^{-1}(U \vee P)$ is open in $f^{-1}(C)$, hence open in B. Applying (EQC) yields that $U \vee P \subseteq C$ is open in A. Because U is the interior of C we conclude first that $U = U \vee P$, and then that $P = 0$ because U and P are disjoint. Since P is dense in H it follows that also $H = 0$, and hence that $C - U = 0$. So we can conclude that $C = U$ is open, i.e. f satisfies (WQ). □

Originally, Li had posed the question of whether weak quotient maps are necessarily regular epimorphisms [9]. Since closed surjections are weak quotient maps, 3.6 answers Li's question in the negative. But this answer doesn't give quite as much insight into surjections in **Loc** as one might have hoped for. The slightly modified question of whether weak quotient maps are necessarily extremal epimorphisms remains open. This question is perhaps the most fundamental unanswered question about quotient maps of locales; a positive answer would lend quite a bit of support to the suggestion that extremal epimorphisms are the "correct" notion of quotient maps in **Loc**, while a negative answer would leave the question of which class of surjections in **Loc** should be taken as quotient maps wide open.

For simple coverings this question has a simple answer. The next proposition shows that for simple coverings the notions of weak quotient map, extremal epimorphism and regular epimorphism all coincide.

PROPOSITION 4.5. *A simple covering* $B \xrightarrow{f} A$ *is a regular epimorphism if and only if it satisfies* (WQ).

Proof. Since extremal epimorphisms always satisfy (WQ) it suffices to show that (WQ) implies (RE). If H is any closed sublocale of A, then $(A - H) \vee (H_o - H_{\bar{m}})$ is a complemented sublocale of A with complement $H_c \vee H_{\bar{m}}$. Since

$$\begin{aligned} & f^{-1}((A - H) \vee (H_o - H_{\bar{m}})) \\ &= (B - f^{-1}(H)) \vee (f^{-1}(H_o) - f^{-1}(H_{\bar{m}})) \\ &= (B - f^{-1}(H)) \vee ((f^{-1}(H) - f^{-1}(H_{\bar{m}})) \wedge U) \\ &= (B - f^{-1}(H)) \vee ((B - f^{-1}(H_{\bar{m}})) \wedge U) \end{aligned}$$

is open in B, (WQ) implies that $(A - H) \vee (H_o - H_{\bar{m}})$ is open in A, and hence that $H_o - H_{\bar{m}}$ is open in H. □

Finally, we introduce one last notion, (RE1) below, which weakens (QO) just enough to be able to deal with the fact that open sublocales equipped with descent data do not necessarily descend to a unique sublocale along regular epimorphisms.

DEFINITION 4.6. Let $B \xrightarrow{f} A$ be a surjection of locales. We say that f satisfies condition (RE1) if

$$\forall S \in S(A)\colon\ f^{-1}(S) \text{ is open } \Rightarrow f^{-1}(S) = f^{-1}(int_A(S)).$$

We have (QO) $\Rightarrow$ (RE1) $\Rightarrow$ (WQ), and neither of these two implications is reversible. (RE1) is strictly stronger than (WQ). It is stronger because if C is any complemented sublocale of A for which C and $int(C)$ have isomorphic pullbacks along some surjection then $C = int(C)$ (1.3), and hence C is open. It is strictly stronger because it is not satisfied by the closed surjection constructed in 3.6 ($f^{-1}(pl(\mathbb{Q}))$ is open and nonzero, but $f^{-1}(int(pl(\mathbb{Q}))) = f^{-1}(0) = 0$). That (RE1) is strictly weaker than (QO) follows because it is satisfied by the map

$X \xrightarrow{f} \mathbb{Q}$ in 4.2. If the map f is a semi-stable epimorphism, for instance if f is a simple covering, then (RE1) is equivalent to (QO). This follows because the map $f^{-1}\colon S(A) \to S(B)$ is injective for semi-stable epimorphisms f.

The other condition we need in order to characterize regular epimorphisms is the following:

DEFINITION 4.7. Let $B \xrightarrow{f} A$ be a surjection of locales and let π_1, π_2 be the kernel pair of f. We say that f satisfies condition (RE2) if it satisfies one (hence all) of the following three equivalent conditions:

(1) open sublocales U of B equipped with descent data relative to f descend;
(2) for all open sublocales U of $B : \pi_1^{-1}(U) = \pi_2^{-1}(U) \Rightarrow \exists S \in S(A) : f^{-1}(S) = U$;
(3) for all open sublocales U of $B : U \times_A (B - U) = 0 \Rightarrow \exists S \in S(A) : f^{-1}(S) = U$.

That (2) and (3) are equivalent is straightforward. (1) and (2) are equivalent because descent data relative to f for any sublocale S of B just amount to the requirement that $\pi_1^{-1}(S) = \pi_2^{-1}(S)$, because the unit and the cocycle condition (see [10] for the definitions) automatically hold for monic structure maps. The reason why I keep mentioning descent theory (or at least use its terminology), is that I believe that an internal characterization of those maps which satisfy (RE2) should provide a first step towards characterizing effective descent morphisms in **Loc**, a problem which looks rather intractable at the moment.

Now we can give a provisional characterization of regular epimorphisms in **Loc** (provisional, because of the missing internal characterization of surjections satisfying (RE2)).

PROPOSITION 4.8. *A surjection $B \xrightarrow{f} A$ in* **Loc** *is a regular epimorphism if and only if it satisfies* (RE1) *and* (RE2).

Proof. If f is a regular epimorphism then it clearly satisfies (RE2). To show that f satisfies (RE1) let S be any sublocale of A for which $f^{-1}(S)$ is open in B. Since f is a regular epimorphism there exists an open sublocale V of A such that $f^{-1}(V) = f^{-1}(S)$ because $\pi_1^{-1}(f^{-1}(S)) = \pi_2^{-1}(f^{-1}(S))$. Now $S = \bigwedge\{C \geq S \mid C \text{ complemented in } A\}$, and for any such C we have that $f^{-1}(C) \geq f^{-1}(S) = f^{-1}(V)$. This implies that $C \geq V$ because the restriction of the map $f^{-1}\colon S(A) \to S(B)$ to complemented sublocales is injective and reflects the ordering. Therefore $S \geq V$, and hence $S \geq int_A(S) \geq V$. Because $f^{-1}(S) \geq f^{-1}(int_A(S)) \geq f^{-1}(V) = f^{-1}(S)$, and $int_A(S)$ and V are complemented we conclude that $int_A(S) = V$. Conversely, if f satisfies (RE1) and (RE2), then f is the coequalizer of its kernel pair because whenever $\pi_1^{-1}(U) = \pi_2^{-1}(U)$ for some open sublocale U of B, then (RE2) provides a sublocale $S \in S(A)$ whose pullback along f is U, and (RE1) then says that S may be replaced by $V = int(S)$. Uniqueness of V follows because f is a surjection. □

It might be worthwhile to point out that the implications (QO) + (RE2) $\Rightarrow$ (RE1) + (RE2) $\Rightarrow$ (WQ) + (RE2) are also irreversible. That the first implication cannot be reversed follows again from 4.2. For the second implication we can also use 3.6 because each open sublocale U of $pl(\mathbb{Q}) +_\bullet X$ equipped with descent data is the union of an open sublocale U_1 of $pl(\mathbb{Q})$ and an open sublocale U_2 of X equipped with descent data relative to f. U_2 descends to an open sublocale S_2 of $\mathbb{Q}$. A straightforward calculation now shows that $h^{-1}(\iota(U_1) \vee S_2) = U$.

Using almost the same arguments as in 4.8 we get the following analogous characterization of regular epimorphisms in terms of closed sublocales:

PROPOSITION 4.9. *A surjection $B \xrightarrow{f} A$ in* **Loc** *is a regular epimorphism if and only if it satisfies the following two conditions*:

(C1) $\forall S \in S(A)$: $f^{-1}(S)$ *is closed* $\Rightarrow f^{-1}(S) = f^{-1}(A - int(sup(S)))$;
(C2) *closed sublocales of B equipped with descent data descend.*

Condition (C1) is equivalent to the requirement that, for all $S \in S(A)$, if $f^{-1}(S)$ is closed, then so is $sup(sup(S))$ and $f^{-1}(S) = f^{-1}(sup(sup(S)))$. Similarly, f satisfies (RE1) if and only if for all $S \in S(A)$, if $f^{-1}(S)$ is open, then so is $sup(sup(S))$ and $f^{-1}(S) = f^{-1}(sup(sup(S)))$. The final example shows that (C1) and (RE1) together are also not sufficient to imply that a map is a regular epimorphism.

EXAMPLE 4.10. A (closed) surjection $Y \xrightarrow{h} \mathbb{Q}$ of locales which is not a regular epimorphism, but which satisfies (C1) and (RE1).

Let $X \xrightarrow{f} \mathbb{Q}$ be again the closed surjection constructed in 3.4, let $\mathbb{P}$ denote the irrationals, and let $\mathbb{P} \xrightarrow{g} \mathbb{Q}$ be any quotient map from the irrationals to the rationals (such maps exist, see for instance [11]). Let $Y \xrightarrow{h} \mathbb{Q}$ be the map $pl(\mathbb{P}) +_\bullet X \xrightarrow{\langle g\iota, f\rangle_\bullet} \mathbb{Q}$ where ι is the inclusion $pl(\mathbb{P}) \xrightarrow{\iota} \mathbb{P}$. The same reasoning which shows that $pl(\mathbb{Q}) +_\bullet X \xrightarrow{\langle\iota, f\rangle_\bullet} \mathbb{Q}$ in 3.6 is not a regular epimorphism applies to h. To show that h satisfies (C1) as well as (RE1), it suffices to show that (‡) $h^{-1}(S) = h^{-1}(pt(S))$ for all $S \in S(\mathbb{Q})$ because f is a quotient map of topological spaces and $sup(sup(S)) = pt(S)$ in any spatial T_D locale. To show that h satisfies (‡) it suffices to show that both f and g satisfy it. But that follows from the fact that $f^{-1}(pl(\mathbb{Q})) = 0$ and $g^{-1}(pl(\mathbb{Q})) = 0$, since both pullbacks have to be pointless and spatial (the second, because O_δ's in completely metrizable spaces are spatial).

Acknowledgement

I would like to thank Peter Johnstone for sharing his ideas about this topic, pointing out some interesting open questions, and also for presenting the main results of this paper at the BB Fest.

References

1. Banaschewski, B.: Singly generated frame extensions, *J. Pure Appl. Algebra* **83** (1990), 1–21.
2. Chen, X.: Closed frame homomorphisms, doctoral dissertation, McMaster University, Hamilton, 1991.
3. van Douwen, E. K.: The integers and topology, in K. Kunen and J. E. Vaughan (eds), *Handbook of Set-theoretic Topology*, North-Holland, 1984, pp. 111–167.
4. Isbell, J. R.: Atomless parts of spaces, *Math. Scand.* **31** (1972), 5–32.
5. Isbell, J. R.: Graduation and dimension in locales, *London Math. Soc. Lecture Note Ser.* **93** (1985), 195–210.
6. Isbell, J. R.: First steps in descriptive theory of locales, *Trans. Amer. Math. Soc.* **327** (1991), 353–371; *Corrections* **341** (1994), 467–468.
7. Johnstone, P. T.: *Stone Spaces*, Cambridge Univ. Press, Cambridge, 1982.
8. Joyal, A. and Tierney, M.: An extension of the Galois theory of Grothendieck, *Mem. Amer. Math. Soc.* **309**, vol. 51, 1984.
9. Li, Y.: Weak locale quotient morphisms and locally connected frames, *J. Pure Appl. Algebra* **110** (1996), 101–107.
10. Moerdijk, I.: Descent theory for toposes, *Bull. Soc. Math. Belgique* **41** (1989), 373–391.
11. Michael, E. and Stone, A. H.: Quotients of the space of irrationals, *Pacific J. Math.* **28**(3) (1969), 629–633.
12. Plewe, T.: Countable products of absolute C_δ spaces, *Topology Appl.* **74** (1996), 39–50.
13. Plewe, T.: Triquotient maps are effective descent maps, *Math. Proc. Cambridge Philos. Soc.* **122** (1997), 17–43.
14. Pultr, A. and Tozzi, A.: Equationally closed subframes and representation of quotient spaces, *Cahiers Topologie Géom. Différentielle Catégoriques* **34**(3) (1993), 167–183.
15. Vermeulen, J. J. C.: Proper maps of locales, *J. Pure Appl. Algebra* **92** (1994), 79–107.

Applied Categorical Structures **8:** 45–66, 2000.
G. Brümmer & C. Gilmour (eds), Papers in Honour of Bernhard Banaschewski.

Some Applications of the Ultrapower Theorem to the Theory of Compacta

Dedicated to Bernhard Banaschewski on the occasion of his 70th Birthday

PAUL BANKSTON
Department of Mathematics, Statistics and Computer Science, Marquette University, Milwaukee, WI 53201-1881, U.S.A. e-mail: paulb@mscs.mu.edu

(Received: 4 September 1996; accepted: 28 May 1998)

Abstract. The ultrapower theorem of Keisler and Shelah allows such model-theoretic notions as elementary equivalence, elementary embedding and existential embedding to be couched in the language of categories (limits, morphism diagrams). This in turn allows analogs of these (and related) notions to be transported into unusual settings, chiefly those of Banach spaces and of compacta. Our interest here is the enrichment of the theory of compacta, especially the theory of continua, brought about by the importation of model-theoretic ideas and techniques.

Mathematics Subject Classifications (2000): 03C20, 54B35, 54C10, 54D05, 54D30, 54D80, 54F15, 54F45, 54F55.

Key words: ultraproduct, ultracoproduct, compactum, continuum, co-elementary map, co-existential map.

1. Introduction

In the ten years since the celebration of Professor Banaschewski's 60th birthday (and the writing of [4] in commemoration), there has been a fair amount of development in the theory of ultracoproducts of compacta (i.e., compact Hausdorff spaces). Papers [3, 5, 7, 6] have been written by this author; also there is the paper [13] by the late R. Gurevič. In addition to this, there has been a parallel development in the first-order theory of Banach spaces/algebras begun by C. W. Henson in [16]. There are important links between Banach space theory and our work (mainly through Gel'fand–Naĭmark duality), and we present here two applications of Banach techniques to the theory of compacta. (See Theorems 3.1 and 4.2 below.) Since Gel'fand–Naĭmark duality is a two-way street, much of the 'dualized' model theory developed for compacta may be directly translated into the Banach model theory of commutative B^*-algebras. (See the results in Sections 5 and 6 below.) It seems likely that the future will see much in the way of progress in these two streams of research, as topological issues stimulate the analytic and *vice versa*.

We begin with a quick review of the topological ultracoproduct construction; detailed accounts may be found in [2–5, 7, 6, 13].

We let **CH** denote the category of compacta and continuous maps. In model theory, it is well known that ultraproducts (and reduced products in general, but we restrict ourselves to maximal filters on the index set) may be described in the language of category theory; i.e., as direct limits of (Cartesian) products, where the directed set is the ultrafilter with reverse inclusion, and the system of products consists of Cartesian products taken over the various sets in the ultrafilter. (Bonding maps are just the obvious restriction maps.) When we transport this framework to **CH**, the result is somewhat less than spectacular: If $\langle X_i : i \in I\rangle$ is a family of nonempty compacta and $\mathcal{D}$ is a nonprincipal ultrafilter on I, then the **CH**-ultraproduct is degenerate (i.e., has only one point). What turns out to be vastly more fruitful, however, is the ultraproduct construction in the category-opposite of **CH**; i.e., take an inverse limit of coproducts. The result is the **topological ultracoproduct**, and may be concretely (if opaquely) described as follows: Given $\langle X_i : i \in I\rangle$ and $\mathcal{D}$, let Y be the disjoint union $\bigcup_{i\in I}(X_i \times \{i\})$ (a locally compact space). With $q: Y \to I$ the natural projection onto the second coördinate (where I has the discrete topology), we then have the Stone–Čech lifting $q^\beta: \beta(Y) \to \beta(I)$. Now the ultrafilter $\mathcal{D}$ may be naturally viewed as an element of $\beta(I)$, and it is not hard to show that the topological ultracoproduct $\sum_{\mathcal{D}} X_i$ is the pre-image $(q^\beta)^{-1}[\mathcal{D}]$. (The reader may be familiar with the Banach ultraproduct [10]. This construction is indeed the ultraproduct in the category of Banach spaces and nonexpansive linear maps, and may be telegraphically described using the recipe: take the usual ultraproduct, throw away the infinite elements, and mod out by the subspace of infinitesimals. Letting $C(X)$ denote the Banach space of continuous real-valued (or complex-valued) continuous functions with X as domain, the Banach ultraproduct of $\langle C(X_i) : i \in I\rangle$ via $\mathcal{D}$ is just $C(\sum_{\mathcal{D}} X_i)$.)

If $X_i = X$ for all $i \in I$, then we have the **topological ultracopower** $XI\backslash\mathcal{D}$, a subspace of $\beta(X \times I)$. In this case there is the Stone–Čech lifting p^β of the natural first-coördinate map $p: X \times I \to X$. Its restriction to the ultracopower is a continuous surjection, called the **codiagonal map**, and is officially denoted $p_{X,\mathcal{D}}$ (with the occasional notation-shortening alias possible). This map is dual to the natural diagonal map from a relational structure to an ultrapower of that structure, and is not unlike the standard part map from nonstandard analysis.)

Many notions from classical first-order model theory, principally elementary equivalence, elementary embedding and existential embedding, may be phrased in terms of mapping conditions involving the ultraproduct construction. Because of the (Keisler and Shelah) ultrapower theorem (see, e.g., [9]), two relational structures are elementarily equivalent if and only if some ultrapower of one is isomorphic to some ultrapower of the other; a function from one relational structure to another is an elementary embedding if and only if there is an ultrapower isomorphism so that the obvious square mapping diagram commutes. A function $f: A \to B$ between relational structures is an existential embedding (i.e., making the image under f a substructure of B that is existentially closed in B) if and only if there are embeddings $g: A \to C$, $h: B \to C$ such that g is elementary and equal to the composition hf. (C may be taken to be an ultrapower of A, with g the natural diagonal.) (See also, e.g., [2, 5, 11].)

In **CH** one then constructs ultracoproducts, and talks of co-elementary equivalence, co-elementary maps and co-existential maps. Co-elementary equivalence is known [2, 5, 13] to preserve important properties of topological spaces, such as being infinite, being a continuum (i.e., connected), being Boolean (i.e., totally disconnected), having (Lebesgue) covering dimension n, and being a decomposable continuum. If $f: X \to Y$ is a co-elementary map in **CH**, then of course X and Y are co-elementarily equivalent ($X \equiv Y$). Moreover, since f is a continuous surjection (see [2]), additional information about X is transferred to Y. For instance, continuous surjections in **CH** cannot raise **weight** (i.e., the smallest cardinality of a possible topological base, and for many reasons the right cardinal invariant to replace cardinality in the dualized model-theoretic setting), so metrizability (i.e., being of countable weight in the compact Hausdorff context) is preserved. Also local connectedness is preserved, since continuous surjections in **CH** are quotient maps. Neither of these properties is an invariant of co-elementary equivalence alone.

When attention is restricted to the full subcategory of Boolean spaces, the dualized model theory matches perfectly with the model theory of Boolean algebras because of Stone duality. In the larger category there is no such match [1, 24], however, and one is forced to look for other (less direct) model-theoretic aids. Fortunately there is a finitely axiomatizable AE Horn class of bounded distributive lattices, the so-called **normal disjunctive** lattices [6] (also called Wallman lattices in [5]), comprising precisely the (isomorphic copies of) **lattice bases**, those lattices that serve as bases for the closed sets of compacta. (To be more specific: The normal disjunctive lattices are precisely those bounded lattices A such that there exists a compactum X and a meet-dense sublattice $\mathcal{A}$ of the closed set lattice $F(X)$ of X such that A is isomorphic to $\mathcal{A}$.) We go from lattices to spaces, as in the case of Stone duality, via the **maximal spectrum** $S(\ )$, pioneered by H. Wallman [27]. $S(A)$ is the space of maximal proper filters of A; a typical basic closed set in $S(A)$ is the set $a^\sharp$ of elements of $S(A)$ containing a given element $a \in A$. $S(A)$ is generally compact with this topology. Normality, the condition that if a and b are disjoint ($a \sqcap b = \bot$), then there are a', b' such that $a \sqcap a' = b \sqcap b' = \bot$ and $a' \sqcup b' = \top$, ensures that the maximal spectrum topology is Hausdorff. Disjunctivity, which says that for any two distinct lattice elements there is a nonbottom element that is below one of the first two elements and disjoint from the other, ensures that the map $a \mapsto a^\sharp$ takes A isomorphically onto the canonical closed set base for $S(A)$. $S(\ )$ is contravariantly functorial: If $f: A \to B$ is a homomorphism of normal disjunctive lattices and $M \in S(B)$, then $f^S(M)$ is the unique maximal filter extending the prime filter $f^{-1}[M]$. (For normal lattices, each prime filter is contained in a unique maximal one.) It is a fairly straightforward task to show, then, that $S(\ )$ converts ultraproducts to ultracoproducts, elementarily equivalent lattices to co-elementarily equivalent compacta, and elementary (resp., existential) embeddings to co-elementary (resp., co-existential) maps. Furthermore, if $f: A \to B$ is a **separative** embedding; i.e., an embedding such that if $b \sqcap c = \bot$ in B, then there exists $a \in A$ such that $f(a) \geq b$ and $f(a) \sqcap c = \bot$, then f^S is a homeomorphism (see [2, 4–6, 13]). Because of this, there is much flexibility in how we may obtain $\sum_{\mathcal{D}} X_i$: Simply choose a lattice base $\mathcal{A}_i$ for each X_i and apply $S(\ )$ to the ultraproduct $\prod_{\mathcal{D}} \mathcal{A}_i$.

2. The Topological Behavior of Co-existential Maps

Recall that a first-order formula in prenex form is an **existential formula** if all its quantifiers are existential. A function $f: A \to B$ between structures is an **elementary** (resp., **existential**) **embedding** if for every formula (resp. existential formula) $\varphi(x_1, \ldots, x_n)$ and every n-tuple $\langle a_1, \ldots, a_n \rangle$ from A, $A \models \varphi[a_1, \ldots, a_n]$ if and only if $B \models \varphi[f(a_1), \ldots, f(a_n)]$.

The ultrapower theorem states that a function $f: A \to B$ is an elementary embedding if and only if there is an isomorphism of ultrapowers $h: A^I/\mathcal{D} \to B^J/\mathcal{E}$ such that the obvious mapping square commutes; i.e., such that $d_{\mathcal{E}} f = h d_{\mathcal{D}}$, where $d_{\mathcal{D}}$ and $d_{\mathcal{E}}$ are the natural diagonal embeddings. There is also a characterization of existential embeddings along similar lines: $f: A \to B$ is an existential embedding if and only if there are embeddings $g: A \to C$ and $h: B \to C$ such that g is elementary and $g = hf$. (By the ultrapower theorem, we may take C to be an ultrapower $A^I/\mathcal{D}$ and $g = d_{\mathcal{D}}$.)

These characterizations have inspired the *definition* of the notions of co-elementary map and co-existential map in **CH** in terms of analogous mapping diagrams involving ultracopowers. Co-elementary (indeed, co-existential) maps are clearly continuous surjections. However, since specific situations involving co-elementary and co-existential maps are not guaranteed to correspond to analogous situations in elementary classes of relational structures, one may not take too much for granted. For example, there is no assurance *a priori* that the classes of co-elementary and co-existential maps are closed under composition or terminal factors (i.e., if both f and gf have a particular property, then g has the property). As it happens, it is shown in [2] that closure under composition and terminal factors holds for co-elementary maps; also in [7] it is shown that there is an amalgamation property for co-elementary maps: If $f: Y \to X$ and $g: Z \to X$ are co-elementary, then there exist co-elementary maps $u: W \to Y$ and $v: W \to Z$ such that $fu = gv$. We do not know whether there is amalgamation for co-existential maps; however we can show closure under composition and terminal factors. We first prove a useful lemma.

LEMMA 2.1. *Let $\{f_\delta: X_\delta \to Y_\delta : \delta \in \Delta\}$ be a family of co-elementary and co-existential maps between compacta. Then there is a single ultrafilter witness to the fact. More precisely, there is an ultrafilter $\mathcal{D}$ on a set I such that:*

(i) *if* f_δ *is co-elementary, then there is a homeomorphism* $h_\delta: X_\delta I\backslash\mathcal{D} \to Y_\delta I\backslash\mathcal{D}$ *such that* $f_\delta p_{X_\delta,\mathcal{D}} = p_{Y_\delta,\mathcal{D}} h_\delta$*; and*
(ii) *if* f_δ *is co-existential, then there is a continuous surjection* $g_\delta: Y_\delta I\backslash\mathcal{D} \to X_\delta$ *such that* $f_\delta g_\delta = p_{Y_\delta,\mathcal{D}}$.

Proof. In S. Shelah's (GCH-free) proof of the ultrapower theorem (see [25]), he proves that if λ is an infinite cardinal and $\mu = \min\{\kappa : \lambda^\kappa > \lambda\}$, then there is an ultrafilter $\mathcal{D}$ on λ such that whenever A and B are elementarily equivalent relational structures of cardinality $< \mu$, then $A^\lambda/\mathcal{D}$ and $B^\lambda/\mathcal{D}$ are isomorphic. (We may easily extend this theorem to cover the situation involving elementary embeddings by adding constants to name domain elements. This way we get commuting square diagrams.)

Assume the maps f_δ are all co-elementary. (Co-existential maps are handled in a similar way.) Then we have a collection of homeomorphisms $k_\delta: X_\delta I_\delta\backslash\mathcal{D}_\delta \to Y_\delta J_\delta\backslash\mathcal{E}_\delta$ making the obvious mapping squares commute. Now, using Shelah's theorem, find an ultrafilter $\mathcal{D}$, on a very large index set I, so that the elementarity of the diagonal embeddings $d_\delta: F(X_\delta) \to F(X_\delta)^{I_\delta}/\mathcal{D}_\delta$ and $e_\delta: F(Y_\delta) \to F(Y_\delta)^{J_\delta}/\mathcal{E}_\delta$ (between normal disjunctive lattices), $\delta \in \Delta$, is simultaneously witnessed. When we apply the spectrum functor $S(\ )$ to these squares, we then have witnesses to the co-elementarity of the co-diagonal maps $p_\delta: X_\delta I_\delta\backslash\mathcal{D}_\delta \to X_\delta$ and $q_\delta: Y_\delta I_\delta\backslash\mathcal{E}_\delta \to Y_\delta$. (Recall that $S(\ )$ converts canonical ultrapower diagrams to canonical ultracopower diagrams, and that there is an iteration theorem for ultracopowers: 'ultracopowers of ultracopowers are ultracopowers,' see [2].) Now $(\ )I\backslash\mathcal{D}$, as an operator on compacta, is functorial; it is covariant, and preserves (reflects as well) many interesting properties of continuous maps. In particular, as is straightforward to check, it preserves and reflects the properties of being surjective, of being one-one, of being co-elementary, and of being co-existential. Thus, when we apply $(\ )I\backslash\mathcal{D}$ to the homeomorphisms k_δ, we have a new collection of homeomorphisms; hence we have a homeomorphism linkage between $X_\delta I\backslash\mathcal{D}$ and $Y_\delta I\backslash\mathcal{D}$, witnessing the co-elementarity of f_δ. □

PROPOSITION 2.2. *Let* $f: X \to Y$ *and* $g: Y \to Z$ *be functions between compacta. If* f *and* g *are co-existential maps, then so is* gf*; if* gf *is a co-existential map and* f *is a continuous surjection, then* g *is a co-existential map.*

Proof. Note that the second assertion above says more than just closure under terminal factors; we do not assume the co-existentiality of f. The proof of this is immediate from the definition, so we now consider the issue of closure under composition. Using Lemma 2.1, we have $p_{Y,\mathcal{D}}: YI\backslash\mathcal{D} \to Y$, $u: YI\backslash\mathcal{D} \to X$, $p_{Z,\mathcal{D}}: ZI\backslash\mathcal{D} \to Z$, and $v: ZI\backslash\mathcal{D} \to Y$, witnessing the co-existentiality of f and g. Since g is co-existential, so is the ultracopower map $gI\backslash\mathcal{D}$. Let $r: W \to ZI\backslash\mathcal{D}$ and $s: W \to YI\backslash\mathcal{D}$ witness the fact. (So r is co-elementary and s is a continuous surjection.) Then, because compositions of co-elementary maps are co-elementary, we have a witness to the conclusion that gf is a co-existential map. □

Before proceeding, some notation is in order. If $\langle X_i : i \in I\rangle$ is a family of compacta and $\mathcal{D}$ is an ultrafilter on I, then, as noted before, $\sum_{\mathcal{D}} X_i$ may be viewed as the space of maximal filters of the normal disjunctive lattice $\prod_{\mathcal{D}} F(X_i)$. Let $A_i \subseteq X_i$ for each $i \in I$. Then we extend the 'sharp' notation mentioned above by defining $(\prod_{\mathcal{D}} A_i)^\sharp := \{m \in \sum_{\mathcal{D}} X_i : \prod_{\mathcal{D}} A_i$ contains a member of $m\}$. $(\prod_{\mathcal{D}} A_i)^\sharp$ is closed (resp., open) in the ultracopower if and only if $\{i \in I : A_i$ is closed (resp., open) in $X_i\} \in \mathcal{D}$. Of course, if each A_i is closed in X_i, and therefore a compactum in its own right, there is a natural homeomorphism between $\sum_{\mathcal{D}} A_i$ and the subspace $(\prod_{\mathcal{D}} A_i)^\sharp$.

The following lemma tells us what happens to pre-images of open sets under codiagonal maps. We use an overline to indicate topological closure.

LEMMA 2.3. *Let U be an open subset of the compactum X. Then $p_{X,\mathcal{D}}^{-1}[U] = \bigcup\{(\overline{V}^I/\mathcal{D})^\sharp : \overline{V} \subseteq U, V$ open in $X\} \subseteq (U^I/\mathcal{D})^\sharp$.*

Proof. $p_{X,\mathcal{D}}(m) = x$ if and only if for each open neighborhood W of x, $m \in (W^I/\mathcal{D})^\sharp$. Suppose $m \in p_{X,\mathcal{D}}^{-1}[U]$, say $p_{X,\mathcal{D}}(m) = x$. Let V be any open set with $x \in V \subseteq \overline{V} \subseteq U$. Then $m \in (V^I/\mathcal{D})^\sharp \subseteq (\overline{V}^I/\mathcal{D})^\sharp$.

For the reverse inclusion, suppose $x = p_{X,\mathcal{D}}(m) \notin U$. Pick V open with $\overline{V} \subseteq U$, and let W be an open neighborhood of x such that $\overline{W} \cap \overline{V} = \emptyset$. Then $\overline{W}^I/\mathcal{D} \in m$. But $\overline{W}^I/\mathcal{D} \cap \overline{V}^I/\mathcal{D} = (\overline{W} \cap \overline{V})^I/\mathcal{D} = \emptyset$, so $\overline{V}^I/\mathcal{D} \notin m$; i.e., $m \notin (\overline{V}^I/\mathcal{D})^\sharp$. □

THEOREM 2.4. *Let $f: X \to Y$ be a co-existential map between compacta. Then there exists a $\cup$-semilattice homomorphism f^* from the subcompacta of Y to the subcompacta of X such that for each subcompactum K of Y:*

(i) $f[f^*(K)] = K$;
(ii) $f^{-1}[U] \subseteq f^*(K)$ *whenever U is open and $U \subseteq K$;*
(iii) *the restriction $f|f^*(K)$ is a co-existential map from $f^*(K)$ to K; and*
(iv) $f^*(K) \in \mathbf{K}$ *whenever $K \in \mathbf{K}$ and $\mathbf{K} \subseteq \mathbf{CH}$ is closed under ultracopowers and continuous images.*

Proof. Let $g: YI\backslash\mathcal{D} \to X$ witness the fact that f is a co-existential map (i.e., $fg = p_{\mathcal{D}}$), and let K be a subcompactum of Y. Then $(K^I/\mathcal{D})^\sharp$ is a subcompactum of $YI\backslash\mathcal{D}$, naturally homeomorphic to $KI\backslash\mathcal{D}$, and $p_{\mathcal{D}}|(K^I/\mathcal{D})^\sharp$ may be viewed as the natural codiagonal from $KI\backslash\mathcal{D}$ to K. Thus $f^*(K) := g[(K^I/\mathcal{D})^\sharp]$ is a subcompactum of X, and $f|f^*(K)$ is a co-existential map onto its image K. That condition (iv) is true is now apparent.

Suppose U is open in Y, with $U \subseteq K$. Then $f^{-1}[U] = g[p_{\mathcal{D}}^{-1}[U]]$. But by Lemma 2.3, $p_{\mathcal{D}}^{-1}[U] \subseteq (U^I/\mathcal{D})^\sharp \subseteq (K^I/\mathcal{D})^\sharp$. Thus $f^{-1}[U] \subseteq f^*(K)$.

It remains to show that the mapping f^* is a $\cup$-semilattice homomorphism. If K_1 and K_2 are subcompacta of Y, then $f^*(K_1 \cup K_2) = g[((K_1 \cup K_2)^I/\mathcal{D})^\sharp] = g[(K_1^I/\mathcal{D})^\sharp \cup (K_2^I/\mathcal{D})^\sharp] = f^*(K_1) \cup f^*(K_2)$. Clearly $f^*(\emptyset) = \emptyset$ and $f^*(Y) = X$. □

A number of properties, not generally preserved by continuous surjections between compacta, are now easily seen to be preserved by co-existential maps.

PROPOSITION 2.5. *The following properties are preserved by co-existential maps*:

(i) *being infinite*;
(ii) *being disconnected*;
(iii) *being a Boolean space*;
(iv) *being an indecomposable continuum*; *and*
(v) *being a hereditarily indecomposable continuum.*

Proof. Let f, X, Y, g, $\mathfrak{D}$ and K be as in the proof of Theorem 2.4. If Y is finite (resp., connected), then so is $Y I\backslash\mathfrak{D}$, and hence X. If Y is not Boolean, then we may choose K to be an infinite subcontinuum of Y. This forces $f^*(K)$ to be an infinite subcontinuum of X.

Suppose X is an indecomposable continuum (so there is no way to write X as the union of two proper subcontinua). Then Y is a continuum. Suppose $Y = K_1 \cup K_2$, where each K_n is a proper subcontinuum. Then each $f^*(K_n)$ is also a proper subcontinuum of X, and $f^*(K_1) \cup f^*(K_2) = f^*(K_1 \cup K_2) = f^*(Y) = f^{-1}[Y] = X$. This contradicts the assumption that X is indecomposable.

Suppose X is a hereditarily indecomposable continuum (so no subcontinuum of X is decomposable). If K is a decomposable subcontinuum of Y, then $f^*(K)$ is a decomposable subcontinuum of X, since $f|f^*(K)$ is a co-existential map onto K. (Use the result of the last paragraph.) □

We can actually improve on (iii) of Proposition 2.5. Noting that being a Boolean space means being of covering dimension zero, it is tempting to conjecture that co-existential maps cannot raise dimension. This turns out to be the case, but it seems we need more than just Theorem 2.4 for a proof. For us, the handiest version of 'being of dimension $\leq n$' is due to E. Hemmingsen (see [12]): A normal space Y is of covering dimension $\leq n$, $n < \omega$, ($\dim(Y) \leq n$) if whenever $B_1, \ldots, B_{n+2}$ is a family of closed subsets of Y and $B_1 \cap \cdots \cap B_{n+2} = \emptyset$, then there is a family $F_1, \ldots, F_{n+2}$ of closed subsets of Y such that $F_m \supseteq B_m$ for each $0 \leq m \leq n+2$, $F_1 \cap \cdots \cap F_{n+2} = \emptyset$, and $F_1 \cup \cdots \cup F_{n+2} = Y$.

THEOREM 2.6. *Suppose $f: X \to Y$ is a co-existential map between compacta, $n < \omega$. If $\dim(X) \leq n$, then $\dim(Y) \leq n$.*

Proof. Let $B_1, \ldots, B_{n+2} \subseteq Y$ be given, as per the characterization given above. Let $g: Y I\backslash\mathfrak{D} \to X$ witness the fact that f is a co-existential map, and set $A_m := f^{-1}[B_m]$, $1 \leq m \leq n+2$. Each A_m is closed in X, and $A_1 \cap \cdots \cap A_{n+2} = \emptyset$. Since $\dim(X) \leq n$, we find appropriate closed sets $E_m \supseteq A_m$, $1 \leq m \leq n+2$, witnessing the fact. Pull these sets back to the ultrapower via g. We get a closed cover whose intersection is empty. Using compactness, we can obtain basic closed sets $\sum_{\mathfrak{D}} F_{m,i} \supseteq g^{-1}[E_m] \supseteq B_m I\backslash\mathfrak{D}$, $1 \leq m \leq n+2$, such that

$\bigcap_{m=1}^{n+2} \sum_{\mathcal{D}} F_{m,i} = \emptyset$. By elementary ultraproduct considerations, we then have $\{i \in I : F_{m,i} \supseteq B_m$ for all $1 \leq m \leq n+2$ and $\bigcap_{m=1}^{n+2} F_{m,i} = \emptyset$ and $\bigcup_{m=1}^{n+2} F_{m,i} = Y\} \in \mathcal{D}$. For any i in this set, then, we have witnesses for the fact that $\dim(Y) \leq n$. □

A continuous map between compacta is called **weakly confluent** in the literature (see [22]) if every subcontinuum in the range is the image of a subcontinuum in the domain. It follows from Theorem 2.4, plus the fact that connectedness is preserved by ultracopowers and continuous images, that co-existential maps are weakly confluent. There are many related notions discussed in [22], and weak confluency is the most general. For example, **confluency** in a mapping says that each component of the pre-image of a subcontinuum maps onto the subcontinuum. We do not know whether there is a direct relation between co-existential maps and confluent maps, but suspect not.

The strongest notion along these lines is monotonicity. A continuous map is called **monotone** if the pre-images of points under that map are connected sets. In [6] it is shown that co-elementary maps onto locally connected compacta are monotone; actually this remains true for co-existential maps.

THEOREM 2.7. *Let $f: X \to Y$ be a co-existential map between compacta. If Y is locally connected, then f is monotone.*

Proof. Let $y \in Y$ be given, and let $\mathcal{U}$ be a neighborhood basis for y consisting of connected open sets. For each subcontinuum $\overline{U}$, $U \in \mathcal{U}$, let $f^*(\overline{U})$ be the subcontinuum of X guaranteed to exist by Theorem 2.4 ($f^*(\overline{U})$ being connected by virtue of being a continuous image of an ultracopower of $\overline{U}$). Since $f^{-1}[U] \subseteq f^*(\overline{U})$, we have $f^{-1}[\{y\}] \subseteq f^*(\overline{U})$. Now $\mathcal{U}$ is a directed set under the ordering of reverse inclusion. Moreover, by Theorem 2.4, f^* is a $\cup$-semilattice homomorphism, hence order preserving. Thus the family $\mathcal{V} := \{f^*(\overline{U}) : U \in \mathcal{U}\}$ is also directed under reverse inclusion. Since $\bigcap \mathcal{U} = \{y\}$, it follows that $\bigcap \mathcal{V} = f^{-1}[\{y\}]$. Directed intersections of subcontinua are subcontinua (see [28]); consequently $f^{-1}[\{y\}]$ is connected. □

The next result, an easy application of our methods, deals with isolated points.

PROPOSITION 2.8. *Let $f: X \to Y$ be a co-existential map between compacta.*

(i) *If y is an isolated point of Y, then $f^{-1}[\{y\}]$ is an open singleton in X.*
(ii) *If x is an isolated point of X and Y is locally connected, then $f(x)$ is an isolated point of Y.*

Proof. To prove (i), let $y \in Y$ be isolated. Letting K be $\{y\}$, the corresponding set $f^*(K)$ guaranteed by Theorem 2.4 must be a singleton open set, and must therefore be the pre-image under f of y.

To prove (ii), note that $f^{-1}[\{f(x)\}]$ is connected by Theorem 2.7. Since $\{x\}$ is open in X, we have $f^{-1}[\{f(x)\}] = \{x\}$. $\{f(x)\}$ is therefore open since continuous surjections between compacta are quotient maps. □

The question naturally arises as to the relationship between co-existential maps and co-elementary maps; as yet we have no examples of co-existential maps that are not co-elementary. To remedy the situation, let us define an **arc** to be any homeomorphic copy of the closed unit interval in the real line. Arcs are of central importance in the study of compacta (especially continua); they can be characterized as being metrizable continua with exactly two noncut points (a classic result of R. L. Moore). In [3] it is shown (using another classic result of Moore) that any Peano continuum (i.e., locally connected metrizable continuum) is an arc, as long as it is co-elementarily equivalent to an arc. It follows immediately that co-elementary images of arcs are arcs. We can even do a little better.

PROPOSITION 2.9. *Co-existential images of arcs are arcs.*

Proof. Let $f: X \to Y$ be a co-existential map between continua, where X is an arc. X is locally connected; hence so is Y. By Theorem 2.7, then, f is monotone. Now Y is an infinite metrizable continuum. So the monotonicity of f, together with the fact that X is an arc, tells us that Y has exactly two noncut points. Hence Y is an arc. □

The main result of [6] is that any monotone continuous surjection between two arcs is a co-elementary map. This gives us the following.

THEOREM 2.10. *If $f: X \to Y$ is a co-existential map between compacta, and if X is an arc, then f is co-elementary.*

Proof. Y is an arc by Proposition 2.9, f is monotone by Theorem 2.7. f is therefore co-elementary by the main result (Proposition 2.7) of [6]. □

PROPOSITION 2.11. *If $f: X \to Y$ is a co-elementary map between compacta, and if X is an arc, then for any subcontinuum C of X, $f|C$ is co-elementary onto its image if and only if C and $f[C]$ have the same cardinality.*

Proof. In the case C is a singleton, there is no problem. So suppose C is nondegenerate. Then C is an arc. If $f[C]$ is a singleton, then co-elementarity fails of course. Otherwise, $f[C]$ is an arc too. $f|C: C \to f[C]$ is clearly a monotone continuous surjection, and is therefore co-elementary by Proposition 2.7 of [6]. □

Co-elementary equivalence preserves covering dimension by results of [2]. Thus any example of a co-existential map that changes dimension is an example of a co-existential map that is not co-elementary. (Of course, by Theorem 2.6, a co-existential map that changes dimension must necessarily lower it.)

EXAMPLE 2.12. We construct a dimension-lowering co-existential map $f: X \to Y$ between locally connected metrizable continua as follows.

Set $X := ([0, 1/2] \times \{0\}) \cup [1/2, 1]^2$, $Y := [0, 1/2] \times \{0\}$, $Z := [0, 1] \times \{0\}$, all subsets of the euclidean plane. Y and Z are arcs; X is a 'kite-with-tail.' Y is a subcontinuum of X (resp., Z), and we simply let f (resp., g) retract X (resp., Z)

onto Y, collapsing the square (resp., the interval $[1/2, 1]\times\{0\}$) to the point $\langle 1/2, 0\rangle$. Since Y and Z are arcs, and g is a monotone continuous surjection, we know that g is co-elementary by Proposition 2.7 of [6]. We now let h map Z to X by leaving each element of $[0, 1/2] \times \{0\}$ fixed and taking $[1/2, 1] \times \{0\}$ continuously onto $[1/2, 1]^2$ in such a way that $\langle 1/2, 0\rangle$ remains fixed. Clearly $fh = g$, and f is therefore a co-existential map.

3. An Analog of the Löwenheim–Skolem Theorem (Sharper Version)

Of the many assertions that lie under the rubric 'Löwenheim–Skolem,' the one with the most 'algebraic' phrasing takes on the form of a factorization theorem: Let $\mathcal{L}$ be a first-order lexicon, with $f\colon A \to B$ an embedding of $\mathcal{L}$-structures. If κ is an infinite cardinal number such that $|A| + |\mathcal{L}| \leq \kappa \leq |B|$, then there exists an $\mathcal{L}$-structure C and embeddings $g\colon A \to C$, $h\colon C \to B$ such that $|C| = \kappa$, h is an elementary embedding and $f = hg$.

For a dualized version of this in the compact Hausdorff setting, we must eliminate reference to a first-order lexicon, as well as decide what cardinal invariant of compacta is to take the place of the underlying-set cardinality. If we take a clue from Stone duality, noting that the weight of a Boolean space equals the cardinality of its clopen algebra, then we must decide upon weight as the invariant of choice. (There are, of course, other reasons to choose weight, but the one given above is the most accessible.) We denote the weight of X by $w(X)$.

THEOREM 3.1. *Let $f\colon X \to Y$ be a continuous surjection between compacta, with κ an infinite cardinal such that $w(Y) \leq \kappa \leq w(X)$. Then there is a compactum Z and continuous surjections $g\colon X \to Z$, $h\colon Z \to Y$ such that $w(Z) = \kappa$, g is a co-elementary map, and $f = hg$.*

Proof. By way of a preliminary comment, this is a sharper version of the Löwenheim–Skolem theorem proved in [4]. In that result we used the model theory of lattices to obtain Z such that $w(Z) \leq \kappa$. There is a bit more work involved in making sure that $w(Z)$ may be any prescribed cardinal in the interval of possibilities.

There is no loss of generality in assuming that Y is infinite; so first assume that $\kappa = w(Y)$, and let $\mathcal{B} \subseteq F(Y)$ be a lattice base of cardinality κ. Let $\varphi := f^F|\mathcal{B}$. Then we may treat $S(\mathcal{B})$ as Y and φ^S as f. By the usual Löwenheim–Skolem theorem, there is a normal disjunctive lattice A and lattice embeddings $\psi\colon \mathcal{B} \to A$, $\theta\colon A \to F(X)$ such that $|A| = \kappa$, θ is an elementary embedding, and $\varphi = \theta\psi$. Set $Z := S(A)$, $g := \theta^S$, and $h := \psi^S$. Then g is a co-elementary map, $w(Z) \leq |A| = \kappa$, and $hg = (\theta\psi)^S = \varphi^S = f$. (Note: This much was proved in [3].) Now h maps Z onto Y; and, since continuous surjections between compacta cannot raise weight, we infer that $w(Z) = \kappa$.

The rest of the argument is not model-theoretic at all. Suppose we could factor f into continuous surjections $g\colon X \to Z$, $h\colon Z \to Y$, where $w(Z) = \kappa$. Then we could apply the argument in the last paragraph to factor g, and we would be done.

To effect this factorization ($f = hg$), we use the Gel'fand–Naĭmark duality theorem between **CH** and the category **CBA** of commutative (Banach) B^*-algebras and nonexpansive linear maps. (Recall [26] that a B^*-algebra is a complex Banach algebra with a unary operation $(\ )^*$ satisfying: $(a+b)^* = a^* + b^*$; $(ab)^* = b^*a^*$; $(\lambda a)^* = \overline{\lambda}a^*$ (λ is a complex scalar, $\overline{\lambda}$ is the complex conjugate of λ); $\|a^*\| = \|a\|$; and $a^{**} = a$.) For $X \in$ **CH**, $C(X)$ is the B^*-algebra of continuous complex-valued functions on X; the norm is the supremum norm, and the involution $(\ )^*$ is defined pointwise by $\varphi^*(x) := \overline{\varphi(x)}$. The other half of the duality is the maximal ideal space construction $M(\ )$, restricted to objects in **CBA**.

Claim 1. $w(X) = d(C(X))$, where d is the **density**, the least cardinality of a dense subset of a topological space.

To see this, let $\mathcal{U}$ be the collection of open disks with rational centers and rational radii in the complex plane, and suppose $\Phi \subseteq C(X)$ is a dense subset. Set $\mathcal{V} := \{\varphi^{-1}[U] : \varphi \in \Phi, U \in \mathcal{U}\}$. $\mathcal{V}$ is an open base for X. Indeed, if W is an open neighborhood of x in X, let $\psi : X \to [0,1]$ take x to 0 and $X \setminus W$ to 1. Let $\varphi \in \Phi$ be such that $\|\varphi - \psi\| < 1/4$. Then $\|\varphi(0)\| < 1/4$ and $\|\varphi(y)\| > 3/4$ for $y \in X \setminus W$; hence if $U \in \mathcal{U}$ is the open disk of radius 1/2 centered at the origin, then $\varphi^{-1}[U]$ is a set in $\mathcal{V}$ containing x and contained in W. Since $|\mathcal{V}| \leq |\Phi|$, we have $w(X) \leq d(C(X))$. Now let $\mathcal{V}$ be an open base for X. By Weierstrass approximation, we can get a dense $\Phi \subseteq C(X)$ of cardinality $\leq |\mathcal{V}|$. Thus we have $d(C(X)) \leq w(X)$.

Claim 2. There is a subset Φ of $C(X)$, consisting of maps into the unit interval, such that:

(i) $|\Phi| = w(X)$; and
(ii) $\|\varphi - \psi\| \geq 1/2$ for all $\varphi, \psi \in \Phi$.

To see this, first note that any set Φ satisfying (ii) must have cardinality at most $d(C(X))$ (which is $w(X)$, by Claim 1). Now let Φ be maximal with regard to satisfying (ii), let U be the open disk of radius 1/2 and centered at the origin in the complex plane, and set $\mathcal{V} := \{\varphi^{-1}[U] : \varphi \in \Phi\}$. We need to show $\mathcal{V}$ is an open base for X. Assuming the contrary, there exist $x \in W \subseteq X$, W open in X, such that for all $\varphi \in \Phi$, if $\varphi(x) \in U$ (so $\varphi(x) \in [0, 1/2)$), then there is some $y \notin W$ with $\varphi(y) \in [0, 1/2)$. Let $\psi : X \to [0,1]$ take x to 0 and $X \setminus W$ to 1. If $\varphi \in \Phi$ and $\varphi(x) \in [1/2, 1]$, then $\|\psi - \varphi\| \geq |\psi(x) - \varphi(x)| \geq 1/2$. If $\varphi(x) \in [0, 1/2)$, let $y \in X \setminus W$ be such that $\varphi(y) \in [0, 1/2)$. Then $\|\psi - \varphi\| \geq |\psi(y) - \varphi(y)| \geq 1/2$. Thus $\Phi \cup \{\psi\}$ properly contains Φ and satisfies (ii), contradicting the maximality of Φ. Since $\mathcal{V}$ is an open base for X, and $|\mathcal{V}| \leq \Phi$, we have $w(X) \leq |\Phi|$. Thus $|\Phi| = w(X)$.

Now we have an embedding $f^C : C(Y) \to C(X)$. Let κ be such that $w(Y) \leq \kappa \leq w(X)$, and let $\Phi \subseteq C(X)$ have cardinality κ and satisfy (ii) in Claim 2.

Let $\Psi \subseteq C(Y)$ be dense of cardinality $w(Y)$. Then $\Phi \cup f^C[\Psi]$ is a subset of $C(X)$ of cardinality κ; closure of this set under the Banach algebra operations with scalar multiplication restricted to the rational complex numbers then results in a subring A, also of cardinality κ. The topological closure $\overline{A}$ of A is therefore a B^*-subalgebra of $C(X)$. Because $|A| = \kappa$, $d(\overline{A}) \leq \kappa$. Because $\Phi \subseteq \overline{A}$, $d(\overline{A}) \geq \kappa$. Because Ψ is dense in $C(Y)$, $f^C[C(Y)] \subseteq \overline{A}$. Thus, by applying the maximal ideal functor $M(\ )$ (i.e., $Z := M(\overline{A})$), we get the factorization we want. This completes the proof. □

Let us return, for the moment, to the model-theoretic setting; for simplicity, assume the underlying lexicon $\mathcal{L}$ is countable. A nice application of the Löwenheim–Skolem factorization theorem mentioned in the lead paragraph of this section is the following.

PROPOSITION 3.2. *Let $f: A \to B$ be an embedding between infinite $\mathcal{L}$-structures, with $\kappa \leq |A|$ an infinite cardinal. Suppose that for each $\mathcal{L}$-structure C of cardinality κ, and each elementary embedding $g: C \to A$, the composition fg is elementary* (*resp., existential*). *Then f is elementary* (*resp., existential*).

Proof. Let $\varphi(x_1, \ldots, x_n)$ be any (existential) formula, with $\langle a_1, \ldots, a_n\rangle$ an n-tuple from A. By Löwenheim and Skolem, there is an $\mathcal{L}$-structure C of cardinality κ and an elementary embedding $g: C \to A$ such that $g[C]$ contains each a_m. The desired conclusion is now immediate. □

The proof above, while extremely simple, suffers the major failing of being non-portable. The proposition has an obvious restatement in the language of compacta, and this proof sheds very little light on how to establish the dualized version. There is, however, an argument, based solely on ultraproduct considerations (inspired by C. C. Chang's ultraproduct proof of the compactness theorem, see [9, 11]), which can be easily transported to the topological context. Its drawback is that it is relatively cumbersome. We dualize that argument in the following.

THEOREM 3.3. *Let $f: X \to Y$ be a continuous surjection between infinite compacta, with $\kappa \leq w(Y)$ an infinite cardinal. Suppose that for each compactum Z of weight κ, and each co-elementary map $g: Y \to Z$, the composition gf is co-elementary* (*resp., co-existential*). *Then f is co-elementary* (*resp., co-existential*).

Proof. In model theory, the obstruction to being able to show, with mapping diagrams, that the class of elementary (resp., existential) embeddings is closed under terminal factors, resides in a failure of surjectivity: one cannot carry out a successful diagram chase. On the topological side, the obstruction to closure under initial factors resides in a failure of injectivity. Both these failures can be remedied somewhat with the use of ultra(co)products.

We establish our result for the co-elementary case; the co-existential case is handled in like fashion. First let Δ be the set of finite subsets of Y. For each $\delta \in \Delta$ there is a continuous mapping r_δ from Y to the closed unit interval, such that the restriction $r_\delta|\delta$ is one-one. Let $W_\delta := r_\delta[Y]$. By Theorem 3.1, there is a compactum Z_δ of weight κ, and continuous surjections $g_\delta: Y \to Z_\delta$, $t_\delta: Z_\delta \to W_\delta$, such that g_δ is co-elementary and $r_\delta = t_\delta g_\delta$. So each g_δ is a co-elementary map onto a compactum of weight κ, and $g_\delta|\delta$ is one-one.

Now, with the aid of Lemma 2.1, we build a diagram $\mathbf{D}_\delta$ that witnesses the co-elementarity above, for each $\delta \in \Delta$. More precisely, we have a single ultrafilter $\mathcal{D}$ on a set I, and homeomorphisms $h_\delta: XI\backslash\mathcal{D} \to Z_\delta I\backslash\mathcal{D}$, $k_\delta: YI\backslash\mathcal{D} \to Z_\delta I\backslash\mathcal{D}$ witnessing the co-elementarity of $g_\delta f$ and g_δ respectively. Letting $p_\delta: Z_\delta I\backslash\mathcal{D} \to Z_\delta$, $p_X: XI\backslash\mathcal{D} \to X$, and $p_Y: YI\backslash\mathcal{D} \to Y$ be the canonical codiagonal maps, we then have $p_\delta k_\delta = g_\delta p_Y$ and $p_\delta h_\delta = g_\delta f p_X$.

In general we cannot expect $fp_X = p_Y k_\delta^{-1} h_\delta$; however it is true that $g_\delta f p_X = g_\delta p_Y k_\delta^{-1} h_\delta$. We now proceed to take an 'ultracoproduct' of the diagrams $\mathbf{D}_\delta$.

For each $\delta \in \Delta$, let $\hat{\delta} := \{\gamma \in \Delta : \delta \subseteq \gamma\}$. Then the set $\{\hat{\delta} : \delta \in \Delta\}$ clearly satisfies the finite intersection property, and hence extends to an ultrafilter $\mathcal{H}$ on Δ. Form the '$\mathcal{H}$-ultracoproduct' diagram $\mathbf{D}$ in the obvious way. Then we have the canonical codiagonal maps $u_X: X\Delta\backslash\mathcal{H} \to X$, $v_X: (XI\backslash\mathcal{D})\Delta\backslash\mathcal{H} \to XI\backslash\mathcal{D}$, $u_Y: Y\Delta\backslash\mathcal{H} \to Y$, and $v_Y: (YI\backslash\mathcal{D})\Delta\backslash\mathcal{H} \to YI\backslash\mathcal{D}$. Moreover, $p_X v_X = u_X(p_X\Delta\backslash\mathcal{H})$ and $p_Y v_Y = u_Y(p_Y\Delta\backslash\mathcal{H})$ are (essentially) codiagonal maps from iterated ultracopowers, and these ultracopowers are isomorphic via

$$\left(\sum_{\mathcal{H}} k_\delta\right)^{-1}\left(\sum_{\mathcal{H}} h_\delta\right) = \sum_{\mathcal{H}} k_\delta^{-1} h_\delta.$$

We need to show that $fp_X v_X = p_Y v_Y (\sum_{\mathcal{H}} k_\delta)^{-1}(\sum_{\mathcal{H}} h_\delta)$. Suppose otherwise. Then we have some $x \in (XI\backslash\mathcal{D})\Delta\backslash\mathcal{H}$ with $[fp_X v_X](x) = y_1 \neq y_2 = [p_Y v_Y(\sum_{\mathcal{H}} k_\delta)^{-1}(\sum_{\mathcal{H}} h_\delta)](x)$. Because $\mathcal{H}$ is an ultrafilter, we have $\{\gamma \in \Delta : v_Y(\sum_{\mathcal{H}} k_\delta)^{-1}(\sum_{\mathcal{H}} h_\delta) = k_\gamma^{-1} h_\gamma v_X\} \in \mathcal{H}$; hence $\{\gamma \in \Delta : y_2 = [p_Y k_\gamma^{-1} h_\gamma v_X](x)\} \in \mathcal{H}$. Now $[g_\gamma p_Y k_\gamma^{-1} h_\gamma v_X](x) = [g_\gamma f p_X v_X](x)$ for all $\gamma \in \Delta$, so $\{\gamma \in \Delta : g_\gamma(y_2) = g_\gamma(y_1)\} \in \mathcal{H}$. However, $\{\gamma \in \Delta : g_\gamma(y_2) \neq g_\gamma(y_1)\} \supseteq \widehat{\{y_1, y_2\}} \in \mathcal{H}$. This contradiction completes the proof. □

4. An Analog of the Elementary Chains Theorem

The simplest version of the elementary chains theorem, due jointly to A. Tarski and R. L. Vaught, says that the union of an ω-indexed elementary chain of relational structures is an elementary extension of each of its summands. The obvious translation into the realm of compacta, what we refer to here as the **co-elementary chains hypothesis** (CECH) is the assertion that the inverse limit of an ω-indexed co-elementary chain of compacta is a co-elementary cover of each of its factors. More precisely, if $\langle X_n \overset{f_n}{\leftarrow} X_{n+1} : n < \omega\rangle$ is an ω-indexed family of co-elementary maps, and if $X := \varprojlim X_n$ is the inverse limit space with natural connecting maps

$g_n: X \to X_n, n < \omega$, then each g_n is a co-elementary map. (Recall that X is defined to be the subspace $\{\langle x_0, x_1, \ldots\rangle \in \prod_{n<\omega} X_n : x_n = f_{n+1}(x_{n+1})$ for all $n < \omega\}$, and g_n is projection to the n th factor, restricted to this subspace.)

The way one normally goes about proving the usual elementary chains theorem is to use induction on the complexity of formulas. We know of no other way; in particular, we know of no way to prove the result using the ultrapower theorem and chains of isomorphisms of ultrapowers. This perhaps speaks to an inherent lack of 'fine structure' in ultraproduct methods and definitely presents difficulties when one tries to find a direct proof of the CECH. While the CECH is true, the only proof we know of uses Gel'fand–Naĭmark duality and Banach model theory (see below).

The following weak version of the CECH *can* be proved by direct methods (avoiding Banach model theory, at any rate), and is worth while exploring. The difference between the weak and strong versions of the CECH may be compared to the difference between the amalgamation property for elementary embeddings (resp., co-elementary maps) and the existence of elementary pushouts (resp., co-elementary pullbacks).

PROPOSITION 4.1. *Let* $\langle X_n \overset{f_n}{\leftarrow} X_{n+1} : n < \omega\rangle$ *be a co-elementary chain of compacta. Then there exists a compactum X and co-elementary maps* $g_n: X \to X_n$, $n < \omega$, *such that for all* $n < \omega$, $g_n = f_n g_{n+1}$.

Proof. By Lemma 2.1, there is an ultrafilter $\mathcal{D}$ on a set I and homeomorphisms $X_n I\backslash\mathcal{D} \overset{h_n}{\leftarrow} X_{n+1} I\backslash\mathcal{D}$, $n < \omega$, such that for each n, $p_{X_n,\mathcal{D}} h_n = f_n p_{X_{n+1},\mathcal{D}}$. Let X be the inverse limit of this chain, with natural maps $k_n: X \to X_n I\backslash\mathcal{D}$. Since each h_n is a homeomorphism, so is each k_n, and we set $g_n := p_{X_n,\mathcal{D}} k_n$. □

Of course the weakness of our weak version of the CECH in Proposition 4.1 lies in the fact that the compactum X is not generally $\varprojlim X_n$. This would be remedied if we could show that the natural map $h: X \to \varprojlim X_n$, a continuous surjection, is actually co-elementary; but that seems to be beyond our methods.

The only way we know of to prove the CECH is to use Banach model theory. For reasons of space, our argument is far from self-contained; the interested reader is referred to the appropriate literature (e.g., [14 – 18]) for details. The skeleton of the approach is this: The basic lexicon of Banach model theory uses the usual symbols from the theory of abelian groups, plus countably many unary operation symbols to allow for scalar multiplication. (The scalar field can be either the field of real rationals or complex rationals.) Also there is a unary predicate symbol whose intended interpretation is the closed unit ball in a Banach space. (The theory is flexible enough to allow for additional operation symbols; e.g., lattice symbols, multiplication and involution.)

Banach model theory does not use all first-order formulas, just the positive-bounded ones (i.e., conjunction, disjunction, and quantification restricted to the unit ball). Also the relation of satisfaction is weakened to what is called *approximate satisfaction*. Once this is all laid out, the notions of elementary equivalence and elementary embedding make sense in the Banach setting. Finally there is an analogous ultrapower theorem, where 'ultrapower' means Banach ultrapower and 'isomorphic' means isometrically isomorphic. This tells us that the Banach notion of elementary embedding is the exact analog of our notion of co-elementary map.

We now come to the main bridge connecting the Banach world and the world of compacta, namely the Gel'fand–Naĭmark duality, a genuine category-theoretic duality between **CH** and the category **CBA** of commutative B^*-algebras and non-expansive linear maps (see the proof of Theorem 3.1 above). Using a result called the perturbation lemma [17], the obvious analog of the classic elementary chains theorem is proved in the Banach setting [18]. (Connecting maps are elementary embeddings in the Banach sense, and the full direct limit is the direct limit in the category **CBA** when the summands happen to be commutative B^*-algebras.) So the CECH is nothing but the Gel'fand–Naĭmark dual of the Banach elementary chains theorem restricted to objects in **CBA**. We thus have what is essentially a proof of the CECH.

THEOREM 4.2. *In Proposition* 4.1, *X may be taken to be the inverse limit* $\varprojlim X_n$.

REMARK 4.3. The technique for proving the elementary chains theorem allows for generalization to directed systems of elementary embeddings. This then gives us a corresponding generalized CECH.

5. An Analog of Robinson's Test

The notion of model completeness in first-order model theory was invented by A. Robinson, who was inspired by classical (Nineteenth-Century) algebra; in particular, Hilbert's *Nullstellensatz* concerning the location of solutions of systems of equations and inequations. (See, e.g., [19, 21].) One of many equivalent formulations of this notion is to say that a first-order theory is **model complete** if any embedding between two models of the theory is elementary. Since elementary embeddings are a great deal rarer than embeddings in general, it is very interesting when a first-order theory is discovered to be model complete. (Examples include (see [9]): dense linear orderings without endpoints; atomless Boolean algebras; algebraically closed fields; real closed fields; real closed ordered fields.) It is not surprising, then, that much effort has been expended in the study of this phenomenon; especially in the search for readily applicable tests to detect its presence.

Robinson's test says that a theory is model complete just in case every embedding between models of the theory is an existential embedding. Since testing for existentiality is ostensibly easier than testing for elementarity, this result is very important in the general study (as well as being a strikingly elegant application of the elementary chains theorem).

To give Robinson's test a proper phrasing in the compact Hausdorff setting, we first define a class of compacta to be **co-elementary** if it is closed under co-elementary equivalence and the taking of ultracoproducts. (This corresponds precisely to a class of relational structures being the class of models of a first-order theory; examples include (see [2]): Boolean spaces without isolated points; continua; (in)decomposable continua; compacta of covering dimension n, $n < \omega$; infinite-dimensional compacta. The class of locally connected compacta is *not* co-elementary [13, 5].) A co-elementary class is **model cocomplete** if any continuous surjection between two of its members is a co-elementary map. (Aside: recalling what completeness means in model theory, the right criterion for a co-elementary class to be **cocomplete** is that it consist of just one co-elementary equivalence class. It is not entirely trivial to show that co-elementary equivalence classes are indeed co-elementary classes, but true nonetheless. (See [7]. Lemma 2.1 greatly facilitates the proof that co-elementary equivalence classes are closed under ultracoproducts.) The notions of (co)completeness and model (co)completeness, while cognate, are not directly related logically. An obvious analog of the prime model test, also obviously true in the topological setting, is that a model cocomplete co-elementary class $\mathbf{K}$ is cocomplete if there is some $X \in \mathbf{K}$ such that every member of $\mathbf{K}$ continuously surjects onto X.) It is our belief that the reasons for studying model completeness in model theory remain just as compelling when we consider the topological analog; the present work is just a beginning of a process of discovering interesting co-elementary classes that are model cocomplete.

THEOREM 5.1. *A co-elementary class of compacta is model cocomplete if and only if every continuous surjection in the class is a co-existential map.*

Proof. Our proof is an exact dualization of the well-known proof of Robinson's test using elementary chains (see, e.g., [21]); we present it here for the sake of thematic completeness.

Only one direction is nontrivial. Assume that $\mathbf{K}$ is a co-elementary class with the property that every continuous surjection in $\mathbf{K}$ is a co-existential map, and let $f: X \to Y$ be a continuous surjection in $\mathbf{K}$. f is co-existential, so we get a compactum Z_1 and continuous surjections $g_0: Z_1 \to Y$, $h_0: Z_1 \to X$ with g_0 co-elementary and $g_0 = fh_0$. Z_1 is thus in $\mathbf{K}$, and we conclude that h_0 is co-existential. Repeat the procedure. We get a compactum Z_2 and continuous surjections $g_1: Z_2 \to X$, $h_1: Z_2 \to Z_1$ with g_1 co-elementary and $g_1 = h_0h_1$. $Z_2 \in \mathbf{K}$, and h_1 is thus a co-existential map. Proceeding in this way, we obtain compacta $Z_n \in \mathbf{K}$ and continuous surjections g_n, h_n, $n < \omega$. For $n \geq 1$, $h_n: Z_{n+1} \to Z_n$, $g_{n+1}: Z_{n+2} \to Z_n$, $g_{n+1} = h_nh_{n+1}$. Each h_n is co-existential, each g_n is co-elementary. The inverse limit Z of the maps h_n, $n < \omega$, is also the inverse limit of the co-elementary maps g_n, n odd (resp., n even). By the CECH (Theorem 4.2), there are co-elementary maps $u: Z \to X$ and $v: Z \to Y$ such that $v = fu$. Since the class of co-elementary maps is closed under terminal factors, it follows that f is co-elementary. □

There is a sharper version of Robinson's test (see, e.g., [9]). One assumes that the theory has no finite models; and that, for some infinite cardinal $\kappa \geq |\mathcal{L}|$, any embedding between models of cardinality κ is an existential embedding. Then the theory is model complete. The following is the topological analog of this fact.

THEOREM 5.2. *Let* **K** *be a co-elementary class containing no finite compacta. Then* **K** *is model cocomplete if (and only if) there is some infinite cardinal* κ *such that each continuous surjection between compacta of weight* κ *in* **K** *is a co-existential map.*

Proof. Let $f: X \to Y$ be a continuous surjection in **K**. By Theorem 5.1, it suffices to show that f is co-existential. Assume first that $\kappa \leq w(Y)$. By Theorem 3.3, it suffices to show that gf is co-existential whenever $g: Y \to Z$ is co-elementary and $w(Z) = \kappa$. So let g be given. By Theorem 3.1, we can obtain a factorization $u: X \to W$, $v: W \to Z$, where u is co-elementary, $w(W) = w(Z) = \kappa$, and $gf = vu$. Of course both W and Z are in **K**, so v is co-existential. Then vu is co-existential by Proposition 2.2.

If $\kappa > w(Y)$, then find an ultrafilter $\mathcal{D}$ on a set I such that $w(YI\backslash\mathcal{D}) \geq \kappa$ (see [2]). By the argument above, we conclude that $fI\backslash\mathcal{D}$ is co-existential. Immediately we infer (see Proposition 2.2) that f is co-existential. □

6. Co-inductive Co-elementary Classes

The usual elementary chains theorem shows that any model complete theory is **inductive**; i.e., closed under chain unions (indeed, direct limits of directed systems of embeddings). Being inductive, by the Chang–Łoś–Suszko theorem, is equivalent to being AE axiomatizable. A co-elementary class **K** is **co-inductive** if it is closed under inverse limits of directed systems of continuous surjections. Because of Theorem 4.2, we know that model cocomplete co-elementary classes are co-inductive.

The following notion dualizes that of being existentially closed relative to an elementary class of relational structures. Let **K** be a co-elementary class, with $X \in$ **K**. X is **co-existentially closed** in **K** if whenever $f: Y \to X$ is a continuous surjection and $Y \in$ **K**, then f is a co-existential map.

THEOREM 6.1. *Let* **K** *be a co-inductive co-elementary class, with* $X \in$ **K** *infinite. Then* X *is the continuous image of some* Y *that is co-existentially closed in* **K**, *such that* $w(Y) = w(X)$.

Proof. This is the dualization of a well-known property of inductive elementary classes, so we supply just the transition steps.

The class of normal disjunctive lattices whose maximal spectra lie in **K** is suitably denoted S^{-1}[**K**]. The class **NDL** of normal disjunctive lattices is itself an inductive elementary class. Since $S(\ )$ converts direct limits to inverse limits, we conclude that S^{-1}[**K**] is an inductive elementary class whenever **K** is a co-inductive co-elementary class. Suppose $X \in \mathbf{K}$ is infinite, and pick a lattice base $\mathcal{A}$ for X of infinite cardinality $w(X)$. Then $\mathcal{A} \in S^{-1}[\mathbf{K}]$, and there is an embedding $f\colon \mathcal{A} \to B$ for some existentially closed $B \in S^{-1}[\mathbf{K}]$ such that $|B| = |\mathcal{A}|$. (The proof of this is elementary model theory, involving unions of chains of embeddings (see [9, 19]).) Set $Y := S(B)$. Then f^S is (essentially) a continuous surjection from Y to X, so $w(X) \leq w(Y)$. But $w(Y) \leq |B| = |\mathcal{A}| = w(X)$, so the weights are equal. Finally, if $g\colon Z \to Y$ is a continuous surjection, $Z \in \mathbf{K}$, then $g^F\colon F(Y) \to F(Z)$ is an embedding in $S^{-1}[\mathbf{K}]$. Let $u\colon B \to F(Y)$ be the obvious separative embedding. Since B is existentially closed, $g^F u$ is an existential embedding. Since $S(\ )$ converts existential embeddings to co-existential maps (and separative embeddings to homeomorphisms), we infer that g is a co-existential map. □

By way of a linguistic aside, suppose 'blob' is a noun given to name a class of compacta that is a co-inductive co-elementary class. Then we may abbreviate 'X is co-existentially closed in the class of blobs' as 'X is a co-existentially closed blob.' Note, however, that this does not mean that X is a blob that is co-existentially closed in the class of compacta, any more than 'G is a free abelian group' means that G is a free group that happens to be Abelian. (Or, more prosaically, any more than 'George is a small elephant' means that George is a small entity that happens to be an elephant.) (W. V. O. Quine [23] applies the word *syncategorematic* to such adjectives as 'co-existentially closed,' 'free,' and 'small' used in this way.) Because of Theorem 6.1, we know that there exist co-existentially closed compacta in all infinite weights. The same can be said for the existence of co-existentially closed continua, since the class of continua is co-inductive co-elementary. Characterizing the co-existentially closed blobs, then, is a new problem each time we change the meaning of the word 'blob.' When this meaning is least restrictive, the problem turns out to have an easy solution.

PROPOSITION 6.2. *The co-existentially closed compacta are precisely the Boolean spaces without isolated points.*

Proof. We call a space without isolated points **self-dense**. Suppose X is a co-existentially closed compactum. Then X is a co-existential image of a Stone–Čech compactification $\beta(I)$ for a suitably large discrete space I. By Proposition 2.5(iii), X must therefore be Boolean. By Stone duality, its clopen algebra $B(X)$ must be an existentially closed Boolean algebra. Now the class of such algebras comprises the atomless Boolean algebras (see [21]); hence X is a self-dense Boolean space.

In the other direction, suppose X is a self-dense Boolean space, with $f\colon Y \to X$ a continuous surjection between compacta. Let I be a discrete space of cardinality $|Y|$, and set $Z := \beta(I) \times W$, where W is any self-dense Boolean space. Then Z is a self-dense Boolean space, and there is a continuous surjection $g\colon Z \to Y$. Let $h := fg$. Then h is a continuous surjection between self-dense Boolean spaces; hence $h^B\colon B(X) \to B(Z)$ is an embedding between atomless Boolean algebras. The class of such algebras is model complete; therefore h^B is elementary. This says that h is a co-elementary map, proving that f is co-existential. □

The problem of identifying the co-existentially closed continua is still open, and seems to be quite difficult. What little we know so far, besides the fact that they exist in all infinite weights, is the following.

PROPOSITION 6.3. *Every co-existentially closed continuum is indecomposable. (Consequently, every nondegenerate continuum is a continuous image of an indecomposable continuum of the same weight.)*

Proof. Suppose X is a decomposable continuum. Then (see [20, 28]) X has a proper subcontinuum K and a nonempty open set $U \subseteq K$. Pick $x_0 \in X \setminus K$, and let $Y := (X \times \{0, 1\})/\sim$ be the quotient space where the only nontrivial identification is $\langle x_0, 0\rangle \sim \langle x_0, 1\rangle$. Let $f: Y \to X$ be induced by projection onto the first factor. Then f is a continuous surjection between continua, and $f^{-1}[U]$ intersects both components of $f^{-1}[K]$. By Theorem 2.4, then, f cannot be a co-existential map and X is therefore not a co-existentially closed continuum. The parenthetical assertion is now an instant corollary of Theorem 6.1 and the above. □

In model theory, there is an important connection between model completeness and categoricity. In particular, there is Lindström's test (see [19, 21]), which says that if a consistent AE theory T has no finite models and is κ-categorical for some infinite cardinal $\kappa \geq |\mathcal{L}(T)|$, then T is model complete. The techniques developed here allow us to establish a topological analog of this. First define a co-elementary class **K** to be **κ-categorical**, where κ is an infinite cardinal, if: (a) there are compacta of weight κ in **K**; and (b) any two compacta of weight κ in **K** are homeomorphic. (Clearly, by the Löwenheim–Skolem theorem (Theorem 3.1), if **K** is a co-elementary class containing no finite compacta, and if **K** is κ-categorical for some cardinal κ, then **K** is cocomplete.)

THEOREM 6.4. *Let* **K** *be a co-elementary class containing no finite compacta. If* **K** *is co-inductive and κ-categorical for some infinite cardinal κ, then* **K** *is model cocomplete.*

Proof. By Theorem 5.2, it suffices to show that every continuous surjection between compacta of weight κ in **K** is co-existential. But Theorem 6.1 and κ-categoricity tell us that all compacta in **K** of weight κ are co-existentially closed in **K**. □

REMARKS AND QUESTIONS 6.5. (i) Concrete examples of co-elementary and co-existential maps are hard to find. One simple question we would like to settle is the following: Are monotone continuous surjections between nondegenerate locally connected continua necessarily co-existential maps? In particular, is the projection map from the closed unit square onto its first coördinate a co-existential map?

(ii) If **K** is a co-elementary class containing at least one infinite compactum, then $S^{-1}[\mathbf{K}]$ is *never* model complete. By Proposition 2.8 in [7], $f^F: F(Y) \to F(X)$ is an elementary embedding just in case $f: X \to Y$ is a homeomorphism.

The existence of codiagonal maps in **K** that are not homeomorphisms then provides embeddings in $S^{-1}[\mathbf{K}]$ that are not elementary.

(iii) Define a compactum X to be κ**-categorical** if: (a) the co-elementary equivalence class of X is κ-categorical; and (b) $w(X) = \kappa$. X is **categorical** if X is $w(X)$-categorical. The only categorical compacta we know of are Boolean (the Cantor discontinuum, for example, being $\aleph_0$-categorical). By the main result of [3], arcs are 'categorical' in a restricted sense: one may look only among the locally connected compacta. Indeed, every infinite compactum is co-elementarily equivalent to a compactum (of any prescribed infinite weight) that is not locally connected [5], so any examples of infinite categorical compacta must fail to be locally connected. In fact, if X is κ-categorical and $Y \equiv X$ has weight $\geq \kappa$, then Y co-elementarily surjects onto some compactum Z of weight κ. Z is then homeomorphic to X, which is not locally connected. It follows that Y is not locally connected either. In particular, no compactum co-elementarily equivalent to an $\aleph_0$-categorical compactum can be locally connected. For any cardinal κ (finite too), define a compactum X to be κ**-wide** if for each cardinal $\lambda < \kappa$, X contains a family of λ pairwise disjoint proper subcontinua with nonempty interiors. Clearly an infinite locally connected compactum is $\aleph_1$-wide; and decomposability for continua is equivalent to being 2-wide. By techniques similar to those used to show that the classes of (in)decomposable continua are co-elementary (see [5, 13]), one can prove that the class of $\aleph_0$-wide compacta is co-elementary (but not its complement). (Indeed, for $n < \omega$, $\sum_{\mathcal{D}} X_i$ is n-wide if and only if $\{i \in I : X_i \text{ is } n\text{-wide}\} \in \mathcal{D}$.) So if X is co-elementarily equivalent to an infinite locally connected compactum, then X is $\aleph_0$-wide. We do not think that the converse is true. (In a private conversation, C. W. Henson has told us that an $\aleph_0$-categorical compactum must fail to be $\aleph_0$-wide. His method uses a Banach version of the Ryll–Nardzewski theorem.)

(iv) We still lack examples of model cocomplete co-elementary classes that are not subclasses of the Boolean spaces.

(v) We lack examples of pairs of co-elementarily equivalent Peano continua that are not homeomorphic.

(vi) What, besides Proposition 6.3, can we infer about co-existentially closed continua? (In view of Theorem 2.6, for example, we suspect that co-existentially closed continua are **curves**; i.e., one-dimensional.) Dare we hope for a classically topological classification? Are any of the familiar examples of indecomposable continua (e.g., pseudo-arcs, solenoids) co-existentially closed?

(vii) Let **CCH** be the class of continua (considered as a full subcategory of **CH**). There is a single AE sentence γ in the first-order language of bounded lattices such that a normal disjunctive lattice A satisfies γ if and only if $S(A) \in \mathbf{CCH}$ (see [7]). Let **CNDL**:$= \{A \in \mathbf{NDL} : A \models \gamma\} = S^{-1}[\mathbf{CCH}]$. Then, as in the proof of Theorem 6.1, each existentially closed member of **CNDL** gives rise to a co-existentially closed member of **CCH**. Is this assignment surjective? Is the class of existentially closed members of **CNDL** an elementary class? (This deals with the existence of model completions/companions; see [21].) What about the dual question for the class of co-existentially closed members of **CCH**? (For example, if we could characterize the co-existentially closed members of **CCH** as the indecomposable curves, a co-elementary class, then the answer to the last question would be yes, and we would have a nice example of a model cocomplete co-elementary class of continua.)

(viii) Two properties closely related to being co-existentially closed in the class of continua are class(W) and class(C). A continuum X is **in class(W)** (resp., **in class(C)**) if whenever $f: Y \to X$ is a continuous surjection and Y is a continuum, then f is weakly confluent (resp., confluent). (These notions are due to A. Lelek [22].) Clearly if X is a co-existentially closed continuum, then X is in class(W). Class(W) is definitely broader, since it contains all arcs. Spaces in class(W) are **unicoherent**; i.e., possessed of the feature that the intersection of any two subcontinua whose union is everything must be connected. There are some nice characterizations of class(W), but they go beyond our scope. An elegant characterization of class(C) is simply being hereditarily indecomposable.

(ix) We would like to see a direct ultracoproduct proof of Theorem 4.2, one that doesn't rely on Banach model theory.

[Added in Proof: Re Question 6.5(vi), co-existentially closed continua are one-dimensional. Re Question 6.5(ix), there is now a proof of Theorem 4.2 that involves no Banach model theory. These results appear in paper [8].]

References

1. Banaschewski, B.: More on compact Hausdorff spaces and finitary duality, *Canad. J. Math.* **36** (1984), 1113–1118.
2. Bankston, P.: Reduced coproducts of compact Hausdorff spaces, *J. Symbolic Logic* **52** (1987), 404–424.
3. Bankston, P.: Model-theoretic characterizations of arcs and simple closed curves, *Proc. Amer. Math. Soc.* **104** (1988), 898–904.
4. Bankston, P.: Co-elementary equivalence for compact Hausdorff spaces and compact Abelian groups, *J. Pure Appl. Algebra* **68** (1990), 11–26.
5. Bankston, P.: Taxonomies of model-theoretically defined topological properties, *J. Symbolic Logic* **55** (1990), 589–603.
6. Bankston, P.: Co-elementary equivalence, co-elementary maps, and generalized arcs, *Proc. Amer. Math. Soc.* **125** (1997), 3715–3720.
7. Bankston, P.: On the topological model theory of normal disjunctive lattices (submitted).
8. Bankston, P.: A hierarchy of maps between compacta, *J. Symbolic Logic* (to appear).
9. Chang, C. C. and Keisler, H. J.: *Model Theory*, Third edn., North-Holland, Amsterdam, 1989.
10. Dacunha-Castelle, D. and Krivine, J.: Applications des ultraproduits à l'étude des espaces et des algèbres de Banach, *Studia Math.* **41** (1972), 315–334.
11. Eklof, P.: Ultraproducts for algebraists, in *Handbook of Mathematical Logic*, North-Holland, Amsterdam, 1977, pp. 105–137.
12. Engelking, R.: *Outline of General Topology*, North-Holland, Amsterdam, 1968.
13. Gurevič, R.: On ultracoproducts of compact Hausdorff spaces, *J. Symbolic Logic* **53** (1988), 294–300.

14. Heinrich, S. and Henson, C. W.: Model theory of Banach spaces, II: Isomorphic equivalence, *Math. Nachr.* **125** (1986), 301–317.
15. Heinrich, S., Henson, C. W. and Moore, L. C., Jr.: A note on elementary equivalence of $C(K)$ spaces, *J. Symbolic Logic* **52** (1987), 368–373.
16. Henson, C. W.: Nonstandard hulls of Banach spaces, *Israel J. Math.* **25** (1976), 108–144.
17. Henson, C. W.: Nonstandard analysis and the theory of Banach spaces, Lecture Notes in Math. 983, Springer, 1983, pp. 27–112.
18. Henson, C. W. and Iovino, J.: *Banach Space Model Theory*, I (lecture notes monograph, in preparation).
19. Hodges, W.: *Model Theory*, Cambridge University Press, Cambridge, 1993.
20. Kuratowski, K.: *Topology*, Vol. II, Academic Press, New York, 1968.
21. Macintyre, A.: Model completeness, in *Handbook of Mathematical Logic*, North-Holland, Amsterdam, 1977, pp. 139–180.
22. Nadler, S. B., Jr.: *Continuum Theory, An Introduction*, Marcel Dekker, New York, 1992.
23. Quine, W. V. O.: *Word and Object*, The MIT Press, Cambridge, MA, 1960.
24. Rosický, J.: Categories of models, *Seminarberichte Mathematik Informatik Fernuniversität* **19** (1984), 337–413.
25. Shelah, S.: Every two elementarily equivalent models have isomorphic ultrapowers, *Israel J. Math.* **10** (1971), 224–233.
26. Simmons, G. F.: *Introduction to Topology and Modern Analysis*, McGraw-Hill, New York, 1963.
27. Wallman, H.: Lattices and topological spaces, *Ann. Math. (2)* **39** (1938), 112–126.
28. Willard, S.: *General Topology*, Addison-Wesley, Reading, MA, 1970.

Applied Categorical Structures **8:** 67–80, 2000.
G. Brümmer & C. Gilmour (eds), Papers in Honour of Bernhard Banaschewski.

Nearness, Subfitness and Sequential Regularity

In honour of Bernhard Banaschewski on the occasion of his 70th birthday

H. HERRLICH
Fachbereich 3, Universität Bremen, Bibliothekstr. 1, D 28359 Bremen, Germany

A. PULTR*
Department of Applied Mathematics, Charles University, Malostranské nám. 25, CZ 11800 Praha 1, Czech Republic

(Received: 21 July 1996; accepted: 11 April 1998)

Abstract. In the point-free context, the structure of nearness has been so far studied in the regular case only. Here we answer the question as to how far beyond that one can go. It turns out that a frame (locale) (quasi-)admits a nearness iff it is subfit. Unlike in the case of spaces, where admitting nearness is a hereditary property, subfitness is not; therefore, also the hereditary subfitness (here called *sequential regularity* for reasons obvious from the properties presented) is studied. It is weaker than regularity and seems to be of some interest also in the spatial case.

Mathematics Subject Classifications (2000): 54A05, 06D10, 54E17, 54D10.

Key words: nearness space, nearnes frame (locale), subfit frame, regular, sequentially regular.

Introduction

Nearness structures (see, e.g., [5, 6], T-uniformities in [11]) have been studied in the point-free context so far only in the regular case (see, e.g., [3]), where it has particularly natural properties. In this paper we are concerned with the question as to how far beyond the regularily one can go, that is, what is the weakest separation property under which an admissible nearness makes a reasonable sense. The answer is the *subfitness* (also studied under the name of *conjunctivity* by Simmons in [15], see also [12]) as we show in Section 2 below.

Unlike the separation axiom of symmetry ($x \in \overline{\{y\}} \Rightarrow y \in \overline{\{x\}}$), necessary for admitting a nearness in spaces, subfitness is not hereditary. Thus, the question naturally arises as to which locales (point-free spaces) have the property that all their sublocales admit a nearness structure. This question is treated in Sections 3 and 4, and a necessary and sufficient condition is found (Theorem 4.5), in a form of a 'separation axiom' which we call *sequential regularity*. Finally, in Section 5 we compare the sequential regularity with more common separation axioms, and add a few further remarks.

* The support of the Alexander von Humboldt-Stiftung and the Grant Agency of the Czech Republic under Grant 201/93/950 is gratefully acknowledged.

1. Preliminaries

1.1. A *frame* is a complete lattice L satisfying the distributive law

$$a \wedge \bigvee S = \bigvee \{a \wedge s \mid s \in S\}$$

for all $a \in L$ and $S \subseteq L$, and a *frame homomorphism* $\varphi\colon L \to M$ is a map preserving finite meets including the top (usually denoted by 1), and arbitrary joins including the bottom (usually denoted by 0). The resulting category will be denoted by **Frm**.

EXAMPLE. The frame $\mathfrak{O}(X)$ of open sets of a topological space X, the homomorphism $\mathfrak{O}(f)\colon \mathfrak{O}(Y) \to \mathfrak{O}(X)$ obtained from a continuous map $f\colon X \to Y$ by putting $\mathfrak{O}(f)(u) = f^{-1}(u)$. Thus a contravariant functor

$$\mathfrak{O}\colon \mathbf{Top} \to \mathbf{Frm}$$

is obtained; the dual of **Frm** is called the *category of locales*, usually denoted by **Loc**, and $\mathfrak{O}$ becomes a covariant functor **Top** $\to$ **Loc**.

For details on frames see [8, 16].

1.2. For $x, y \in L$ write

$$x \prec y$$

if there is a $z \in L$ such that $x \wedge z = 0$ and $z \vee y = 1$. L is said to be *regular* if

$$a = \bigvee \{x \mid x \prec a\}$$

for each $a \in L$. Note that $\mathfrak{O}(X)$ is regular iff X is a regular space.

1.3. A *sublocale* (*part* in [7]) of a frame L is a surjective frame homomorphism $\varphi\colon L \to M$. We write

$$\varphi \sqsubseteq \psi$$

if there is a homomorphism γ such that $\gamma \circ \psi = \varphi$. Sublocales φ, ψ are considered to be equivalent if $\varphi \sqsubseteq \psi$ and $\psi \sqsubseteq \varphi$. The (equivalence classes of) sublocales can be represented by the associated congruences

$$x E_\varphi y \equiv \varphi(x) = \varphi(y)$$

and one has $\varphi \sqsubseteq \psi$ iff $E_\varphi \supseteq E_\psi$. In consequence, the preorder $\sqsubseteq$ is complete; the suprema and infima will be denoted by

$$\bigsqcup \varphi_i, \varphi_1 \sqcup \varphi_2, \varphi_1 \sqcap \varphi_2, \quad \text{etc.}$$

In particular, the congruence associated with $\bigsqcup \varphi_i$ is $\bigcap E_{\varphi_i}$.

The elements $a \in L$ induce the *open*, resp. *closed*, sublocales

$$\hat{a} = (x \longmapsto a \wedge x)\colon L \longrightarrow {\downarrow}a, \quad \text{resp.} \quad \check{a} = (x \longmapsto a \vee x)\colon L \longrightarrow {\uparrow}a$$

(where ${\downarrow}a = \{x \mid x \leq a\}$, ${\uparrow}a = \{x \mid x \geq a\}$), complementing each other.

The closure of a sublocale $\varphi\colon L \to M$ is

$$\overline{\varphi} = \check{c}\colon L \longrightarrow {\uparrow}c$$

where $c = \bigvee\{x \mid \varphi(x) = 0\}$, and it is easy to see that this is the smallest closed $\check{a}$ such that $\varphi \sqsubseteq \check{a}$.

The meet $\varphi \sqcap \hat{a}$ is obtained from the pushout

$$\begin{array}{ccc} L & \xrightarrow{\ \varphi\ } & M \\ {\scriptstyle \hat{a}}\big\downarrow & & \big\downarrow{\scriptstyle \widehat{\varphi(a)}} \\ {\downarrow}a & \longrightarrow & {\downarrow}\varphi(a). \end{array}$$

In particular,

$$\begin{aligned} &\varphi \sqcap \hat{a} \neq 0 \quad \text{iff} \quad \varphi(a) \neq 0, \\ &\check{b} \sqcap \hat{a} \neq 0 \quad \text{iff} \quad a \nleq b. \end{aligned} \tag{1.3.1}$$

Another formula we will use is that

$$\check{b} \sqsubseteq \hat{a} \quad \text{iff} \quad a \vee b = 1. \tag{1.3.2}$$

1.4. A *cover* of a frame L is a subset $A \subseteq L$ such that $\bigvee A = 1$. We say that A refines B and write $A \leq B$ if

$$\forall a \in A \quad \exists\, b \in B,\ a \leq b.$$

For $x \in L$ and a cover A we write

$$Ax = \bigvee\{a \in A \mid a \wedge x \neq 0\}$$

(see, e.g., [13, 3]). Obviously,

$$A \leq B \quad \& \quad x \leq y \Longrightarrow Ax \leq By$$

and

$$Ax \leq y \Longrightarrow x \prec y. \tag{1.4.1}$$

More generally, for a sublocale $\varphi\colon L \to M$,

$$A\varphi = \bigvee\{a \in A \mid \hat{a} \sqcap \varphi \neq 0\} = \bigvee\{a \in A \mid \varphi(a) \neq 0\}.$$

In particular (recall (1.3.1))

$$A\hat{x} = Ax, \qquad A\check{x} = \bigvee\{a \in A \mid a \nleq x\}. \tag{1.4.2}$$

It is easy to check that

$$A\varphi = A\overline{\varphi}. \tag{1.4.3}$$

1.5. CONGRUENCES, NUCLEI AND PRENUCLEI

If $R \subseteq L \times L$ is a relation on L, an $s \in L$ is called *saturated* if

$$\forall a, b, c \in L \quad aRb \Longrightarrow (a \wedge c \leq s \text{ iff } b \wedge c \leq s).$$

The system of saturated elements is closed under meets. If we define $\kappa(a)$ as the least saturated $s \geq a$ we obtain the so called *nucleus* (associated with R) $\kappa: L \to L$, a monotone function satisfying the formulas $a \leq \kappa(a), \kappa\kappa(a) = \kappa(a)$ and $\kappa(a \wedge b) = \kappa(a) \wedge \kappa(b)$. This κ can be viewed as a sublocale

$$\kappa: L \longrightarrow L/R = \{a \mid \kappa(a) = a\}$$

and $\{(a, b) \mid \kappa(a) = \kappa(b)\}$ is the least congruence containing R.

If $aRb \Rightarrow a \leq b$, the nucleus κ can be constructed from a *prenucleus*

$$\kappa_1(u) = u \vee \bigvee \{x \wedge b \mid x \wedge a \leq u \text{ for some } (a, b) \in R\}$$

by setting

$$\kappa_{\alpha+1}(u) = \kappa_1(\kappa_\alpha(u)), \quad \kappa_\alpha(u) = \bigvee_{\beta<\alpha} \kappa_\beta(u) \quad \text{for limit } \alpha,$$

and $\kappa(u) = \kappa_\alpha(u)$ where α is such that $\kappa_{\alpha+1}(u) = \kappa_\alpha(u)$. (See [2, 8, 9, 10].)

1.6. CONVENTION

To unify the notation in the spatial and in the general point-free case, we will denote open sets of spaces by small case letters (a, u, x, etc.), open covers by capitals (A, B, etc.) and points by greek letters (ξ, η, etc.).

1.7. A *nearness on a topological space* X (this has appeared, in essence, as T-*uniformity* [11], for a detailed study of this structure see, e.g., [5, 6]) is a nonvoid system $\mathfrak{N}$ of open covers such that

(a) if $A \in \mathfrak{N}$ and A refines B then $B \in \mathfrak{N}$, and
(b) for any $A, B \in \mathfrak{N}$ there is a common refinement $C \in \mathfrak{N}$.

Similarly, a *nearness on a frame* L (cf., e.g., [3]) is a nonvoid system $\mathfrak{N}$ of covers of L such that

(a) $A \in \mathfrak{N}$ & $A \leq B \Rightarrow B \in \mathfrak{N}$,
(b) $A, B \in \mathfrak{N} \Rightarrow A \wedge B = \{a \wedge b \mid a \in A,\ b \in B\} \in \mathfrak{N}$.

A nearness on a space induces a new topology by declaring u to be an $\mathfrak{N}$-neighbourhood of a point ξ if $A\xi = \bigcup\{a \in A \mid \xi \in a\} \subseteq u$ for some $A \in \mathfrak{N}$, and u to be open if it is an $\mathfrak{N}$-neighbourhood of each of its points (it should be noted that the $\mathfrak{N}$-neighbourhood $A\xi$ of ξ itself need not be open, indeed not even a neighbourhood of

ξ, in this new topology). We say that the given $\mathfrak{N}$ is *admissible* if this new topology coincides with the original one. The admissibility (also in the point-free case) will be discussed in Section 2; so far we will only prepare the notation. We write $a \lhd_{\mathfrak{N}} b$ if there is an $A \in \mathfrak{N}$ such that $Aa \leq b$ and, more generally, for sublocales, $\varphi \lhd_{\mathfrak{N}} \psi$ if $\widehat{A\varphi} \sqsubseteq \psi$ for an $A \in \mathfrak{N}$. If clear from the context, the subscript in $\lhd_{\mathfrak{N}}$ may be omitted.

Obviously (recall (1.4.1))

$$a \lhd b \Longrightarrow a \prec b$$

and in case of an $\mathfrak{N}$ containing all the finite covers,

$$a \lhd b \quad \text{iff} \quad a \prec b$$

(as $\{x, b\}a \leq b$ when $x \wedge a = 0$ and $x \vee b = 1$).

1.8. SOBRIETY AND T_D

A space X is said to be *sober* if

(Sob) X is T_0 and each meet-irreducible open $u \neq X$ (i.e., a u such that $u = u_1 \cap u_2 \Rightarrow u = u_j$ for some j) is of the form $X \setminus \overline{\{\xi\}}$ for some $\xi \in X$.

In particular,

if Y is sober then each frame homomorphism φ: $\mathfrak{O}(Y) \to \mathfrak{O}(X)$ is of the form $\mathfrak{O}(f)$ for a uniquely defined continuous map f: $X \to Y$

(see, e.g., [8]).

A space X is said to be a *T_D-space* (first defined in [1]) if

(T_D) for each $\xi \in X$ there is an open $u \subseteq X$ such that $\xi \in u$ and $u \setminus \{\xi\}$ is open.

In particular,

a space is T_D iff distinct subspaces $Y \subseteq X$ generate nonequivalent sublocales

$$(u \longmapsto u \cap Y)\colon \mathfrak{O}(X) \longrightarrow \mathfrak{O}(Y)$$

(see [4, 14]).

(Sob) is between T_0 and T_2 and incomparable with T_1. T_D is between T_0 and T_1 and incomparable with (Sob).

1.9. SUBFIT FRAMES

A frame is said to be *subfit* (see [7]) if the following condition holds:

(Sf1) *each open sublocale is a join of closed ones.*

This is well-known to be equivalent to the statement

(Sf) *whenever* $a \nleq b$ *there is a* c *such that*

$$a \vee c = 1 \neq b \vee c.$$

(The equivalence (Sf1) $\equiv$ (Sf) is very easy to see: in view of (1.3.2), (Sf1) can be rewritten as

(Sf2) $(x \wedge a \neq y \wedge a) \Longrightarrow (\exists c,\ a \vee c = 1$ and $x \vee c \neq y \vee c)$.

Now if $a \nleq b$ consider in (Sf2) $x = a$, $y = b$ to obtain (Sf). On the other hand, if $x \wedge a \neq y \wedge a$ consider in (Sf) $x \wedge a \nleq x \wedge y \wedge a$ or the other case to obtain (Sf2).)

It should be noted that (Sf) was called *conjunctivity* in [15] for frames and *weak regularity* for spaces [11].

2. (Quasi-)Admissible Nearnesses

2.1. Recall 1.7. Explicitly, a nearness $\mathfrak{N}$ is admissible on a space X if for each open $u \subseteq X$, $u = \{\xi \mid A\xi \subseteq u$ for an $A \in \mathfrak{N}\}$, or, in a reformulation more handy for a pointfree treatment:

$$\text{for each open } u \subseteq X, \quad u = \bigcup\{\varphi \subseteq X \mid A\varphi \subseteq u \text{ for an } A \in \mathfrak{N}\}. \tag{2.1.1}$$

The *admissibility in the point-free case* has been so far studied as the condition (see [3], cf. [13])

$$\forall u \in L, \quad u = \bigvee\{x \in L \mid Ax \leq u \text{ for an } A \in \mathfrak{N}\}. \tag{$*$}$$

This is correct – that is, in accordance with the admissibility in spaces – for regular L. As the formula implies regularity (see (1.4.1)), however, it cannot be used in a more general case. Thus, we will use, extending the formula (2.1.1) above, the following

DEFINITION. A nearness $\mathfrak{N}$ is *quasi-admissible on a frame* L if for each $u \in L$,

$$\hat{u} = \bigsqcup\{\varphi \mid \varphi \text{ sublocale of } L,\ A\varphi \leq u \text{ for an } A \in \mathfrak{N}\}. \tag{2.1.2}$$

NOTE. A nearness can be quasi-admissible on L without being admissible in the sense of the $(*)$ above: for instance, no quasi-admissible nearness on a nonregular frame is admissible. See, however, 2.4 below.

2.2. As $A\varphi = A\overline{\varphi}$ (recall (1.4.3)), the formula (2.1.2) can be rewritten to

$$\forall\, u \in L, \quad \hat{u} = \bigsqcup\{\breve{a} \mid \exists\, A \in \mathfrak{N},\ A\breve{a} \leq u\}, \tag{2.2.1}$$

and since the inclusion $\sqsupseteq$ is trivial, the inclusion $\sqsubseteq$ is what has to be checked in a concrete case.

More explicitly, by 1.3, the condition in (2.2.1) can be expressed by the requirement that

if for all $c \in L$ such that there is an $A \in \mathfrak{N}$ with $A\check{c} = \bigvee\{a \in A \mid a \nleq c\} \leq u$ one has $c \vee x = c \vee y$, then $x \wedge u = y \wedge u$.

We have a trivial

OBSERVATION. If $\mathfrak{N}$ is quasi-admissible and $\mathfrak{N} \subseteq \mathfrak{M}$ then $\mathfrak{M}$ is quasi-admissible.

Thus, L quasi-admits a nearness iff the conditions above hold for $\mathfrak{N} = \{\text{all covers}\}$.

2.3. THEOREM. *A frame L quasi-admits a nearness iff it is subfit.*

Proof. ($\Rightarrow$) follows immediately from (2.2.1).

($\Leftarrow$) We will use the formula (Sf) from 1.9 to prove (2.2.1) for the system of all covers. Thus, let $x \vee a = y \vee a$ for any a such that there is a cover A with $A\breve{a} = \bigvee\{b \in A \mid b \nleq a\} \leq u$. We want to prove that then $x \wedge u = y \wedge u$. Let, say, $x \wedge u \nleq y \wedge u$. By (Sf) there is a c with $1 = (x \wedge u) \vee c \neq (y \wedge u) \vee c$ (in particular, $u \vee c = 1$). But $\{x \wedge u, c\}\check{c} \leq x \wedge u \leq u$ and hence $x \vee c = y \vee c$ and we get a contradiction

$$(y \wedge u) \vee c = y \vee c = x \vee c = (x \vee c) \wedge (u \vee c) = (x \wedge u) \vee c = 1. \qquad \square$$

2.4. Recall that a nearness $\mathfrak{N}$ is said to be *regular* ([6]; *strong* in [3]) if for each $A \in \mathfrak{N}$ also

$$\widetilde{A} = \{a \mid a \lhd_{\mathfrak{N}} b \text{ for some } b \in A\} \in \mathfrak{N}.$$

We have

THEOREM. *A regular nearness is quasi-admissible if and only if it is admissible. Consequently, L admits a regular nearness if and only if it is regular.*

Proof. We have

$$\widetilde{A}\varphi = \bigvee\{a \in \widetilde{A} \mid \varphi(a) \neq 0\} \leq \bigvee\{a \mid a \lhd_{\mathfrak{N}} A\varphi\}$$

and hence

$$u = \bigvee\{\widetilde{A}\varphi \mid A \in \mathfrak{N},\ A\varphi \leq u\} \leq \bigvee\{a \mid a \lhd_{\mathfrak{N}} u\}.$$

For the second statement it now suffices to realize that the system of all covers of a regular frame is a regular nearness. $\square$

2.5. For the spatial case, the equivalence from 2.3 is analoguous to that from [11], Theorem 1. Another well-known necessary and sufficient condition for a space X to admit a nearness (see [5, 6]) is that it satisfied the symmetry axiom

$$\xi \in \overline{\{\eta\}} \quad \text{iff} \quad \eta \in \overline{\{\xi\}}$$

(which is the same as saying that the T_0-modification of X is T_1, and hence we will refer to this property as T_{1-0}).

Now in view of 2.3 one might expect that this T_{1-0} is equivalent to the subfitness of $\mathfrak{O}(X)$. But this is false, and the discrepancy should be explained.

Recall 1.8: If X were not T_D, two subsets may be distinct although the associated congruence coincides. In particular, this can happen with u and $\bigcup\{\varphi$ closed $\mid \varphi \subseteq u\}$. On the other hand, under T_D, T_{1-0} is indeed equivalent to subfitness. More exactly, one has

$$T_1 \equiv \text{(Sf)} \mathbin{\&} T_D, \quad \text{and}$$
$$T_{1-0} \equiv \text{(Sf)} \mathbin{\&} T_{D-0},$$

where T_{D-0} stands for the condition that the T_0-modification of the space is T_D.

3. Sobriety as a Partly Hereditary Subfitness

3.1. EXAMPLE. Consider the following spaces:

X: the ordinal ω with the cofinite topology,
Y: $\omega + 1$ with open $\emptyset$ and $Y \setminus K$ where K is finite and $\subseteq \omega$,
S: the Sierpiński space represented as the subspace of Y generated by the subset $\{0, \omega\}$,

and the maps:

j: the embedding $S \subseteq Y$,

k: the embedding $X \subseteq Y$,

f, f': the maps $f\colon X \to X$, resp. $f'\colon Y \to Y$, sending 0 to 0, $2n+1$ to $2n+2$, $2n+2$ to $2n+1$, and ω to ω,

$$\iota = \mathfrak{O}(j), \kappa = \mathfrak{O}(k), \ \varphi = \mathfrak{O}(f).$$

Obviously

(3.1.1) κ is an isomorphism,
(3.1.2) j is the equalizer of f' and id,
(3.1.3) $\iota \circ \kappa^{-1}$ is the coequalizer of φ and id.

((3.1.3) does not follow from (3.1.2) but it is easy to check.)

Now one sees that:

Subfitness is not inherited by subspaces (see $S \subseteq Y$), *and not even by equalizers with subfit range* (see (3.1.2)). *A T_1-space* (although all of its subspaces are T_1, and hence subfit) *can have a spatial sublocale that is not subfit and, again, even a coequalizer with a subfit domain* (see (3.1.3)).

In this section we will investigate the question of subfitness of spaces hereditary with respect to the spatial sublocales. In the next section, the general hereditarity question will be treated.

3.2. The following is certainly folklore and can be left to the reader:

LEMMA. *A space X is T_0 iff the embeddings $f\colon X \to Y$ are exactly those continuous maps f for which $\mathfrak{O}(f)\colon \mathfrak{O}(Y) \to \mathfrak{O}(X)$ are onto.*

3.3. Recall that an element $a \in L$ is *prime* if the following holds:

$$a \neq 1 \quad \text{and} \quad (x \wedge y \leq a \Longrightarrow x \leq a \text{ or } y \leq a).$$

LEMMA. *Let $a, b \in L$ be prime and let $a \lneqq b$. Then the Sierpiński space is a sublocale of L.*

Proof. Represent the Sierpiński frame ($\mathfrak{O}(S)$ of a Sierpiński space S) as $\mathbb{S} = \{0 < \frac{1}{2} < 1\}$.

Define $\varphi\colon L \to \mathbb{S}$ by setting

$$\varphi(x) = \begin{cases} 0 & \text{if } x \leq a, \\ \frac{1}{2} & \text{if } x \nleq a \text{ and } x \leq b, \\ 1 & \text{if } x \nleq b. \end{cases}$$

It is easy to check that φ is a frame homomorphism. □

3.4. THEOREM. *Let X be a T_1-space. Then the following statements are equivalent:*

(i) *X is sober,*
(ii) *each T_0-spatial sublocale is T_1,*
(iii) *each spatial sublocale is subfit,*
(iv) *X has no Sierpiński sublocale.*

Proof. (i) $\Rightarrow$ (ii) Let X be sober. Suppose we have a surjective homomorphism $\varphi\colon \mathfrak{O}(X) \to \mathfrak{O}(Y)$, with Y a T_0-space. Then, by 3.2, Y is a subspace and hence it is T_1.

(ii) $\Rightarrow$ (iii) Consider a surjective $\varphi\colon \mathfrak{O}(X) \to \mathfrak{O}(Y)$ and the T_0-modification $p\colon Y \to \widetilde{Y}$ of Y. Then $\mathfrak{O}(p)$ is an isomorphism. Considering $\mathfrak{O}(p)^{-1} \circ \varphi$ we see that $\widetilde{Y}$ is T_1, hence $\mathfrak{O}(\widetilde{Y})$ is subfit, and hence $\mathfrak{O}(Y)$ is subfit.

(iii) $\Rightarrow$ (iv) is trivial.

(iv) $\Rightarrow$ (i) Let X not be sober. Then there is a prime $u \in \mathfrak{O}(X)$ unequal to $X \setminus \{\xi\}$ for any ξ. Choosing $\xi \in X \setminus u$ we obtain prime

$$u \lneqq v = X \setminus \{\xi\}$$

and hence there is a Sierpiński sublocale by 3.3. □

4. Hereditary Subfitness

4.1. In this section we will characterize the frames L such that each sublocale of L quasi-admits a nearness. In view of 2.2, this amounts to the *hereditary subfitness*, that is,

(HSf) for every sublocale φ: $L \to M$, M is subfit.

4.2. LEMMA. *A frame L is hereditarily subfit iff for each sublocale* $\varphi : L \to M$ *the M satisfies*

(WSf) *if* $a \neq 0$ *in M then there is* $x \in M$ *such that* $x \neq 1$ *and* $x \vee a = 1$.

Proof. Indeed, if L has the property indicated then, by force of the sublocales $\check{b} = (x \mapsto x \vee b)$: $L \to \uparrow b$ it is, first, subfit, and as the property is hereditary, it is hereditarily subfit. □

NOTE. Obviously, (WSf) *can be reformulated as the implication*

$$(x \vee a = 1 \Longleftrightarrow x = 1) \Longrightarrow a = 0. \tag{4.2.1}$$

4.3. SEQUENTIAL REGULARITY

Let α be an ordinal. We say that L is α*-sequentially regular* if

(SRα) for each $a \in L$ there is a nondecreasing sequence $(a_\beta)_{\beta\leq\alpha}$ such that $a_0 = 0$, $a_{\beta+1} \leq \bigvee\{u \mid \exists\, x,\ x \vee a = 1 \text{ and } u \wedge x \leq a_\beta\}$, $a_\beta = \bigvee_{\gamma<\beta} a_\gamma$ for limit ordinals β, and $a = a_\alpha$.

L is *sequentially regular* if

(SR) L is α-sequentially regular for some α.

NOTE. One has $a_1 \leq \bigvee\{u \mid \exists x,\ x \vee a = 1 \text{ and } u \wedge x = 0\}$. Thus, 1-regularity is the same as regularity.

4.4. LEMMA. (SRα) *is hereditary and implies* (*hereditary*) *subfitness.*

Proof. Consider a sublocale φ: $L \to M$ and a $\varphi(a) \in M$. Let $(a_\beta)_{\beta\leq\alpha}$ be the sequence from (SRα). Then it is easy to check that $(\varphi(a_\beta))_{\beta\leq\alpha}$ is the required sequence for $\varphi(a)$.

As for the second statement, by 4.2 it suffices to show that (SRα) implies (WSf). Let $a \neq 0$ and let $(a_\beta)_{\beta\leq\alpha}$ be the associated sequence. Then there is a β such that

$$a_\beta \lneqq a_{\beta+1} \leq \bigvee\{u \mid \exists\, x,\ x \vee a = 1 \text{ and } u \wedge x \leq a_\beta\}.$$

In particular there has to be a u and an x such that

$$x \vee a = 1, \quad u \wedge x \leq a_\beta \quad \text{and} \quad u \nleq a_\beta.$$

Then x cannot be 1. □

4.5. THEOREM. *The following statements on a frame L are equivalent:*

(i) *Each sublocale of L quasi-admits a nearness,*
(ii) *L is hereditarily subfit,*
(iii) *L is sequentially regular,*
(iv) *each congruence on L is uniquely determined by the class containing* 1.

Proof. (i) ≡ (ii) by 2.2.

(ii) ⇒ (iii) In view of 4.2 and 4.3 it suffices to prove that if each sublocale of L satisfies (WSf) then L is (SR). Consider an $a \in L$ and the sublocale $L/_\sim$ where $\sim$ is the congruence generated by

$$R = \{(x, 1) \mid x \vee a = 1\}.$$

Recall 1.5. $L/_\sim$ consists of all the R-saturated elements of L.

CLAIM. *Each $y \geq a$ is saturated.*

(Indeed, if $x \vee a = 1$ and $u \wedge x \leq y$ we have

$$u = (x \vee a) \wedge u = (x \wedge u) \vee (a \wedge u) \leq y \vee a = y.)$$

By (4.2.1), if $L/_\sim$ satisfies (WSf) we have $a \sim 0$ and since by the above claim a is saturated, this amounts to

$$\kappa(0) = a$$

where κ is the nucleus associated with R. The prenucleus of R, κ_1, is defined by

$$\kappa_1(b) = \bigvee\{u \mid \exists\, x, \quad x \vee a = 1,\ u \wedge x \leq b\}.$$

Set $a_0 = 0$, $a_\beta = \kappa_\beta(0)$ for $\beta > 1$.

(iii) ⇒ (iv) Let E_1, E_2 be congruences on L, let $\varphi_j\colon L \to L/E_j$ be the associated sublocale homomorphisms. Let $xE_1 1$ iff $xE_2 1$ and let $aE_1 b$. We have $\varphi_2(a_0) = 0 \leq \varphi_2(b)$. Let $\varphi_2(a_\beta) \leq \varphi_2(b)$. If u is such that there is an x with $x \vee a = 1$ and $u \wedge x \leq a_\beta$, we have $\varphi_1(x \vee b) = \varphi_1(x \vee a) = 1$, hence $\varphi_2(x \vee b) = 1$ and

$$\varphi_2(u) = \varphi_2(u \wedge (x \vee b)) \leq \varphi_2(a_\beta) \vee \varphi_2(b) \leq \varphi_2(b).$$

Thus, $\varphi_2(a_{\beta+1}) \leq \varphi_2(b)$ and since the limit step is trivial, $\varphi_2(a) \leq \varphi_2(b)$. Similarly, $\varphi_2(b) \leq \varphi_2(a)$ and hence $\varphi_2(b) = \varphi_2(a)$.

(iv) ⇒ (ii) It is easy to see that the property in (iv) is hereditary and hence, by 4.2, it suffices to prove that (iv) ⇒ (WSf). Recall (4.2.1). If $x \vee b = 1 \equiv x = 1$, $\check{b} = (x \mapsto x \vee b)$ is the identity and hence $b = 0$. □

NOTE. It was pointed out by P. Johnstone and J. Vermeulen that the condition 4.5(iv) is, in turn, equivalent to *fitness* of [7]. This raises several new questions which will be analyzed in a separate paper.

5. More on Sequential Regularity

5.1. In this section all the concrete examples will be spaces. In these we will forget the convention 1.6 and use the more common notation (small case roman letters for points, capitals for subsets).

The topologies in the examples will be indicated as τ, and the closures by overline. $\mathbb{R}$ (resp. $\mathbb{Q}$, resp. $\mathbb{N}$) is the *set* of real (resp. rational, resp. natural) numbers, and the intervals in $\mathbb{R}$ will be denoted by

$$[a, b], \quad]a, b[, \quad \text{etc.}$$

The Urysohn separation axiom (requiring separation of distinct points by *closed* neighbourhoods) will be indicated by (U), and regularity by (R).

5.2. Let $(a_\beta)_{\beta\le\alpha}$ be the sequence from the definition of (SRα). Define

$$a'_0 = 0, \quad a'_{\beta+1} = \bigvee\{u \mid \exists x,\ x \vee a = 1,\ u \wedge x \le a'_\beta\},$$
$$a'_\beta = \bigvee_{\gamma<\beta} a'_\gamma \text{ in the limit case.}$$

Thus $(a'_\beta)_{\beta\le\alpha}$ has again the required property. Thus we can reformulate (SRα) by *defining*, for any $a \in L$, by induction

$$a_0 = 0, \quad a_{\beta+1} = \bigvee\{u \mid \exists\, x,\ x \vee a = 1,\ u \wedge x \le a_\beta\}, \text{ etc.,}$$

and stating that L has (SRα) if for each $a \in L$, $a = a_\alpha$.

In a space X, the formula (for $A \subseteq X$ open)

$$A_{\beta+1} = \bigcup\{U \mid \exists\, V,\ A \cup V = X \text{ and } V \cap U \subseteq A_\beta\}$$

is equivalent to a handier

$$A_{\beta+1} = \bigcup\{U \mid \overline{U \setminus A_\beta} \subseteq A\} \tag{5.2.1}$$

(think of $V = X \setminus (\overline{U \setminus A_\beta})$), and this way it will be viewed in the constructions to follow.

5.3. THEOREM. *If $\alpha < \beta$ then* (SRα) *is strictly stronger than* (SRβ).

Proof. We will produce spaces (X_α, τ_α) and open sets $A \in \tau_\alpha$ such that for $\beta < \alpha$, $A_\beta \ne A$, and $A_\alpha = A$.

Set

$$X_\alpha = \{(\beta, n) \mid \beta < \alpha,\ n < \omega\}$$

with $B \subseteq X_\alpha$ in τ_α iff

$$(\beta, n) \in B \Longrightarrow \exists m \in \omega,\ \{(\gamma, k) \mid \gamma < \beta,\ k > m\} \subseteq B.$$

Thus in particular any neighbourhood of (β, n) meets any $\{(\gamma, m) \mid m \in \omega\}$ with $\gamma < \beta$ while it does not need to meet $\{(\beta, m) \mid m \in \omega\}$ in anything but the (β, n) itself. Hence, if we set $A = \{(\beta, n) \in X_\alpha \mid n \neq 0\}$ we get $A_\gamma = \{(\beta, n) \mid \beta < \gamma,\ n \in \omega\}$ for $\gamma < \alpha$, and $A_\alpha = A$. □

5.4. THEOREM. *Between the mentioned separation axioms the following implications and (besides compositions) no others hold:*

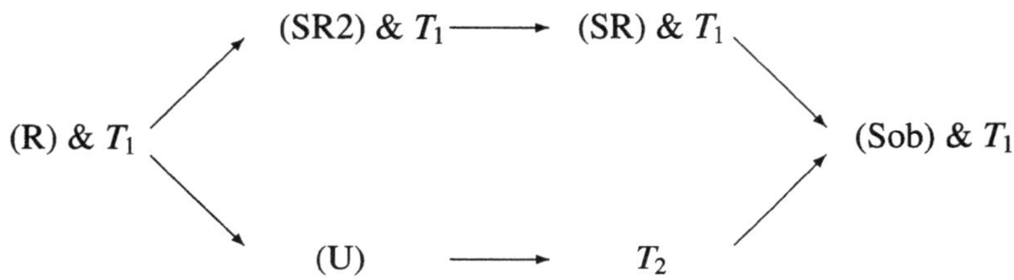

Proof. The only less standard implication in the scheme (SR) & $T_1 \Rightarrow$ (Sob) & T_1 follows from 4.5 and 3.4. Thus, it remains to be proved that (SR2) & T_1 does not imply T_2 and that (U) does not imply (SR).

(SR2) & $T_1 \not\Rightarrow$ (T_2): Consider $X = \mathbb{N} \cup \{a, b\}$ with $a \neq b,\ a, b \notin \mathbb{N}$, and the topology τ,

$$A \in \tau \quad \text{iff} \quad (A \cap \{a, b\} \neq 0 \Longrightarrow \exists n \in \mathbb{N},\ \{m \mid m \geq n\} \subseteq A).$$

Obviously, (X, τ) is not Hausdorff. Using the trivial observation that the $\{n\}$ with $n \in \mathbb{N}$ are clopen we easily deduce that

(1) if $A \subseteq \mathbb{N}$ or $\{a, b\} \subseteq A$, $A = A_1$,
(2) if $A \cap \{a, b\} = \{x\}$ ($x = a$ or b) then $A_1 = A \setminus \{x\}$ and $A_2 = A$.

Thus, (X, τ) satisfies (SR2).

(U) $\not\Rightarrow$ (SR): Consider $(\mathbb{R}, \tau)$ with

$$A \in \tau \quad \text{iff} \quad \begin{cases} \forall a \in A \cap (\mathbb{R} \setminus \mathbb{Q}) \quad \exists \varepsilon > 0 \text{ such that }]a - \varepsilon, a + \varepsilon[\subseteq A, \\ \qquad\qquad \text{and} \\ \forall a \in A \cap \mathbb{Q} \quad \exists \varepsilon > 0 \text{ such that } Q \cap]a - \varepsilon, a + \varepsilon[\subseteq A. \end{cases}$$

Obviously $(\mathbb{R}, \tau)$ is Urysohn; but it is not (SR) as the closure of any $\mathbb{Q} \cap\ ]a, b[$ equals $[a, b]$ and hence, in particular, $\mathbb{Q}_\alpha = \emptyset$ for all α. □

Acknowledgement

We are indebted to the referee for helpful comments.

References

1. Aull, C. E. and Thron, W. J.: Separation axioms between T_0 and T_1, *Indag. Math.* **24** (1962), 26–37.
2. Banaschewski, B.: Compactification of frames, *Math. Nachr.* **149** (1990), 105–116.
3. Banaschewski, B. and Pultr, A.: Cauchy points in uniform and nearness frames, *Quaestiones Math.* **19** (1996), 101–127.
4. Banaschewski, B. and Pultr, A.: Variants of openness, *Appl. Categ. Structures* **2** (1994), 331–350.
5. Herrlich, H.: A concept of nearness, *Gen. Topology Appl.* **5** (1974), 191–212.
6. Herrlich, H.: *Topologie II: Uniforme Räume*, Heldermann Verlag, Berlin, 1988.
7. Isbell, J. R.: Atomless parts of spaces, *Math. Scand.* **31** (1972), 5–32.
8. Johnstone, P. T.: *Stone Spaces*, Cambridge University Press, Cambridge, 1982.
9. Joyal, A. and Tierney, M.: An extension of the Galois theory of Grothendieck, *Mem. Amer. Math. Soc.* **51**(309) (September 1984).
10. Kříž, I.: A direct description of uniform completion in locales and a characterization of *LT*-groups, *Cahiers Topologie Géom. Différentielle Catégoriques* **27** (1986), 19–34.
11. Morita, K.: On the simple extension of a space with respect to a uniformity I, *Proc. Japan Acad.* **27** (1951), 65–72.
12. Paseka, J. and Šmarda, B.: T_2-frames and almost compact frames, *Czech. Math. J.* **42**(117) (1992), 385–402.
13. Pultr, A.: Pointless uniformities I, *Comment. Math. Univ. Carolin.* **25** (1984), 91–104.
14. Pultr, A. and Tozzi, A.: Separation axioms and frame representation of some topological facts, *Appl. Categ. Structures* **2** (1994), 107–118.
15. Simmons, H.: The lattice theoretic part of topological separation properties, *Proc. Edinburgh Math. Soc. (2)* **21** (1978), 41–48.
16. Vickers, S. J.: *Topology via Logic*, Cambridge Tracts Theoret. Comput. Sci. 5, Cambridge University Press, Cambridge, 1989.

Applied Categorical Structures **8:** 81–113, 2000.
G. Brümmer & C. Gilmour (eds), Papers in Honour of Bernhard Banaschewski.

The Structural Nature of the Nerve Functor for n-Groupoids

DOMINIQUE BOURN
LANGAL – Université du Littoral – 1, Quai Freycinet, BP 5526, F-59379, Dunkerque Cedex, France

(Received: 4 October 1996; accepted: 28 May 1998)

Abstract. Given a left exact category $\mathbb{B}$, the construction of the nerve functor ν_n for n-groupoids in $\mathbb{B}$ is related to a certain property of the category $S\text{-}Simpl_{n-1}\mathbb{B}$ of the split $(n-1)$-truncated simplicial objects in $\mathbb{B}$, which allows us to define the split n-truncated simplicial objects in $\mathbb{B}$ completely internally to $S\text{-}Simpl_{n-1}\mathbb{B}$ and thus to construct intrisincally from it the category $S\text{-}Simpl_n\mathbb{B}$.

Mathematics Subject Classifications (2000): Primary: 18G30; Secondary: 18D05, 18D30, 55U10.

Key words: simplicial object, nerve, n-groupoid, fibration.

Introduction

The notion of the nerve of a category has been introduced by Grothendieck around 1960. Since that time, the question of the nerve for n-categories has been studied by various authors, mainly at the Australian [16, 17, 13, 15] and the Welsh [8, 10, 2] schools, but see also [14, 1, 9]. It was always presented in a geometrico-combinatoric way which is descriptively growing in rather a discouraging complexity as the level n grows up, and always defined only in the set-theoretical context.

A recent result about a monadicity theorem for n-groupoids [4, 5], using a kind of polyhedral presentation led me here to an intrinsic and synthetic definition of the nerve, in the case of n-groupoids. It is, in that context, determined inductively by the same construction at each level and makes sense in any internal situation.

The work is thus mainly devoted to establish that the categories $S\text{-}Simpl_1\mathbb{B}$, $S\text{-}Simpl_2\mathbb{B}, \ldots, S\text{-}Simpl_n\mathbb{B}$ of the split $1, 2, \ldots, n$-truncated simplicial objects in $\mathbb{B}$ can be obtained successively one from the previous one by a unique iterative construction $(\mathbb{B}_1, T_1), (\mathbb{B}_2, T_2), \ldots, (\mathbb{B}_n, T_n)$ from a category $\mathbb{B}_i$ with a monad T_i to the category $\mathbb{B}_{i+1}$ with the monad T_{i+1}, the starting point being $\mathbb{B}$ with the identity monad.

Moreover, this construction, which makes sense in a very general context, determines at each step a comparison functor $\nu_T\colon Grd_\tau AlgT_n \to AlgT_{n+1}$, where $Grd_\tau AlgT_n$ is a certain full subcategory of the category $GrdAlgT_n$ of the internal groupoids in $AlgT_n$. This comparison functor is the heart of the nerve functor for the $(n+1)$-groupoids and enlights its structural nature.

More precisely, starting from a left exact fibration $\tau\colon \mathbb{E} \to \mathbb{B}$ and a monad (T, λ, μ) on $\mathbb{E}$, no relationship between τ and (T, λ, μ) being required, we construct the category $T\mathbb{E}$ of the T-elements of $\mathbb{E}$, whose objects $\underline{X_1}$ are defined as triples (X_0, d, l) where X_0 is an object of $\mathbb{E}$, $d\colon X_1 \to TX_0$ is a map in the fibre of TX_0 and $l\colon X_0 \to X_1$ is a map in $\mathbb{E}$ such that $d.l = \lambda X_0$:

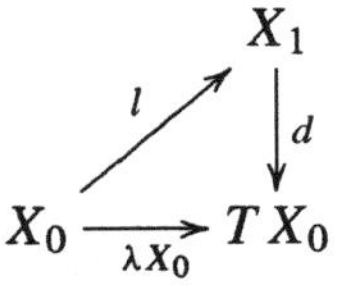

Such a T-element must be thought of as a map in the fibres of τ which is split 'modulo the monad T'.

Moreover, there is on $T\mathbb{E}$ a monad (T_1, λ_1, μ_1) with $T_1\underline{X_1} = (X_1, \bar{d}, \bar{l})$, where the map $\bar{d}$ is defined by the following pullback in $\mathbb{E}$:

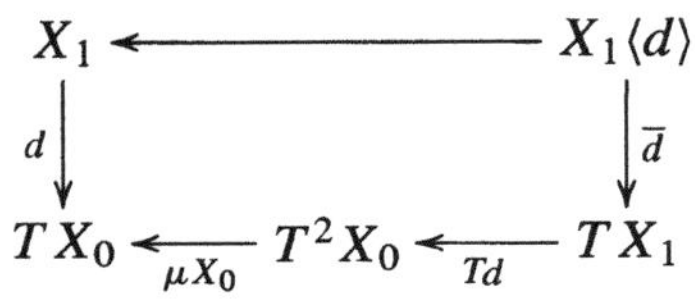

and the map $\bar{l}$ is $[1_{X_1}, \lambda X_1]$.

If the monad (T, λ, μ) is left exact, then the functor $\tau_0\colon T\mathbb{E} \to \mathbb{E}$ defined by $\tau_0(\underline{X_1}) = X_0$ is again a left exact fibration and the monad (T_1, λ_1, μ_1) is again left exact. It is, therefore, the beginning of an inductive construction.

On the other hand, if $F\colon S\text{-}Simpl\mathbb{B} \to Simpl\mathbb{B}$ denotes the functor from the split simplicial objects in $\mathbb{B}$ to the simplicial objects forgetting the splittings and U denotes its right adjoint, then it is well known ([11], for example) that U is monadic. If (T, λ, μ) denotes the monad on $S\text{-}Simpl\mathbb{B}$ determined by the adjoint pair (U, F), then, for every $n \in \mathbb{N}$, it restricts to a monad (T_n, λ_n, μ_n) on the category $S\text{-}Simpl_n\mathbb{B}$ of the split n-truncated simplicial objects (from now on we shall say split n-simplicial objects for the sake of simplicity). Moreover, the inclusion $j\colon S\text{-}Simpl_n\mathbb{B} \hookrightarrow S\text{-}Simpl\mathbb{B}$ is a morphism of monads and produces a comparison $AlgT_n \hookrightarrow AlgT = Simpl\mathbb{B}$ at the level of the categories of algebras.

Our first important result is that $S\text{-}Simpl_{n+1}\mathbb{B} = T_n(S\text{-}Simpl_n\mathbb{B})$ the category of the T_n-elements of $S\text{-}Simpl_n\mathbb{B}$, and that the monad induced on the T_n-elements is just $(T_{n+1}, \lambda_{n+1}, \mu_{n+1})$. Thus, the category $S\text{-}Simpl_{n+1}\mathbb{B}$ of the split $(n+1)$-simplicial objects is nothing but the category of the iterated T-elements on the final fibration $\mathbb{B} \to 1$ equipped with the identity monad.

The main technical theorem is that given $\mathbb{E}$, τ, (T, λ, μ), there is always a comparison functor $\tilde{U}_\tau\colon Pt_\tau AlgT \to T\mathbb{E}$, where $Pt_\tau AlgT$ is the category whose objects are the split epimorphisms (d, s) in the category $AlgT$ of the algebras of the monad (T, λ, μ) such that the underlying map d in $\mathbb{E}$ is in a fibre of τ, the morphisms being the commutative squares between such data. This functor $\tilde{U}_\tau$ is actually a lax morphism of monads which consequently extends to the categories of algebras as a functor ν_T: $Grd_\tau AlgT \to AlgT_1$ where $Grd_\tau AlgT$ is the full subcategory of the category $GrdAlgT$ of internal groupoids in $AlgT$ whose objects are the groupoids in the fibres of τ. So a category of the form $T\mathbb{E}$ always carries intrinsically a comparison from a certain subcategory of the category of the internal groupoids in $AlgT$ to $AlgT_1$ and, in a way, is the place of the 'nerve' of something.

Now starting from the previous result following which $S\text{-}Simpl_{n+1}\mathbb{B} = T_n(S\text{-}Simpl_n\mathbb{B})$, we can define inductively a left exact functor $\nu_n = \nu_{T_{n-1}}.Grd\nu_{n-1}$:

$$n\text{-}Grd\mathbb{B} \overset{Grd\nu_{n-1}}{\longrightarrow} Grd_\tau AlgT_{n-2} \overset{\nu_{T_{n-1}}}{\longrightarrow} AlgT_{n-1},$$

where $Grd\,\nu_{n-1}$ is the extension of ν_{n-1}: $(n-1)\text{-}Grd\mathbb{B} \to AlgT_{n-2}$ to the internal groupoids. This functor ν_n, thanks to the fact that $AlgT_{n-1}$, as we recall, is a subcategory of $Simpl\mathbb{B}$, is the nerve functor for the n-groupoids in $\mathbb{B}$.

Thus, it is the fact that the category $S\text{-}Simpl_n\mathbb{B}$ posseses the structure of the iterated T-elements which determines entirely the existence of the nerve functor in the case of n-groupoids.

Clearly, the role, in this work, of the monad (T_n, λ_n, μ_n) on $S\text{-}Simpl_n\mathbb{B}$ is crucial and a great part of this paper will be devoted to its study and to the related notion of simplicial kernel which is necessary to define it. Although well known, these notions will be dealt with at length, in a way which perhaps will be felt unnecessarily pedantic. But it is unavoidable, on the one hand, in order to fix the notation (and in mathematics the notation is part of the structure), and, on the other hand, to describe precisely the algebras of this monad (the Kan operations) which are particular simplicial objects the nerves of n-groupoids belong to.

This paper is organized along the following line:

(1) Review on the simplicial and n-simplicial objects.

(2) The algebra of simplicial kernels: preliminary works for the study of monads (T_n, λ_n, μ_n).

(3) The point on the pointed objects: crucial presentation of the monad (T_0, λ_0, μ_0) whose algebras are groupoids.

(4) Kan operation on n-simplicial objects: characterization of the category $K_{n+1}\mathbb{B} = AlgT_n$ of algebras of the monad (T_n, λ_n, μ_n) on $S\text{-}Simpl_n\mathbb{B}$ with a special glance at the additive case. If A is additive, then $K_{n+1}A = Simpl_{n+1}A$.

(5) The nerve of a Kan operation: characterization of $K_{n+1}\mathbb{B}$ as a subcategory of $Simpl\mathbb{B}$.

(6) $S\text{-}Simpl_n\mathbb{B}$ from $Pt\mathbb{B}$: an inductive construction. This is the introduction of T-elements and the proof that $T_n(S\text{-}Simpl_n\mathbb{B}) = S\text{-}Simpl_{n+1}\mathbb{B}$.

(7) The canonical factorization induced by $T\mathbb{E}$: the functor ν_T: $Grd_\tau AlgT \to AlgT_1$ is described.

(8) The nerve functor for n-groupoids with a special glance at the case $n = 2$.

The last remark: once observed the underlying structure of T-elements for $S\text{-}Simpl_n\mathbb{B}$, the nerve functor for n-groupoids is easily defined, but, of course, as expected from the case of the nerve of n-categories, the actual calculation and geometric description in the last paragraph are quite difficult technically. The real benefit of this presentation of the nerve functor will be effective when this geometric description will be, perhaps frustratingly, left aside for the profit of a characteristic algebraic property: this will be shown in a further paper [7].

1. Review on Simplicial Objects and n-Simplicial Objects in a Category

Let $\mathbb{B}$ be a category.

A simplicial object $\underline{X}$ in $\mathbb{B}$ is a collection of objects X_k ($k \geq 0$) in $\mathbb{B}$ together with maps: $X_{n-1} \xleftarrow{d_i} X_n \xrightarrow{s_i} X_{n+1}$ ($0 \leq i \leq n$) which satisfy the well known identities:

$$\begin{aligned} d_i d_j &= d_{j-1} d_i \quad \text{for } i < j, & s_i s_j &= s_{j+1} s_i \quad \text{for } i \leq j,\\ d_i s_j &= s_{j-1} d_i \quad \text{for } i < j, & d_i s_j &= s_j d_{i-1} \quad \text{for } i \geq j+2,\\ d_{i+1} s_i &= d_i s_i = 1. \end{aligned}$$

A simplicial map $f: \underline{X} \to \underline{Y}$ is a collection of maps f_k: $X_k \to Y_k$ ($k \geq 0$) which commute with the maps d_i and s_i. Hence a category denoted by $Simpl\mathbb{B}$.

A n-simplicial object $\underline{X}_n$, when $n \geq 1$, has a similar definition except that the collection of objects X_k is defined only for $0 \leq k \leq n$. Hence a category denoted by $Simpl_n\mathbb{B}$.

Thus a 1-simplicial object $\underline{X}_1$ is just a reflexive graph: $\underline{X}_1$: $X_0 \underset{d_1}{\overset{d_0}{\leftleftarrows}}\!\!\xrightarrow{s_0} X_1$.

A simplicial object $\underline{X}$ is split when there is an extra family of maps, namely, the splittings s_{n+1}: $X_n \to X_{n+1}$, satisfying the previous equations at their own level, when $n > 1$ with, moreover:

$$s_1 d_1 = 1, \qquad d_0 s_1 d_0 = d_0 s_1 d_1 \quad \text{and} \quad s_0 d_0 s_1 = s_1 d_0 s_1.$$

A morphism of split simplicial objects is a simplicial map which commutes with the splittings. Hence again a category denoted by $S\text{-}Simpl\mathbb{B}$.

REMARK. When $\mathbb{B}$ is left exact, the third last equations are equivalent to the data of augmentation: $X_{-1} \underset{s_0}{\overset{d_0}{\rightleftarrows}} X_0$ which also satisfies the simplicial equations. The object X_{-1} is at the same time the kernel of s_0 and s_1, the cokernel of d_0 and d_1, and the splitting of the idempotent map $d_0.s_1$: $X_0 \to X_0$, as well.

Again, a split n-simplicial object $\underline{X}_n$ ($n \geq 1$) has a similar definition except that X_k is defined only for $0 \leq k \leq n$. Hence again a category denoted by $S\text{-}Simpl_n\mathbb{B}$.

We are now going to review the organization of all these categories, the ones in relation to the others.

There is an obvious left exact forgetful functor F: $S\text{-}Simpl\mathbb{B} \to Simpl\mathbb{B}$ which forgets the splittings, which clearly restricts itself to a functor

$$F_n\colon S\text{-}Simpl_n\mathbb{B} \longrightarrow Simpl_n\mathbb{B}.$$

This functor F admits as a right adjoint the functor

$$U\colon Simpl\mathbb{B} \longrightarrow S\text{-}Simpl\mathbb{B}$$

which is defined by $(UX)_k = X_{k+1}$ and deletes the last face operator $d_{k+1}\colon X_{k+1} \to X_k$ at each level.

Apparently, its 'restriction' $\underline{U}_n$ to $Simpl_n\mathbb{B}$, because of shifting of indexation, takes its values in $S\text{-}Simpl_{n-1}\mathbb{B}$ and cannot be a right adjoint to F_n.

Nevertheless, this functor $\underline{U}_n$ has an extremely important property: not only does it reflect isomorphisms but also reflects identity maps.

1.1. THE STRUCTURE OF THE PAIR (U, F)

As soon as $\mathbb{B}$ is left exact, what we shall assume from now on, the functor F is always comonadic since:

(1) the natural transformation $\lambda\colon 1 \to UF$ is the kernel of λUF and $UF\lambda$ (since, at each level, $s_{n+1}\colon X_n \to X_{n+1}$is the split equalizer of s_{n+1} and s_{n+2}), this fact being equivalent to the fact that the comparison functor with the category of coalgebras of the comonad generated by FU on $Simpl\mathbb{B}$ is fully faithful.

(2) any parallel pair $S\text{-}Simpl\mathbb{B}$ whose image by F has a split equalizer has an equalizer ($\mathbb{B}$ left exact) which is preserved by F since F is left exact.

(3) the functor F is clearly conservative (= reflects the isomorphisms).

The functor U itself is monadic [11] since

(1) the natural transformation $\varepsilon\colon FU \to 1$ is the cokernel of εFU and $FU\varepsilon$ since, at each level, $d_{n+1}\colon X_{n+1} \to X_n$ is the split coequalizer of d_{n+1} and d_{n+2}.

(2) any parallel pair in $Simpl\mathbb{B}$ whose image by U has a split coequalizer has a coequalizer which is preserved by U, since a split coequalizer in $S\text{-}Simpl\mathbb{B}$ is pointwise and thus determines a pointwise coequalizer in $Simpl\mathbb{B}$.

(3) the functor U is clearly conservative.

1.2. THE STRUCTURE OF THE PAIR (U_n, F_n)

Actually the functor F_n: $S\text{-}Simpl_n \to Simpl_n$ itself has a right adjoint U_n [11].

Obviously, U_n and $\underline{U}_n$ coincide on their components as far as level $(n-1)$. The component at level n needs the classical notion of $(n+1)$-simplicial kernel [11]. But, before going into details (very important, as we shall see) of this construction, let us emphasize that:

(1) F_n is comonadic for exactly the same reasons as F.

(2) U_n is no longer monadic:

(α) again, the natural transformation $\varepsilon_n\colon F_n.U_n \to 1$ is the cokernel of $\varepsilon_n F_n U_n$ and $F_n U_n \varepsilon_n$ for equational reasons and thus the comparison functor $\overline{k_{n+1}}$ from

$Simpl_n\mathbb{B}$ to the categories of algebras $K_{n+1}\mathbb{B}$ of the monad (T_n, λ_n, U_n) generated by (U_n, F_n) is fully faithful.

(β) again, when $\mathbb{B}$ is left exact, any parallel pair in $Simpl_n\mathbb{B}$ whose image by U_n has a split coequalizer admits itself a coequalizer which is also pointwise as in $Simpl\mathbb{B}$.
However, it is not preserved by U_n, since, in general, the coequalizers in $\mathbb{B}$ do not commute with the left exact limits and, in particular, with the simplicial kernels. Nevertheless, the existence of such coequalizers yields a left adjoint $\overline{\tau_n}$ to the fully faithful comparison functor $\overline{k_{n+1}}$: $Simpl_n\mathbb{B} \to AlgT_n = K_{n+1}\mathbb{B}$.

(γ) Nevertheless, the functor U_n is also conservative.

On the other hand, the iterated simplicial kernels allow us to construct full embeddings j_n:

$$\begin{array}{ccc} Simpl_n\mathbb{B} & \xrightarrow{j_n} & Simpl\mathbb{B} \\ U_n \downarrow \uparrow F_n & & U \downarrow \uparrow F \\ S\text{-}Simpl_n\mathbb{B} & \xrightarrow[j_n]{} & S\text{-}Simpl\mathbb{B} \end{array}$$

which commute up to isomorphisms with the pairs (U, F). Consequently, the monad (T_n, λ_n, μ_n) generated by (U_n, F_n) on S-$Simpl_n\mathbb{B}$ is the restriction of the monad (T, λ, μ) generated by (U, F) on S-$Simpl\mathbb{B}$. Thus, any algebra of the monad T_n 'is' an algebra of T and therefore is somewhere in $Simpl\mathbb{B}$, since U: $Simpl\mathbb{B} \to S$-$Simpl\mathbb{B}$ is monadic, and $Simpl\mathbb{B} = AlgT$.

The aim of the next paragraphs is to explicate the nature of the algebras of T_n and to determine where exactly they lie in $Simpl\mathbb{B}$.

2. The Algebra of Simplicial Kernels

2.1. SIMPLICIAL KERNELS

Let us recall first the well known definition of a simplicial kernel, see [11] for instance:

DEFINITION 1. Let f_i: $X \to Y$, $0 \leq i \leq n-1$, be a n-family of maps in $\mathbb{B}$, the n-simplicial kernel of this family is an object $X[f_0, \dots, f_{n-1}]$ together with a $(n+1)$-family of maps p_i: $X[f_0, \dots, f_{n-1}] \to X$, $0 \leq i \leq n$, which is universal with respect to the identities: $f_i.p_j = f_{j-1}p_i$ for $0 \leq i < j \leq n$.

Clearly, the 1-simplicial kernel $X[f_0]$ of a map f_0: $X \to Y$ is just given by its kernel pair.

If f_i': $X' \to Y'$, $0 \leq i \leq n-1$, is another n-family of maps and if ψ: $X \to X'$and φ: $Y \to Y'$ are a pair such that $f_i'.\psi = \varphi.f_i$, $0 \leq i \leq n-1$, then we shall denote by $\psi[\varphi]$: $X[f_0, \ldots, f_{n-1}] \to X'[f'_0, \ldots, f'_{n-1}]$ the unique factorization such that $p_i.\psi[\varphi] = \psi.p_i$, $0 \le i \le n$.

The following proposition will give us an inductive construction of simplicial kernels:

PROPOSITION 1. *The following diagram is a pullback:*

$$\begin{array}{ccc} X & \xleftarrow{\quad p_n \quad} & X[f_0, \ldots, f_{n-1}] \\ \Big\downarrow{\scriptstyle [f_0,\ldots,f_{n-1}]} & & \Big\downarrow{\scriptstyle j} \\ Y[\varphi_y, \ldots, \varphi_y] = Y^n & \xleftarrow[f_{n-1}[\varphi_x]]{} & X[f_0, \ldots, f_{n-2}] \end{array}$$

where φ_x and φ_y denote the terminal maps.

Proof. Let U be an object of $\mathbb{B}$ and g_i: $U \to X, 0 \le i \le n$, a $(n+1)$-family of maps such that $f_i.g_j = f_{j-1}.g_i$, $0 \le i < j \le n$.

The n-family g_i, $0 \le i \le n-1$ produces a unique factorization γ: $U \to X[f_0, \ldots, f_{n-2}]$. Now $[f_0, \ldots, f_{n-1}].g_n = f_{n-1}[\varphi_x].\gamma$ means exactly that $f_i.g_n = f_n.p_i.\gamma = f_n.g_i$, $0 \le i \le n-1$. Whence the result.

2.2. ITERATED SIMPLICIAL KERNELS

Let g_i: $Y \to Z$, $0 \le i \le n-1$, be an n-family of maps and f_j: $X \to Y$, $0 \le j \le n$, be a $(n+1)$-family of maps satisfying $g_i f_j = g_{j-1} f_i$, $0 \le i < j \le n$. Then there is a unique factorization τ: $X \to Y[g_0, \ldots, g_{n-1}]$ such that $p_j.\tau = f_j$, $0 \le j \le n$.

Furthermore, the map f_n and the equalities $g_i.f_n = g_{n-1}.f_i$, $0 \le i \le n-1$, determine a map

$$f_n[g_{n-1}]\colon X[f_0, \ldots, f_{n-1}] \longrightarrow Y[g_0, \ldots, g_{n-1}].$$

PROPOSITION 2. *The following square is a pullback:*

$$\begin{array}{ccc} X & \xleftarrow{\quad p_{n+1} \quad} & X[f_0, \ldots, f_n] \\ \Big\downarrow{\scriptstyle \tau=[f_0,\ldots,f_n]} & & \Big\downarrow{\scriptstyle \pi=[p_0,\ldots,p_n]} \\ Y[g_0, \ldots, g_{n-1}] & \xleftarrow[f_n[g_{n-1}]]{} & X[f_0, \ldots, f_{n-1}] \end{array}$$

Proof. The previous square commutes since, for all i, $0 \le i \le n$,

$$\begin{aligned} p_i.\tau.p_{n+1} &= f_i.p_{n+1} = f_n.p_i = f_n.p_i.\pi \\ &= p_i.f_n[g_{n-1}].\pi. \end{aligned}$$

Now let (k, k_{n+1}) be a pair of maps $k\colon C \to X[f_0, \dots, f_{n-1}]$, $k_{n+1}\colon C \to X$ such that $f_n[g_{n-1}].k = \tau.k_{n+1}$. The map k determines a $(n+1)$-family $k_j\colon C \to X$, $0 \le j \le n$, such that $f_i.k_j = f_{j-1}.k_i$, $0 < i \le j \le n$.

On the other hand, the equality $f_n[g_{n-1}].k = \tau.k_{n+1}$ means that $f_n.k_i = f_i.k_{n+1}$, $0 \le i \le n$. Consequently, the $(n+2)$-family k_i, $0 \le i \le n+1$, satisfies the desired identities for factorization through $X[f_0, \dots, f_n]$.

This result has an important corollary.

COROLLARY 1. *Given $g_i\colon Y \to Z$, $0 \le i \le n-1$, an n-family of maps, then $Y[g_0, \dots, g_{n-1}][p_0, \dots, p_n]$ is isomorphic to the iterated simplicial kernel $Y[g_0, \dots, g_{n-1}][p_0, \dots, p_{n-1}]$ which we shall denote as $Y^2[g_0, \dots, g_{n-1}]$ from now on, where the maps p_i are the projection $p_i\colon Y[g_0, \dots, g_{n-1}] \to Y$, $0 \le i \le n$.*

Proof. If τ is an isomorphism, then π is an isomorphism. And then the map $p_{n+1}\colon Y^2[g_0, \dots, g_{n-1}] \to Y[g_0, \dots, g_{n-1}]$ is nothing but $p_n[g_{n-1}]$.

2.3. SIMPLICIAL KERNELS OF SIMPLICIAL OBJECTS

When the $(n+1)$-family $d_i\colon X_n \to X_{n-1}$ comes from the last level of an n-simplicial object $\underline{X}_n$, the identities satisfied by the maps $s_i\colon X_{n-1} \to X_n$, $0 \le i \le n-1$, produce maps $s_i\colon X_n \to X_n[d_0, \dots, d_n]$, $0 \le i \le n$, so that they complete the n-simplicial object $\underline{X}_n$ into a $(n+1)$-simplicial object denoted by $k_{n+1}(\underline{X}_n)$[11]. This construction determines a functor $k_{n+1}\colon Simpl_n\mathbb{B} \to Simpl_{n+1}\mathbb{B}$ which is the right adjoint right inverse for the left exact functor $\tau_n\colon Simpl_{n+1}\mathbb{B} \to Simpl_n\mathbb{B}$ which truncates the last level of an $(n+1)$-simplicial object. Thus, this functor τ_n is a fibration.

Furthermore, when $\underline{X}_n$ is a split n-simplicial object, $k_{n+1}(\underline{X}_n)$ is split. We shall again denote by $k_{n+1}\colon S\text{-}Simpl_n\mathbb{B} \to S\text{-}Simpl_{n+1}\mathbb{B}$ the functor defined in this way. Again, k_{n+1} is a right adjoint right inverse for the truncation $\tau_n\colon S\text{-}Simpl_{n+1}\mathbb{B} \to S\text{-}Simpl_n\mathbb{B}$, making this last functor a fibration.

Furthermore, the following diagram commutes:

$$\begin{array}{ccc}
S\text{-}Simpl_{n+1}\mathbb{B} & \xrightarrow{F_{n+1}} & Simpl_{n+1}\mathbb{B} \\
\tau_n \downarrow \uparrow k_{n+1} & & \tau_n \downarrow \uparrow k_{n+1} \\
S\text{-}Simpl_n\mathbb{B} & \xrightarrow[F_n]{} & Simpl_n\mathbb{B}
\end{array}$$

Since F_{n+1} preserves and reflects the pullbacks, F_{n+1} preserves and reflects the cartesian maps. Moreover, since F_n is conservative, the functor F_{n+1} also reflects the τ_n-invertibles maps.

Now, coming back to our functor U_n: $Simpl_n\mathbb{B} \to S\text{-}Simpl_n\mathbb{B}$, $U_n(\underline{X}n)$ is given by the following diagram:

$$X_1 \;\underset{d_1}{\overset{d_0}{\leftleftarrows}}\; X_2 \;\cdots\; X_{n-1} \;\underset{d_{n-1}}{\overset{d_0}{\underset{\vdots}{\leftarrow}}}\; X_n \;\underset{p_n}{\overset{p_0}{\underset{\vdots}{\leftarrow}}}\; X_n[d_0, \ldots, d_n] \qquad (s_0, s_1, \ldots, s_{n-1}, s_n)$$

So by definition of U_n, we have $\underline{U}_n = \tau_{n-1}.U_n$: $Simpl_n\mathbb{B} \to S\text{-}Simpl_{n-1}\mathbb{B}$, where, as we saw, $\underline{U}_n$ just deletes the last face operator.

PROPOSITION 3. *The functors $U_{n+1}.k_{n+1}$ and $k_{n+1}.U_n$ are naturally isomorphic.*

Proof. The last level of $k_{n+1}.U_n(\underline{X}_n)$ is given by $X_n[d_0, \ldots, d_n][p_0, \ldots, p_n] = X_n^2[d_0, \ldots, d_n]$ and the last level of $U_{n+1}.k_{n+1}(\underline{X}_n)$ is given by $X_n[d_0, \ldots, d_n][p_0, \ldots, p_{n+1}]$. But these two objects are isomorphic by Corollary 1. □

Consequences:

(1) The adjunctions (U_i, F_i) commute with the inclusions k_i, and thus the monads (T_i, λ_i, μ_i) on $S\text{-}Simpl_i\mathbb{B}$ commute with the inclusions k_i.

(2) Let us denote by ω the natural tranformation $1 \to k_{n+1}.\tau_n$. In $S\text{-}Simpl_{n+1}\mathbb{B}$, we have $(\omega\underline{X}_{n+1})_k = 1_{X_k}$ $\forall k$, $0 \le k \le n$, and $(\omega\underline{X}_{n+1})_{n+1} = [d_0, \ldots, d_{n+1}]$: $X_{n+1} \to X_n[d_0, \ldots, d_n]$.

Then, $\tau_n T_{n+1}\omega$ determines a lax cartesian morphism of monads:

$$\begin{array}{ccc}
S\text{-}Simpl_{n+1}\mathbb{B} & \xrightarrow{T_{n+1}} & S\text{-}Simpl_{n+1}\mathbb{B} \\
\downarrow{\scriptstyle \tau_n} & \tau_n T_{n+1}\omega \Downarrow & \downarrow{\scriptstyle \tau_n} \\
S\text{-}Simpl_n\mathbb{B} & \xrightarrow[T_n]{} & S\text{-}Simpl_n\mathbb{B}
\end{array}$$

PROPOSITION 4. *For any object $\underline{X}_{n+1}$ in $S\text{-}Simpl_{n+1}\mathbb{B}$, the following diagram is a pullback:*

$$\begin{array}{ccccc}
\tau_n T_{n+1}\underline{X}_{n+1} & & \xleftarrow{\tau_n\mu_{n+1}\underline{X}_{n+1}} & & \tau_n T_{n+1}^2\underline{X}_{n+1} \\
\downarrow{\scriptstyle \tau_n T_{n+1}\omega\underline{X}_{n+1}} & & & & \downarrow{\scriptstyle \tau_n T_{n+1}\omega T_{n+1}\underline{X}_{n+1}} \\
T_n\tau_n\underline{X}_{n+1} & \xleftarrow[\mu_n\tau_n\underline{X}_{n+1}]{} & T_n^2\tau_n\underline{X}_{n+1} & \xleftarrow[T_n\tau_n T_{n+1}\omega\underline{X}_{n+1}]{} & T_n\tau_n T_{n+1}\underline{X}_{n+1}
\end{array}$$

Proof. The maps $\tau_n T_{n+1}\omega\underline{Z}_{n+1}$ being τ_{n-1}-inversible, it is sufficient to prove this at the last level. So, let us consider the following square:

$$\begin{array}{ccccc}
X_{n+1} & & \xleftarrow{p_{n+2}} & & X_{n+1}[d_0, \ldots, d_{n+1}] \\
\downarrow{\scriptstyle [d_0,\ldots,d_{n+1}]} & & & & \downarrow{\scriptstyle [p_0,\ldots,p_{n+1}]} \\
X_n[d_0, \ldots, d_n] & \xleftarrow[p_{n+1}[d_n]]{} & X_n^2[d_0, \ldots, d_n] & \xleftarrow[[d_0,\ldots,d_{n+1}][1_{X_n}]]{} & X_{n+1}[d_0, \ldots, d_n]
\end{array}$$

But $p_{n+1}[d_n].[d_0, \dots, d_{n+1}][1_{X_n}] = d_{n+1}[d_n]$. Then, according to Proposition 2, this square is a pullback. □

(3) We also need the following observation:

Let us denote by $M\mathbb{B}$ the category whose objects are the maps in $\mathbb{B}$ and maps are the commutative squares in $\mathbb{B}$. Let us denote by p: $M\mathbb{B} \to \mathbb{B}$ a functor associating its codomain with each map. If $\mathbb{B}$ is left exact, then p is a fibration.

Given a fibration τ: $\mathbb{E} \to \mathbb{B}$, let us denote by $M_\tau\mathbb{E}$ the full subcategory of $M\mathbb{E}$ whose objects are the τ-identity maps, which are those maps in $\mathbb{E}$ whose image by τ are identity maps. When τ is left exact, the restriction p: $M_\tau\mathbb{E} \to \mathbb{E}$ is again a fibration.

PROPOSITION 5. *The following square is a pullback and the functor j is an embedding:*

$$\begin{array}{ccc} Simpl_{n+1}\mathbb{B} & \xrightarrow{\ j\ } & M_\tau(S\text{-}Simpl_n\mathbb{B}) \\ {\scriptstyle \tau_n}\downarrow & & \downarrow{\scriptstyle p} \\ Simpl_n\mathbb{B} & \xrightarrow[U_n]{} & S\text{-}Simpl_n\mathbb{B} \end{array}$$

Proof. In other words, we must prove that a $(n+1)$-simplicial object is just a τ_{n-1}-identity map δ: $\underline{Z}_n \to U_n\underline{X}_n$ in $S\text{-}Simpl_n\mathbb{B}$.

Let us define $j\underline{X}_{n+1}$: $\tau_n U_{n+1}\underline{X}_{n+1} \to U_n\tau_n\underline{X}_{n+1}$ by $(j\underline{X}_{n+1})_k = 1_{X_k}$ $\forall k$, $0 \leq k \leq n$, and $(j\underline{X}_{n+1})_{n+1} = [d_0, \dots, d_{n+1}]$: $X_{n+1} \to X_n[d_0, \dots, d_n]$.

Then, clearly, the previous diagram commutes.

Now, let $\underline{X}_n$ be an n-simplicial object and δ: $\underline{Z}_n \to U_n\underline{X}_n$ a τ_{n-1} identity map in $S\text{-}Simpl_n\mathbb{B}$. Its last level is of the following kind.

$$\begin{array}{ccc} Z_n & \xrightarrow{\ \delta\ } & X_n[d_0, \dots, d_n] \\ {\scriptstyle d_0}\downarrow \cdots \downarrow{\scriptstyle d_n} \uparrow{\scriptstyle s_n} & & {\scriptstyle p_0}\downarrow \cdots \downarrow{\scriptstyle p_n} \uparrow{\scriptstyle s_n} \\ Z_{n-1} & = & X_n \end{array}$$

It produces a unique $(n+1)$-simplicial object $\underline{X}_{n+1}$ such that $\tau_n\underline{X}_{n+1} = \underline{X}_n$ and $X_{n+1} = Z_n$ with d_i: $X_{n+1} = Z_n \to X_n = Z_{n-1}$, $0 \leq i \leq n$, and $d_{n+1} = p_{n+1}.\delta$. Then, clearly, $\delta = j\underline{X}_{n+1}$. The functor j is an embedding since U_n is an embedding (since it is left exact + conservative).

3. The Point on the Pointed Objects (the case $n = 0$)

Until now we have deliberately neglected the case $n = 0$.

We shall denote by $Pt\mathbb{B}$ the category whose objects are the split epimorphisms with a given splitting and morphisms are the commutative squares between such data. Of course, we must think of $Pt\mathbb{B}$ as $S\text{-}Simpl_0\mathbb{B}$, but we do prefer this heterogeneous notation because this category $Pt\mathbb{B}$ makes sense outside the simplicial context and has by itself some strong classifying properties [6]. Let us denote by τ_0: $S\text{-}Simpl_1\mathbb{B} \to Pt\mathbb{B}$ the functor associating with each object $\underline{X}_1$ its split augmentation. It is a left exact fibration with a right adjoint right inverse k_1, associating with each split epimorphism the underlying graph of its kernel relation which is obviously split as a 1-simplicial object.

Let us denote by p: $Pt\mathbb{B} \to \mathbb{B}$ a forgetful functor associating its codomain with each epimorphism. It has a right adjoint right inverse assigning to each object Y in $\mathbb{B}$ the split epimorphism 1_Y and, thus, it is a fibration.

DEFINITION 2. The fibration p is called the fibration of pointed objects of $\mathbb{E}$ [3, 4, 6].

On $Pt\mathbb{B}$ there is a left exact monad which will be one of our main tools. Let F_0: $Pt\mathbb{B} \to M\mathbb{B}$ be the functor forgetting the splitting. It has a right adjoint U_0 which assigns to each morphism f: $X \to Y$ the splitting (p_0, s_0) determined by its kernel pair. We shall denote by (T_0, λ_0, μ_0) the monad on $Pt\mathbb{B}$ generated by the adjunction (U_0, F_0). It is described by the following diagram:

$$\begin{array}{ccccc}
X & \xrightarrow{s_1=[s.d,1]} & X_{X_Y}X & \xleftarrow{p_2} & X_{X_Y}X_{X_Y}X \\
d\downarrow\uparrow s & & p_0\downarrow\uparrow s_0=[1,1] & & p_0\downarrow\uparrow s_0 \\
Y & \xrightarrow[s]{} & X & \xleftarrow[p_1]{} & X_{X_Y}X
\end{array}$$

$$\frac{\mathbb{B}}{Pt\mathbb{B}}$$

$$(d,s) \xrightarrow[\lambda_0(d,s)]{} T_0(d,s) \xleftarrow[\mu_0(d,s)]{} T_0^2(d,s)$$

The category of algebras of this monad is the category $Grd\mathbb{B}$ of internal groupoids in $\mathbb{B}$ ([4]).

Following this result, an internal groupoid X_1 in $\mathbb{B}$ is determined by the following underlying diagram in $\mathbb{B}$:

$$X_1 : X_0 \underset{d_1}{\overset{d_0}{\leftleftarrows}}\!\!\xrightarrow{s_0} mX_1 \underset{d_2}{\overset{p_0}{\leftleftarrows}}\!\!\xleftarrow{p_1} mX_1[d_0]$$

where X_0 is called the object of objects and mX_1 the object of morphisms of X_1.

The functor $(\;)_0$: $Grd\mathbb{B} \to \mathbb{B}$ associating X_0 to X_1 has a right adjoint right inverse gr where $gr\ X$ is the kernel groupoid of the terminal map $X \to 1$ and a left adjoint right inverse dis, where $dis\ X$ is the kernel groupoid of 1_X: $X \to X$.

4. Kan Operation on n-Simplicial Objects

Given an n-simplicial object $\underline{X}_n$, its initial simplicial $(n+1)$-horn $\Lambda\underline{X}_n$ is just the n-simplicial kernel $X_n[d_0, \ldots, d_{n-1}]$, $d_i\colon X_n \to X_{n-1}$ (the map d_n is excluded) [12]. It is the universal open $(n+1)$-simplex whose last face is missing.

DEFINITION 3. A Kan operation on an n-simplicial object $\underline{X}_n$ $(n \geq 1)$ is a $(n+1)$-ary partial operation $t\colon \Lambda X_n \to X_n$ satisfying the following axioms:
(1) contiguity axioms:

$$\begin{aligned} d_i.t &= d_n.p_i, & 0 \leq i \leq n, \\ t.s_i &= s_i.d_n, & 0 \leq i \leq n-1, \end{aligned}$$

(2) 'unitary' axiom:

$$t.s_n = 1_{X_n},$$

(3) 'associativity' axiom:

$$t.t[d_n] = t.p_n[d_{n-1}] \; (= t.p_{n+1}).$$

If $\mathbb{B}$ is the category Set of sets, given $(x_0, x_1, \ldots, x_n)$ an initial horn of $\underline{X}_n$, then $t(x_0, x_1, \ldots, x_n)$ is a top face of this horn such that:
(1) for every n-simplex x:

$$t(s_{n-1}.d_0 x, s_{n-1}.d_1 x, \ldots, s_{n-1}.d_{n-1} x, x) = x,$$

(2) for every family of n-simplexes satisfying the desired contiguity conditions

$$t(t(x_{00}, \ldots, x_{0n}), t(x_{10}, \ldots, x_{1n}) \ldots t(x_{n0}, \ldots, x_{nn})) = t(x_{0n}, x_{1n}, \ldots, x_{nn}).$$

In other words, the operation satisfies the axioms of the hypergroupoid [12], except that the operation is only defined at the last level and uniquely for initial horns. Hence, a considerable saving of data here. The morphism of Kan operation is an n-simplicial map which preserves the $(n+1)$-ary operation. Hence a category denoted by $K_n\mathbb{B}$. We have a clearly forgetful functor $w_n\colon K_n\mathbb{B} \to Simpl_n\mathbb{B}$.

EXAMPLES. (1) A 1-simplicial object is a reflexive graph. The Kan operation on a reflexive graph is a groupoid structure on this graph [4], where $t(f, g) = g.f^{-1}$. Therefore, $K_1\mathbb{B}$ is $Grd\mathbb{B}$ the category of internal groupoid in $\mathbb{B}$. At this level the notion of hypergroupoid [12], the notion of T-simplex [10], and the notion of Kan operation are the same.

(2) Let X_2 be a 2-groupoid in the category of sets. It has an associated 2-simplicial object:

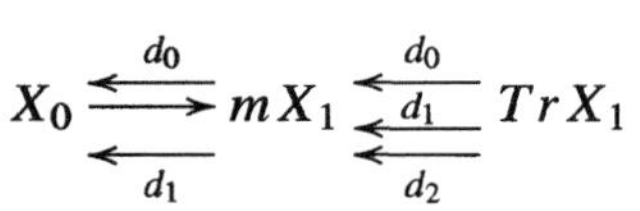

where X_0 is the set of objects, mX_1 is the set of 1-morphisms, $Tr\,X_2$ is the set of the oriented triangles of X_2 (see [16, 5], for instance):

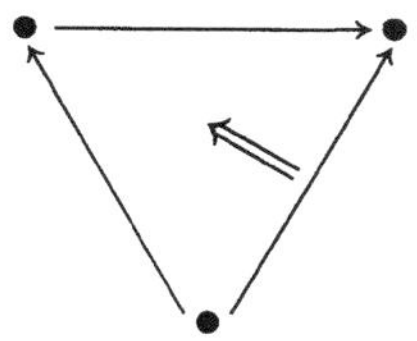

There is a canonical Kan operation on this 2-simplicial object, namely, that which, from any triangle of oriented triangles:

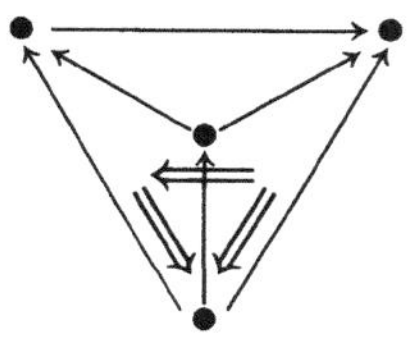

produces a unique 'pasting' filling the upper face, on account of the 2-groupoidal nature of X_2 [5]. This 2-simplicial Kan operation is actually the 'nerve' of this 2-groupoid, we shall it denote by $\nu_2\,X_2$.

(3) Let $\underline{X}_{n-1}$ be a $(n-1)$-simplicial object in $\mathbb{B}$. Then the n-simplicial object $k_n\underline{X}_{n-1}$ admits a canonical Kan operation. Indeed, $(k_n\underline{X}_{n-1})_n$ is $X_{n-1}[d_0, \ldots, d_{n-1}]$ and $\Lambda(k_n\underline{X}_{n-1})$ is $X^2_{n-1}[d_0, \ldots, d_{n-1}]$.

The map $p_{n+1} = p_n[d_{n-1}]\colon X^2_{n-1}[d_0, \ldots, d_{n-1}] \to X_{n-1}[d_0, \ldots, d_{n-1}]$ as in the corollary of Proposition 2 determines a Kan operation on $k_n\underline{X}_{n-1}$ and it is actually the unique one possible. Let us denote by $\overline{k}_n\colon Simpl_{n-1}\mathbb{B} \to K_n\mathbb{B}$ the left exact functor determined by this construction. We have then $w_n.\overline{k}_n = k_n$.

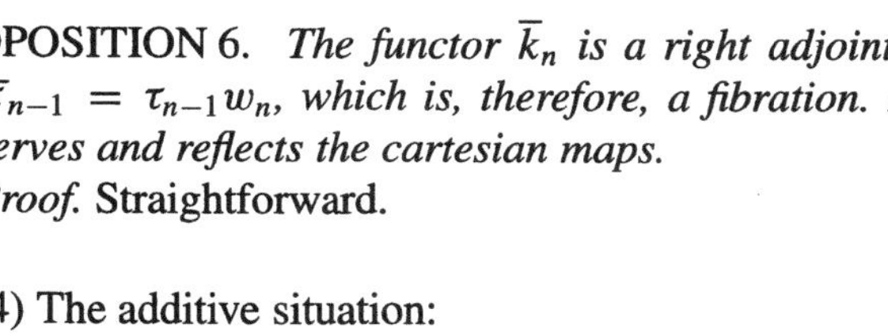

PROPOSITION 6. *The functor $\overline{k}_n$ is a right adjoint right inverse for the functor $\overline{\tau}_{n-1} = \tau_{n-1}w_n$, which is, therefore, a fibration. Furthemore, the functor w_n preserves and reflects the cartesian maps.*

Proof. Straightforward. □

(4) The additive situation:

Let us suppose that $\mathbb{B} = A$ is an additive category. It is well known that the category $Simpl_1A$ of internal reflexive graphs in A is isomorphic to the category $GrdA$ of internal groupoids in A (see [3], for instance).

But $GrdA = K_1A$ and then $Simpl_1A \cong K_1A$, which means that the functor w_1 is an isomorphism.

Let us consider the following reflexive graph in A:

$$A_0 \underset{d_1}{\overset{d_0}{\leftleftarrows}} \xrightarrow{s_0} A_1$$

The associated Kan operation t: $A_1[d_0] \to A_1$ is defined by

$$t = p_1 - p_0 + s_0d_1p_0 = p_1 + (s_0d_1 - 1)p_0.$$

We are going to show now that w_n: $K_nA \to Simpl_nA$ is an isomorphism for for each n, $n \geq 2$.

PROPOSITION 7. *Given $\underline{A}_n$ a split n-simplicial object in an additive category A, then the simplicial kernel $A_n[d_0, \ldots, d_n]$ is the simplicial cokernel $A_n]s_0, \ldots, s_n[$.*

Proof. Let us recall first that, in any additive category A the downward square of a commutative square of split epimorphisms:

$$\begin{array}{ccc} \bullet & \longrightarrow & \bullet \\ \downarrow\uparrow & & \downarrow\uparrow \\ \bullet & \longrightarrow & \bullet \end{array}$$

is a pullback if and only if the upward square is a pushout [3].

Consequently, given any split epimorphism (d_0, s_0); d_0: $A_0 \to A_{-1}$, the following upward square is a pushout:

$$\begin{array}{ccc} A_0 & \xrightarrow{s_1} & A_0[d_0] \\ {\scriptstyle d_0}\downarrow\uparrow{\scriptstyle s_0} & & {\scriptstyle p_0}\downarrow\uparrow{\scriptstyle s_0} \\ A_{-1} & \xrightarrow{s_0} & A_0 \end{array}$$

and $A_0[d_0]$ is $A_0]s_0[$. Let us suppose now that the result holds up to the level $(n-1)$. Let $\underline{A}_n$ be a split n-simplicial object. Then, according to Proposition 2 the following square is a pullback

$$\begin{array}{ccc} A_n[d_0, \ldots, d_n] & \xrightarrow{p_{n+1}} & A_n \\ {\scriptstyle [p_0,\ldots,p_n]}\downarrow & & \downarrow{\scriptstyle [d_0,\ldots,d_n]} \\ A_n[d_0, \ldots, d_{n-1}] & \xrightarrow[d_n[d_{n-1}]]{} & A_{n-1}[d_0, \ldots, d_{n-1}] \end{array}$$

Consequently, the following downward square is a pullback:

$$\begin{array}{ccc} A_n & \xrightarrow{s_{n+1}} & A_n[d_0, \ldots, d_n] \\ {\scriptstyle [p_0,\ldots,p_n]}\downarrow\uparrow{\scriptstyle]s_0,\ldots,s_n[} & & {\scriptstyle [d_0,\ldots,d_n]}\downarrow\uparrow{\scriptstyle]s_0,\ldots,s_n[} \\ A_{n-1}[d_0, \ldots, d_{n-1}] & \xrightarrow[s_n[s_{n-1}]]{} & A_n[d_0, \ldots, d_{n-1}] \end{array}$$

But the inductive hypothesis says that $A_{n-1}[d_0, \ldots, d_{n-1}] = A_{n-1}]s_0, \ldots, s_{n-1}[$ and, therefore, the maps $[p_0, \ldots, p_n]$ and $[d_0, \ldots, d_n]$ are split by maps $]s_0, \ldots, s_n[$ in a way that makes the upward square commutative. Consequently, it is a pushout and by the duality of Proposition 2,

$$A_n[d_0, \ldots, d_n] = A_n]s_0, \ldots, s_n[.$$

PROPOSITION 8. *In an additive category A, any n-simplicial object has a unique Kan operation. And thus the functor* w_n: $K_n A \to Simpl_n A$ *is an isomorphism.*

Proof. Let us consider the following diagram:

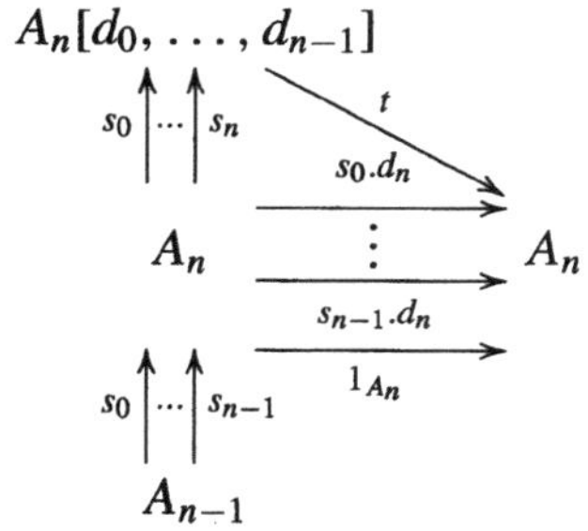

The $(n+1)$-family $\sigma_i, \sigma_i = s_i.d_n$, $0 \leq i \leq n-1$, $\sigma_n = 1_{A_n}$ satisfies the identities which determine a unique factorisation t through the simplicial cokernel $A_n[d_0, \ldots, d_{n-1}] = \Lambda \underline{A}_n$.

Therefore, all the degeneracy contiguity axioms and the unitary axiom are immediately satisfied.

Furthermore, $d_i.t = d_n.p_i$ is verified through the following equalities:

$$d_i.t.s_j = d_n.p_i.s_j, \quad 0 \leq j \leq n.$$

Finally, $t.t[d_n] = t.p_n[d_{n-1}]$ is given by: $t.t[d_n].s_j = t.s_j.t$ since $t[d_n] = t]d_n[$. Now $t.s_j.t = s_j.d_n.t = s_j.d_n.d_n$ and $t.t[d_n].s_j = t.s_j.d_n = s_j.d_n.d_n$.

REMARK. The explicit formula for this Kan operation is

$$t = p_n + (s_{n-1}d_n - 1)p_{n-1} + (s_{n-2}d_{n-1} - 1)(s_{n-1}d_n - 1)p_{n-2} + \cdots + \\ + (s_0 d_1 - 1) \cdots (s_{n-1}d_n - 1)p_0.$$

The Category of Algebras of the Monad $(T_{n-1}, \lambda_{n-1}, \mu_{n-1})$

We are now going to establish our first point, namely, that $K_n\mathbb{B}$ is precisely the category of algebras of the monad $(T_{n-1}, \lambda_{n-1}, \mu_{n-1})$ on S-$Simpl_{n-1}\mathbb{B}$ through the following functor:

$$\overline{U}_{n-1} = \underline{U}_n.w_n\colon K_n\mathbb{B} \to S\text{-}Simpl_{n-1}.$$

PROPOSITION 9. $K_n\mathbb{B}$ *is the category of algebras of the monad* $(T_{n-1}, \lambda_{n-1}, \mu_{n-1})$ *on* S-$Simpl_{n-1}\mathbb{B}$ *where* $n \geq 2$. *The functor* $\overline{k}_n$: $Simpl_{n-1}\mathbb{B} \to K_n\mathbb{B}$ *is the comparison functor induced by the adjunction* (U_{n-1}, F_{n-1}) *and its left adjoint left inverse* $\bar{\tau}_n$ *(Proposition 6) is the one expected in the paragraph about the structure of the pairs* (U_n, F_n).

Proof. The functor $\overline{U}_{n-1} = \underline{U}_n.w_n\colon K_n\mathbb{B} \to S\text{-}Simpl_{n-1}$ admits $\overline{F}_{n-1} = \bar{k}_n.F_{n-1}$ as left adjoint. Moreover,

$$\overline{U}_{n-1}.\overline{F}_{n-1} = \underline{U}_n.w_n.\bar{k}_n.F_{n-1} = \underline{U}_n.k_n.F_{n-1} = U_{n-1}.F_{n-1}$$

and, consequently, the monads generated by (U_{n-1}, F_{n-1}) and $(\overline{U}_{n-1}, \overline{F}_{n-1})$ are the same.

Let us show that $\overline{U}_{n-1}$ is monadic. First, it is clearly conservative.

Now, let $(f, g)\colon (\underline{Y}_n, t) \to (\underline{Z}_n, t)$ be a pair of maps in $K_n\mathbb{B}$ whose image by $\overline{U}_{n-1}$ has a split coequalizer. But $\overline{U}_{n-1}$ is $\underline{U}_n.\ w_n$ and $\underline{U}_n\colon Simpl_n\mathbb{B} \to S\text{-}Simpl_{n-1}\mathbb{B}$ inheriting the property of $U\colon Simpl\mathbb{B} \to S\text{-}Simpl\mathbb{B}$ creates coequalizers for those parallel pairs whose images have a split coequalizer. Consequently, $w_n(f)$ and $w_n(g)$ do have a coequalizer k in $Simpl_n\mathbb{B}$ which is preserved by $\underline{U}_n$.

Now let us consider the following diagram:

$$\begin{array}{ccccc}
Y_n[d_0, \ldots, d_{n-1}] & \overset{f_n[f_{n-1}]}{\underset{g_n[g_{n-1}]}{\rightrightarrows}} & Z_n[d_0, \ldots, d_{n-1}] & \xrightarrow{k_n[k_{n-1}]} & Q_n[d_0, \ldots, d_{n-1}] \\
{\scriptstyle p_0}\downarrow\cdots\downarrow{\scriptstyle p_n}\ \downarrow{\scriptstyle t} & & {\scriptstyle p_0}\downarrow\cdots\downarrow{\scriptstyle p_n}\ \downarrow{\scriptstyle t} & & {\scriptstyle p_0}\downarrow\cdots\downarrow{\scriptstyle p_n} \\
Y_n & \overset{f_n}{\underset{g_n}{\rightrightarrows}} & Z_n & \xrightarrow{k_n} & Q_n \\
{\scriptstyle d_0}\downarrow\cdots\downarrow{\scriptstyle d_{n-1}} & & {\scriptstyle d_0}\downarrow\cdots\downarrow{\scriptstyle d_{n-1}} & & {\scriptstyle d_0}\downarrow\cdots\downarrow{\scriptstyle d_{n-1}} \\
Y_{n-1} & \overset{f_{n-1}}{\underset{g_{n-1}}{\rightrightarrows}} & Z_{n-1} & \xrightarrow{k_{n-1}} & Q_{n-1}
\end{array}$$

Let us show that $k_n[k_{n-1}]$ is the coequalizer of $f_n[f_{n-1}]$ and $g_n[g_{n-1}]$.

The splittings of f and g, lying in $S\text{-}Simpl_{n-1}\mathbb{B}$, commute with all the map d_i except d_n, which is sufficient to determine a splitting at the level of simplicial kernels and to make $k_n[k_{n-1}]$ the desired coequalizer. The maps f and g preserving the Kan action t, determine a Kan action on the coequalizer Q_n. Thus, $\overline{U}_{n-1}$ is monadic and the end of the proposition is straightforward.

REMARKS. (1) The functor $\overline{F}_{n-1}$ being equal to $\bar{k}_n.\ F_{n-1}$ is left exact. Furthermore, the monad $(T_{n-1}, \lambda_{n-1}, \mu_{n-1})$ being such that λ_{n-1} is the kernel of $\lambda_{n-1}T_{n-1}$ and $T_{n-1}\lambda_{n-1}$, the functor $\overline{F}_{n-1}$ is always comonadic exactly for the same reasons as F_{n-1}.

(2) Following Proposition 8, when A is additive, $U_n\colon Simpl_nA \to S\text{-}Simpl_nA$ is monadic.

(3) Let us make explicit the T_{n-1}-algebraic presentation of the Kan operation on an n-simplicial object $\underline{X}_n$ and let us emphasize the shift of the indexation. Given

a Kan operation $(\underline{X}_n, t)$, then the associated algebra in $S\text{-}Simpl_{n-1}\mathbb{B}$ $A(\underline{X}_n, t)$: $T_{n-1}\overline{U}_{n-1}(X_n, t) \to \overline{U}_{n-1}(\underline{X}_n, t)$ is given at its $(n-1)$-st and $(n-2)$-nd level by the following diagram in $\mathbb{B}$:

$$\begin{array}{ccc} X_n[d_0, \ldots, d_{n-1}] & \xrightarrow{\ t\ } & X_n \\ {\scriptstyle p_0}\downarrow \cdots \downarrow{\scriptstyle p_{n-1}} \ \uparrow{\scriptstyle s_{n-1}} & & {\scriptstyle d_0}\downarrow \cdots \downarrow{\scriptstyle d_{n-1}} \ \uparrow{\scriptstyle s_{n-1}} \\ X_n & \xrightarrow[d_{n-1}]{} & X_{n-1} \end{array}$$

In other words, the underlying split $(n-1)$-simplicial object of the Kan operation $(\underline{X}_n, t)$, seen as T_{n-1}-algebras, is the following shifted split $(n-1)$-simplicial object:

$$X_1 \underset{d_1,\ s_1}{\overset{d_0}{\underset{s_0}{\leftrightarrows}}} X_2 \ldots X_{n-1} \underset{d_{n-1},\ s_{n-1}}{\overset{d_0}{\underset{\vdots}{\leftarrow}}} X_n$$

as expected from the analogous underlying normalized $(n-1)$-groupoid associated with an n-groupoid [5].

(4) Therefore, the following diagram commutes:

$$\begin{array}{ccc} K_n\mathbb{B} & \xrightarrow{\ \overline{\tau}_{n-1}\ } & Simpl_{n-1}\mathbb{B} \\ {\scriptstyle \overline{U}_{n-1}}\downarrow & & \downarrow{\scriptstyle \overline{U}_{n-1}} \\ S\text{-}Simpl_{n-1}\mathbb{B} & \xrightarrow[\tau_{n-2}]{} & S\text{-}Simpl_{n-2}\mathbb{B} \end{array}$$

(5) As we observed the functor k_n: $S\text{-}Simpl_{n-1}\mathbb{B} \to S\text{-}Simpl_n\mathbb{B}$ commutes, up to isomorphism with the monads and thus determines an extension $Algk_n$: $K_n\mathbb{B} \to K_{n+1}\mathbb{B}$ to the categories of algebras such that the following diagram commutes:

$$\begin{array}{ccc} K_n\mathbb{B} & \xrightarrow{\ Algk_n\ } & K_{n+1}\mathbb{B} \\ {\scriptstyle \overline{U}_{n-1}}\downarrow & & \downarrow{\scriptstyle \overline{U}_{n}} \\ S\text{-}Simpl_{n-1}\mathbb{B} & \xrightarrow[k_n]{} & S\text{-}Simpl_{n}\mathbb{B} \end{array}$$

so that $[Algk_n(X_n, t)]_k = X_k \ \forall\, k,\ 0 \le k \le n,$

$$[Algk_n(X_n, t)]_{n+1} = X_n[d_0, \ldots, d_{n-1}]$$

with $d_i = p_i, 0 \le i \le n$, and $d_{n+1} = t$. The Kan operation on this $(n+1)$-simplicial object being the map $t[d_n]$: $X_n^2[d_0, \ldots, d_{n-1}] \to X_n[d_0, \ldots, d_{n-1}]$.

5. The Nerve of the Kan Operation

The iterated composite $j_n = \ldots k_{n+1}.k_n\colon S\text{-}Simpl_{n-1}\mathbb{B} \to S\text{-}Simpl\mathbb{B}$ again commutes with the monads up to isomorphisms. Let $Algj_n\colon K_n\mathbb{B} \to Simpl\mathbb{B}$ be its extension to the categories of algebras.

Then

$$(Algj_n(X_n, t))_k = X_k \quad \forall k \leq n,$$

and

$$(Algj_n(X_n, t))_{n+k} = X_n^k[d_0, \ldots, d_{n-1}]$$

with $d_i = p_i \quad \forall i \leq n$, $d_{n+1} = p_n[p_{n-1}]$, $d_{n+j} = p_n^j[p_{n-1}]\ \forall j,\ 1 \leq j \leq k-2$, $d_{n+k-1} = p_n^{k-1}[d_{n-1}]$, $d_{n+k} = t^{k-1}[d_n]$.

THEOREM 10. *The following square is a pullback:*

$$\begin{array}{ccc} K_n\mathbb{B} & \xrightarrow{\quad Algj_n \quad} & Simpl\mathbb{B} \\ {\scriptstyle \overline{U}_{n-1}}\big\downarrow & & \big\downarrow{\scriptstyle U} \\ S\text{-}Simpl_{n-1}\mathbb{B} & \xrightarrow[\quad j_n \quad]{} & S\text{-}Simpl\mathbb{B} \end{array}$$

Proof. Clearly, the diagram commutes. Now, an object of the pullback is nothing but a simplicial object $\underline{X}$ such that $X_{n+k} = X_n^k[d_0, \ldots, d_{n-1}]\ \forall k \geq 1$.

Then, $d_{n+1}\colon X_{n+1} = X_n[d_0, \ldots, d_{n-1}] \to X_n$ is the Kan action on the truncated n-simplicial object $\underline{X}_n$ such that $Algj_n(\underline{X}_n, d_{n+1}) = \underline{X}$. □

We have, then, precisely located the algebras of $(\mathrm{T}_{n-1}, \lambda_{n-1}, \mu_{n-1})$ among the objects of $Simpl\mathbb{B}$ as expected in the paragraph concerning the organization of the adjoint pair (U_n, F_n). A simplicial object $\underline{X}$ is the nerve of the Kan operation on an n-simplicial object if and only if $X_{k+1} = X_k[d_0, \ldots, d_{n-1}]\ \forall k \geq n$.

But we can say much more: the functor $Algj_n$ 'factors' through $Simpl_{n+1}\mathbb{B}$:

PROPOSITION 11. *The functor $Algj_n$ is naturally isomorphic to $j_{n+2}.w_{n+1}.Algk_n$*

$$\begin{array}{ccccccc} K_n\mathbb{B} & & & \xrightarrow{\qquad\qquad Algj_n \qquad\qquad} & & & Simpl\mathbb{B} \\ & \searrow^{Algk_n} & & & & \nearrow^{j_{n+2}} & \\ {\scriptstyle \overline{U}_{n-1}}\big\downarrow & & K_{n+1}\mathbb{B} & \xrightarrow{\ w_{n+1}\ } & Simpl_{n+1}\mathbb{B} & & \big\downarrow{\scriptstyle U} \\ & & {\scriptstyle \overline{U}_n}\big\downarrow & \beta_{n+1}\Downarrow & \big\downarrow{\scriptstyle U_{n+1}} & \cong & \\ S\text{-}Simpl_{n-1}\mathbb{B} & \xrightarrow[k_n]{} & S\text{-}Simpl_n\mathbb{B} & \xrightarrow[k_{n+1}]{} & S\text{-}Simpl_{n+1}\mathbb{B} & \xrightarrow[j_{n+2}]{} & S\text{-}Simpl\mathbb{B} \end{array}$$

Proof. First let us show that, given a Kan operation $(\underline{X}_n, t)$, the canonical factorization

$$\begin{aligned} &[p_0, \ldots, p_n, p_n[d_{n-1}], t[d_n]]\colon X_n^2[d_0, \ldots, d_{n-1}] \\ &\quad \to X_n[d_0, \ldots, d_{n-1}][p_0, \ldots, p_n, t] \end{aligned}$$

obtained from the map: $(Algj_n(\underline{X}_n, t))_{n+2} \to (j_{n+2}.w_{n+1}.Algk_n(X_n, t))_{n+2}$ is an isomorphism.

For that let us consider the following diagram:

$$\begin{array}{ccccc}
X_n[d_0, \ldots, d_n] & \xleftarrow{t[d_n]_{n+1}} & X_n[d_0, \ldots, d_{n-1}][p_0, \ldots, p_n] & = & X_n^2[d_0, \ldots, d_{n-1}] \\
{\scriptstyle [p_0,\ldots,p_n]}\downarrow & & {\scriptstyle [p_0,\ldots,p_n]}\downarrow & & \| \\
X_n[d_0, \ldots, d_{n-1}] & \xleftarrow[t[d_n]]{} & X_n[d_0, \ldots, d_{n-1}][p_0, \ldots, p_{n-1}] & = & X_n^2[d_0, \ldots, d_{n-1}]
\end{array}$$

Following the corollary of Proposition 2, the vertical map on the right-hand side is an identity map. Moreover, the map $[p_0, \ldots, p_n, t]$: $X_n[d_0, \ldots, d_{n-1}] \to X_n[d_0, \ldots, d_n]$ is a splitting of $[p_0, \ldots, p_n]$ on the left-hand side.

We have also: $[p_0, \ldots, p_n, t].t[d_n] = t[d_n]_{n+1}$ since $t.t[d_n] = t.p_n[d_{n-1}] = t.p_{n+1}$ (since t is a Kan operation).

Then the following square is a pullback:

$$\begin{array}{ccc}
X_n[d_0, \ldots, d_{n-1}] & \xleftarrow{t[d_n]} & X_n^2[d_0, \ldots, d_{n-1}] \\
{\scriptstyle [p_0,\ldots,p_n,t]}\downarrow & & \| \\
X_n[d_0, \ldots, d_n] & \xleftarrow[t[d_n]_{n+1}]{} & X_n^2[d_0, \ldots, d_{n-1}]
\end{array}$$

and, following Proposition 2, the map $[p_0, \ldots, p_n, p_n[d_{n-1}], t[d_n]]$ is an isomorphism. More precisely, it is the inverse of

$$[p_0, \ldots, p_n]\colon X_n[d_0, \ldots, d_{n-1}][p_0, \ldots, p_n, t] \to X_n^2[d_0, \ldots, d_{n-1}]$$

which is nothing but $(\beta_{n+1} Algk_n(\underline{X}_n, t))_{n+1}$. Thus, $Algk_n$ makes β_{n+1} invertible.

Now it is easy to check that the factorization

$$\begin{aligned}
&[p_0, \ldots, p_n^{m-1}[d_{n-1}], t^{m-1}[d_n]]\colon X_n^m[d_0, \ldots, d_{n-1}] \\
&\quad \to (X_n[d_0, \ldots, d_{n-1}])^{m-1}[p_0, \ldots, p_n, t]
\end{aligned}$$

has

$$[p_0, \ldots, p_n]\colon (X_n[d_0, \ldots, d_{n-1}])^{m-1}[p_0, \ldots, p_n, t] \to X_n^m[d_0, \ldots, d_{n-1}])$$

as an inverse.

Consequently, the whole information about $Algj_n(\underline{X}_n, t)$ is contained in $w_{n+1}.Algk_n(\underline{X}_n, t)$, whence the following definition.

DEFINITION 4. We shall call $N_n = w_{n+1}.Algk_n$ the nerve functor for n-Kan operations.

PROPOSITION 12. *The following square is a quasi pullback that is the universal square with a natural isomorphism:*

$$\begin{array}{ccc}
K_n\mathbb{B} & \xrightarrow{\quad w_{n+1}Algk_n \quad} & S\text{-}Simpl_{n+1}\mathbb{B} \\
\Big\downarrow{\scriptstyle \overline{U}_{n-1}} & \beta_{n+1}Algk_n \ \Updownarrow & \Big\downarrow{\scriptstyle U_{n+1}} \\
S\text{-}Simpl_{n-1}\mathbb{B} & \xrightarrow[\quad k_{n+1}k_n \quad]{} & S\text{-}Simpl_{n+1}\mathbb{B}
\end{array}$$

Proof. Let $\underline{X}_{n+1}$ be a $(n+1)$-simplicial object such that $U_{n+1}(\underline{X}_n)$ is isomorphic to some $k_{n+1}.k_n(Z_{n-1})$. This means that X_{i+1} is isomorphic to Z_i $\forall i,\ 0 \leq i \leq n-1$, and that X_{n+1} is isomorphic to $Z_{n-1}[d_0, \dots, d_{n-1}]$ which is itself isomorphic to $X_n[d_0, \dots, d_{n-1}]$. Then, the map $d_{n+1}\colon X_{n+1} \to X_n$ determines a Kan operation on $\underline{X}_n$. □

REMARK. This gives us another characterization of a simplicial object which is the nerve of the Kan operation on an n-simplicial object and, probably, the best one:

COROLLARY. A simplicial object X is the nerve of the Kan operation on an n-simplicial object if and only if:

$$X_{n+1} = X_n[d_0, \dots, d_{n-1}], \qquad X_{n+k+1} = X_{n+k}[d_0, \dots, d_{n+k}], \quad k \geq 1.$$

6. $S\text{-}Simpl_n\mathbb{B}$ from $Pt\mathbb{B}$: an Inductive Autonomous Construction

We saw in Proposition 5 that a $(n+1)$-simplicial object in $\mathbb{B}$ is just a τ_{n-1}-identity $\delta\colon \underline{Z}_n \to U_n\underline{X}_n$ in $S\text{-}Simpl_n\mathbb{B}$.

If, furthermore, $\underline{X}_n$ is itself a split n-simplicial object and if we consider a τ_{n-1}-identity of the form $\delta\colon \underline{Z}_n \to U_nF_n\underline{X}_n = T_n\underline{X}_n$ its corresponding $(n+1)$-simplicial object is split as far as the level n.

To have a splitting $s_{n+1}\colon X_n \to X_{n+1}$ at the level $n+1$ is exactly to have a map $l\colon \underline{X}_n \to \underline{Z}_n$ in $S\text{-}Simpl_n\mathbb{B}$ of the following form:

$$\begin{array}{ccc}
X_n & \xrightarrow{\ s_{n+1}\ } & Z_n = X_{n+1} \\
{\scriptstyle d_0}\downarrow \cdots \downarrow{\scriptstyle d_n} \uparrow{\scriptstyle s_n} & & {\scriptstyle d_0}\downarrow \cdots \downarrow{\scriptstyle d_n} \uparrow{\scriptstyle s_n} \\
X_{n-1} & \xrightarrow{\ s_n\ } & Z_{n-1} = X_n
\end{array}$$

that is to say, a map l in $S\text{-}Simpl_n\mathbb{B}$ such that $\delta.l = \lambda_n\underline{X}_n$ and, consequently, a T_n-element in the sense given in the introduction.

That is why we introduce now the following more general context:

Let $\mathbb{E}$ be a category with a class Σ of maps which satisfies the following property: for any map $d\colon Z \to X$ in Σ and any map $f\colon X' \to X$ in $\mathbb{E}$ there is a pullback in $\mathbb{E}$ with d' in Σ:

$$\begin{array}{ccc} Z' & \xrightarrow{k} & Z \\ {\scriptstyle d'}\big\downarrow & & \big\downarrow{\scriptstyle d} \\ X' & \xrightarrow[f]{} & X \end{array}$$

We shall call such a class of maps pullback amenable. It is clear that this property is different from the stability under pullback even if always there are pullbacks in $\mathbb{E}$.

If we denote by $M_\Sigma\mathbb{E}$ the full subcategory of $M\mathbb{E}$ whose objects are the maps in Σ, the previous property is equivalent to the fact that the restriction of p to $M_\Sigma\mathbb{E}$, $p\colon M_\Sigma\mathbb{E} \to \mathbb{E}$ is a fibration.

The main example we have in mind is the following one:

A left exact category $\mathbb{B}$ with a fibration $\tau\colon \mathbb{E} \to \mathbb{B}$ such that each fibre of τ is left exact and each change of the base functor is left exact: we say, then, that τ is a left exact fibration.

In this situation, the category $\mathbb{E}$ and the functor τ are left exact and the pullbacks in the fibres of τ are pullbacks in $\mathbb{E}$.

Now, the class Σ_τ of τ-identities (the maps in the fibres of τ) is pullback amenable. Indeed, let $f = f_2.f_1$ be the decomposition of f with f_1 a τ-identity and f_2 a cartesian map and let us consider the following diagram:

$$\begin{array}{ccccc} Z' & \xrightarrow{\overline{f_1}} & V & \xrightarrow{\overline{f_2}} & Z \\ {\scriptstyle d'}\big\downarrow & & \big\downarrow{\scriptstyle d''} & & \big\downarrow{\scriptstyle d} \\ X' & \xrightarrow[f_1]{} & U & \xrightarrow[f_2]{} & X \end{array}$$

where $\overline{f}_2$ is the Cartesian map above $\tau(f_2)$ with codomain Z and d'' is the image of d by the change of the base along $\tau(f_2)$.

Then, d'' is a τ-identity and the right hand square is a pullback. The pullback of f_1 and d'' in the fibre above $\tau(U)$ will give us the required amenable pullback.

Clearly, for each $n \in \mathbb{N}$ the fibration $\tau_{n-1}\colon S\text{-}Simpl_n\mathbb{B} \to S\text{-}Simpl_{n-1}\mathbb{B}$ is always left exact and, therefore, the class of τ_{n-1}-identities is pullback amenable.

Let us suppose we have a category $\mathbb{E}$ endowed with a pullback amenable class Σ and a monad (T, λ, μ), no relation between them being assumed.

DEFINITION 5. A ΣT-element of $\mathbb{E}$ is a triple $\underline{X}_1 = (X_0, d, l)$ with X_0 an object $\mathbb{E}$, $d\colon X_1 \to TX_0$ a map in Σ and $l\colon X_0 \to X_1$, a map such that $d.l = \lambda X_0$. A morphism of ΣT-elements between $\underline{X}_1$ and $\underline{X}'_1$ is a pair (f_0, f_1), $f_0\colon X_0 \to X'_0$, $f_1\colon X_1 \to X'_1$ of morphisms in $\mathbb{E}$ such that $d'.f_1 = Tf_0.d$ and $f_1.l = l'.f_0$.

We denote by $\Sigma T\mathbb{E}$ the category of the ΣT-elements in $\mathbb{E}$ (and often simply $T\mathbb{E}$ in our main example with the class of τ−identities).

There is a forgetful functor τ_0: $\Sigma T\mathbb{E} \to \mathbb{E}$ associating X_0 to $\underline{X}_1$. The functor τ_0 is a fibration and the morphism (f_0, f_1) is cartesian in $\Sigma T\mathbb{E}$ if and only if the following square is a pullback.

$$\begin{array}{ccc} X_1 & \xrightarrow{f_1} & X'_1 \\ {\scriptstyle d}\downarrow & & \downarrow{\scriptstyle d'} \\ TX_0 & \xrightarrow[Tf_0]{} & TX'_0 \end{array}$$

From now on, for the sake of simplicity and since it is the case in our main example (the fibrations τ_n), we suppose that the category $\mathbb{E}$ and the monad (T, λ, μ) are left exact and that the class Σ fulfils the following conditions:

(1) Σ contains the identity maps,

(2) Σ is left proper: i.e., such that if g is in Σ, then $f.g$ is in Σ if and only if f is in Σ.

Then, the fibration τ_0 is left exact as a fibration and the category $\Sigma T\mathbb{E}$ is left exact. Moreover, τ_0 has the right adjoint right inverse k_1, defined by $k_1(X) = (X, 1_{TX}, \lambda X)$, the projection $\omega(\underline{X}_1)$: $\underline{X}_1 \to k_1 X_0 = k_1\tau_0\underline{X}_1$ being the pair $(1_{X_0}, d)$.

EXAMPLES. (1) If the fibration τ is the terminal functor $\mathbb{B} \to 1$ and the monad on $\mathbb{B}$ is the identity monad, then $T\mathbb{B} = Pt\mathbb{B}$ and the fibration τ_0 is p: $Pt\mathbb{B} \to \mathbb{B}$ with its right adjoint right inverse k.

(2) We just observed that:

PROPOSITION 13. *If the fibration τ is τ_{n-1}: $S\text{-}Simpl_n\mathbb{B} \to S\text{-}Simpl_{n-1}\mathbb{B}$ and the monad (T, λ, μ) is (T_n, λ_n, μ_n), then $T_n(S\text{-}Simpl_n\mathbb{B}) = S\text{-}Simpl_{n-1}\mathbb{B}$ and the associated fibration τ_0 is nothing but τ_n with its right adjoint right inverse k_{n+1}.*

The monad (T, λ, μ) on $\mathbb{E}$ can be extended to $\Sigma T\mathbb{E}$:

THEOREM 14. *There is on $\Sigma T\mathbb{E}$ a monad (T_1, λ_1, μ_1) which is transversely cartesian to τ_0 (i.e., $\lambda_1\underline{X}_1$ and $\mu_1\underline{X}_1$ are always cartesian maps) and such that the following diagram is a pullback:*

$$\begin{array}{ccccc} \tau_0 T_1\underline{X}_1 & & \xleftarrow{\quad\tau_0\mu_1\underline{X}_1\quad} & & \tau_0 T_1^2\underline{X}_1 \\ {\scriptstyle \tau_0 T_1\omega\underline{X}_1}\downarrow & & & & \downarrow{\scriptstyle \tau_0 T_1\omega T_1\underline{X}_1} \\ T\tau_0\underline{X}_1 & \xleftarrow[\mu\tau_0\underline{X}_1]{} & T^2\tau_0\underline{X}_1 & \xleftarrow[T\tau_0T_1\omega\underline{X}_1]{} & T\tau_0T_1\underline{X}_1 \end{array}$$

Proof. Given $\underline{X}_1 = (X_0, d, l)$ a ΣT-element, let us consider the following amenable pullback:

$$\begin{array}{ccccc} X_1 & & \xleftarrow{\quad m\langle d\rangle \quad} & & X_1\langle d\rangle \\ {\scriptstyle d}\downarrow & & & & \downarrow{\scriptstyle d_0} \\ TX_0 & \xleftarrow[\mu X_0]{} & T^2X_0 & \xleftarrow[T_d]{} & TX_1 \end{array}$$

We have $\mu X_0.Td.\lambda X_1 = \mu X_0.\lambda TX_0.d = d$, whence a unique map $l_0\colon X_1 \to X_1\langle d\rangle$ satisfying $d_0.l_0 = \lambda X_1$ and $m\langle d\rangle .l_0 = 1$.

Let us set $T_1\underline{X_1} = (X_1, d_0, l_0)$. This is clearly an endofunctor satisfying $\tau T_1\underline{X_1} = X_1$.

The map Tl is a section of $\mu X_0.Td$ and produces a section $l\langle d\rangle$ of $m\langle d\rangle$. Hence the natural transformation $\lambda_1\underline{X}_1 = (l, l\langle d\rangle)\colon \underline{X}_1 \to T_1\underline{X}_1$, with $\lambda_1 X_1$ cartesian.

Now $T_1^2\underline{X}_1$ is determined by the following pullback:

$$\begin{array}{ccccc} X_1\langle d\rangle & & \xleftarrow{\quad m\langle d_0\rangle \quad} & & X_1^2\langle d\rangle \\ {\scriptstyle d_0}\downarrow & & & & \downarrow{\scriptstyle d_0} \\ TX_1 & \xleftarrow[\mu X_1]{} & T^2X_1 & \xleftarrow[Td_0]{} & T_1X_1\langle d\rangle \end{array}$$

Then the equality

$$\begin{aligned} \mu X_0.Td.\mu X_1.Td_0 &= \mu X_0.\mu TX_0.T^2d.Td_0 = \mu X_0.T\mu X_0.T^2d.Td_0 \\ &= \mu X_0.Td.Tm\langle d\rangle \end{aligned}$$

produces a unique map $m_1\langle d\rangle\colon X_1^2\langle d\rangle \to X_1\langle d\rangle$ such that $d_0.m_1\langle d\rangle$

$$= Tm\langle d\rangle.d_0 \text{ and } m_1\langle d\rangle.m\langle d\rangle = m\langle d\rangle.m\langle d_0\rangle.$$

Hence the natural transformation $\mu_1\underline{X_1} = (m\langle d\rangle, m_1\langle d\rangle)$ with $\mu_1\underline{X}_1$ cartesian. To check if λ_1 and μ_1 satisfy the axioms of a monad is straightforward. □

REMARKS. (1) The monad on $Pt\mathbb{B}$ considered as the category of the Id-elements (see Example (1)) is just the canonical monad (T_0, λ_0, μ_0).

(2) If $\mathbb{E} = S\text{-}Simpl_n\mathbb{B}$ and, $\tau = \tau_{n-1}$ and $(T, \lambda, \mu) = (T_n, \lambda_n, \mu_n)$, then according to Proposition 4 the monad determined on $T_n(S\text{-}Simpl_n\mathbb{B})$ by Theorem 14 is $(T_{n+1}, \lambda_{n+1}, \mu_{n+1})$.

(3) In case (T, λ, μ) is left exact, then (T_1, λ_1, μ_1) is left exact.

(4) The construction of the category $\Sigma T\mathbb{E}$ is the beginning of an inductive process. We started from a left exact category $\mathbb{E}$ with a pullback amenable class Σ which contains the identity maps and is left proper, and with a left exact monad (T, λ, μ). Then we got the category $\Sigma T\mathbb{E}$ with the left exact fibration $\tau_0\colon \Sigma T\mathbb{E} \to$

$\mathbb{E}$. Then the class Σ_1 of the τ_0-identities is pullback amenable, left proper and contains the identity maps. Furthermore, the monad (T_1, λ_1, μ_1) is left exact.

We denote by $T_n\mathbb{E}$ and (T_n, λ_n, μ_n) the category and the monad obtained at the n-th step of this construction.

(5) An example of this inductive process is given by the tower of the split n-simplicial objects, since we observed that starting from $\mathbb{E} = Pt\mathbb{B}$, Σ the class of p-identities and $(T, \lambda, \mu) = (T_0, \lambda_0, \mu_0)$. Then $T_n(S\text{-}Simpl_n\mathbb{B}) = S\text{-}Simpl_{n+1}\mathbb{B}$.

(6) Clearly, $k_1 TX = T_1 k_1 X$ and k_1 is a strict morphism of monads from (T, λ, μ) to (T_1, λ_1, μ_1). Therefore it determines an extension functor $Algk_1\colon AlgT \to AlgT_1$ by $Algk_1(X, x) = (k_1 X, k_1 x)$.

It is exactly the situation of $Algk_{n-1}\colon K_n\mathbb{B} \to K_{n+1}\mathbb{B}$ as described in Proposition 9, Remark 5.

7. The Canonical Factorization Determined by $T\mathbb{E}$

Let Σ be a pullback amenable proper class of maps in $\mathbb{E}$ containing the identities and let us denote by $Pt_\Sigma\mathbb{E}$ the full subcategory of $Pt\mathbb{E}$ whose objects are split epimorphisms in Σ. The restriction functor $p\colon Pt_\Sigma\mathbb{E} \to \mathbb{E}$ is then a left exact fibration with the right adjoint right inverse k. Moreover, the canonical monad (T_0, λ_0, μ_0) makes sense on $Pt_\Sigma\mathbb{E}$.

On the one hand, we denote by $U_\Sigma\colon \Sigma T\mathbb{E} \to Pt\mathbb{E}$ the left exact functor which assigns to $\underline{X}_1 = (X_0, d, l)$ the Σ split epimorphism given by the following amenable pullback:

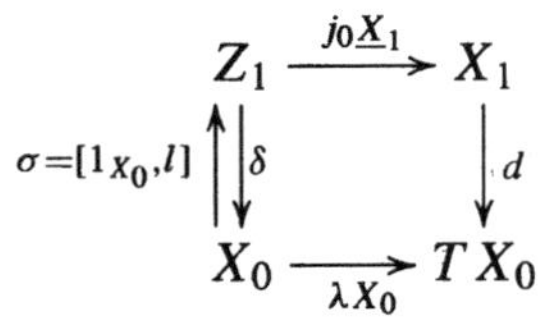

The following diagram commutes:

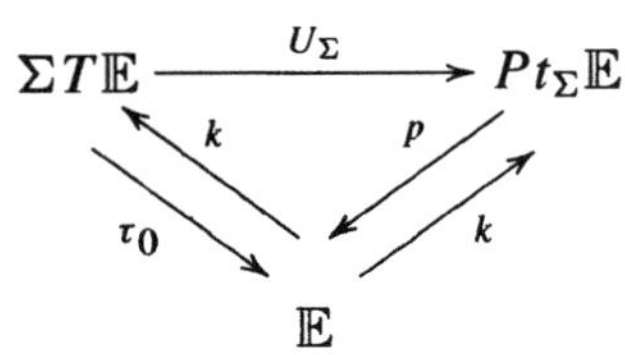

PROPOSITION 15. *The functor U_Σ is left exact and cartesian. There is a natural transformation $j\colon T_0.U_\Sigma \Rightarrow U_\Sigma.T_1$ which determines a lax morphism of monads, i.e.:*

$$j.\lambda_0 U_\Sigma = U_\Sigma\lambda_1, \qquad j.\mu_0 U_\Sigma = U_\Sigma\mu_1.jT_1.T_0 j.$$

Proof. The functor U_Σ is cartesian since the previous triangle commutes and U_Σ is left exact.

Since $\mu X_0.Td.\lambda X_1 = \mu X_0.\lambda TX_0.d = d$, the object $U_\Sigma T_1\underline{X}_1$ is just (p_0, δ_0), p_0: $X_1[d] \to X_1$ and the pair $(\lambda X_0, j_0\underline{X}_1)$ determines a unique factorization $j_1\underline{X}_1$: $Z_1[\delta] \to X_1[d]$.

Let us set $j\underline{X}_1 = (j_0\underline{X}_1, j_1\underline{X}_1)$: $T_0U_\Sigma(\underline{X}_1) \Rightarrow U_\Sigma.T_1(\underline{X}_1)$. It determines a natural transformation j: $T_0.U_\Sigma \Rightarrow U_\Sigma.T_1$. To check the axioms of a lax morphism of monads is straightforward. □

REMARK. As a consequence, there is an extension J of U_Σ to the categories of algebras, J: $AlgT_1 \to AlgT_0 = Grd_\Sigma\mathbb{E}$ where $Grd_\Sigma\mathbb{E}$ is the full subcategory of $Grd\mathbb{E}$ whose objects are the groupoids in Σ. Given $x = (x_0, x_1)$: $T_1\underline{X}_1 \to \underline{X}_1$ a T_1-algebra structure, then $J(\underline{X}_1, x) = (U_\Sigma.\underline{X}_1, U_\Sigma x.j\underline{X}_1)$: $T_0U_\Sigma\underline{X}_1 \to U_\Sigma\underline{X}_1$.

This groupoid is described by the following left-hand side vertical diagram:

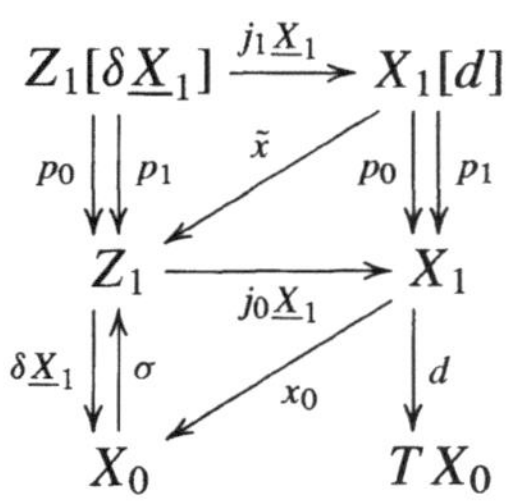

where $\widetilde{x}$ is the unique factorization: $X_1[d] \to Z_1$ such that $j_0\underline{X}_1.\widetilde{x} = x_1.j_0T_1\underline{X}_1$ and $\delta\underline{X}_1.\widetilde{x} = x_0.\delta T_1\underline{X}_1$.

On the other hand, let us consider $AlgT$ the category of algebras of the monad (T, λ, μ) on $\mathbb{E}$ and U: $AlgT \to \mathbb{E}$ the canonical left exact forgetful functor. Let us denote by $Pt_\Sigma AlgT$ the full subcategory of $PtAlgT$ whose objects are those split epimorphims in $AlgT$ whose image by U are in Σ.

Then, again p: $Pt_\Sigma AlgT \to AlgT$ is a left exact fibration and the monad (T_0, λ_0, μ_0) makes sense on $Pt_\Sigma AlgT$.

The functor U being left exact, it clearly defines an extension $Pt_\Sigma AlgT \to Pt_\Sigma\mathbb{E}$ which is a strict morphism of monads.

Let us denote by $\widetilde{U}_\Sigma$: $Pt_\Sigma AlgT \to \Sigma T\mathbb{E}$ the left exact functor assigning to each Σ split epimorphism (d, s) in $AlgT$, d: $(Z, z) \to (X, x)$, the ΣT-element defined by the following pullback in $\mathbb{E}$ (and, thus, in $AlgT$ since d and x are morphisms of algebras):

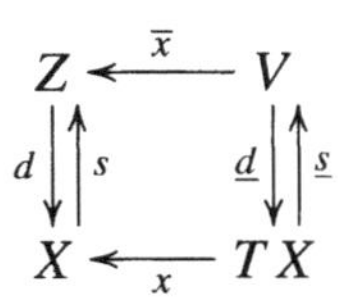

by setting $\widetilde{U}_\Sigma(d, s) = (X, \overline{d}, \overline{s}.\lambda X)$.

Since $x.\lambda X = 1_X$, the functor $U_\Sigma.\widetilde{U}_\Sigma$ is, up to a natural isomorphism, equal to $Pt_\Sigma U$.

This is the canonical factorisation associated with $\Sigma T\mathbb{E}$:

$$\begin{array}{ccc} Pt_\Sigma AlgT & \xrightarrow{Pt_\Sigma U} & Pt_\Sigma\mathbb{E} \\ & \simeq & \\ {\scriptstyle \widetilde{U}_\Sigma}\searrow & & \nearrow{\scriptstyle U_\Sigma} \\ & \Sigma T\mathbb{E} & \end{array}$$

The following square commutes:

$$\begin{array}{ccc} Pt_\Sigma AlgT & \xrightarrow{\widetilde{U}_\Sigma} & \Sigma T\mathbb{E} \\ {\scriptstyle p}\downarrow\uparrow{\scriptstyle k} & & {\scriptstyle \tau_0}\downarrow\uparrow{\scriptstyle k} \\ AlgT & \xrightarrow[U]{} & \mathbb{E} \end{array}$$

PROPOSITION 16. *The functor $\widetilde{U}_\Sigma$ preserves the cartesian maps and reflects them. There is a natural transformation $\gamma\colon T_1.\widetilde{U}_\Sigma \Rightarrow \widetilde{U}_\Sigma.T_0$ which is a lax morphism of monads.*

Proof. The functor $\widetilde{U}_\Sigma$ is cartesian since the previous square commutes and $\widetilde{U}_\Sigma$ is left exact. It reflects the cartesian maps since the algebra x has λX as a section.

Since d and x are morphisms in $AlgT$, there is a T-algebra structure $v\colon TV \to V$ on V which is such that $\overline{x}.v = z.T\overline{x}$ and $\overline{d}.v = \mu X.T\overline{d}$.

The pullback which defines $\widetilde{U}_\Sigma.T_0(d,s)$ is the following one:

$$\begin{array}{ccc} Z[d] & \xleftarrow{\overline{z}} & V \\ {\scriptstyle p_0}\downarrow\uparrow{\scriptstyle s_0} & & {\scriptstyle \overline{p_0}}\downarrow\uparrow{\scriptstyle \overline{s_0}} \\ Z & \xleftarrow[z]{} & TZ \end{array}$$

Since $\overline{d}.v = \mu X.T\overline{d}$, there is a unique factorization $\overline{v}\colon V\langle\overline{d}\rangle \to V[\overline{d}]$ such that $p_0\overline{v} = v.d_0$ and $p_1\overline{v} = m\langle d\rangle$.

Hence follows a unique factorization $w\colon V\langle\overline{d}\rangle \to W$ such that $p_0.w = T\overline{x}.d_0$ and $\overline{x}.w = \overline{x}[x].\overline{v}$.

The map $\gamma(d,s) = (\overline{x}, w)\colon T_1.\widetilde{U}_\Sigma(d,s) \to \widetilde{U}_\Sigma.T_0(d,s)$ determines a natural transformation $\gamma\colon T_1.\widetilde{U}_\Sigma \Rightarrow \widetilde{U}_\Sigma T_0$. To check if it is a lax morphism of monads is straightforward. □

Mainpoint: consequently, there is an extension ν_T of $\widetilde{U}_\Sigma$ to the categories of algebras, $\nu_T\colon Grd_\Sigma AlgT \to AlgT_1$, where $Grd_\Sigma AlgT$ is the full subcategory of $Grd_\Sigma AlgT$ whose objects are those groupoids in $AlgT$ whose image by U are in Σ.

Since $\widetilde{U}_\Sigma$ is left exact, ν_T is left exact, as well.

Before going into details of description of the functor ν_T which is the key tool of the nerve functor for n-groupoids, note that:

(1) The canonical isomorphism: $Pt_\Sigma U \simeq U_\Sigma.\widetilde{U}_\Sigma$ produces a canonical isomorphism $Grd_\Sigma U \simeq \nu_T.J$:

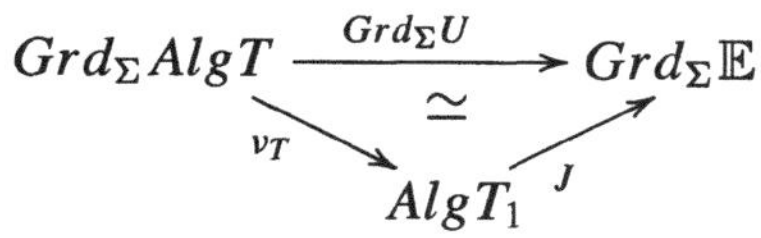

(2) The following outer square commutes since the two inner squares do:

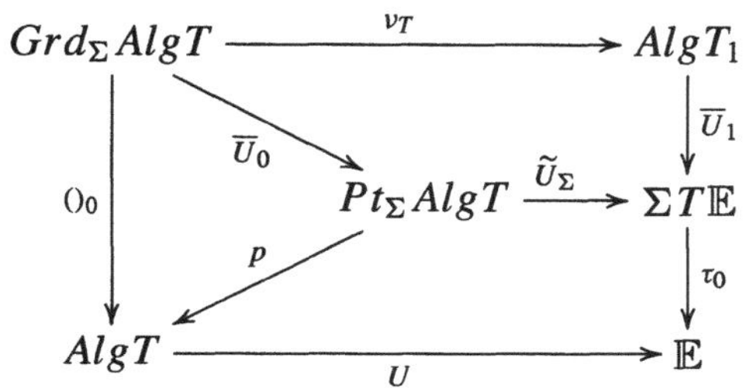

Description of the Functor ν_T

Given X_1 a Σ-groupoid in $AlgT$

$$X_1 : (X_0, x_0) \underset{d_1}{\overset{d_0}{\leftleftarrows}} \!\!\xrightarrow{s_0}\! (mX_1, x_1) \underset{d_2}{\overset{p_0}{\leftleftarrows}}\!\!\xleftarrow{p_1}\! (mX_1[d_0], x),$$

then $\nu_T(X_1)$ has $\widetilde{U}_\Sigma(d_0, s_0)$ as an underlying object in $\Sigma T\mathbb{E}$ and the map $\widetilde{U}_\Sigma(d_1, d_2).\gamma(d_0, s_0)$ as T_1-algebra structure in $\Sigma T\mathbb{E}$.

The first component of this map is $d_1.\bar{x}$ and the second one is the unique factorization α: $V\langle \bar{d}\rangle \to TV$ such that: $\overline{d_0}.\alpha = Td_1.T\overline{x_0}.d_0$ and $\overline{x_0}.\alpha = d_2.\overline{x_0}[x_0].\overline{v}$.

In other words, the map α is determined by the following pair of dotted arrows:

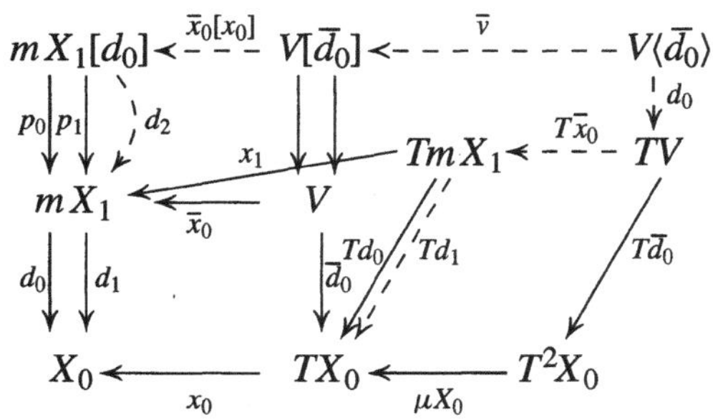

REMARK. We need the following specification: the map $\overline{x_0}[x_0].\overline{v}$ is characterized by the equalities

$$p_0.(\overline{x_0}[x_0].\overline{v}.) = \overline{x_0}.p_0.\overline{v} = \overline{x_0}.v.d_0 = x_1.T\overline{x_0}.d_0,$$
$$p_1.(\overline{x_0}[x_0].\overline{v}.) = \overline{x_0}.p_1.\overline{v} = \overline{x_0}.m\,\langle d\rangle\,.$$

PROPOSITION 17. *The nerve of a discrete groupoid is given by* $Algk_1$.
In other words: $\nu_T.dis = Algk_1$.
Proof. Straightforward from the previous description. □

8. The Nerve Functor for the 2-Groupoids

We observed that $Grd\mathbb{B} = K_1\mathbb{B}$, so that the nerve functor ν_1 for groupoids is just the identity functor.

Now, taking $\mathbb{E} = Pt\mathbb{B}$, Σ the class of the p-identities and the canonical monad (T_0, λ_0, μ_0) on $Pt\mathbb{B}$, then the previous construction determines a functor ν_{T_0}: $Grd_0(Grd\mathbb{B}) \to AlgT_1 = K_2\mathbb{B}$ since the category of algebras of the monad (T_1, λ_1, μ_1) on $T_0(Pt\mathbb{B}) = S\text{-}Simpl_1\mathbb{B}$ is $K_2\mathbb{B}$ (Proposition 9).

On the other hand, $Grd_0(Grd\mathbb{B})$ is the full subcategory of $Grd(Grd\mathbb{B})$ whose objects are groupoids in the fibres of the canonical left exact fibration $()_0$: $Grd\mathbb{B} \to \mathbb{B}$, that is precisely the category $2\text{-}Grd\mathbb{B}$ of the internal 2-groupoids in $\mathbb{B}$.

Therefore, let us set $\nu_2 = \nu_{T_0}$: $2\text{-}Grd\mathbb{B} \to \mathbb{K}_2\mathbb{B}$ and call it the nerve functor for the 2-groupoids.

Description of the Nerve of the 2-Groupoids

An internal 2-groupoid X_2 in $\mathbb{B}$ is, thus, an internal groupoid in $Grd\mathbb{B}$.

$$X_2 : X_1 \underset{\tau_1}{\overset{\tau_0}{\leftleftarrows}} m_1X_2 \underset{\substack{p_1 \\ \longleftarrow \\ \tau_2}}{\overset{p_0}{\leftleftarrows}} m_1X_2[d_0]$$

such that all its structural maps τ_i are $()_0$-identity maps, where $()_0$: $Grd\mathbb{B} \to \mathbb{B}$ assigns to each groupoid X_1 its object of objects X_0.

The groupoid X_1 is called the groupoid of 1-arrows or 1-cells of X_2 and the groupoid m_1X_2 the groupoid of its 2-cells.

Clearly, there is a functor $()_1$: $2\text{-}Grd\mathbb{B} \to Grd\mathbb{B}$ associating X_1 with X_2 which is a left exact fibration. It has the right adjoint right inverse gr_2 where gr_2X_1 is the kernel groupoid of the terminal map $X_1 \to grX_0$ in the fibre of X_1 with respect to $()_0$, and the left adjoint right inverse dis, where $dis\,X_1$ is the kernel groupoid of 1_{X_1}: $X_1 \to X_1$.

We need first the following result:

PROPOSITION 18. *The following square commutes*

$$\begin{array}{ccc} 2\text{-}Grd\mathbb{B} & \xrightarrow{\ \nu_2\ } & K_2\mathbb{B} \\ {\scriptstyle ()_1}\Big\downarrow & & \Big\downarrow{\scriptstyle \overline{\tau}_1} \\ Grd\mathbb{B} & \xrightarrow[w_1]{} & Simpl_1\mathbb{B} \end{array}$$

Proof. For every groupoid X_2, the map i_2: $dis\,X_1 \to X_2$ in $2\text{-}Grd\mathbb{B}$ is a $()_1$-identity map ($dis\,X_1$ is the initial object in the fibre above X_1).

Then $\underline{U}_1.\overline{\tau_1}.\nu_2(i_2) = \tau_0.\overline{U_1}.\nu_2(i_2) = \overline{U_0}.()_1(i_2)$ and, thus, the map $\underline{U}_1.\overline{\tau_1}.\nu_2(i_2)$ is an identity map. But the functor $\underline{U}_1$: $Simpl_1\mathbb{B} \to Pt\mathbb{B}$ reflects the identities and then $\overline{\tau_1}.\nu_2(i_2)$ is an identity. But its domain is $\overline{\tau_1}.\nu_2.dis(X_1) = \overline{\tau_1}.\nu_2.dis.()_1(X_2) = \overline{\tau_1}.\ Algk_1.()_1(X_2) = w_1.()_1(X_2)$ and its codomain is just $\overline{\tau_1}.\nu_2(X_2)$. Therefore, $\overline{\tau_1}.\nu_2 = w_1.()_1$.

REMARK. This result means that, given a 2-groupoid X_2, the underlying 1-simplicial object of its nerve is the underlying 1-simplicial object (= reflexive graph) of its underlying groupoid X_1.

According to this result we have: $(\nu_2 X_2)_0 = X_0$ and $(\nu_2 X_2)_1 = mX_1$.

It remains to describe $(\nu_2 X_2)_2$ and the Kan operation. According to the description of ν_T, $(\nu_2 X_2)_2$ is given by the following pullback of dotted arrows in $\mathbb{B}$, where τ_0 plays the role of d_0 and d_2 the role of the algebra x_0 in the previous description of ν_T, and mX_2 denotes $(m_1 X_2)_0$:

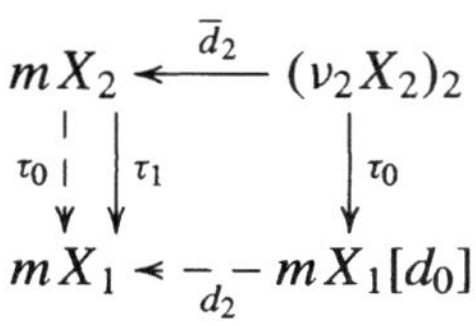

with d_i: $(\upsilon_2 X_2)_2 \to (\upsilon_2 X_2)_1 = mX_1$ given by: $d_i = p_i.\overline{\tau_0}$, $\quad 0 \leq i \leq 1$, and $d_2 = \tau_1.\overline{d_2}$.

The Kan operation t_2: $(\upsilon_2 X_2)_2[d_0, d_1] \to (\upsilon_2 X_2)_2$ is the unique map such that:

$$\overline{\tau_0}.t_2 = \tau_1[1].[\overline{d_2}.p_0, \overline{d_2}.p_1], \overline{d_2}.t_2 = \tau_2.[d_2.[\overline{d_2}.p_0, \overline{d_2}.p_1], \overline{d_2}.p_2].$$

Here t_2 plays the role of α and $\overline{d_2}$ the role of $\overline{x_0}$ in the general description of ν_T.

Therefore, given a 2-groupoid X_2 in the category of sets, $(\upsilon_2 X_2)_0$ is the set of the objects of X_2, $(\upsilon_2 X_2)_1$ is the set of 1-arrows of X_2 and $(\upsilon_2 X_2)_2$ is the set of the triangles of the form:

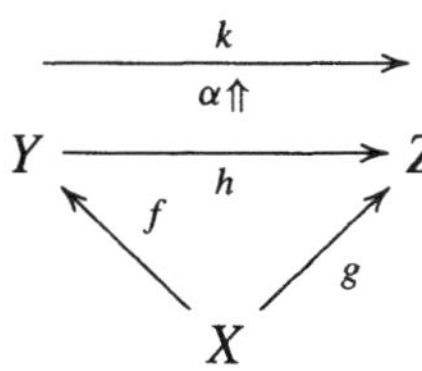

where $h.f = g$ holds and α: $h \Rightarrow k$ is a 2-cell.

Thus, $(\upsilon_2 X_2)_2$ is isomorphic as we expected, to the set of oriented triangles:

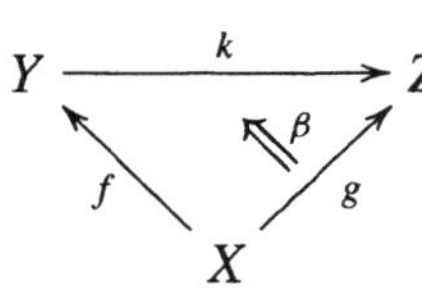

with $\beta = \alpha.f \Longleftrightarrow \alpha = \beta.f^{-1}$.

9. The Nerve Functor for the n-Groupoids in $\mathbb{B}$

We have defined the nerve functor at level 1 and 2. We are now going to define it inductively at any level.

Let us recall first what is an internal n-groupoid in $\mathbb{B}$.

Let us suppose we have defined the category $k\text{-}Grd\mathbb{B}$ of the internal k-groupoids in $\mathbb{B}$ with the left exact fibration $()_{k-1}$: $k\text{-}Grd\mathbb{B} \to (k-1)Grd\mathbb{B}$ with its right adjoint right inverse gr_k and its left adjoint right inverse dis, and this for all k, $0 \leq k \leq n-1$.

Then, the category $n\text{-}Grd\mathbb{B}$ of the internal n-groupoids in $\mathbb{B}$ is the full subcategory of $Grd((n-1)\text{-}Grd\mathbb{B})$ whose objects are the groupoids in the fibres of $()_{n-2}$: $(n-1)\text{-}Grd\mathbb{B} \to (n-2)\text{-}Grd\mathbb{B}$.

An object X_n of $n\text{-}Grd\mathbb{B}$ is, thus, an internal groupoid in $(n-1)\text{-}Grd\mathbb{B}$:

$$X_n : X_{n-1} \underset{d_1}{\overset{d_0}{\underset{s_0}{\leftrightarrows}}} m_{n-1}X_n \underset{d_2}{\overset{p_0}{\underset{p_1}{\leftleftarrows}}} m_{n-1}X_n[d_0]$$

such that any of these structural maps are $()_{n-2}$-identities.

The object X_{n-1} is the $(n-1)$-groupoid of the $(n-1)$ cells and $m_{n-1}X_n$ is the $(n-1)$-groupoid of the n-cells.

We have a forgetful functor $()_{n-1}$: $n\text{-}Grd\mathbb{B} \to (n-1)\text{-}Grd\mathbb{B}$ associating X_{n-1} with X_n. It is a left exact fibration. The right adjoint right inverse gr_n assigns to X_{n-1} the kernel groupoid of the terminal map $X_{n-1} \to gr_{n-1}X_{n-2}$ in the fibre of X_{n-1}, and the left adjoint right inverse dis is just the kernel groupoid of $1_{X_{n-1}}$: $X_{n-1} \to X_{n-1}$.

Now, suppose we have defined the nerve functor as a left exact functor ν_k: $k\text{-}Grd\mathbb{B} \to K_k\mathbb{B}$ for all k, $1 \leq k \leq n-1$, such that the following diagram commutes:

$$\begin{array}{ccccc}
k\text{-}Grd\mathbb{B} & & \xrightarrow{\quad \nu_k \quad} & & K_k\mathbb{B} \\
 & & & & \downarrow{\scriptstyle \overline{U}_{k-1}} \\
\downarrow{\scriptstyle ()_{k-1}} & & & & S\text{-}Simpl_{k-1}\mathbb{B} \\
 & & & & \downarrow{\scriptstyle \tau_{k-2}} \\
(k-1)Grd\mathbb{B} & \xrightarrow[\nu_{k-1}]{} & K_{k-1}\mathbb{B} & \xrightarrow[\overline{U}_{k-2}]{} & S\text{-}Simpl_{k-2}\mathbb{B}
\end{array}$$

The functor ν_{n-1}, being left exact, extends to a functor

$$Grd\nu_{n-1}\colon Grd((n-1)\text{-}Grd\mathbb{B}) \to Grd(K_{n-1}\mathbb{B}).$$

However, because of the commutativity of the previous square for $k = n - 1$, a $()_{n-2}$-identity map has its image by $\tau_{n-3}.\overline{U}_{n-2}.\nu_{n-1}$ an identity map. Then the restriction $Grd_n \nu_{n-1}$ of the functor $Grd\ \nu_{n-1}$ to n-$Grd\mathbb{B}$ takes its values in the full subcategory $Grd_{\tau\overline{U}}(\mathbb{K}_{n-1}\mathbb{B})$ whose objects are the groupoids in $\mathbb{K}_{n-1}\mathbb{B}$ such that any of their structural maps are $\tau_{n-3}.\overline{U}_{n-2}$-identity maps.

On the other hand, Proposition 16 defines the left exact functor

$$\nu_{T_{n-1}}\colon\ Grd_{\tau\overline{U}}(K_{n-1}\mathbb{B}) \to AlgT_{n-1} = K_n\mathbb{B}.$$

Hence a functor $\nu_n = \nu_{T_{n-1}}.Grd_n \nu_{n-1}\colon\ n\text{-}Grd\mathbb{B} \to K_n\mathbb{B}$ which is left exact. Let us call it the nerve functor for the n-groupoids.

Furthermore, the whole following outer square does commute since any of its two components commute:

$$\begin{array}{ccccc}
n\text{-}Grd\mathbb{B} & \xrightarrow{Grd_n\nu_{n-1}} & Grd_{\tau\overline{U}}K_{n-1}k\mathbb{B} & \xrightarrow{\nu_{T_{n-1}}} & \mathbb{K}_n\mathbb{B} \\
 & & & & \downarrow{\scriptstyle \overline{U}_{k-1}} \\
\downarrow{\scriptstyle ()_{n-1}} & & \downarrow{\scriptstyle ()_0} & & S\text{-}Simpl_{n-1}\mathbb{B} \\
 & & & & \downarrow{\scriptstyle \tau_{n-2}} \\
(n-1)Grd\mathbb{B} & \xrightarrow[\nu_{n-1}]{} & K_{n-1}\mathbb{B} & \xrightarrow[\overline{U}_{n-2}]{} & S\text{-}Simpl_{n-2}\mathbb{B}
\end{array}$$

Consequently, the inductive hypothesis still holds.

Therefore, as we claimed, the construction of the nerve functor for n-groupoids is a structural consequence of the fact that the split n-simplicial objects are iterative T-elements.

Description of the Nerve of the n-Groupoids

We need first the following result.

PROPOSITION 19. *The following square commutes:*

$$\begin{array}{ccccc}
n\text{-}Grd\mathbb{B} & & \xrightarrow{\quad\nu_n\quad} & & K_n\mathbb{B} \\
\downarrow{\scriptstyle ()_{n-1}} & & & & \downarrow{\scriptstyle \overline{\tau}_{n-1}} \\
(n-1)Grd\mathbb{B} & \xrightarrow[\nu_{n-1}]{} & K_{n-1}\mathbb{B} & \xrightarrow[w_{n-1}]{} & Simpl_{n-1}\mathbb{B}
\end{array}$$

Proof. For each groupoid X_n, the initial map $i_n\colon\ dis\ X_{n-1} \to X_n$ in the fibre of X_n is a $()_{n-1}$-identity map. But we have:

$$\underline{U}_{n-1}.\overline{\tau}_{n-1}.\nu_{n(i_n)} = \tau_{n-2}.\overline{U}_{n-1}.\nu_n(i_n) = \overline{U}_{n-2}.\nu_{n-1}.()_{n-1}(i_n).$$

Thus, $\underline{U}_{n-1}.\overline{\tau}_{n-1}.\nu_{n(i_n)}$ is an identity map. Now, the functor $\underline{U}_{n-1}\colon\ Simpl_{n-1}\mathbb{B} \to Simpl_{n-2}\mathbb{B}$ (which just deletes the last face operator) reflects the identity maps and then $\overline{\tau}_{n-1}.\nu_{n(i_n)}$ is an identity map. Its domain is:

$$\begin{aligned}\overline{\tau}_{n-1}.\nu_n.dis(X_{n-1}) &= \overline{\tau}_{n-1}.\nu_n.dis.()_{n-1}(X_n)\\ &= \overline{\tau}_{n-1}.\nu_{T_{n-1}}.dis.\nu_{n-1}.()_{n-1}(X_n)\\ &= \overline{\tau}_{n-1}.Algk_{n-1}.\nu_{n-1}.()_{n-1}(X_n)\\ &= w_{n-1}.\nu_{n-1}.()_{n-1}(X_n),\end{aligned}$$

and its codomain is $\overline{\tau}_{n-1}.\nu_n(X_n)$.

Hence $\overline{\tau}_{n-1}.\nu_n = w_{n-1}.\nu_{n-1}()_{n-1}$. □

REMARK. This result means that, given an n-groupoid X_n, the underlying $(n-1)$-simplicial object of its nerve is the underlying $(n-1)$-simplicial object of the nerve of its underlying $(n-1)$-groupoid X_{n-1}.

Thus, $(\nu_n X_n)_k = (\nu_{n-1}X_{n-1})_k$ for all k, $0 \le k \le n-1$.

It remains to describe its n-th component and the Kan operation. Let t_n denote the Kan operation of the nerve $\nu_{n-1}X_{n-1}$ of X_{n-1} and t_n^m the Kan operation of the nerve $\nu_{n-1}(m_{n-1}X_n)$ of $m_{n-1}X_n$.

Then, according to the description of ν_T, the n-th component $(\nu_n X_n)_n$ is defined by the following pullback of dotted arrows in B, where $\nu_{n-1}\tau_0$ plays the role of d_0 and t_n the role of the algebra x_0 in the general description of ν_T.

$$\begin{array}{ccc}(\nu_{n-1}(m_{n-1}X_n))_{n-1} & \xleftarrow{\quad\overline{t}_n\quad} & (\nu_n X_n)_n\\ {\scriptstyle (\nu_{n-1}\tau_0)_{n-1}}\downarrow\,\downarrow{\scriptstyle (\nu_{n-1}\tau_1)_{n-1}} & & \downarrow{\scriptstyle \overline{\tau_0}}\\ (\nu_{n-1}X_{n-1})_{n-1} & \xleftarrow[t_n]{} & (\nu_{n-1}X_{n-1})_{n-1}[d_0,\dots,d_{n-2}]\end{array}$$

with $d_i\colon (\nu_n X_n)_n \to (\nu_n X_n)_{n-1} = (\nu_{n-1}X_{n-1})_{n-1}$ is given by $d_i = p_i.\overline{\tau_0}$ for all i, $0 \le i \le n-1$, $d_n = (\nu_{n-1}\tau_1)_{n-1}.\overline{t}_n$.

According to the same description, the Kan operation

$$t_{n+1}\colon (\nu_n X_n)_n[d_0,\dots,d_{n-1}] \to (\nu_n X_n)_n$$

is the unique map t_{n+1} such that

$$\begin{aligned}\overline{\tau_0}.t_{n+1} &= (\nu_{n-1}\tau_1)_{n-1}[1].[\overline{t_n}.p_0,\dots,\overline{t}_n.p_{n-1}],\\ \overline{t_n}.t_{n+1} &= (\nu_{n-1}.\tau_2)_{n-1}.[t_n^m.[\overline{t}_n.p_0,\dots,\overline{t}_n.p_{n-1}],\overline{t}_n.p_n],\end{aligned}$$

where t_{n+1} plays the role of α, $\overline{t}_n$ the role of $\overline{x_0}$, and t_n^m plays the role of x_1 in the general description of ν_T.

References

1. Agl, F. A. and Steiner, R.: Nerves of multiple categories, *Proc. London Math. Soc. (3)* **66** (1993), 92–128.
2. Ashley, N.: Simplicial T-complexes and crossed complexes: A non-Abelian version of a theorem of Dold and Kan, *Dissertationes Math.* **265** (1988).

3. Bourn, D.: Normalization equivalence, Kernel equivalence and affine categories, Lecture Notes in Math. 1488, Springer, 1991, pp. 43–62.
4. Bourn, D.: Low dimensional geometry of the notion of choice, in *Cat. Theory 1991, CMS Conf. Proc.*, Vol. 13, 1992, pp. 55–73.
5. Bourn, D.: Polyhedral monadicity of n-groupoids and standardized adjunction, *J. Pure Appl. Algebra* **99** (1995), 135–181.
6. Bourn, D.: Mal'cev categories and fibration of pointed objects, *Appl. Categ. Structures* **4** (1996), 307–327.
7. Bourn, D.: n-groupoids from n-truncated simplicial objects as a solution to a universal problem, preprint Univ-Littoral, Cahiers du Langal no. 17, 1997.
8. Brown, R. and Higgins, P. J.: The equivalence of ∞-groupoids and crossed complexes, *Cahiers Topologie Géom. Différentielle* **22**(4) (1981), 371–386.
9. Burroni, A. and Penon, J.: Une construction d'un nerf des ∞-categories, in preprint Univ. Caen: Categories, algèbres, esquisses et neo-esquisses, 1994, pp. 45–55.
10. Dakin, M. K.: Kan complexes and multiple groupoid structures, Ph.D. Thesis, Univ. of Wales, 1974.
11. Duskin, J.: Simplicial methods in the interpretation of Triple Cohomology, *Mem. Amer. Math. Soc.* **3**(163) (1975).
12. Glenn, P.: Realization of cohomology classes in arbitrary exact categories, *J. Pure Appl. Algebra* **25** (1982), 33–105.
13. Johnson, M.: The combinatorics of n-categorical pasting, *J. Pure Appl. Algebra* **62** (1989), 211–225.
14. Kapranov, M. M. and Voevodsky, V. A.: Combinatorial geometric aspects of polycategory theory: pasting schemes and higher Bruhat orders, *Cahiers Topologie Géom. Différentielle* **32**(1) (1991), 11–27.
15. Power, A. J.: An n-categorical pasting theorem, in *Category Theory, Proc. Internat. Conf., Lake Como, 1990*, Lecture Notes in Math. 1488, Springer, 1991, pp. 326–358.
16. Street, R.: The algebra of oriented simplexes, *J. Pure Appl. Algebra* **49** (1987), 283–335.
17. Street, R.: Parity complexes, *Cahiers Topologie Géom. Différentielle* **32**(4) (1991), 315–343.

Applied Categorical Structures **8:** 115–144, 2000.
G. Brümmer & C. Gilmour (eds), Papers in Honour of Bernhard Banaschewski.

Prime Ideal Theory for General Algebras

Dedicated to Bernhard Banaschewski on the occasion of his 70th birthday

MARCEL ERNÉ
Department of Mathematics, University of Hannover, D-30167 Hannover, Germany

(Received: 24 October 1996; accepted: 28 May 1998)

Abstract. We introduce ideals, radicals and prime ideals in arbitrary algebras with at least one binary operation, and we show that various separation lemmas and prime ideal theorems are special instances of one general theorem which, in turn, is equivalent to the Boolean Prime Ideal Theorem (or Ultrafilter Principle).

Mathematics Subject Classifications (2000): Primary: 06F05, 08A30, 20M12, 20N02; Secondary: 03G05, 16D30, 16N80, 17A65.

Key words: closure system, *cm*-lattice, distributive, ideal, locale, prime, quantale, radical, residuated, semiprime.

1. The Concept of Total Separation

The logical framework for our subsequent study is Zermelo–Fraenkel set theory (ZF) or, if classes are involved, von Neumann–Bernays–Gödel set theory (NBG). The Axiom of Choice (AC) will not be used. However, at some places we shall need one or both of the two principles below, which together are still weaker than AC (see Halpern [16] and Pincus [21]):

(UP) The *Ultrafilter Principle*:
Every proper set-theoretical filter can be extended to an ultrafilter.

(CDC) The *Principle of Countable Dependent Choices*:
If ρ is a relation on a set E such that for each $a \in E$ there is a $b \in E$ with $a\rho b$ then for each $a_0 \in E$ there is a sequence $(a_n \mid n \in \omega)$ with $a_n \rho a_{n+1}$ for all n.

The following situation is encountered in many algebraic, topological and order-theoretical contexts: there are given certain collections $\mathcal{A}$ and $\mathcal{B}$ of subsets of a ground set G (endowed with a relational and/or a topological structure), and one wants to 'blow up' any two disjoint sets $A \in \mathcal{A}$ and $B \in \mathcal{B}$ so that A is extended to some member of $\mathcal{A}$ whose complement belongs to $\mathcal{B}$ and contains the set B. If this is always possible, we say that the members of $\mathcal{A}$ are *totally separated* from those of $\mathcal{B}$. Statements of that kind are referred to as *separation lemmas*.

Without loss of relevant applications, we shall assume throughout that $\mathcal{A}$ is a set-theoretical closure system (that is, closed under arbitrary intersections, with $\bigcap \emptyset = G$). In view of classical algebraic examples, a subset of G might be called *prime* if its complement belongs to $\mathcal{B}$, and *semiprime* if its complement is a union of sets in $\mathcal{B}$. It is an immediate consequence of the respective separation lemma that the semiprime members of $\mathcal{A}$ are precisely the intersections of prime members of $\mathcal{A}$. Therefore, in algebraic contexts where $\mathcal{A}$ consists of certain 'ideals', separation lemmas or the correponding intersection theorems are also referred to as *prime ideal theorems*. Denoting by $\mathcal{B}^c$ the collection of all prime sets (i.e., complements of members of $\mathcal{B}$) and by $\Omega(\mathcal{S})$ the closure system generated by $\mathcal{S}$, we may formulate the previous representation theorem as an equation:

$$\mathcal{A} \cap \Omega(\mathcal{B}^c) = \Omega(\mathcal{A} \cap \mathcal{B}^c).$$

Often, the meets of prime ideals or prime elements form a *spatial locale*, that is, a complete lattice isomorphic to a topology (the so-called *hull-kernel topology*). This is then the starting point for a spectral theory, invoking topological tools in the given algebraic setting. On the other hand, such *'spectral theorems'* frequently provide certain subdirect product representations in the sense of universal algebra.

Here are a few typical examples:

(1) A complete lattice is *superalgebraic*, i.e., join-generated by its completely join-prime elements, iff the principal ideals are totally separated from the principal filters (in fact, an element is completely join-prime iff it generates a principal filter whose complement is a principal ideal). The 'prime' principal ideals are here generated by completely meet-prime elements, while all principal ideals are 'semiprime'.

(2) In a *Boolean space* (that is, in a compact Hausdorff space with a base of clopen sets), the closed sets are totally separated from the closed sets. The 'prime' closed sets are here the clopen sets – not to be confused with the meet-prime or join-prime (i.e. irreducible) closed sets. More generally, in any space with a base of compact-open sets, the closed sets are totally separated from the compact sets.

Similarly, in a *Priestley space* (that is, in a compact totally order-disconnected space, see, e.g. [22]), the closed lower (decreasing) sets are totally separated from the closed upper (increasing) sets. Here, 'prime' means 'clopen lower set', while every closed lower set is 'semiprime'.

Needless to say that the previous topological separation lemmas hold in ZF without choice and do not require any kind of algebraic prime ideal theorems.

(3) One form of the *Hahn–Banach Theorem* says that the convex (closed) subsets of Euclidean spaces (or of certain more general topological linear spaces) are totally separated from the convex open sets. In that context, 'prime' means 'closed half-space'. Apparently, the precise position of Hahn–Banach Theorems in ZF set theory without choice is not entirely settled, but some versions are known to be consequences of the Ultrafilter Principle.

(4) In its general noncommutative form, *Krull's Separation Lemma* [18] states that the (two-sided) ideals of any ring are totally separated from the *m-filtered* subsets (i.e. nonempty subsets containing with any two elements a, b the product ab or acb for some c). An ideal is prime iff its complement is m-filtered. As a consequence, the semiprime ideals (i.e. ideals complementary to a union of m-filtered sets) are exactly the intersections of prime ideals. The equivalence of Krull's Separation Lemma with the Ultrafilter Principle was shown in [6]. If one wants to characterize semiprime ideals by the more familiar condition to contain with the square of any ideal that ideal itself, some further principles like CDC seem to be indispensable unless the ring in question is commutative.

(5) *Stone's Separation Lemma or Prime Ideal Theorem for Distributive Lattices* states that a lattice is distributive iff its ideals are totally separated from its filters (or meet-subsemilattices). In fact, the prime ideals are those ideals whose complement is a filter. More generally, Rav [24] calls an ideal S of an arbitrary lattice L *semiprime* if $a \wedge c \in S$ and $b \wedge c \in S$ imply $(a \vee b) \wedge c \in S$, and shows that the semiprime ideals are totally separated from the filters; consequently, they are the intersections of prime ideals. A still more general result on so-called *distributors* in algebras of type (2, 2) was established in [11]. The semiprime ideals in the sense of Rav are in fact the semiprime members of the closure system $\mathcal{A}$ of all ideals if one takes for $\mathcal{B}$ the system of all semiprime filters (see Sections 2 and 5).

(6) The *Boolean Prime Ideal Theorem* (in its strong form) states that the ideals of a Boolean algebra are totally separated from the filters; consequently, every ideal is an intersection of prime ideals, and dually, every filter is an intersection of ultrafilters, i.e. complements of prime ideals (*Boolean Ultrafilter Theorem*). By the one-to-one correspondence between Boolean algebras, Boolean lattices and Boolean rings, the Boolean Prime Ideal Theorem is a common specialization of (4) and (5). Moreover, it is known that each of the previous prime ideal theorems is equivalent to the Ultrafilter Principle (see [3, 6, 23, 24, 26]). As a consequence, the representation of Boolean algebras and distributive lattices as subdirect products of two-element chains, but also Stone's representation of bounded distributive lattices by compact-open sets in sober spaces, as well as their representation by clopen lower sets in Priestley spaces (see [22]), are equivalent to UP.

(7) The *Birkhoff–Pierce Representation Theorem for f-rings* (see [8] or [15]) states that every f-ring is a subdirect product of totally ordered rings. For this, it is necessary and sufficient to ensure that the zero ideal of an f-ring is an intersection of prime l-ideals, because an l-ideal of an f-ring is prime iff the factor ring modulo that ideal is totally ordered. Thus, the above representation theorem arises from a separation lemma, saying that the l-ideals of an f-ring are totally separated from those sets which are closed under the operation $m(a, b) = |a| \wedge |b|$. The equivalence of the Birkhoff–Pierce Theorem with UP was established by Feldman and Henriksen [14] and, in a different way using the so-called *Prime Element Theorem*, by Banaschewski (see [4] and [5]). Similar results hold for l-ideals in lattice-ordered groups or semigroups (cf. Lorenzen [20] and Aubert [1]). For details, see Section 5.

(8) The separation lemmas and prime ideal theorems in (4), (5), (6) and (7) are easy consequences of the *Separation Lemma for Quantales*, stating that the principal ideals (and, more generally, the directed subsets) of a two-sided quantale are totally separated from the Scott-open multiplicatively closed subsets (see below). The equivalence of the Separation Lemma for Quantales with the Ultrafilter Principle was established in [6].

Since Krull's pioneer work on ideals in (noncommutative) rings, quantales and more general types of complete lattices with an extra multiplication have been introduced and studied by many authors with the aim to develop an abstract ideal theory (see Dilworth and Ward [10] for an early reference). Let us briefly recall the basic definitions; for more background, consult Birkhoff [7], Blyth and Janowitz [9], or Rosenthal [25].

Following Keimel [17], we mean by a *complete multiplicative lattice* or *cm-lattice* a complete lattice L endowed with a binary operation (called *multiplication*)

$$m\colon L^2 \longrightarrow L, (a, b) \longmapsto ab$$

such that the *fundamental left translations*

$$\lambda_a\colon L \longrightarrow L, x \longmapsto ax \quad (a \in L),$$

the *fundamental right translations*

$$\rho_b\colon L \longrightarrow L, x \longmapsto xb \quad (b \in L),$$

and, consequently, all *translations* (i.e. compositions of finitely many fundamental translations) are order-preserving. Elements $a \in L$ with $ab \leq a$ and $ba \leq a$ for all $b \in L$ are called *two-sided* or *ideal* elements, and the whole cm-lattice L is said to be *two-sided* if each of its elements has that property. Furthermore, L is called *residuated* or a *prequantale* if for all $a, c \in L$ there exists a *right residual* $c \mathbin{.\!\cdot} a$ and a *left residual* $c \mathbin{\cdot\!.} a$ such that

$$ab \leq c \Leftrightarrow b \leq c \mathbin{.\!\cdot} a \quad \text{and} \quad ba \leq c \Leftrightarrow b \leq c \mathbin{\cdot\!.} a$$

for all $b \in L$. A cm-lattice L is a prequantale iff it satisfies the identities

$$a\Big(\bigvee B\Big) = \bigvee(aB) \text{ and } \Big(\bigvee B\Big)a = \bigvee(Ba) \quad (a \in L, B \subseteq L),$$

where $\bigvee B$ denotes the join (least upper bound) of B, and aB (respectively, Ba) denotes the set of all products ab (respectively, ba) with $b \in B$. In order to prevent confusion, we mention that Birkhoff [7] includes the above identities for finite B in the definition of multiplicative lattices.

A *quantale* is a prequantale with associative multiplication, and a *locale* is a two-sided quantale whose elements are idempotent. Since the multiplication of a locale must coincide with the binary meet, a locale is merely the same thing as a *frame*, that is, a complete lattice enjoying the distributive law $a \wedge \bigvee B = \bigvee(a \wedge B)$.

A subset S of a complete (multiplicative) lattice L is *Scott open* iff it intersects exactly those directed subsets of L whose join belongs to S. Any such set is *increasing* (or an *upper set*), i.e. $b \geq a \in S$ implies $b \in S$. Further, a nonempty set $S \subseteq L$ is *multiplicatively closed* or *m-closed* iff $ab \in S$ for all $a, b \in S$. In accordance with [11] and [13], we mean by a *distributor* of L a nonempty subset S such that

$$a \vee c \in S \text{ and } b \vee c \in S \Leftrightarrow (ab) \vee c \in S.$$

In two-sided prequantales, the distributors are precisely the m-closed upper sets. Now we are ready to formulate the lattice-theoretical fundament for a broad variety of separation lemmas and prime ideal theorems (see [13] for details):

THEOREM 1.1. *In* ZF *or* NBG *set theory, the Ultrafilter Principle is equivalent to*

(SLC) *The Separation Lemma for Complete Multiplicative Lattices:*
The principal ideals of any two-sided cm-lattice are totally separated from the Scott-open distributors.

Otherwise stated, SLC says that any element outside a Scott-open distributor S is dominated by a prime element outside S; in fact, an element p of a cm-lattice is prime (i.e., p is not the top element and $ab \leq p$ implies $a \leq p$ or $b \leq p$) iff p generates a principal ideal whose complement is a (Scott-open) distributor.

The Separation Lemma for Quantales (see Example 8 above) and, in particular, Banaschewski's Prime Element Theorem [4] are special instances of SLC.

2. Ideal Systems

Let $\mathcal{A}$ be a (set-theoretical) closure system on a set G, and suppose there is given a multiplication

$$m\colon \mathcal{A}^2 \longrightarrow \mathcal{A}, \ (A, B) \longmapsto A \cdot B$$

such that $\mathcal{A}$ becomes a cm-lattice (see Section 1). Then we say $\mathcal{A}$ is an *m-closure system*. The two-sided members of $\mathcal{A}$ are referred to as *($\mathcal{A}$-)ideals* (cf. Aubert [2] for a more restricted notion of so-called *x-ideals*). If each member of $\mathcal{A}$ is two-sided, i.e., $A \cdot B \subseteq A \cap B$ for all $A, B \in \mathcal{A}$, then we speak of a *(two-sided) ideal system* $\mathcal{A}$. Notice that every two-sided cm-lattice is isomorphic to the ideal system of its principal ideals.

Recall that a nonempty collection $\mathcal{D}$ of sets is *directed (by inclusion)* iff any two members of it are contained in a common member of $\mathcal{D}$, and that a closure system is *inductive* or *algebraic* iff it is closed under unions of directed subsystems.

At the first glance, the following definitions look a bit artificial. However, they prepare separation lemmas and prime ideal theorems of great generality. Given an m-closure system $\mathcal{A}$ on a set G, we say a subset S of G is *($\mathcal{A}$-)closed* if for any directed subsystem $\mathcal{D}$ of $\mathcal{A}$ with $\bigcup \mathcal{D} \subseteq S$, there is an $A \in \mathcal{A}$ with $\bigcup \mathcal{D} \subseteq A \subseteq S$. The complement of a closed set is called *($\mathcal{A}$-)open*. All members of $\mathcal{A}$ as well as arbitrary intersections of closed sets are closed, and if $\mathcal{A}$ is algebraic then *every* subset of G is closed and open. (Caution: In other contexts, only the members of the given closure system are called closed.) Now, a proper subset S of G is said to be *($\mathcal{A}$-)prime* if it is $\mathcal{A}$-closed and for all $A, B, C \in \mathcal{A}$,

$$(A \cdot B) \vee C \subseteq S \text{ implies } A \vee C \subseteq S \text{ or } B \vee C \subseteq S,$$

where the binary joins are formed in the complete lattice $\mathcal{A}$. The complement of a prime set is said to be *($\mathcal{A}$-)coprime*. The following two closely related definitions are easier to handle (but sometimes too weak for the desired conclusions): call a proper subset S of G *($\mathcal{A}$-)quasiprime* if for all $A, B \in \mathcal{A}$,

$$A \cdot B \subseteq S \text{ implies } A \subseteq S \text{ or } B \subseteq S,$$

and call a nonempty subset F of G *($\mathcal{A}$-)filtered* if for all $A, B \in \mathcal{A}$,

$$A \cap F \neq \emptyset \text{ and } B \cap F \neq \emptyset \text{ imply } (A \cdot B) \cap F \neq \emptyset,$$

in other words, if F is complementary to a quasiprime set.

The following straightforward observations will be helpful:

LEMMA 2.1. *Every prime set is quasiprime. For ideals and, in case of algebraic residuated ideal systems, for arbitrary subsets the properties 'prime' and 'quasiprime' are equivalent.*

For the last claim, use the inclusion

$$(A \vee C) \cdot (B \vee C) = (A \cdot B) \vee (A \cdot C) \vee (C \cdot B) \vee (C \cdot C) \subseteq (A \cdot B) \vee C$$

valid in any residuated ideal system.

An intersection of prime sets is called *($\mathcal{A}$-)semiprime*. Finally, generalizing a well-known construction from ring theory (see, for example, [19]), one may associate with any subset S of G its *($\mathcal{A}$-)radical*, the intersection of all quasiprime sets containing S. By definition, the radical of S consists of all elements a such that every filtered set containing a meets S. We say a set is *($\mathcal{A}$-)radical* if it coincides with its own radical.

Obviously, any intersection of prime ideals is a semiprime ideal, every semiprime ideal is a radical ideal, and it is one of the 'primary' goals of algebraic spectral theory to establish the converse implications in suitable ideal systems. By

Lemma 2.1, 'semiprime' is tantamount to 'radical' whenever we are dealing with an algebraic residuated ideal system.

THEOREM 2.2. *In* ZF *or* NBG *set theory, each of the following three statements is equivalent to the Ultrafilter Principle:*

(SLG) *The Separation Lemma for General Ideals:*
The ideals are totally separated from the coprime sets.

(PIG) *The Prime Ideal Theorem for General Ideals:*
The semiprime ideals are the intersections of prime ideals.

(PRG) *The Prime Radical Theorem for General Ideals:*
For any algebraic residuated ideal system, the radical of an ideal is the intersection of all prime ideals containing it.

Proof. We know that UP implies SLC (see Theorem 1.1), and the latter entails SLG, since those ideals which intersect a given coprime set constitute a Scott-open distributor of the ideal system in question. Obviously, SLG implies PIG and PRG (use Lemma 2.1) which in turn entail the Boolean Prime Ideal Theorem (hence UP), because the ideals of a Boolean algebra (or a distributive lattice) form an algebraic residuated ideal system in which every ideal is semiprime (radical). □

Each of the following separation lemmas is a special instance of SLG:

(SLC) The Separation Lemma for cm-Lattices.
(SLP) The Separation Lemma for Prequantales.
(SLR) The Separation Lemma for (Semi)Rings.
(SLL) The Separation Lemma for Lattices.
(SLB) The Separation Lemma for Boolean Algebras.

For SLC, observe that the principal ideals of a two-sided cm-lattice L form an ideal system which is, in general, neither algebraic nor residuated; but it is isomorphic to L, and a Scott-open distributor of L is just a coprime upper set. Clearly, SLP is the residuated case of SLC.

In SLR, we mean by a semiring an algebra $(G, +, \cdot)$ of type $(2, 2)$ whose multiplication is associative and distributes over the addition (which need not satisfy any further identities). Rings and distributive lattices (but not arbitrary lattices) are special instances of semirings. An ideal is an additively closed subset that is also closed under multiplication with elements from the left and from the right. The resulting ideal system is both algebraic and residuated, and SLR assures that the ideals are totally separated from the m-filtered subsets (see Section 3).

Concerning SLL, we mention here that for the (algebraic but in general not residuated!) system of all ideals in an arbitrary lattice L, the coprime upper sets are just the semiprime filters in the sense of Rav [24], that is, the nonempty subsets S satisfying the equivalence

$$a \vee c \in S \text{ and } b \vee c \in S \Leftrightarrow (a \wedge b) \vee c \in S.$$

Hence, SLG entails Rav's Separation Lemma for Lattices (saying that the ideals are totally separated from the semiprime filters) and, consequently, the representation of semiprime filters as intersections of prime filters. On the other hand, taking for $\mathcal{A}$ the (algebraic and residuated!) system of all semiprime ideals, i.e. subsets S satisfying the dual equivalence

$$a \wedge c \in S \text{ and } b \wedge c \in S \Leftrightarrow (a \vee b) \wedge c \in S,$$

one obtains from PIG the representation of semiprime ideals as intersections of prime ideals (which is, by duality, just the same result as the above prime filter theorem).

It remains to investigate circumstances under which semiprimeness or the radical property may be substituted by the simpler condition of reducedness, where a subset S of G is said to be ($\mathcal{A}$-)*reduced* if for all members A of the given ideal system $\mathcal{A}$,

$$A \cdot A \subseteq S \Rightarrow A \subseteq S.$$

Notice that this is tantamount to the condition

$$(a) \cdot (a) \subseteq S \Rightarrow a \in S,$$

where (a) denotes the smallest $\mathcal{A}$-ideal containing a. Similarly, the implication

$$(a) \cdot (b) \subseteq S \Rightarrow a \in S \text{ or } b \in S$$

together with $S \neq G$ characterizes quasiprime sets.

PROPOSITION 2.3. *Every radical set is reduced, and under hypothesis* CDC, *the converse is also true.*

Proof. The first claim is obvious because quasiprime sets and intersections of reduced sets are reduced. Now let S be a reduced set and define a relation ρ on $B = G \setminus S$ by

$$a\rho b \Leftrightarrow b \in (a) \cdot (a).$$

Since the ideals (a) are two-sided, $a\rho b$ entails $(b) \subseteq (a) \cdot (a) \subseteq (a)$, and CDC provides for any $a \in B$ a sequence $(a_n \mid n \in \omega)$ with $a_0 = a$ and $a_n \rho a_{n+1}$ for all n, whence $F = \{a_n \mid n \in \omega\}$ is a filtered set contained in B: indeed, given $a_m, a_n \in F$, we may assume $m \leq n$; then $a_m \rho \cdots \rho a_n$ yields $(a_n) \subseteq (a_m)$ and consequently $a_{n+1} \in (a_n) \cdot (a_n) \cap F \subseteq (a_m) \cdot (a_n) \cap F$. It follows that S is an intersection of quasiprime sets. □

Let us summarize the previous results concerning an arbitrary ideal system:

COROLLARY 2.4. *Consider the following conditions on an element b and an ideal A:*

(p) *$b \in P$ for each prime set P containing A,*
(q) *$b \in Q$ for each quasiprime set Q containing A,*
(r) *$b \in R$ for each reduced set R containing A,*

and let (n′) *denote the statement obtained from* (n) *when 'set' is replaced with 'ideal'* (n = p, q, r). *Then* (r) $\Rightarrow$ (q) $\Rightarrow$ (p) $\Rightarrow$ (p′) *and* (r) $\Rightarrow$ (r′) $\Rightarrow$ (q′) $\Leftrightarrow$ (p′).

Furthermore,

(p) $\Leftrightarrow$ (p′) *under hypothesis* UP,
(p) $\Leftrightarrow$ (q) *if the given ideal system is algebraic and residuated,*
(q) $\Leftrightarrow$ (r) *under hypothesis* CDC.

Hence, assuming UP *and* CDC, *all six conditions are equivalent for any algebraic residuated ideal system; in particular, the radical of an ideal A is then not only the least semiprime ideal but also the least reduced ideal containing A.*

Under suitable circumstances, certain separation lemmas and prime ideal theorems hold in ZF or NBG without UP, namely, if the ideal system $\mathcal{A}$ in question is *maximized* in the following sense: for each ideal A disjoint from an open set B, there is a maximal ideal among those which contain A and are disjoint from B. This is certainly the case whenever $\mathcal{A}$ is closed under nonempty unions. If one assumes the Axiom of Choice then *every* ideal system is maximized, by Zorn's Lemma.

Although it is our intention to avoid maximal ideals (whose existence usually requires AC), the following property of prime sets is of some interest for a general prime ideal theory, since it reveals the basic reason for the frequently observed phenomenon that under appropriate assumptions, maximal ideals are prime.

PROPOSITION 2.5. *If S is a prime set then every maximal ideal contained in S is a prime ideal. Conversely, any closed set S with that property is prime, provided the ideal system is maximized.*

Proof. Let S be a prime set, and suppose C is maximal among those ideals which are contained in S. If A, B are ideals but neither A nor B is contained in C then $A \vee C \not\subseteq S$ and $B \vee C \not\subseteq S$, hence $(A \cdot B) \vee C \not\subseteq S$ and, in particular, $A \cdot B$ cannot be a subset of C.

Conversely, if $\mathcal{A}$ is a maximized ideal system and S is a closed set such that every maximal member of $\mathcal{A}$ contained in S is prime then S must be a proper subset of G, and for all $A, B, C \in \mathcal{A}$ with $(A \cdot B) \vee C \subseteq S$, we find a maximal, hence prime ideal P with $(A \cdot B) \vee C \subseteq P \subseteq S$. But then $A \vee C \subseteq P \subseteq S$ or $B \vee C \subseteq P \subseteq S$. □

COROLLARY 2.6. *If an ideal system on G is algebraic and maximized* (*in particular, if it is closed under nonempty unions*) *then:*

(1) *A subset of G is prime iff every maximal ideal contained in it is prime.*
(2) *The ideals are totally separated from the coprime sets.*
(3) *The semiprime ideals are the intersections of prime ideals.*

Our Separation Lemma for General Ideals heavily leans on the somewhat technical notion of (co)prime sets. If one prefers to work with the simpler notion of filtered sets, it may be necessary to restrict the class of 'ideals' under consideration, by extracting suitable residuated ideal systems from a given m-closure system $\mathcal{A}$ (whose multiplication is denoted by $A \cdot B$). Thus, we say a closure system $\mathcal{C} \subseteq \mathcal{A}$ is *residuated in* $\mathcal{A}$ if for all $A \in \mathcal{A}$ and $C \in \mathcal{C}$, there exists a *right residual* $C \therefore A \in \mathcal{C}$ and a *left residual* $C \because A \in \mathcal{C}$ such that for all $B \in \mathcal{A}$,

$$B \subseteq C \therefore A \Leftrightarrow A \cdot B \subseteq C \quad \text{and} \quad B \subseteq C \because A \Leftrightarrow B \cdot A \subseteq C.$$

$\mathcal{C}$ is then a residuated m-closure system in its own right, hence a prequantale with respect to the modified multiplication

$$A \cdot_{\mathcal{C}} B = \bigcap \{C \in \mathcal{C} \mid A \cdot B \subseteq C\} \quad (A, B \in \mathcal{C}).$$

If a closure system $\mathcal{C}$ is residuated in $\mathcal{A}$ and $A \cdot B \subseteq A \cap B$ for all $A, B \in \mathcal{C}$ then we say $\mathcal{C}$ is an *ideal prequantale in* $\mathcal{A}$. If, moreover, the multiplication of $\mathcal{C}$ is associative then we say $\mathcal{C}$ is an *ideal quantale in* $\mathcal{A}$, and if the multiplication of $\mathcal{C}$ is the binary intersection, $\mathcal{C}$ is an *ideal locale in* $\mathcal{A}$.

For later purposes, we note the following set-theoretical variants of some basic results on cm-lattices from [13]:

PROPOSITION 2.7. *If $\mathcal{C}$ is an ideal prequantale in an m-closure system $\mathcal{A}$ then $\mathcal{C}$ is a residuated ideal system with respect to the modified multiplication $\cdot_{\mathcal{C}}$, and the closure operator J of $\mathcal{C}$ restricts to a homomorphism from $\mathcal{A}$ onto $\mathcal{C}$:*

$$J(A \cdot B) = J(A) \cdot_{\mathcal{C}} J(B) \quad \textit{for all } A, B \in \mathcal{A}.$$

The reduced $\mathcal{C}$-ideals form the largest ideal locale in $\mathcal{C}$ and also an ideal locale in $\mathcal{A}$. Hence, the corresponding closure operator R satisfies the identity

$$R(A \cdot B) = R(A) \cap R(B) \quad \textit{for all } A, B \in \mathcal{A}.$$

Thus, a $\mathcal{C}$-ideal is $\mathcal{C}$-prime iff it is $\mathcal{A}$-(quasi)prime, and a $\mathcal{C}$-ideal is $\mathcal{C}$-reduced iff it is $\mathcal{A}$-reduced. Moreover, under the hypotheses UP *and* CDC, *the $\mathcal{C}$-radical of a $\mathcal{C}$-ideal C agrees with the $\mathcal{A}$-radical of C, being the intersection of all prime $\mathcal{C}$-ideals containing C.*

On account of these facts, we hope that there will be no risk of confusion in omitting the prefix $\mathcal{A}$ or $\mathcal{C}$ when we speak of prime, reduced or radical ideals in the above context.

Observe that an ideal prequantale $\mathcal{C}$ is an ideal locale (in itself) iff every $\mathcal{C}$-ideal is reduced. In that case, the $\mathcal{C}$-filtered sets are just the filtered (= down-directed) sets with respect to the *specialization quasiorder* $\leq$ defined by $a \leq b$ iff a belongs to each member of $\mathcal{C}$ containing b. We conclude this section with the remark that within isomorphism, the two-sided (pre)quantales are just the ideal (pre)quantales, and the locales are just the ideal locales.

3. Ideals of m-Algebras

In what follows, we shall apply the previous results to certain ideal systems arising in universal algebra. We admit general algebras with a possibly class-indexed family of operations (some of which may be infinitary). Thus, by a *type*, we mean an arbitrary function ν assigning to each member k of a class K a cardinal number $\nu_k = \nu(k)$, and an *m-algebra of type* ν is a set G together with a family of operations $m_k\colon G^{\nu_k} \to G$ $(k \in K)$ such that $0 \in K$ and

$$m = m_0\colon G^2 \longrightarrow G, (a, b) \longmapsto ab$$

is a binary operation (i.e. $\nu_0 = 2$), called the *multiplication* of the algebra. It will be convenient to denote by G_0 the groupoid reduct $(G; m)$ and by G_1 the reduct $(G;\ m_k \mid k \in K\backslash\{0\})$, while G will also denote the whole algebra $(G;\ m_k \mid k \in K)$.

Given any function $\mathcal{Z}$ assigning to each m-algebra G a certain m-closure system $\mathcal{Z}G$, it will be convenient to write $A \cdot_{\mathcal{Z}} B$ for $A \cdot_{\mathcal{Z}G} B$, *$\mathcal{Z}$-closed* for *$\mathcal{Z}G$-closed* etc.

The power set $\mathcal{P}G$ endowed with the multiplication

$$AB = \{ab \mid a \in A, b \in B\} \quad (A, B \subseteq G)$$

is a residuated m-closure system but not an ideal system unless G has at most one element. A (*two-sided*) *ideal* of the groupoid G_0 is the same thing as a $\mathcal{P}$-ideal, that is, a subset A of G with $GA \cup AG \subseteq A$. The (*two-sided*) *ideals* of the m-algebra G are those $\mathcal{P}$-ideals which are closed under all operations of G. Sometimes, one will have to exclude one or more of these operations from the above closedness requirement for ideals; for example, if a ring has a unit, regarded as a nullary operation, then certainly it will make no sense to postulate that an ideal should be closed under that operation, i.e., contain the unit element. However, the adaption to such situations is straightforward.

We denote by $\mathcal{I}_0G$ the set of all ideals of the groupoid G_0 and by $\mathcal{I}G$ that of all ideals of the m-algebra G (thus $\mathcal{I}G_0 = \mathcal{I}_0G$). Any closure system $\mathfrak{C} \subseteq \mathcal{I}_0G$ is referred to as an *ideal system of* G, being in fact an ideal system in the sense of Section 2 with respect to the multiplication

$$A \cdot B = A \cdot_{\mathfrak{C}} B = \bigcap\{C \in \mathfrak{C} \mid AB \subseteq C\}.$$

In particular, this holds for the closure systems $\mathcal{I}_0G$ and $\mathcal{I}G$. Whereas $\mathcal{I}_0G$ is a superalgebraic ideal prequantale in $\mathcal{P}G$, being closed under arbitary unions and intersections (see Proposition 4.10), the ideal system $\mathcal{I}G$ is not even a prequantale, in general. For example, if L is a lattice, regarded as an m-algebra of type $(2, 2)$ with the binary meet as multiplication, then the ideal system $\mathcal{I}L$ is a (pre)quantale if and only if L is distributive. Hence, we must impose suitable restrictions on the ideals in question in order to obtain suitable ideal (pre)quantales. One effective approach will be described immediately below, another one in Section 4.

First, adjoin, if necessary, a multiplicative unit u to G so that

$$ua = a = au \quad \text{for all } a \in G_u = G \cup \{u\}$$

and call a subset A *associating* if

$$\text{(A.)} \quad (a(bc))d \in A \Leftrightarrow ((ab)c)d \in A$$

for all $a, b, c, d \in G_u$. In particular, (A.) implies

$$\text{(A)} \quad a(bc) \in A \Leftrightarrow (ab)c \in A,$$

and if (A) holds, we may write unambiguously $abc \in A$ for the equivalent expressions $a(bc) \in A$ and $(ab)c \in A$, although the elements $a(bc)$ and $(ab)c$ may be different.

A *zero* (*element*) of G is an element 0 such that $\{0\}$ is an ideal. Notice that G can have at most one such element. An example of a groupoid whose zero element constitutes an ideal satisfying (A) but not (A.) is given in [12], where also the subsequent equivalences are proven:

LEMMA 3.1. *The following conditions on a subset A of an m-algebra G are equivalent:*

(a) *A is associating.*
(b) $\Theta_A = \{(b, b') \in G^2 \mid a(bc) \in A \Leftrightarrow a(b'c) \in A$ *for all* $a, c \in G_u\}$ *is a congruence on G_0, and the quotient G_0/Θ_A is a semigroup.*
(c) *There is a homomorphism ϕ from G_0 onto a semigroup with $A = \phi^{-1}[\phi[A]]$.*
(d) *If an equation $s = t$ in n variables is valid in the variety of semigroups then* $s_G(b) \in A \Leftrightarrow t_G(b) \in A$ *for all* $b \in G^n$.

Moreover, A is a nonempty associating ideal iff G_0/Θ_A is a semigroup with zero A.

The associating ideals form an ideal system $\mathcal{A}G$ but not a quantale, in general. In fact, the above remark on nondistributive lattices shows that even the system $\mathcal{R}\mathcal{A}G$ of all reduced associating ideals may fail to be a prequantale (whereas the reduced associating ideals of a *groupoid* do always form a superalgebraic locale; see Corollary 4.11). Therefore, we introduce now the class of *quantic* ideals. These are ideals A satisfying (A.) and

$$\text{(D.)} \quad ab_i c \in A (i \in \nu_k) \Rightarrow a\, m_k(b)c \in A$$

for all $k \in K \setminus \{0\}$, $a, c \in G_u$ and $b = (b_i \mid i < \nu_k) \in G^{\nu_k}$.

If m_k is a binary operation, say $m_k(b_0, b_1) = b_0 + b_1$, then (D.) amounts to the distribution law

$$ab_0c \in A \text{ and } ab_1c \in A \Rightarrow a(b_0 + b_1)c \in A.$$

LEMMA 3.2. *If C is a quantic ideal of an m-algebra G and A is a subset of G with $AG \subseteq A$ then the right residual*

$$C \mathrel{.\!\cdot} A = \{b \in G \mid Ab \subseteq C\}$$

is again a quantic ideal. The same holds for the left residual

$$C \mathrel{\cdot\!.} B = \{a \in G \mid aB \subseteq C\}$$

provided $GB \subseteq B \subseteq G$.

Proof. If $b \in R = C \mathrel{.\!\cdot} A$ then $(Ac)b \subseteq Ab \subseteq C$ and $(Ab)c \subseteq C$ for all $c \in G$. As C is associating, it follows that $A(cb) \subseteq C$ and $A(bc) \subseteq C$, that is, $cb \in R$ and $bc \in R$. It is easy to verify that R is associating if so is C. Further, the hypothesis $ab_ic \in R$ for $i \in \nu_k$ yields $A(a(b_ic)) \subseteq C$, hence $(Aa)(b_ic) \subseteq C$ for all $i < \nu_k$, and therefore $(Aa)(m_k(b)c) \subseteq C$ and $A(a(m_k(b)c)) \subseteq C$, i.e. $am_k(b)c \in R$. In all, we see that R is in fact a quantic ideal. □

Now we are in a position to prove an important structural property of the m-closure system of all quantic ideals, motivating the nomenclature:

PROPOSITION 3.3. *The quantic ideals of any m-algebra G form an ideal quantale $\mathcal{Q}G$ in $\mathcal{I}_0G$ (and in $\mathcal{I}G$) whose closure operator Q satisfies the identity*

$$Q(AB) = Q(A \cdot B) = Q(A) \cdot_{\mathcal{Q}} Q(B) \quad \textit{for all } A, B \in \mathcal{I}_0G.$$

If G is a finitary algebra then $\mathcal{Q}G$ is an algebraic residuated ideal system.

Proof. Clearly, $\mathcal{Q}G$ is a closure system contained in $\mathcal{I}G$, and if each operation of G is finitary then $\mathcal{Q}G$ is algebraic. Let $A, B \in \mathcal{I}_0G$ and $C \in \mathcal{Q}G$. By Lemma 3.2, $C \mathrel{.\!\cdot} A$ and $C \mathrel{\cdot\!.} B$ are quantic ideals with

$$A \subseteq C \mathrel{\cdot\!.} B \Leftrightarrow AB \subseteq C \Leftrightarrow A \cdot B \subseteq C \Leftrightarrow B \subseteq C \mathrel{.\!\cdot} A.$$

Hence $\mathcal{Q}G$ is an ideal prequantale in $\mathcal{I}_0G$ and in $\mathcal{I}G$, so that Proposition 2.7 applies. It remains to prove the associative law in $\mathcal{Q}G$. For $A, B, C \in \mathcal{Q}G$, we obtain:

$$A \cdot_{\mathcal{Q}} (B \cdot_{\mathcal{Q}} C) = Q(A(BC)) = Q((AB)C) = (A \cdot_{\mathcal{Q}} B) \cdot_{\mathcal{Q}} C. \qquad □$$

By Proposition 2.7, a quantic ideal C is reduced iff $AA \subseteq C$ implies $A \subseteq C$ for each (quantic) ideal A, and we have:

PROPOSITION 3.4. *The reduced quantic ideals form the largest ideal locale $\mathcal{RQ}G$ in $\mathcal{Q}G$, and its closure operator RQ satisfies the identity*

$$RQ(AB) = RQ(A) \cap RQ(B) \textit{ for all } A, B \in \mathcal{I}_0G.$$

In particular, a quantic ideal C is prime iff $C \neq G$ and

$$AB \subseteq C \text{ implies } A \subseteq C \text{ or } B \subseteq C \quad \text{for all } A, B \in \mathcal{I}_0G\ (\mathcal{I}G, \mathcal{Q}G, \mathcal{RQ}G).$$

The prime quantic ideals admit an elementary characterization which is well known from ring theory but requires some care in the general case of m-algebras. Call a nonempty subset B of an m-algebra G *m-closed* if $a, b \in B$ implies $ab \in B$, and *m-filtered* (also *weakly m-closed* or an *m-system*; cf. [19]) if $a, b \in B$ implies $a(cb) \in B$ or $(ac)b \in B$ for some $c \in G_u$. Notice that an m-filtered subset of an m-algebra G is $\mathcal{C}$-filtered for each ideal system $\mathcal{C}$ of G. In meet-semilattices, the m-filtered subsets are precisely the filtered (= down-directed) subsets in the usual order-theoretical sense.

LEMMA 3.5. *A quantic ideal of an m-algebra G is prime iff it has an m-filtered complement.*

Proof. Let $T_r(G_0)$ denote the monoid of all right translations (i.e. compositions of finitely many fundamental right translations ρ_b: $x \mapsto xb$) of G_0. For $a \in G$, the set $A = \{\rho(a) \mid \rho \in T_r(G_0)\}$ contains a and satisfies $AG \subseteq A$. From Lemma 3.1 we conclude that for each $C \in \mathcal{Q}G$ and each $\rho \in T_r(G_0)$, there is a $c \in G_u$ with $\rho(a)b \in C \Leftrightarrow acb \in C$ (because C is associating). Now, assuming that certain elements $a, b \in G$ satisfy $acb \in C$ for all $c \in G_u$, we deduce the inclusion $Ab \subseteq C$. From Lemma 3.2, we know that $C \mathbin{.\!\cdot} A$ is a quantic ideal containing b, and it follows that $(b) = \bigcap\{D \in \mathcal{Q}G \mid b \in D\} \subseteq C \mathbin{.\!\cdot} A$, i.e. $A \subseteq C \mathbin{\cdot\!.} (b)$. Again by Lemma 3.2, that residual is a quantic ideal; consequently, $(a) \subseteq C \mathbin{\cdot\!.} (b)$ and so $(a) \cdot (b) \subseteq C$. If C is prime then the latter entails $a \in C$ or $b \in C$. By contraposition, we conclude that $G \setminus C$ is m-filtered. □

In the context of an m-algebra G, we mean by the *radical* $\sqrt{S}$ of a set $S \subseteq G$ the collection of all elements $a \in G$ such that every $\mathcal{I}_0$-filtered set containing a meets S. Although this is not identical with the notion of *$\mathcal{Q}$-radical* (which involves $\mathcal{Q}$-filtered instead of $\mathcal{I}_0$-filtered sets), we know from Propositions 2.7 and 3.3 that UP together with CDC collapses both types of radicals for quantic ideals. In associative structures, the present definition of radical reduces to a simpler one known from ring theory (cf. [19]).

LEMMA 3.6. *An element belongs to the radical of an associating ideal A iff every m-filtered set containing that element meets A.*

Proof. Consider first the more specific case of a semigroup H with zero 0. We claim that for each $\mathcal{I}_0$-filtered set $F \subseteq H \setminus \{0\}$,

$$E = \{b \in H \mid abc \in F \text{ for some } a, c \in H_u\}$$

is an m-filtered set with $F \subseteq E \subseteq H \setminus \{0\}$. This is an easy consequence of the fact that the least ideal of H containing a is

$$(a) = \{dae \mid d, e \in H_u\},$$

and the product $(a) \cdot (b)$ in $\mathcal{I}_0 H$ consists of all elements $daebf$ with $d, e, f \in H_u$.

The general case of an m-algebra is reduced to the previous one with the help of Lemma 3.1: we know that $G' = G_0/\Theta_A$ is a semigroup with zero A, and for the natural epimorphism ϕ from G_0 onto G', we have $a \in A \Leftrightarrow \phi(a) = A$.

Now, for any element $b \in G \setminus \sqrt{A}$, we find an $\mathcal{I}_0$-filtered set F with $b \in F \subseteq G \setminus A$ and $A \notin F' = \phi[F]$ (otherwise $A = \phi(a)$ for some $a \in F \cap A$). Since ϕ commutes with ideal generation and F is $\mathcal{I}_0$-filtered, so is F', and as G' is a semigroup, the above consideration yields an m-filtered set E' containing F' but not the zero element A. Then $E := \phi^{-1}[E']$ is m-filtered, too, and we obtain $b \in F \subseteq E \subseteq G \setminus A$. By contraposition, it follows that every element b of G with the property that every m-filtered set containing b meets A must belong to the radical of A.

The converse inclusion is clear since m-filtered sets are always $\mathcal{I}_0$-filtered. □

In accordance with our convention on ideal systems (see Section 2), we call a subset B of an m-algebra $\mathfrak{Q}$-*open* if every set that is a directed union of quantic ideals and disjoint from B is contained in a quantic ideal disjoint from B. In a finitary m-algebra, *every* subset is $\mathfrak{Q}$-open because the quantic ideals of such an algebra form an algebraic ideal system.

LEMMA 3.7. *Consider the following statements on a quantic ideal C of an m-algebra G:*

(a) *C is an intersection of prime quantic ideals.*
(b) *$G \setminus C$ is a union of $\mathfrak{Q}$-open m-filtered sets.*
(c) *C is semiprime, i.e., $G \setminus C$ is a union of $\mathfrak{Q}$-open $\mathfrak{Q}$-filtered sets.*
(d) *C is radical, i.e., $G \setminus C$ is a union of m-filtered sets.*
(e) *C is $\mathfrak{Q}$-radical, i.e., $G \setminus C$ is a union of $\mathfrak{Q}$-filtered sets.*
(f) *C is reduced.*
(g) *$aca \in C$ for all $c \in G_u$ implies $a \in C$.*

Without any further assumptions, one has the implications

$$(a) \Rightarrow (b) \Rightarrow (c) \Rightarrow (e) \quad \textit{and} \quad (b) \Rightarrow (d) \Rightarrow (e) \Rightarrow (f) \Leftrightarrow (g).$$

Under hypothesis UP, *the first three conditions are equivalent, while* CDC *entails the equivalence of the last four conditions. Hence, assuming* UP *and* CDC, *all seven conditions are equivalent for quantic ideals of finitary m-algebras.*

Proof. For (a) ⇒ (b), see Lemma 3.5. The implications (b) ⇒ (c) ⇒ (e) ⇒ (f) and (b) ⇒ (d) are straightforward (cf. Proposition 2.3).

For the equivalence of (f) and (g), proceed as in the proof of Lemma 3.5 (with $a = b$).

The Prime Ideal Theorem PIG (see Theorem 2.2), applied to the ideal system of all quantic ideals, yields the implication (c) ⇒ (a).

It remains to show that (g) in combination with CDC entails (d). First, we note the following more special result, known from semigroup and ring theory:

(s) *Let H be a semigroup with zero* 0 *such that for each $a \neq 0$, there is some $c \in H$ with $aca \neq 0$ or $aa \neq 0$. Then* CDC *implies that each $a \neq 0$ lies in an m-filtered subset of H not containing* 0.

For the proof of (s), define a relation ρ on $H \setminus \{0\}$ by $a\rho b$ iff there is some $c \in H_u$ with $b = aca$. Then, for any $a \in H \setminus \{0\}$, CDC provides a sequence $(a_n \mid n \in \omega)$ in $H \setminus \{0\}$ such that $a_0 = a$ and $a_n \rho a_{n+1}$ for all n. Now the set $F = \{a_n \mid n \in \omega\}$ turns out to be m-filtered: in fact, given $a_m, a_n \in F$ with $m \leq n$, we find $b, c \in H_u$ with $a_n = a_m b$ and $a_{n+1} = a_n c a_n = a_m b c a_n \in F$; analogously for $m \geq n$.

In the general case of an m-algebra G, use the canonical homomorphism ϕ from G_0 onto $G' = G_0/\Theta_C$ (see Lemma 3.1). On account of our assumption (g), the zero element C of the semigroup G' satisfies the same premise as in (s). For $b \in G \setminus C$, the image $\phi(b)$ is distinct from C, so we find an m-filtered subset F' of G' with $\phi(b) \in F'$ but $C \notin F'$. Thus $F = \phi^{-1}[F']$ is an m-filtered subset of G containing b and disjoint from C. □

By Proposition 3.3, our results on residuated ideal systems apply to $\mathcal{Q}G$, and we arrive at

THEOREM 3.8. *In* NBG *set theory, the Ultrafilter Principle is equivalent to each of the following statements:*

(SLQ) *The Separation Lemma for Quantic Ideals:*
The quantic ideals of an m-algebra are totally separated from the $\mathcal{Q}$-open m-filtered (or $\mathcal{Q}$-filtered) sets.

(PIQ) *The Prime Ideal Theorem for Quantic Ideals:*
The semiprime quantic ideals of an m-algebra are the meets of prime quantic ideals.

(PRQ) *The Prime Radical Theorem for Quantic Ideals:*
In any finitary m-algebra, the radical of a quantic ideal is the meet of all prime quantic ideals containing it.

COROLLARY 3.9. *Under hypothesis* CDC, *the semiprime quantic ideals of a finitary m-algebra agree with the reduced quantic ideals, and* UP *together with* CDC *guarantees that the radical of a quantic ideal C is not only the least semiprime but also the least reduced quantic ideal containing C.*

Now let us have a look at a larger class of ideals which often have similar properties as quantic ideals. Call a subset A of an m-algebra G *distributing* if it satisfies the distribution law

(D) $ab_i \in A \ (i < \nu_k) \Rightarrow am_k(b) \in A$

and the corresponding right-sided law for all $k \in K \setminus \{0\}$ and $a, b_i \in G$ $(i < \nu_k)$. Obviously, every quantic ideal is distributing, but the converse only holds under additional assumptions (like associativity and commutativity of the multiplication). By definition, an ideal C is distributing iff the 'point residuals'

$$C \mathbin{.\!\cdot} a = \{b \in G \mid ab \in C\} \text{ and } C \mathbin{\cdot\!.} a = \{b \in G \mid ba \in C\}$$

belong to the closure system SubG_1 of all subalgebras of the reduct G_1. Although we shall see in Section 5 that the ideal system $\mathcal{D}G$ of all distributing ideals may fail to be residuated, rather mild assumptions on the algebra G make $\mathcal{D}G$ a prequantale.

PROPOSITION 3.10. *If $\mathcal{C}$ is an ideal system contained in $\mathcal{D}G$ and join-closed in* SubG_1 *(i.e., if for each $\mathcal{B} \subseteq \mathcal{C}$, the subalgebra of G_1 generated by $\bigcup \mathcal{B}$ belongs to $\mathcal{C}$) then $\mathcal{C}$ is a residuated ideal system, and Theorem* 2.2 *applies.*

Proof. For $A, C \in \mathcal{C}$, the set-theoretical right residual

$$C \mathbin{.\!\cdot} A = \{b \in G \mid Ab \subseteq C\} = \bigcap \{C \mathbin{.\!\cdot} a \mid a \in A\}$$

is a subalgebra of G_1, and consequently, the join

$$C \mathbin{\underset{\mathcal{C}}{.\!\cdot}} A = \bigvee \{B \in \mathcal{C} \mid AB \subseteq C\} = \bigvee \{B \in \mathcal{C} \mid B \subseteq C \mathbin{.\!\cdot} A\}$$

is a member of $\mathcal{C}$ such that for each $B \in \mathcal{C}$,

$$A \cdot_{\mathcal{C}} B \subseteq C \Leftrightarrow AB \subseteq C \Leftrightarrow B \subseteq C \mathbin{.\!\cdot} A \Leftrightarrow B \subseteq C \mathbin{\underset{\mathcal{C}}{.\!\cdot}} A.$$

Analogously, one treats the case of left residuals. □

In many important algebraic structures, every ideal is quantic, and consequently, the system $\mathcal{I}G$ of all ideals is a quantale. Indeed, this is the case in any *associative and distributive m-algebra*, that is, in any m-algebra G whose multiplication is associative and whose fundamental translations

$$\lambda_a \colon G \longrightarrow G,\ x \longmapsto ax \quad \text{and} \quad \rho_b \colon G \longrightarrow G,\ x \longmapsto xb \quad (a, b \in G)$$

are homomorphisms of the reduct G_1. Examples of such algebras of type (2, 2) are all semirings, in particular, all rings and all distributive lattices.

Notice that the ideals of an m-algebra G are just the subalgebras of the algebra G_2 whose underlying set is G and whose operations are those of G_1 together with the fundamental left and right translations of G_0. Often, a weaker notion of distributivity suffices in order to ensure that the ideals form a prequantale. Let us call an m-algebra G *weakly distributive* provided for each term function $t\colon G^n \to G$ of the reduct G_1 and all $a, b_i \in G$ $(i < n)$, there are term functions s_i $(i < n)$ of the algebra G_2 and a term function $t'\colon G^n \to G$ of G_1 such that

$$\text{(wd)} \quad at(b_i \mid i < n) = t'(s_i(ab_i) \mid i < n),$$

and analogously for multiplications from the right. Of course, every distributive m-algebra is weakly distributive. Interesting classes of weakly distributive (but often not distributive) m-algebras occur in commutator theory (see Section 5 and [7], Ch. XIV-8).

LEMMA 3.11. *A finitary m-algebra G is weakly distributive iff every ideal of G is distributing and $\mathcal{I}G$ is join-closed in* SubG_1.

Proof. Suppose G is weakly distributive. For $A \in \mathcal{I}G, t = m_k$ ($k \in K \setminus \{0\}$) and $a, b_i \in G$ with $ab_i \in A$ for all $i < \nu_k = n$, choose term functions s_i and t' according to the above definition. We get $s_i(ab_i) \in A$ and $at(b_i \mid i < n) = t'(s_i(ab_i) \mid i < n) \in A$. This and a corresponding right-hand argument show that A is distributing. Furthermore, if $\mathcal{B}$ is a collection of ideals and C is the subalgebra of G_1 generated by $\bigcup \mathcal{B}$ then each element $c \in C$ is of the form $t(b_i \mid i < n)$ where t is a term function of G_1 and $b_i \in B_i$ for suitable $B_i \in \mathcal{B}$. Choose s_i and t' as before. Then $s_i(ab_i) \in B_i$, and by (wd), it follows that $ac = at(b_i \mid i < n) \in C$; analogously, $ca \in C$.

Conversely, assume $\mathcal{I}G = \mathcal{D}G$ is join-closed in SubG_1. Consider a term function t: $G^n \to G$ of G_1 and elements $a \in G, b = (b_i \mid i < n) \in G^n$. By hypothesis, the join C of the principal ideals (ab_i) $(i < n)$ in SubG_1 is a distributing ideal C of G with $b_i \in C \cdot\!\!\cdot\, a$, and it follows that $t(b) \in C \cdot\!\!\cdot\, a$, i.e. $at(b) \in C$. As G is a finitary algebra, there are suitable term functions satisfying (wd). □

Combining Proposition 3.10 with Lemma 3.11, we arrive at

PROPOSITION 3.12. *The ideals of a weakly distributive finitary m-algebra G form an algebraic residuated ideal system.*

COROLLARY 3.13. *In* NBG *set theory, the Ultrafilter Principle is equivalent to the following Separation Lemma for Weakly Distributive Algebras:*

> *The ideals of any weakly distributive finitary m-algebra G are totally separated from the $\mathcal{I}$-filtered sets. Hence, the $\mathcal{I}$-radical of any ideal $A \in \mathcal{I}G$ is the least semiprime ideal and the intersection of all prime ideals containing A.*

4. Strong Ideals and Radicals

It is well known that in commutative rings and similar structures, prime ideals may be characterized as proper ideals A satisfying

(P) $ab \in A \Rightarrow a \in A$ or $b \in A$,

and reduced (or radical) ideals A by the condition

(R) $aa \in A \Rightarrow a \in A$.

In fact, many authors use the term 'reduced' for the latter property even in non-commutative ring theory (see, for example, [19]).

In the sequel, we shall study the above two properties in the more general setting of an m-algebra G. Generalizing an approach proposed by Aubert [2] for the case of associative multiplications, we consider the monoid $T(G_0)$ of all translations (i.e. compositions of finitely many fundamental translations) of the groupoid reduct G_0 and call an ideal A of G

all-sided if $\tau^{-1}[A]$ is an ideal of G_0 for each $\tau \in T(G_0)$,
prequantic if $\tau^{-1}[A]$ is a subalgebra of G_1 for each $\tau \in T(G_0)$,
strong if $\tau^{-1}[A]$ is an ideal of G for each $\tau \in T(G_0)$.

As the name indicates, an all-sided ideal A is not only invariant under multiplication from the left and from the right, but also under 'intermediate' multiplications, for example:

$$ac \in A \Rightarrow a(bc) \in A \quad \text{and} \quad (ab)c \in A.$$

By definition, an ideal is strong iff it is both all-sided and prequantic. On the other hand, it is easy to see that the quantic ideals are just the associating prequantic ideals. In any associative, commutative and distributive m-algebra, *all* ideals are strong. More generally, calling an associating set A *commuting* if it satisfies

$$\text{(C)} \quad abc \in A \Leftrightarrow acb \in A \quad \text{for all } a, b, c \in G_u,$$

one can show (cf. [12]):

LEMMA 4.1. *For any commuting ideal C of an m-algebra G, the quotient G_0/Θ_C is a commutative semigroup with zero C. Hence, every commuting ideal C is all-sided, and the properties 'quantic' and 'strong' are equivalent for C.*

Moreover, by similar methods as in Proposition 3.3, one proves the following three facts (see [13] for more general results on cm-lattices):

PROPOSITION 4.2. *Let G be any m-algebra.*

(1) *The strong ideals form an ideal prequantale $\mathcal{S}G$ in $\mathcal{P}G$ (and in $\mathcal{I}G$).*
(2) *The associating strong ideals form an ideal quantale $\mathcal{AS}G$ in $\mathcal{P}G$.*
(3) *The commuting strong ideals form a commutative ideal quantale $\mathcal{CS}G$ in $\mathcal{P}G$.*

In all three cases, the residuals are the set-theoretical ones, i.e.

$$C \mathbin{.\!\cdot} A = \{b \in G \mid Ab \subseteq C\} \quad \textit{and} \quad C \mathbin{\cdot\!.} A = \{b \in G \mid bA \subseteq C\}.$$

Hence, the corresponding closure operators J satisfy the identity

$$J(AB) = J(A) \cdot J(B) \quad \textit{for all } A, B \subseteq G.$$

In accordance with the above remark on prime ideals and reduced ideals in commutative rings, we call a proper ideal of an arbitrary m-algebra a *strong prime ideal* if it satisfies (P), i.e., if its complement is m-closed. An ideal with the property (R) is said to be *m-reduced*, and an ideal with the three properties (R), (A), (D) (cf. Section 3) will be referred to as a *strong radical ideal*. Every strong radical ideal is in fact a strong ideal, as we shall see in Proposition 4.5. The subsequent remarks on strong prime ideals and strong radical ideals are easily verified (use Propositions 2.7 and 4.2):

LEMMA 4.3. *Let $\mathcal{C}$ be any closure system of strong ideals of an m-algebra G containing at least all strong radical ideals. Then the strong prime ideals are precisely the prime $\mathcal{C}$-ideals, and the strong radical ideals are precisely the reduced $\mathcal{C}$-ideals.*

On account of that coincidence, we need not distinguish between 'strong prime ideals' and 'prime strong ideals'. Now, invoking Proposition 2.7 once more, we may supplement Proposition 4.2 by

PROPOSITION 4.4. *The strong radical ideals of any m-algebra G form the largest ideal locale $\mathcal{RS}G$ in $\mathcal{S}G$, and its closure operator RS satisfies the identity*

$$RS(AB) = RS(A) \cap RS(B) \; (A, B \subseteq G).$$

It is straightforward to check that arbitrary intersections of strong prime ideals are commuting strong ideals and, moreover, strong radical ideals. In order to prove the converse in finitary m-algebras with the help of UP, we need some alternative characterizations of such ideals.

PROPOSITION 4.5. *For a subset A of an m-algebra G, the following conditions are all equivalent:*

(a) *A is a strong radical ideal.*
(b) *A is an m-reduced commuting ideal.*
(c) *A is an m-reduced strong ideal.*
(d) *A is a strong ideal, and G_0/Θ_A is a semilattice with zero A.*
(e) *A is a strong ideal, and each congruence class of Θ_A is m-closed.*
(f) *A is a strong ideal whose complement is a union of m-closed sets.*
(g) *A is a radical $\mathcal{C}$-ideal for one (each) ideal system $\mathcal{C}$ with $\mathcal{RS}G \subseteq \mathcal{C} \subseteq \mathcal{S}G$.*

Proof. In [12], the equivalence of the first six conditions is established for subsets of groupoids (the most involved part is to show that an ideal A satisfying (R) and (A) is already associating). The extension to m-algebras is easily obtained (see also Lemma 4.1).

For (a) $\Rightarrow$ (g), observe that by (a) $\Rightarrow$ (f), A is a radical $\mathcal{C}$-ideal because m-closed sets are $\mathcal{C}$-filtered. Conversely, (g) $\Rightarrow$ (a) follows from Lemma 4.3 and the fact that radical $\mathcal{C}$-ideals are always reduced (cf. Proposition 2.3). □

It is worth mentioning that, in contrast to the corresponding statements in Proposition 2.3 and Lemma 3.7, the previous proof for the coincidence of the terms 'radical $\mathfrak{C}$-ideal' and 'reduced $\mathfrak{C}$-ideal' did not require CDC.

The radical (as defined in Section 3) of a strong ideal admits a surprisingly simple description (which is, of course, known in the special setting of commutative rings but requires considerably more involved computations in the general case of m-algebras). Define inductively certain 'binary' powers of an element $a \in G$ by

$$a^{2^0} = a,$$
$$a^{2^{n+1}} = (a^{2^n})(a^{2^n}).$$

By induction, it follows that

$$a^{2^{m+n}} = (a^{2^m})^{2^n}.$$

THEOREM 4.6. *The radical $\sqrt{A}$ of a strong ideal A in an arbitrary m-algebra G consists of all $a \in G$ such that $a^{2^n} \in A$ for some $n \in \omega$. If G is finitary then $\sqrt{A}$ is the least*

- *strong radical ideal containing A,*
- *semiprime quantic ideal containing A,*
- *radical quantic ideal containing A,*
- *reduced quantic ideal containing A.*

Proof. Given $a \in G$, define inductively sets A_n by

$$A_0 = \{a\}, \quad A_{n+1} = \{\tau(\rho(b)\sigma(b)) \mid b \in A_n,\ \rho, \sigma, \tau \in T(G_0)\}.$$

First, let us prove the implication

$$\text{(i)} \quad \tau(a^{2^n}) \in A \Rightarrow \tau[A_n] \subseteq A \quad \text{for all } \tau \in T(G_0).$$

Proceeding by induction, put $d = a^{2^n}$ and assume $\tau(a^{2^{n+1}}) \in A$. Then $\tau \circ \rho_d(a^{2^n}) \in A$, and the induction hypothesis gives for each $b \in A_n$ the following consequences: $\tau \circ \rho_d(b) = \tau \circ \lambda_b(d) \in A$, hence $\tau \circ \lambda_b[A_n] \subseteq A$, in particular $\tau(bb) \in A$. Since A is an all-sided ideal, we conclude that $\tau(\tau'(\rho(b)\sigma(b))) \in A$ for all $\rho, \sigma, \tau' \in T(G_0)$, so that $\tau[A_{n+1}] \subseteq A$, completing the induction step.

By definition, the radical of A is the intersection of all sets C containing A and having an $\mathcal{I}_0$-filtered complement B. Consider such sets B and $C = G \setminus B$. Then $A \cap B = \emptyset$. Assuming $a \in B$, a further induction gives

$$A_n \cap B \neq \emptyset.$$

Indeed, if $b \in A_n \cap B$ then the $\mathcal{I}_0$-ideal (b) generated by b satisfies $(b) \cdot (b) \cap B \neq \emptyset$, and we find translations $\rho, \sigma, \tau \in T(G_0)$ with $\tau(\rho(b)\sigma(b)) \in A_{n+1} \cap B$. On the other hand, by (i), $a^{2^n} \in A$ implies $A_n \subseteq A \subseteq C$ and so $A_n \cap B = \emptyset$. Hence every element of the set

$$R = \{a \in G \mid a^{2^n} \in A \text{ for some } n \in \omega\}$$

must belong to C, and R is contained in $\sqrt{A}$.

Conversely, if $a^{2^n} \in A$ for no n then $F = \{a^{2^n} \mid n \in \omega\}$ is an $\mathcal{I}_0$-filtered set containing a and disjoint from A. Indeed, given $m \leq n$, we find a $\rho \in T(G_0)$ such that $\rho(a^{2^m}) = a^{2^n}$, whence $\rho(a^{2^m})a^{2^n} = a^{2^{n+1}} \in F$. Analogously, if $m \geq n$ then there is a $\sigma \in T(G_0)$ with $a^{2^m}\sigma(a^{2^n}) \in F$. Consequently, a does not belong to $\sqrt{A}$.

In case of a finitary algebra G, we can show that R is a strong radical ideal. Clearly, $a^2 \in R$ implies $a^{2^{n+1}} = (a^2)^{2^n} \in A$ for some n, hence $a \in R$.

Suppose $\tau \in T(G_0)$ and $a \in \tau^{-1}[R]$. Then $\tau(a) \in R$, so $(\tau(a))^{2^n} \in A$ for some n, and as A is all-sided, an easy induction yields $(\tau(ab))^{2^n} \in A$, i.e. $ab \in \tau^{-1}[R]$ for each $b \in G$. Analogously, we get $ba \in \tau^{-1}[R]$.

More difficult is the proof that $\tau^{-1}[R]$ is closed under all operations m_k of G. Suppose $b_i \in \tau^{-1}[R]$ for $i < \nu_k$ and put $d = \tau(m_k(b_i \mid i < \nu_k))$. We must show that d belongs to R. Using the hypothesis that G is a *finitary* algebra, we may assume that ν_k is a natural number q and there is an exponent n such that $\tau(b_i)^{2^n} \in A$ for all $i < q$.

We need some auxiliary definitions and facts. Generalizing the above definition of powers, define products $\prod_{r=1}^{2^n} c_r$ of elements $c_r \in G$ by

$$\prod_{r=1}^{2^0} c_r = c_1, \qquad \prod_{r=1}^{2^{n+1}} c_r = \left(\prod_{r=1}^{2^n} c_r\right)\left(\prod_{r=1}^{2^n} c_{r+2^n}\right).$$

The specialization quasiorder $\leq$ is defined by $a \leq b$ iff a 'specializes' b, i.e., a belongs to the strong ideal (b) generated by b. On account of Proposition 4.2, we have $(ab) = (a) \cdot (b)$; therefore, $c \in (a)$ implies $cb \in (ab)$ and $bc \in (ba)$; in other words, all translations are monotone with respect to $\leq$.

Let s be a natural number such that $q < 2^s$, and let C be any subset of G with at most q elements. We claim that

(∗) For all $c_1, \ldots, c_{2^{ns}} \in C$, there is a $c \in C$ with $\prod_{r=1}^{2^{ns}} c_r \leq c^{2^n}$.

The case $n = 0$ is clear. Assume (∗) has been shown for some n.

For any $c_1, \ldots, c_{2^{(n+1)s}} \in C$, the element $\prod_{r=1}^{2^{(n+1)s}} c_r$ is a product of 2^s subproducts with 2^{ns} factors (suitably parenthesized). By induction hypothesis, each of these subproducts specializes c^{2^n} for certain $c \in C$, and by the assumption $q < 2^s$, this occurs at least twice for the same c. Hence the whole product specializes $(c^{2^n})^2 = c^{2^{n+1}}$.

Now we may complete the proof that R is a strong ideal. The set $C = \{\tau(b_i) \mid i < q\}$ has at most q elements, so by $(*)$ and the hypothesis $c^{2^n} \in A$ for $c \in C$, we get

$$\prod_{r=1}^{2^{ns}} c_r \in A \text{ for arbitrary choices } c_r \in C \quad (1 \leq r \leq 2^{ns}).$$

Assume we have already shown for some $p \leq 2^{ns}$ that $\prod_{r=1}^{2^{ns}} c_r$ lies in A for all choices c_r with $c_r = d$ in case $r < p$ and $c_r \in C$ if $r \geq p$. Then, using the fact that A is prequantic, we conclude that the same is true for $p+1$ instead of p, and we arrive at $d^{2^{ns}} \in A$, i.e. $d \in R$.

Summarizing our results, we see that R is a strong radical ideal, and any m-reduced ideal C containing A must also contain R, since $a^{2^n} \in A \subseteq C$ implies $a \in C$.

Similarly, if C is a reduced quantic ideal with $A \subseteq C$ then, by Proposition 4.2, $a^{2^n} \in A$ implies $(a)^{2^n} \subseteq A \subseteq C$ for the strong ideal (a) generated by a. By Proposition 3.3, we get $Q((a))^{2^n} = Q((a)^{2^n}) \subseteq C$, and it follows that $a \in Q((a)) \subseteq C$. This proves the inclusion $R \subseteq C$, and R itself is a reduced quantic ideal by Proposition 4.5.

Finally, R is a semiprime quantic ideal since each element $a \in G \setminus R$ is contained in the m-filtered, hence $\mathcal{Q}$-coprime set $\{a^{2^n} \mid n \in \omega\}$ which is disjoint from A.

The remark that every semiprime quantic ideal is radical and every radical quantic ideal is reduced concludes the proof. □

Now, we infer from Theorem 2.2 (using Proposition 4.5 and Theorem 4.6) the following result:

THEOREM 4.7. *In* NBG *set theory, the Ultrafilter Principle is equivalent to each of the following statements:*

(SLS) *The Separation Lemma for Strong Ideals:*
The strong ideals of any m*-algebra are totally separated from the* $\mathcal{S}$*-open and* m*-closed sets.*

(PIS) *The Prime Ideal Theorem for Strong Ideals:*
The strong radical ideals of any finitary m*-algebra are exactly the intersections of strong prime ideals.*

(PRS) *The Prime Radical Theorem for Strong Ideals:*
In a finitary m*-algebra, the radical of any strong ideal is the intersection of all strong prime ideals containing that ideal.*

In case of an associative, commutative and distributive multiplication, the prefix 'strong' may be omitted in Theorem 4.7. Notice that every subset of a finitary m-algebra is $\mathcal{S}$-open. Whereas it was easy to see that the system $\mathcal{S}G$ of all strong ideals is residuated, the corresponding result on prequantic ideals is more involved.

PROPOSITION 4.8. *The prequantic ideals of any m-algebra G form an ideal prequantale $\mathcal{PQ}G$ in $\mathcal{I}_0G$, hence a residuated ideal system.*

Proof. In contrast to the case of strong ideals, the set-theoretic residuals

$$C \therefore A = \{b \in G \mid Ab \subseteq C\}$$

of a prequantic ideal C need not even be ideals. But we have

$$\tau^{-1}[C \therefore A] = \bigcap\{(\lambda_a \circ \tau)^{-1}[C] \mid a \in A\} \in \mathrm{Sub}G_1$$

for each translation $\tau \in T(G_0)$. In order to construct a right residual that belongs to $\mathcal{PQ}G$, consider the set D of all $d \in G$ such that there is a tuple $b = (b_i \mid i < n) \in G^n$, a term function $t\colon G^n \to G$ of G_1 and a translation $\sigma \in T(G_0)$ with

$$d = \sigma \circ t(b) \text{ and}$$
$$\rho \circ \sigma(b_i) \in C \therefore A \text{ for all } \rho \in T(G_0) \text{ and } i < n.$$

(1) $D \subseteq C \therefore A$: Indeed, $d \in D$ implies $d = \sigma \circ t(b)$ for suitable b, t, σ as above, and so $d \in C \therefore A$ because $b_i \in \sigma^{-1}[C \therefore A] \in \mathrm{Sub}G_1$ for $i < n$ entails $t(b) \in \sigma^{-1}[C \therefore A]$.

(2) $B \in \mathcal{I}_0G$ and $B \subseteq C \therefore A$ imply $B \subseteq D$: For $b \in B$ and all $\rho \in T(G_0)$, we have $\rho(b) \in B \subseteq C \therefore A$, and taking for σ and t the identity maps, we get $b = \sigma \circ t(b) \in D$. Combining (1) with (2), we obtain

$$AB \subseteq C \Leftrightarrow B \subseteq C \therefore A \Leftrightarrow B \subseteq D$$
$$\text{for all } B \in \mathcal{I}_0G.$$

(3) $D \in \mathcal{I}_0G$: Given $d \in D$ and $\tau \in T(G_0)$, choose b, t and σ according to the definition of D. Then we have $\rho \circ \tau \circ \sigma(b_i) \in C \therefore A$ for all $\rho \in T(G_0)$ and $i < n$, whence $\tau(d) = \tau \circ \sigma \circ t(b) \in D$.

(4) $D \in \mathcal{PQ}G$: Let $k \in K \setminus \{0\}$, $b = (b_i \mid i < \nu_k) \in G^{\nu_k}$, $\sigma \in T(G_0)$, and suppose $\sigma(b_i) \in D$ for $i < \nu_k$. Then, by (3) and (1), we obtain $\rho \circ \sigma(b_i) \in D \subseteq C \therefore A$ for all $\rho \in T(G_0)$, and the choice $t = m_k$ yields $\sigma \circ m_k(b) \in D$. Left residuals are constructed analogously. □

On account of Proposition 4.8, our general results on residuated ideal systems apply to $\mathcal{PQ}G$ as well. In particular, UP is equivalent to the Separation Lemma and to the Prime Ideal Theorem for Prequantic Ideals.

DIAGRAM 4.9. *Ideal systems of an m-algebra*

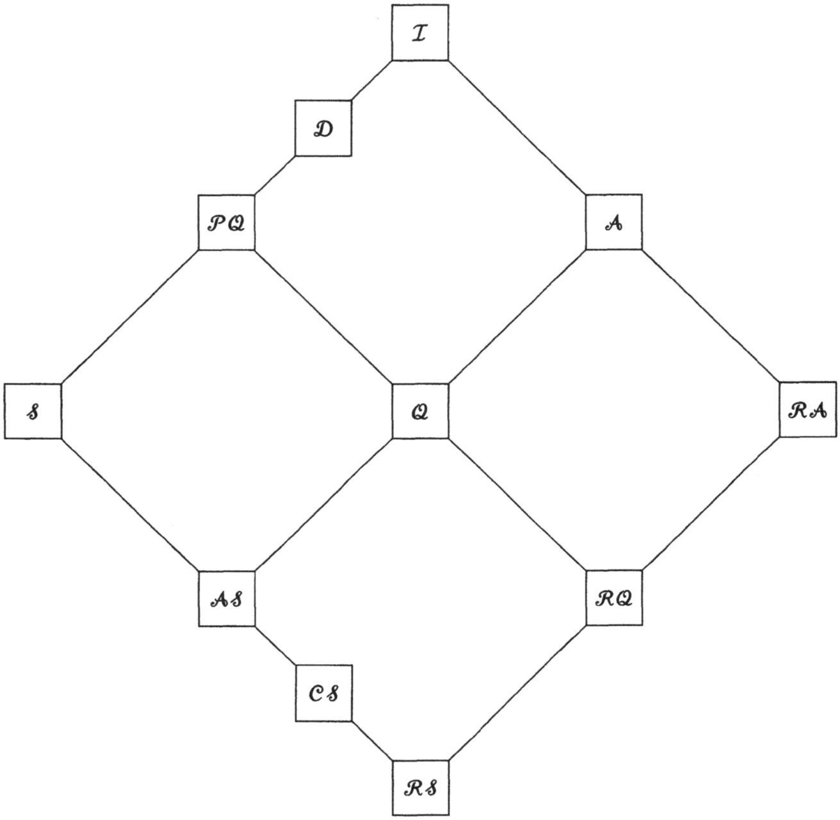

Finally, a large class of ideal systems of m-algebras has to be mentioned where total separation works in NBG without UP. From Proposition 2.3, Corollary 2.6 and Proposition 3.10, we infer:

PROPOSITION 4.10. *If an ideal system $\mathcal{C}$ of an m-algebra G is closed under non-empty unions then it is a superalgebraic residuated ideal system whose members are totally separated from the $\mathcal{C}$-filtered sets. Under hypothesis* CDC, *the reduced (= radical = semiprime) $\mathcal{C}$-ideals are the intersections of prime $\mathcal{C}$-ideals.*

COROLLARY 4.11. *Let G be an m-algebra with no operations of arity > 1 except m. Then each of the ideal systems in Diagram* 4.9 *is closed under nonempty unions, hence superalgebraic and residuated. Moreover:*

(1) *The ideals are totally separated from the $\mathcal{I}$-filtered sets.*
(2) *The quantic ideals are totally separated from the m-filtered sets.*
(3) *The strong ideals are totally separated from the m-closed subsets.*

(1′) *The semiprime ideals are the meets of prime ideals.*
(2′) *The semiprime quantic ideals are the meets of prime quantic ideals.*
(3′) *The strong radical ideals are the meets of strong prime ideals.*

Under hypothesis CDC, *'semiprime' may be replaced with 'reduced'.*

5. Applications and Examples

The simplest examples of m-algebras are *groupoids*, i.e. algebras G with one binary operation m and no further operations. Here, the system of all ideals and the system of all quantic (= associating) ideals are not only closed under arbitrary intersections but also under arbitrary unions. As a consequence, the respective separation lemma and the representation of strong radical ideals as intersections of strong prime ideals holds in ZF or NBG without UP. The corresponding representation theorem for reduced associating ideals as intersections of prime associating ideals probably requires CDC. For details, including various (counter)examples, see [12]. Here we only point out that in the regular semigroup of all selfmaps on a fixed set, all ideals are trivially quantic, but the ideal of constant maps fails to be strong if the underlying set has more than two elements.

Every group $G_1 = (G; +, -, 0)$ (written additively but not necessarily abelian) gives rise to a (frequently nonassociative) m-algebra $(G;\ m, +, -, 0)$ where $m(a, b)$ is the commutator $a + b - a - b$. The ideals of that algebra are just the normal subgroups of G_1, and the product in $\mathcal{I}G$ is the usual commutator product [A, B]. Using the identities

$$m(a, 0) = 0, \qquad m(b, a) = -m(a, b),$$
$$m(a, -b) = m(-b, m(b, a)) + m(b, a),$$
$$m(a, b + c) = m(a, b) + m(b, m(a, c)) + m(a, c),$$

one can show easily that the ideals of G are distributing, and by Lemma 3.11, G is weakly distributive (but rarely distributive). Hence, UP together with CDC entails the result (usually derived from AC) that the reduced normal subgroups R (satisfying $[A, A] \subseteq R \Rightarrow A \subseteq R$) are the intersections of prime normal subgroups P (satisfying $[A, B] \subseteq P \Rightarrow A \subseteq P$ or $B \subseteq P$).

The only strong radical ideal in the above m-algebra is the whole group. Hence, there are no strong prime ideals at all.

In the situation of Corollary 4.11, the closure systems $\mathcal{I}G$ and $\mathcal{D}G$ are closed under nonempty unions, hence certainly join-closed in SubG_1. However, $\mathcal{I}G$ may be distinct from $\mathcal{D}G$ – in other words, G need not be weakly distributive (see Lemma 3.11). For example, let G denote the set of all complex numbers z with $|z| > 1$, and consider the m-algebra $(G; m, ^*)$ where m means complex multiplication and * complex conjugation. Then the set A of all numbers $z \in G$ with $|z| \geq 2$ and $z \neq 2$ turns out to be an ideal which is not distributing (since $(1 + i)(1 + i) = 2i \in A$ but $(1 + i)(1 + i)^* = 2 \notin A$).

On the other hand, there are m-algebras G in which every ideal is distributing, but $\mathcal{I}G$ fails to be a prequantale (this cannot happen in case of an associative and commutative multiplication). Consider, for example, the algebra $(G; \cdot, +, 0)$ of type $(2, 2, 0)$, where $G = \{0, 1, 2, 3, 4\}$,

$$1 \cdot a = a \cdot 1 = a \cdot a = a \ (a \neq 1, 4), \ \ 1 \cdot 1 = 4 \cdot 4 = 4,$$

and $a \cdot b = 0$ otherwise. The operation $+$ is defined by $2 + 3 = 3 + 2 = 1$ and $a + b = 0$ in all other cases. An easy computation leads to the ideal system

$$\mathcal{I}G = \mathcal{D}G = \{\{0\}, \{0, 2\}, \{0, 3\}, \{0, 4\}, \{0, 2, 4\}, \{0, 3, 4\}, G\}.$$

The zero ideal is distributing but not prequantic: for $\tau = \lambda_4 \circ \lambda_1$, we have $\tau(2) = \tau(3) = 0$ but $\tau(2 + 3) = \tau(1) = 4$. For the ideals $A = \{0, 2\}$, $B = \{0, 3\}$ and $C = \{0, 4\}$, the inequality

$$(A \cdot C) \vee (B \cdot C) = \{0\} \neq C = G \cdot C = (A \vee B) \cdot C$$

shows that $\mathcal{I}G = \mathcal{D}G$ cannot be a prequantale, and $\mathcal{I}G$ fails to be join-closed in $G_1 = (G; +, 0)$: indeed, the join of A and B in SubG_1 is $G \setminus \{4\}$. Modifying the above multiplication slightly by setting $4 \cdot 4 = 0$ instead of $4 \cdot 4 = 4$, one obtains an m-algebra with the same (distributing) ideals, but this time the system of all ideals becomes a quantale!

The following similar example shows that, in contrast to the case of quantic ideals and of strong ideals, the set-theoretical residual $C .\!\cdot A = \{b \in G \mid Ab \subseteq C\}$ of two prequantic ideals A, C need not even be an ideal. Define on $G = \{0, 1, 2, 3, 4, 5\}$ an operation $+$ as before, and define a commutative multiplication $m(a, b) = a \cdot b$ by

$$1 \cdot a = a \cdot 1 = a \cdot a = a \ (a \leq 3), \qquad 3 \cdot 4 = 4 \cdot 3 = 4,$$
$$4 \cdot 5 = 5 \cdot 4 = 5,$$

and $a \cdot b = 0$ in all other cases. Obviously, $A = \{0, 5\}$ and $C = \{0\}$ are ideals of $(G; \cdot, +, 0)$, but $C .\!\cdot A = G \setminus \{4\}$ fails to be an ideal. Moreover, all ideals of the present algebra are prequantic (and consequently, $\mathcal{I}G$ is a prequantale). In fact, an easy induction yields $\tau(a + b) \in \{0, \tau(a), \tau(b), \tau(a) + \tau(b)\}$ for all $\tau \in T(G_0)$.

We have seen in Section 2 that various separation lemmas for ideal systems are consequences of the Separation Lemma for Prequantales (SLP). On the other hand, SLP is a special instance of SLS (see Theorem 4.7). Indeed, a two-sided cm-lattice L may be interpreted as an (infinitary) m-algebra G with the given multiplication m, joins of arbitrary arity and unary meet operations $x \mapsto a \wedge x$; the ideals of such an m-algebra are then just the principal ideals of the original cm-lattice, so the system of all (principal) ideals is isomorphic to L. It is easy to see that every (principal) ideal is all-sided, and that L is a prequantale iff all ideals are prequantic (strong). In that case, the $\mathcal{I}$-open and m-closed upper sets are the Scott-open distributors of L. Furthermore, L is a quantale iff every (principal) ideal is quantic.

The following class of examples shows that neither PRQ nor Corollaries 3.9 and 3.13, nor the 'strong versions' PIS and PRS extend to infinitary m-algebras. Consider any locale without prime elements, for instance, any atomless complete Boolean algebra. Taking the binary meet as multiplication and arbitrary joins as further operations, one obtains an associative, commutative and distributive m-algebra G whose ideals are the principal ideals of the original locale. In that algebra G, all ideals are reduced, quantic and strong, but there are no prime ideals. The m-filtered subsets are the down-directed ones. Hence every (principal) ideal agrees with its radical, whereas by PIQ, the only semiprime quantic ideal is the whole locale. Without invoking the Prime Ideal Theorem, we can only say that a principal ideal is semiprime iff its complement is a union of Scott-open filters (which, by SLP, is equivalent to saying that the generating element of that ideal is a meet of primes).

In the classical case of rings, all ideals are quantic, but in noncommutative rings, ideals need not be strong. Here, we may apply SLQ but not SLS in order to obtain Krull's Separation Lemma. On the other hand, the Prime Radical Theorem PRS (see Theorem 4.7) easily yields the subdirect product representation of reduced rings by domains (cf. [19]).

Lattice ideals are certainly associating, commuting and m-reduced; for such ideals, the attributes *(pre)quantic, strong, semiprime* and *radical* are all equivalent. Hence, the main results in [24] are immediate consequences of Theorem 4.7.

More sophisticated applications occur in the theory of ordered algebraic structures (cf. Birkhoff [7] and Fuchs [15]). Recall that a *lattice-ordered group* or *l-group* is an algebra $(G;\ \wedge, +, -, 0)$ such that $(G;\ \wedge)$ is a semilattice and $(G;\ +, -, 0)$ is a group (not necessarily Abelian) satisfying the distributive laws

$$(a \wedge b) + c = (a + c) \wedge (a + c) \text{ and } c + (a \wedge b) = (c + a) \wedge (c + b).$$

Then $(G;\ \wedge, \vee)$ with $a \vee b = -((-a) \wedge (-b))$ turns out to be a distributive lattice. An l-*ideal* is a normal subgroup of $(G;\ +, -, 0)$ that is, at the same time, a convex sublattice of $(G;\ \wedge, \vee)$. Introducing the associative and commutative operation

$$m : G^2 \longrightarrow G, (a, b) \longmapsto |a| \wedge |b| \quad \text{with} \quad |a| = (0 \vee a) + (0 \vee -a),$$

it is now an elementary exercise to show that the l-ideals are precisely the ideals of the m-algebra $(G;\ m, +, -, 0, m_b\ (b \in G))$, where m_b is the inner automorphism mapping a to $b + a - b$. More challenging is the proof that those l-groups in which all l-ideals are quantic are characterized by the implication

$$a \wedge (b + a - b) = 0 \Rightarrow a = 0$$

(which trivially holds in abelian l-groups). It was shown by Lorenzen [20] that the l-groups with that property are precisely those which admit a subdirect product representation by totally (= fully) ordered l-groups. His proof uses the Axiom of Choice, while our approach provides a deduction from the Ultrafilter Principle.

Similar facts hold for *lattice-ordered rings* (*l-rings*), that is, for algebras $(G;\ \wedge, \cdot, +, -, 0)$ such that $(G;\ \wedge, +, -, 0)$ is an l-group, $(G;\ \cdot, +, -, 0)$ is a ring, and multiplication by positive elements is monotone. The l-ideals of such an l-ring are both l-ideals of the l-group $(G;\ \wedge, +, -, 0)$ and ideals of the ring $(G;\ \cdot, +, -, 0)$. Again, they may be interpreted as the ideals of a suitable m-algebra, namely of the algebra

$$(G;\ m, +, -, 0, \lambda_c, \rho_c\ (c \in G)),$$

where m is the same operation as in l-groups, and λ_c and ρ_c denote the left and the right multiplication by c, respectively. It turns out that those l-rings in which all l-ideals are quantic are just the *f-rings* mentioned in Example 7 of Section 1. These f-rings may be characterized by the distributive laws

$$(a \wedge b)c = ac \wedge bc \quad \text{and} \quad c(a \wedge b) = ca \wedge cb \quad \text{for all } c > 0.$$

The Birkhoff–Pierce representation of f-rings as subdirect products of fully ordered l-rings is a rather direct consequence of the Prime Ideal Theorem PIS (see Theorem 4.7). These and more general results on the spectral theory of ordered algebras (cf. Aubert [1]) will be discussed in a forthcoming note.

Acknowledgements

Parts of this study have been elaborated while the author was a guest at McMaster University, Hamilton, Ontario. The great hospitality of Professor Banaschewski and his wife during that time, as well as the financial assistance by the Natural Sciences and Engineering Research Council of Canada in the form of a research grant, are gratefully acknowledged.

References

1. Aubert, K. E.: Géneralisation de la théorie des r-idéaux de Prüfer–Lorenzen, *C.R. Acad. Sci. Paris* **238** (1954), 214–216.
2. Aubert, K. E.: Theory of x-ideals, *Acta Math.* **107** (1962), 1–52.
3. Banaschewski, B.: The power of the ultrafilter theorem, *J. London Math. Soc. (2)* **27** (1983), 193–202.
4. Banaschewski, B.: Prime elements from prime ideals, *Order* **2** (1985), 211–213.
5. Banaschewski, B.: The prime ideal theorem and representation of F-rings, *Algebra Universalis* **25** (1988), 384–387.
6. Banaschewski, B. and Erné, M.: On Krull's separation lemma, *Order* **10** (1993), 253–260.
7. Birkhoff, G.: *Lattice Theory*, 3rd edn, Amer. Math. Soc. Coll. Publ. 25, Providence, RI, 1973.
8. Birkhoff, G. and Pierce, R. S.: Lattice-ordered rings, *An. Acad. Brasil. Ci.* **28** (1956), 41–69.
9. Blyth, T. S. and Janowitz, M. F.: *Residuation Theory*, Pergamon Press, Oxford, 1972.
10. Dilworth, R. P. and Ward, M.: Residuated lattices, *Trans. Amer. Math. Soc.* **45** (1939), 335–354.
11. Erné, M.: Distributors and wallmanufacture, Preprint No. 204/204a, Hannover, 1986 (see also J. *Pure Appl. Algebra* **68** (1990), 109–125).
12. Erné, M.: Ideal theory in groupoids, Preprint, Hannover, 1996.

13. Erné, M.: Closure operators on cm-lattices, Preprint, Hannover, 1996.
14. Feldman, D. and Henriksen, M.: f-rings, subdirect products of totally ordered rings, and the Prime Ideal Theorem, Preprint, 1987.
15. Fuchs, L.: *Partially Ordered Algebraic Systems*, Pergamon Press, Oxford, 1963.
16. Halpern, J.: The independence of the Boolean prime ideal theorem from the axiom of choice, *Fund. Math.* **55** (1964), 57–66.
17. Keimel, K.: A unified theory of minimal prime ideals, *Acta Math. Acad. Sci. Hungar.* **23** (1972), 51–69.
18. Krull, W.: Zur Theorie der zweiseitigen Ideale in nichtkommutativen Bereichen, *Math. Z.* **28** (1927), 481–503.
19. Lam, T. Y.: *A First Course in Noncommutative Rings*, Springer-Verlag, New York, 1991.
20. Lorenzen, P.: Über halbgeordnete Gruppen, *Math. Z.* **52** (1949), 483–526.
21. Pincus, D.: *Adding Dependent Choice to the Prime Ideal Theorem*, Logic Colloquium 76, North-Holland, 1977.
22. Priestley, H. A.: Ordered topological spaces and the representation of distributive lattices, *Proc. London Math. Soc. (3)* **24** (1972), 507–530.
23. Rav, Y.: Variants of Rado's selection lemma and their applications, *Math. Nachr.* **79** (1977), 145–165.
24. Rav, Y.: Semiprime ideals in general lattices, *J. Pure Appl. Algebra* **56** (1989), 105–118.
25. Rosenthal, K. I.: *Quantales and Their Applications*, Pitman Research Notes in Math. 234, Longman Scientific & Technical, 1990.
26. Scott, D. S.: Prime ideal theorems for rings, lattices and Boolean algebras, *Bull. Amer. Math. Soc.* **60** (1954), 390 (Abstract).

Applied Categorical Structures **8:** 145–160, 2000.
G. Brümmer & C. Gilmour (eds), Papers in Honour of Bernhard Banaschewski.

An Illustration of the Power of Structure Theory

Dedicated to Bernhard Banaschewski

KARL H. HOFMANN
Fachbereich Mathematik, Technische Universität Darmstadt, Schlossgartenstr. 7, D-64289 Darmstadt, Germany

(Received: 18 September 1996; accepted: 28 May 1998)

Abstract. There is a hierarchy of concepts of compact semigroups according to the degree of continuity of their multiplications: One-sided continuity, separate continuity, continuity. Each level of generality has its own fields of application and its own peculiarities. This is illustrated in a discussion of three seemingly rather unrelated examples from different branches of mathematics: number theory, functional analysis, logic.

Mathematics Subject Classifications (2000): Primary: 20M10, 22A15; Secondary: 11B25, 11P81, 22A25, 47D03, 68Q60, 06B35.

Key words: semigroup, topological semigroup, operator semigroup, Ramsey theory, van der Waerden's Theorem on arithmetic progressions, λ-calculus, models of the untyped λ-calculus.

One common thread running through the mathematical work of Bernhard Banaschewski is his persistent desire to approach a mathematical question (or the development of a mathematical concept) by discovering and developing a general framework. In this way he became a pioneer of structure theory. A general structure theory often permits us to see things in a wide context and to find not only a solution to the original problem, but often answers to a whole class of questions, indeed questions which may have seemed quite unrelated at first glance.

1. Sampling a List of Problems

We attempt an illustration of the unifying power of structure theory in this essay. Therefore, without further ado, we begin by considering a variety of rather diverse problems arising in different areas of mathematics.

VAN DER WAERDEN'S THEOREM ON ARITHMETIC PROGRESSIONS

An *arithmetic progression* is a finite sequence of natural numbers of the form $a + nq$, $n = 0, \dots, \ell - 1$, and ℓ is called its *length.* A famous theorem due to van der Waerden states

THEOREM 1. *If* $\mathbb{N} = A_1 \dot{\cup} \cdots \dot{\cup} A_p$ *denotes a finite partition of* $\mathbb{N}$, *then at least one of the sets* A_j *contains arithmetic progressions of arbitrary length.*

The theorem was discovered and proved in the twenties. The proof was long, hard and technically complicated. Only in recent years did systematic proofs become available, notably through contributions by Fürstenberg and later Hindman (see, e.g., [3]). A monograph on such matters by Neil Hindman and Dona Papert has just appeared [4].

OPERATORS ON HILBERT SPACE

We consider a linear contraction $T: \mathcal{H} \to \mathcal{H}$ of a Hilbert space. The following result is very helpful.

THEOREM 2. *There exists an orthogonal projection* P *of* $\mathcal{H}$ *which commutes with* T *and which has the following properties:*

(i) $\mathcal{H}$ *decomposes into an orthogonal direct sum* $\mathcal{H} = \mathcal{H}_1 \oplus \mathcal{H}_2$, $\mathcal{H}_1 = (1-P)\mathcal{H}$ *and* $\mathcal{H}_2 = P\mathcal{H}$ *of* T*-invariant subspaces.*
(ii) *If one writes* $T = T_1 \oplus T_2$ *with* $T_1 = (1-P)T \mid \mathcal{H}_1$ *and* $T_2 = PT \mid \mathcal{H}_2$, *then* $G \stackrel{\text{def}}{=} \overline{\{T_2, T_2^2, \ldots\}}$ *is a compact Abelian group of unitary operators on* $\mathcal{H}_2$, *while the closure* $N \stackrel{\text{def}}{=} \overline{\{T_1, T_1^2 \ldots\}}$ *with respect to the weak operator topology is a compact multiplicatively closed set of operators on* $\mathcal{H}_1$ *containing the zero operator* 0.
(iii) *The function* $\pi: N \times G \to L(\mathcal{H})$ *into the set of bounded operators, defined by* $\pi(s, g)(x) = s(1-P)x \oplus gPx \in \mathcal{H}_1 \oplus \mathcal{H}_2$ *is a faithful representation on* $\mathcal{H}$ *whose image contains* $\{T, T^2, \ldots\}$.

The vectors in $\mathcal{H}_2 = P\mathcal{H}$ are called *almost periodic*, the ones in $\mathcal{H}_1 = (1-P)\mathcal{H}$ *flight vectors*. We note that the compactness of G permits the orthogonal decomposition of $\mathcal{H}_2$ in a Hilbert space orthogonal direct sum of finite dimensional T- and T_2-invariant subspaces.

Decompositions of this sort play an important role in ergodic theorems which describe the behavior of the sequence $M_n \stackrel{\text{def}}{=} \frac{1}{n}(T + T^2 + \cdots + T^n)$. Then $PM_n \mid \mathcal{H}_2$ converges to the operator Q, $Qx = \int_G \pi(1, g)(x)\,dg$ with normalized Haar measure dg on G.

By way of an example let (Ω, μ) be a measure space with a probability measure μ and $\tau: \Omega \to \Omega$ a measure preserving self-map. Then we set $\mathcal{H} = L^2(\Omega; \mu)$ and define $T: \mathcal{H} \to \mathcal{H}$ by $(Tf)(x) = f((\tau(x))$. We have

$$\|Tf\|^2 = \int |(Tf)(x)|^2\,d\mu(x) = \int |f(\tau(x))|^2\,d\mu(x) = \|f\|^2,$$

and thus T is a unitary operator. The preceding discussion then yields a decomposition of $\mathcal{H}$ in the way which we specified.

SEMIGROUPS AND SEMANTICS

The previous examples were from number theory and analysis; the following one belongs to logic and is likely to require some extra exposition. We outline the λ-calculus of Church which originally was an algebraic calculus for basing a foundation of mathematics on the concept of a 'function' rather than that of an 'element' and of a 'set.' So we start with a set V of *variables* and a set C of *constants*. The so-called *type free λ-calculus* has a syntax given by the Backus–Naur form:

$$p ::= c|v|pp|\lambda v.p,$$

where $c \in C$ and $v \in V$.

Let us write Λ for the set of all λ-terms. Mathematicians would think of Λ as a free algebra with a binary operation $(p, q) \mapsto pq$ and a set of unary operations $x \mapsto \lambda v.x$, one for each variable $v \in V$ on which a congruence relation will be defined and factored by the 'reductions' we define below. The following convention saves parentheses:

$$m_1m_2m_3 \cdots m_k = m_1(m_2(m_3(\cdots m_k)) \quad \text{and}$$
$$\lambda v_1 \cdots v_n.m = \lambda v_1.(\lambda v_2.(\cdots \lambda v_n.m) \cdots)).$$

The binary operation $(m, n) \mapsto mn: \Lambda \times \Lambda \to \Lambda$ is called *application*.

The *substitution* of a term q for a variable v occurring freely in the term m will be denoted by $m[q/v]$. We shall apply, as needed, the following *reduction rules*:

(α) $\lambda v.m = \lambda u.m[u/v]$ for u not free in m,
(β) $(\lambda v.m)q = m[q/v]$, and
(η) $\lambda v.(mv) = m$ for v not free in m.

If x and y are variables which do not occur freely in the lambda-terms m and n, then we have $\lambda x.mnx \overset{\alpha}{=} \lambda y.(mnx)[y/x] = \lambda y.mny$. Likewise we have $\lambda x.x \overset{\alpha}{=} \lambda y.x[y/x] = \lambda y.y$.

Therefore α-reduction unambiguously allows the following definition:

DEFINITION 3. The function $\circ: \Lambda \times \Lambda \to \Lambda$, called *composition*, is defined by $m \circ n \overset{\text{def}}{=} \lambda x.mnx$, where x must not occur freely in either m nor in n. The lambda term $I = \lambda x.x$ is independent of the choice of the variables x.

One can show without difficulty that $(\Lambda, \circ)$ is a monoid with identity element I and application is a monoid action on Λ.

$$\text{appl}: \Lambda_\circ \times \Lambda \to \Lambda, \qquad \text{appl}(m, n) = mn.$$

$$\begin{array}{ccc}
\Lambda_\circ \times \Lambda_\circ \times \Lambda & \xrightarrow{\circ \times \mathrm{id}_\Lambda} & \Lambda_\circ \times \Lambda \\
{\scriptstyle \mathrm{id}_{\Lambda_\circ} \times \mathrm{appl}} \downarrow & & \downarrow {\scriptstyle \mathrm{appl}} \\
\Lambda_\circ \times \Lambda & \xrightarrow[\mathrm{appl}]{} & \Lambda.
\end{array}$$

Models

We shall explain somewhat informally what we mean when we say that we have *a model of the type free λ-calculus* on a structure X (e.g., a topological space or an ordered set etc.).* The set of endomorphisms $\mathcal{C}(X, X)$, i.e. the monoid of structure preserving self maps of X yields, in appropriate circumstances, again a structure of the kind we consider; we shall call it X^X.

The monoid $\mathcal{C}(X, X)$ operates on the set $|X|$ underlying the structure X via $(f, x) \mapsto |f|(x)$, where $|f|$ simply denotes the function underlying the endomorphism f.

A 'model' of the λ-calculus will have to reflect the idea of identification of functions and elements. In order to implement such an identification we consider morphisms $p: X \to X^X$. For two sets A and B let $\mathbb{S}(A, B)$ denote the set of all functions $f: A \to B$.

DEFINITION 4. Let X be a structure. We say that

$$(\phi, \psi, p) \in \mathbb{S}\big(\Lambda, \mathcal{C}(X, X)\big) \times \mathbb{S}(\Lambda, |X|) \times \mathcal{C}(X, X^X)$$

is *a basic model of* Λ on X if the following conditions are satsified:

(a) α-β-η reduction applies and

$$(\phi, \psi): \big((\Lambda, \circ), \Lambda\big) \to \big(\mathcal{C}(X, X), |X|\big).$$

is an equivariant morphism of monoid actions, i.e. satisfies

$$\psi(fx) = \phi(f)(\psi(x)).$$

(b) $|p|(\psi(m)) = \phi(m)$ in $\mathcal{C}(X, X)$.

A basic model is *degenerate* if ϕ is constant. (Then (b) implies that $|p|$ is constant, too.)

Nondegenerate models of the untyped λ-calculus have been known for about 25 years. They were found by Dana S. Scott on certain *injective topological spaces* or, stated equivalently, on certain *continuous lattices*. (See, e.g., [2].) The specific concept of a basic model appears in articles by Mislove and myself [5, 6].

All models which have been as observed since then are variants or generalisations of Scott's models. There seems to be a vague impression that the existence of models is a rare occurrence. In classical logic, as for instance in Wittgenstein's *Tractatus Logico-Philosophicus* the logical possibility of forming expressions of the form $f(f)$ was rejected as impossible. In the λ-calculus it is certainly syntactically possible. The models show that it is indeed also possible in mathematical

* In a formally correct description 'a structure' is an object in a Cartesian closed category. Then one can talk precisely about 'endomorphisms' of a structure. For details cf. [5, 6, 9].

universes with functions and elements. Scott's models are all compact T_0-spaces. Abramsky raised the question of whether models on T_2-spaces exist.

THEOREM 5. (Hofmann and Mislove [5]). *There are no nondegenerate compact Hausdorff models of the untyped λ-calculus.*

2. The Common Thread

We have surveyed some very different mathematical problems. Now we claim that there is an intrinsic bond which ties them all together, and that this happens in a hierarchical way that needs explanation.

We claim that the synthesis is achieved by *semigroup* theory in conjunction with *topology*. We shall explain this in detail and give a complete and self-contained treatment in each case.

RIGHT TOPOLOGICAL SEMIGROUPS AND VAN DER WAERDEN'S THEOREM

Let S be a semigroup. We consider S as a discrete space and consider the Stone–Čech compactification $\beta(S)$. We shall write $S \subseteq \beta(S)$. Each function $\lambda_s: S \to S$ extends uniquely to a continuous function $\beta(\lambda_s): \beta(S) \to \beta(S)$. The function $s \mapsto \beta(\lambda_s)(\eta): S \to \beta(S)$ extends uniquely to a continuous function $\rho_\eta^*: \beta(S) \to \beta(S)$. If we set $\xi * \eta = \rho_\eta^*(\xi)$ then we have defined a multiplication $*: \beta(S) \times \beta(S) \to \beta(S)$ for which all right translation ρ_η^* are continuous and on which a left translation with an element s from S is of the form $\beta(\lambda_s)$ and is, therefore, continuous, too. Note $s * t = st$ for $s, t \in S$. From $\beta(\lambda_s) \circ \beta(\lambda_t) = \beta(\lambda_s \circ \lambda_t) = \beta(\lambda_{st})$ we get $s * (t * \eta) = (st) * \eta = (s * t) * \eta = \beta(\lambda_s)(t) * \eta$. Consequently, $s * (\tau * \eta) = \beta(\lambda_s)(\tau) * \eta = (s * \tau) * \eta$, and finally $\sigma * (\tau * \eta) = (\sigma * \tau) * \eta$. Hence $\beta(S)$ is a semigroup and a compact space such that right translations are continuous. This calls for a definition.

DEFINITION 6. A semigroup S with a topology is called *right topological* if $\rho_a: S \to S$, $\rho_a(x) = xa$ is continuous for all $a \in S$.

(Warning: The set of semigroup researchers decomposes into two classes. The members of the other class call these semigroups left topological.)

THEOREM 7. (R. Ellis, [1, Lemma 1]). *Every compact right topological semigroup S contains an idempotent element.*

Proof. Set $\mathcal{C} = \{T \subseteq S : TT \subseteq T,\ \overline{T} = T \neq \emptyset\}$. Then $S \in \mathcal{C}$. With respect to $\supseteq$, the set $\mathcal{C}$ is inductive. Let T be minimal in $\mathcal{C}$ and $e \in T$. Then $Te = \rho_e(T)$ is compact, hence closed and $(Te)(Te) = (TeT)e \subseteq Te$. Thus $Te = T$ by minimality. Now the set $T' \stackrel{\text{def}}{=} \rho_e^{-1}(e) \cap T = \{t \in T : te = e\}$ is closed, and if $te = e$ and $t'e = e$ hold, then $(tt')e = t(t'e) = te = e$ follows. Thus $T' \in \mathcal{C}$, whence $T' \subseteq T$ implies $T' = T$ by the minimality of T. Hence $T = Te = \{e\}$. Thus $ee = e$. □

In this proof we used decisively:

ρ_e is continuous.

In the set $E(S)$ of all idempotents of S we set $e \leq f$ iff $ef = fe = e$.

A subset L of S is called *left ideal* if $SL \subseteq L \neq \emptyset$. A semigroup is called a *left group* if it is of the form $X \times G$ with a set X and a group G such that $(x, g)(y, h) = (x, gh)$.

COROLLARY 8. *In a compact right topological semigroup S there is a minimal left ideal, and each such is a left group.*

Proof. Let $\mathcal{L}$ be the set of all compact left ideals of S. Note $S \in \mathcal{L}$. Let $\mathcal{B}$ denote a filter basis in $\mathcal{L}$. Since all $L \in \mathcal{B}$ are compact $B \overset{\text{def}}{=} \bigcap \mathcal{B}$ is not empty. For $s \in S$ we have $sB \subseteq \bigcap_{L \in \mathcal{B}} sL \subseteq \bigcap_{L \in \mathcal{B}} L = B$. Thus $B \in \mathcal{L}$. Therefore $\mathcal{L}$ is $\supseteq$-inductive and thus possesses a minimal element L. For $y \in L$ the set $\rho_y(L) = Ly \subseteq L$ is a compact left ideal since $SLy \subseteq Ly$. Hence $Ly = L$ by the minimality of L. Thus $L = Ly \subseteq Sy \subseteq L$, whence $Sy = L$.

By Theorem 7 the set $X \overset{\text{def}}{=} E(L)$ is not empty. Let $e \in E(S)$. Then $Se = L$. Thus for $x \in L$ there is an $s \in S$ such that $x = se$, i.e., $xe = see = se = x$. In particular this holds for each $x \in X$. Hence $xy = x$ for $x, y \in X$. Set $G = eSe$ and note $eSe = eL$. Let $g \in G$. Then $Lg = L$ implies $Gg = eLg = eL = G$, i.e., $\rho_g^{(G)}: G \to G$ is surjective. Thus there is an $h \in G$ with $hg = e$. On account of $e(e\ell) = e\ell$ we have that G is a semigroup with a two-sided identity and one-sided inverses and thus is a group. Since $(xg)(yh) = [(xg)y]h = [xg]h$ the function $\mu: X \times G \to L$, $\mu(x, g) = xg$ is a homomorphism if on $X \times G$ the multiplication of a left group is considered.

Moreover, let $\ell \in L$; set $g = e\ell \in eL = G$. Let g' denote the inverse of g in G. Set $x = \ell g'$. Then $x^2 = \ell g'\ell g' = \ell(g'e)\ell g' = \ell g'gg' = \ell g'e = \ell g' = x$. Thus $x \in E(L) = X$ and the function $\nu: L \to X \times G$ is well defined by $\nu(\ell) = (\ell(e\ell)', e\ell)$. Because $\mu\nu(\ell) = \ell(e\ell)'(e\ell) = \ell e = \ell$ and $\nu\mu(x, g) = \nu(xg) = (xg(exg)', exg) = (xg(eg)', eg) = (x(gg'), g) = (x, g)$, the function μ is an isomorphism whose inverse is ν. □

Let us mention in passing (although we shall not use this in the sequel), that the conclusions of Corollary 8 imply the existence of a minimial ideal which is a disjoint union of isomorphic left ideals and thus, in particular, of isomorphic groups.

LEMMA 9. *Let $S = X \times G$ be any left group and C a subsemigroup of S which is a compact right topological semigroup. Then there is a subset $X' \subseteq X$ and a* *subgroup $H \subseteq G$ such that $C = X' \times H$. In particular, if e is an idempotent of S, then $e \in eC$.*

Proof. From the preceding corollary, C has a minimal left ideal L which is a left group. The set $E(L)$ is necessarily of the form $X' \times \{1\}$ with a set $X' \subseteq X$. Let $f = (x', 1) \in E(C)$. Then $fC = fCf$ is the maximal subgroup of L containing L and is of the form $\{x'\} \times H$ with a subgroup H of G. Then $L = E(L)fC = X' \times H$. Let $c = (x, g) \in C$. Then $(x, g) = (x, g)(x', 1) = cf \in L$, whence $x \in X'$. Also $(x', g) = fc \in fC = \{x'\} \times H$. Thus $g \in H$. Therefore $c = (x, g) \in X' \times H = L$. Thus $C = L$.

If now $e \in E(S)$, then $e = (x, 1)$ for some $x \in X$. Then $e = (x, 1) = (x, 1)(x', 1) \in eC$.

We are now ready for a semigroup theoretical proof of van der Waerden's Theorem. We present a purely semigroup theoretical version due to J. D. Lawson [8].

By Corollary 8 we find at least one minimal idempotent e of $\beta\mathbb{N}$. We take an arbitrary natural number $\ell \in \mathbb{N}$ and let $T = \{(a, a + q, a + 2q, \ldots, a + (\ell - 1)q) : a, q \in \mathbb{N}\} \subseteq (\beta\mathbb{N})^\ell$ and $\Delta = \{(x, \ldots, x) \in (\beta N)^\ell : x \in \mathbb{N}\}$. Then $\overline{\Delta} = \{(x, \ldots, x) \in (\beta N)^\ell : x \in \beta\mathbb{N}\}$, whence $\mathbf{e} \stackrel{\text{def}}{=} (e, \ldots, e)$ is contained in $\overline{\Delta}$. We quickly verify that T is a subsemigroup and $T + \Delta \subseteq T$. Since left translations with elements from $\mathbb{N}$ are continuous we have $T * \mathbf{e} \subseteq T * \overline{\Delta} \subseteq \overline{T}$ and $\Delta * \overline{T} \subseteq \overline{T}$. Since right translations are continuous we have $\overline{T} * \mathbf{e} \subseteq \overline{T}$ and $\mathbf{e} * \overline{T} \subseteq \overline{\Delta} * \overline{T} \subseteq \overline{T}$. It then follows that $\mathbf{e} * \overline{T} * \mathbf{e} \subseteq \overline{T} * \mathbf{e} \subseteq \overline{T}$.

Now $\overline{T} * \mathbf{e}$ is a compact subsemigroup of the minimal left ideal $(\beta N)^\ell * \mathbf{e}$ of βN^ℓ, which is a left group by Corollary 8. From Lemma 9 we conclude that $\mathbf{e} \in \mathbf{e} * \overline{T} * \mathbf{e} \subseteq \overline{T}$.

Now suppose that $A_1 \dot{\cup} \cdots \dot{\cup} A_p = \mathbb{N}$ is a partition of $\mathbb{N}$. Then $\beta\mathbb{N} = \overline{A_1} \cup \cdots \cup \overline{A}_p$. Hence there is an $A = A_j$ for some $j = j(e)$ such that $e \in \overline{A}$. Then $\overline{A}$ is an open closed neighborhood of e, and thus $\overline{A}^\ell$ is a neighborhood in $(\beta\mathbb{N})^\ell$ of $\mathbf{e} \in \overline{T}$. Hence $\overline{A}^\ell \cap T \neq \emptyset$. I.e., there is an $(a, a + q, \ldots, a + (\ell - 1)q) \in \overline{A}^\ell \cap T \subseteq \overline{A}^\ell \cap \mathbb{N}^\ell = A^\ell$. Thus we have found a j such that A_j contains arithmetic progressions of length ℓ. $\square$

Note that j depends only on e, which we picked in the beginning. No doubt, the axiom of choice is involved, beginning with Theorem 7 and Corollary 8. Let us conclude this section with the passing remark that no one appears to know anything very precise about the structure of the groups in the minimal ideal of $\beta(\mathbb{N})$.

An excellent survey of semigroups and Ramsey theory appears in the prize winning Master's Thesis of Markus Haase [10].

SEMITOPOLOGICAL SEMIGROUPS AND OPERATORS

EXAMPLE 10. Let $\mathcal{H}$ be a Hilbert space and X its closed unit ball. Then X is weakly compact and the semigroup $S(X)$ of all self maps of X is compact and right topological with the topology of pointwise convergence. We endow the multiplicative semigroup $L(\mathcal{H})$ of all bounded operators with the topology of pointwise convergence on $\mathcal{H}$, equipped with the weak topology. Let S_1 denote the submonoid $\{T \in L(\mathcal{H}) : \|T\| \leq 1\}$. (On the group $U(\mathcal{H}) \subseteq S$ of isometries the weak operator topology agrees with the strong operator topology!) The function $s \mapsto s|X: S_1 \to S(X)$ is a semigroup injection and a homeomorphism onto the image which is closed. Thus S_1 is a right topological semigroup. The function $t \mapsto (stx|y) = (tx|s^*y): S_1 \to \mathbb{C}$ is continuous for all $s \in S_1$, $x, y \in \mathcal{H}$. Hence $\lambda_s = (t \mapsto st): S_1 \to S_1$ is continuous for all $s \in S_1$.

DEFINITION 11. A semigroup S with a topology is called *semitopological* if both λ_s and ρ_t are continuous for all $s, t \in S$.

In a right topological semigroup S the *topological center* $Z_{\rm top}(S)$ is the set of all $s \in S$ such that the left translations λ_s are continuous, too. Thus S is semitopological if and only if $Z_{\rm top}(S) = S$. Note that $Z_{\rm top}(S)$ is always a subsemigroup.

If, e.g., X is a compact Hausdorff space and $C(X, X)$ is the subsemigroup of X^X of all continuous selfmaps of X, then $C(X, X) \subseteq Z_{\rm top}(S)$. Indeed, for $f \in C(X, X)$ and for all $x \in X$ the function $g \mapsto f(g(x)): X^X \to X$ is continuous.

For $A \subseteq S$ we call $Z(A) = \{s \in S : (\forall a \in A) as = sa\}$ the *centralizer* of A in S. It is always a subsemigroup. A subset $A \subseteq S$ is called *commutative* if $A \subseteq Z(A)$.

In semitopological semigroups circumstances are much improved in comparison with right topological semigroups.

LEMMA 12. *Let S be a right topological semigroup and $A \subseteq S$.*

(i) *If A is a subsemigroup and is contained in $Z_{\rm top}(S)$, then $\overline{A}$ is a subsemigroup.*
(ii) *If $\overline{A} \subseteq Z_{\rm top}(S)$ and A is a commutative set, then $\overline{A}$ is also a commutative set.*
(iii) *In a* semitopological *semigroup the closure of a subsemigroup is a subsemigroup and the closure of a commutative set is a commutative set. In particular, the closure of a commutative subsemigroup is again a commutative semigroup.*

Proof. (i) From $AA \subseteq A$ we get $A \subseteq \lambda_a^{-1}(A) \subseteq \lambda_a^{-1}(\overline{A})$ for all $a \in A$. Since $a \in Z_{\rm top}(A)$, the set $\lambda_a^{-1}(\overline{A})$ is closed, and thus $\overline{A} \subseteq \lambda_a^{-1}(\overline{A})$ for all $a \in A$, i.e., $A\overline{A} \subseteq \overline{A}$. In other words, $A \subseteq \rho_a^{-1}(\overline{A})$ for all $a \in \overline{A}$. Since S is right topological, $\overline{A} \subseteq \rho_a^{-1}(\overline{A})$ for all $a \in \overline{A}$. Hence $\overline{A}.\overline{A} \subseteq \overline{A}$, which we had to show.

(ii) For all $s \in Z_{\rm top}(S)$ the function $m_s: S \to S \times S$, $m_s(t) = (st, ts) = (\lambda_s(t), \rho_s(t))$ is continuous. Since S is Hausdorff, the diagonal Δ_S of $S \times S$ is closed. Hence $m_s^{-1}\Delta_S$ is closed. But this set is the centralizer $Z(\{s\})$ of s. Thus the intersection $Z(A) = \bigcap_{s \in A} m_s^{-1}\Delta_S$, i.e., the centralizer of any set A in $Z_{\rm top}(S)$ is closed. If A is Abelian, then $A \subseteq Z(A)$; consequently, $\overline{A} \subseteq Z(A)$ and thus also $A \subseteq Z(\overline{A})$. Then $\overline{A} \subseteq Z_{\rm top}(S)$ implies $\overline{A} \subseteq Z(\overline{A})$ and $\overline{A}$ is Abelian.

(iii) follows from (i) and (ii). □

The proof of (iii) used essentially

λ_s & ρ_s are continuous.

For example, the additive semigroup $\mathbb{N}$ is commutative while the right topological semigroup $\beta(\mathbb{N})$ is horribly noncommutative.

Now let $T: \mathcal{H} \to \mathcal{H}$ be a linear contraction of a Hilbert space $\mathcal{H}$, i.e. T is a member of the compact semitopological semigroup $\{\phi \in L(\mathcal{H}) \mid \|\phi\| \leq 1\}$. By Lemma 12(iii) above, $\langle T \rangle \stackrel{\text{def}}{=} \overline{\{T, T^2, T^3, \ldots\}}$ is a compact commutative subsemigroup of S. By Corollary 8, there is minimal left ideal $L \subseteq \langle T \rangle$ and a projection $P \in E(L)$ such that L is isomorphic to $E(L) \times P\langle T \rangle P$. The commutativity of $\langle T \rangle$ implies that the left zero semigroup $E(L)$ is singleton, and $L = P\langle T \rangle = \langle T \rangle P = P\langle T \rangle P$ is the minimal ideal. Furthermore, $G = P\langle T \rangle = \langle T \rangle P$ is a semitopological compact group since $s \mapsto \rho_P(s) = sP$ is continuous. A Baire category argument shows that G is a topological group. As a linear contraction, the projection P is orthogonal: Indeed let e be a vector from $((1-P)\mathcal{H})^{\perp}$. Then $Pe = e \oplus (P-1)e$. Since this is an orthogonal decomposition and P is a contraction, we have $\|e\|^2 \geq \|Pe\|^2 = \|e\|^2 + \|(P-1)e\|^2$. This implies $(P-1)e = 0$, hence $Pe = e$. Thus $((1-P)\mathcal{H})^{\perp} \subseteq P\mathcal{H}$. We write $P\mathcal{H} = \mathcal{K} \oplus ((1-P)\mathcal{H})^{\perp}$ with $\mathcal{K} \perp (1-P)\mathcal{H}$. From $\mathcal{H} = P\mathcal{H} \oplus (1-P)\mathcal{H} = \mathcal{K} \oplus ((1-P)\mathcal{H})^{\perp} \oplus (1-P)\mathcal{H}) = \mathcal{K} \oplus \mathcal{H}$ we get $\mathcal{K} = \{0\}$. Hence $(1-P)\mathcal{H} \perp \mathcal{H}$, as asserted.

TOPOLOGICAL SEMIGROUPS AND THE LAMBDA CALCULUS

The first part of the following discourse is pure semigroup algebra.

The Bicyclic Semigroup

In a semigroup with 1 consider elements p and q with $pq = 1$. Are p and q inverses of each other? Let $\mathcal{H}$ denote the Hilbert space ℓ^2 of all square summable sequences $x = (x_n)_{n \in \mathbb{N}}$ and let the operators $P, Q: \mathcal{H} \to \mathcal{H}$ be given by $Px = (x_2, x_3, \ldots)$ und $Qx = (0, x_1, x_2, \ldots)$ (*unilateral shift operators*). Then $PQx = x$ and $QPx = (0, x_2, x_3, \ldots)$: Furthermore $(Px \mid y) = x_2y_1 + x_3y_2 + \cdots = (x \mid Qy)$, i.e., $Q = P^*$. Thus there are very natural circumstances in which we observe semigroup elements p und q which satisfy $pq = 1$ but which are not inverses of each other.

The monoid which is generated freely by elements p and q subject only to the relation $pq = 1$ consists of the elements $q^m p^n$, $m, n \in \{0, 1, 2, 3, \ldots\}$ with a formula for the multiplication which results from the cancellation rule $pq = 1$. There are concrete manifestations of this semigroup. Let $\mathbb{Z}$ be the additive group of integers and $\mathbf{Z}$ the semigroup of integers with the multiplication $m, n \mapsto \max\{m, n\}$. The group $\mathbb{Z}$ operates on $\mathbf{Z}$ as a group of automorphisms via $(g, m) \mapsto -g +$

$m: \mathbb{Z} \times \mathbf{Z} \to \mathbf{Z}$. Hence there is a semidirect product $\mathbf{Z} \rtimes \mathbb{Z}$ with a semigroup multiplication

$$(m, g)(n, h) = (\max\{m, (-g + n)\}, g + h).$$

The subsemigroup $B(p, q)$ generated by $q = (1, -1)$ and $p = (0, 1)$ consisting of all (m, g) with $m \geq -g$ has exactly the desired properties. The elements $\mathbf{1} = (0, 0), \ldots, (n, 0) = q^n p^n$, $n = 1, 2, \ldots$ are the idempotents, which form a descending chain in the sense of the order on idempotents. The map $(m, g) \mapsto g: B(p, q) \to \mathbb{Z}$ is a surjective homomorphism.

DEFINITION 13. The semigroup $B(p, q)$ (and any semigroup isomorphic to it) is called *bicyclic.*

LEMMA 14. *The only* proper *homomorphic images of the bicyclic semigroups are* $\mathbb{Z}$ *and its homomorphic images. In particular, they are all groups.*

Proof. Let $\phi: B \to H$ denote a surjective homomorphism failing to be injective. We set $x = \phi(p)$ and $y = \phi(q)$. Then H is generated by x and y and $xy = \phi(1) = 1$. Now let $y^a x^b = \phi(q^a p^b) = \phi(q^u p^v) = y^u x^v$ with $a \neq u$ or $b \neq v$. We shall show that H is a group. For this purpose it suffices to show that x (respectively, y) is invertible; then x and y are necessarily inverses of each other. W.l.o.g. let $u = a + h$ with $0 < h$. Then $x^b = x^a y^a x^b = x^a y^u x^v = y^h x^v$.

If $v = b + k$, then multiplication on the right by y^b yields the relation $1 = y^h x^k$. Then y^h has an inverse on the left x^h, and an inverse on the right x^k. Hence y^h has an inverse z in H. Thus $(zy^{h-1})y = 1 = y(y^{h-1}x)$. Therefore y is invertible and the assertion is proved in this case.

If $b = v + k$, the multiplication on the right by y^v yields $x^k = y^h$. Now multiplication on the left by x^h produces $x^{k+h} = 1$. Hence x is contained in a subgroup and again this suffices for a proof of the assertion.

But if H is a group, then it is generated by $x = y^{-1}$ and thus it is cyclic. The proof of the lemma is complete. □

LEMMA 15. *If* $B(p, q)$ *is a submonoid of a monoid* S, *then* $Sp = S$.

Proof. The right translation ρ_p is surjective because of $\rho_p \circ \rho_q = \rho_{pq} = \rho_1 = \mathrm{id}_S$. Hence $S = \rho_p(S) = Sp$. □

Since the work of Lyapin and his school in St. Petersburg in the forties a highly developed algebraic theory exists around the bicyclic semigroup. The bicyclic semigroup is an important building block in semigroup theory. We have observed that it occurs, e.g., as a submonoid of the unit ball of bounded operators on a Hilbert space, a compact semitopological semigroup.

If P and Q are self maps of a set X and if $PQ = \mathrm{id}_X$, then P is surjective and Q is injective. Hence the semigroup of endomorphisms of a finite dimensional vector space does not contain a genuine bicyclic semigroup, since here injective and surjective endomorphisms are automorphisms.

There is, however, another general circumstance which excludes the existence of bicyclic subsemigroups, and we shall address this case presently.

Compact Topological Semigroups

DEFINITION 16. A semigroup S with a topology is said to be *topological* if multiplication $S \times S \to S$ is continuous.

In a semigroup with 1 we set $R(1) = \{x \in S : (\exists y \in S) \quad xy = 1\}$ and $H(1) = H_S(1) = \{x \in S : (\exists y \in S) \quad xy = 1 = yx\}$. Then $H_S(1) = H(1)$ is the group of all invertible elements.

LEMMA 17. *Let S be a compact topological semigroup with* 1. *Then*

(i) $R(1)$ *is closed, and*
(ii) $R(1) = H(1)$.

Proof. (i) Set $R = \{(x, y) \in S \times S : xy = 1\}$. The projection $\mathrm{pr}_1 : S \times S \to S$ onto the left factor maps R onto $R(1)$. For a proof of (i) it is sufficient to verify the closedness of R in $S \times S$. For then R is compact and then so is its continuous image $R(1)$. The function $m: S \times S \to S, m(x, z) = xy$ is continuous. As a consequence, $R = m^{-1}(\{1\})$ is closed.

(ii) Let $x \in R(1)$. Then there is a y such that $(x, y) \in R$. In $S \times S$ we set $A = \{(x, y), (x, y)^2, (x, y)^3, \ldots\}$. Then A is a compact topological Abelian semigroup by Lemma 13. By Corollary 8 there exists a minimal idempotent (e, e') in A. Since $xy = 1$ implies $x^n y^n = 1$ we have $A \subseteq R$, as R is closed. Thus $ee' = m(e, e') = 1$. Then $e = e1 = eee' = ee' = ee'e' = 1e' = e'$ and we conclude $e = ee = ee' = 1$. Now $A = A(1, 1) = A(e, e)$ is a group, therefore contained in $H_{S \times S}(1, 1) = H_S(1) \times H_S(1)$. Hence $x = \mathrm{pr}_1(x, y)$ contained in $H(1)$. □

The crucial property of this proof was

$$\boxed{m: S \times S \to S \text{ is continuous.}}$$

THEOREM 18. *If $f: B(p, q) \to S$ is an algebraic homomorphism of a bicyclic semigroup in a compact topological semigroup, then* $\mathrm{im} f$ *and* $\overline{\mathrm{im} f}$ *are groups.*

Proof. By Lemma 17(i), the set $\overline{f(B(p, q))}$ is a compact subsemigroup; we may assume $S = \overline{f(B(p, q))}$ and $f(1) = 1$. Let $x = f(p)$ and $y = f(q)$. Then $xy = f(pq) = 1$. Now $x \in R(1) = H(1)$ by Lemma 17. The relation $xx^{-1} = 1 = xy$ implies upon multiplication with x^{-1} on the left side that $y = x^{-1} \in H(1)$. Hence $f(B(p, q))$ is a dense subgroup of $H(1)$. The assertion follows. □

This result is classical and has been known since the fifties (cf. [7]). We will apply it to a proof of the impossibility of certain models of the untyped λ-calculus. But before we do this we need a crash course in some basic aspects of the λ-calculus and its models.

Back to the λ-Calculus

For a basic model (ϕ, ψ, p) on a structure X the set $\phi(\Lambda_\circ)$ is a submonoid of $\mathcal{C}(X, X)$ and therefore acts in a natural way on $|X|$. Further, $\phi(\Lambda) = p(\psi(\Lambda)) \subseteq p(|X|)$, and equality holds if ψ is surjective. In particular we have $\text{card}(\phi(\Lambda)) \leq \text{card}|X|$. If the model is *extensional*, i.e., the function $\phi\colon \Lambda \to X^X$ is surjective, then we claim that $\psi\colon \Lambda \to |X|$ is surjective, and and, consequently,

$$\text{card}(\mathcal{C}(X, X)) \leq \text{card}|X| \leq \text{card}(\Lambda).$$

Indeed if $x \in |X|$ and $\kappa\colon X \to X^X$ is the morphism, which is defined by $|\kappa|(x)(y) = x$, then the surjectivity of ϕ entails the existence of a term $m \in \Lambda$ such that $\phi(m) = |\kappa|(x)$. Thus $\psi(mI) = \phi(m)(\psi(I)) = |\kappa|(x)(\psi(I)) = x$.

If one considers a set X without *additional* structure as a 'structure' then according to Cantor's Lemma there cannot be any surjection $X \to 2^X$. Hence on a *set* with at least two elements there is no surjective function $X \to X^X$, and thus no extensional basic model. In the domain of pure set theory we have justified the vague prejudice that we have against the use of expressions of the form $f(f)$.

We need two particular lambda terms without constants, so-called *combinators*.

The Constant-Combinator

With the aid of α-reduction we define unambiguously the following special lambda-term:

$$K = \lambda xy.x, \quad x, y \text{ variable}.$$

The following reduction is easily verified:

$$(Km)n \overset{\beta}{=} m. \tag{\#}$$

The combinator K is called *K-combinator* or *constant combinator*. The combinator $I = \lambda x.x$ is the *unit combinator*.

For the following arguments we again assume $\alpha - \beta - \eta$-reduction.

THEOREM 19. *$\Lambda_\circ$ has minimal two-sided ideal $M(\Lambda_\circ)$ consisting of left zeros. The function $m \mapsto Km\colon \Lambda_\circ \to \Lambda_\circ$ satisfies*

$$(\forall m, n \in \Lambda) \quad m \circ Kn = K(mn) \tag{1}$$

and has image $K\Lambda = M(\Lambda_\circ) = \Lambda \circ (KI)$.

Proof. (i) $(Km \circ n)x \overset{\alpha\beta}{=} (Km)nx = m = (Km)x$ by (#). Then $\lambda x.(Km \circ n)x = \lambda x.(Km)x$ and η-reduction yields $Km \circ n = Km$ for all $m, n \in \Lambda$. Thus all elements of Km are left zeros in $\Lambda_\circ$. If a semigroup S has at least one left zero ℓ, then it also has a unique minimal left ideal $S\ell$, which happens to be the minimal (two sided) ideal $M(S)$ of S, and each element of $M(S)$ is a left zero.

(ii) We have $(m \circ Kn)x \overset{\alpha\beta}{=} m(Kn)x \overset{\beta}{=} mn \overset{\beta}{=} K(mn)x$. Hence $\lambda x.(m \circ Kn)x = \lambda x.(Kmn)x$, and η-reduction yields $m \circ Kn \overset{\alpha\beta\eta}{=} Kmn$. The unique minimal left ideal $M(\Lambda_\circ)$, which has the form $\Lambda \circ Kn$ for any lambda-term n, e.g., for $n = I$, is contained in the set $K\Lambda$, which consists of left zeros of $\Lambda_\circ$ by (i). Hence it is contained in $M(\Lambda_\circ)$. Thus $K(\Lambda) = M(\Lambda_\circ)$. □

The Paradoxical Combinator

Using α-reduction we define the following combinator:

$$Y \overset{\text{def}}{=} \lambda f.(\lambda v.f(vv))(\lambda v.f(vv)), \quad v,\ f \text{ variable.}$$

The lambda-term $Y \in \Lambda$ is called *the paradoxical combinator*. It satisfies the reduction

$$mYm = Ym, \tag{\#\#}$$

which is verified with the aid of β-reduction.

A Bicyclic Monoid in the λ-Calculus

LEMMA 20. *Via β-reduction we obtain*

$$(\forall m \in \Lambda) \quad m \overset{\beta}{=} YKm, \tag{2}$$

and if α-η-reduction is admitted, too,

$$Y \circ K \overset{\alpha\beta\eta}{=} I. \tag{3}$$

Proof. By (#) and (##) we compute

$$Im \overset{\beta}{=} m \overset{\beta}{=} (Km)(YKm) \overset{\beta}{=} YKm, \tag{†}$$

hence (2). Further, (†) and (1) imply the reduction $I \overset{\eta}{=} \lambda x.Ix = \lambda x.YKx \overset{\alpha\beta}{=} \lambda x.(Y \circ K)x \overset{\eta}{=} Y \circ K$. □

Now let $\langle Y, K\rangle$ denote the monoid which is generated in $\Lambda_\circ$ by Y und K. By (2) we have a surjective morphism $\pi\colon B(p, q) \to \langle Y, K\rangle$ such that $\pi(p) = Y$ and $\pi(q) = K$.

The following is a purely algebraic property of the λ-calculus.

THEOREM 21. *Assume that α-β-reduction applies. Then π is an isomorphism. I.e., $\langle Y, K\rangle$ is a genuinely bicyclic submonoid of $\Lambda_\circ$.*

Proof. By Lemma 14 it suffices to show $K \circ Y \neq I$. Suppose $K \circ Y = I$. Then we would have $m \overset{\beta}{=} Im = (K \circ Y)m \overset{\alpha\beta}{=} KYm$ for all m. Thus $n \overset{\beta}{=} In = [K(YI)]n \overset{\beta}{=} YI$ for all n, and this yields a contradition for any λ-term $n \neq YI$. □

Equivariant Morphisms of Monoid Actions

If X a set then the set X^X of all self-maps of X is a monoid with respect to composition.

DEFINITION 22. Now let (S_j, X_j), $j = 1, 2$ be two pairs of monoids S_j which operate on the sets X_j. A pair (ϕ, ψ) of maps $\phi: S_1 \to S_2$ and $\psi: X_1 \to X_2$ is called an *equivariant morphism*

$$(\phi, \psi): (S_1, X_1) \to (S_2, X_2), \quad \text{if} \quad \psi(sx) = \phi(s)\psi(x)$$

holds and $\phi: S_1 \to S_2$ is a monoid morphism. An equivariant morphism (ϕ, ψ): $(S, X_1) \to ((X_2)^{X_2}, X_2)$ is called an equivariant *representation* of (S, X_1) on X_2.

Let us now assume that $(\phi, \psi): (S, X_1) \to ((X_2)^{X_2}, X_2)$ is an equivariant representation. We set $X' = \psi(X_1)$ and $S' = \phi(S)$. Then $S'(X') \subseteq X'$ and the functions $\psi': X_1 \to X'$, $\psi'(x) = \psi(x)$ and $\phi': S \to X'^{X'}$, $\phi'(s) = \phi(s)|X'$ define an equivariant morphism $(\phi', \psi'): (S, X_1) \to (S', X')$ with a surjective component ψ' and $\phi'(S) = S'$. For an equivariant represention $(\phi, \psi): (S, X_1) \to ((X_2)^{X_2}, X_2)$ it is therefore no serious loss of generality to assume that ψ is surjective.

THEOREM 23. *We assume α-β-η-reduction and assume further that*

$$(\phi, \psi): (\Lambda_\circ, \Lambda) \to \big((X^X, \circ), X\big)$$

is an equivariant representation on X. Then

(i) $\phi(\lambda x.m)\psi(n) = \psi(m[n/x])$,

(ii) $\phi(Km)(x) = \psi(m)$ *for all $x \in \psi(\Lambda)$, i.e., the self-map $\phi(Km)$ of X has, at any rate on $\psi(\Lambda)$, the constant value $\psi(m)$.*

(iii) $\phi(m)(\psi(Ym)) = \psi(Ym)$ *for all $m \in \Lambda$, i.e., $\psi(Ym)$ is a fixed point of the self-map $\phi(m)$ of X.*

(iv) $\phi(Y)\phi(K) = \mathrm{id}_X$. *In particular, $\phi(Y)$ is surjective and $\phi(K)$ is injective.*

(v) *If ϕ is not injective on $\langle Y, K\rangle$, then $\psi(\Lambda) = \psi(M(\Lambda_\circ))$.*

Proof. (i) The equivariance implies at once

$$\phi(\lambda x.m)\psi(n) = \psi((\lambda x.m)n) \stackrel{\beta}{=} \psi(m[n/x]).$$

(ii) $\phi(Km)(\psi(n))) = \psi((Km)n) = \psi(n)$ because of equivariance and of (#).

(iii) $\phi(m)(\psi(Ym)) = \psi(m(Ym)) = \psi(Ym)$ because of equivariance and of (##).

(iv) This is clear from Lemma 20(3) because ϕ is a monoid morphism.

(v) The monoid $\langle K, Y\rangle$ is bicyclic by Theorem 21. If ϕ is not injective on $\langle Y, K\rangle$, then $\phi(\langle Y, K\rangle)$ is a group with identityby $\phi(I)$ by Lemma 14, while $\phi(K)$ and $\phi(Y)$ are inverses of each other. Hence $\phi(Y)^{-1}\psi(\Lambda) = \psi(\Lambda)$. For each λ-term m there is therefore a λ-term m' such that $\phi(Y)^{-1}\psi(m') = \psi(m)$. Hence

$\psi(Km') = \phi(K)(\psi(m')) = \phi(K)\phi(Y)(\psi(m)) = \psi(m)$. In view of Theorem 19, this is the claim. □

We now recognize the bicyclic semigroup as one typical obstruction to the construction of models. This statement is made more precise in the following lemma.

LEMMA 24. *If (ϕ, ψ, p) is a basic model on a structure X and if $\phi \mid \langle Y, K\rangle$ is not injective, then ϕ and p are constant.*

Proof. By Theorem 23(v) we have $\psi(\Lambda) = \psi(M(\Lambda_\circ))$. The monoid $\phi(\Lambda)$ therefore is now equal to $p\psi(\Lambda) = p\psi(M(\Lambda_\circ)) = \phi(M(\Lambda_\circ))$. By Theorem 19 consists of left zeros. A monoid consisting of left zeros is singleton. □

We shall now implement this observation in a proof of the impossibility of compact Hausdorff models of the λ-calculus.

Impossibility of Hausdorff Models

Let X be a compact Hausdorff space. Then $\mathcal{C}(X, X)$, the set of all continuous self-maps gives a topological monoid X^X when endowed with the topology of uniform convergence. A basic model on X therefore consists of a semigroup morphism $\phi: \Lambda_\circ \to \mathcal{C}(X, X)$, of a function $\psi: \Lambda \to |X|$ and a continuous map $p: X \to X^X$ with $\psi(mn) = \phi(m)\psi(n)$ and $p(\psi(m)) = \phi(m)$. Because of the compactness of X and the continuity of p, the set $p(X)$ is a compact subspace of X^X and thus $S \stackrel{\text{def}}{=} \overline{\phi(\Lambda_\circ)}$ is a compact topological semigroup.

With the semigroup theoretical tools we now have the means for the following proof for the impossibility of models living on compact Hausdorff spaces:

THEOREM 25. (Impossibility of compact Hausdorff λ-models). *Assume α-β-reduction. If (ϕ, ψ, p) is a basic model on a compact Hausdorff space, then ϕ and ψ are constant.*

Proof. We noted that $S = \overline{\phi(\Lambda_\circ)}$ is compact topological semigroup. The monoid $\langle K, Y\rangle$ is bicyclic by Theorem 21. Then $\phi \mid \langle K, Y\rangle$ cannot be injective by Theorem 18. By Lemma 24, the functions ϕ and p are constant. □

One conjectures that basic models even on Hausdorff topological spaces with no further restriction are impossible. For normal spaces containing a nondegenerate arc this was shown by Mislove and myself [6]. Mislove has shown [9] that the proof works to exclude the possibility of models of more general forms of the untyped lambda calculus in which, e.g., the constant combinator K is not available. At the moment it is not clear whether semigroup theory provides the right tools to show that all Hausdorff basic models are degenerate.

References

1. Ellis, R.: Distal transformation groups, *Pacific J. Math.* **8** (1958), 401–405.
2. Gierz, G., Hofmann, K. H., Keimel, K., Lawson, J. D., Mislove, M. W., and Scott, D. S.: *A Compendium of Continuous Lattices*, Springer-Verlag, Heidelberg, 1980.
3. Hindman, N.: The semigroup $\beta\mathbb{N}$ and its applications to number theory, in K. H. Hofmann, J. D. Lawson, and J. S. Pym (eds), *The Analytical and Topological Theory of Semigroups*, De Gruyter, Berlin, 1990, pp. 347–360.
4. Hindman, N. and Papert, D.: *Algebra in the Stone–Čech Compactification*, De Gruyter, Berlin, 1998.
5. Hofmann, K. H. and Mislove, M. W.: All compact Hausdorff models are degenerate, *Fund. Inform.* **22** (1995), 23–52.
6. Hofmann, K. H. and Mislove, M. W.: Principles underlying the degeneracy of topological models of the untyped lambda calculus, in K. H. Hofmann and M. W. Mislove (eds), *Semigroup Theory and Its Applications*, Cambridge Univ. Press, 1996, pp. 123–155.
7. Hofmann, K. H. and Mostert, P. S.: *Elements of Compact Semigroups*, Charles E. Merrill Publishers, Columbus, OH, 1966.
8. Lawson, J. D.: A purely semigroup theoretical proof of van der Waerden's Theorem, Personal Communication, December 1995.
9. Mislove, M. W.: Topology, domain theory, and theoretical computer science, Elsevier Preprint, September 1996.
10. Haase, M.: Ramsey Theorie und die Stone-Čech-Kompaktifizierung diskreter Halbgruppen, Diplom Thesis, University of Tübingen, 1997.

Applied Categorical Structures **8:** 161–173, 2000.
G. Brümmer & C. Gilmour (eds), Papers in Honour of Bernhard Banaschewski.

Coreflectivity of E-Monads and Algebraic Hulls

Dedicated to Bernhard Banaschewski on his 70th birthday

GÜNTHER RICHTER
Fakultät für Mathematik, Universität Bielefeld, Universitätsstrasse 25, D-33615 Bielefeld, Germany
e-mail: heidrich@mathematik.uni-bielefeld.de

(Received: 7 October 1996; accepted: 26 November 1997)

Abstract. Let $\boldsymbol{X}$ be a category with a given $(\boldsymbol{E}, \boldsymbol{M})$-factorization structure for morphisms, $\boldsymbol{M} \subseteq Mono\boldsymbol{X}$. In general, an arbitrary endofunctor T of $\boldsymbol{X}$ fails badly to preserve the $\boldsymbol{E}$-class. If T carries a monad structure, then $T(\boldsymbol{E}) \subseteq \boldsymbol{E}$ implies that the corresponding category of Eilenberg–Moore-algebras admits $(\boldsymbol{E}, \boldsymbol{M})$-factorizations and vice versa. In order to get T as close as possible to this nice algebraic behaviour, a couniversal modification $\widehat{T} \hookrightarrow T$ with $\widehat{T}(\boldsymbol{E}) \subseteq \boldsymbol{E}$ is constructed in two different ways using mild and natural assumptions on $\boldsymbol{E}$ and $\boldsymbol{M}$, respectively. $\widehat{T}$ inherits its monad structure from T. In case of $T = UF$, $F \dashv U$, the Eilenberg–Moore-category of $\widehat{T}$ contains a universal $(\boldsymbol{E}, \mathcal{M})$-algebraic hull (completion) of U [2, 3]. There are further applications to varietal hulls [4] and to function spaces.

Mathematics Subject Classifications (2000): 18A22, 18A32, 18C05, 18C15, 54B30, 54C35.

Key words: monads, factorizations of (homo)morphisms, varietal and algebraic hulls, function spaces.

Introduction

The induced multiplication on a quotient of a topological semigroup may fail to be continuous with respect to the quotient topology [15]. Consequently, the monad T: **Top** $\to$ **Top** of topological semigroups on the category **Top** of topological spaces doesn't preserve quotients. The image of a functor may fail to be a subcategory. Consequently, the monad T: **Grph** $\to$ **Grph** of small categories (given by the underlying graph functor and the categories of paths defining its left adjoint) doesn't preserve surjections. Multiplication by a fixed space X defines an endofunctor T of **Top** which fails to preserve quotients, in general [6]. Nevertheless, there is always a couniversal modification $\widehat{T}$ of T which behaves well with respect to the class of epimorphisms in question.

To this end consider an arbitrary endofunctor T of a category $\boldsymbol{X}$ with a given $(\boldsymbol{E}, \boldsymbol{M})$-factorization structure for morphisms [1], $\boldsymbol{M} \subseteq Mono\boldsymbol{X}$, together with a natural transformation $\kappa\colon \widehat{T} \to T$ such that $\widehat{T}(\boldsymbol{E}) \subseteq \boldsymbol{E}$ and every natural transformation $\nu\colon \widetilde{T} \to T$, $\widetilde{T}(\boldsymbol{E}) \subseteq \boldsymbol{E}$, factorizes uniquely via κ

$$\begin{array}{ccc} T & \xleftarrow{\ \nu\ } & \widetilde{T} \\ {\scriptstyle\kappa}\uparrow & \swarrow{\scriptstyle \exists!\,\widehat{\nu}} & \\ \widehat{T} & & \end{array} \quad , \qquad \kappa \circ \widehat{\nu} = \nu.$$

Then κ is necessarily pointwise in $\boldsymbol{M}$, because the $(\boldsymbol{E}, \boldsymbol{M})$-factorization of κ would yield some $\widetilde{T}$ as above.

PROPOSITION 0.1 *If (T, η, μ) is a monad on $\boldsymbol{X}$ then $\widehat{T}$ carries a unique monad structure $\widehat{\eta}$, $\widehat{\mu}$ such that κ becomes a monad morphism. Moreover, if ν as above is a monad morphism then the same holds for $\widehat{\nu}$.*

Proof. The identity $\mathrm{id}\boldsymbol{X}\colon \boldsymbol{X} \to \boldsymbol{X}$ preserves $\boldsymbol{E}$. Consequently, $\eta\colon \mathrm{id}\boldsymbol{X} \to T$ defines $\widehat{\eta}\colon \mathrm{id}\boldsymbol{X} \to \widehat{T}$ with $\eta = \kappa \circ \widehat{\eta}$. Similarly, $\widehat{\mu}$ is given by the following commutative diagram

$$\begin{array}{ccccc} T & \xleftarrow{\ \mu\ } & T^2 & \xleftarrow{\ \kappa^2\ } & \widehat{T}^2 \\ {\scriptstyle\kappa}\uparrow & & & \swarrow{\scriptstyle\widehat{\mu}} & \\ \widehat{T} & & & & \end{array} \quad ,$$

where $\kappa^2 = (\kappa T) \circ (\widehat{T}\kappa) = (T\kappa) \circ (\kappa\widehat{T})$ denotes the so called 'vertical composition' of κ with itself [16] (or [1, 6A]). The monad properties of $(\widehat{T}, \widehat{\eta}, \widehat{\mu})$ follow immediately from the given ones on (T, η, μ) using $\kappa X \in \boldsymbol{M} \subseteq \mathit{Mono}\boldsymbol{X}$ for any object X. The same applies for $\widehat{\nu}$ being a monad morphism if ν does. □

Using the wellknown contravariant bijection from monad morphisms to concrete functors between the corresponding Eilenberg–Moore-categories [16], κ yields a universal concrete functor

$$K\colon \boldsymbol{X}^T \to \boldsymbol{X}^{\widehat{T}}$$

and vice versa. Therefore, the universal $(\boldsymbol{E}, \mathcal{M})$-algebraic completion (hull) constructed by H. Bargenda [2, 3] (with respect to a conglomerate $\mathcal{M}$ of mono-sources) in the presence of enough $\boldsymbol{E}$-projectives in $\boldsymbol{X}$ yields a couniversal modification κ as above. The converse will be exposed in the last section 4 of this paper. Categorical terminology follows [1].

The construction of κ in Section 1 needs a remainder of Bargenda's projectivity assumption, the *square condition* on $\boldsymbol{E}$, i.e. for any $\boldsymbol{X}$-morphism f and $e \in \boldsymbol{E}$ with common codomain there is a commutative square

$$\begin{array}{ccc} \cdot & \xrightarrow{\ g\ } & \cdot \\ {\scriptstyle \boldsymbol{E} \ni d}\downarrow & & \downarrow{\scriptstyle e} \\ \cdot & \xrightarrow[\ f\ ]{} & \cdot \end{array} .$$

This is much more general but it still excludes interesting topological categories $\boldsymbol{X}$ where $\boldsymbol{E}$ is the class of quotient maps.

This gap will be filled in Section 2 where existence of κ is proved without any additional assumption on $\boldsymbol{E}$. Consequently there are topological examples including function spaces and varietal hulls [4] for categories of topological algebras like topological semigroups in Section 3.

1. An Internal Construction

Every $e \in \boldsymbol{E}$ with an $\boldsymbol{E}$-projective codomain P is a retract, hence $Te \in \boldsymbol{E}$ in this case, because $\boldsymbol{E}$ contains all retracts. First let us assume that $\boldsymbol{X}$ has enough $\boldsymbol{E}$-projectives P. Then the above mentioned square condition holds and there are enough P with $T(\{e \in \boldsymbol{E} \mid codomain\ e = P\}) \subseteq \boldsymbol{E}$. Choose such a $P = P_X$ and $e_X\colon P_X \twoheadrightarrow X$ in $\boldsymbol{E}$ for every object X and let $\kappa X\colon \widehat{T}X \hookrightarrow TX$ be the $\boldsymbol{M}$-part in the $(\boldsymbol{E}, \boldsymbol{M})$-factorization of Te_X. For every morphism $f\colon X \to Y$ in $\boldsymbol{X}$ there is a commutative 'square'

$$\begin{array}{ccc}
Q & \xrightarrow{\ g\ } & P_Y \\
{\scriptstyle \boldsymbol{E}\ni e}\downarrow & & \Big\| \\
P_X & & \Big\downarrow {\scriptstyle e_Y \in \boldsymbol{E}} \\
{\scriptstyle \boldsymbol{E}\ni e_X}\downarrow & & \Big\| \\
X & \xrightarrow{\ f\ } & Y
\end{array} \qquad . \qquad (*)$$

Applying T one gets $\widehat{T}f$ as indicated

$$\begin{array}{ccc}
TQ & \xrightarrow{\ Tg\ } & TP_Y \\
{\scriptstyle \boldsymbol{E}\ni Te}\downarrow & & \Big\| \\
TP_X & & \Big\downarrow {\scriptstyle \in \boldsymbol{E}} \\
{\scriptstyle \boldsymbol{E}\ni}\downarrow & & \Big\| \\
\widehat{T}X & \overset{\widehat{T}f}{- - \to} & \widehat{T}Y \\
{\scriptstyle \boldsymbol{M}\ni\kappa X}\big\downarrow & & \big\downarrow{\scriptstyle \kappa Y \in \boldsymbol{M}} \\
TX & \xrightarrow{\ Tf\ } & TY
\end{array} \qquad . \qquad (**)$$

PROPOSITION 1.1 *$\widehat{T}(\boldsymbol{E}) \subseteq \boldsymbol{E}$ and $\kappa\colon \widehat{T} \hookrightarrow T$ is couniversal among all natural transformations $\nu\colon \widetilde{T} \to T$ with $\widetilde{T}(\boldsymbol{E}) \subseteq \boldsymbol{E}$.*

Proof. If $f \in \boldsymbol{E}$ consider $(*)$ with g being in $\boldsymbol{E}$ instead of e. Then the outer and the lower rectangle in $(**)$ commute as well, hence the upper one, because κY is a monomorphism. Using $Tg \in \boldsymbol{E}$ one gets $\widehat{T}f \in \boldsymbol{E}$. The following commutative diagram establishes the factorization of ν via κ

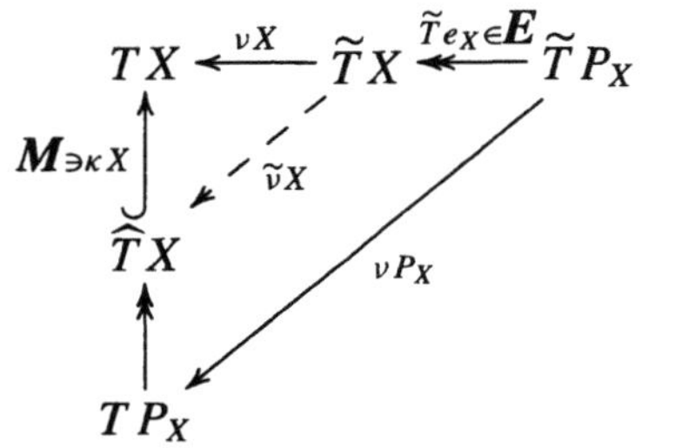

□

REMARK In [18, 1.2] a related condition is assumed for the construction of $\kappa\colon \widehat{T} \to T$ which needs a monad structure on T. This condition implies the existence of enough P as above, and it is implied by that in the presence of the square condition. The latter simplifies everything a lot up to easy computations of $\widehat{T}$. For instance, if T assignes to every graph X its graph of paths (including the empty ones), then $\widehat{T}X$ is nothing but the original graph equipped with the empty path at each vertex or identities, respectively [3, 6.2].

What can be done, if there are not enough P? Consider simply all morphisms $e\colon Y \twoheadrightarrow X$ in $\boldsymbol{E}$ instead of the distinguished one e_X, take the $(\boldsymbol{E}, \boldsymbol{M})$-factorizations $Te = s_e m_e$, $m_e\colon U_e \hookrightarrow TX$, and the intersection of the $\boldsymbol{M}$-parts assuming existence

$$\bigcap_{Y \overset{e \in \boldsymbol{E}}{\twoheadrightarrow} X} U_e =: T_1 X \overset{\kappa_1 X}{\hookrightarrow} TX.$$

Now, assuming the square condition, this construction turns out to be functorial, because for any morphism $f\colon X' \to X$ in $\boldsymbol{X}$ and any e there is some commutative square $ef' = fe'$ with $e'\colon Y' \twoheadrightarrow X'$ in $\boldsymbol{E}$ and the following diagram commutes

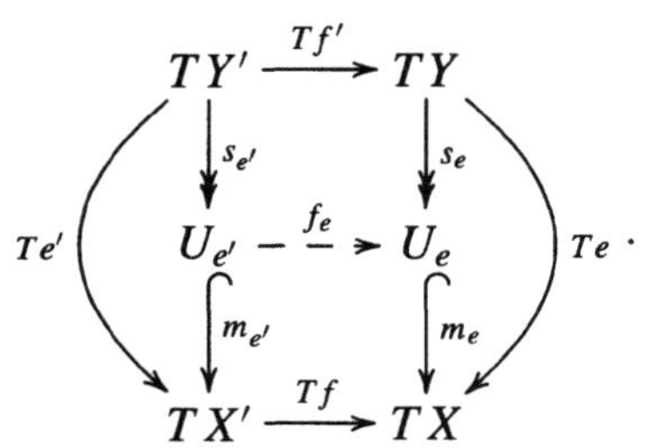

This yields $T_1 f$ as composition $T_1 X' \hookrightarrow \bigcap_{e'} U_{e'} \overset{\bigcap f_e}{\to} T_1 X$.

LEMMA 1.2 *Every natural transformation* $\nu\colon \widetilde{T} \to T$ *with* $\widetilde{T}(\boldsymbol{E}) \subseteq \boldsymbol{E}$ *factorizes (uniquely) via* κ_1.

Proof. Consider for every $e\colon Y \twoheadrightarrow X$ in $\boldsymbol{E}$ the following commutative diagram

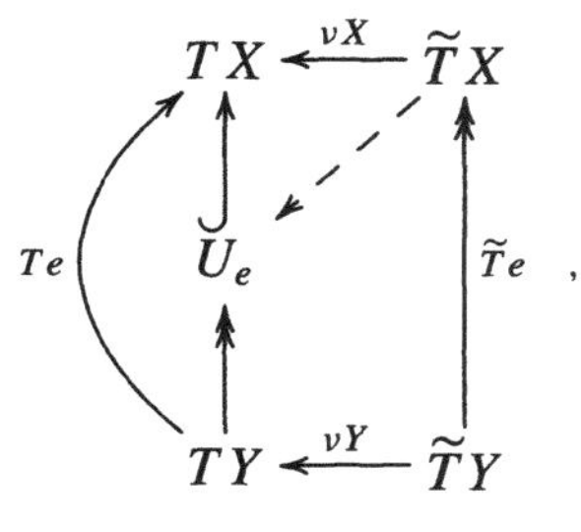

which induces a morphism $\nu_1 X\colon \widetilde{T}X \to T_1X$ with $\kappa_1 X \nu_1 X = \nu X$. □

Unfortunately there is no evidence for T_1 to preserve $\boldsymbol{E}$ in general, because $\boldsymbol{E}$-morphisms and intersections are not compatible at all (although it may happen in special cases as above). By Proposition 1.1 and Lemma 1.2, $\widehat{T} \hookrightarrow T_1$ and $\widehat{T} = \widehat{T_1}$ assuming existence of κ. Therefore try transfinite recursion

$$T_0 := T, \qquad T_{\alpha+1} := (T_\alpha)_1, \qquad T_\beta := \bigcap_{\alpha<\beta} T_\alpha$$

for a limit ordinal β. This yields a *descending chain*, i.e. a functorial image of the category of all ordinals

$$T_0 \hookleftarrow T_1 \ldots T_\alpha \hookleftarrow T_{\alpha+1} \ldots \hookleftarrow T_\beta \hookleftarrow \ldots .$$

This chain is pointwise in $\boldsymbol{M}$, because $\boldsymbol{M}$ is closed under the formation of intersections [1, 14.15]. The following is straightforward.

LEMMA 1.3 *Let $\boldsymbol{X}$ be $\boldsymbol{M}$-wellpowered. Then every descending chain in $\boldsymbol{M}$*

$$U_0 \hookleftarrow U_1 \ldots U_\alpha \hookleftarrow U_{\alpha+1} \ldots \hookleftarrow U_\beta \hookleftarrow \ldots$$

terminates, i.e. there is an ordinal γ such that $U_\gamma = U_\alpha$ for all $\alpha > \gamma$.

THEOREM 1.4 *If $\boldsymbol{X}$ has intersections of $\boldsymbol{M}$-subobjects, is $\boldsymbol{M}$-wellpowered, and fulfils the square condition, then $\widehat{T} := \bigcap_{\alpha \text{ ordinal}} T_\alpha$ preserves $\boldsymbol{E}$ and $\widehat{T} \overset{\kappa}{\hookrightarrow} T$ is couniversal among all natural transformations $\nu\colon \widetilde{T} \to T$ with $\widetilde{T}(\boldsymbol{E}) \subseteq \boldsymbol{E}$.*

Proof. Lemma 1.2 together with the couniversal property of intersections shows that every $\kappa_\alpha\colon T_\alpha \hookrightarrow T$ and hence $\kappa\colon \widehat{T} \hookrightarrow T$ behaves couniversal with respect to the ν's in question.

Now fix some $e\colon Y \twoheadrightarrow X$ in $\boldsymbol{E}$. Then there is a common ordinal γ such that T_γ, $(T_\gamma)_1 = T_{\gamma+1}$, and $\widehat{T}$ coincide at X and Y. But the $\boldsymbol{M}$-part of $T_\gamma e = \widehat{T}e$ contains $(T_\gamma)_1 X = \widehat{T}X$ as $\boldsymbol{M}$-subobject by construction. This shows $\widehat{T}e \in \boldsymbol{E}$. □

2. A Fairly General but External Approach

The advantage of the above construction of $\kappa: \widehat{T} \to T$ is that it works with the given functor T. Hence there is at least a chance to compute $\widehat{T}$. Its disadvantage is the use of the square condition which is not at hand in many topological examples. The following approach avoids the square condition but refers to the conglomerate of **all** natural transformations $\nu: \widetilde{T} \to T$ with $\widetilde{T}(\boldsymbol{E}) \subseteq \boldsymbol{E}$.

Observe that the conglomerate of all ν's is pointwise a class. This legitimates a usual construction of $\boldsymbol{M}$-coreflections by $(\mathcal{E}, \boldsymbol{M})$-factoriazations (for a suitable conglomerate $\mathcal{E}$ of sinks) pointwise in case of κ. The following is implied by the dual of Theorem 15.14 in [1] for the $\boldsymbol{M}$-part of the given $(\boldsymbol{E}, \boldsymbol{M})$-factorization structure on $\boldsymbol{X}$ with $\boldsymbol{M} \subseteq Mono\boldsymbol{X}$. Note that $\boldsymbol{M}$ is preserved under intersections and inverse images in this case.

THEOREM 2.1 *$\boldsymbol{X}$ is an $(\mathcal{E}, \boldsymbol{M})$-category for some conglomerate $\mathcal{E}$ of sinks iff $\boldsymbol{X}$ has arbitrary large intersections of $\boldsymbol{M}$-subobjects as well as inverse images of $\boldsymbol{M}$-subobjects along arbitrary $\boldsymbol{X}$-morphisms.*

For the rest of this section assume one of the equivalent conditions in Theorem 2.1. The $\boldsymbol{E}$-class is uniquely determined by $\boldsymbol{M}$ [1, 14.6], thus $\boldsymbol{E} = \mathcal{E} \cap Mor\boldsymbol{X}$.

PROPOSITION 2.2 *The full subcategory $\boldsymbol{C}$ of all objects C for which the empty sink (ϕ, C) belongs to $\mathcal{E}$ is $\boldsymbol{M}$-coreflective in $\boldsymbol{X}$* [1, 15.2(4)]. *Moreover, every morphism in $\boldsymbol{C}$ belongs to $\boldsymbol{E}$.*

Proof. The dual of the first part is known for arbitrary factorization structures of sources. Moreover, every $\boldsymbol{C}$-object is $\boldsymbol{M}$-projective [1, 15G]. Consequently, every $\boldsymbol{C}$-morphism in $\boldsymbol{M}$ is a retract, hence an isomorphism according to $\boldsymbol{M} \subseteq Mono\boldsymbol{X}$ (dual of [1, 15.4]). Now consider the $(\boldsymbol{E}, \boldsymbol{M})$-factorization $f = me$ of some $\boldsymbol{C}$-morphism f, which belongs to $\boldsymbol{C}$, because the $\boldsymbol{M}$-coreflective $\boldsymbol{C}$ is closed under $\boldsymbol{E}$-morphisms. Consequently, m is an isomorphism and f belongs to $\boldsymbol{E}$. □

COROLLARY 2.3 *There is at least one natural transformation $\nu: \widetilde{T} \to T$ with $\widetilde{T}(\boldsymbol{E}) \subseteq \boldsymbol{E}$.*

Proof. Consider the coreflector $C: \boldsymbol{X} \to \boldsymbol{X}$ of $\boldsymbol{C}$ as above and the coreflection $\gamma: C \hookrightarrow \mathrm{id}\boldsymbol{X}$. Composition with T from the right yields

$$\nu := \gamma T: CT \to T, \quad \widetilde{T} := CT. \qquad \square$$

LEMMA 2.4 *Let $(e_i)_{i \in I}$ be a conglomerate-indexed sink in $\boldsymbol{X}$ such that the self-indexed class $\{e_i \mid i \in I\}$ belongs to $\mathcal{E}$. Then $(e_i)_{i \in I}$ has the unique diagonalization property with respect to $\boldsymbol{M}$.*

Proof. Consider commutative diagrams

$$\begin{array}{ccc} \cdot & \xrightarrow{e_i} & \cdot \\ {\scriptstyle f_i}\big\downarrow & & \big\downarrow{\scriptstyle g} \\ \cdot & \overset{m}{\hookrightarrow} & \cdot \end{array}, \quad i \in I.$$

Then $f_i = f_j$ if $e_i = e_j$, because m is a monomorphism. Therefore $(e_i)_{i \in I}$ inherits the unique diagonalization property from $\{e_i \mid i \in I\}$. □

THEOREM 2.5 *There is a couniversal natural transformation $\kappa\colon \widehat{T} \to T$ among all $\nu\colon \widetilde{T} \to T$ with $\widetilde{T}(\boldsymbol{E}) \subseteq \boldsymbol{E}$.*

Proof. For fixed X the $(\mathcal{E}, \boldsymbol{M})$-factorization of $\{\nu X \mid \nu\colon \widetilde{T} \to T, \widetilde{T}(\boldsymbol{E}) \subseteq \boldsymbol{E}\}$ yields

$$\nu X = \kappa X \widehat{\nu} X$$

with $\kappa X\colon \widehat{T}X \hookrightarrow TX$ in $\boldsymbol{M}$ and $(\widehat{\nu}X)_\nu \ldots$ such that $\{\widehat{\nu}X \mid \nu \ldots\}$ belongs to $\mathcal{E}$. For any morphism $f\colon X \to Y$ one gets $\widehat{T}f$ as unique 'diagonal' in

$$\begin{array}{ccccc} TX & \overset{\kappa X}{\hookleftarrow} & \widehat{T}X & \xleftarrow{\widehat{\nu}X} & \widetilde{T}X \\ {\scriptstyle Tf}\big\downarrow & & {\scriptstyle \widehat{T}f}\big\downarrow & & \big\downarrow{\scriptstyle \widetilde{T}f} \\ TY & \overset{\kappa Y}{\hookleftarrow} & \widehat{T}Y & \xleftarrow{\widehat{\nu}Y} & \widetilde{T}Y \end{array}.$$

Obviously, $\widetilde{T}f \in \boldsymbol{E}$ for all ν implies $\widehat{T}f \in \boldsymbol{E}$ by the dual of [1, 15.5], thus $\widehat{T}(\boldsymbol{E}) \subseteq \boldsymbol{E}$. □

3. Topological Examples and Varietal Hulls

Among the following topological examples are full reflective subcategories $\boldsymbol{C}$ of **Top** and T is the reflector composed with the embedding in this case. Does the same hold for $\widehat{T}$?

PROPOSITION 3.1 *Let $\boldsymbol{C} \subseteq \widehat{\boldsymbol{C}} \subseteq \boldsymbol{X}$ be isomorphism-closed, full, reflective subcategories with corresponding monads (T, η, μ) and $(\widehat{T}, \widehat{\eta}, \widehat{\mu})$ such that for every X in $\boldsymbol{X}$ there is some $e\colon C \twoheadrightarrow X$ in $\boldsymbol{E}$ with C in $\boldsymbol{C}$. Then the following are equivalent.*

(i) *$\eta\widehat{T}\colon \widehat{T} \to T\widehat{T} = T$ is couniversal among all natural transformations $\nu\colon \widetilde{T} \to T$ with $\widetilde{T}(\boldsymbol{E}) \subseteq \boldsymbol{E}$.*
(ii) *$\widehat{T}(\boldsymbol{E}) \subseteq \boldsymbol{E}$ and $\eta\widehat{T}$ is pointwise in $\boldsymbol{M}$.*

Proof. Clearly, (i) $\Rightarrow$ (ii). Assuming (ii) consider $\nu\colon \widetilde{T} \to T$, $e\colon C \twoheadrightarrow X$ as above, and the following commutative diagram using $TC = \widehat{T}C$ and $Te = \eta\widehat{T}X\widehat{T}e$

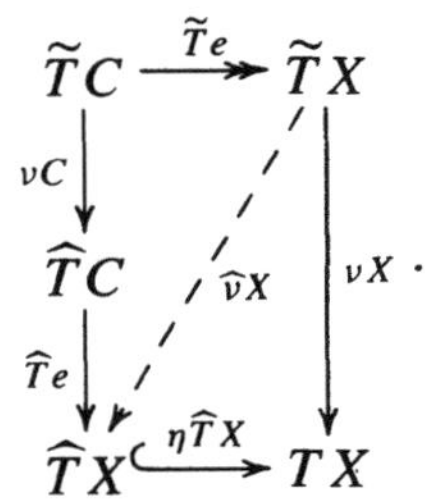

□

Note that $\widehat{T}(\boldsymbol{E}) \subseteq \boldsymbol{E}$ if $\widehat{\boldsymbol{C}}$ is $\boldsymbol{E}$-reflective. J. R. Isbell has shown that every topological space X appears as a quotient of some zero-dimensional T_1-space C which is even paracompact [12]. M. Katětov [13] proved that a paracompact Hausdorff space C is realcompact iff every closed discrete subspace of C is realcompact. The latter holds if there is no measurable cardinal. Therefore, Proposition 3.1 proves the quotient preserving monad $(\widehat{T}, \widehat{\eta}, \widehat{\mu})$ of the quotient-reflective subcategory $\widehat{\boldsymbol{C}} = \boldsymbol{F}$**Haus** $\subseteq$ **Top** of functionally Hausdorff spaces to be couniversal with respect to $\boldsymbol{C} =$ **Tych**, the category of Tychonoff spaces ($T_{3\frac{1}{2}}$ and T_1), and the category $\boldsymbol{C}$ of all realcompact spaces, assuming non-existence of measurable cardinals for the latter.

This implies that the **Tych**-reflection α as well as the Hewitt-realcompactification υ fail to preserve quotients even when restricted to $\boldsymbol{F}$**Haus**. Another example of this kind is the category $\boldsymbol{C}$ of zero-dimensional T_0-spaces and $\widehat{\boldsymbol{C}}$ the category of all spaces in which pairs of distinct points can be separated by a clopen set.

Next consider the functor D: **Top** $\to$ **Top** which maps each space to its underlying discrete space. The underlying identities define a natural transformation $\varepsilon\colon D \to \mathrm{id}$**Top**. If T: **Top** $\to$ **Top** maps quotients to surjections, then $\widetilde{T} := DT$ preserves quotients and $\nu := \varepsilon T\colon \widetilde{T} \to T$ is a natural transformation with pointwise underlying identities. This shows that $\kappa\colon \widehat{T} \hookrightarrow T$ is pointwise bijective according to $\kappa\widehat{\nu} = \nu$. Therefore assume without any loss of generality, that κ has pointwise underlying identities.

This observation applies to $T := X \times -$ for any space X. In general, T preserves surjections but fails to preserve quotients except X is *quasi locally compact* (see [11] for details and further references). $\widehat{T}Y$ is nothing but the cartesian product $X \times Y$ equipped with a certain refinement of the product topology as it is the case for any bicoreflection of the product space. Especially, this holds for $X = T(1)$ and $\widehat{X} := \widehat{T}(1)$. But there is a refinement $\widetilde{T}$ of T given by the weak product that preserves quotients and fulfils $\widetilde{T}(1) = X$ [14]. Therefore $\widehat{X} = X$.

THEOREM 3.2 *$\widehat{T}$ has a right adjoint given by the function spaces $C(X, Y)$ equipped with a suitable topology such that the evaluation maps* ev: $\widehat{T}(C(X, Y)) \to Y$, $\mathrm{ev}(x, f) := f(x)$, *are couniversal.*

Proof. By construction, $\widehat{T}$ preserves quotients. Moreover, $\widehat{T}$ preserves coproducts $\coprod_I Y_i$, because this holds for underlying sets, $\widehat{T}(Y_i)$ is an open subset of $\widehat{T}(\coprod_I Y_i)$ (being a refinement of $X \times \coprod_I Y_i$), and $\widehat{T}$ maps the coproduct injections (being coretractions in case of nonempty domains) to embeddings. Consequently, the same procedure as in the proof of the theorem in [10] applies.

$\widehat{T}$ preserves final epi-sinks. Supply the set $C(X, Y)$ with the final topology with respect to all maps $f\colon Z \to C(X, Y)$ for which $\mathrm{ev} \circ (\mathrm{id}\, X \times f)\colon \widehat{T}Z \to Y$ is continuous. This is an epi-sink, because for $Z = 1$ the composition $\mathrm{ev} \circ (\mathrm{id}\, X \times f)$ is just the continuous map $X = \widehat{T}1 \to Y$ picked by f. Consequently, $\widehat{T}(C(X, Y))$ carries the final topology with respect to all $\widehat{T}(f) = \mathrm{id}\, X \times f$ for which $\mathrm{ev} \circ (\mathrm{id}\, X \times f)$ is continuous, hence ev is continuous.

Now consider some continuous map $g\colon \widehat{T}Z \to Y$ and $z\colon 1 \to Z$ defined by $z \in Z$. Then $g_z := g \circ \widehat{T}z\colon X \to Y$ is continuous and $f\colon Z \to C(X, Y)$ with $f(z) := g_z$ is a map such that $\mathrm{ev} \circ (\mathrm{id}\, X \times f) = g\colon \widehat{T}Z \to Y$ is continuous. Hence f is continuous with $ev \circ \widehat{T}f = g$. Moreover, f is uniquely determined by g. This proves the couniversal property for ev. □

COROLLARY 3.3 *The projections $p_x\colon C(X, Y) \to Y$, $p_x(f) := f(x)$, $x \in X$, are continuous as well as the canonical map*

$$p\colon X \to \mathrm{hom}(C(X, \mathbb{R}), \mathbb{R}),$$

$x \mapsto p_x$, *where* $\mathrm{hom}(C(X, \mathbb{R}), \mathbb{R})$ *denotes the set of all continuous $\mathbb{R}$-algebra homomorphisms equipped with the topology of pointwise convergence. $C(X, \mathbb{R})$ is a topological $\mathbb{R}$-Algebra.*

Proof. Every $x \in X$ defines a map $x\colon 1 \to X$ and a natural transformation $1 \times - \to T = X \times -$ which factorizes via $\kappa\colon \widehat{T} \hookrightarrow T$. Therefore, the map

$$1 \times C(X, Y) = C(X, Y) \ni f \mapsto (x, f) \in \widehat{T}(C(X, Y))$$

is continuous. Its composition with the continuous evaluation is just p_x.

The composition of p with the projection at $f \in C(X, \mathbb{R})$ is just f, hence continuous. Using right adjointness of the functor $C(X, -)$, addition and multiplication on $C(X, \mathbb{R})$ are given by $C(X, +)$ and $C(X, \cdot)$ where $+, \cdot\colon \mathbb{R} \times \mathbb{R} \to \mathbb{R}$ are continuous. Multiplication by scalars is the restriction to constant functions as left factors. Observe furthermore that $\mathbb{R}$ is isomorphic to $C(1, \mathbb{R})$ and the canonical map $C(1, \mathbb{R}) \to C(X, \mathbb{R})$ is continuous, because it is given by the conjugate $C(1, -) \to C(X, -)$ of the natural transformation $\widehat{T} \to 1 \times -$, defined by $X \to 1$. □

Recall from [5] that there is (in general) no $\mathbb{R}$-vector space topology on $C(X) = C(X, \mathbb{R})$ for which the evaluation map ev is continuous with respect to the usual product topology on $X \times C(X)$, even for $X \in$ **Tych**. The first part of Corollary 3.3

shows that convergence in $C(X, Y)$ implies pointwise convergence. Besides all advantages the product $\widehat{X \times -}$ may have the disadvantage of being not symmetric [7]. Related product constructions have already been introduced by R. Brown [6].

Next consider the power functor $T = -^n$: **Top** $\rightarrow$ **Top**, $X \mapsto X^n$, n some cardinal. Its couniversal quotient preserving modification $\widehat{T}$ has the same underlying sets (using the functor D as above). The diagonal functor Δ: id **Top** $\rightarrow T$ factorizes via $\widehat{T}$ showing that $\widehat{T}X = \widehat{X^n}$ is not discrete if this holds for X. Therefore these modified powers are not trivial. They may be convenient for topological universal algebra. If R is a congruence on a topological algebra A of some not necessarily finitary type Ω [17] then the algebraic and topological quotient q: $A \rightarrow A/R$ yields a *generalized topological algebra*, because the induced algebraic operations on A/R turn out to be continuous with respect to modified powers.

$$\begin{array}{ccc} \widehat{A^n} & \xrightarrow[\text{quotient}]{\widehat{q^n}} & \widehat{(A/R)^n} \\ {\scriptstyle \omega A}\downarrow & & \downarrow{\scriptstyle \omega A/R} \\ A & \xrightarrow{\quad q \quad} & A/R \end{array}, \qquad \omega \in \Omega n.$$

The underlying space functor $\widetilde{U}$ of the category $\widetilde{\boldsymbol{Alg}}(\Omega|\Gamma, \mathbf{Top})$ of all generalized topological algebras of type Ω, which fulfil all equations in Γ, has a left adjoint $\widetilde{F}$. It is given by the algebraically free algebra $F|X|$ on the underlying set $|X|$ of a space X equipped with the initial topology with respect to all algebraic extensions $\overline{f}$: $F|X| \rightarrow A$ of all continuous maps f: $X \rightarrow A \in \widetilde{\boldsymbol{Alg}}(\Omega|\Gamma, \mathbf{Top})$. The diagram above shows that $\widetilde{U}$ preserves quotients. Moreover $\widetilde{U}$ reflects quotients and congruences. Altogether $\widetilde{U}$ is a *varietal* [1, 17] extension of the underlying space functor U of the category $\boldsymbol{Alg}(\Omega|\Gamma, \mathbf{Top})$ of all topological Ω-algebras which fulfil the equations in Γ. In general, U is not varietal, even not algebraic [15, 17] but monadic [17]. Among all varietal extensions of U there is a best one, the *varietal hull.*

THEOREM 3.4 *Let T be the monad on* **Top** *given by* $\boldsymbol{Alg}(\Omega|\Gamma, \mathbf{Top})$, κ: $\widehat{T} \hookrightarrow T$ *its couniversal quotient-preserving modification, and* K: $\boldsymbol{Alg}(\Omega|\Gamma, \mathbf{Top}) \rightarrow \mathbf{Top}^{\widehat{T}}$ *defined by* κ [16]. *Then* $U^{\widehat{T}}$: $\mathbf{Top}^{\widehat{T}} \rightarrow$ **Top** *is varietal. K is a full embedding and universal among all concrete functors* H: $\boldsymbol{Alg}(\Omega|\Gamma, \mathbf{Top}) \rightarrow (\boldsymbol{A}, V)$, V: $\boldsymbol{A} \rightarrow$ **Top** *varietal.*

Proof. K inherits its universal property from the couniversal one of κ even with respect to all quotient preserving monads on **Top** [16]. Consider $\widetilde{T} := \widetilde{U}\widetilde{F}$ as above. There is a natural transformation ν: $\widetilde{T} \rightarrow T$ with pointwise underlying identities. The same can be assumed for κ and the unique factorization $\widehat{\nu}$ of ν via

κ. As pointwise epimorphisms $\widehat{\nu}$ and κ induce full embeddings $\widehat{N}$ and K (as above) such that the following diagram commutes.

$$\begin{array}{ccccc} \boldsymbol{Alg}(\Omega|\Gamma, \mathbf{Top}) & \xrightarrow{K} & \mathbf{Top}^{\widehat{T}} & \xrightarrow{\widehat{N}} & \widetilde{\boldsymbol{Alg}}(\Omega|\Gamma, \mathbf{Top}) \\ & {\scriptstyle U}\searrow & \downarrow{\scriptstyle U^{\widehat{T}}} & \swarrow{\scriptstyle \widetilde{U}} & \\ & & \mathbf{Top} & & \end{array}$$

It remains to show that $U^{\widehat{T}}$ reflects congruences. But this holds for $\widetilde{U}$. Therefore a $U^{\widehat{T}}$-congruence may be considered as a kernel pair of some morphism p in $\widetilde{\boldsymbol{Alg}}(\Omega|\Gamma, \mathbf{Top})$, the codomain of which carrying the indiscrete topology, thus being in $\boldsymbol{Alg}(\Omega|\Gamma, \mathbf{Top})$. This proves $p = \widehat{N}p'$ and the given $U^{\widehat{T}}$-congruence to be the kernel pair of p'. □

REMARK There are obvious generalizations to topological categories $\boldsymbol{X}$ over some (RegEpi, Mono)-category $\boldsymbol{S}$ with regular epimorphism preserving powers instead of $\boldsymbol{X} = \mathbf{Top}$ and $\boldsymbol{S} = \mathbf{Set}$.

4. Arbitrary Algebraic Hulls

H. Bargenda introduced $(\boldsymbol{E}, \mathcal{M})$-algebraic functors ranging in an $(\boldsymbol{E}, \mathcal{M})$-category $\boldsymbol{X}$, $\mathcal{M}$ a conglomerate of mono-sources, as a natural generalization of H. Herrlich's regular functors [2, 3, 9]. Moreover H. Bargenda constructed a universal $(\boldsymbol{E}, \mathcal{M})$-algebraic completion (hull) for any right adjoint functor $U: \boldsymbol{A} \to \boldsymbol{X}$ provided that $\boldsymbol{X}$ has enough $\boldsymbol{E}$-projectives. This implies the existence of a couniversal $\boldsymbol{E}$-preserving modification $\kappa: \widehat{T} \to T$ of the monad T given by U. Now assume the existence of κ and nothing special for $\boldsymbol{X}$. Then the following holds with respect to $K: \boldsymbol{X}^T \to \boldsymbol{X}^{\widehat{T}}$ induced by κ and the comparison functor $H: \boldsymbol{A} \to \boldsymbol{X}^T$.

THEOREM 4.1 *U has an $(\boldsymbol{E}, \mathcal{M})$-algebraic hull given by the restriction $\widehat{U}$ of $U^{\widehat{T}}$ to the $\boldsymbol{E}$-reflective hull $\widehat{\boldsymbol{A}}$ of the image of KH in $\boldsymbol{X}^{\widehat{T}}$ and the respective corestriction G of KH.*

Proof. By construction, $\widehat{U}$ is right adjoint and $\widehat{\boldsymbol{A}}$ is closed under the formation of (underlying) $(\boldsymbol{E}, \mathcal{M})$-factorizations of $\widehat{\boldsymbol{A}}$-sources as well as $\boldsymbol{X}^{\widehat{T}}$ is. This proves $\widehat{U}$ to be $(\boldsymbol{E}, \mathcal{M})$-algebraic.

Now consider an $(\boldsymbol{E}, \mathcal{M})$-algebraic functor $\widetilde{U}: \widetilde{\boldsymbol{A}} \to \boldsymbol{X}$ and $D: \boldsymbol{A} \to \widetilde{\boldsymbol{A}}$ with $\widetilde{U}D = U$. Recall that the comparison functor H is claimed [8] and easily seen to be universal among all concrete functors from $(\boldsymbol{A}, U)$ to some category of Eilenberg–Moore-algebras over $\boldsymbol{X}$. Applying this to the composition of D with the comparison functor $\widetilde{H}: \widetilde{\boldsymbol{A}} \to \boldsymbol{X}^{\widetilde{T}}$ given by $\widetilde{U}$ yields a concrete functor $E: \boldsymbol{X}^T \to \boldsymbol{X}^{\widetilde{T}}$ hence

$\widehat{E}\colon \boldsymbol{X}^{\widehat{T}} \to \boldsymbol{X}^{\widetilde{T}}$ by the universal property of K such that the following diagram commutes.

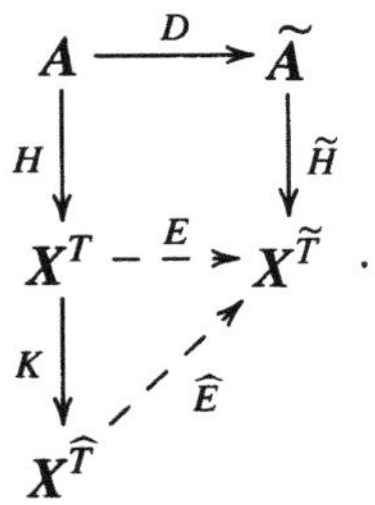

$\widehat{\boldsymbol{A}}$ contains the free $\widehat{T}$-algebras $F^{\widehat{T}}X = (\widehat{T}X, \widehat{\mu}X)$, because the following diagram commutes (according to $\kappa\widehat{\mu} = \mu\kappa^2$), κ is pointwise in $\mathcal{M}$, and $KHFX = K(TX, \mu X) = (TX, \mu X \kappa TX)$

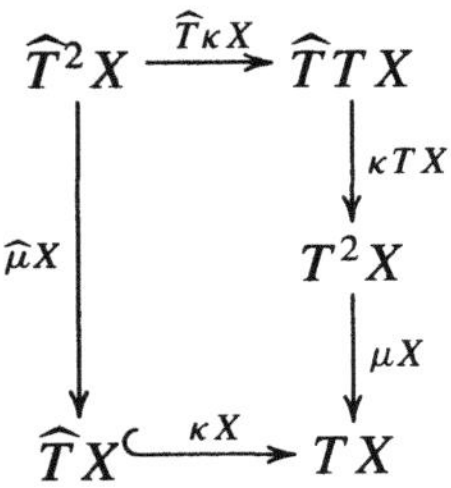

Consequently, the monad defined by $\widehat{U}$ is just $\widehat{T}$ and the embedding $\widehat{H}\colon \widehat{\boldsymbol{A}} \hookrightarrow \boldsymbol{X}^{\widehat{T}}$ is nothing but the corresponding comparison functor which shares the universal property mentioned above. Moreover, $\widetilde{H}$ is a full, isomorphism-closed, and $\boldsymbol{E}$-reflective embedding [2] the image of which is closed under the formation of $\mathcal{M}$-sources in $\boldsymbol{X}^{\widetilde{T}}$ and contains $\widehat{E}KH(\boldsymbol{A}) = \widetilde{H}D(\boldsymbol{A})$. Furthermore $\widehat{E}$ preserves $\mathcal{M}$-sources, hence $\widehat{E}(\widehat{\boldsymbol{A}}) \subseteq \widetilde{H}(\widetilde{\boldsymbol{A}})$, $\widehat{E}\widehat{H} = \widetilde{H}\widetilde{D}$ for some concrete functor $\widetilde{D}\colon \widehat{\boldsymbol{A}} \to \widetilde{\boldsymbol{A}}$, and $\widetilde{H}\widetilde{D}G = \widehat{E}\widehat{H}G = \widehat{E}KH = \widetilde{H}D$ implies $\widetilde{D}G = D$. Uniqueness of $\widetilde{D}$ follows from uniqueness of $\widehat{E}$ and the universal property of $\widehat{H}$. □

References

1. Adámek, J., Herrlich, H. and Strecker, G. E.: *Abstract and Concrete Categories*, Wiley, 1990.
2. Bargenda, H.: Algebraische Hüllen rechtsadjungierter Funktoren, Thesis, Universität Bremen, 1987.
3. Bargenda, H.: Universal algebraic completions of right adjoint functors, in *Category Theory at Work*, Heldermann, Berlin, 1991, pp. 289–306.
4. Bargenda, H. and Richter, G.: Varietal hulls of functors, *Quaestiones Math.* **4** (1980), 121–158.
5. Binz, E.: *Continuous Convergence on* $C(X)$, Lecture Notes in Math. 469, Springer-Verlag, Berlin, Heidelberg, New York, 1975.
6. Brown, R.: Ten topologies for $X \times Y$, *Quart. J. Math. Oxford (2)* **14** (1963), 303–319.

7. Činčura, J.: Tensor products in the category of topological spaces, *Comment. Math. Univ. Carol.* **20** (1979), 431–445.
8. Eilenberg, S. and Moore, J.C.: Adjoint functors and triples, *Illinois J. Math.* **9** (1965), 381–398.
9. Herrlich, H.: Regular categories and regular functors, *Canad. J. Math.* **26** (1974), 709–720.
10. Herrlich, H.: Cartesian closed topological categories, *Math. Colloq. Univ. Cape Town* **9** (1974), 1–16.
11. Herrlich, H.: Categorical topology 1971–1981, in *Proc. 5th Prague Top. Symp., Prague 1981*, Heldermann, Berlin, 1982, pp. 279–383.
12. Isbell, J.: A note on complete closure algebras, *Math. Systems Theory* **3** (1969), 310–312.
13. Katětov, M.: Measures in fully normal spaces, *Fund. Math.* **38** (1951), 73–84.
14. Knight, C. J., Moran, W. and Pym, J. S.: The topologies of separate continuity. I, *Proc. Camb. Phil. Soc.* **68** (1970), 663–671.
15. Lawson, J. and Madison, B.: On congruences and cones, *Math. Z.* **120** (1971), 18–24.
16. Manes, E. G.: *Algebraic Theories*, Springer-Verlag, Berlin, Heidelberg, New York, 1976.
17. Richter, G.: *Kategorielle Algebra*, Akademie-Verlag, Berlin, 1979.
18. Richter, G.: Algebra $\subset$ Topology?!, in *Category Theory at Work*, Heldermann, Berlin, 1991, pp. 261–273.

Applied Categorical Structures **8:** 175–207, 2000.
G. Brümmer & C. Gilmour (eds), Papers in Honour of Bernhard Banaschewski.

Separation and Epimorphisms in Quasi-Uniform Spaces

Dedicated to Prof. B. Banaschewski on his seventieth birthday

DIKRAN DIKRANJAN
Dipartimento di Matematica e Informatica, Università di Udine, Via delle Scienze 206, 33100 Udine, Italy, e-mail: dikranja@dimi.uniud.it

HANS-PETER KÜNZI
Department of Mathematics, University of Bern, Sidlerstrasse 5, 3012 Bern, Switzerland e-mail: kunzi@math-stat.unibe.ch

(Received: 30 October 1996; accepted: 14 December 1999)

Abstract. We study some categorical aspects of quasi-uniform spaces (mainly separation and epimorphisms) via closure operators in the sense of Dikranjan, Giuli, and Tholen. In order to exploit better the corresponding properties known for topological spaces we describe the behaviour of closure operators under the lifting along the forgetful functor T from quasi-uniform spaces to topological spaces. By means of appropriate closure operators we compute the epimorphisms of many categories of quasi-uniform spaces defined by means of separation axioms and study the preservation (reflection) of epimorphisms under the functor T.

Mathematics Subject Classifications (2000): Primary 18A20, 54E15, 54B30. Secondary 18B30, 54B17.

Key words: quasi-uniform space, uniform space, (regular, semiregular) closure operator, θ-closure, sequential closure, $S(\alpha)$-space, ϱ-separated space, epimorphism.

1. Introduction

Different densities (or closures) in the category **QUnif** of quasi-uniform spaces have already been used in various contexts. If T: **QUnif** $\to$ **Top** is the forgetful functor, then one speaks of T*-density* and T*-closedness* and this can be extended to the other two functorial topologies lying on a quasi-uniform space (co-dense, supdense). But one also has *doubly dense* sets ([28]) when sets, both T-dense and T^{ι}-dense, are concerned. This requires a well organized approach to "density" and "closure" in **QUnif** which goes much beyond the three closure operators $\mathrm{cl}_{T(X)}(\text{-})$, $\mathrm{cl}_{T^{\iota}(X)}(\text{-})$ and $\mathrm{cl}_{T^{*}(X)}(\text{-})$. An appropriate tool for this seems to provide the closure operators introduced in a quite general setting [9] and working especially well in topological categories [7]. We illustrate here the potential power of closure operators in **QUnif** by describing separation axioms and epimorphisms of subcategories of **QUnif**. Closure operators of the category **Unif** of uniform spaces are studied in

([2, 14]), where they are shown to be "few". We show here that there are many closure operators in **QUnif** which coincide, when "restricted" to **Unif**.

A morphism $f: X \to Y$ in a category $\mathcal{X}$ is an *epimorphism* if for each pair of morphisms $g, h: Y \to Z$ in $\mathcal{X}$ such that $gf = hf$ we have $g = h$. In this paper we consider subcategories of the category **QUnif** of quasi-uniform spaces and quasi-uniformly continuous maps.

In Section 2 we give the necessary categorical background on **QUnif** (the forgetful functors T, T^ι, T^*: **QUnif** $\to$ **Top**, etc.). In Section 3.1 we recall the definition of a closure operator in the context of **QUnif** and the three leading examples $\alpha = \mathrm{cl}_{T(X)}$, $\beta = \mathrm{cl}_{T^*(X)}$ and $\gamma = \mathrm{cl}_{T^\iota(X)}$ (Example 3.2). In Section 3.2 we offer two alternative ways of generating closure operators in **QUnif** based on functors: one of them just lifts closure operators of **Top** along the forgetful functors T, T^*, T^ι (as in the case of α, β, γ, cf. Section 3.3), the other gives additive closure operators which are more intrinsically related to the quasi-uniform structure (Section 3.2). In this direction we define a series of closure operators $\overline{\theta}_n$ in analogy to the θ-closure of Velichko [34] and the closure operators θ_n of **Top** defined in [10].

We dedicate Section 4 to the amalgamation in **QUnif** and its underlying topology which need not coincide with the topology of the amalgamation, carried out in **Top**, of the underlying topological spaces (Section 4.1). Since the amalgamation is a very special instance of quotient, let us recall that even for uniform spaces the topology of the quotient uniformity need not coincide with the quotient topology of the underlying topology (Himmelberg [21]). This feature of the amalgamation in **QUnif** prevents an easy "lifting" of the known results in **Top** along the functors T and T^* and makes it necessary to study problems of epimorphisms and cowellpoweredness in **QUnif** separately.

The concept of closure operator was gradually developed in [7 – 12, 15] on the base of the important model of regular closure operators, introduced by Salbany in the context of topological spaces [31]. In Section 5 we recall the definition of a regular closure operator associated to a subcategory of **QUnif** and the simplest separation axioms. By means of separation properties of the amalgamation from Section 4.1 we compute the regular closure operators of the respective subcategories.

In Section 6 we define two types of separation axioms $\Delta(\varrho)$ and $\mathcal{T}_\varrho$ depending on a closure operator ϱ. The separations axiom $\mathcal{T}_\varrho$ is built on the intuitive notion of neighbourhood and works well in topological categories whereas $\Delta(\varrho)$ makes no recourse to points and neighbourhoods (see [4] for separation in abstract categories). The axioms $\Delta(\varrho)$ and $\mathcal{T}_\varrho$ coincide in the case $\varrho = \overline{\theta}_n$ (see Lemma 6.4). We compare then the subcategory $\mathcal{T}_n$ of spaces satisfying that axiom with the category $S(n)$ in **Top** introduced by Viglino [36] (see Section 6.2 for the precise definition), by showing that $T(\mathcal{T}_n) = S(n)$, but $\mathcal{T}_n \neq T^{-1}(S(n))$ (Corollary 6.10 and Example 6.11). In Section 6.3 we discuss subcategories of **QUnif** defined in a different way (totally bounded spaces, transitive spaces, etc.).

In Section 7 we recall the semiregular closure operators introduced in [7] in order to compute the categorically defined regular closure operators by means of iterations. The main advantage of semiregular closure operators is that they are naturally defined and easier to deal with.

The epimorphisms of a subcategory $\mathcal{A}$ of **QUnif** are described in Section 8 as the dense maps with respect to the regular closure operator of $\mathcal{A}$. Since most of the closure operators under consideration are semiregular, the description we obtain is quite simple, e.g., the epimorphisms of $\mathcal{T}_{\alpha\wedge\gamma}$ are the doubly dense maps (cf. Proposition 8.4) and the epimorphisms of $\mathcal{T}_n$ are the $(\overline{\theta}_n)^\infty$-dense maps, for $1 \leq n \leq \omega$ (see Example 8.3 where other examples can be found). We show also that the epimorphisms of the category of transitive (resp., totally bounded) T_0-spaces are the T^*-dense maps (8.5).

The results of this paper are applied in a forthcoming paper [13] dealing with cowellpoweredness in **QUnif**.

This project was initiated during the first author's visit at the Department of Mathematics of the University of Berne in July 1994. This author takes the opportunity to thank his hosts for the excellent hospitality.

2. Categorical Background on Quasi-Uniform Spaces

A *quasi-uniform space* is a pair $(X, \mathcal{U})$, where X is a set and $\mathcal{U}$ is a filter of reflexive relations on X such that for every $U \in \mathcal{U}$, there exists $V \in \mathcal{U}$ with $V \circ V \subseteq U$. A *quasi-uniformly continuous map* $f: (X, \mathcal{U}) \to (Y, \mathcal{V})$ is a map such that for each $V \in \mathcal{V}$ there exists $U \in \mathcal{U}$ with $(f \times f)(U) \subseteq V$. This defines the category **QUnif** of quasi-uniform spaces which contains the category **Unif** of uniform spaces (and uniformly continuous maps) as a full subcategory. The forgetful functor G: **QUnif** $\to$ **Set** is topological (in the sense of [1]).

For a quasi-uniform space $(X, \mathcal{U})$ one can define a topology $T(\mathcal{U})$ on X by taking as a base of neighbourhoods at $x \in X$ the filter base $\{U(x)\}_{U\in\mathcal{U}}$, where $U(x) = \{y \in X : (x, y) \in U\}$. Every quasi-uniformly continuous map $f: (X, \mathcal{U}) \to (Y, \mathcal{V})$ gives a continuous map with respect to the induced topologies. Hence one has a concrete functor T: **QUnif** $\to$ **Top** which composed with the forgetful functor V: **Top** $\to$ **Set** gives $G = V \circ T$. The functor T preserves initial sources, but it is not topological, since a topological space need not admit a coarsest quasi-uniformity inducing the given topology.

The functor T admits many sections, i.e., functorially chosen quasi-uniformities on the underlying set X of a topological space (X, τ) inducing the given topology τ (see [27]), of which the following two are the most relevant ones. The *finest quasi-uniformity* $F(X, \tau)$ is defined as the finest quasi-uniformity $\mathcal{U}$ on X with $T(\mathcal{U}) = \tau$. The *Pervin quasi-uniformity* $P(X, \tau)$ of a topological space (X, τ) is generated by the filter subbase of binary relations $\{S_G\}$, with $S_G := (G \times G) \cup ((X \setminus G) \times X)$ and G running through the family of all open subsets of X. In this way one obtains two functors F, P: **Top** $\to$ **QUnif**, of which F is left adjoint to T. Then the counit

ε_X: $FTX \to X$ is the identity map, where FTX is the set X equipped with the finest quasi-uniformity of the topological space $T(X)$.

Consider the natural involutive endofunctor ι: **QUnif** $\to$ **QUnif** defined by $(X, \mathcal{U}) \overset{\iota}{\mapsto} (X, \mathcal{U}^{-1})$, where $\mathcal{U}^{-1}$ denotes the filter $\{U^{-1} : U \in \mathcal{U}\}$. The composite $T^\iota := T \circ \iota$: **QUnif** $\to$ **Top** has $F^\iota := \iota \circ F$ as a left adjoint with counit $\varepsilon^\iota_X := \iota(\varepsilon_{\iota X})$: $F^\iota T^\iota X \to X$. The category **Unif** of uniform spaces is simply the full subcategory of **QUnif** of all quasi-uniform spaces left fixed by ι. Actually, **Unif** is coreflective in **QUnif**, with the coreflector defined by $(X, \mathcal{U}) \mapsto (X, \mathcal{U}^*)$, with $\mathcal{U}^* = \mathcal{U} \vee \mathcal{U}^{-1}$ and the join taken in the lattice of all quasi-uniformities on X. The restriction of the functor T: **QUnif** $\to$ **Top** to **Unif** coincides with the forgetful functor W: **Unif** $\to$ **Top** defined in analogy with T. The composition of the coreflector $*$: **QUnif** $\to$ **Unif** with W is usually denoted by T^*: **QUnif** $\to$ **Top**. Hence one has the equation $W \circ (*) = T^*$.

Recall that a quasi-uniformity $\mathcal{U}$ is said to be totally bounded provided that $\mathcal{U}^*$ is a precompact uniformity (see [17, p. 12]). A quasi-uniform space is *transitive* if it has a base consisting of transitive entourages. Products and subspaces preserve total boundedness (see Fletcher and Lindgren [17, p. 12]). The same is true for transitivity: this follows directly from the fact that $(f \times f)^{-1}U$ is transitive for an arbitrary map f: $X \to Y$ and a transitive entourage U of Y; hence transitivity of quasi-uniform spaces is preserved by taking initial structures (for the definition of the quasi-uniform product see [17, p. 7]).

3. Closure Operators in QUnif

3.1. BASIC DEFINITIONS

DEFINITION 3.1. A collection of functions $\{\varrho_X\colon 2^X \to 2^X\}_{X \in \mathbf{QUnif}}$ is a *closure operator of* **QUnif** if

(1) $(\forall A \subseteq X)\varrho_X(A) \supseteq A$,
(2) $(\forall A \subseteq X)(\forall B \subseteq A)\varrho_X(B) \subseteq \varrho_X(A)$,
(3) $\varrho_X(\emptyset) = \emptyset$,
(4) $f(\varrho_X(A)) \subseteq \varrho_Y(f(A))$ for each quasi-uniformly continuous map $f\colon X \to Y$ and $A \subseteq X$.

We write simply $\varrho(A)$ instead of $\varrho_X(A)$ when no confusion is possible.

As any topological category over **Set**, **QUnif** is $\mathcal{M}$-complete (cf. [15, p. 18]), where $\mathcal{M}$ is the class of all embeddings. Hence, in the terminology of [15] (see also [9] and [12]) these are *grounded* closure operators of **QUnif** with respect to $\mathcal{M}$. In fact, one can relax the requirement $\varrho_X(\emptyset) = \emptyset$ as in [15]. Since this gives only one more *trivial* closure operator (defined by $\varrho(M) = X$ for all M, including $M = \emptyset$) we prefer to impose the natural constraint $\varrho_X(\emptyset) = \emptyset$.

Closure operators of **Top** are defined analogously (cf. [8] or [10]).

The closure operator ϱ is *additive* if $\varrho_X(A \cup B) = \varrho_X(A) \cup \varrho_X(B)$ holds for all $A, B \subseteq X$. Note that a closure operator need not be *idempotent*, i.e. $\varrho(\varrho(A)) = \varrho(A)$ need not always be true.

We say that a map $f: X \to Y$ in **QUnif** is *ϱ-dense* (resp. *ϱ-closed*) if $\varrho_Y(f(X)) = Y$ (resp. $\varrho_Y(f(X)) = f(X)$).

The *discrete* closure operator δ is defined by setting $\delta_X(M) = M$ for each $X \in$ **QUnif** and $M \subseteq X$. A partial order between closure operators is defined by setting $\varrho \leq \varrho'$ whenever $\varrho(M) \subseteq \varrho'(M)$ always holds. Composition $\varrho\varrho'$ between closure operators ϱ and ϱ' is defined by $\varrho\varrho'(M) = \varrho(\varrho'(M))$. For every ordinal α we define the αth iteration of ϱ as the closure operator ϱ^α given recursively by $\varrho^1 = \varrho, \varrho^{\alpha+1} = \varrho\varrho^\alpha$ and $\varrho^\alpha(M) = \bigcup\{\varrho^\beta(M) : \beta < \alpha\}$ for limit α. The *idempotent hull* ϱ^∞ of a closure operator ϱ is the finest idempotent closure operator coarser than ϱ; it can be obtained by $\varrho^\infty(M) = \bigcup\{\varrho^\alpha(M) : \alpha \in \mathbf{Ord}\}$.

EXAMPLE 3.2. The ordinary *topological* closure α in **QUnif** is defined by $\alpha_X(M) = \bigcap\{V^{-1}[M] : V \in \mathcal{U}\} = \overline{M}$, where the closure is taken in $T(X)$. Define also β by $\beta_X(M) = \bigcap\{(V \cap V^{-1})[M] : V \in \mathcal{U}\}$ and γ by $\gamma_X(M) = \bigcap\{V[M] : V \in \mathcal{U}\}$. These are additive and idempotent closure operators in the above sense. In order to keep the standard notation in **QUnif**, we shall sometimes refer to them also as *T-closure* (resp. *T^*-closure*, *T^ι-closure*) and write $\mathrm{cl}_{T(X)}(M)$ (resp. $\mathrm{cl}_{T^*(X)}(M)$, $\mathrm{cl}_{T^\iota(X)}(M)$) instead of $\alpha_X(M)$ (resp. $\beta_X(M)$, $\gamma_X(M)$).

The meet $\varrho \wedge \varrho'$ of two closure operators is defined as the intersection of the ϱ-closure and the ϱ'-closure, i.e. $(\varrho \wedge \varrho')_X(M) = \varrho_X(M) \cap \varrho'_X(M)$. Note that in general this intersection *may fail to be additive* even if both ϱ and ϱ' are additive (note that $\mathrm{cl}_T \wedge \mathrm{cl}_{T^\iota}$-dense sets are usually called *doubly dense*). To get an additive closure operator one has to take the biggest closure operator $E = (\varrho \wedge \varrho')^+$ such that for each $M \subseteq X \in$ **QUnif**, $e_X(M)$ is contained in $\varrho_X(M) \cap \varrho'_X(M)$. E is called *additive core* of ϱ and ϱ' (for example $\mathrm{cl}_{T^*} = (\mathrm{cl}_T \wedge \mathrm{cl}_{T^\iota})^+$). Here is an example of a space X where the meet $\varrho \wedge \varrho'$ of two additive closure operators ϱ, ϱ' fails to be additive. Let $\mathbf{S} \in$ **QUnif** be the Sorgenfrey's line, i.e., the reals $\mathbf{R}$ equipped with the quasi-uniformity $\mathrm{fil}\{U_\varepsilon\}_{\varepsilon>0}$, where $U_\varepsilon := \{(x, y) \in \mathbf{R}^2 : x \leq y < x + \varepsilon\}$. Then with $\varrho = \mathrm{cl}_T$ and $\varrho' = \mathrm{cl}_{T^\iota}$ the additive core $\mathrm{cl}_{T^*} = (\mathrm{cl}_T \wedge \mathrm{cl}_{T^\iota})^+$ is discrete, while $\varrho \wedge \varrho'$ is not discrete (note that $\mathbf{Q}$ is doubly dense in $\mathbf{S}$).

3.2. ADDITIVE CLOSURE OPERATORS VIA NEIGHBOURHOODS

Now we give an equivalent approach to additive closure operators based on the notion of neighbourhood.

DEFINITION 3.3. Let ϱ be a closure operator of **QUnif**. For $X \in$ **QUnif** and a point $x \in X$ a subset U of X is said to be a *ϱ-neighbourhood* of x if $x \notin \varrho(X \setminus U)$.

The next proposition follows directly from Definitions 3.1 and 3.3.

PROPOSITION 3.2.1. *Let ϱ be a closure operator of* **QUnif**, *$f: X \to Y$ be a map in* **QUnif**, *$x \in X$ and let U be a ϱ-neighbourhood of $f(x)$ in Y. Then:*

(a) *the preimage $f^{-1}(U)$ is a ϱ-neighbourhood of x in X,*
(b) *if ϱ is additive, then the ϱ-neighbourhoods of x form a filter.*

It is easy to show that if each space X is equipped axiomatically with families (filters) of ϱ-neighbourhoods of points, such that property (a) from the above proposition holds, then one can define a (additive) closure operator ϱ by setting for each $M \subseteq X$

$$\varrho_X(M) = \{x \in X : (\forall \varrho\text{-neighbourhood } U \text{ of } x)\ U \cap M \neq \emptyset\}.$$

In the sequel we often define only one of these two notions (and use the same notation), in the sense that giving one of them determines automatically the other one. The neighbourhood-structure corresponding, in this sense, to the additive closure operator α of the above example at $x \in X$ is defined by the filter $\{V(x) : V \in \mathcal{U}\}$, β is defined by the filter $\{(V \cap V^{-1})(x) : V \in \mathcal{U}\}$, γ is defined by the filter $\{V^{-1}(x) : V \in \mathcal{U}\}$. A more precise form to speak about "neighbourhood-structure" on **QUnif** is the following.

A *pretopological* (or *Čech*) space is a set X equipped with an additive closure operator ϱ_X (i.e., such that conditions (1)–(3) from Definition 3.1 hold (cf. [5, 18])). Let **PrTop** be the category of pretopological spaces and continuous maps between them (a map f: $X \to Y$ between two pretopological spaces is continuous if (4) in Definition 3.1 holds). Let H: **PrTop** $\to$ **Set** be the obvious forgetful functor. Consider a concrete functor ρ: **QUnif** $\to$ **PrTop** (i.e. commuting with both forgetful functors G: **QUnif** $\to$ **Set** and H). Now $\rho(X)$ has the "ρ-neighbourhood-structure" in the sense discussed above so we have a closure operator on X defined by $M \mapsto \varrho_{\rho(X)}(M)$. The most natural examples (the functors T, T^ι and T^*) were considered in Example 3.2. From this last point of view we can consider together with any additive closure operator ϱ: **QUnif** $\to$ **PrTop** also the composition $\varrho^\iota := \varrho \circ \iota$: **QUnif** $\to$ **PrTop**. Let us note here that the modification $\varrho \mapsto \varrho^\iota$ can be defined in a way to make no recourse to any functor **QUnif** $\to$ **PrTop**, so that it can be extended to the general case of non-necessarily additive closure operator ϱ. Indeed, let ϱ^ι be the closure operator defined by $(\varrho^\iota)_X(M) = \varrho_{\iota(X)}(M)$ for every $X \in$ **QUnif** and every subset M of X.

As an application of this approach to closure operators now we define a series of additive closure operators of **QUnif** which fully exploit the specific features of **QUnif**. Let $n \in \mathbf{N}$ and $\mathcal{U}$ be a quasi-uniformity in X. For $V \in \mathcal{U}$ consider the relation $w_n(V) = V_n \circ \cdots \circ V_2 \circ V_1$, on X, where $V_1 = V$, $V_2 = V^{-1}, \ldots, V_n = V^{(-1)^{n-1}}$. Then $\{w_n(V)(x) : V \in \mathcal{U}\}_{x \in X}$ is a neighbourhood system of a pretopological space $\overline{\boldsymbol{\theta}}_n(X)$ satisfying the condition from Proposition 3.2.1. Hence $\overline{\boldsymbol{\theta}}_n$: **QUnif** $\to$ **PrTop** is a concrete functor, which defines an additive closure operator $\overline{\theta}_n$ for every $n \in \mathbf{N}$. Finally, set $\overline{\theta}_\omega = \bigvee_n \overline{\theta}_n$. Obviously, $\overline{\theta}_1 = \mathrm{cl}_T = \alpha$, while the closure operator $\overline{\theta}_2$

is the quasi-uniform counterpart of Velichko's θ-closure in **Top** (see [34]). More properties of these closure operators will be given in Section 6.2.

3.3. CLOSURE OPERATORS VIA THE FUNCTORS T, T^ι AND T^*

Section 3.2 offered construction of additive closure operators of **QUnif** via concrete functors ρ: **QUnif** $\to$ **PrTop**. In this way the closure operators of **QUnif** are obtained when we *vary the functor* ρ that each time defines an additive closure operator of **QUnif** generated by the usual closure in **PrTop** lifted to **QUnif** along the functor ρ. Here we consider a different way to generate closure operators following the procedure from Example 3.2. Namely, we shall keep fixed one of the three functors T, T^ι, T^*: **QUnif** $\to$ **Top** and we *vary the closure operator* ϱ of **Top**. With the help of the adjunction $F \dashv T$ we "lift" the closure operator ϱ of **Top** along the functor T to obtain a closure operator ${}^T\!\varrho$ of **QUnif** (defined by ${}^T\!\varrho_X(M) = \varrho_{T(X)}(M)$). Analogously, the functors T^ι, T^* can be used to produce closure operators of **QUnif** out of closure operators of **Top**. For the Kuratowski operator K of **Top** we obtain the closure operators ${}^T\!K = \alpha$, ${}^{T^\iota}\!K = \gamma$ and ${}^{T^*}\!K = \beta$ defined in Example 3.2 (since $\alpha_X(M) = \mathrm{cl}_{T(X)}(M)$, $\gamma_X(M) = \mathrm{cl}_{T^\iota(X)}(M)$ and $\beta_X(M) = \mathrm{cl}_{T^*(X)}(M)$). It should be mentioned that the restrictions of the closure operators γ, β and α to **Unif** coincide; in particular, $\beta|_{\mathbf{Unif}}$ coincides with the Kuratowski closure lifted from **Top** to **Unif**.

Here is a formal definition of this procedure of "lifting" closure operators. Let ξ: **QUnif** $\to$ **Top** be a concrete functor that preserves embeddings, e.g., one of the three functors T, T^ι and T^*: **QUnif** $\to$ **Top** (we consider almost exclusively the case $\xi = T$). For every closure operator ϱ of **Top** we shall denote by ${}^\xi\varrho$ the closure operator of **QUnif** defined by ${}^\xi\varrho_X(M) := \varrho_{\xi(X)}(M)$. In general, the correspondence $\varrho \mapsto {}^\xi\varrho$ preserves all operations between closure operators: composition, meets and joins, idempotent hulls. Consequently, it preserves idempotency and additivity of closure operators.

Let us mention that the notation ${}^\xi\varrho$ we choose here for the sake of simplicity diverges from that used in [15, Section 5.13] where the counit of the adjunction appears instead of the functor: in the case of $\xi = T$ that leads to ${}^\varepsilon\varrho$ instead of ${}^T\!\varrho$ used in this paper. Accordingly, the term "pushing closure operators along the counit ε" is replaced here by "lifting closure operators along T".

4. Amalgamation in QUnif

4.1. CONSTRUCTION OF THE AMALGAMATION

For $(X, \mathcal{U}) \in$ **QUnif** and a subspace $M \subseteq X$, one defines the amalgamation $X \amalg_M X$ in **QUnif** as follows. The underlying set of $X \amalg_M X$ is the quotient set of the disjoint union $X \amalg X = X \times \{1, 2\}$ with respect to the equivalence relation $\sim$: $(x, \nu) \sim (y, \mu)$ iff $x = y$ and either $\nu = \mu$ or $x = y \in M$. Let q: $X \amalg X \to X \amalg_M X$ be the natural quotient map. Define the canonical embeddings k_ν: $X \to$

$X \amalg_M X$ by $k_\nu(x) = q(x, \nu)$ for $\nu = 1, 2$. Consider also the canonical embeddings $j_\nu\colon X \to X \amalg X$ for $\nu = 1, 2$ such that j_1 sends X identically to the first copy of X in $X \amalg X$ and j_2 sends X identically to the second copy of X in $X \amalg X$. Finally, define the canonical projection $p\colon X \amalg_M X \to X$ by $p(k_i(x)) = x$ for $x \in X$ and $i = 1, 2$. This gives the commutative diagram

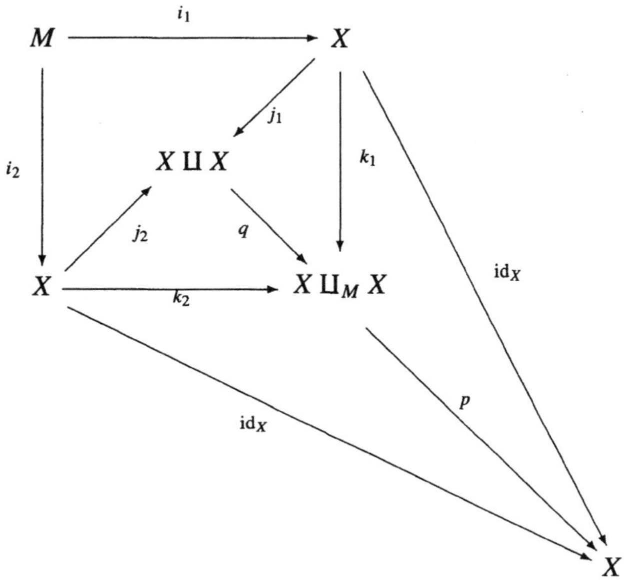

Diagram 1.

The involution (symmetry) $s\colon X \amalg_M X \to X \amalg_M X$, defined by $s(k_1(x)) = k_2(x)$ and $s(k_2(x)) = k_1(x)$ for $x \in X$, is another natural map related to this diagram.

Now we define explicitly a quasi-uniform structure $\mathcal{U} \amalg_M \mathcal{U}$ on $X \amalg_M X$. For $V \in \mathcal{U}$ set

$$\begin{aligned} R_{M,V} &= \{(x, y) \in X \times X : V^{-1}(y) \cap V(x) \cap M \neq \emptyset\} \\ &= \{(x, y) \in X \times X : y \in V(V(x) \cap M)\}, \end{aligned}$$

and note that $R_{M,V} \subseteq V^2$. Define W_V to be the following subset of the Cartesian square $(X \amalg_M X) \times (X \amalg_M X)$:

$$\begin{aligned} W_V := {} & (k_1 \times k_1)(V^2) \cup (k_2 \times k_2)(V^2) \\ & \cup (k_1 \times k_2)(R_{M,V}) \cup (k_2 \times k_1)(R_{M,V}). \end{aligned}$$

Set $\mathcal{U} \amalg_M \mathcal{U} = \mathrm{fil}\{W_V\}_{V \in \mathcal{U}}$. We prove now that with this quasi-uniformity $(k_1, k_2, X \sqcup_M X)$ is the pushout of (M, i_1, i_2).

THEOREM 4.1. *$\mathcal{U} \amalg_M \mathcal{U}$ is a quasi-uniformity on $X \sqcup_M X$ that makes the maps q, k_1, k_2, and p quasi-uniform and the map q a quotient map in* **QUnif**. *Moreover,*

(1) *$\mathcal{U} \amalg_M \mathcal{U}$ is a uniformity iff $\mathcal{U}$ is a uniformity.*
(2) *$\mathcal{U} \amalg_M \mathcal{U}$ is totally bounded iff $\mathcal{U}$ is totally bounded.*
(3) *$\mathcal{U} \amalg_M \mathcal{U}$ is transitive iff $\mathcal{U}$ is transitive.*
(4) *The symmetry s: $X \amalg_M X \to X \amalg_M X$ is a uniform homeomorphism with respect to this quasi-uniformity.*

A direct proof of this theorem can be obtained from the relations

$$W_V^{-1} = W_{V^{-1}},\ (p \times p)(W_V) \subseteq V^2 \quad \text{and} \\ (k_i \times k_i)(V^2) = (k_i(X) \times k_i(X)) \cap W_V \quad (i = 1, 2) \tag{1}$$

for $V \in \mathcal{U}$ (to check the last equality note that the inclusion "$\subseteq$" is trivial, the inclusion "$\supseteq$" follows easily from the inclusion $(p \times p)(W_V) \subseteq V^2$). Since the verification is technically hard to write down, we prefer to give here a more general and rather transparent proof of this theorem based on the following general fact about quotients in **QUnif**.

Let $(X, \mathcal{U}) \in$ **QUnif**, let $\sim$ be an equivalence relation on X and let q: $X \to X/{\sim}$ be the quotient map. We define now

$$H^* = \{(q(x), q(y)) : (\exists z, z' \in X) q(z) = q(z'), (x, z) \in H, (z', y) \in H\}$$

for every $H \in \mathcal{U}$.

LEMMA 4.2. *If the quasi-uniformity $\mathcal{U}$ of X has a base $\mathcal{B}$ of entourages H with the property:*

$$\textit{if } m \sim m', n \sim n', m \neq m', n \neq n', \quad \textit{then } (m, n) \in H^2 \Rightarrow (m', n') \in H^4, \tag{2}$$

then $\mathcal{U}/{\sim} = \mathrm{fil}\{H^* : H \in \mathcal{B}\}$ is the quotient quasi-uniformity of $X/{\sim}$. Moreover, the passage $\mathcal{U} \mapsto \mathcal{U}/{\sim}$ preserves each of the following properties:

(a) *$\mathcal{U}$ is a uniformity,*
(b) *$\mathcal{U}$ is totally bounded,*
(c) *$\mathcal{U}$ is transitive.*

Proof. We first show that $\mathcal{U}/{\sim}$ is a quasi-uniformity. Let $V, H \in \mathcal{B}$ and $H^6 \subseteq V$. We show that $(H^*)^2 \subseteq V^*$. Our proof will fully exploit the monotonicity of the assignment $H \mapsto H^*$ with respect to inclusion.

Suppose that $(q(x), q(y)) \in H^*$ and $(q(y'), q(z)) \in H^*$ where $q(y) = q(y')$. Then there are $m, m', n, n' \in X$ such that $q(m) = q(m')$, $q(n) = q(n')$, $(x, m) \in H$, $(m', y) \in H$, $(y', n) \in H$ and $(n', z) \in H$.

Case 1. At most one of the equalities $m = m'$, $y = y'$ and $n = n'$ fails.

(a) $m = m'$ and $n = n'$. Now $(x, y), (y', z) \in H^2$, so that $(qx, qz) \in (H^2)^* \subseteq V^*$ by monotonicity of *.

(b) $y = y'$, but one of the equalities $m = m'$ and $n = n'$ fails, say $n \neq n'$. Now $(x, y) \in H^2$, $(y, n) \in H$ yield $(x, n) \in H^3$. Therefore $(n', z) \in H \subseteq H^3$ with $n \sim n'$ gives $(qx, qz) \in (H^3)^* \subseteq V^*$.

Case 2. Suppose next that at least two adjacent ones of these equalities fail, say, e.g., $m \neq m'$ and $y \neq y'$. Thus by $(m', y) \in H$ and (2) we conclude $(m, y') \in H^4$ and hence $(x, n) \in H^6$. Therefore $(q(x), q(z)) \in V^*$. The second subcase can be dealt with analogously, using that then $(y, n') \in H^4$ and consequently $(m', z) \in H^6$.

Case 3. We have two nonadjacent inequalities, i.e. $m \neq m'$, $n \neq n'$, but $y = y'$. Then by (2) we conclude that $(m', n) \in H^2$ implies $(m, n') \in H^4$. Thus $(x, n') \in H^5$ and consequently $(qx, qz) \in (H^5)^* \subseteq V^*$.

We deduce that $\mathcal{U}/\sim$ is a quasi-uniformity. In the case when $\mathcal{U}$ is transitive it suffices to note that according to our proof when H is transitive (so $H^6 \subseteq H$) one has $(H^*)^2 \subseteq H^*$, i.e., H^* is transitive as well. This proves that $\mathcal{U}/\sim$ is transitive.

To see that $\mathcal{U}/\sim$ is the quotient quasi-uniformity on $X/\sim$ with respect to q note that for every entourage H of X we have $(q \times q)(H^2) \subseteq H^*$. Hence q is quasi-uniformly continuous with respect to $\mathcal{U}/\sim$. This resolves item (b) of the second assertion of the lemma since total boundedness is preserved by quasi-uniformly continuous maps.

Suppose that $\mathcal{Q}$ is a quasi-uniformity on $X/\sim$ such that q is quasi-uniformly continuous and let $V, U \in \mathcal{Q}$ such that $V^2 \subseteq U$. Then there exists an entourage H of X such that $(q \times q)(H) \subseteq V$. Now this obviously leads to $H^* \subseteq V^2$ by the definition of H^*. Hence $H^* \subseteq U$. Thus $\mathcal{Q} \subseteq \mathcal{U}/\sim$. We conclude that $\mathcal{U}/\sim$ is the finest quasi-uniformity such that q is quasi-uniformly continuous. Hence it is the quotient quasi-uniformity on $X/\sim$ with respect to q.

Finally, to verify item (a) of the second assertion of the lemma let us check that H^* is symmetric, provided that $H \in \mathcal{U}$ is symmetric. Indeed, $(x, y) \in H$ and $(y', z) \in H$ and $q(y) = q(y')$ gives then $(z, y') \in H$ and $(y, x) \in H$ and $q(y) = q(y')$. Therefore, $\mathcal{U}/\sim$ is a uniformity whenever $\mathcal{U}$ is a uniformity. □

Let us mention here that the quasi-uniformity $\mathcal{U}/\sim$ on $X/\sim$ is different from the quasi-uniformity $q(\mathcal{U}) = \text{fil}\{(q \times q)(U) : U \in \mathcal{U}\}$ considered by Himmelberg [21] even in the case of the amalgamation. In fact, according to his Theorem 2 the topology of $(X/\sim, q(\mathcal{U}))$ coincides with the quotient topology of $T(X)/\sim$ when q has compact fibers. We see in the next section that this may fail for $\mathcal{U}/\sim$ even in the case of the amalgamation (cf. Corollary 4.7).

Proof of Theorem 4.1. Our plan is to apply the previous lemma to the pointwise identification $q\colon X \amalg X \to X \amalg_M X$ on $M \subseteq X$. To this end we have to verify that the equivalence relation $\sim$ associated with q satisfies (2), so that the above lemma can be applied. Let us recall that the coproduct quasi-uniformity $\mathcal{U} \amalg \mathcal{U}$ of $X \amalg X$ has a fundamental system of entourages of the form $V \amalg V := (j_1 \times j_1)(V) \cup (j_2 \times j_2)(V)$. In other words, $V \amalg V$ is the entourage of the sum $X \amalg X$ formed by those pairs $(x, y) \in X \amalg X$ such that either x and y belong to the first copy of X and

$(x, y) \in V$ or x and y belong to the second copy of X and $(x, y) \in V$. Now to check (2) for $H = V \amalg V$ note first that

$$(V \amalg V) \circ (V \amalg V) = (V \circ V) \amalg (V \circ V). \qquad (*)$$

Now suppose $m \sim m'$, $n \sim n'$, $m \neq m'$, $n \neq n'$ and $(m, n) \in H^2$. Hence we can assume without loss of generality that $m = j_1(u)$ and $m' = j_2(u)$ for some $u \in M$. Since $(m, n) \in H^2 = (V^2) \amalg (V^2)$ by (*), we have $n = j_1(v)$ for some $v \in X$ with $(u, v) \in V^2$. By $n \sim n'$, $n \neq n'$ we have $n' = j_2(v)$. Hence $(m', n') \in V^2 \amalg V^2 = H^2 \subseteq H^4$.

Let V be an entourage of X. We shall show that $W_V = (V \amalg V)^*$ as defined in the lemma. This will show that $\mathcal{U} \amalg_M \mathcal{U}$ coincides with the quotient quasi-uniformity $(\mathcal{U} \amalg \mathcal{U})/\sim$ of $\mathcal{U} \amalg \mathcal{U}$ with respect to q. The inclusion $W_V \subseteq (V \amalg V)^*$ follows directly from the definitions of W_V and $(V \amalg V)^*$. Assume $(q(x), q(y)) \in (V \amalg V)^*$, then there exist $z, z' \in X \amalg X$ such that $q(z) = q(z')$ and $(x, z) \in V \amalg V$, $(z', y) \in V \amalg V$. Let $q(x) = k_\nu(u)$, $q(y) = k_\mu(v)$, so that $x = j_\nu(u)$, $y = j_\mu(v)$. Now $(x, z) \in V \amalg V$, $(z', y) \in V \amalg V$ imply that $z \in k_\nu(X)$ and $z' \in k_\mu(X)$. From $p(q(z)) = p(q(z'))$ we conclude that $z = k_\nu(w)$ and $z' = k_\mu(w)$ for some $w \in X$. This immediately gives $(u, v) \in V^2$. Hence $\nu = \mu$ gives $(q(x), q(y)) \in (k_\nu \times k_\nu)(V^2) \subseteq W_V$. Assume that $\nu \neq \mu$, say $\nu = 1$, $\mu = 2$, so that $x = j_1(u)$, $y = j_2(v)$. Hence $z \neq z'$. Now $q(z) = q(z')$ implies $w \in M$ that yields $v \in V(V(u) \cap M)$, hence $(u, v) \in R_{M,V}$ and $(q(x), q(y)) \in (k_1 \times k_2)(R_{M,V}) \subseteq W_V$. This finishes the proof of $W_V = (V \amalg V)^*$.

Finally, $\mathcal{U} \amalg_M \mathcal{U}$ is a uniformity whenever $\mathcal{U}$ is a uniformity by the last part of the lemma. The assertions about total boundedness and transitivity follow analogously from Lemma 4.2. Note that all three properties (1)–(3) are preserved under taking subspaces and k_i: $X \to X \amalg_M X$ are embeddings in **QUnif**. Therefore total boundedness (transitivity, uniformity) of the amalgamation entails the same property for X. The assertion about s follows immediately from the categorical property of the amalgamation. □

Note that $m \neq m'$ and $n \neq n'$ is essential constraint in (2) to make it valid for the amalgamation map q: $X \amalg X \to X \amalg_M X$ of the coproduct (take for example $n = n' = m = j_1(u)$, $m' = j_2(u)$ with $u \in M$ and the entourage $H = (X \times X) \amalg (X \times X)$).

4.2. COMPARISON OF THE TOPOLOGIES ON THE AMALGAMATION

Our next aim is to compare the T-topology τ of the so defined amalgamation with the topology τ_t of the amalgamation $TX \amalg_M TX$ defined for the *topological space* TX. The categorical properties of the last amalgamation, and the application of the functor T to the commutative diagram in Diag. 1 yield the continuity of the identity $TX \amalg_M TX \to T(X \amalg_M X)$, i.e., $\tau_t \geq \tau$. We show that when the subspace M of X

is T-closed both topologies coincide. For a uniform T_1-space the converse is also true: if both topologies coincide then M is closed (cf. Corollary 4.9).

Since $s\colon TX \amalg_M TX \to TX \amalg_M TX$ and $s\colon T(X \amalg_M X) \to T(X \amalg_M X)$ are both homeomorphisms, it is clear that for $x \in X$, both topologies coincide at $k_1(x)$ iff they coincide at $k_2(x)$. So that from now on we shall study the set

$$E(X) = \{x \in X : \text{both topologies coincide at } k_1(x)\} \quad \text{and}$$
$$D(X) = X \setminus E(X).$$

Our plan to distinguish τ and τ_t will consist of detecting pairs of points of the amalgamation having the same τ_t-neighbourhoods. Indeed, we show in Lemma 4.4 that $E(X)$ contains the set

$$A_t(X) = \{x \in X : k_1(x) \text{ and } k_2(x) \text{ have the same } \tau_t\text{-neighbourhoods}\}.$$

Clearly, $M \subseteq A_t(X)$. The particular choice of the pairs $k_1(x), k_2(x)$ is explained by the fact that if the space $T(X)$ is T_0 then two distinct points $k_i(x)$ and $k_j(y)$ of $X \amalg_M X$ have the same τ_t-neighbourhoods iff $i \neq j$ and $x = y \in A_t(X)$. Analogous argument shows that two distinct points $k_i(x)$ and $k_j(y)$ of $X \amalg_M X$ have the same τ-neighbourhoods iff $i \neq j$ and $x = y$ belongs to the following counterpart of $A_t(X)$:

$$A(X) = \{x \in X : k_1(x) \text{ and } k_2(x) \text{ have the same } \tau\text{-neighbourhoods}\}.$$

The sets $A_t(X)$ and $A(X)$ admit natural and easy characterizations (cf. 4.4 and 4.5); in particular we show that $A_t(X) \subseteq A(X) \subseteq \overline{M}$. This helps to find an approximation of $E(X)$ in terms of $A_t(X)$, $A(X)$ and $\overline{M}$ (cf. Corollary 4.6).

The next lemma shows that $A(X) \subseteq \overline{M}$ and $D(X) \subseteq \overline{M}$.

LEMMA 4.3. *If x does not belong to the T-closure of M, then $k_1(X \setminus M)$ is a T-neighbourhood of $k_1(x)$. In particular $A(X) \subseteq \overline{M}$ and $D(X) \subseteq \overline{M}$.*

Proof. In fact, if $x \notin \overline{M}$ we can find a V such that $V^2(x)$ misses M, so that $W_V(k_1(x)) = k_1(V^2(x))$. Since $V^2(x)$ is a neighbourhood of x in $T(X)$ that misses M, $k_1(V(x))$ is a neighbourhood of $k_1(x)$ in $TX \amalg_M TX$ contained in $k_1(X \setminus M)$. Hence $k_1(X \setminus M)$ is a T-neighbourhood of $k_1(x)$. Since $k_1(V^2(x))$ and $k_2(V^2(x))$ are disjoint τ-neighbourhoods of $k_1(x)$ and $k_2(x)$ respectively, we have $x \notin A(X)$. This proves $A(X) \subseteq \overline{M}$.

To prove the other inclusion we have to show that $x \in E(X)$. Take an arbitrary τ_t-neighbourhood of $k_1(x)$. It is not restrictive to assume that $O \subseteq k_1(X \setminus M)$, i.e., $O = k_1(U(x))$ for some entourage U such that $U(x)$ does not meet M. If $H^2 \subseteq U$ then $W_H(k_1(x)) = k_1(H^2(x)) \subseteq O$ is a typical T-neighbourhood of $k_1(x)$. This proves $x \in E(X)$, consequently $D(X) \subseteq \overline{M}$. □

One can prove the following for the amalgamation in **Top**. We denote by b (and call it *b-closure*) the Skula-closure in **Top**. Since $A(X)$ (and, obviously, $b_{T(X)}(M)$)

is contained in $\mathrm{cl}_{T(X)}(M)$ by Lemma 4.3, our next considerations occur essentially in $\overline{M}$ $(= \mathrm{cl}_{T(X)}(M))$.

LEMMA 4.4. $A_t(X) = b_{T(X)}(M) \subseteq E(X)$.

Proof. Assume $x \notin b_{T(X)}(M)$. Then there exists a $T(X)$-open neighbourhood U of x in X such that $U \cap \overline{\{x\}} \cap M = \emptyset$. Then for the $T(X)$-open set $V = U \setminus \overline{\{x\}}$ one has $V \cap M = U \cap M$, thus $W = k_1(U) \cup k_2(V)$ is a τ_t-open neighbourhood of $k_1(x)$ which obviously avoids $k_2(x)$. Analogously, $W' = k_1(V) \cup k_2(U)$ is a τ_t-open neighbourhood of $k_2(x)$ which obviously avoids $k_1(x)$. This proves that the points $k_1(x) \neq k_2(x)$ are T_1-separated in $T(X) \amalg_M T(X)$, so that they cannot have the same neighbourhoods in τ_t, i.e., $x \notin A_t(X)$.

Assume that $x \notin A_t(X)$ and let W be a τ_t-open neighbourhood of $k_1(x)$ which avoids $k_2(x)$. Set $U = k_1^{-1}(W)$, $V = k_2^{-1}(W)$. Then U is a $T(X)$-open neighbourhood of x such that $V \cap M = U \cap M$ for the $T(X)$-open set V and $x \notin V$. Then obviously also $V \cap \overline{\{x\}} = \emptyset$, hence $U \cap \overline{\{x\}} \cap M = \emptyset$. Therefore, $x \notin b_{T(X)}(M)$. This finishes the proof of the equality $A_t(X) = b_{T(X)}(M)$.

Let $x \in A_t(X)$. Then $k_1(x)$ and $k_2(x)$ have the same τ_t-neighbourhoods. Take as before a τ_t-open neighbourhood $W = k_1(U) \cup k_2(V)$ of $k_1(x)$, so that U is a $T(X)$-open neighbourhood of x in X and V is a $T(X)$-open set of X with $V \cap M = U \cap M$. By our hypothesis W is a neighbourhood of $k_2(x)$ as well. Hence $k_2(x) \in W$ and consequently $x \in V$. Since $U \cap V$ is a $T(X)$-open neighbourhood of x in X we can choose an entourage H such that $H^2(x) \subseteq U \cap V$. Then $W_H(k_1(x)) = k_1(H^2(x)) \cup k_2(H(H(x) \cap M)) \subseteq W$ is a τ-neighbourhood of $k_1(x)$. Hence W is also a τ-neighbourhood of $k_1(x)$. This proves $x \in E(X)$. □

On the other hand, we have the following situation regarding the topology of the amalgamation in **QUnif**.

LEMMA 4.5. $A(X) = \mathrm{cl}_{T^*(X)}(M)$.

Proof. Every τ-open neighbourhood of $k_1(x)$ contains $k_2(V(V(x) \cap M))$ for some $V \in \mathcal{U}$. Hence the points $k_1(x)$ and $k_2(x)$ of $X \amalg_M X$ have the same τ-neighbourhoods iff $x \in V(V(x) \cap M)$ for each $V \in \mathcal{U}$. This means that $(V \cap V^{-1})(x)$ meets M for each $V \in \mathcal{U}$. The latter property is obviously equivalent to $x \in \mathrm{cl}_{T^*(X)}(M)$. □

Note that for $x \in A(X) \setminus A_t(X)$ the points $k_1(x) \neq k_2(x)$ have the same τ-neighbourhoods, while they do not have the same τ_t-neighbourhoods. Along with the equalities in Lemmas 4.4 and 4.5 this proves the first inclusion in the following corollary, the second one follows from the inclusion in Lemma 4.4.

COROLLARY 4.6. *Let* $D_0 = \mathrm{cl}_{T^*(X)}(M) \setminus b_{T(X)}(M)$ *and* $D_1 = \mathrm{cl}_{T(X)}(M) \setminus b_{T(X)}(M)$. *Then* $D_0 \subseteq D(X) \subseteq D_1$.

This corollary says that the topologies τ and τ_t differ at each point $k_1(x)$ for $x \in D_0$ and coincide at each point $k_1(x)$ when x is in the complement of D_1.

Now we can isolate a sufficient condition ($D_1 = \emptyset$) for the equality $\tau = \tau_t$, as well as a necessary one ($D_0 = \emptyset$). We see below that if X is uniform, then these conditions coincide.

COROLLARY 4.7. (1) *Both topologies coincide when* $\overline{M} = b_{T(X)}(M)$, *in particular when* M *is* T*-closed.*

(2) *If* $\tau = \tau_t$ *then* $\mathrm{cl}_{T^*(X)}(M) = b_{T(X)}(M)$.

COROLLARY 4.8. *Let* $X \in$ **Unif** *and* $M \subseteq X$.

(i) *If* $x \in \overline{M}$, *then* $k_1(x)$ *and* $k_2(x)$ *have the same* $T(\mathcal{U} \amalg_M \mathcal{U})$*-neighbourhoods.*
(ii) *Both topologies on* $X \amalg_M X$ *coincide iff* $\overline{M} = b_{T(X)}(M)$, *i.e.,* $D(X) = \overline{M} \setminus b_{T(X)}(M)$.
(iii) *If* X *is* T_0, *both topologies coincide iff* M *is closed in* TX.

Proof. Part (i) follows from Lemma 4.5 since $T^*(X) = T(X)$. Part (ii) follows from Corollary 4.7. Part (iii) follows from part (ii) and the fact that the Skula topology of the Tychonoff space $T(X)$ is discrete. □

PROPOSITION 4.9. *Let* $X \in$ **Top** *be equipped with its Pervin quasi-uniformity and* $M \subseteq X$. *Then both topologies of* $X \amalg_M X$ *coincide.*

Proof. The assertion immediately follows from the following:

CLAIM. *Let* $X \in$ **Top** *and* $f\colon (X, P(X)) \to Y$ *a surjection onto a set* Y. *Then the quotient quasi-uniformity on* Y *is the Pervin quasi-uniformity of the quotient topology.*

Proof of the claim. Let τ be the quotient topology on Y and let $\mathcal{V}$ be the quotient quasi-uniformity on Y. Then $T(\mathcal{V}) \subseteq \tau$, since $f\colon X \to (X, T(\mathcal{V}))$ is continuous. Since P is a section, $f\colon (Y, P(X)) \to (Y, P(\tau))$ is quasi-uniformly continuous. Thus $P(\tau) \subseteq \mathcal{V}$ and $\tau \subseteq T(\mathcal{V})$. Consequently $T(\mathcal{V}) = \tau$. If $f\colon (X, P(X)) \to (Y, \mathcal{U})$ is quasi-uniformly continuous for some quasi-uniformity $\mathcal{U}$ on Y, then $\mathcal{U}$ is totally bounded, because $P(X)$ is totally bounded. Since the Pervin quasi-uniformity on any topological space is the finest compatible totally bounded quasi-uniformity (see [17, p. 28]), we conclude that $P(\tau) = \mathcal{V}$.

The main assertion follows now by considering the quotient map $q\colon X \amalg X \to X \amalg_M X$. To this end note that $X \amalg X$ carries the Pervin quasi-uniformity of its topology (by an argument similar to the one just given, taking into account that the union of two totally bounded quasi-uniform spaces is totally bounded) provided that X carries the Pervin quasi-uniformity. □

5. Regular Closure Operators

A subcategory $\mathcal{A}$ of **QUnif** is *reflective* if for every $X \in$ **QUnif** there exists a morphism $r_X\colon X \to rX$ (the $\mathcal{A}$*-reflexion* of X) with $rX \in \mathcal{A}$ such that for each morphism $f\colon X \to A$ with $A \in \mathcal{A}$ there exists a unique morphism $f'\colon rX \to A$ such that $f = f'r_X$. In case each r_X is surjective (bijective, a quotient map) the reflective subcategory $\mathcal{A}$ is *epireflective* (resp., *bireflective, strongly epireflective*). The name comes from the fact that surjective (bijective, quotient) maps are precisely the epimorphisms (resp., bimorphism, strong [= extremal] epimorphisms) in **QUnif**. A subcategory $\mathcal{A}$ of **QUnif** closed under **QUnif**-isomorphisms is:

- epireflective iff $\mathcal{A}$ is closed with respect to taking products and subspaces,
- strongly epireflective iff $\mathcal{A}$ is closed with respect to taking products, subspaces and finer quasi-uniformities;
- bireflective iff $\mathcal{A}$ is closed with respect to taking products and subspaces and contains the full subcategory $\mathcal{I}nd$ of **QUnif** having as object all indiscrete spaces.

Consequently, intersections of epireflective (bireflective, strongly epireflective) subcategories have the same property, thus every subcategory $\mathcal{A}$ of **QUnif** admits an *epireflective* (*bireflective, strongly epireflective*) *hull* $E(\mathcal{A})$ ($B(\mathcal{A})$, $S(\mathcal{A})$ resp.) – the smallest epireflective (resp., bireflective, strongly epireflective) subcategory of **QUnif** containing $\mathcal{A}$. Clearly, the bireflective hull of $\mathcal{A}$ is $B(\mathcal{A}) = E(\mathcal{A}_0 \cup \mathcal{I}nd)$. It turns out that a morphism f in $\mathcal{A}$ is epi in $\mathcal{A}$ iff f is epi in $E(\mathcal{A})$ iff f is epi in $S(\mathcal{A})$, hence the study of epimorphisms and cowellpoweredness of subcategories of **QUnif** is reduced to strongly epireflective ones (see Section 8).

The following easy to check lemma determines the largest proper strongly epireflective subcategory of **QUnif**.

LEMMA 5.1. *For a* $(X, \mathcal{U}) \in$ **QUnif** *the following are equivalent:*

(i) $\bigcap\{(V \cap V^{-1}) : V \in \mathcal{U}\} = \Delta_X$;
(ii) $(X, T(\mathcal{U})) \in \mathbf{Top}_0$;
(iii) $(X, T(\mathcal{U}^{-1})) \in \mathbf{Top}_0$;
(iv) *the topological space* $(X, T(\mathcal{U}^*))$ *is Hausdorff.*

The full subcategory $\mathbf{QUnif}_0$ of **QUnif** consisting of spaces with these equivalent properties is strongly epireflective. The reflector $\mathbf{QUnif} \to \mathbf{QUnif}_0$ is defined as follows: for $(X, \mathcal{U}) \in$ **QUnif** set $L_X := \bigcap\{(V \cap V^{-1}) : V \in \mathcal{U}\}$. Clearly L_X is an equivalence relation on X, such that the quotient map $\rho\colon X \to RX = (X/L_X, \rho(\mathcal{U}))$ is the $\mathbf{QUnif}_0$-reflexion of X. Indeed, the map ρ satisfies (2) for every $H \in \mathcal{U}$ since $\rho(m) = \rho(m')$, $\rho(n) = \rho(n')$ along with $(m, n) \in H^2$ imply $(m', n') \in H \circ H^2 \circ H \subseteq H^4$ (as $(m', m), (n, n') \in H$). Therefore the quotient quasi-uniformity $\mathcal{U}/{\sim}$ of RX with base $\{V^* : V \in \mathcal{U}\}$ is defined as in Lemma 4.2 and it is easy to check that $RX \in \mathbf{QUnif}_0$, so that RX is the $\mathbf{QUnif}_1$-reflexion of X. Let us verify that ρ sends β-closed sets of X to β-closed sets of RX. Indeed, let $x \in X$ and assume that $\rho(x) \in \beta_{RX}\rho(M)$. Then for every $V \in \mathcal{U}$ the typical β-neighbourhood $W = (V^* \cap (V^*)^{-1})(\rho(x))$ meets $\rho(M)$. Since V^* is contained in $(\rho \times \rho)(V^3)$, we see that $\rho^{-1}(W) \subseteq V^3(x)$ and therefore

$V^3(x)$ meets M. Hence $x \in \beta_X(M)$ since this occurs for every V. This proves that $\rho(\beta_X(M)) = \beta_Y(\rho(M))$.*

To see that $\mathbf{QUnif}_0$ is the largest proper strongly epireflective subcategory of **QUnif** note that if $X \notin \mathbf{QUnif}_0$ then X contains a two-point indiscrete subspace. Hence any epireflective subcategory $\mathcal{A}$ of **QUnif** that contains X must contain also the subcategory $\mathcal{I}nd$. If $\mathcal{A}$ is also strongly epireflective then $\mathcal{A} \supseteq S(\mathcal{I}nd) =$ **QUnif**. The fact that $\mathbf{QUnif}_0$ is the largest proper strongly epireflective subcategory of **QUnif** along with the initiality of the $\mathbf{QUnif}_0$-reflexions of every space in **QUnif** permit to apply a result of Marny [29, Proposition 3] that describes the bireflective subcategories of **QUnif** as follows. For every bireflective subcategory $\mathcal{B}$ consider the subcategory $\mathcal{B}_0 = \mathcal{B} \cap \mathbf{QUnif}_0$. Then one can recover $\mathcal{B}$ from $\mathcal{B}_0$ as $\mathcal{B} = E(\mathcal{B}_0 \cup \mathcal{I}nd)$. In this way one obtains a bijection between the disjoint (big) lattices of bireflective subcategories of **QUnif** from one hand and epireflective subcategories of **QUnif** contained in $\mathbf{QUnif}_0$ on the other hand.

Each subcategory $\mathcal{A}$ of **QUnif** induces a *regular closure operator* $\mathrm{reg}^{\mathcal{A}}$ on **QUnif** in the following way. For $X \in \mathbf{QUnif}$ and $M \in 2^X$ set

$$\mathrm{reg}_X^{\mathcal{A}}(M) = \bigcap\{\mathrm{Eq}(f, g) : f, g\colon X \to A \in \mathcal{A},\ f|_M = g|_M\}, \qquad (*)$$

where $\mathrm{Eq}(f, g) = \{x \in X : f(x) = g(x)\}$. If $\mathcal{A}$ is reflective, then in $(*)$ only one pair of maps can be taken. Namely consider the reflection in $\mathcal{A}$

$$\rho = r_{X \amalg_M X}\colon X \amalg_M X \to r(X \amalg_M X).$$

Then

$$\mathrm{reg}_X^{\mathcal{A}}(M) = \mathrm{Eq}(\rho k_1, \rho k_2) = \{x \in X : \rho k_1(x) = \rho k_2(x)\} \qquad (3)$$

([12], 3.1]). The regular closure is idempotent, it is also called $\mathcal{A}$-*closure* and we abbreviate $\mathrm{reg}^{\mathcal{A}}$-closed ($\mathrm{reg}^{\mathcal{A}}$-dense) by $\mathcal{A}$-*closed* ($\mathcal{A}$-*dense*). In particular, (3) yields that M is $\mathcal{A}$-closed whenever $X \amalg_M X \in \mathcal{A}$ (cf. Lemma 5.3).

For a subcategory $\mathcal{B}$ of **Top** a regular closure operator $\mathrm{reg}^{\mathcal{B}}$ of **Top** is defined analogously.

We show below that β is regular, while neither α nor γ are regular closure operators of **QUnif** (see Examples 5.4, 5.7). To this end we need some technique to compute the regular closure.

The following lemma permits to recover the regular closure of an epireflective subcategory $\mathcal{A}$ of **QUnif** from its restriction on $\mathcal{A}$, i.e., the $\mathcal{A}$-closure is determined by its values on $\mathcal{A}$-spaces (this was already contained in the above comments).

LEMMA 5.2 ([7, Lemma 3.4]). *Let $\mathcal{A}$ be an epireflective subcategory of* **QUnif**, *$X \in$* **QUnif** *and $r\colon X \to rX$ be the $\mathcal{A}$-reflexion of X. Then*

$$\mathrm{reg}_X^{\mathcal{A}}(M) = r^{-1}(\mathrm{reg}_{rX}^{\mathcal{A}}(r(M)))$$

* and yields also that the map ρ is T^*-open.

for every $M \subseteq X$.

The next lemma is of great help. It was isolated by Frolík from various concrete cases in **Top**. It is crucial for the computation of the $\mathcal{A}$-closure on $\mathcal{A}$-spaces by means of the amalgamation.

LEMMA 5.3 ([12, Corollary 3.10]). *Let* $\mathcal{A}$ *be a strongly epireflective subcategory of* **QUnif**, $X \in$ **QUnif** *and* $M \subseteq X$. *If* $X \in \mathcal{A}$ *then* M *is* $\mathcal{A}$*-closed iff* $X \amalg_M X \in \mathcal{A}$.

As an application we show how Frolík's lemma can be used to compute the regular closure of $\mathbf{QUnif}_0$. It seems that the essential part of this result was known to Brümmer (for the validity of $\beta = \mathrm{reg}^{\mathbf{QUnif}_0}$ on $\mathbf{QUnif}_0$ we found a citation to [3] in Holgate's thesis [23, Lemma 3.2.15]), but we offer a proof here since we have no access to [3].

EXAMPLE 5.4 ([23, Theorem 3.2.16]). $\mathrm{reg}^{\mathbf{QUnif}_0} = \beta$.

Proof. To prove the inequality $\beta \leq \mathrm{reg}^{\mathbf{QUnif}_0}$, note that both closure operators are idempotent, hence it suffices to show that each equalizer $\mathrm{Eq}(f, g)$ with $f, g\colon X \to Y \in \mathbf{QUnif}_0$ is β-closed. In fact, $\mathrm{Eq}(f, g) = h^{-1}(\Delta_Y)$ for $h = \langle f, g\rangle\colon X \to Y \times Y$, and the diagonal Δ_Y is β-closed in $Y \times Y$ by (i) of Lemma 5.1. To prove the opposite inequality, consider a β-closed subset M of X. By Lemma 4.5 $k_1(x)$ and $k_2(x)$ are separated in $X \amalg_M X$. In case $X \in \mathbf{QUnif}_0$ the other pairs of distinct points of $X \amalg_M X$ can be separated via the projection $X \amalg_M X \to X$, so that by Frolík's lemma, M is $\mathbf{QUnif}_0$-closed. In the general case one has to observe that the $\mathbf{QUnif}_0$-reflexion $\rho\colon X \to rX$ sends β-closed sets to β-closed sets. □

We have seen above that $\mathbf{QUnif}_0 = T^{-1}(\mathbf{Top}_0) = (T^{\iota})^{-1}(\mathbf{Top}_0)$. This suggests to examine the categories $T^{-1}(\mathbf{Top}_1)$ and $(T^{\iota})^{-1}(\mathbf{Top}_1)$ and see if they coincide as well.

LEMMA 5.5. *For a* $(X, \mathcal{U}) \in$ **QUnif** *the following are equivalent:*

(i) $\bigcap\{V : V \in \mathcal{U}\} = \Delta_X$;
(ii) $\bigcap\{V^{-1} : V \in \mathcal{U}\} = \Delta_X$;
(iii) $(X, T(\mathcal{U})) \in \mathbf{Top}_1$;
(iv) $(X, T(\mathcal{U}^{-1})) \in \mathbf{Top}_1$.

Let $\mathbf{QUnif}_1$ be the full subcategory of **QUnif** containing all spaces $(X, \mathcal{U})$ with this property. Then $\mathbf{QUnif}_1 = T^{-1}(\mathbf{Top}_1) = (T^{\iota})^{-1}(\mathbf{Top}_1)$. Now we compute the regular closure operator of $\mathbf{QUnif}_1$.

EXAMPLE 5.6. $\beta \leq \mathrm{reg}^{\mathbf{QUnif}_1}$ and they coincide on $\mathbf{QUnif}_1$.

Proof. The inequality follows from Example 5.4 and the fact that the assignment $\mathcal{A} \mapsto \mathrm{reg}^{\mathcal{A}}$ is order-reversing, so that $\beta = \mathrm{reg}^{\mathbf{QUnif}_0} \leq \mathrm{reg}^{\mathbf{QUnif}_1}$. To prove the opposite inequality on $\mathbf{QUnif}_1$ take a T^*-closed subset M of $X \in \mathbf{QUnif}_1$. If $x \notin M$ then by Lemma 4.5 $k_1(x)$ and $k_2(x)$ are T_1-separated in $X \amalg_M X$. Moreover, points $z, z' \in X \amalg_M X$ with distinct projection $p(z) \neq p(z')$ can be T_1-separated (resp. T_0-separated) by means of inverse images under p of appropriate neighbourhoods of $p(z) \neq p(z')$ in $T(X)$ since $T(X)$ is T_1 by assumption. □

We shall see later, that $\mathrm{reg}^{\mathbf{QUnif}_1}$ and β cannot coincide out of $\mathbf{QUnif}_1$.

A similar argument gives the following:

EXAMPLE 5.7. The full subcategory $\mathbf{QUnif}_2$ of $\mathbf{QUnif}$ consisting of all spaces $X \in \mathbf{QUnif}$ with $\bigcap\{V \circ V^{-1} : V \in \mathcal{U}\} = \Delta_X$ coincides with $T^{-1}(\mathbf{Haus})$, i.e., these are the T-Hausdorff quasi-uniform spaces (this is proved in a more general case in Lemma 6.4). The regular closure operator of $\mathbf{QUnif}_2$ coincides on $\mathbf{QUnif}_2$ with α (this follows from the description of the amalgamation in Section 4.1, Lemma 4.5). However, they do not coincide out of $\mathbf{QUnif}_2$ since, unlike β, neither α nor γ are regular closure operators ([23, Proposition 3.2.2]).

Unlike the previous examples, $\mathbf{QUnif}_2 = T^{-1}(\mathbf{Haus}) \neq (T^{\iota})^{-1}(\mathbf{Haus}) = \iota(\mathbf{QUnif}_2)$. In fact, there exists an $X \in \mathbf{QUnif}_2$ (actually, $T(X)$ is discrete), such that $T^{\iota}(X) \notin \mathbf{Haus}$ (see Example 6.11).

The following example shows that the computation of the regular closure operator of some strongly epireflective subcategory $\mathcal{A}$ in $\mathbf{QUnif}$ is not possible by "lifting" along the forgetful functor T: $\mathbf{QUnif} \to \mathbf{Top}$ the regular closure operator of $T(\mathcal{A})$ even in the case when $\mathcal{A}$ is T-saturated, i.e. $\mathcal{A} = T^{-1}(T(\mathcal{A}))$. One example to this effect is $\mathbf{QUnif}_0 = T^{-1}(\mathbf{Top}_0)$ (see Example 5.4 and note that $\beta \neq {}^{T}b$). Here we give another example.

EXAMPLE 5.8. Take $\mathcal{A} = \mathbf{QUnif}_1 = T^{-1}(\mathbf{Top}_1)$. Any space $X \in \mathcal{A}$ such that $T^*(X)$ is not discrete will work as the $\mathbf{QUnif}_1$-closure coincides on X with $\beta_X = \mathrm{cl}_{T^*(X)}$ because of Example 5.6, while the $\mathbf{Top}_1$-closure in $T(X)$ is discrete since $\mathrm{reg}^{\mathbf{Top}_1}$ is discrete on $\mathbf{Top}_1$ ([8]) and $T(X) \in \mathbf{Top}_1$. For an example of a space X with this property take the reals $\mathbf{R}$ with the metric uniformity.

In the following example we show that a regular closure operator need not be additive. Even if such an example is known in $\mathbf{Top}$, and essentially used here, we cannot argue directly by lifting it along T and exploit preservation of (non-) additivity, since regularity need not be preserved by the lifting as shown in the previous example.

EXAMPLE 5.9. Let $\mathcal{X} = \{X_i\}_{i \in I}$ be a strongly rigid family in $\mathbf{QUnif}$, i.e., the only morphisms $X_i \to X_{i'}$, $i, i' \in I$, are either constants or identities. Then for the epireflective subcategory $\mathcal{A}$ of $\mathbf{QUnif}$ generated by $\mathcal{X}$ the regular closure operator $\mathrm{reg}^{\mathcal{A}}$ is not additive. In fact, for $X = X_i \in \mathcal{X}$ with $|X| > 2$ every singleton $\{x\}$, $x \in X$, is $\mathcal{A}$-closed, while every doubleton is $\mathcal{A}$-dense. To get such a strongly rigid family in **QUnif** one can take a strongly rigid family $\mathcal{Y}$ in **Top** (see [25] or [6] for the existence of such families or just single spaces – a Cook continuum, for example) and take its image $\mathcal{X} = P(\mathcal{Y})$ under Pervin's functor P in **QUnif**. Since T is faithful, we easily conclude that $\mathcal{X}$ is a strongly rigid family in **QUnif**. The importance of the condition $|X| > 2$ is discussed in Remark 6.15.

6. Correspondences between Subcategories and Closure Operators of QUnif

Here we consider subcategories of **QUnif** defined by means of closure operators.

6.1. TWO WAYS OF POINT-SEPARATION

Every closure operator ϱ of **QUnif** gives rise to a strongly epireflective subcategory $\Delta(\varrho)$ of **QUnif** (the *Delta-subcategory* of ϱ) having as objects all spaces $X \in$ **QUnif** such that the diagonal Δ_X is ϱ-closed in $X \times X$.

LEMMA 6.1 ([15, Corollary 7.1]). *For any family* $\{\varrho_i\}$ *of closure operators of* **QUnif** $\Delta(\bigvee_i \varrho_i) = \bigcap_i \Delta(\varrho_i)$.

EXAMPLE 6.2. (1) $\mathbf{QUnif}_0 = \Delta({}^Tb) = \Delta(\beta)$ (by Lemma 5.1), $\mathbf{QUnif}_2 = \Delta(\alpha)$ (by Example 5.7) and $\mathbf{QUnif}_1 = \Delta({}^T\pi)$, where π is the closure operator of **Top** defined as follows: $x \in \pi_X(M)$ in $X \in$ **Top** iff $\overline{\{x\}} \cap M \neq \emptyset$, i.e., $\overline{\{x\}}$ is the smallest π-neighbourhood of x. Then π is an idempotent and additive closure operator of **Top**, hence ${}^T\pi$ is an idempotent and additive closure operator of **QUnif**. An alternative way to describe ${}^T\pi$ is to take, for a space $X \in$ **QUnif** and $M \subseteq X$, the T-closure of M computed in the space $(PT)^\iota(X)$.

(2) The Delta-subcategory helps to see how big is the difference between $\beta = (\alpha \wedge \gamma)^+$ and $\alpha \wedge \gamma$. To this end we compare $\mathbf{QUnif}_0 = \Delta(\beta)$ with $\Delta(\alpha \wedge \gamma) \subset \mathbf{QUnif}_1$.

Let us prove the inclusion $\Delta(\alpha \wedge \gamma) \subseteq \mathbf{QUnif}_1$. Note that $X \in \Delta(\alpha \wedge \gamma)$ iff $\forall x \in X, \forall y \in X$ with $x \neq y$ there exists $V \in \mathcal{U}$ such that either $V(x) \cap V(y) = \emptyset$ or $V^{-1}(x) \cap V^{-1}(y) = \emptyset$. Then $y \notin V(x) \cup V^{-1}(x)$. By Lemma 5.6 this proves that $X \in \mathbf{QUnif}_1$. To see that the inclusion is proper take X to be the Sorgenfrey line (cf. Section 3.1) and $M = \mathbf{Q}$ the set of rationals, then M is β-closed, so that the amalgamation $Y = X \amalg_M X$ belongs to $\mathbf{QUnif}_1$ by Example 5.6, while for each irrational point $x \in X$ the pair $(k_1(x), k_2(x))$ is in the $(\alpha \wedge \gamma)$-closure of the diagonal of $Y \times Y$, but not in the diagonal Δ_Y, so $Y \in \mathbf{QUnif}_1 \setminus \Delta(\alpha \wedge \gamma)$.

Let us note that $\Delta(\alpha \wedge \gamma)$ properly contains both $\mathbf{QUnif}_2$ and $\iota(\mathbf{QUnif}_2)$. In fact, as $\alpha \geq \alpha \wedge \gamma$, $\mathbf{QUnif}_2 \subseteq \Delta(\alpha \wedge \gamma)$. Since $\Delta(\alpha \wedge \gamma)$ is obviously invariant under ι, we also get $\iota(\mathbf{QUnif}_2) \subseteq \Delta(\alpha \wedge \gamma)$. As $\mathbf{QUnif}_2 \neq \iota(\mathbf{QUnif}_2)$ (see Example 6.11), the above inclusions are proper.

If ϱ is a closure operator on **QUnif**, a space X of **QUnif** is *ϱ-separated* if every pair of distinct points of X can be separated by disjoint ϱ-neighbourhoods in X. We denote by $\mathcal{T}_\varrho$ the full subcategory of ϱ-separated spaces of **QUnif**. According to [7], Proposition 5.3, $\mathcal{T}_\varrho$ is a strongly epireflective subcategory of **QUnif**. Obviously, X is ϱ-separated iff X equipped with the ϱ-neighbourhood structure is a separated pretopological (i.e. Čech) space in the usual sense. Hence $\mathcal{T}_\varrho \subseteq \Delta(\varrho)$, but this inclusion may be strict (see Example 6.7).

REMARK 6.3. A sufficient condition for the equality $\mathcal{T}_\varrho = \Delta(\varrho)$ is ϱ to be *finitely multiplicative*: for every $X \in$ **QUnif** the ϱ-neighbourhood structure of $X \times X$ is just the product of the ϱ-neighbourhood structures of X, i.e. for $(x, y) \in X \times X$ a typical ϱ-neighbourhood of (x, y) in $X \times X$ has the form $U \times V$, where U is a ϱ-neighbourhood of x in X and V is a ϱ-neighbourhood of y in Y.

We shall use the following abbreviations: $\mathcal{T}_n := \mathcal{T}_{\overline{\theta}_n}$ and $\mathcal{T}_\omega := \mathcal{T}_{\overline{\theta}_\omega}$. Obviously, $\mathcal{T}_1 = \mathbf{QUnif}_2$, as $\overline{\theta}_1 = \alpha$. Here we show that this can be extended to all $n \in \mathbf{N}$.

LEMMA 6.4. $\Delta(\alpha \wedge \gamma) = \mathcal{T}_{\alpha\wedge\gamma}$, $\Delta(\alpha \vee \gamma) = \mathcal{T}_{\alpha\vee\gamma}$ *and* $\Delta(\overline{\theta}_n) = \mathcal{T}_n = \{X : \bigcap_{V\in\mathcal{U}} w_n(V) \circ w_n(V)^{-1} = \Delta_X\}$ *for every natural number* $n \geq 1$.

Proof. Assume that $(X, \mathcal{U}) \in \Delta(\alpha \wedge \gamma)$ and x, y are distinct points of X. Then there exists an entourage $V \in \mathcal{U}$ such that either $V(x) \cap V(y) = \emptyset$ or $V^{-1}(x) \cap V^{-1}(y) = \emptyset$. Note that both $V(x)$ and $V^{-1}(x)$ are $\alpha \wedge \gamma$-neighbourhoods of x, while $V(y)$ and $V^{-1}(y)$ have the analogous property with respect to y. Hence x and y have disjoint $\alpha \wedge \gamma$-neighbourhoods. This proves $X \in \mathcal{T}_{\alpha\wedge\gamma}$ and the first statement.

Now assume that $(X, \mathcal{U}) \in \Delta(\alpha \vee \gamma)$ and x, y are distinct points of X. Then there exists an entourage $V \in \mathcal{U}$ such that $V(x) \cap V(y) = \emptyset$, $V^{-1}(x) \cap V^{-1}(y) = \emptyset$. Since $\alpha \vee \gamma \geq \alpha$ and $\alpha \vee \gamma \geq \gamma$, we can also assume that $y \notin V^2(x)$ and $x \notin V^2(y)$. Then $V(x) \cup V^{-1}(x)$ and $V(y) \cup V^{-1}(y)$ are disjoint $\alpha \vee \gamma$-neighbourhoods of x and y. This proves $X \in \mathcal{T}_{\alpha\vee\gamma}$ and the second statement.

By Remark 6.3 to prove the equality $\Delta(\overline{\theta}_n) = \mathcal{T}_n$ it suffices to note that the $\overline{\theta}_n$-closure is finitely multiplicative, i.e. the $\overline{\theta}_n$-neighbourhood structure of $X \times X$ is just the Cartesian product of the $\overline{\theta}_n$-neighbourhood structures of X. Indeed, the quasi-uniform structure $\mathcal{U}_{X\times X}$ of the Cartesian product $X \times X$ has as basic entourages of the diagonal in $(X \times X) \times (X \times X)$ the sets of the form $U_V := \{((x, y), (x', y')) : (x, x'), (y, y') \in V\}$, where $V \in \mathcal{U}$. Moreover, the equalities $U_V(x, y) = V(x) \times V(y)$, $U_V^{-1} = U_{V^{-1}}$ and $U_V \circ U_{V'} = U_{V\circ V'}$ hold for $x, y \in X$ and $V, V' \in \mathcal{U}$. This yields $w_n(U_V)(x, y) = U_{w_n(V)}(x, y) = w_n(V)(x) \times w_n(V)(y)$.

Finally, to prove the last equality we show that $\overline{\theta}_n(\Delta_X) = \bigcap_{V\in\mathcal{U}} w_n(V) \circ w_n(V)^{-1}$, the $\overline{\theta}_n$-closure being taken in $X \times X$. Indeed, as shown above, a typical $\overline{\theta}_n$-neighbourhood for $(x, y) \in X \times X$ has the form $U = w_n(V)(x) \times w_n(V)(y)$, where $V \in \mathcal{U}$. It remains to note now that U meets Δ_X iff $(x, y) \in w_n(V) \circ w_n(V)^{-1}$. □

Here is the relation (known as "Diagonal Theorem") between strongly epireflective subcategories of **QUnif** and regular closure operators.

THEOREM 6.5 ([19]). *The correspondence $C \mapsto \Delta(C)$, $\mathcal{A} \mapsto \mathrm{reg}^{\mathcal{A}}$ is a Galois connection between closure operators of* **QUnif** *and subcategories of* **QUnif**, *which restricted to regular closure operators and strongly epireflective subcategories of* **QUnif** *becomes a Galois equivalence.*

Following [12, 4.1], we shall refer to the above Galois connection as *Salbany correspondence*. It follows from this theorem that for every subcategory $\mathcal{A}$ of **QUnif** $S(\mathcal{A}) = \Delta(\mathrm{reg}^{\mathcal{A}})$ and for every closure operator C the Δ-closure is the smallest regular closure operator containing C.

LEMMA 6.6. *If ϱ is a closure operator of* **Top**, *then for its T-lifting ${}^{T}\!\varrho$ we have:*

$$\Delta({}^{T}\!\varrho) = T^{-1}(\Delta(\varrho)) \quad \textit{and} \quad \mathcal{T}_{{}^{T}\!\varrho} = T^{-1}(\mathcal{T}_{\varrho}).$$

EXAMPLE 6.7. There exists a closure operator ϱ of **Top** such that the inclusion $\mathcal{T}_{\varrho} \subseteq \Delta(\varrho)$ is proper in **Top** ([7, Example 6.7]). Now apply Lemma 6.6 to "lift" it in **QUnif**.

6.2. SEPARATION BY ENTOURAGES

Here we compare $\mathcal{T}_n$ with Viglino's [36] axioms $S(n)$ in **Top**. In order to do it we need the notion of θ_n-neighbourhood in **Top**. Our definition differs from the one given in [10].

DEFINITION 6.8. Let n be a natural number and $k = n/2$ in case n is even, otherwise $k = (n+1)/2$. If X is a topological space and $U \subseteq X$ and $x \in X$ then we say that U is a *θ_n-neighbourhood* of x if there is a family of open sets $\{U_s : 1 \leq s \leq k\}$ such that

(i) $x \in U_1$ and $\overline{U_s} \subseteq U_{s+1}$ for $1 \leq s < k$;
(ii) $U_k \subseteq U$ if n is odd, otherwise $\overline{U_k} \subseteq U$.

The closure operator θ_1 is the usual Kuratowski closure, while θ_2 is Velichko's θ-closure. A space $X \in$ **Top** is an $S(n)$-space if every pair of distinct points have disjoint θ_n-neighbourhoods. Obviously, $S(1)$-spaces coincide with Hausdorff spaces, while $S(2)$-spaces are known also as *Urysohn spaces*. We denote by $S(n)$ the category of topological $S(n)$-spaces. For examples distinguishing all these categories see Porter and Votaw [30].

LEMMA 6.9. *For every n one has $\overline{\theta}_n \geq {}^{T}\theta_n$. If $(X, \mathcal{U})$ satisfies also $\mathcal{U} \supseteq P(T(\mathcal{U}))$, then $\overline{\theta}_n = {}^{T}\theta_n$ on X.*

Proof. To show $\overline{\theta}_n \geq {}^T\theta_n$ take $X \in \mathbf{QUnif}$, $x \in X$ and a $\overline{\theta}_n$-neighbourhood U of x. Then there exists V such that $w_n(V)(x) \subseteq U$. Let U_1 be an open neighbourhood of x contained in $V(x)$. Then $\overline{U_1} \subseteq V^{-1} \circ V(x)$. Since $V \circ V^{-1} \circ V(x)$ is a neighbourhood of every $y \in V^{-1} \circ V(x)$, we can find an open set U_2 in $V \circ V^{-1} \circ V(x)$ containing $\overline{U_1}$. We continue in this way until we build the chain $U_1 \subseteq U_2 \subseteq \cdots$ in U that defines a θ_n-neighbourhood as in Definition 6.8.

Now, for the rest of the proof assume that the Pervin quasi-uniformity of $T(X)$ is contained in the quasi-uniformity $\mathcal{U}$ of X. For $x \in X$ fix a ${}^T\theta_n$-neighbourhood U of x. Let $\{U\}_{i=1}^k$ be a chain of neighbourhoods of x in $T(X)$ witnessing, as in Definition 6.8, that U is a θ_n-neighbourhood of x. Now set $V_1 = S_{U_1}$, $V_2 = S_{X\setminus\overline{U_1}}, \ldots, V_{2i-1} = S_{U_i}$, $V_{2i} = S_{X\setminus\overline{U_i}}, \ldots$ for $1 \leq 2i-1 < 2i \leq k$. Then $V_i \in P(T(\mathcal{U}))$, so that $V = \bigcap_{i=1}^n V_i \in \mathcal{U}$ by our assumption. On the other hand, $\underbrace{(V_n^{(-1)^{n-1}} \circ \cdots \circ V_2^{-1} \circ V_1)}_{n}(x)$ coincides with U_k or $\overline{U_k}$ depending on whether k is odd or even. Hence $w_n(V)(x) \subseteq U$, so that U is a $\overline{\theta}_n$-neighbourhood of x as well. □

These two closure operators do not coincide in general (see Example 6.11 below).

COROLLARY 6.10. *$T(\mathcal{T}_n) = S(n)$ for $n \geq 1$. Moreover, for spaces X having the Pervin quasi-uniformity $X \in \mathcal{T}_n$ is equivalent to $X \in T^{-1}(S(n))$.*

Now we establish the inequality $\mathcal{T}_n \neq T^{-1}(S(n))$. Let us note that $\mathcal{T}_{n+1} \subseteq \iota(\mathcal{T}_n) \subseteq \mathcal{T}_{n-1}$. In particular, $\mathcal{T}_2 \subseteq \iota(\mathcal{T}_1)$.

EXAMPLE 6.11. There exists a space $X \in \mathbf{QUnif}$ such that $T(X)$ is discrete and $X \notin \iota(\mathcal{T}_1)$. Hence, for $n > 1$ the category $T^{-1}(S(n))$ properly contains $\mathcal{T}_n$.

Proof. Set $X = \mathbf{N} \cup \{\infty\}$, and for each n set $L_n(\infty) = \{\infty\}$, $L_n(k) = \{k\}$ if $k \leq n$, and $L_n(k) = X$ otherwise. Then the sets $L_n = \bigcup_{x\in X}\{x\} \times L_n(x)$ generate a quasi-uniformity $\mathcal{U}$ on X whose induced topology $T(\mathcal{U})$ is discrete, so that $TX \in S(n)$ with $n \leq \omega$. On the other hand, for each n obviously $n+1 \in L_n^{-1}(1) \cap L_n^{-1}(\infty)$, so that $X \notin \iota(\mathcal{T}_1)$. □

REMARK 6.12. (1) On the space X in the above example the closure operator ${}^T\theta_n$ is discrete for every n, while $\overline{\theta}_2$ (and a fortiori $\overline{\theta}_n$ for $n > 1$) is nondiscrete.

(2) By (1) we get $X \in \mathbf{QUnif}_2 = T^{-1}(S(1))$, but $\iota(X) \notin \mathbf{QUnif}_2$.

(3) The above example shows that for $n > 1$ the closure operator $\overline{\theta}_n$ does not coincide with the lifting along T of any closure operator ϱ of **Top**. In fact, assume that $\overline{\theta}_n = {}^T\varrho$. Then for every space $X \in \mathbf{QUnif}$ satisfying $\mathcal{U} \supseteq P(T(\mathcal{U}))$ one also would have $\overline{\theta}_n = {}^T\theta_n$ on X. Consequently, $\varrho = S(n)$ in **Top**, so that (1) applies.

(4) In contrast with $\mathcal{T}_n = \Delta(\overline{\theta}_n)$, one can easily check that for $n > 2$ the inclusions $S(n) \subset \Delta(\theta_n)$ in **Top** are proper, hence the corresponding inclusions

$T^{-1}(S(n)) \subseteq T^{-1}(\Delta(\theta_n))$ in **QUnif** are proper as well (e.g., for every topological space $X \in S(2n) \setminus S(2n+1)$ the diagonal of $X \times X$ is θ_{2n+1}-closed, so $X \in \Delta(\theta_{2n+1}) \setminus S(2n+1)$; the existence of such a space X for every $n > 0$ was proved in [30]).

(5) Note that $\alpha = \gamma$ on every uniform space. Actually, it is easy to see that $\overline{\theta}_n = \alpha$ for $n \leq \omega$ on **Unif**. Consequently, all closure operators $\alpha, \beta, \gamma, \overline{\theta}_n$ $(n \leq \omega)$ coincide on **Unif**.

6.3. SUBCATEGORIES OF QUnif THAT DO NOT ARISE FROM SEPARATION

The smallest epireflective subsubcategory of **QUnif** is the category $\mathcal{S}ngl$ of all singleton spaces. It is easy to see that the $\mathcal{S}ngl$-closure is indiscrete, i.e., $\mathrm{reg}_X^{\mathcal{S}ngl}(M) = X$ for each $X \in$ **QUnif** and $\emptyset \neq M \subseteq X$. On the other hand, it follows easily from (3) that if $\mathrm{reg}^{\mathcal{A}}$ is indiscrete then $\mathcal{A} = \mathcal{S}ngl$.

We start by computation of the regular closure of bireflective subcategories of **QUnif**. We shall need the Sierpinski dyad **2** that admits a unique quasi-uniformity $\mathcal{U}$, which is generated by the entourage $E = [\{0\} \times 2] \cup [\{1\} \times \{1\}]$. In general bireflective subcategories are not closed under taking coproducts (take the category $\mathcal{I}nd$ of all indiscrete spaces). But if $\mathcal{B}$ has at least one space that is not indiscrete, then $\mathcal{B}$ contains the discrete dyad (since $\mathcal{B}$ is closed under taking products and subspaces, and either $\mathbf{2} \in \mathcal{B}$ or $\{0, 1\} \in \mathcal{B}$ anyway). Again the productivity of $\mathcal{B}$ will yield $B \in \mathcal{B} \Rightarrow B \amalg B \in \mathcal{B}$. However, the amalgamation $B \amalg_M B$ need not belong to $\mathcal{B}$ in general. If this occurs whenever $B \in \mathcal{B}$ and $M \subseteq B$ we say that $\mathcal{B}$ *is stable under amalgamation.*

THEOREM 6.13. *Let $\mathcal{B}$ be a bireflective subcategory of* **QUnif**. *Then*

(1) $\mathrm{reg}^{\mathcal{B}}$ *is discrete;*
(2) *if $\mathcal{B}$ is stable under amalgamation, then* $\mathrm{reg}^{\mathcal{B}_0}$ *coincides with β when restricted on $\mathcal{B}_0$;*
(3) $\mathrm{reg}^{\mathcal{B}_0}$ *is additive when $\mathcal{B}$ is stable under amalgamation.*

Proof. Let $X \in$ **QUnif** and $M \subseteq X$. With $Y = X \amalg_M X$ the $\mathcal{A}$-reflexion $r_Y \colon Y \to rY$ is injective, hence (3) gives $M = \mathrm{reg}_X^{\mathcal{A}}(M)$.

Since $\mathcal{B}_0 \subseteq \mathbf{QUnif}_0$ we have $\mathrm{reg}^{\mathcal{B}_0} \geq \mathrm{reg}^{\mathbf{QUnif}_0} = \beta$. On the other hand, if $X \in \mathcal{B}_0$ and M is β-closed then the amalgamation $X \amalg_M X \in \mathbf{QUnif}_0$. Moreover, $X \amalg_M X \in \mathcal{B}$ by the stability of $\mathcal{B}$ under amalgamation. Thus $X \amalg_M X \in \mathcal{B}_0$, hence again (3) gives $M = \mathrm{reg}_X^{\mathcal{B}_0}(M)$.

Item 3 follows from item 2, the additivity of β and Lemma 5.2. □

This theorem shows that many epireflective subcategories may happen to have the same regular closure operator. Hence "strongly epireflective" is a necessary constraint in the Diagonal Theorem.

In the sequel we denote by $\mathcal{T}ot\mathcal{B}$ the full subcategory of **QUnif** having as objects all totally bounded quasi-uniform spaces. $\mathcal{T}ot\mathcal{B}$ is epireflective and obviously contains $\mathcal{I}nd$, hence $\mathcal{T}ot\mathcal{B}$ is a bireflective subcategory of **QUnif** (i.e., every quasi-uniform space admits a finest coarser totally bounded quasi-uniformity). Hence, the $\mathcal{T}ot\mathcal{B}$-closure is discrete by Theorem 6.13.

The full subcategory $\mathcal{T}r$ of **QUnif** having as objects all transitive spaces is epireflective and obviously contains $\mathcal{I}nd$. Hence $\mathcal{T}r$ is a bireflective subcategory of **QUnif**, i.e., every quasi-uniform space $(X, \mathcal{U})$ admits a finest coarser transitive quasi-uniformity $\mathcal{U}_t$, but in this case $T(\mathcal{U}_t)$ need not coincide with $T(\mathcal{U})$. The $\mathcal{T}r$-closure must be discrete on $\mathcal{T}r$ by Theorem 6.13.

Now we shall compute the $\mathcal{T}ot\mathcal{B}_0$-closure and the $\mathcal{T}r_0$-closure. To this end we note that both $\mathcal{T}ot\mathcal{B}$ and $\mathcal{T}r$ are stable under amalgamation. Indeed, by Theorem 4.1 the amalgamation $X \amalg_M X$ is transitive (totally bounded) whenever X is transitive (totally bounded). By Theorem 6.13 the $\mathcal{T}ot\mathcal{B}_0$-closure coincides with β on $\mathcal{T}ot\mathcal{B}_0$ and the $\mathcal{T}r_0$-closure coincides with β on $\mathcal{T}r_0$. This implies that the $\mathcal{T}ot\mathcal{B}_0$-closure and the $\mathcal{T}r_0$-closure are additive.

As above, one could consider the smaller subcategory $\mathcal{T}ot\mathcal{B} \cap \mathcal{T}r$ having analogous properties. We see now that it is possible to encounter this subcategory and the respective subcategory $(\mathcal{T}ot\mathcal{B} \cap \mathcal{T}r)_0$ starting from a different point. Following the notation of Section 5 we denote by $E(\mathbf{2})$ the category of subspaces of powers of the Sierpinski dyad $\mathbf{2}$. Now we compute the $E(\mathbf{2})$-closure and show that it is additive. To this end we consider the category $(\mathcal{T}ot\mathcal{B} \cap \mathcal{T}r)_0$ of all totally bounded transitive quasi-uniform T_0-spaces. While $E(\mathbf{2})$ is a proper subcategory of both $\mathcal{T}ot\mathcal{B}_0$ and $\mathcal{T}r_0$, we have:

THEOREM 6.14. $E(\mathbf{2}) = (\mathcal{T}ot\mathcal{B} \cap \mathcal{T}r)_0$. *Consequently, the* $E(\mathbf{2})$*-closure is additive and coincides with* β *on* $E(\mathbf{2})$.

Proof. By Theorem 6.13 the $(\mathcal{T}ot\mathcal{B} \cap \mathcal{T}r)_0$-closure is additive and coincides with β on $(\mathcal{T}ot\mathcal{B} \cap \mathcal{T}r)_0$. Hence we are left with the proof of the equality $E(\mathbf{2}) = (\mathcal{T}ot\mathcal{B} \cap \mathcal{T}r)_0$. Clearly $(\mathbf{2}, \mathcal{U})$ is a totally bounded T_0-space that is transitive. An arbitrary quasi-uniform subspace of a product of copies of $(\mathbf{2}, \mathcal{U})$ is a totally bounded transitive T_0-space, because the three properties of transitivity, total boundedness and T_0-separation are preserved under products and subspaces.

On the other hand, let $(X, \mathcal{U})$ be a transitive totally bounded quasi-uniform T_0-space. Set

$$\mathcal{B} = \{G \subseteq X : \text{there is } U \in \mathcal{U} \text{ such that } U(G) = G\}.$$

Note that $\mathcal{B}$ is a base for the topology of $(X, \mathcal{U})$, since $\mathcal{U}$ is a transitive quasi-uniformity.

Furthermore, for each $G \in \mathcal{B}$ the set $S_G = [(X \setminus G) \times X] \cup [G \times G]$ belongs to the quasi-uniformity $\mathcal{U}$ since for $G = V(G)$ with transitive V one obviously has $V \subseteq S_G$. Moreover, these sets form a subbase for the quasi-uniformity $\mathcal{U}$. Indeed, being totally bounded, the quasi-uniformity $\mathcal{U}$, is generated by all sets of the form (in the notation of [17])

$$T(A, C) := (X \times X) \setminus (A \times C) = [(X \setminus A) \times X] \cup [X \times (X \setminus C)],$$

where $A, C \subseteq X$ and A is far from C ([17, Theorem 1.33]). Now it suffices to note that if A is far from C, then there is a transitive entourage V such that $V(A) \cap C = \emptyset$. Then $V(A)$ belongs to $\mathcal{B}$ and $S_{V(A)}$ is obviously a subset of the set $T(A, C)$. Hence the quasi-uniformity $\mathcal{U}$ is generated by all sets of the form S_G where G belongs to $\mathcal{B}$.

For each $G \in \mathcal{B}$ consider a copy $\mathbf{2}_G$ of the Sierpinski dyad and define a map $f_G\colon X \to \mathbf{2}_G$ by $f_G(x) = 1$ iff $x \in G$. Then clearly each f_G is quasi-uniformly continuous. Thus the product map $f := \Pi_{G\in\mathcal{B}} f_G\colon X \to \Pi_{G\in\mathcal{B}}\mathbf{2}_G$ is quasi-uniformly continuous. It is injective, since $(X, \mathcal{U})$ is a T_0-space.

For each $G \in \mathcal{B}$ we have $((\mathrm{pr}_G \times \mathrm{pr}_G)^{-1}E) \cap (f(X) \times f(X)) \subseteq (f \times f)(S_G)$. Since f is injective and $\mathcal{U}$ is generated by all sets of the form S_G, $G \in \mathcal{B}$, this suffices to see that f is a quasi-uniform embedding. Therefore $X \in E(\mathbf{2})$. □

REMARK 6.15.

- We saw in Example 5.9 that the $E(X)$-closure is not additive whenever X is a strongly rigid space with $|X| > 2$. Since $\mathbf{2}$ is obviously the only strongly rigid two-point space, the above theorem shows that the condition $|X| > 2$ in Example 5.9 cannot be removed.
- We see now that the strongly epireflective hull $S(\mathbf{2})$ of $E(\mathbf{2})$ does not coincide with $\mathbf{QUnif}_0$. To get a counterexample take the metric uniformity $\mathcal{U}$ on any connected metric space (X, d) (say, the unit interval I). Suppose $f\colon (X, U) \to \mathbf{2}$ is quasi-uniformly continuous and nonconstant. Then the proper open set $A = f^{-1}(1)$ would be far from its complement in X. Indeed, the quasi uniform continuity of f entails the existence of an $\varepsilon > 0$ such that $(f(x), f(y)) \in E$ whenever $d(x, y) < \varepsilon$, so that $x \in A$ and $d(x, y) < \varepsilon$ entails $y \in A$ as well, i.e., $d(A, X \setminus A) \geq \varepsilon$. Thus A is a proper clopen set of X – a contradiction. Note that this example shows actually $S(\mathbf{2}) \not\supseteq \mathbf{QUnif}_2$ or even $S(\mathbf{2}) \not\supseteq \mathcal{T}_\omega$ (cf. Remark 6.12(5)). With $X = I$ we get a space in $\mathcal{T}_\omega \cap \mathcal{T}ot\mathcal{B}$ with only constant maps to $\mathbf{2}$.

7. Semiregular Closure Operators

The following property of closure operators, introduced in [7, Definition 4.2], will be useful in describing the epimorphisms (see Fact 8.2 below).

DEFINITION 7.1. A closure operator ϱ of $\mathbf{QUnif}$ is *semiregular*, if for every $X \in \Delta(\varrho)$ and ϱ-closed subspace M of X the embeddings k_1 and k_2 of X into $X \amalg_M X$ are ϱ-closed.

Note that the semiregular closure operators were called essentially strong in [15, Definition 7.4]. Since the functor **QUnif** $\to$ **Set** is topological, we can apply [7, Lemma 4.1] and claim:

FACT 7.2. *Regular closure operators of* **QUnif** *are semiregular.*

The following lemma makes clear the advantage of semiregular closure operators for the explicit computation of regular closure operators.

LEMMA 7.3. *If ϱ is a semiregular closure operator of* **QUnif**, *then* $\mathrm{reg}^{\Delta(\varrho)}$ *coincides on $\Delta(\varrho)$ with the idempotent hull of ϱ.*

Proof. Since the functor **QUnif** $\to$ **Set** is topological, we can apply [7, Corollary 4.9]. □

It is easy to see that the assignment $\varrho \mapsto \varrho^\iota$ preserves (and reflects) semiregularity.

Here we discuss semiregularity of some closure operators.

LEMMA 7.4. *Semiregularity is stable under joins of closure operators.*

Proof. In fact, if $\varrho = \bigvee \varrho_\lambda$ and $X \in \Delta(\varrho)$, then $X \in \Delta(\varrho_\lambda)$ for every λ. Now every ϱ-closed $M \subseteq X$ is also ϱ_λ-closed, hence $k_1(X)$ is ϱ_λ-closed in $X \amalg_M X$. But then it is also ϱ-closed. □

EXAMPLE 7.5. One can easily see that α is semiregular: if $M \subseteq X \in \mathbf{QUnif}_2 = \Delta(\alpha)$ then $k_2(X)$ is closed in $T(X \amalg_M X)$ whenever M is closed in $T(X)$ by Lemma 4.3 (this is proved in a more general setting in Lemma 7.7, note that $\alpha = \overline{\theta}_1$). By the above remark also $\gamma = \alpha\iota$ is semiregular. By the above lemma also $\alpha \vee \gamma$ is semiregular. A direct verification shows that β is semiregular (this follows also by Fact 7.2 and Example 5.4). On the other hand, if one knows that β is semiregular, then regularity of β follows from the fact that β is also additive and initial (see [7, Theorem 4.12]).

We also have the following more general fact:

LEMMA 7.6. *The closure operators $\overline{\theta}_n$ and $\overline{\theta}_\omega$ in* **QUnif** *are semiregular.*

Proof. In view of Lemma 7.4 the semiregularity of the closure operator $\overline{\theta}_\omega$ will follow from the semiregularity of the closure operators $\overline{\theta}_n$. To check that $\overline{\theta}_n$ is semiregular take $(X, \mathcal{U}) \in$ **QUnif**, $M \subseteq X$ and assume that $k_2(X)$ is not $\overline{\theta}_n$-closed in $X \amalg_M X$. Then there exists some $x \in X$ such that $z = k_1(x) \in \overline{\theta}_n(k_2(X))$, i.e. for every $V \in \mathcal{U}$ the neighbourhood $w_n(W_V)(z)$ meets $k_2(X)$, with $w_n(U) = \underbrace{(\cdots \circ U^{-1} \circ U)}_{n}$ as defined in Section 3.2. Fix $z' \in k_2(X) \cap w_n(W_V)(z)$. Then $(z, z') \in w_n(W_V)$ implies (by (1)) that $(p(z), p(z')) \in w_n(V^2)$. Moreover, since $z' \in k_2(X)$ and since $w_n(W_V) = \cdots \circ W_V^{-1} \circ W_V$, there exists $k \leq n$ such that the neighbourhood $w_k(V^2)(x)$ hits M. Since this can be achieved for every $V \in \mathcal{U}$, we conclude that $x \in \overline{\theta}_n(M)$. Hence M is not $\overline{\theta}_n$-closed in X. □

LEMMA 7.7. *For $n \leq \omega$ $\mathrm{reg}^{\mathcal{T}_n}$ coincides on $\mathcal{T}_n$ with the idempotent hull of $\overline{\theta}_n$.*

Proof. By the above lemma, $\overline{\theta}_n$ is semiregular for every $n \leq \omega$. Now Lemma 7.3 applies. □

In the following lemma we show that the assignment $\varrho \mapsto {}^T\varrho$ reflects semiregularity. This will help us to find examples of non-semiregular closure operators of **QUnif** in Example 7.9.

LEMMA 7.8. *A closure operator ϱ of* **Top** *is semiregular, whenever the closure operator ${}^T\varrho$ of* **QUnif** *is semiregular.*

Proof. In fact, assume that the closure operator ϱ is not semiregular and let $M \subseteq X \in \Delta(\varrho)$ witness non-semiregularity of ϱ. Then M is ϱ-closed in X, while $k_1(X)$ is not ϱ-closed in $X \amalg_M X$. Now provide X with the Pervin quasi-uniformity; by Lemma 6.6 $X \in \Delta({}^T\varrho)$. Since the topology of $X \amalg_M X$ is coarser than that of the topological amalgamation, $k_1(X)$ is not ${}^T\varrho$-closed in $X \amalg_M X$. □

Now we give four examples of nonsemiregular closure operators of **QUnif**. The first two are obtained by means of nonsemiregular closure operators ϱ of **Top**.

EXAMPLE 7.9. According to the above lemma, it suffices to provide nonsemiregular closure operators ϱ of **Top**, then the closure operator ${}^T\varrho$ of **QUnif** is not semiregular. One such example is the sequential closure in **Top**. Another possibility for a nonsemiregular closure operator of **Top** is the additive core of the regular closure of a proper strongly rigid class as in [7] (p. 136, Corollary 4.15).

We show now that Lemma 7.8 cannot be inverted. This produces two other examples of nonsemiregular closure operators of **QUnif**.

EXAMPLE 7.10. Next we show that the assignment $\varrho \mapsto {}^T\varrho$ does not preserve semiregularity.

(1) Let π be the closure operator of **Top** defined in Example 6.2. Then π is an idempotent, additive and semiregular closure operator of **Top**. On the other hand, $\Delta({}^T\pi) = \mathbf{QUnif}_1$. Hence, if ${}^T\pi$ were regular, then ${}^T\pi$ would necessarily be discrete by Lemma 5.2, since ${}^T\pi$ is discrete on $\mathbf{QUnif}_1$.
(2) Another example to this effect is $\varrho = b$. Indeed, b is a regular closure operator of **Top**, while ${}^Tb$ is not semiregular. In fact, by $\Delta({}^Tb) = \Delta(\beta) = \mathbf{QUnif}_0$ we can see first that ${}^Tb$ cannot be regular in view of Theorem 6.5. Since ${}^Tb$ is idempotent and initial, by Theorem 4.12 of [7], ${}^Tb$ is not even semiregular.

8. Epimorphisms

It follows immediately from the definition of regular closure operator that the epimorphisms in a subcategory $\mathcal{A}$ of **QUnif** are precisely the $\mathcal{A}$-dense maps. Hence we are led to computing the regular closure operator in all cases of interest. We start with the following easy example.

EXAMPLE 8.1. It follows from Theorem 6.13(1) that for any bireflective subcategory $\mathcal{B}$ of **QUnif** the epimorphisms are surjective. In particular, the epimorphisms in $\mathcal{T}ot\mathcal{B}$, $\mathcal{T}r$ and $\mathcal{T}ot\mathcal{B} \cap \mathcal{T}r$ are surjective.

The following fact explains the importance of semiregular closure operators in detecting epimorphisms.

FACT 8.2. *Let ϱ be an additive semiregular closure operator of* **QUnif**. *Then the epimorphisms in $\mathcal{T}_\varrho$ and $\Delta(\varrho)$ are precisely the ϱ^∞-dense maps.*

Proof. Since the functor **QUnif** $\to$ **Set** is topological, we can apply [7, Corollary 5.9]. □

EXAMPLE 8.3. The closure operators α, β, γ, $\alpha \vee \gamma$ and $\overline{\theta}_n$ $(1 \leq n \leq \omega)$ are semiregular (see Example 7.5 and Lemma 7.6).

(1) The epimorphisms of $\mathbf{QUnif}_0$ and $\mathbf{QUnif}_1$ are the β-dense maps.
(2) The epimorphisms of $\mathbf{QUnif}_2$ are the T-dense maps.
(3) The epimorphisms of $\mathbf{QUnif}_2 \cap \iota(\mathbf{QUnif}_2) = \Delta(\alpha \vee \gamma)$ are the $(\alpha \vee \gamma)^\infty$-dense maps.
(4) The epimorphisms of $\mathcal{T}_n$ are the $(\overline{\theta}_n)^\infty$-dense maps, for $1 < n \leq \omega$.

We isolate the following case:

PROPOSITION 8.4. *The epimorphisms of $\mathcal{A} = \Delta(\alpha \wedge \gamma)$ are the $\alpha \wedge \gamma$-dense (i.e., doubly dense) maps. Moreover, $\mathcal{A}$ is the biggest full subcategory of* **QUnif** *with this property.*

Proof. We shall show that $\alpha \wedge \gamma = \mathrm{reg}^{\mathcal{A}}$ on spaces of $\mathcal{A}$. Since the closure operator $\alpha \wedge \gamma$ is idempotent and $\alpha \wedge \gamma \leq \mathrm{reg}^{\mathcal{A}}$, it suffices to see that if $M \subseteq X \in \mathcal{A}$ is $\alpha \wedge \gamma$-closed then M is also $\mathcal{A}$-closed. By Lemma 5.3 it suffices to show that $X \amalg_M X \in \mathcal{A}$. Let D be the diagonal of the Cartesian square $(X \amalg_M X)^2$. Then $p \times p$: $(X \amalg_M X)^2 \to X^2$ sends D into $\Delta_X \subseteq X^2$. By our hypothesis Δ_X is $\alpha \wedge \gamma$-closed in X^2, hence the $\alpha \wedge \gamma$-closure of D is contained in $(p \times p)^{-1}(\Delta_X)$, i.e., consists of points $(k_1(x), k_2(x))$ and $(k_2(x), k_1(x))$, with $x \in X$. Now assume that $x \notin M$. Then by our assumption on M there exists an entourage V of X such that either $V(x) \cap M = \emptyset$ or $V^{-1}(x) \cap M = \emptyset$. The first case entails $W_V(k_1(x)) = k_1(V(x))$, hence if $(k_1(x), z) \in W_V$ and $(z, k_1(x)) \in W_V^{-1}$ for some $z = k_i(y)$, $y \in X$, then $i = 1$. Now $(k_1(y), k_2(x)) \in W_V^{-1}$ yields $y \in V(V(x) \cap$

M) – this contradicts $V(x) \cap M = \emptyset$. The nonexistence of such z proves that $(k_1(x), k_2(x)) \notin W_V \circ W_V^{-1}$. Analogously one shows that $V^{-1}(x) \cap M = \emptyset$ yields $(k_1(x), k_2(x)) \notin W_V^{-1} \circ W_V$. Hence $(k_1(x), k_2(x)) \notin (W_V \circ W_V^{-1}) \cap (W_V^{-1} \circ W_V)$. As $\bigcap_V [(W_V \circ W_V^{-1}) \cap (W_V^{-1} \circ W_V)]$ coincides with the $\alpha \wedge \gamma$-closure of D in $(X \amalg_M X)^2$, we have proved that D is $\alpha \wedge \gamma$-closed.

To prove the second assertion assume that $\mathcal{B} \supset \mathcal{A}$ and let $X \in \mathcal{B} \setminus \mathcal{A}$. Then Δ_X is not $\alpha \wedge \gamma$-closed in $X \times X$. Let C be the $\alpha \wedge \gamma$-closure of Δ_X in $X \times X$. Then the inclusion $j\colon \Delta \hookrightarrow C$ is $\alpha \wedge \gamma$-dense. On the other hand, Δ_X is obviously $\mathcal{B}$-closed in C since $\Delta_X = \mathrm{Eq}(p_1, p_2)$, where $p_i\colon X \times X \to X \in \mathcal{B}$, $i = 1, 2$, are the projections. By our assumption $C \neq \Delta_X$, so that j cannot be $\mathcal{B}$-dense. Hence j is not an epimorphism in $\mathcal{B}$. □

Now a direct application of Theorem 6.13 gives

THEOREM 8.5. *For any bireflective subcategory $\mathcal{B}$ of* **QUnif** *stable under amalgamation, the epimorphisms in $\mathcal{B}_0$ are the β-dense maps. In particular, the epimorphisms in $\mathcal{T}ot\mathcal{B}_0$, $\mathcal{T}r_0$ and $E(\mathbf{2}) = (\mathcal{T}ot\mathcal{B} \cap \mathcal{T}r)_0$ are the β-dense maps.*

It is possible to put this result differently, starting from the part $\mathcal{A} = \mathcal{B}_0$. To this end we consider the following condition that is weaker than stability of $\mathcal{B} = E(\mathcal{A} \cup \mathcal{I}nd)$ under amalgamation:

$$X \amalg_M X \in \mathcal{A} \text{ for every } X \in \mathcal{A} \text{ and for every } \beta\text{-closed subset } M \subseteq X. \qquad (4)$$

It turns out that (4) characterizes those strongly epireflective $\mathcal{A}$ such that the epimorphisms in $\mathcal{A}$ are the β-dense maps and the $\mathcal{A}$-closure is weakly hereditary ($\mathrm{reg}^{\mathcal{A}}$ is *weakly hereditary* if for $M \subseteq X \in \mathcal{A}$ one can always assume M is $\mathcal{A}$-dense in $\mathrm{reg}_X^{\mathcal{A}}(M)$).

COROLLARY 8.6. *Let $\mathcal{A}$ be an epireflective subcategory of* **QUnif** *contained in* $\mathbf{QUnif}_0$. *If* (4) *holds then the epimorphisms in $\mathcal{A}$ are the β-dense maps. If $\mathcal{A}$ is strongly epireflective and* $\mathrm{reg}^{\mathcal{A}}$ *is weakly hereditary, then the converse is also true, i.e., the epimorphisms in $\mathcal{A}$ are the β-dense maps iff $\mathcal{A}$ satisfies* (4).

Proof. Obviously (4) implies that for $X \in \mathcal{A}$ every β-closed subset M is also $\mathcal{A}$-closed. Since $\mathcal{A} \subseteq \mathbf{QUnif}_0$, $\mathrm{reg}^{\mathcal{A}} \geq \beta$, so that the $\mathcal{A}$-closure coincides with β on X. Consequently, the epimorphisms in $\mathcal{A}$ (the $\mathcal{A}$-dense maps) coincide with the β-dense maps.

Now assume that $\mathcal{A}$ is strongly epireflective, $\mathrm{reg}^{\mathcal{A}}$ is weakly hereditary and the epimorphisms in $\mathcal{A}$ are the β-dense maps. This means that the $\mathcal{A}$-dense maps coincide with the β-dense maps in $\mathcal{A}$. Take $X \in \mathcal{A}$ and a β-closed subset M of X. Since M is $\mathcal{A}$-dense in $\mathrm{reg}_X^{\mathcal{A}}(M)$ by the weak hereditariness of $\mathrm{reg}^{\mathcal{A}}$, we conclude that M will be β-dense in $\mathrm{reg}_X^{\mathcal{A}}(M)$. By the β-closedness of M in X we get $\mathrm{reg}_X^{\mathcal{A}}(M) = M$. Hence by Frolík's Lemma 5.3 $X \amalg_M X \in \mathcal{A}$. □

It follows from this corollary and Example 8.3 that the strongly epireflective subcategory $\mathbf{QUnif}_1$ of **QUnif** satisfies (4).

8.1. FUNCTORS THAT PRESERVE EPIMORPHISMS

Inclusion functors between categories of topological spaces that preserve epimorphisms were studied in [22, 11]. According to Example 8.1 for any bireflective subcategory $\mathcal{B}$ of **QUnif** the inclusion functor $\mathcal{B} \hookrightarrow$ **QUnif** preserves epimorphisms (since in both categories they are surjective). This can be stated more generally for any pair $\mathcal{B} \subseteq \mathcal{A}$ of bireflective subcategories of **QUnif**, or follows from the next more general result.

PROPOSITION 8.7. *Let $\mathcal{A}$ be a full subcategory of* **QUnif**. *Then the inclusion $\mathcal{A} \hookrightarrow S(\mathcal{A})$ preserves epimorphisms.*

Proof. This is well known (cf. [22]), but we include a proof for reader's convenience. Let $e\colon X \to Y$ be an epi in $\mathcal{A}$. To check that it is an epi in $S(\mathcal{A})$, too, take two maps $f, g\colon Y \to Z \in S(\mathcal{A})$ such that $f \circ e = g \circ e$. Then there exists a mono-source $(p_i\colon Z \to A_i)_{i \in I}$ with $A_i \in \mathcal{A}$ for every $i \in I$. Our assumption gives $p_i \circ f = p_i \circ g$ for every $i \in I$. Therefore, $f = g$. □

In view of Theorem 8.5 we can say that for any bireflective subcategory $\mathcal{B}$ of **QUnif** stable under amalgamation the inclusion functor $\mathcal{B}_0 \hookrightarrow \mathbf{QUnif}_0$ preserves epimorphisms. Here we prefer to put it as in Corollary 8.6.

THEOREM 8.8. *Let $\mathcal{A}$ be an epireflective subcategory of* **QUnif** *contained in* $\mathbf{QUnif}_0$. *If* (4) *holds then the inclusion functor $i\colon \mathcal{A} \hookrightarrow \mathbf{QUnif}_0$ preserves epimorphisms. If $\mathcal{A}$ is strongly epireflective and* $\mathrm{reg}^{\mathcal{A}}$ *is weakly hereditary, then i preserves epimorphisms iff $\mathcal{A}$ satisfies* (4).

Now we are interested when epimorphisms are preserved by the functor T. Let us note that reflection of epimorphisms is automatically available in the case of arbitrary faithful functors (in particular, inclusion functors $\mathcal{A} \hookrightarrow \mathcal{B}$ or the functor T and its restrictions).

THEOREM 8.9. *Let $\mathcal{A}$ be an epireflective subcategory of* **QUnif** *and let $\mathcal{B}$ be an epireflective subcategory of* **Top** *such that $T(\mathcal{A}) \subseteq \mathcal{B}$. Then, $\mathrm{reg}^{\mathcal{A}} \geq {}^{T}\mathrm{reg}^{\mathcal{B}}$.*

Proof. Since regular closure operators are idempotent, in order to prove $\mathrm{reg}^{\mathcal{A}} \geq {}^{T}\mathrm{reg}^{\mathcal{B}}$, it suffices to show that any $\mathcal{A}$-closed subspace $M \subseteq X \in$ **QUnif** is $\mathcal{B}$-closed in $T(X)$. Let $x \in X$ and assume that $x \notin M$. Then by our assumption $x \notin \mathrm{reg}_X^{\mathcal{A}}(M)$. Hence there exists a pair of maps $f, g\colon X \to A \in \mathcal{A}$ in **QUnif** such that $f(x) \neq g(x)$ and $f|_M = g|_M$. Applying the forgetful functor T we get a pair of continuous maps $f, g\colon T(X) \to T(A) \in \mathcal{B}$ in **Top** that witness $x \notin \mathrm{reg}_{T(X)}^{\mathcal{B}}(M)$.

To prove the second assertion assume that for the map $f\colon X \to Y$ in $\mathcal{A}$ the map $f\colon TX \to TY$ is an epimorphism in $\mathcal{B}$. Then $f(X)$ is $\mathcal{B}$-dense in $T(Y)$. By

$\mathrm{reg}^{\mathcal{A}} \geq {}^{T}\mathrm{reg}^{\mathcal{B}}$, this yields that $f(X)$ is $\mathcal{A}$-dense in Y. Hence f is an epimorphism in $\mathcal{A}$. □

In case that the subcategories $\mathcal{A}$ and $\mathcal{B}$ are strongly epireflective one can prove the first part by a double application of Frolík's lemma.

THEOREM 8.10. *Let $\mathcal{B}$ be an epireflective subcategory of* **Top**. *Then the subcategory $\mathcal{A} = T^{-1}(\mathcal{B})$ of* **QUnif** *is epireflective and* $\mathrm{reg}^{\mathcal{A}} \geq {}^{T}\mathrm{reg}^{\mathcal{B}}$. *If $\mathcal{B} \subseteq$* **Haus** *is strongly epireflective, then* $\mathrm{reg}^{\mathcal{A}} = {}^{T}\mathrm{reg}^{\mathcal{B}}$ *on $\mathcal{A}$, so that the restriction $T|_{\mathcal{A}}\colon \mathcal{A} \to \mathcal{B}$ preserves epimorphisms.*

Proof. The first assertion follows from Theorem 8.9. To prove the last assertion, assume $\mathcal{B} \subseteq$ **Haus** and proceed with the proof of the inequality $\mathrm{reg}^{\mathcal{A}} \leq {}^{T}\mathrm{reg}^{\mathcal{B}}$, which will imply $\mathrm{reg}^{\mathcal{A}} = {}^{T}\mathrm{reg}^{\mathcal{B}}$. In view of the idempotency of the regular closure operators it suffices to see that if $M \subseteq X \in \mathcal{A}$ is $\mathcal{B}$-closed in $T(X)$, then it is also $\mathcal{A}$-closed. By Frolík's lemma, $\mathcal{B}$-closedness of M gives $T(X) \amalg_M T(X) \in \mathcal{B}$. Since $\mathcal{B} \subseteq$ **Haus**, $\mathcal{B}$-closedness of M yields also closedness of M in TX. Now Corollary 4.7 guarantees that both topologies on the amalgamation $X \amalg_M X$ coincide, so that also $T(X \amalg_M X) \in \mathcal{B}$. Thus $X \amalg_M X \in \mathcal{A}$, consequently M is $\mathcal{A}$-closed by Frolík's lemma. □

REMARK 8.11. Here we comment on the extent to which the condition $\mathcal{B} \subseteq$ **Haus** in Theorem 8.10 is necessary. Items (1) and (2) show that the functor T need not preserve epimorphisms if $\mathcal{B}$ is not contained in **Haus**. On the other hand, item (4) gives an example where the functor T preserves epimorphisms even if $\mathcal{B} \not\subseteq$ **Haus**.

(1) While the epimorphisms in $\mathbf{QUnif}_0$ are the β-dense maps (Example 5.4), the epimorphisms in $\mathbf{Top}_0 = T(\mathbf{QUnif}_0)$ are the b-dense maps (it is easy to find a β-dense map f such that Tf is not b-dense).
(2) Another example is provided by the smaller subcategory $\mathbf{QUnif}_1$. Its epimorphisms are (as in (1)) the β-dense maps, while the epimorphisms in $\mathbf{Top}_1 = T(\mathbf{QUnif}_1)$ are *surjective*. Here is an example to this effect: the subset $\mathbf{Q}$ of the Sorgenfrey line $\mathbf{S} \in \mathbf{QUnif}_1$ is $\mathbf{QUnif}_1$-dense, while $\mathbf{Q}$ is $\mathbf{Top}_1$-closed in $T(\mathbf{S})$, hence ${}^{T}\mathrm{reg}^{\mathbf{Top}_1}$ does not coincide with $\mathrm{reg}^{\mathbf{QUnif}_1}$ on $\mathbf{S}$, cf. Example 5.8.
(3) Making use of Proposition 4.10 one can easily show that for $\mathcal{B}$ and $\mathcal{A} = T^{-1}(\mathcal{B})$ in Theorem 8.10 (without requiring $\mathcal{B} \subseteq$ **Haus**) the equality $\mathrm{reg}^{\mathcal{A}} = {}^{T}\mathrm{reg}^{\mathcal{B}}$ remains valid on all spaces X such that the original quasi-uniformity of X contains $P(TX)$.
(4) Let **SUS** be the full subcategory of **Top** consisting of those topological spaces in which convergent sequences have a unique accumulation point. Then **SUS**-closed subspaces of **SUS**-spaces are precisely the sequentially closed ones. Now with $\mathcal{A} = T^{-1}(\mathbf{SUS})$, the $\mathcal{A}$ closed subsets of a space $X \in \mathcal{A}$ are precisely those which are sequentially closed in $T(X)$, i.e., $\mathrm{reg}^{\mathcal{A}} = {}^{T}\mathrm{reg}^{\mathbf{SUS}}$ on $\mathcal{A}$.

Acknowledgments

This work was partially supported by the research projects 60% and 40% of the Italian Ministero dell' Università e della Ricerca Scientifica e Tecnologica and by the Swiss National Science Foundation under grant 21-32382.91.

This paper enjoyed a very careful and rigorous reading by a referee. His numerous critical suggestions improved essentially the exposition, corrected an erroneous Example 5.8 and simplified the proof of Theorem 8.9.

References

1. Adámek, J., Herrlich, H. and Strecker, G. E.: Abstract and Concrete Categories, *Pure and Applied Mathematics*, Wiley, New York, 1990.
2. Burke, M.: Characterizing uniform continuity with closure operators, *Topology Appl.* **59** (1994), 245–259.
3. Brümmer, G. C. L.: Uniform topology versus bitopology, Seminar on General Topology, University of l'Aquila, unpublished lecture notes.
4. Clementino, M. and Tholen, W.: Separation versus connectedness, *Topology Appl.* **75** (1997), 143–181.
5. Čech, E.: *Topologickě Prostory*, Prague, 1959.
6. Cook, H.: Continua which admit only the identity mapping onto non-degenerate subcontinua, *Fund. Math.* **60** (1967), 241–249.
7. Dikranjan, D.: Semiregular closure operators and epimorphisms in topological categories, *Suppl. Rend. Circ. Mat. Palermo, Ser. II* **29** (1992), 105–160.
8. Dikranjan, D. and Giuli, E.: Ordinal invariants and epimorphisms in some categories of weak Hausdorff spaces, *Comm. Math. Univ. Carolin.* **27** (1986), 395–417.
9. Dikranjan, D. and Giuli, E.: Closure operators I, *Topology Appl.* **27** (1987), 129–143.
10. Dikranjan, D. and Giuli, E.: $S(n)$-θ-closed spaces, *Topology Appl.* **28** (1988), 59–74.
11. Dikranjan, D. and Giuli, E.: Inclusions preserving epimorphisms (English), in *Baku International Topological Conference* (Russian) (Baku, 1987), "Èlm", Baku, 1989, pp. 220–230.
12. Dikranjan, A., Giuli, E. and Tholen, W.: Closure operators II, in *Proceedings of the International Conference on Categorical Topology*, Prague, 1988, World Scientific, Singapore, 1989, pp. 297–335.
13. Dikranjan, D. and Künzi, H.-P.: Cowellpoweredness in quasi-uniform spaces, work in progress.
14. Dikranjan, D. and Pelant, J.: The impact of closure operators on the structure of a concrete category, *Quaestiones Math.* **18** (1995), 381–396.
15. Dikranjan, D. and Tholen, W.: *Categorical Structure of Closure operators with Applications to Topology, Algebra and Discrete Mathematics*, Mathematics and its Applications 346, Kluwer Academic Publishers, Dordrecht, 1995.
16. Engelking, R.: *General Topology*, Warszawa, 1977.
17. Fletcher, P. and Lindgren, W. F.: *Quasi-Uniform Spaces*, Lecture Notes in Pure Appl. Math. 77, Marcel-Dekker, Inc., New York, 1982.
18. W. Gähler, *Grundstrukturen der Analysis I*, Akademie-Verlag, Berlin, 1977.
19. Giuli, E. and Hušek, M.: A diagonal theorem for epireflective subcategories of Top and cowellpoweredness, *Ann. Mat. Pura Appl.* **145** (1986), 337–346.
20. Herrlich, H., Salicrup, G. and Strecker, G.: Factorization, denseness, separation and relatively compact objects, *Topology Appl.* **27** (1987), 157–169.
21. Himmelberg, C.: Quotient uniformities, *Proc. Amer. Math. Soc.* **17** (1966), 1385–1388.

22. Hoffmann, R.: Topological functors admitting a generalized Cauchy-completion, in *Proc. Conf. Categorical Topology* (Mannheim 1975), Lecture Notes in Math. 540, Springer-Verlag, Berlin, 1976, pp. 286–344.
23. Holgate, D.: Closure operators in categories, M.Sci. Thesis, Department of Mathematics, University of Cape Town, 1992.
24. Hušek, M. and Pumplün, D.: Disconnectedness, *Quaestiones Math.* **13** (1990), 449–459.
25. Kannan, V. and Rajagopalan, M.: Construction and application of rigid spaces I, *Adv. in Math.* **29** (1978), 1139–1172.
26. Künzi, H. P.: Quasi-uniform spaces – eleven years later, *Topology Proc.* **18** (1993), 143–171.
27. Künzi, H. P.: Functorial admissible quasi-uniformities on topological spaces, *Topology Appl.* **43** (1992), 27–36.
28. Künzi, H. P. and Lüthy, A.: Dense subspaces of quasi-uniform spaces, *Stud. Sci. Math. Hungar.* **30** (1995), 289–301.
29. Marny, Th.: On epireflective subcategories of topological categories, *Gen. Topology Appl.* **10** (1979), 175–181.
30. Porter, J. and Votaw, C.: $S(\alpha)$ spaces and regular Hausdorff extensions, *Pacific J. Math.* **45**(1) (1973), 327–345.
31. Salbany, S.: *Reflective Subcategories and Closure Operators*, Lecture Notes in Math. 540, Springer-Verlag, Berlin, 1976, pp. 548–565.
32. Schröder, J.: Epi und extremer Mono in $T_{2,5}$, *Arch. Math.* **XXV** (1974), 561–565.
33. Schröder, J.: The category of Urysohn spaces is not cowellpowered, *Topology Appl.* **16** (1983), 237–241.
34. Velichko, H.: *H-closed topological spaces*, *Mat. Sb. (N.S.)* **70**(112) (1966), 98–112. (*Amer. Math. Soc. Transl. Ser. 2* **78** (1969), 103–118.)
35. Tozzi, A.: **US**-spaces and closure operators, *Suppl. Rend. Circ. Mat. Palermo Ser. II* **12** (1986), 291–300.
36. Viglino, G.: $\overline{T}_n$-spaces, *Notices Amer. Math. Soc.* **16** (1969), 846.

Applied Categorical Structures **8**: 209–234, 2000.
G. Brümmer & C. Gilmour (eds), Papers in Honour of Bernhard Banaschewski.

A Homotopy 2-Groupoid of a Hausdorff Space

Dedicated to Bernhard Banaschewski on his 70th birthday

K. A. HARDIE
Department of Mathematics, University of Cape Town, 7700 Rondebosch, South Africa

K. H. KAMPS
Fachbereich Mathematik, Fernuniversität, Postfach 940, D-58084 Hagen, Germany

R. W. KIEBOOM
Department of Mathematics, Vrije Universiteit Brussel, Pleinlaan 2, F 10, B-1050 Brussels, Belgium

(Received: 17 October 1996; accepted: 4 January 1999)

Abstract. If X is a Hausdorff space we construct a 2-groupoid $\mathbf{G}_2X$ with the following properties. The underlying category of $\mathbf{G}_2X$ is the 'path groupoid' of X whose objects are the points of X and whose morphisms are equivalence classes $\langle f\rangle$, $\langle g\rangle$ of paths f, g in X under a relation of *thin* relative homotopy. The groupoid of 2-morphisms of $\mathbf{G}_2X$ is a quotient groupoid $\Pi X/NX$, where ΠX is the groupoid whose objects are paths and whose morphisms are relative homotopy classes of homotopies between paths. NX is a normal subgroupoid of ΠX determined by the thin relative homotopies. There is an isomorphism $\mathbf{G}_2X(\langle f\rangle, \langle f\rangle) \approx \pi_2(X, f(0))$ between the 2-endomorphism group of $\langle f\rangle$ and the second homotopy group of X based at the initial point of the path f. The 2-groupoids of function spaces yield a 2-groupoid enrichment of a (convenient) category of pointed spaces.

We show how the 2-morphisms may be regarded as *2-tracks*. We make precise how cubical diagrams inhabited by 2-tracks can be pasted.

Mathematics Subject Classifications (2000): 18D05, 20L05, 55Q05, 55Q35.

Key words: 2-groupoid, 2-track, track, homotopy, higher homotopy structures, tree, fundamental groupoid, pasting, piecewise linear map, Gray tensor product, interchange 2-track, folding map.

Introduction

The investigation and application of higher dimensional van Kampen type theorems to calculations in homotopy theory has led to a rich variety of higher homotopy structures. (Motivation for the identification and construction of higher homotopy structures has been given by Brown in several papers, most recently [11].) For example, for a filtered space the equivalent homotopy structures of *ω-groupoid, crossed complex*, and *simplicial T-complex* are available (cf. [14]). More general forms of higher homotopy groupoids were introduced, under the name of cat^n-*groups*, by Loday in [25]. These are defined on n-cubes of pointed spaces, rather than on filtered spaces.

In this paper we are concerned with homotopy structures in dimension 2. The notion of *crossed module* (the terminology is due to J. H. C. Whitehead [39]) has been a landmark in the development of 2-dimensional homotopical algebra. Whitehead exploited the notion when investigating the structure of the second homotopy group of a space in a series of papers [37 – 39]. The major geometric example [12] of crossed module is given by the boundary map

$$\partial\colon \pi_2(X, A, x) \longrightarrow \pi_1(A, x)$$

from the second relative homotopy group of a based pair of spaces (X, A, x) to the fundamental group together with the standard action of $\pi_1(A, x)$ on $\pi_2(X, A, x)$ (for an exposition see [23]). The resulting crossed module might be called the Whitehead crossed module. Whitehead's major result was the algebraic description of the second relative homotopy group $\pi_2(A \cup \{e^2_\lambda\}_{\lambda\in\Lambda}, A, x)$ as the *free crossed* $\pi_1(A, x)$*-module* on the characteristic maps of the 2-cells e^2_λ [39].

Different needs led to different developments. Brown and Higgins [13, 14] when investigating and proving a 2-dimensional van Kampen theorem having Whitehead's theorem as a corollary were led to use as an essential tool what is now called a *double groupoid with thin structure* (for an explicit definition we refer to [8, 33, 12]) and constructed a homotopy double groupoid of a triple of spaces [13].

In category theory 2-categories [24] became a subject of investigation on their own right. The notion of 2-*groupoid* is a canonical specialisation. This notion has been exploited by Moerdijk and Svensson [28] in order to classify equivariant homotopy 2-types. They give, as one of their examples of 2-groupoids, a construction for what they call the Whitehead 2-groupoid $W(X, Y, S)$ of a space X with subspace Y and basepoint set $S \subseteq Y$.

The main observation is that the different approaches described are essentially equivalent. More precisely, there are equivalences of categories between crossed modules over groupoids (generalising the notion of crossed module to the groupoid context), 2-groupoids, and double groupoids with thin structure [15, 33, 34, 14, 29]. Likewise, under mild restrictions, the Brown–Higgins double groupoid and the Moerdijk–Svensson 2-groupoid are equivalent to each other and, in the special case of a based pair of spaces, to the Whitehead crossed module.

Note that a common feature of all these constructions is that they apply to a relative situation of a pointed pair resp. a triple of spaces.

The aim of this paper is to present a new contribution to 2-dimensional homotopical algebra which applies to the absolute case. We associate with a Hausdorff space X a 2-dimensional structure, $\mathbf{G}_2X$, which will be called the *homotopy* 2-*groupoid* of X. We proceed in analogy with the construction of the fundamental groupoid, $\mathbf{G}_1X$, of X in dimension 1.

Poincaré's notion of fundamental group based at a point of a topological space X has been a spur of conceptual and technical development since the earliest days of topology. The fundamental groupoid $\mathbf{G}_1 X$ was well known by the 1950's. However, the utility of the fundamental groupoid was not recognised in full until the 1960's ([32, 16, 5, 6, 22]). The elements, $\{f\}$, of $\mathbf{G}_1 X$ which we shall call *tracks*, are classes of paths, f, in X, two paths f and f' belonging to the same class if the initial (resp. final) points of f and f' coincide and if there is a relative homotopy (i.e., a homotopy rel endpoints), f_t, between f and f'. However no distinction is made between different relative homotopies. This means that information is cut down at a very early stage of the story. Thus it seems natural to retain information at least one dimension up and compare relative homotopies between paths. Then one is led to what we shall call 2-*tracks*, i.e., relative homotopy classes, $\{f_t\}$, of relative homotopies.

NOTE ON TERMINOLOGY. Even today there does not seem to be a universally accepted term for the elements (i.e., morphisms) of $\mathbf{G}_1 X$. Brown [5, 6] has called them 'roads' but many authors do not name them. Yet for the special case $X = B^A$ of a function space of pointed maps an established term is at hand, in the sense that the endomorphism groups of $\mathbf{G}_1 B^A$ coincide (up to isomorphism subject to mild restrictions on A) with the track groups introduced by Barratt [2]. It may therefore be permissible to call the elements of $\mathbf{G}_1 X$ *tracks*, or rather 1-*tracks* as opposed to 2-tracks.

In practice 2-tracks arise in coherent homotopy theory. In topology one often encounters diagrams of spaces and maps that are commutative only up to homotopy. In some cases (as discussed, for example, in [27]) it is necessary to add further data to the diagram such as to specify that the 2-cells are inhabited by specific 1-tracks (i.e., that the homotopies by which 2-cells commute belong to the relevant relative homotopy classes). Then one may still regard the diagram as homotopy commutative if there exist appropriate higher homotopies. One might hope to add further data to the diagram, specifying that the higher homotopies belong to specific 2-tracks. However one also wishes to paste such 2-tracks (in a manner as far as possible choice free) and to compute with them. Attempting to provide such facilities one is led to the problem of defining the *homotopy* 2-*groupoid* of X.

In order to be able to compute with 2-tracks, a convenient algebraic structure should be available. We note that 2-tracks can be pasted vertically and horizontally in the obvious way. Vertical pasting is well behaved and gives rise to a groupoid structure, ΠX (cf. Section 1). Horizontal pasting, like concatenation of paths, is neither strictly associative nor do we have strict identities and inverses. However, there are canonical relative homotopies which we will call *rescale, dilation* and *cancellation* to the effect that associativity does hold and identities and inverses do exist up to coherent isomorphisms. This gives rise to a *bigroupoid*, a structure which specialises the notion of a bicategory in the sense of Bénabou [4]. This kind of structure will be investigated in a separate paper [21]. In the present paper we proceed in a different direction (cf. Section 2). We enforce strict associativity, etc. by taking the quotient of ΠX by a suitable normal subgroupoid, NX. The elements of NX are the 2-tracks of so called *thin* homotopies. Thin homotopies are roughly relative homotopies which factor through a tree (for the precise definition see 2.1 below) and include rescale, dilation and cancellation. The quotient $\Pi X/NX$ is a 2-*groupoid* which we will call the *homotopy* 2-*groupoid* (cf. [28]), $\mathbf{G}_2X$, of the Hausdorff space X . The 1-morphisms of $\mathbf{G}_2X$ are the equivalence classes of paths (in X) under the relation of thin homotopy: these will be called *semitracks*. Restriction to Hausdorff spaces seems to be necessary in order to establish transitivity of the thin homotopy relation. Note that, under the construction described, 2-tracks are essentially retained (see 2.8 below). By analogy with the fundamental groupoid the 2-endomorphism groups of the homotopy 2-groupoid coincide (up to isomorphism) with the second homotopy groups of X based at appropriate points of X.

As mentioned above, the category of 2-groupoids is equivalent to the category of double groupoids with thin structure. In the present paper we have given preference to exhibiting the 2-groupoid structure arising in a natural way when working with 2-tracks. We hope to look elsewhere at the double groupoid approach and generalisation to higher dimensions.

With a view to applications we consider in Section 3 the special case of the 2-groupoid $\mathbf{G}_2X^V$ associated with the (unpointed) function space of pointed continuous maps from V to X. Here we work in the category of compactly generated Hausdorff spaces and take adjoints systematically to arrive at a 2-groupoid $\mathbf{G}_2(V, X)$ isomorphic to $\mathbf{G}_2X^V$. This construction is functorial in V and X. We investigate the resulting algebraic structure on the associated category of pointed spaces, $\mathbf{T}^*$. In this context *interchange* 2-tracks play a rather significant role and we examine their properties in some detail using, in particular, a homotopy addition lemma for the homotopy 2-groupoid (cf. Section 6). The algebra developed in Section 3 gives rise to a 2-groupoid enrichment of $\mathbf{T}^*$ with respect to the closed monoidal structure on the category of 2-groupoids given by the Gray tensor product. Pasting of cubical diagrams of spaces inhabited by 2-tracks is described in Section 4. It turns out that the 2-track inhabiting the pasted cube depends not only on the 2-tracks inside the given cubes but also on interchange 2-tracks determined by semitracks in their faces. Further applications are envisaged for subsequent papers. In particular we have in mind (i) a 2-track theoretic definition of the third order homotopy composition operation (the quaternary Toda bracket) from which new properties can be derived and (ii) the construction of certain coherent homotopy categories whose morphisms involve cubical diagrams.

1. Paths, 1-Tracks, 2-Tracks

If X is a topological space then a *path*, $f : p \simeq q$, in X from p to q, where p and q are points of X, is a map f from the unit interval $I = [0, 1]$ into X such that $f(0) = p$ and $f(1) = q$. If $f : p \simeq q$ and $g : q \simeq r$ are paths in X we denote by

$$g \bullet f : p \simeq r$$

their *concatenation*, i.e.,

$$(g \bullet f)(s) = f(2s), \quad 0 \le s \le 1/2;$$

$$(g \bullet f)(s) = g(2s - 1), \quad 1/2 \le s \le 1.$$

The *constant path* at $x \in X$ will be denoted c_x. If f is a path in X, we denote by f^{-1} the path *reverse* to f, i.e., $f^{-1}(s) = f(1 - s)$.

Let $f, f' : p \simeq q$ be paths. We say f and f' are *equivalent* if there exists a *relative homotopy* between f and f', i.e., a homotopy $f_t : f \simeq f'$ such that the initial and final points remain fixed during the homotopy. In that case we will use the notation $f_t : f \simeq f' : p \simeq q$. The equivalence class of f, denoted $\{f\}$, will be called the *track* of f. In the following we refer to these tracks as 1-*tracks* since we wish to consider 2-tracks as well.

If we regard the points of X as objects and the 1-tracks $\{f\}$ as morphisms from $f(0)$ to $f(1)$ with composition of morphisms induced by concatenation of paths we obtain a groupoid $\mathbf{G}_1 X$, the *fundamental groupoid* of X. We note that in order to verify the groupoid properties of $\mathbf{G}_1 X$ (associativity, existence of identities and inverses) certain relative homotopies

$$h \bullet (g \bullet f) \simeq (h \bullet g) \bullet f; \qquad c_{f(1)} \bullet f \simeq f \simeq f \bullet c_{f(0)};$$

$$f^{-1} \bullet f \simeq c_{f(0)} \tag{1.1}$$

are used which we will call *rescale, dilation* and *cancellation*. Explicit formulae are given in [32, pp. 47–48].

Relative homotopies can be concatenated (pasted) vertically and horizontally. More precisely, let $f, f', f'' : p \simeq q$ be paths, let $f_t : f \simeq f'$, $f'_t : f' \simeq f''$ be relative homotopies we define the *vertical pasting*

$$f_t + f'_t : f \simeq f'' : p \simeq q$$

to be the relative homotopy h_t such that

$$h_t = f_{2t}, \quad 0 \le t \le 1/2; \qquad h_t = f'_{2t-1}, \quad 1/2 \le t \le 1.$$

If $f_t : f \simeq f' : p \simeq q$ and $g_t : g \simeq g' : q \simeq r$ are relative homotopies, then we obtain the *horizontal pasting*

$$g_t \bullet f_t : g \bullet f \simeq g' \bullet f' : p \simeq r$$

by concatenation of the respective paths at each stage of the homotopy.

Vertical and horizontal pasting are interrelated by the following *interchange formula*. The proof is a simple verification.

PROPOSITION 1.2. *Let*

$$f_t : f \simeq f' : p \simeq q, \qquad f'_t : f' \simeq f'' : p \simeq q,$$

$$g_t : g \simeq g' : q \simeq r, \qquad g'_t : g' \simeq g'' : q \simeq r$$

be relative homotopies. Then the following equality holds:

$$(g_t + g'_t) \bullet (f_t + f'_t) = (g_t \bullet f_t) + (g'_t \bullet f'_t) : g \bullet f \simeq g'' \bullet f'' : p \simeq r.$$

Let $f_t, f'_t : f \simeq f' : p \simeq q$ be two relative homotopies. We may consider f_t and f'_t themselves to be *relatively homotopic*, if they are homotopic via a homotopy $I \times I \times I \to X$ which is constant on the boundary of $I \times I$. We call the relative homotopy class $\{f_t\}$ a 2-*track*. We now obtain a category ΠX whose objects are the paths f, f' etc. in X and whose morphisms $\{f_t\} : f \Rightarrow f'$ are the relative homotopy classes of relative homotopies between paths. Composition in ΠX is induced by vertical pasting. Since relative homotopies are reversible ΠX is a groupoid.

2. The Homotopy 2-Groupoid

In the next step it seems natural to try to integrate horizontal pasting in order to obtain a 2-dimensional categorical structure (cf. [24, 4]) with objects (0-cells) the points of X and 2-morphisms (2-cells) the 2-tracks. The question arises: what are the appropriate 1-morphisms (1-cells). One might first think the 1-cells should be the paths. This would give rise to the structure of a bigroupoid as indicated in the introduction. Here we proceed in a different way and construct a 2-groupoid (cf. [28]) as a quotient of ΠX by a suitable normal subgroupoid NX.

In order to construct the 1-cells of the desired 2-groupoid a relation, $\sim$, among paths in X based on what we call 'thin homotopies' will be used. Thin homotopies are relative homotopies and include rescale, dilation and cancellation but the relation is finer than equivalence under arbitrary relative homotopies. In order to establish transitivity of the relation $\sim$ we seem to need the space X to satisfy the Hausdorff separation property.

The following definition is crucial.

DEFINITION 2.1. Let f and f' be paths in X with the same initial and final points.

(1) A relative homotopy $F : f \simeq f' : I \times I \to X$ is *thin* if there is a factorisation

$$F = I \times I \xrightarrow{\Phi_F} J_F \xrightarrow{p_F} X, \tag{2.1.1}$$

where J_F is a tree, $\Phi_F : \phi \simeq \phi'$ is a relative homotopy, ϕ and ϕ' are paths in J_F which

(i) satisfy $\phi(0) = \phi'(0)$, $\phi(1) = \phi'(1)$,
(ii) are piecewise linear (PWL, see below) and
(iii) satisfy $p_F\phi = f$, $p_F\phi' = f'$.

Here, by a *tree*, we mean the underlying space $|K|$ of a finite 1-connected 1-dimensional simplicial complex K. A map $\phi : |K| \to |L|$, where K and L are (finite) simplicial complexes, is PWL (*piecewise linear*) if there exist subdivisions of K and L relative to which ϕ is simplicial.

(2) A 2-track $\alpha : f \Rightarrow f'$ is *thin* if it has a thin representative.
(3) The path f is *equivalent* to f', denoted $f \sim f'$, if there is a thin homotopy between f and f'.

EXAMPLE. Let $f : I \to X$ be a path and suppose that $0 < \alpha < 1$. Let $f_{\alpha,c}$ denote the path in X which agrees with $f(t\alpha^{-1})$ on the interval $[0, \alpha]$ and which is constant at $f(1)$ on $[\alpha, 1]$. Then we claim that $f \sim f_{\alpha,c}$. It is not a problem to construct a relative homotopy $F : f \simeq f_{\alpha,c}$, but we have to describe J_F, p_F and Φ_F. We take $J_F = I$ (regarded as a 1-simplex), $p_F = f$ and we choose Φ_F to be the corresponding relative homotopy from 1 to $1_{\alpha,c}$, where 1 refers to the identity map of I. Note that if $\alpha = 1/2$ then we recover the dilation relative homotopy.

PROPOSITION 2.2. *The class $\mathcal{T}$ of thin relative homotopies into a Hausdorff space X satisfies the following.*

$\mathcal{T}$(i) *Dilation, cancellation and rescale are thin. Any constant relative homotopy is thin.*
$\mathcal{T}$(ii) *$\mathcal{T}$ is closed under vertical pasting of relative homotopies.*
$\mathcal{T}$(iii) *$\mathcal{T}$ is closed under vertical reversion of relative homotopies: if $\sigma_t : f \simeq f' : p \simeq q$ is thin, then*
$$\sigma_{1-t} : f' \simeq f : p \simeq q$$
is thin.
$\mathcal{T}$(iv) *$\mathcal{T}$ is closed under horizontal pasting of relative homotopies.*

Proof. To check that cancellation is thin we may take $J_F = I$ as before, $p_F = f$ and choose Φ_F to be the corresponding relative homotopy from $1^{-1} \bullet 1$ to c_0. Note that $1^{-1} \bullet 1$ is PWL. It is easy to check similarly that rescale is thin. To verify the property (iv), suppose that J_F, p_F, $\Phi_F : \phi_f \simeq \phi_{f'}$ realise $f \sim f'$ and that J_G, p_G, $\Phi_G : \phi_g \simeq \phi_{g'}$ realise $g \sim g'$. Then we may take $J_H = J_F \vee J_G$, the *graft* of the two trees at their common vertex $\phi_f(1) = \phi_g(0)$. Since the graft is actually a pushout and since p_F and p_G agree at the common point, these maps induce p_H. It is now clear how we should define Φ_H, $\phi_{g\bullet f}$ and $\phi_{g'\bullet f'}$ so that the desired properties are satisfied. □

It is easy to verify property (iii) but (ii) is nontrivial and requires the Hausdorff separation property on the space X. Suppose that we have $f \sim f'$, $f' \sim f''$ and that these relations are realised by J_F, p_F, $\Phi_F : \phi_f \simeq \phi_{f',F}$ and by $J_{F'}$, $p_{F'}$, $\Phi_{F'} : \phi_{f',F'} \simeq \phi_{f''}$. The following diagram is relevant.

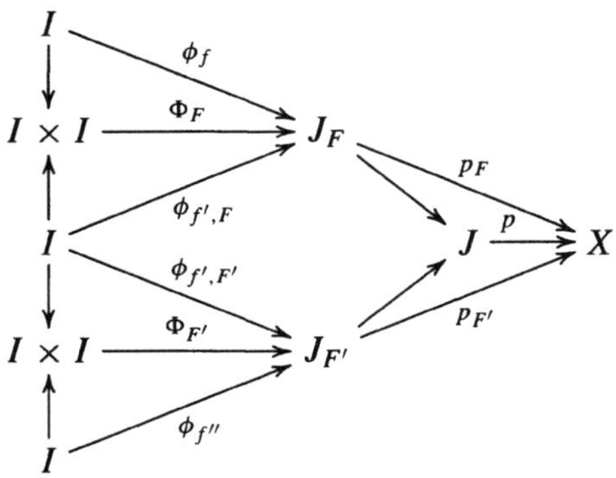

(2.2.1)

The idea of the proof is to construct the central quadrilateral with sink J as a pushout with J a tree. The argument depends on the following lemma, a proof of which is given in Section 5, and the well-known fact that composites of PWL maps are PWL [17, Chapter 1].

LEMMA 2.3. *Suppose, in the diagram*

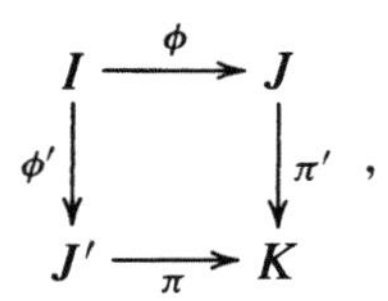

that J, J' are trees and that ϕ, ϕ' are PWL. *Then K, the pushout of ϕ and ϕ' in the category of Hausdorff spaces is a tree and π, π' are* PWL.

For the remainder of this paper, unless otherwise stated, all spaces are assumed to belong to **Haus**, the category of Hausdorff spaces.

COROLLARY 2.4. *The relation $\sim$ is an equivalence.*

Let NX denote the subgroupoid of ΠX whose morphisms are those 2-tracks which are thin. Then the following is a central result.

THEOREM 2.5. *The sets $NX(f, f')$ are either singletons or empty.*

REMARK. Theorem 2.5 can be expressed as follows in a language which is commonly used in connection with thin structures.

The 2-track of a relative homotopy which is thin is completely determined by its boundary.

Proof of Theorem 2.5. It is sufficient to consider the case $f = f'$. (If $\alpha, \beta \in NX(f, f')$ are distinct elements then $\alpha - \beta$ and 0 are distinct elements of $NX(f, f)$.)

Suppose that F and G are thin representatives of two elements of $NX(f, f)$. Then there exist factorisations

$$F = I \times I \xrightarrow{\Phi} J_F \xrightarrow{p_F} X,$$

$$G = I \times I \xrightarrow{\Psi} J_G \xrightarrow{p_G} X$$

of type (2.1.1), where $\Phi : \phi \simeq \phi'$, $\Psi : \psi \simeq \psi'$ and $p_F\phi = p_F\phi' = f = p_G\psi = p_G\psi'$. Applying Lemma 2.3, suppose that K is the pushout of ϕ and ϕ' and that K' is the pushout of ψ and ψ'. Then the corresponding composite maps $I \to K$ and $I \to K'$ are PWL and we may construct the pushout of these to obtain a tree L and factorisations

$$F = I \times I \longrightarrow L \xrightarrow{p} X, \qquad G = I \times I \longrightarrow L \xrightarrow{p} X$$

through a common tree. Now between any two points of a tree there is a unique shortest path that varies continuously with the given points. It follows that any two relative homotopies between maps that factor through a tree are themselves relatively homotopic, completing the proof of Theorem 2.5. □

COROLLARY 2.6. *NX is a normal subgroupoid of ΠX.*

Proof. Recall [6, p. 278], [7, p. 104], [22, p. 85] that a subgroupoid N of a groupoid A is *normal* if

(i) N contains all the identity elements of A,
(ii) $x \in N(f, f), a \in A(f, f')$ implies $axa^{-1} \in N(f', f')$.

In the case of NX since, by Theorem 2.5, $x \in NX(f, f)$ is unique, x is the identity element, so (ii) (and certainly also (i)) is satisfied. □

Since NX is a normal subgroupoid of ΠX we have the *quotient groupoid* $\Pi X/NX$ (cf. [7, p. 104], [22, p. 86]). The objects of $\Pi X/NX$ are the equivalence classes $\langle f\rangle$ of objects f of ΠX, i.e., of paths f in X, under the relation $\sim$ of Definition 2.1(3).

The morphisms $\langle f\rangle \Rightarrow \langle f'\rangle$ of $\Pi X/NX$ are the equivalence classes of morphisms $\{f_t\} : f \Rightarrow f'$ of $\Pi(X)$ under the relation $\{f_t\} \cong \{\bar{f}_t\}$, where $f_t : f \simeq f'$ and $\bar{f}_t : \bar{f} \simeq \bar{f}'$, which holds if there are thin homotopies $\sigma_t : f \simeq \bar{f}$ and $\sigma_t' : f' \simeq \bar{f}'$ such that

$$\{\bar{f}_t\} = \{\sigma_{1-t}\} + \{f_t\} + \{\sigma_t'\}.$$

Composition in $\Pi X/NX$, denoted $+$, is induced by composition in ΠX, i.e., by vertical pasting of relative homotopies. Note that the identity element

$$0_{\langle f\rangle} : \langle f\rangle \Longrightarrow \langle f\rangle$$

is represented by the constant relative homotopy at f.

There is a natural projection functor $\Pi X \to \Pi X/NX$. By 2.5 we have:

COROLLARY 2.7. *The natural projections*

$$\Pi X(f, f') \longrightarrow \Pi X/NX(\langle f\rangle, \langle f'\rangle)$$

are bijections.

We are now in a position to define a 2-groupoid $\mathbf{G}_2X$, i.e., a 2-category (cf. [24]) such that all 1-cells and all 2-cells are invertible. The objects (0-cells) of $\mathbf{G}_2X$ are the points of X, the 1-morphisms (1-cells) are the objects of $\Pi X/NX$, and the 2-morphisms (2-cells) are the morphisms of $\Pi X/NX$. Note that by 2.7 the 2-cells of $\mathbf{G}_2X$ can be interpreted as 2-tracks.

We have a composition of 1-morphisms induced by concatenation of paths:

$$\langle g\rangle \bullet \langle f\rangle = \langle g \bullet f\rangle,$$

where $f : p \simeq q, \quad g : q \simeq r$ are paths in X. By $\mathcal{T}$(iv) composition of 1-morphisms is well defined. By $\mathcal{T}$(i) the objects and the morphisms of $\mathbf{G}_2X$ form a category, $\mathbf{G}X$, the underlying category of $\mathbf{G}_2X$, in such a way that any 1-morphism is invertible. Thus $\mathbf{G}X$ is a groupoid, the *path groupoid* of X. Its morphisms, i.e., the classes $\langle f\rangle$, will be called *semitracks*.

Vertical composition of 2-cells, denoted $+$, is given by composition in the category $\Pi X/NX$. Since $\Pi X/NX$ is a groupoid, any 2-cell is invertible.

Horizontal composition of 2-cells, denoted $\bullet$, is defined by the formula

$$\{g_t\} \bullet \{f_t\} = \{g_t \bullet f_t\},$$

where $f_t : f \simeq f' : p \simeq q$ and $g_t : g \simeq g' : q \simeq r$ are relative homotopies. Using $\mathcal{T}$(i) and $\mathcal{T}$(iv) it is not difficult to prove that horizontal composition of 2-cells is well defined and is associative.

It follows from 1.2 that vertical and horizontal composition of 2-cells are interrelated by the obvious *interchange law*

$$(\beta + \beta') \bullet (\alpha + \alpha') = (\beta \bullet \alpha) + (\beta' \bullet \alpha'). \tag{2.8}$$

Finally, we observe that $\bullet$ has horizontal identity elements of form

$$0_{\langle c_x\rangle} : \langle c_x\rangle \Longrightarrow \langle c_x\rangle,$$

where x is a point of X.

This completes the construction of the 2-groupoid $\mathbf{G}_2X$ of the Hausdorff space X which will be called the *homotopy 2-groupoid* of X.

REMARK. The elements of $\mathbf{G}_2X$ have been constructed by a two stage process, part geometric (definition of the groupoid ΠX) and part algebraic (take the quotient by the normal subgroupoid NX). We note that the elements of $\mathbf{G}_2X$ can also be obtained in a one stage process as classes with respect to the following equivalence relation between relative homotopies.

Let $f_t : f \simeq f' : p \simeq q$ and $\bar{f}_t : \bar{f} \simeq \bar{f}' : p \simeq q$ be relative homotopies. Then f_t is said to be *equivalent* to $\bar{f}_t$ if there exists a homotopy $H : I^2 \times I \to X$ between f_t and $\bar{f}_t$ such that

(i) $H \mid \{0\} \times I \times I$ resp. $H \mid \{1\} \times I \times I$ are constant and
(ii) $H \mid I \times \{0\} \times I$ and $H \mid I \times \{1\} \times I$ are thin.

HISTORICAL REMARK. Teleman [36] has described a generalisation of the fundamental group of a space X. When X is Hausdorff the groupoid version of Teleman's construction seems to yield a structure closely related to the path groupoid. Another similar structure is that of the semi-fundamental groupoid of X. (See [6, 6.2, Exercise 7].)

Basic properties of $\mathbf{G}_2X$ are summarised in the following proposition. If f is a path in X, we write $\mathbf{G}_2(X, f)$ for the group of 2-endomorphisms of $\langle f \rangle$.

PROPOSITION 2.9. (i) *For each point x of X, the group $\mathbf{G}_2(X, c_x)$ is Abelian and isomorphic to the second homotopy group $\pi_2(X, x)$.* (ii) *For each path f in X, the map*

$$\Gamma : \mathbf{G}_2(X, c_{f(0)}) \longrightarrow \mathbf{G}_2(X, f)$$

defined by

$$\Gamma(\alpha) = 0_{\langle f \rangle} \bullet \alpha$$

is an isomorphism of groups.

Proof. In order to interpret $\mathbf{G}_2(X, c_x)$ as the second homotopy group $\pi_2(X, x)$ we observe that by 2.7 an element of $\mathbf{G}_2(X, c_x)$ is the relative homotopy class of a homotopy $I \times I \to X$ which maps the boundary of $I \times I$ to the point x.

Routine manipulations based on the interchange law 2.9 show that $\mathbf{G}_2(X, c_x)$ is Abelian and Γ is a homomorphism. Finally the map

$$\Gamma' : \mathbf{G}_2(X, f) \longrightarrow \mathbf{G}_2(X, c_{f(0)})$$

defined by

$$\Gamma'(\alpha') = 0_{\langle f^{-1} \rangle} \bullet \alpha'$$

is inverse to Γ (cf. [31]). The verification is left to the reader. □

We remark that the operators Γ and Γ' used in the proof of Proposition 2.9 are special cases of the 'star operations'

$$\langle f \rangle_* \alpha = 0_{\langle f \rangle} \bullet \alpha \text{ and } \langle g \rangle^* \alpha = \alpha \bullet 0_{\langle g \rangle} \tag{2.9.1}$$

which are defined whenever $\alpha \in \mathbf{G}_2X(\langle h\rangle, \langle h'\rangle)$ and $g(1) = h(0) = h'(0)$, $h(1) = h'(1) = f(0)$.

COROLLARY 2.10. *For each path f in X, the group $\mathbf{G}_2(X, f)$ is Abelian and isomorphic to the second homotopy group $\pi_2(X, f(0))$.*

REMARK. It is an open problem to generalise the methods of this section so as to construct a homotopy n-groupoid $\mathbf{G}_nX$ for $n > 2$.

The connection of our construction of the homotopy 2-groupoid with the structures described in the introduction may best be seen by comparing it with the Moerdijk–Svensson 2-groupoid. If X is a CW-complex, Y its 1-skeleton and S its set of 0-cells the Moerdijk–Svensson 2-groupoid $W(X, Y, S)$ is weakly equivalent to the homotopy 2-groupoid $\mathbf{G}_2X$ in the sense that in each case

(i) the 2-endomorphism groups are isomorphic with the second homotopy groups of X,
(ii) the associated Poincaré groupoid is isomorphic with the fundamental groupoid of X.

(Recall that the Poincaré groupoid (cf. [4]) of a 2-groupoid is obtained from the 1-morphism sets by factoring out by the relation of 'homotopy' where a homotopic to b means there exists a 2-morphism from a to b.)

However, there seems to be no obvious relation between the path groupoid, $\mathbf{G}X$, of X and the underlying groupoid of $W(X, Y, S)$ which is isomorphic to the fundamental groupoid of Y on the set S.

3. Homotopy 2-Groupoids of Function Spaces

In this section we work in the convenient category $\mathbf{T}$ of compactly generated Hausdorff spaces or the corresponding pointed category $\mathbf{T}^*$. As usual, the base point of a space in $\mathbf{T}^*$ will be denoted by $*$. We are particularly concerned with homotopy 2-groupoids of the form $\mathbf{G}_2X^V$. Here V and X are spaces in $\mathbf{T}^*$ and X^V denotes the (unpointed) space of pointed continuous maps from V to X considered as subspace of the space of *all* continuous maps from V to X equipped with the compactly generated coreflection of the compact-open topology. If Z is a space in $\mathbf{T}$, by adjunction we have a bijection between $\mathbf{T}(Z, X^V)$ and the subset of $\mathbf{T}(Z \times V, X)$ consisting of those continuous functions from $Z \times V$ to X sending $Z \times *$ to the base point of X. Specialising to $Z = I^n$, $n = 0, 1, 2, 3$, the bijection described gives rise to a 2-groupoid $\mathbf{G}_2(V, X)$, isomorphic to $\mathbf{G}_2X^V$, whose objects are pointed maps from V to X, 1-morphisms are semitracks of pointed homotopies and whose

2-morphisms (also called 2-tracks) are (equivalence classes of) relative homotopy classes of 2-homotopies. We have the following table of correspondences.

$$
\begin{array}{l}
\mathbf{G}_2X^V \leftrightarrow \mathbf{G}_2(V,X)\\
f: I \to X^V \leftrightarrow f_t : V \to X\\
\langle f\rangle \leftrightarrow \langle f_t\rangle\\
f_s : f \simeq f' \leftrightarrow f_{t,s} : f_t \simeq f'_t\\
\{f_s\} + \{g_s\} \leftrightarrow \{f_{t,s}\} + \{g_{t,s}\}\\
\{f_s\} \bullet \{h_s\} \leftrightarrow \{f_{t,s}\} \bullet \{h_{t,s}\}\\
\langle f\rangle_* \leftrightarrow \langle f_t\rangle_*\\
\langle g\rangle^* \leftrightarrow \langle g_t\rangle^*\\
\Pi_2X^V \leftrightarrow \Pi_2(V,X)\\
NX^V \leftrightarrow N(V,X)
\end{array}
$$

Note that for each 1-morphism $\langle f_t\rangle$, with f_0 homotopic to the constant map, there is an isomorphism

$$\sigma_{\langle f_t\rangle} : \mathbf{G}_2(V,X)(\langle f_t\rangle) \xrightarrow{\approx} \pi(\Sigma^2 V, X). \tag{3.1}$$

The elements (2-morphisms) of $\mathbf{G}_2(V,X)(\langle f_t\rangle, \langle f'_t\rangle)$ are generalised cosets of $N(V,X)$ determined by 2-tracks. We shall refer to them also (with some abuse) as 2-tracks and denote them by capital letters. In the following list of properties, c_t refers to a constant homotopy.

$$
\begin{array}{l}
H + (G + F) = (H + G) + F\\
H + 0 = 0 + H = H\\
H + (-H) = 0 = (-H) + H\\
\langle h_t\rangle^*(-H) = -\langle h_t\rangle^* H, \quad \langle g_t\rangle_*(-H) = -\langle g_t\rangle_* H\\
\langle h_t\rangle^*(H + G) = \langle h_t\rangle^* H + \langle h_t\rangle^* G,\\
\langle g_t\rangle_*(H + G) = \langle g_t\rangle_* H + \langle g_t\rangle_* G\\
\big(\langle g_t\rangle \bullet \langle f_t\rangle\big)_* = \langle g_t\rangle_*\langle f_t\rangle_*, \quad \langle c_t\rangle_* = \text{identity}\\
\big(\langle f_t\rangle \bullet \langle h_t\rangle\big)^* = \langle h_t\rangle^*\langle f_t\rangle^*, \quad \langle c_t\rangle^* = \text{identity}, \quad \langle g_t\rangle_*\langle h_t\rangle^* = \langle h_t\rangle^*\langle g_t\rangle_*\\
G \bullet H = \langle f_t\rangle^* G + \langle g'_t\rangle_* H = \langle g_t\rangle_* H + \langle f'_t\rangle^* G,
\end{array}
\tag{3.2.1}
$$

in the situation

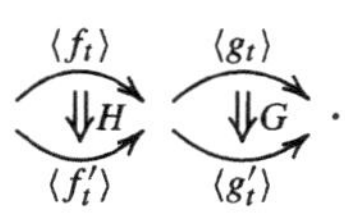

Each 2-track $H : \langle f_t\rangle \Rightarrow \langle f'_t\rangle$ has an inverse with respect to the horizontal composition denoted $H^{-1} : \langle f_{1-t}\rangle \Rightarrow \langle f'_{1-t}\rangle$. Via the Equations (3.2.1) we obtain

$$H^{-1} = -\langle f'_{1-t}\rangle^*\langle f_{1-t}\rangle_* H = -\langle f_{1-t}\rangle^*\langle f'_{1-t}\rangle_* H. \tag{3.2.2}$$

These properties are all standard in a 2-groupoid. They indicate that $\mathbf{G}_2(V,X)$ has the structure of a *track category* in the sense of [3].

Moreover, $\mathbf{G}_2(V, X)$ is a bifunctor, covariant in X and contravariant in V: given pointed maps $k' : V' \to V$ and $f' : X \to X'$ there are associated 2-groupoid morphisms

$$(k')^{\cdot} : \mathbf{G}_2(V, X) \to \mathbf{G}_2(V', X) \quad \text{and} \quad f. : \mathbf{G}_2(V, X) \to \mathbf{G}_2(V, X').$$

Furthermore, the 2-dimensional structure comprising pointed spaces, pointed maps and semitracks admits two horizontal composition operations $\blacktriangleleft$ and $\blacktriangleright$. Semitracks $\langle g_t\rangle \blacktriangleleft \langle f_t\rangle$ and $\langle g_t\rangle \blacktriangleright \langle f_t\rangle$ are defined in the situation

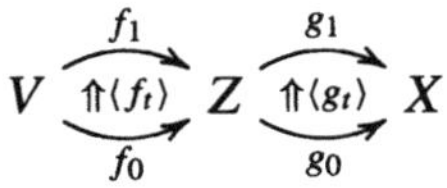

as follows.

$$\langle g_t\rangle \blacktriangleleft \langle f_t\rangle = g_1.\langle f_t\rangle \bullet f_0^{\cdot}\langle g_t\rangle \quad \text{and} \quad \langle g_t\rangle \blacktriangleright \langle f_t\rangle = f_1^{\cdot}\langle g_t\rangle \bullet g_0.\langle f_t\rangle. \tag{3.3}$$

These compositions are associative. They endow the spaces, maps and semitracks with the structure of a sesquicategory [35]. The interchange law not being valid in general for semitracks, the compositions are distinct. Hence they do not give rise to another 2-category structure. However, the compositions are linked by a natural isomorphism

$$[g_t, f_t] : \langle g_t\rangle \blacktriangleleft \langle f_t\rangle \Rightarrow \langle g_t\rangle \blacktriangleright \langle f_t\rangle$$

obtained by considering the interchange 2-track.

Suppose that $f_t : V \to Z$ and $g_t : Z \to X$ are homotopies. Then there is an associated 2-track, an *interchange*, depending only on $\langle f_t\rangle$ and $\langle g_t\rangle$ and inhabiting the diagram

$$\begin{array}{ccccc} g_0 f_0 & \xrightarrow{g_t f_0} & g_1 f_0 & \xrightarrow{g_1 f_t} & g_1 f_1 \\ \| & & [g_t, f_t] \Downarrow & & \| \\ g_0 f_0 & \xrightarrow[g_0 f_t]{} & g_0 f_1 & \xrightarrow[g_t f_1]{} & g_1 f_1 \end{array} \quad . \tag{3.4}$$

Various essentially equivalent versions of the interchange can be found in the literature but a rather convenient definition uses the 'lens collapse' construction described by Marco Grandis [18]. Let $f_{t,s}$ be a *square* in a Hausdorff space W, i.e., a map $I \times I \to W$. Then a representative of the *lens collapse* 2-track

$$\{\hat{f}_{t,s}\} : \langle f_{1,t}\rangle \bullet \langle f_{t,0}\rangle \Longrightarrow \langle f_{t,1}\rangle \bullet \langle f_{0,t}\rangle \tag{3.4.1}$$

of $f_{t,s}$ in $\mathbf{G}_2 W$ is suggested by the following sketch.

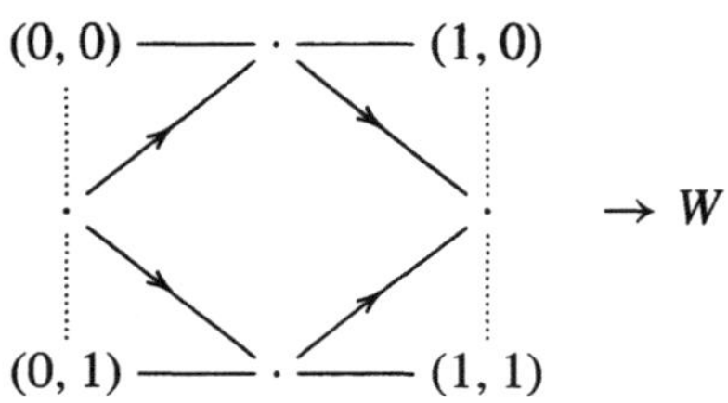

The outer square is first retracted on to the inner square by collapsing the triangles vertically on to its edges. The inner square is then mapped to W by applying $\hat{f}_{t,s}$.

The *interchange* 2-*track* $[g_t, f_t]$ is then the adjoint of the special case of $\{\hat{f}_{t,s}\}$ in which $W = X^V$ and $f_{t,s}$ is the adjoint of the composite 2-homotopy $g_t f_s : V \to X$.

PROPOSITION 3.5. *The 2-track* $[g_t, f_t] \in \mathbf{G}_2(V, X)(\langle g_1 f_t \bullet g_t f_0 \rangle, \langle g_t f_1 \bullet g_0 f_t \rangle)$ *depends only on the classes* $\langle g_t \rangle$ *and* $\langle f_t \rangle$. *The following properties hold.*

(i) *For a constant homotopy* c_t, $[g_t, c_t] = 0 = [c_t, f_t]$;
(ii) $[h_t \bullet g_t, f_t] = \langle g_t f_0 \rangle^* [h_t, f_t] + \langle h_t f_1 \rangle_* [g_t, f_t]$;
(iii) $[g_t, k_t \bullet f_t] = \langle g_1 k_t \rangle_* [g_t, f_t] + \langle g_0 f_t \rangle^* [g_t, k_t]$;
(iv) $[g_{1-t}, f_t] = -\langle g_{1-t} f_0 \rangle^* \langle g_{1-t} f_1 \rangle_* [g_t, f_t]$;
(v) $[g_t, f_{1-t}] = -\langle g_0 f_{1-t} \rangle^* \langle g_1 f_{1-t} \rangle_* [g_t, f_t]$;
(vi) *Given* $k : V' \to V$, $f_t : V \to Z$, $g_t : Z \to X$ *then* $[g_t, f_t k] = k^\cdot [g_t, f_t]$;
(vii) *Given* $f_t : V \to Z$, $g_t : Z \to X$, $h : X \to X'$ *then* $[h g_t, f_t] = h.[g_t, f_t]$;
(viii) *Given* $f_t : V \to Z$, $g : Z \to X$, $h_t : X \to W$ *then* $[h_t g, f_t] = [h_t, g f_t]$;
(ix) $[g_t, f_t]^{-1} = -[g_{1-t}, f_{1-t}]$;
(x) *If* $f_0 = f_1 = *$ *and* $g_0 = g_1 = *$ *are trivial maps then* $[g_t, f_t] = 0$;
(xi) *Given 2-tracks*

$$\mu : \langle f_t \rangle \Longrightarrow \langle f_t' \rangle : f \longrightarrow f' : V \longrightarrow Z,$$
$$\nu : \langle g_t \rangle \Longrightarrow \langle g_t' \rangle : g \longrightarrow g' : Z \longrightarrow X$$

then

$$\begin{aligned} &\langle g_t f \rangle^* (g'.\mu) + \langle g' f_t' \rangle_* (f^\cdot \nu) + [g_t', f_t'] \\ &\quad = [g_t, f_t] + \langle g f_t \rangle^* (f'^\cdot \nu) + \langle g_t' f' \rangle_* (g.\mu) \end{aligned}$$

in $\mathbf{G}_2(V, X)(\langle g' f_t \bullet g_t f \rangle, \langle g_t' f' \bullet g f_t' \rangle)$.

Proof. Let $f_{t,s} : f_t \simeq f_t' : V \to Z$ be a 2-homotopy representing an element $\alpha \in N(V, Z)(f_t, f_t')$ then the cube

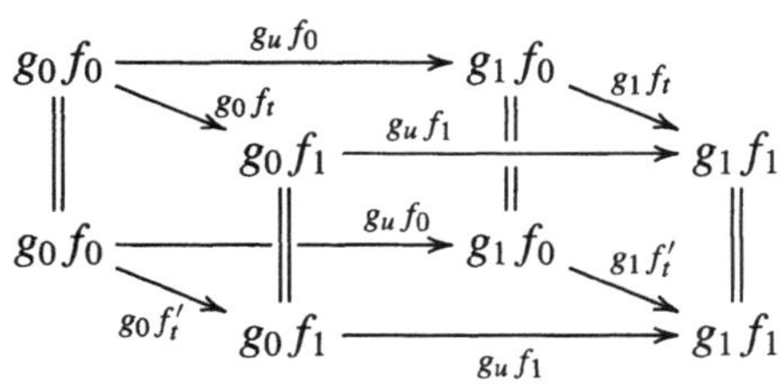

(3.5.1)

commutes via the 3-homotopy $g_u f_{t,s}$. Note that the 2-track represented by the top face is $[g_t, f_t]$ and that represented by the bottom face is $[g_t, f_t']$. Since the sides represent elements of $N(V, X)$, these two are equal. In a similar way we can show that $[g_t, f_t]$ depends only on the class $\langle g_t \rangle$.

To verify (i), note that $g_t c_s$ is a constant 2-homotopy.

The proofs of (ii), (iii) and (xi) are postponed until section 6, where they will be deduced from Propositions 6.1 and 6.2.

(iv) This can be deduced from (i) and (ii), similarly (v) can be deduced from (i) and (iii).

(vi), (vii) and (viii) follow directly from the definition of the interchange 2-track; (ix) can be checked using (3.2.2) together with (iv) and (v) and the properties of the star operators.

(x) Consider the paths $f : I \to Z^V$, $g : I \to X^Z$ adjoint to f_t and g_t. Note that the map $g \times f : I \times I \to X^Z \times Z^V$ is trivial on the boundary of $I \times I$ and hence represents an element of $\pi_2(X^Z \times Z^V)$. Since any such element lies in the image of the homomorphism induced by the inclusion map $X^Z \vee Z^V \to X^Z \times Z^V$, the desired equality follows. □

REMARK. The algebraic structure exhibited above can be interpreted by means of the Gray tensor product for 2-categories, resp. 2-groupoids, in the following way. Let V, Z, X be pointed spaces. Then by Proposition 3.5(i), (ii), (xi) a *quasi-functor of two variables* in the sense of [20, 1.1], resp. [19, I, 4.1(i)]

$$\mathbf{G}_2(V, Z) \times \mathbf{G}_2(Z, X) \xrightarrow{\cdot\cdot} \mathbf{G}_2(V, X)$$

is given by the following data:

(i) the families of 2-functors

$$f^{\cdot} : \mathbf{G}_2(Z, X) \longrightarrow \mathbf{G}_2(V, X),$$
$$g_{\cdot} : \mathbf{G}_2(V, Z) \longrightarrow \mathbf{G}_2(V, X)$$

for all pointed maps $f : V \to Z$, resp. $g : Z \to X$,

(ii) the collection of interchange 2-tracks

$$[g_t, f_t] : \langle g_t \rangle \blacktriangleleft \langle f_t \rangle \Longrightarrow \langle g_t \rangle \blacktriangleright \langle f_t \rangle$$

for each pair of semitracks $\langle f_t \rangle$, $\langle g_t \rangle$ in $\mathbf{G}_2(V, Z)$, resp. $\mathbf{G}_2(Z, X)$.

By [19, p. 73] a quasi-functor of two variables is equivalent to a 2-functor from the (Gray) tensor product

$$\mathbf{G}_2(V, Z) \otimes \mathbf{G}_2(Z, X) \longrightarrow \mathbf{G}_2(V, X).$$

In this way the category $\mathbf{T}^*$ of pointed spaces admits a 2-groupoid enrichment with respect to the closed monoidal structure on the category of 2-groupoids given by the Gray tensor product, i.e., $\mathbf{T}^*$ is a Gray-category in the sense of [35].

Alternatively, the algebra described in this section should fit into a 3-dimensional structure à la Baues [3, p. 330] with the following elements. Objects: pointed spaces; 1-morphisms: pointed maps; 2-morphisms: semitracks; 3-morphisms: 2-tracks.

4. Cubical Diagrams

Homotopy commutative cubes have been dealt with, amongst others, by [27] and, in an abstract setting, by [34]. For our purposes, in contrast with [27] and [34], we will have to consider homotopy commutative cubes inhabited by specific 2-tracks realising homotopy commutativity. The task is to keep track of the 2-tracks when pasting such cubes.

Suppose we have a cubical diagram

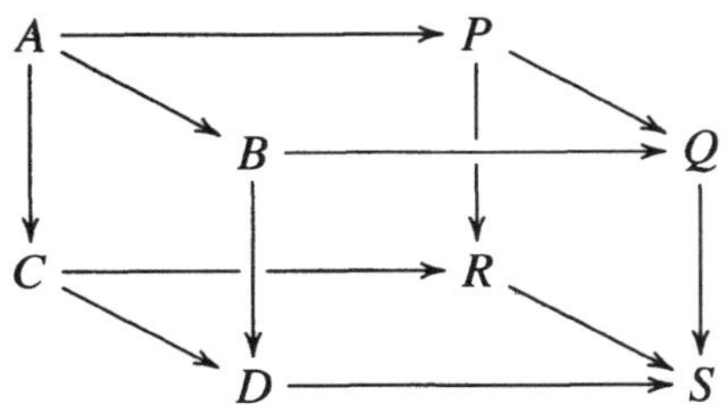

(4.1)

of spaces, maps and homotopies (occupying the faces). It will be convenient to use the notation AB for the map from A to B and the notation ABQ for the composition of AB and BQ etc. Similarly we use the convention that, for example, $\langle CB\rangle$ refers to the semitrack from ACD to ABD. Then the following hexagonal diagram (4.2), in which each vertex indicates a composite map from A to S and each arrow a semitrack between such composites, can be regarded as a representation in $\mathbf{G}_2(A, S)$ of the cube (4.1).

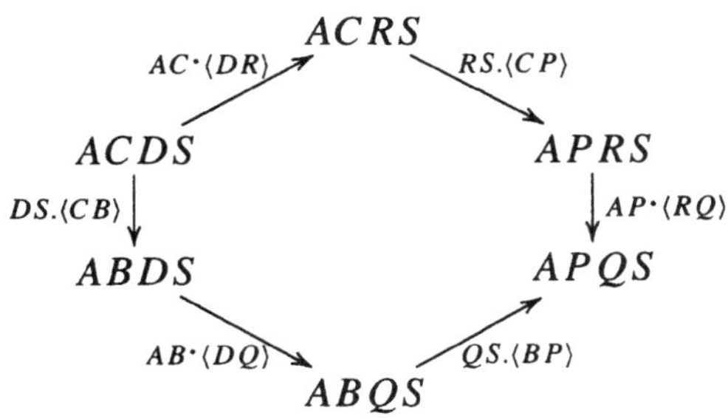

(4.2)

Consider the semitracks

$$\phi = AP^{\cdot}\langle RQ\rangle \bullet RS.\langle CP\rangle \bullet AC^{\cdot}\langle DR\rangle,$$

$$\psi = QS.\langle BP\rangle \bullet AB^{\cdot}\langle DQ\rangle \bullet DS.\langle CB\rangle.$$

If the 1-tracks corresponding to these are equal then the cube (4.1) is homotopy commutative in the sense of Mather [27] and there will exist a 2-track

$$\alpha \in \mathbf{G}_2(A, S)(\phi, \psi) \tag{4.3}$$

inhabiting the cube.

If the two cubes, inhabited by α and β respectively,

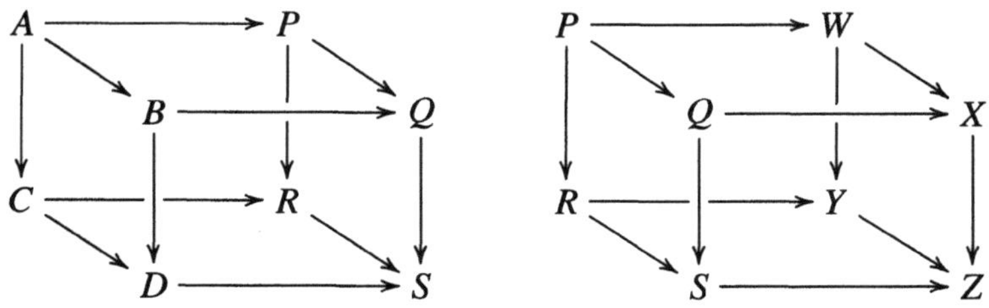

(4.4)

carry the same semitrack on their common face then their 2-tracks can be pasted in the following manner. Note that the hexagon

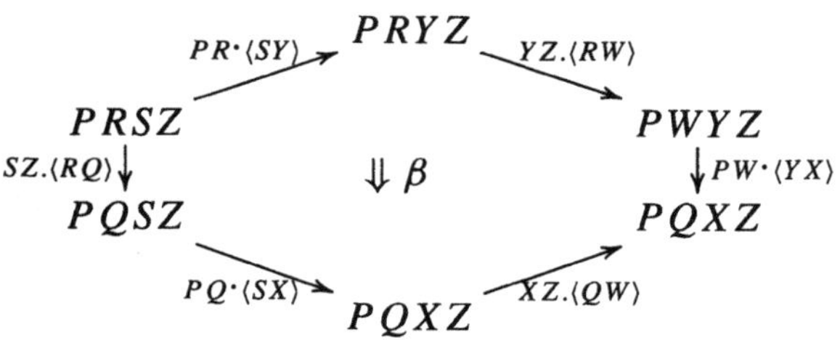

(4.5)

represents, in $\mathbf{G}_2(P, Z)$, the righthand cube of (4.4). The hexagons can be adjoined to form a composite diagram in $\mathbf{G}_2(A, Z)$ by first composing each entry in (4.2) with SZ and (precomposing) each entry in (4.5) with AP. Doing so we obtain the following diagram in which the rectangular cells (needed to recover a hexagon) are occupied by interchange 2-tracks. The outer (larger) hexagon then represents the cube obtained by pasting the individual cubes in (4.4) along the common face.

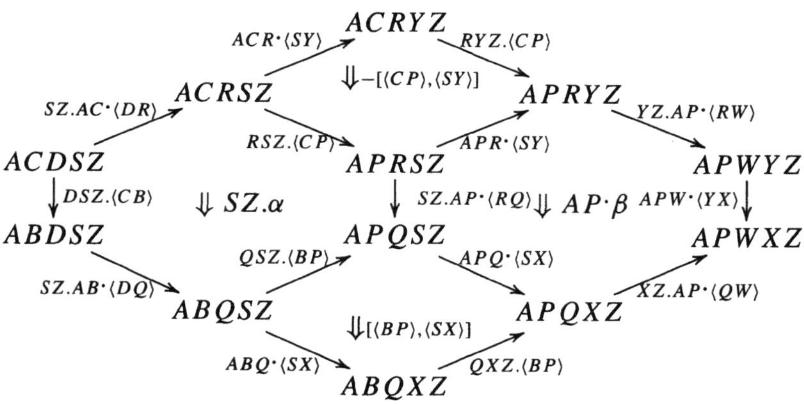

(4.6)

Note that composing the pair of arrows in each edge of the larger hexagon amounts to pasting corresponding faces of the cubes (4.4).

Similar definitions can be given for other pastings of cubes provided that their orientations are coherent, the common faces carry the same homotopy and provided that we always paste a 'sink' face to a 'source' face. For example, in the case of the cubes

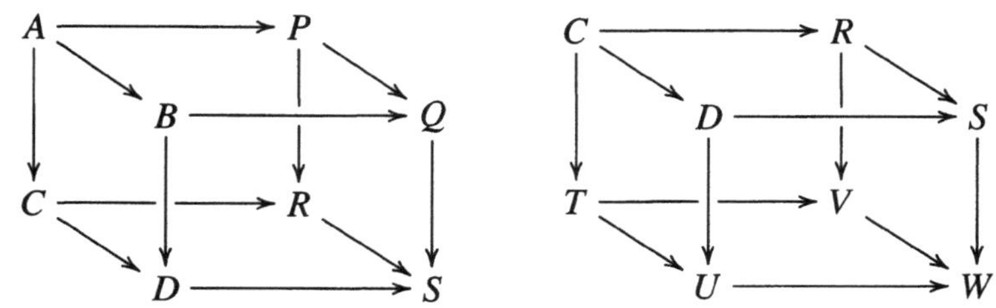

(4.7)

we obtain the diagram

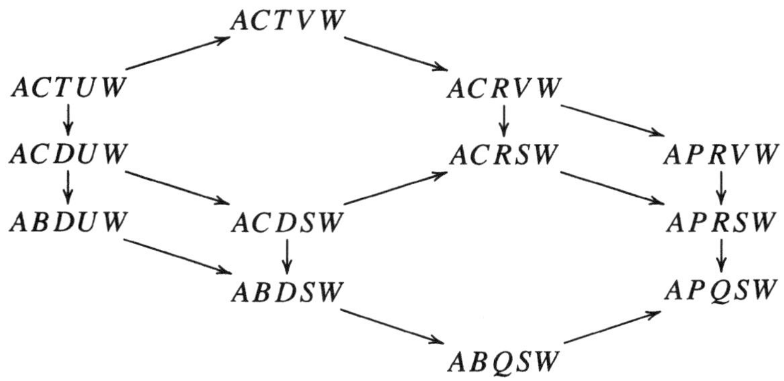

. (4.8)

5. Folding Maps of Trees

In this section we study the pushout of folding maps of trees in order to prove the Lemma 2.3. We begin by recalling a definition and a result concerning piecewise linear maps. For a proof of the result, the reader may consult [17, Chapter 1].

DEFINITION 5.1. A map $f : |K| \to |L|$, where K and L are finite simplicial complexes is *piecewise linear* (PWL) if there exist subdivisions of K and L with respect to which f is simplicial.

PROPOSITION 5.2. *The collection of all finite simplicial complexes and* PWL *maps constitutes a category.*

Suppose that J is a 1-dimensional, 1-connected finite simplicial complex, and let $\phi : I \to |J|$ be a PWL map. Then there is a subdivision J' of J and a subdivision of I with vertices

$$0 = t_0 < t_1 < t_2 < \cdots < t_n = 1 \qquad (5.2.1)$$

on which $\phi : I \to J'$ is simplicial. Since the image of ϕ in $|J'|$ is a subtree, we may assume without loss of generality that ϕ is surjective. We describe such a ϕ as a *folding map*. Note that the points $\phi(0)$ and $\phi(1)$, which need not be distinct, may be regarded as *roots* of $|J'|$ so that $|J'|$ is a *doubly rooted* tree.

DEFINITION 5.3. Let P be a vertex of J, where $|J|$ is a tree. A vertex Q of J is *adjacent* to P if PQ is a (possibly degenerate) simplex of J. Suppose that the vertex $\bar{Q}$ of J is also adjacent to P, then there is a simplicial map $\psi : J \to \bar{J}$ which identifies the simplexes PQ and $P\bar{Q}$ (and the vertices Q and $\bar{Q}$) but leaves unchanged all other vertices and simplexes of J. Note that $|\bar{J}|$ is also a tree. We describe ψ as a *simple folding map*. Note that if just one of PQ and $P\bar{Q}$ is degenerate then ψ shrinks the nondegenerate one to a single vertex of $\bar{J}$.

LEMMA 5.4. *Any folding map $\phi : I \to J'$ is the composite of a finite sequence of simple folding maps:*

$$\phi = I = J_0 \xrightarrow{\phi_1} J_1 \xrightarrow{\phi_2} J_2 \xrightarrow{\phi_3} J_3 \cdots \xrightarrow{\phi_n} J_n = J'.$$

Proof. Suppose that the vertices of J' are labelled by letters A, B, C etc. Then the successive vertices $\phi(t_0), \phi(t_1), \ldots$ visited by the simplicial path ϕ are given by a finite sequence of form

$$A,\ B,\ C,\ B',\ D,\ B'',\ C'\ \ldots,$$

where the number of primes attached to any given letter indicates the number of times that vertex has previously been visited by ϕ. Note that the first letter in the sequence is always the root vertex $\phi(0)$ and the last is always $\phi(1)$. For brevity we rather denote the sequence as the symbolic product $ABCB'DB''C'\ldots$. For example the product $0ABCB'A'DA''B''1$ specifies a path obtained by following arrows in the diagram

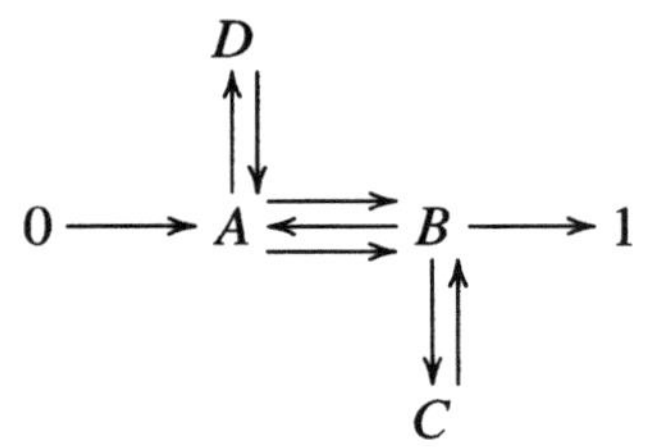

from which also the associated tree can be recovered by forgetting the directions of the arrows and identifying all arrows between each pair of letters.

We now consider an operation ρ on formal products which acts as follows. If a product has no primed letters then ρ leaves it unchanged. Otherwise ρ singles out the first letter (from the left) that later occurs with (at least one) prime. Then ρ replaces that letter by its barred version at its first occurrence (e.g., $B \to \bar{B}$) and removes exactly one prime at each subsequent occurrence, ρ leaving all other letters unchanged. When iterating ρ, a barred letter is regarded as distinct from its unbarred version and multiple bars may be needed. Thus, for example,

$$\rho(0ABCB'A'DA''B''1) = 0\bar{A}BCB'ADA'B''1$$

and the successive iterates (until the product stabilises) are

$$0ABCB'A'DA''B''1 \to 0\bar{A}BCB'ADA'B''1 \to 0\bar{A}\bar{B}CBADA'B'1$$
$$\to 0\bar{A}\bar{B}C\bar{\bar{B}}ADA'B1 \to 0\bar{A}\bar{B}C\bar{\bar{B}}\bar{\bar{A}}DAB1.$$

The products obtained by applying ρ represent successive unfoldings of the original path and tree. Thus $0\bar{A}BCB'ADA'B''1$ represents

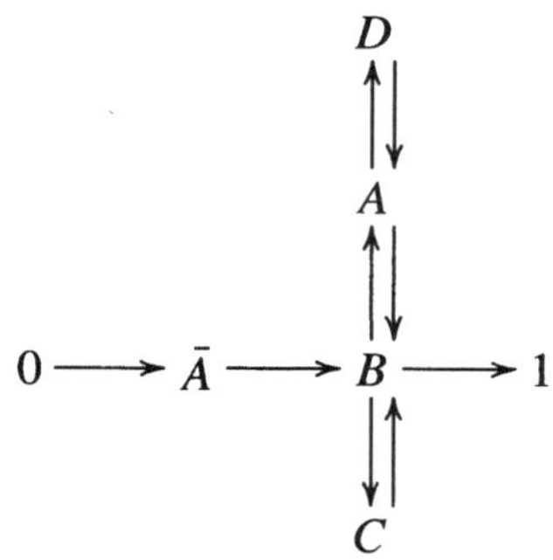

Every stable product represents a decomposition of I and the associated identity path. The unfoldings produced by application of ρ are simple (i.e., they can be reversed by simple foldings): to see this, suppose that A is the letter replaced by $\bar{A}$ during an application of ρ and that $B \neq A$ is the next vertex visited by ϕ then, when ϕ next revisits A it has to do so from a revisit to B, for otherwise the path would determine a nontrivial cycle of J (not possible since $|J|$ is a tree). Then the relevant unfolding is reversed by folding $\bar{A}B$ against AB. If, on the other hand, it happens that A *is* the next vertex visited by ϕ then the unfolding is reversed by shrinking $\bar{A}A$. This completes the proof of Lemma 5.4. □

It is now evident, in view of Lemma 5.4, that Lemma 2.3 is a consequence of the following proposition.

PROPOSITION 5.5. *Suppose that the diagram*

$$\begin{array}{ccc} |J| & \xrightarrow{|\phi_1|} & |J_1| \\ {\scriptstyle |\phi_2|}\downarrow & & \downarrow{\scriptstyle \pi_2} \\ |J_2| & \xrightarrow{\pi_1} & K \end{array}$$

is the pushout in **Haus** *of the simple folding maps ϕ_1 and ϕ_2. Then K is a tree and the maps π_1 and π_2 are finite composites of simple folding maps.*

Proof. We first consider the case $|J| = I$. Then ϕ_1 is a simplicial map from a subdivision

$$A_1 \leq A_2 \leq A_3 \cdots \leq A_n$$

of I that folds the simplexes $A_{i-1}A_i$ and $A_{i+1}A_i$, for some i. Similarly ϕ_2 is a simplicial map from some other subdivision with vertices (say) $B_1, B_2, \ldots, B_m$ that folds $B_{j-1}B_j$ and $B_{j+1}B_j$ for some j. If $A_{i+1} \leq B_{j-1}$ (or if $B_{j+1} \leq A_{i-1}$) then the foldings do not interact and there is no problem, K is simply the tree produced by making both foldings and π_1 and π_2, the induced maps, are both simple foldings. □

More interesting cases occur when there is overlap of the intervals. For example if we have the situation

$$0 \leq A < M_A < B < \breve{A} < M_B < \breve{B} \leq 1$$

in which AM_A, $\breve{A}M_A$ and BM_B, $\breve{B}M_B$ are to be folded then J_1 and J_2 have the form

$$J_1 = \begin{array}{ccccccccc} & & M_A & & & & & & \\ & & | & & & & & & \\ & & (B) & & & & & & \\ & & | & & & & & & \\ 0 & — & A & — & (M_B) & — & (\breve{B}) & — & 1 \end{array} \qquad J_2 = \begin{array}{ccccccccc} & & & & & & M_B & & \\ & & & & & & | & & \\ & & & & & & (\breve{A}) & & \\ & & & & & & | & & \\ 0 & — & (A) & — & (M_A) & — & B & — & 1 \end{array}$$

and each of these fold to

$$K = \begin{array}{ccccccc} & & M_B & & M_A & & \\ & & | & & | & & \\ 0 & — & A & — & B & — & 1 \end{array} .$$

Note that (B) is not a vertex of J_1, but simply the image of the point B under the folding map ϕ_1. In this example it is possible to make ϕ_1 and ϕ_2 simplicial on a common simplicial subdivision of I, namely

$$0 < A < B_1 < M_A < B < \breve{A} < M_B < A_1 < \breve{B} < 1, \tag{5.5.1}$$

where B_1 (respectively A_1) refers to the *reflection* of B in M_A (respectively of A in M_B): specifically B_1 is the point of I identified with B when AM_A is folded against $\breve{A}M_A$. Then $|\phi_1| = |\phi_1'|$, where ϕ_1' is the simplicial folding to

$$J_1' = \begin{array}{ccccccccccc} & & M_A & & & & & & & & \\ & & | & & & & & & & & \\ & & B & & & & & & & & \\ & & | & & & & & & & & \\ 0 & — & A & — & M_B & — & A_1 & — & \breve{B} & — & 1 \end{array} .$$

Similarly we may modify ϕ_2 and J_2 to ϕ_2' and J_2'. The maps π_1 and π_2 are then each composites of two simple foldings. The composite folding map into K is specified by the formal product $0ABM_AB'A'M_BA''B''1$, which can be recovered from the successive entries of (5.5.1).

Problems arise with situations such as

$$0 \leq A = B < M_A < M_B < A' = B' \leq 1. \tag{5.5.2}$$

If both identifications are made then the resulting equivalence relation on I has the following property. Let x be a point in the interior of the interval $[A, A']$. The equivalence class $\langle x \rangle$ of x has infinitely many members and there are infinitely many distinct equivalence classes of interior points. Moreover each neighborhood of A contains a member of $\langle x \rangle$. The quotient space L, the pushout of the two folding maps in the category of all topological spaces, is not Hausdorff. Indeed it is not T_0, since $\langle A \rangle$ (which is distinct from $\langle x \rangle$) belongs to the closure of $(\langle x \rangle)$. Instead we have to consider K, the pushout in **Haus** of the folding maps. Then K is the largest Hausdorff quotient of L. Let $q : L \to K$ be the quotient map. Then q is a closed map. (See [26, p. 153].) Hence

$$q\big(\overline{\langle x \rangle}\big) = \overline{q(\langle x \rangle)} = q\big(\langle x \rangle\big),$$

it follows that $q(\langle x \rangle) = q(\langle A \rangle)$, so that K is simply the space obtained from I by shrinking the interval $[A, A']$ to a single point. Hence the assertion of Proposition 5.5 holds in this case also.

Collapsing of intervals occurs in cases in which equivalence classes of points contain infinitely many members. This certainly happens when one of the intervals $[A, M_A]$, $[M_A, A]$, $[B, M_B]$, $[M_B, B]$ is degenerate. However suppose none of these is degenerate and let $R_A : [A, A'] \to [A, A']$ (respectively $R_B : [B, B'] \to [B, B']$) be the map which sends each point into the other point with which it would be identified under the respective folding. Then R_A and R_B are involutions and the only way in which infinitely many points are identified with a given point x of I is if infinitely many iterates of form

$$(R_A R_B)^n x \quad \text{or} \quad (R_B R_A)^n x$$

are defined and distinct. This occurs when there is a closed subinterval of I which is sent into itself by one of the two composite mappings $R_A R_B$, $R_B R_A$ and that mapping is contractive.

In the general case of a pair of simple foldings of a tree $|J|$ deriving from different simplicial decompositions, there is again no problem if the pairs of simplexes to be folded are disjoint, or if there is overlap of only one pair. However if there is overlap of both pairs of relevant simplexes, the subset of $|J|$ consisting of the line segments to be identified is necessarily homeomorphic to I. The identifications on this subset yield foldings as have been discussed. The remainder of the tree $|J|$ is unaffected by the foldings but is carried along at the places where it is attached. This completes the proof of Proposition 5.5 and the results of Section 2. □

6. Lens Collapse and the Homotopy Addition Lemma

In this section we are concerned with basic algebraic properties of the lens collapse construction and the homotopy addition lemma for the homotopy 2-groupoid of a Hausdorff space. These properties are required for the proof of the main part of Proposition 3.5. Throughout this section let X denote a Hausdorff space.

First we describe how the lens collapse construction $\{\hat{f}_{t,s}\}$ associated to a square in X, $f_{t,s} : I \times I \to X$, behaves with respect to pasting of squares.

PROPOSITION 6.1. (i) *Suppose that $f_{t,s}$ and $g_{t,s}$ are squares in X such that $f_{1,s} = g_{0,s}$. Let $h_{t,s} = g_{t,s} \bullet f_{t,s}$ denote the square obtained by pasting $f_{t,s}$ and $g_{t,s}$ horizontally in the canonical way. Then in $\mathbf{G}_2X$ we have*

$$\{\hat{h}_{t,s}\} = \langle f_{t,0}\rangle^*\{\hat{g}_{t,s}\} + \langle g_{t,1}\rangle_*\{\hat{f}_{t,s}\}.$$

(ii) *Similarly, if $f_{t,s}$ and $g_{t,s}$ are squares in X such that $f_{t,1} = g_{t,0}$, let $k_{t,s} = f_{t,s} + g_{t,s}$ denote the square obtained by pasting $f_{t,s}$ and $g_{t,s}$ vertically in the canonical way. Then in $\mathbf{G}_2X$ we have*

$$\{\hat{k}_{t,s}\} = \langle g_{1,t}\rangle_*\{\hat{f}_{t,s}\} + \langle f_{0,t}\rangle^*\{\hat{g}_{t,s}\}.$$

Sketch of proof. (i) Apply lens collapse to the composite square and deform as suggested by the following picture.

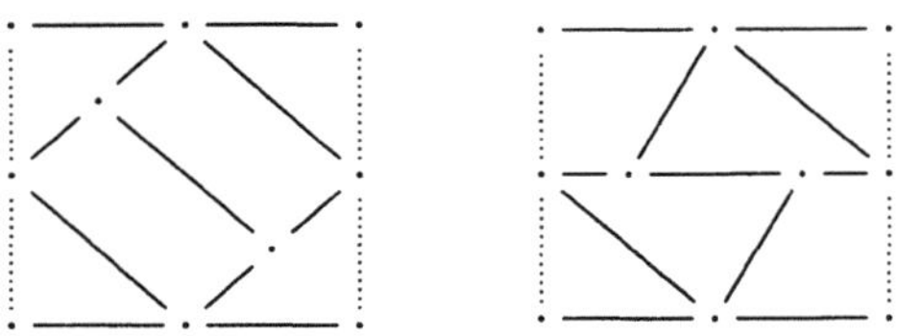

(ii) is obtained from (i) by interchanging the role of t and s. □

Next we state the homotopy addition lemma for the homotopy 2-groupoid.

PROPOSITION 6.2 (Homotopy Addition Lemma). *Let $f_{t,s,r}$ be a cube in X, i.e., a map $I \times I \times I \to X$, such that*

$$f_{0,0,r}, \quad f_{0,1,r}, \quad f_{1,0,r}, \quad f_{1,1,r}$$

are constant maps. Then in $\mathbf{G}_2X(\langle f_{1,t,0} \bullet f_{t,0,0}\rangle, \langle f_{t,1,1} \bullet f_{0,t,1}\rangle)$ we have

$$\{\hat{f}_{t,s,0}\} + \{f_{t,1,s}\} \bullet \{f_{0,t,s}\} = \{f_{1,t,s}\} \bullet \{f_{t,0,s}\} + \{\hat{f}_{t,s,1}\}.$$

Sketch of proof. By means of lens collapse the situation can be reduced to the special case of a cube with a pair of opposite faces which are mapped constantly. Lens collapse applied in a similar way to the special case yields the result.

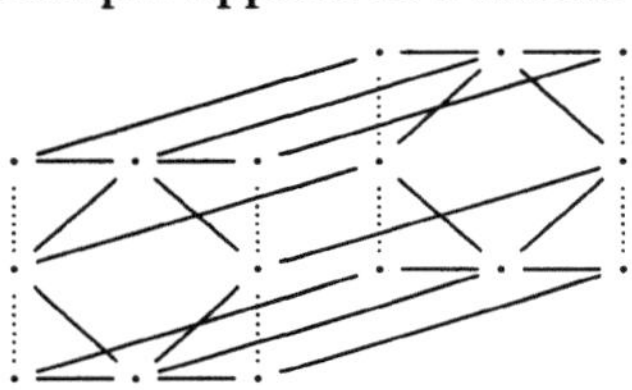

□

REMARK. Due to the interchange law, the homotopy addition lemma can be rephrased in several ways. We shall make use of the following variant.

$$\begin{aligned}&\{\hat{f}_{t,s,0}\} + \langle f_{t,0,0}\rangle^*\{f_{1,t,s}\} + \langle f_{1,t,1}\rangle_*\{f_{t,0,s}\}\\ &\quad = \langle f_{0,t,0}\rangle^*\{f_{t,1,s}\} + \langle f_{t,1,1}\rangle_*\{f_{0,t,s}\} + \{\hat{f}_{t,s,1}\}\end{aligned} \tag{6.3}$$

Proof of Proposition 3.5. Proposition 6.1(i), applied to $(h_t f_s) \bullet (g_t f_s)$, yields the assertion in (ii).

Part (iii) follows from Proposition 6.1(ii), applied to $g_t f_s + g_t k_s$.

(xi) Let μ resp. ν be represented by $m_{t,s}$, resp. $n_{t,s}$. Then (xi) follows from (6.3), applied to $f_{t,s,r} = n_{s,r} m_{t,r}$. □

REMARK. A double groupoid version of the homotopy addition lemma has been given by Brown and Higgins [13, p. 199]. Al-Agl [1] has a version of the homotopy addition lemma in the context of a triple category with connection which is slightly more general than Proposition 6.2 in the sense that one does not have to assume that some of the edges are constant.

Acknowledgements

The authors gratefully acknowledge suggestions by R. Brown, M. Bullejos, I. Moerdijk, T. Porter, and J. J. C. Vermeulen.

The first author acknowledges support from the Foundation for Research Development and the University of Cape Town via the Topology and Category Theory Group.

References

1. Al-Agl, F. A.: Aspects of multiple categories, Ph.D. Thesis, University of Wales, Bangor, 1989.
2. Barratt, M. G.: Track groups I, *Proc. London Math. Soc.* (*3*) **5** (1955), 71–106.
3. Baues, H. J.: *Combinatorial Homotopy and 4-Dimensional Complexes*, W. de Gruyter, Berlin, New York, 1981.
4. Bénabou, J.: Introduction to bicategories, in *Reports of the Midwest Category Seminar*, Lecture Notes in Math. 47, Springer-Verlag, Berlin, 1967, pp. 1–77.
5. Brown, R.: Groupoids and van Kampen's theorem, *Proc. London Math. Soc.* (*3*) **17** (1967), 385–401.
6. Brown, R.: *Elements of Modern Topology*, McGraw-Hill, Maidenhead, 1968.
7. Brown, R.: Fibrations of groupoids, *J. Algebra* **15** (1970), 103–132.
8. Brown, R.: Higher dimensional group theory, in *Proceedings of the Conference on Topology in Low Dimension, Bangor 1979*, London Math. Soc. Lecture Notes Series 48, Cambridge University Press, 1982, pp. 215–238.
9. Brown, R.: From groups to groupoids: A brief survey, *Bull. London Math. Soc.* **19** (1987), 113–134.
10. Brown, R.: *Topology: A Geometric Account of General Topology, Homotopy Types and the Fundamental Groupoid,* Ellis Horwood, Chichester, 1988.
11. Brown, R.: Homotopy theory, and change of base for groupoids and multiple groupoids, *Appl. Cat. Structures* **4** (1996), 175–193.
12. Brown, R.: Groupoids and crossed objects in algebraic topology, Lectures Given at the Summer School on the Foundations of Algebraic Topology, Grenoble, 1997.

13. Brown, R. and Higgins, P. J.: On the connection between the second relative homotopy groups of some related spaces, *Proc. Lond. Math. Soc. (3)* **36** (1978), 193–212.
14. Brown, R. and Higgins, P. J.: On the algebra of cubes, *J. Pure Appl. Algebra* **21** (1981), 233–260.
15. Brown, R. and Spencer, C. B.: Double groupoids and crossed modules, *Cahiers Topologie Géom. Différentielle Catégoriques* **17** (1976), 343–362.
16. Gabriel, P. and Zisman, M.: *Calculus of Fractions and Homotopy Theory*, Ergebnisse der Mathematik und ihrer Grenzgebiete 35, Springer-Verlag, 1967.
17. Glaser, L. C.: *Geometrical Combinatorial Topology, Vol. 1*, Van Nostrand Reinhold Math. Stud. 27, Van Nostrand Reinhold Company, New York, 1970.
18. Grandis, M.: Categorically algebraic foundations for homotopical algebra, *Appl. Cat. Structures* **5** (1997), 363–413.
19. Gray, J. W.: *Formal Category Theory: Adjointness for 2-Categories*, Lecture Notes in Math. 391, Springer-Verlag, New York, 1974.
20. Gray, J. W.: Coherence for the tensor product of 2-categories, and braid groups, in *Algebra, Topology, and Category Theory (a collection of papers in honour of Samuel Eilenberg)*, Academic Press, New York, 1976, pp. 63–76.
21. Hardie, K. A., Kamps, K. H. and Kieboom, R. W.: A homotopy bigroupoid of a topological space, *Appl. Cat. Structures* (to appear).
22. Higgins, P. J.: *Notes on Categories and Groupoids*, Van Nostrand Math. Stud. 32, Van Nostrand Reinhold Company, London, 1971.
23. Hilton, P. J.: *An Introduction to Homotopy Theory*, Cambridge Tract in Math. and Math. Phys. 43, Cambridge University Press, 1966.
24. Kelly, G. M. and Street, R.: Review of the elements of 2-categories, in *Category Seminar Sydney 1972/73*, Lecture Notes in Math. 420, Springer-Verlag, Berlin, 1974, pp. 75–103.
25. Loday, J.-L.: Spaces with finitely many non-trivial homotopy groups, *J. Pure Appl. Algebra* **24** (1982), 179–202.
26. Mac Lane, S.: *Categories for the Working Mathematician*, Springer-Verlag, Berlin, 1971.
27. Mather, M.: Pull-backs in homotopy theory, *Canad. J. Math.* **28** (1976), 225–263.
28. Moerdijk, I. and Svensson, J.-A.: Algebraic classification of equivariant homotopy 2-types, I, *J. Pure Appl. Algebra* **89** (1993), 187–216.
29. Porter, T.: Crossed modules in *Cat* and a Brown–Spencer theorem for 2-categories, *Cahiers Topologie Géom. Différentielle Catégoriques* **26** (1985), 381–388.
30. Rourke, C. P. and Sanderson, B. J.: *Introduction to Piecewise-Linear Topology*, Springer-Verlag, Berlin, 1982.
31. Rutter, J. W.: A homotopy classification of maps into an induced fibre space, *Topology* **6** (1967), 379–403.
32. Spanier, E. H.: *Algebraic Topology*, McGraw-Hill, 1966.
33. Spencer, C. B.: An abstract setting for homotopy pushouts and pullbacks, *Cahiers Topologie Géom. Différentielle Catégoriques* **18** (1977), 409–430.
34. Spencer, C. B. and Wong, Y. L.: Pullback and pushout squares in a special double category with connection, *Cahiers Topologie Géom. Différentielle Catégoriques* **24** (1983), 161–192.
35. Street, R.: Categorical structures, in M. Hazewinkel (ed.), *Handbook of Algebra*, Vol. 1, Elsevier, Amsterdam, 1996, pp. 529–577.
36. Teleman, C.: Généralisation du groupe fondamental, *Ann. Scient. Éc. Norm. Sup. (3)* **77** (1960), 195–234.
37. Whitehead, J.H.C.: On adding relations to homotopy groups, *Ann. Math.* **42** (1941), 409–428.
38. Whitehead, J.H.C.: Note on a previous paper entitled 'On adding relations to homotopy groups', *Ann. Math.* **47** (1946), 806–810.
39. Whitehead, J.H.C.: Combinatorial homotopy II, *Bull. Amer. Math. Soc.* **55** (1949), 453–496.

Applied Categorical Structures **8:** 235–245, 2000.
G. Brümmer & C. Gilmour (eds), Papers in Honour of Bernhard Banaschewski.

A Complete Classification of the Simultaneously Bireflective and Bicoreflective Subconstructs of Approach Spaces

Dedicated to our friend and colleague, Bernhard Banaschewski on the occasion of his 70th birthday

R. LOWEN and P. WUYTS
Universiteit Antwerpen, Departement Wiskunde en Informatica, Groenenborgerlaan 171, 2020 Antwerpen, Belgium, e-mail: rlow@ruca.ua.ac.be

(Received: 24 October 1996; accepted: 15 December 1999)

Abstract. In this paper, using a technique developed by Herrlich and Lowen [3], we classify and describe all subconstructs of AP, the construct of approach spaces, which are at the same time bireflective and bicoreflective in AP.

Mathematics Subject Classifications (2000): 18A35, 18A40, 54B30, 54E99.

Key words: approach space, distance, regular function frame, bireflective, bicoreflective, monoidal subalgebra, subconstruct.

1. Introduction

It is a well-known fact that in many familiar topological constructs, such as, e.g., TOP, $PrTOP$, $UNIF$ and BOR there do not exist non-trivial simultaneously bireflective and bicoreflective subconstructs, see, e.g., Giuli [2], Hušek [4] and Kannan [5] among others.

The situation is quite different in AP, the construct of approach spaces and contractions, [6, 7], since already in [6] it was shown that TOP is bireflectively bicoreflectively embedded in AP. In the present paper, not only do we prove that TOP is not the only such subconstruct, but we are also able to completely classify all such subconstructs. Moreover we can give an extremely simple internal characterization of the objects of each of these subconstructs.

Basic to our considerations are the results of Herrlich and Lowen [3], where a technique for determining simultaneously bireflective and bicoreflective subconstructs of specific topological constructs was given. The fundamental result of that paper determines a relationship between simultaneously bireflective and bicoreflective subconstructs on the one hand and particular function algebras on the other hand. In [3] it was already pointed out that AP is a construct of the required type. In the present paper, for the particular case of the construct of approach spaces, we refine this relationship, and we completely describe all possible particular function algebras which can arise. Via this description we then also give some internal characterizations of the objects of the associated simultaneously bireflective and bicoreflective subconstructs.

2. Preliminaries

In [7] several equivalent descriptions of approach spaces are given. We recall two of them, namely the notions of distance and of regular function frame, since these play a central role in our paper. In what follows $\vee$ stands for supremum and $\wedge$ stands for infimum.

2.1. DEFINITIONS AND NOTATIONS

An approach space is a pair (X, δ), where X is a set and where $\delta\colon X \times 2^X \to [0, \infty]$ is a function, called a *distance*, which satisfies the following properties:

(D1) $\forall x \in X : \delta(x, \{x\}) = 0$,
(D2) $\forall x \in X : \delta(x, \emptyset) = \infty$,
(D3) $\forall x \in X, \forall A, B \in 2^X : \delta(x, A \cup B) = \delta(x, A) \wedge \delta(x, B)$,
(D4) $\forall x \in X, \forall A \in 2^X, \forall \epsilon \in [0, \infty] : \delta(x, A) \leq \delta(x, A^{(\epsilon)}) + \epsilon$ where $A^{(\epsilon)} := \{y \in X \mid \delta(y, A) \leq \epsilon\}$.

Equivalently, an approach space can also be characterized as a pair $(X, \mathcal{R})$, where X is a set and where the collection $\mathcal{R} \subset [0, \infty]^X$, called a regular function frame, satisfies the following properties:

(R1) $\forall \mathcal{S} \subset \mathcal{R} : \vee \mathcal{S} \in \mathcal{R}$,
(R2) $\forall \mu, \nu \in \mathcal{R} : \mu \wedge \nu \in \mathcal{R}$,
(R3) $\forall \mu \in \mathcal{R}, \forall \alpha \in [0, \infty] : \mu + \alpha \in \mathcal{R}$,
(R4) $\forall \mu \in \mathcal{R}, \forall \alpha \in [0, \inf_{x \in X} \mu(x)] : \mu - \alpha \in \mathcal{R}$.

For details on the transition formulas for going from a distance to a regular function frame and vice versa we refer the reader to [7]. In the last section we will however briefly recall that part of these formulas which is required in 5.1.

Approach spaces form the objects of a topological construct, denoted AP. The morphisms in this construct, given objects (X, δ) and (X', δ'), where δ and δ' are distances, are functions $f\colon X \to X'$, called contractions, such that

$$\forall x \in X, \forall A \in 2^X : \delta'(f(x), f(A)) \leq \delta(x, A).$$

If $\mathcal{R}$ and $\mathcal{R}$' are the corresponding regular function frames on X, respectively X', then the foregoing condition is equivalent to:

$$\forall \mu' \in \mathcal{R}' : \mu' \circ f \in \mathcal{R}.$$

We also need to recall some concepts and results from Herrlich and Lowen [3]. An algebra [3] basically is a set, equipped with a class of operations, each of arbitrary arity. Although some of the results in the present paper remain valid for general algebras, we shall only be dealing with the algebra $\mathcal{A}$, which consists of the set $[0, \infty]$ endowed with the following class of operations:

1. for arbitrary sets $S : \sigma_S : \mathcal{A}^S \to \mathcal{A} : (a_s)_{s \in S} \mapsto \sup_{s \in S} a_s$,
2. for finite sets $S : \iota_S : \mathcal{A}^S \to \mathcal{A} : (a_s)_{s \in S} \mapsto \min_{s \in S} a_s$,
3. for $a \in [0, \infty] : \alpha_a : \mathcal{A} \to \mathcal{A} : b \mapsto b + a$,
4. for $a \in [0, \infty] : \beta_a : \mathcal{A} \to \mathcal{A} : b \mapsto (b - a) \vee 0$.

Sometimes we may use the notation $\mathcal{A}$ also simply for the underlying set. Given any set X, the algebra $\mathcal{A}^X$ is defined, as usual, by pointwise extension of the operations.

By a subalgebra of a given algebra is meant a subset which is closed under all operations of the given class. Hence, it follows that in our case a subalgebra of $\mathcal{A}^X$ is nothing else than a regular function frame on X. It is this fact, as noted in [3], that ensures that AP is a construct of the type required in order to be able to apply the results of [3]. More precisely, it was shown that AP is isomorphic to $\mathbf{T}(\mathcal{A})$ for the above algebra. Basically, $\mathbf{T}(\mathcal{A})$ is the topological category where the objects are all sets structured by a subalgebra of $\mathcal{A}^X$ and where the morphisms preserve these subalgebras under composition. We refer the reader to [3] for details.

If $\Gamma \subset [0, \infty]^{[0,\infty]}$, then we shall denote by $\tilde{\Gamma}$ the smallest *monoidal subalgebra* of $\mathcal{A}^{\mathcal{A}}$ containing Γ, i.e. the smallest subset of $[0, \infty]^{[0,\infty]}$ which contains Γ and which is simultaneously a submonoid of $([0, \infty]^{[0,\infty]}, \circ, \mathrm{id}_{\mathcal{A}})$ and a subalgebra of $\mathcal{A}^{\mathcal{A}}$. A monoidal subalgebra necessarily contains all constant maps and $\mathrm{id}_{\mathcal{A}}$ (see 2.2).

Characterizing approach spaces by regular function frames, and given a subconstruct $\mathcal{C}$ of AP, we define

$$\Omega(\mathcal{C}) := \{\gamma \in [0, \infty]^{[0,\infty]} \mid \forall (X, \mathcal{R}) \in |\mathcal{C}|, \forall \mu \in \mathcal{R} : \gamma \circ \mu \in \mathcal{R}\}.$$

If $\Gamma \subset [0, \infty]^{[0,\infty]}$, we say that $\mathcal{C}$ is Γ-invariant if $\Gamma \subset \Omega(\mathcal{C})$ or explicitly if

$$\forall (X, \mathcal{R}) \in |\mathcal{C}|, \forall \gamma \in \Gamma, \forall \mu \in \mathcal{R} : \gamma \circ \mu \in \mathcal{R}.$$

Further we recall from [3] the following full subconstructs of $\mathbf{T}(\mathcal{A})$ (or equivalently of AP), determined by an arbitrary subset $\Gamma \subset \mathcal{A}^{\mathcal{A}}$:

$$\mathbf{T}(\mathcal{A}, \Gamma) := \{(X, \mathcal{R}) \in |AP| \mid (\gamma \in \Gamma \text{ and } \mu \in \mathcal{R}) \Rightarrow \gamma \circ \mu \in \mathcal{R}\}.$$

Since in our case $\mathcal{A}$ is fixed, we will write $\mathbf{T}(\Gamma)$ instead of $\mathbf{T}(\mathcal{A}, \Gamma)$. We call $\mathcal{C}$ maximal Γ-invariant if $\mathcal{C} = \mathbf{T}(\Gamma)$.

We recall from [3] that

$$\Gamma_{AP} := \left\{\mu \in [0, \infty]^{[0,\infty]} \mid \mu \text{ nonexpansive, increasing, } \mu(\infty) = \sup_{x<\infty} \mu(x)\right\},$$

and

$$\begin{aligned}\Gamma_{TOP} &:= \{\mu \in [0,\infty]^{[0,\infty]} \mid \mu \text{ l.s.c. and increasing}\} \\ &:= \{\mu \in [0,\infty]^{[0,\infty]} \mid \mu \text{ leftcontinuous and increasing}\}\end{aligned}$$

are monoidal subalgebras of $\mathcal{A}^{\mathcal{A}}$, and thus regular function frames on $[0,\infty]$ and that we have

$$AP = \mathbf{T}(\Gamma_{AP}) \quad \text{and} \quad TOP = \mathbf{T}(\Gamma_{TOP}).$$

2.2. REMARKS

(1) If $\mathcal{R} \subset [0,\infty]^X$ is a regular function frame, then the following holds:

$$\forall \mu \in \mathcal{R}, \forall a \in [0,\infty], \forall b \in [-a,\infty] : \mu \vee a + b \in \mathcal{R}.$$

(2) If $\mathcal{C}$ is a subconstruct of AP and if $\Gamma \subset [0,\infty]^{[0,\infty]}$, then $\Gamma \subset \Omega(\mathcal{C})$ if and only if $|\mathcal{C}| \subset |\mathbf{T}(\Gamma)|$.

3. Simultaneously Bireflective and Bicoreflective Subconstructs of AP

In this section, making use of the results of [3], we will tag all the simultaneously bireflective and bicoreflective subconstructs of AP. More precisely, we will prove that there exists a bijection between the class of these subconstructs on the one hand and a particular set of monoidal subalgebras of $\mathcal{A}^{\mathcal{A}}$ on the other hand. In particular this will already show that there is only a proper set of simultaneously bireflective and bicoreflective subconstructs of AP. In the following section we will determine this set precisely.

PROPOSITION 3.1. (1) *If $\mathcal{C}$ is a subconstruct of AP, then $\Omega(\mathcal{C})$ is the final regular function frame on $\mathcal{A}$ for the sink $(\mu\colon (X,\mathcal{R}) \to \mathcal{A})_{(X,\mathcal{R})\in|\mathcal{C}|,\mu\in\mathcal{R}}$, and moreover it is a monoidal subalgebra of $\mathcal{A}^{\mathcal{A}}$.*

(2) *If $(\mathcal{C}_j)_{j\in J}$ is a family of subconstructs of AP, then $\Omega(\bigcup_{j\in J}\mathcal{C}_j) = \bigcap_{j\in J}\Omega(\mathcal{C}_j)$.*

Proof. That $\Omega(\mathcal{C})$ is a subalgebra of $\mathcal{A}^{\mathcal{A}}$ follows easily by making use of the remark in 2.2.1. The remaining claims are immediate. □

PROPOSITION 3.2. (1) *If $\Gamma \subset \Lambda \subset \mathcal{A}^{\mathcal{A}}$, then $\mathbf{T}(\Lambda)$ is a subconstruct of $\mathbf{T}(\Gamma)$.*

(2) *If $(\Gamma_j)_{j\in J}$ is a family of subsets of $\mathcal{A}^{\mathcal{A}}$, then $|\mathbf{T}(\bigcup_{j\in J}\Gamma_j)| = \bigcap_{j\in J}|\mathbf{T}(\Gamma_j)|$.*

Proof. This is trivial. □

PROPOSITION 3.3. *For any $\Gamma \subset [0,\infty]^{[0,\infty]}$ we have*

(1) $\mathbf{T}(\Gamma) = \mathbf{T}(\tilde{\Gamma})$ *and*

(2) $\tilde{\Gamma} = \Omega(\mathbf{T}(\Gamma))$.

Proof. To prove (1): since $\Gamma \subset \tilde{\Gamma}$, we have $|\mathbf{T}(\tilde{\Gamma})| \subset |\mathbf{T}(\Gamma)|$. On the other hand, it follows from 2.2(2) that $\Gamma \subset \Omega(\mathbf{T}(\Gamma))$, and since the right hand side is a monoidal subalgebra of $\mathcal{A}^{\mathcal{A}}$ (see 3.1(1)), it also follows that $\tilde{\Gamma} \subset \Omega(\mathbf{T}(\Gamma))$. Applying 2.2(2) once more, we obtain $|\mathbf{T}(\Gamma)| \subset |\mathbf{T}(\tilde{\Gamma})|$. To prove (2): it follows from the first part and 2.2(2) that $\tilde{\Gamma} \subset \Omega(\mathbf{T}(\Gamma))$. Conversely, it follows from 3.1(1) that the sink

$$(\mu\colon (X, \mathcal{R}) \longrightarrow (\mathcal{A}, \Omega(\mathbf{T}(\Gamma))))_{(X,\mathcal{R})\in|\mathbf{T}(\Gamma)|,\mu\in\mathcal{R}}$$

is final, and thus, since clearly $(\mathcal{A}, \tilde{\Gamma}) \in |\mathbf{T}(\Gamma)|$, we furthermore have $\Omega(\mathbf{T}(\Gamma)) \subset \tilde{\Gamma}$. □

PROPOSITION 3.4. *For any subconstruct $\mathcal{C}$ of AP the following properties hold:*

(1) *$\mathcal{C}$ is a subconstruct of* $\mathbf{T}(\Omega(\mathcal{C}))$,
(2) *$\mathcal{C}$ is maximal Γ-invariant for some Γ if and only if $\mathcal{C} = \mathbf{T}(\Omega(\mathcal{C}))$,*
(3) $\Omega(\mathbf{T}(\Omega(\mathcal{C}))) = \Omega(\mathcal{C})$.

Proof. The first claim follows immediately from the definitions. To prove the second claim, let $\mathcal{C} := \mathbf{T}(\Gamma)$, then since $\Gamma \subset \Omega(\mathcal{C})$, it follows from 3.2 that $\mathbf{T}(\Omega(\mathcal{C}))$ is a subconstruct of $\mathcal{C}$. Finally, the third claim follows from 3.3(2) and the fact that $\Omega(\mathcal{C})$ is a monoidal subalgebra of $\mathcal{A}^{\mathcal{A}}$. □

THEOREM 3.5. *There exists a one-to-one correspondence between the collection $\mathfrak{G}$ of simultaneously bireflective and bicoreflective subconstructs of AP and the set $\mathfrak{M}_t(\mathcal{A})$ of monoidal subalgebras of $\mathcal{A}^{\mathcal{A}}$ which are contained in Γ_{TOP}. This correspondence is given by the bijection*

$$\mathbf{T} : \mathfrak{M}_t(\mathcal{A}) \to \mathfrak{G} : \Gamma \mapsto \mathbf{T}(\Gamma)$$

the inverse of which is given by

$$\Omega : \mathfrak{G} \to \mathfrak{M}_t(\mathcal{A}) : \mathcal{C} \mapsto \Omega(\mathcal{C}).$$

Proof. If $\mathcal{C}$ is a simultaneously bireflective and bicoreflective subconstruct of AP, then it follows that $|TOP| \subset |\mathcal{C}|$, since TOP itself has no such subconstructs. The remainder now follows at once from 3.3 and 3.4. □

4. The Main Result

The results of the previous section have reduced the task of finding all simultaneously bireflective and bicoreflective subconstructs of AP to that of determining all monoidal subalgebras of $\mathcal{A}^{\mathcal{A}}$ which are contained in Γ_{TOP}. It is an easy exercise to prove that $\Gamma_{AP} \subset \Gamma$ for any monoidal subalgebra Γ of $\mathcal{A}^{\mathcal{A}}$. For any $m \in [0, \infty]$, we now define Γ_m to be the set of all functions $\varphi \in \mathcal{A}^{\mathcal{A}}$ which are both leftcontinuous and increasing and which fulfill the property

$$\forall x < y : \varphi(y) - \varphi(x) \leq (y - x) \vee m.$$

LEMMA 4.1. *For any $m \in [0, \infty]$, Γ_m is a monoidal subalgebra of $\mathcal{A}^{\mathcal{A}}$, such that $m\mathbf{1}_{]0,\infty]} \in \Gamma_m$ and $\Gamma_m \subset \Gamma_{TOP}$.*

Proof. That Γ_m contains $\mathrm{id}_{\mathcal{A}}$, that it is invariant under translations and stable for arbitrary suprema and for finite infima follows by standard verification. If now $\varphi \in \Gamma_m$ and $\psi \in \Gamma_m$, then it follows trivially that $\psi \circ \varphi$ is increasing and leftcontinuous. If $x < y$, then either $\varphi(x) = \varphi(y)$, and it follows that $\psi \circ \varphi(x) = \psi \circ \varphi(y)$, or $\varphi(x) < \varphi(y)$, and we obtain $\psi \circ \varphi(x) - \psi \circ \varphi(y) \leq (\varphi(x) - \varphi(y)) \vee m \leq (y - x) \vee m$. □

LEMMA 4.2. *If Γ is a monoidal subalgebra of $\mathcal{A}^{\mathcal{A}}$, satisfying $\Gamma_{AP} \neq \Gamma \subset \Gamma_{TOP}$, then there exists an $m \in]0, \infty]$ such that $m\mathbf{1}_{]0,\infty]} \in \Gamma$.*

Proof. Since $\Gamma_{AP} \neq \Gamma$, there exists $\varphi \in \Gamma$ and $x, y \in [0, \infty]$ such that $x < y$ and such that $\varphi(y) - \varphi(x) > y - x$. Since $\alpha_x \in \Gamma_{AP}$ we also have that $\theta := (\varphi \circ \alpha_x - \varphi(x)) \wedge p \in \Gamma$, where $p := \varphi(y) - \varphi(x)$. Clearly we then have

$$\theta(0) = 0 \quad \text{and} \quad \theta_{|[y-x,\infty]} = p. \tag{1}$$

Furthermore, by hypothesis $a := y - x < p$.

CLAIM. *If Γ contains an element θ satisfying* (1), *then there exist $\tau \in \Gamma$ and $b \in]0, a[$ such that*

$$\tau(0) = 0 \quad \textit{and} \quad \tau_{|[b,\infty]} = p.$$

Indeed, since $\theta \in \Gamma$ is leftcontinuous, there exists $b \in]0, a[$ such that

$$b \leq t < a \Rightarrow a < \theta(t) \leq p.$$

So, if we put $\tau := \theta \circ \theta$, then we have that $\tau \in \Gamma$, $\tau(0) = \theta(0) = 0$ and since $\theta(b) > a$, also that

$$t \geq b \Rightarrow \tau(t) = \theta \circ \theta(t) \geq \theta(\theta(b)) = p.$$

Now, let

$$\sigma := \sup\{\varphi \in \Gamma \mid \varphi(0) = 0, \exists z \in]0, \infty[: \varphi_{|[z,\infty]} = p\},$$

and let

$$c := \inf\{t \mid \sigma(t) = p\}.$$

The above claim shows that $c > 0$ and $\sigma(c) = p$ is impossible. Two other cases can now occur. Either $c = 0$, or $c > 0$ and $p > q := \sigma(c)$. In the first case

$\sigma = p\mathbf{1}_{]0,\infty]}$ and we can take $m := p$. In the second case, since $\sigma(c+) = p$ and since $\rho := \sigma \circ \alpha_c - \sigma(c)$ is a member of Γ we can take $m := p - q$. □

LEMMA 4.3. *If Γ is a monoidal subalgebra of $\mathcal{A}^{\mathcal{A}}$ which contains the function $\theta := m\mathbf{1}_{]0,\infty]}$ for some $m \in]0, \infty]$, then it also contains the functions*

$$\varphi_{p,q,r} := p \vee (p+q)\mathbf{1}_{]r,\infty]},$$
$$\rho_{p,q,r,b} := \varphi_{p,q,r} \wedge \alpha_b \quad \textit{and}$$
$$\sigma_{p,q,r,b} := \varphi_{p,q,r} \wedge \beta_b$$

for all $p \in [0, \infty]$, $q \in [0, m]$, $r \in [0, \infty]$ and $b \in [0, \infty]$.

Proof. Since $\beta_r = (\mathrm{id}_{\mathcal{A}} - r) \vee 0 \in \Gamma_{AP}$, the result for the first function follows from the fact that $\varphi_{p,q,r} = (\theta \circ \beta_r + p) \wedge (p+q)$. For the remaining functions the result then follows from the facts that $\alpha_b = \mathrm{id}_{\mathcal{A}} + b \in \Gamma_{AP}$ and $\beta_b = (\mathrm{id}_{\mathcal{A}} - b) \vee 0 \in \Gamma_{AP}$. □

LEMMA 4.4. *If Γ is a monoidal subalgebra of $\mathcal{A}^{\mathcal{A}}$ which contains the function $\theta := m\mathbf{1}_{]0,\infty]}$ for some $m \in]0, \infty]$, then for any $\gamma \in \Gamma_m$, $\varepsilon > 0$ and $x \in]0, \infty[$ there exists a function $\theta_{x,\varepsilon} \in \Gamma$ such that $\theta_{x,\varepsilon} \leq \gamma$ and $\gamma(x) - \varepsilon \leq \theta_{x,\varepsilon}(x)$.*

Proof. We use the notations of the foregoing lemma. It is clear that we can always suppose $0 < \varepsilon < \gamma(x) \wedge m$. Since γ is leftcontinuous, we can find $\delta_0 \in]0, x[$ such that

$$x - \delta_0 \leq t \leq x \Rightarrow \gamma(x) - \frac{\varepsilon}{2} < \gamma(t) \leq \gamma(x).$$

We now distinguish four cases and in all cases we take $\delta \in]0, \delta_0 \wedge \frac{\varepsilon}{2}]$.

Case 1. $\gamma(x) \geq x \geq m$. We define

$$\theta_{x,\varepsilon} := \rho_{\gamma(x)-m,m-\varepsilon/2,x-\delta,\gamma(x)-x}.$$

One can easily verify that $\theta_{x,\varepsilon} \leq \gamma$ by successively considering the intervals $[0, x-m]$, $]x-m, x-\delta]$, $]x-\delta, x]$ and $]x, \infty]$.

Case 2. $\gamma(x) \geq x$, $x \leq m$. We now define

$$\theta_{x,\varepsilon} := \varphi_{(\gamma(x)-m)\vee 0,(\gamma(x)-\varepsilon/2)\wedge(m-\varepsilon/2),x-\delta}.$$

In this case we verify that $\theta_{x,\varepsilon} \leq \gamma$ by successively considering the intervals $[0, x-\delta]$, $]x-\delta, x]$ and $]x, \infty]$.

Case 3. $m \leq \gamma(x) < x$. We now define

$$\theta_{x,\varepsilon} := \sigma_{\gamma(x)-m,m-\varepsilon/2,x-\delta,x-\gamma(x)}.$$

In order to verify that $\theta_{x,\varepsilon} \leq \gamma$ one now has to consider the intervals $[0, x-\gamma(x)]$, $]x-\gamma(x), x-m]$, $]x-m, x-\delta]$, $]x-\delta, x]$ and $]x, \infty]$.

Case 4. $\gamma(x) < x$, $\gamma(x) < m$. For this last case we define

$$\theta_{x,\varepsilon} := \varphi_{0,\gamma(x)-\frac{\varepsilon}{2},x-\delta}.$$

Once more the inequality $\theta_{x,\varepsilon} \leq \gamma$ is verified by successively considering the intervals $[0, x - \delta]$, $]x - \delta, x]$ and $]x, \infty]$.

Finally, in all four cases, it is immediately seen that $\theta_{x,\varepsilon}(x) = \gamma(x) - \varepsilon/2 > \gamma(x) - \varepsilon$. □

PROPOSITION 4.5. *If Γ is a monoidal subalgebra of $\mathcal{A}^{\mathcal{A}}$ which contains the function $\theta := m\mathbf{1}_{]0,\infty]}$ for some $m \in]0, \infty]$, then $\Gamma_m \subset \Gamma$.*

Proof. If $\gamma \in \Gamma_m$, then, making use of 4.4, for each $x \in]0, \infty[$ and each $\varepsilon > 0$, we can find $\theta_{x,\varepsilon} \in \Gamma$ such that $\theta_{x,\varepsilon} \leq \gamma$ and $\gamma(x) - \varepsilon < \theta_{x,\varepsilon}(x)$. It then follows that

$$\gamma = \gamma(0) \vee \sup_{x \in]0,\infty[} \sup_{\varepsilon > 0} \theta_{x,\varepsilon},$$

and thus $\gamma \in \Gamma$. □

COROLLARY 4.6. *For any $m \in]0, \infty[$ we have that $\widetilde{\{m\mathbf{1}_{]0,\infty]}\}} = \Gamma_m \subset \Gamma_{TOP}$.*

Proof. Immediate from the foregoing result. □

We are now in a position to prove the main result of our paper.

THEOREM 4.7. *The only monoidal subalgebras Γ of $\mathcal{A}^{\mathcal{A}}$ satisfying $\Gamma \subset \Gamma_{TOP}$ are the subalgebras Γ_m for $m \in [0, \infty]$. Furthermore, the only simultaneously bireflective and bicoreflective subconstructs of AP are the subconstructs $AP_m := \mathbf{T}(\Gamma_m)$, with $AP_0 = AP$ and $AP_\infty = TOP$. They form a set which is linearly ordered by inclusion in the sense that $p < q$ implies $|AP_q| \subset |AP_p|$.*

Proof. Since each Γ_m is a monoidal subalgebra of $\mathcal{A}$ satisfying $\Gamma_m \subset \Gamma_{TOP}$, and since clearly $m\mathbf{1}_{]0,\infty]} \in \Gamma_p$ if $m \leq p$, the following definition makes sense for any Γ such that $\Gamma \subset \Gamma_{TOP}$:

$$m_\Gamma := \sup\{p \mid p\mathbf{1}_{]0,\infty]} \in \Gamma\}.$$

Since $m\mathbf{1}_{]0,\infty]} = \sup_{p<m} p\mathbf{1}_{]0,\infty]}$, it follows that the supremum is actually a maximum and hence

$$m_\Gamma := \max\{p \mid p\mathbf{1}_{]0,\infty]} \in \Gamma\}.$$

Furthermore it is easily seen that $\Gamma_0 = \Gamma_{AP}$ and $\Gamma_\infty = \Gamma_{TOP}$. The result hence follows from the foregoing results and the fact that $\Gamma_{m_\Gamma} = \Gamma$. □

PROPOSITION 4.8. *If (X, δ) is an approach space then its AP_m-coreflection is given by*

$$c_m\colon (X, \delta_m) \to (X, \delta),$$

where $\delta_m(x, A) = 0$ if $\delta(x, A) = 0$, and $\delta_m(x, A) = \delta(x, A) \wedge m$ if $\delta(x, A) \neq 0$.

Proof. In order to verify that δ_m is a distance, noting that the verification of (D1)–(D3) is trivial, we only have to show that (D4) holds. We shall put $A_m^{(\varepsilon)} := \{x \mid \delta_m(x, A) \leq \varepsilon\}$. If $\delta(x, A) = 0$ then $\delta_m(x, A) = 0$ and there is nothing to show. Thus we may suppose that $\delta(x, A) \neq 0$ and then two cases can occur. First, if $\varepsilon < m$, then $A_m^{(\varepsilon)} = A^{(0)}$ and

$$\delta(x, A) \leq \delta(x, A^{(\varepsilon)}) + \varepsilon \leq \delta(x, A^{(0)}) + \varepsilon$$

and consequently

$$\delta(x, A) \vee m \leq \delta(x, A^{(0)}) \vee m + \varepsilon.$$

Since $\delta(x, A^{(0)}) = \delta(x, A) \neq 0$, this implies that

$$\delta_m(x, A) \leq \delta_m(x, A^{(0)}) + \varepsilon = \delta_m(x, A_m^{(\varepsilon)}) + \varepsilon.$$

Second, if $m \leq \varepsilon$, then $A_m^{(\varepsilon)} = A^{(\varepsilon)}$. If $\delta(x, A^{(\varepsilon)}) = 0$, then it follows that $\delta(x, A) \leq \varepsilon$, so either $\delta_m(x, A) = 0$, or

$$\delta_m(x, A) = \delta(x, A) \vee m \leq \varepsilon \vee m = \varepsilon \leq \delta_m(x, A_m^{(\varepsilon)}) + \varepsilon.$$

If $\delta(x, A^{(\varepsilon)}) \neq 0$, then we obtain

$$\begin{aligned} \delta_m(x, A) &= \delta(x, A) \vee m \leq \delta(x, A^{(\varepsilon)}) \vee m + \varepsilon \\ &= \delta_m(x, A^{(\varepsilon)}) + \varepsilon = \delta_m(x, A_m^{(\varepsilon)}) + \varepsilon. \end{aligned}$$

That (X, δ_m) is indeed the AP_m-coreflection of (X, δ) is a standard verification. □

PROPOSITION 4.9. *If $(X, \mathcal{R})$ is an approach space then its AP_m-reflection is given by*

$$r_m\colon (X, \mathcal{R}) \to (X, \mathcal{R}_m),$$

where $\mathcal{R}_m := \{\mu \in \mathcal{R} \mid \forall \alpha \in [0, \infty[\colon m\mathbf{1}_{\{\mu > \alpha\}} \in \mathcal{R}\}$.

Proof. We leave the straightforward verification to the reader. □

5. Internal Description of the Objects of AP_m

For detailed information concerning the appropriate transition formulas between the two structures which we shall use in the following proposition we refer the reader to [7]. We recall however that, given a distance δ and given a bounded function ρ for which there exists a finite partitioning $(M_i)_{i=1}^n$ of X and an associated set of real numbers $(m_i)_{i=1}^n$ such that $\rho_{|M_i} = m_i$, for all $i \in \{1, \ldots, n\}$, ρ is a member of the associated regular function frame $\mathcal{R}$ if and only if

$$\forall x \in X\colon \inf_{i=1,\ldots,n} (m_i + \delta(x, M_i)) \geq \rho(x).$$

Conversely, given a regular function frame $\mathcal{R}$ the associated distance is determined by the formula

$$\forall x \in X, \forall A \in 2^X : \delta(x, A) = \sup\{\mu(x) \mid \mu \in \mathcal{R}, \mu|_A = 0\}.$$

We also recall that the topology underlying a distance δ is characterized by the closure operator $\mathrm{cl}(A) := A^{(0)}$ for all $A \subset X$, and that a member ρ of the associated regular function frame $\mathcal{R}$ is lower semicontinuous with respect to this topology.

PROPOSITION 5.1. *For an approach space with support X and for $m \in [0, \infty]$ the following properties of respectively the distance and the regular function frame are equivalent:*

(1) $\delta(X \times 2^X) \subset \{0\} \cup [m, \infty]$.
(2) $\forall \mu \in \mathcal{R}$ *such that* $\{\mu = 0\} \neq \emptyset : m\mathbf{1}_{\{\mu > 0\}} \in \mathcal{R}$.

Proof. The cases $m = 0$ and $m = \infty$, i.e. respectively AP and TOP, are trivial. So from now on we suppose that $m \in]0, \infty[$

(1) $\Rightarrow$ (2). Let $\mu \in \mathcal{R}$ be such that $\{\mu = 0\} \neq \emptyset$. The function $\mu^* := m\mathbf{1}_{\{\mu>0\}}$ is then bounded and for the partitioning $\{X, X\backslash M\}$, where $M := \{\mu > 0\}$, it fulfils $\mu^*_{|M} = m$ and $\mu^*|_{X\backslash M} = 0$. Hence, taking into account the remarks preceding this proposition, we will have that $\mu^* \in \mathcal{R}$ if and only if

$$\forall x \in X : (m + \delta(x, M)) \wedge \delta(x, X\backslash M) \geq \mu^*(x).$$

Obviously it suffices to consider the case where $x \in M$. If $\delta(x, X\backslash M) = 0$, then, since $X\backslash M$ is closed in the topology underlying δ it follows that $x \in X\backslash M$, which is a contradiction. Hence, $\delta(x, X\backslash M) > 0$, which by our supposition implies that $\delta(x, X\backslash M) \geq m$ and we are done.

(2) $\Rightarrow$ (1). If $A \subset X$ and $x \in X$ are such that $\delta(x, A) > 0$, then it follows from the formula given in the remarks preceding this proposition that there exists a $\mu \in \mathcal{R}$ such that $\mu(x) \neq 0$ and $\mu_{|A} = 0$. Taking $\mu^* := m\mathbf{1}_{\{\mu>0\}}$, it follows that $\mu^*(x) = m$ and $\mu^*|_A = 0$, and so we immediately obtain that $\delta(x, A) \geq m$. □

THEOREM 5.2. *For any $m \in [0, \infty]$ we have that $\mathbf{T}(\Gamma_m)$ is the full subconstruct of AP with objects those approach spaces satisfying one of the conditions stated in the foregoing proposition.*

Proof. Making use of 5.1(2) this follows at once from the fact that $\mathbf{T}(\Gamma_m) = \mathbf{T}(\{m\mathbf{1}_{]0,\infty]}\})$. □

PROPOSITION 5.3. *If $r \in]0, \infty[$, then the functor $AP \to AP$ which turns an approach space (X, δ) into $(X, r\delta)$ and which leaves morphisms unchanged is an isomorphism. In particular for any $p, q \in]0, \infty[$ the subconstructs AP_p and AP_q are isomorphic.*

Proof. Immediate from the foregoing results. □

It follows from 5.1 that none of the subconstructs AP_m with $m \in]0, \infty]$ contains all metric spaces. So no simultaneously bireflective and bicoreflective subconstruct of AP, except for AP itself, contains all of TOP and MET. In this sense, AP is the smallest "reasonable" extension of TOP and MET.

References

1. Adámek, J., Herrlich, H., and Strecker, G. E.: *Abstract and Concrete Categories*, Wiley, 1990.
2. Giuli, E.: Two families of Cartesian closed topological categories, *Quaestiones Math.* **6**'(1983), 353–362.
3. Herrlich, H. and Lowen, R.: On simultaneously reflective and coreflective subconstructs, in *Proceedings Symposium on Categorical Topology UCT 1994*, 1999, pp. 121–130.
4. Hušek, M.: Reflective and coreflective subcategories of $UNIF$ and TOP, in *Sem. Unif. Spaces 1973–74*, 1975, pp. 113–126.
5. Kannan, V.: Reflexive cum coreflexive subcategories in topology, *Math. Ann.* **195** (1972), 168–174.
6. Lowen, R.: Approach spaces: A common supercategory of TOP and MET, *Math. Nachr.* **141** (1989), 183–226.
7. Lowen, R.: *Approach Spaces: The Missing Link in the Topology-Metric-Uniformity Triad*, Oxford Mathematical Monographs, Oxford University Press, 1997.

Applied Categorical Structures **8:** 247–256, 2000.
G. Brümmer & C. Gilmour (eds), Papers in Honour of Bernhard Banaschewski.

Exponential Objects in Coreflective or Quotient Reflective Subconstructs: A Comparison

Dedicated to Professor Bernhard Banaschewski on the occasion of his 70th birthday

E. LOWEN-COLEBUNDERS
Departement Wiskunde, Vrije Universiteit Brussel, Pleinlaan 2, 1050 Brussel, Belgium
e-mail: evacoleb@vub.ac.be

C. VERBEECK
Departement Wiskunde-Informatica, University of Antwerp, U.I.A., Universiteitsplein 1, 2610 Wilrijk, Belgium, e-mail: verbeeck@uia.ua.ac.be

(Received: 22 October 1996; accepted: 28 October 1997)

Abstract. We prove that in the construct *PRAP* of pre-approach spaces the class of exponential objects completely determines the exponential objects in certain subconstructs. We show that $\mathrm{Exp}\,\mathcal{B} \subset \mathrm{Exp}\,PRAP$ for every coreflective subconstruct $\mathcal{B}$ and from this inclusion we deduce the equality $\mathrm{Exp}\,\mathcal{B} = \mathcal{B} \cap \mathrm{Exp}\,PRAP$ for every subconstruct $\mathcal{B}$ that is coreflective and finitely productive. We prove that the same equality holds for non-trivial quotient reflective subconstructs. These results induce well known answers to similar questions on the construct of pretopological spaces and are compared to the topological situation.

Mathematics Subject Classifications (2000): 18B99, 18D15, 54A05, 54B30, 54C35, 54E35.

Key words: exponential object, pre-approach space, pretopological space, coreflective subconstruct, quotient reflective subconstruct.

1. Introduction

We will consider several examples of well fibred topological constructs $\mathcal{C}$ and in $\mathcal{C}$ some full and isomorphism closed subconstructs $\mathcal{B}$ that are either coreflective in $\mathcal{C}$ or quotient reflective in $\mathcal{C}$. We investigate how the class $\mathrm{Exp}\,\mathcal{B}$ of all exponential objects in $\mathcal{B}$ is related to the class $\mathrm{Exp}\,\mathcal{C}$ of all exponential objects in $\mathcal{C}$.

In case $\mathcal{C}$ is a cartesian closed topological construct the answer to the previous question is well known [11]. If $\mathcal{B}$ is coreflective and finitely productive in $\mathcal{C}$ or if $\mathcal{B}$ is quotient reflective in $\mathcal{C}$ then $\mathcal{B}$ inherits cartesian closedness. So in both cases the formula

$$\mathrm{Exp}\,\mathcal{B} = \mathcal{B} \cap \mathrm{Exp}\,\mathcal{C}$$

trivially holds.

If $\mathcal{C}$ is a well fibred topological construct (but not necessarily Cartesian closed) and $\mathcal{B}$ is coreflective and finitely productive or quotient reflective, the inclusion

$$\operatorname{Exp}\mathcal{B} \supset \mathcal{B} \cap \operatorname{Exp}\mathcal{C}$$

still holds. For the coreflective and finitely productive case the statement follows immediately from the characterization that X is exponential if and only if $X \times -$ preserves final epi-sinks. For the quotient reflective case it suffices to observe that for $X \in |\mathcal{B} \cap \operatorname{Exp}\mathcal{C}|$, $Y \in |\mathcal{B}|$, and Y^X the power object in $\mathcal{C}$, the map

$$Y^X \longrightarrow \prod_X Y$$

is a monomorphism. Therefore the power object belongs to $\mathcal{B}$.

The validity of the equality above is investigated in various situations. We start from the discussion of the equality $\operatorname{Exp}\mathcal{B} = \mathcal{B} \cap \operatorname{Exp}\mathcal{C}$ in the setting of the topological construct *PRAP*.

This construct contains both the construct *PRTOP* of pretopological (= closure-) spaces and continuous maps and the construct *TOP* of topological spaces and continuous maps as nicely embedded full subconstructs. In [10] an internal characterization of the objects in Exp *PRAP* was given. Here we prove that for every coreflective subconstruct $\mathcal{B}$ of *PRAP*,
$\operatorname{Exp}\mathcal{B} \subset \operatorname{Exp} PRAP$ holds.

From these results we deduce the equality $\operatorname{Exp}\mathcal{B} = \mathcal{B} \cap \operatorname{Exp} PRAP$ for coreflective and finitely productive subconstructs of *PRAP*. In particular for *PRTOP*, which is both coreflectively and bireflectively embedded in *PRAP*, we deduce three known results, namely

1. $\operatorname{Exp} PRTOP = FING$ [7, 12].
2. $\operatorname{Exp}\mathcal{B} \subset FING$ for every $\mathcal{B}$ coreflective in *PRTOP* [8, 12].
3. $\operatorname{Exp}\mathcal{B} = \mathcal{B} \cap FING$ for every $\mathcal{B}$ coreflective and finitely productive in *PRTOP* [8, 12].

We compare these answers to the topological situation studied by Činčura in [2].

The equality $\operatorname{Exp}\mathcal{B} = \mathcal{B} \cap \operatorname{Exp}\mathcal{C}$ for non-trivial quotient reflective subconstructs $\mathcal{B}$ of $\mathcal{C}$ is investigated in each of the examples $\mathcal{C} = PRAP$ and $\mathcal{C} = PRTOP$ and the results are compared to the situation in *TOP*, investigated in [9]. The positive answer for $\mathcal{C} = PRTOP$ recently appeared in [12] and combining Richter's ideas with results in [10] we deduce a positive answer for *PRAP* too. The following table summarizes the results and compares them to the topological situation.

$\mathcal{C}$	*PRAP*	*PRTOP*	*TOP*
Exp $\mathcal{C}$	∞-pqs-metric spaces	finitely gen. spaces	core compact spaces
Exp $\mathcal{B} \subset$ Exp $\mathcal{C}$ $\forall \mathcal{B}$ coreflective	+	+	−
Exp $\mathcal{B} = \mathcal{B} \cap$ Exp $\mathcal{C}$ $\forall \mathcal{B}$ corefl. + fin. prod.	+	+	−
Exp $\mathcal{B} = \mathcal{B} \cap$ Exp $\mathcal{C}$ $\forall \mathcal{B}$ quotient reflective	+	+	+

2. Preliminaries

Given a set X we denote its power set by 2^X and the set of its finite subsets by $2^{(X)}$. A set $\mathcal{I} \subset [0, \infty]^X$ is called an *ideal* in $[0, \infty]^X$ if and only if it satisfies

1. The constant function ∞ does not belong to $\mathcal{I}$.
2. If $\varphi \in \mathcal{I}$, $\psi \in [0, \infty]^X$, and $\psi \leqslant \varphi$, then $\psi \in \mathcal{I}$.
3. If $\varphi, \psi \in \mathcal{I}$, then $\varphi \vee \psi \in \mathcal{I}$.

A collection $(\mathcal{A}(x))_{x \in X}$ of ideals in $[0, \infty]^X$ is called a *pre-approach system* on X if it fulfils

(A1) $\forall x \in X, \forall \varphi \in \mathcal{A}(x) : \varphi(x) = 0$.

(A2) $\forall x \in X, \forall \varphi \in [0, \infty]^X$:

$$(\forall \varepsilon, N \in]0, \infty[, \exists \varphi_N^\varepsilon \in \mathcal{A}(x) \text{ s.t. } \varphi \wedge N \leqslant \varphi_N^\varepsilon + \varepsilon) \Rightarrow \varphi \in \mathcal{A}(x).$$

The members of $\mathcal{A}(x)$ are called *local pre-distances* in x. For ease in notation we shall, whenever convenient denote a pre-approach system $(\mathcal{A}(x))_{x \in X}$ also simply by $\mathcal{A}$. $(X, \mathcal{A})$ is called a *pre-approach space.* If $(X, \mathcal{A})$ and $(X', \mathcal{A}')$ are pre-approach spaces and if $f\colon (X, \mathcal{A}) \to (X', \mathcal{A}')$ is a function, then f is a *contraction* if

$$\forall x \in X, \forall \varphi' \in \mathcal{A}'(f(x))\colon \varphi' \circ f \in \mathcal{A}(x).$$

The construct with as objects 'pre-approach spaces' and as morphisms 'contractions' is denoted by *PRAP*. This construct was introduced in [4] and [5]. *PRAP* is related to the construct *AP* of approach spaces which has been extensively studied in [6]. One of the motivations for introducing *PRAP* is that it provides a unifying theory for both pretopological spaces (= closure spaces) and generalized metric spaces.

If X is endowed with a pretopological structure $(\mathcal{V}(x))_{x\in X}$, where $\mathcal{V}(x)$ is the neighborhood filter in x (which need not satisfy the open kernel condition) then there is a related pre-approach system $(\mathcal{A}(x))_{x\in X}$ where

$$\mathcal{A}(x) = \{\varphi \in [0,\infty]^X \mid \varphi(x) = 0 \text{ and } \{\varphi < \varepsilon\} \in \mathcal{V}(x) \text{ for every } \varepsilon > 0\},$$

and $\{\varphi < \varepsilon\} := \{y \in X \mid \varphi(y) < \varepsilon\}$. Through this embedding continuous maps between pretopological spaces correspond exactly with contractions.

An extended pseudo-quasi-semimetric on X is a map $d\colon X \times X \to [0,\infty]$ which is zero on the diagonal. If X is endowed with an extended pseudo-quasi-semimetric, then a natural pre-approach system $(\mathcal{A}(x))_{x\in X}$ is given by $\mathcal{A}(x) = \{\varphi \in [0,\infty]^X \mid \varphi \leqslant d(x,\cdot)\}$.

Through this embedding non-expansive maps correspond exactly with contractions. So both the constructs *PRTOP* of pretopological spaces and continuous maps and *pqs-MET*$^\infty$ of extended pseudo-quasi-semimetric spaces are fully embedded in *PRAP*.

PRTOP is both coreflectively and bireflectively embedded whereas *pqs-MET*$^\infty$ is a coreflective and finitely productive subconstruct. The intersection *PRTOP* $\cap$ *pqs-MET*$^\infty$ coincides with *FING*, the construct of those pretopological spaces in which every point has a smallest neighborhood. The following picture clarifies the previous relations.

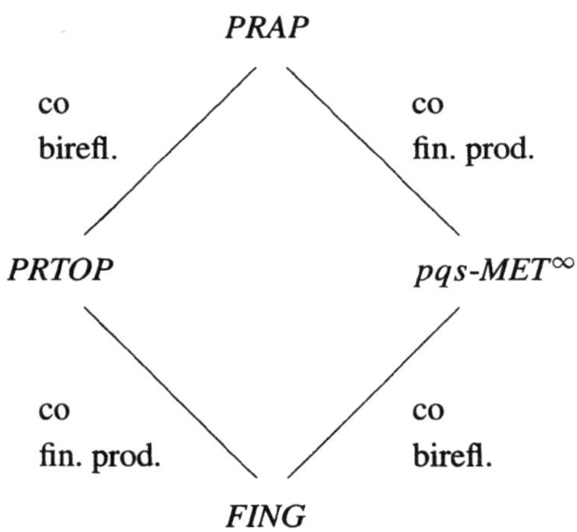

If $(X,\mathcal{A})$ is a pre-approach space then $(\Lambda(x))_{x\in X}$ is a *basis* for $\mathcal{A}$ if it fulfils $\mathcal{A}(x) = \hat{\Lambda}(x)$ for every $x \in X$, where

$$\hat{\Lambda}(x) = \{\varphi \in [0,\infty]^X \mid \forall \varepsilon, N \in]0,\infty[, \exists \psi \in \Lambda(x)\colon \varphi \wedge N \leqslant \psi + \varepsilon\}.$$

PRAP is a well fibred topological construct. Initial and final structures are described as follows:

If $f\colon X \to Y$ is a map and $\varphi \in [0,\infty]^X$ then we denote $f(\varphi)$ the function defined as $f(\varphi)\colon Y \to [0,\infty]\colon y \mapsto \inf\{\varphi(x) \mid f(x) = y\}$. Let $((X_i,\mathcal{A}_i))_{i\in I}$ be a class of *PRAP*-objects. If $(f_i\colon X \to (X_i, \mathcal{A}_i))_{i\in I}$ is a source then the collection $(\Lambda(x))_{x\in X}$ is a basis for the initial pre-approach system, where for all $x \in X$

$$\Lambda(x) = \left\{ \sup_{j\in J} \varphi_j \circ f_j \mid J \in 2^{(I)}, \varphi_j \in \mathcal{A}_j(f_j(x)), \forall j \in J \right\}.$$

If $(f_i\colon (X_i, \mathcal{A}_i) \to X)_{i\in I}$ is a sink then the collection $(\Lambda(x))_{x\in X}$ is a basis for the final pre-approach system, where for all $x \in X$

$$\Lambda(x) = \begin{cases} \bigcap_{\substack{i\in I \text{ s.t.} \\ f_i^{-1}(x)\neq\emptyset}} \bigcap_{z\in f_i^{-1}(x)} \{f_i(\varphi) \mid \varphi \in \mathcal{A}_i(z)\} & \text{if } \bigcup_{i\in I} f_i^{-1}(x) \neq \emptyset, \\ \{\varphi \in [0,\infty]^X \mid \varphi(x) = 0\} & \text{if } \bigcup_{i\in I} f_i^{-1}(x) = \emptyset. \end{cases}$$

All constructs under study in this paper are (isomorphic to) subconstructs of *PRAP*. The term 'subconstruct' implies full, non-empty, and isomorphism closed subconstruct. All subconstructs in this paper are initially structured [11].

An object X in $\mathcal{C}$ is exponential in $\mathcal{C}$ if $X \times -$ has a right adjoint. In the setting considered in this paper X is exponential in $\mathcal{C}$ if and only if $X \times -$ preserves final epi-sinks [11].

For categorical notions and terminology we refer to [1]. For previous investigations on *PRAP* we refer to [4] and [5].

3. Exponential Objects in Coreflective Subconstructs of *PRAP*

In [10] it was shown that the class Exp*PRAP* of all exponential objects in *PRAP* coincides with the subconstruct $pqs\text{-}MET^{\infty}$. In this section we investigate the class of exponential objects in some coreflective subconstruct $\mathcal{B}$ of *PRAP*.

In order to do so we consider a special final sink as described in the next result. The construction of this sink is inspired by the pretopological situation studied by Richter in [12].

PROPOSITION 3.1 *Let $(X, \mathcal{A}_X)$ be a pre-approach space. For $x \in X$, $Y_x = \coprod_{y\neq x} X_y$, where X_y is a copy of X and Y_x carries the coproduct structure in PRAP. Let the function $s_x\colon Y_x \to X$ be defined by $s_x(x, y) = x$ and $s_x(z, y) = y$ if $z \neq x$. Let $(s_x\colon Y_x \to (X, \mathcal{A}_{s_x}))$ for every element x of X and $(s_x\colon Y_x \to (X, \mathcal{A}_s))_{x\in X}$ be final sinks in PRAP. Then the identity map $1_X\colon (X, \mathcal{A}_X) \to (X, \mathcal{A}_s)$ is a contraction.*

Proof. It suffices to prove that for $x \in X$, $\mathcal{A}_{s_x}(x) \subset \mathcal{A}_X(x)$. We describe basic local pre-distances in $\mathcal{A}_{s_x}(x)$. Let $\psi = \inf_{y\neq x} s_x(\psi_y^{\nu_y})$ be such a local pre-distance, where $\psi_y^{\nu_y}\colon Y_x \longrightarrow [0,\infty]$ is defined by

$$\psi_y^{\nu_y}(u, y') = \begin{cases} \nu_y(u) & \text{if } y' = y, \\ \infty & \text{if } y' \neq y \end{cases}$$

for $\nu_y \in \mathcal{A}_X(x)$, $y \neq x$. If $z \neq x$ then straightforward calculations prove that

$$\psi(z) = \inf_{u\neq x} \nu_z(u).$$

In [10] Lemma 3.4 and 3.5, it is shown that there exists a positive real number ε_x with $\inf_{u\neq x}\varphi(u) \leqslant \varepsilon_x$ for every $\varphi \in \mathcal{A}_X(x)$ and such that the function $\theta_x^{\varepsilon_x}$ defined by

$$\begin{cases} \theta_x^{\varepsilon_x}(u) = \varepsilon_x & \text{if } u \neq x, \\ \theta_x^{\varepsilon_x}(x) = 0 \end{cases}$$

belongs to $\mathcal{A}_X(x)$. It follows that $\psi \leqslant \theta_x^{\varepsilon_x}$ and therefore ψ too belongs to $\mathcal{A}_X(x)$. □

PROPOSITION 3.2 *With the notations of the previous proposition consider* $(1_X \times s_x\colon X \times Y_x \to X \times X)_{x\in X}$. *The final PRAP-structure* $\mathcal{A}_{1\times s}$ *on* $|X \times X|$ *induces a pqs-MET*$^\infty$*-structure on the diagonal* Δ *where the largest local pre-distance in* (t, t) *is* $\psi_t \circ \mathrm{pr}_1 \vee \psi_t \circ \mathrm{pr}_2$ *with* $\psi_t = \sup\{\varphi \mid \varphi \in \mathcal{A}_X(t)\}$.

Proof. Let $(t, t) \in \Delta$ be fixed. First we consider the quotient on $|X \times X|$ determined by $1_X \times s_t\colon X \times Y_t \to |X \times X|$. We construct basic local pre-distances for this quotient on $|X \times X|$ in (t, t) in the following way. Let $\varepsilon > 0$ and $N > 0$. For $y \neq t$: if $\psi_t(y) < \infty$, choose $\mu_y \in \mathcal{A}_X(t)$ such that $\mu_y(y) + \varepsilon \geqslant \psi_t(y)$; if $\psi_t(y) = \infty$, choose $\mu_y \in \mathcal{A}_X(t)$ such that $\mu_y(y) \geqslant N$. We form a basic local pre-distance

$$\eta_t^{\varepsilon,N} = \inf_{y\neq t} 1_X \times s_t\big(\psi_y^{\mu_y\circ\mathrm{pr}_1\vee\mu_y\circ\mathrm{pr}_2}\big)$$

and consider the induced pre-distance on the diagonal. Evaluation of this function in (r, r) for $r \neq t$ gives

$$\eta_t^{\varepsilon,N}(r, r) = \mu_r(r).$$

From this equality we can deduce that

$$\psi_t(r) \wedge N \leqslant \eta_t^{\varepsilon,N}(r, r) + \varepsilon \quad \text{for } r \neq t.$$

Next we consider $1_X \times s_x$ for $x \neq t$ and again we evaluate the contribution of the quotient induced on Δ in (t, t). Let $\varepsilon > 0$ and $N > 0$ and choose $\mu_t \in \mathcal{A}_X(t)$ and $\mu_y \in \mathcal{A}_X(y)$, $y \neq x$ such that $\psi_t(y) < \infty \Rightarrow \mu_t(y) + \varepsilon \geqslant \psi_t(y)$ and $\psi_t(y) = \infty \Rightarrow \mu_t(y) \geqslant N$. Then

$$\eta_x^{\varepsilon,N} = \inf_{y\neq x} 1_X \times s_x\big(\psi_t^{\mu_t\circ\mathrm{pr}_1\vee\mu_y\circ\mathrm{pr}_2}\big)$$

is a basic local function. Evaluation of $\eta_x^{\varepsilon,N}(r, r)$ with $r \neq t$ gives

$$\begin{cases} \eta_x^{\varepsilon,N}(r, r) = \infty & \text{if } r \neq x, \\ \eta_x^{\varepsilon,N}(x, x) = \mu_t(x). \end{cases}$$

Then again we have

$$\psi_t(r) \wedge N \leqslant \eta_x^{\varepsilon,N}(r, r) + \varepsilon \quad \text{for } r \neq t.$$

Let $\mathcal{A}_{1\times s}(t,t)$ be the pre-approach system in (t,t) of the final structure for the sink $(1_X \times s_x\colon X \times Y_x \to |X \times X|)_{x\in X}$ and let $\mathcal{A}_\Delta(t,t)$ be the induced system on Δ. A basis for $\mathcal{A}_\Delta(t,t)$ is formed by the functions of the form

$$\varphi^{(\varepsilon),(N)} = \inf_{x\in X} \eta_x^{\varepsilon^x,N^x},$$

where $(\varepsilon) = (\varepsilon^x)_{x\in X}$ and $(N) = (N^x)_{x\in X}$ are collections of positive real numbers and $\eta_x^{\varepsilon^x,N^x}$, $x \neq t$ and $\eta_t^{\varepsilon^t,N^t}$ are constructed as before. Since $\varphi^{(\varepsilon),(N)}(r,r) \leqslant \eta_t^{\varepsilon^t,N^t}(r,r) \leqslant \psi_t(r)$ we can conclude that $\varphi^{(\varepsilon),(N)}$ belongs to the ∞-*pqs*-metric pre-approach structure on Δ with largest local pre-distance $\psi_t \circ \mathrm{pr}_1 \vee \psi_t \circ \mathrm{pr}_2$. On the other hand also $\psi_t \circ \mathrm{pr}_1 \vee \psi_t \circ \mathrm{pr}_2$ belongs to $\mathcal{A}_\Delta(t,t)$ since

$$(\psi_t \circ \mathrm{pr}_1 \vee \psi_t \circ \mathrm{pr}_2) \wedge N \leqslant \varphi^{\varepsilon,N} + \varepsilon$$

for $\varepsilon > 0$ and $N > 0$, where $\varphi^{\varepsilon,N} = \inf_{x\in X} \eta_x^{\varepsilon,N}$. □

THEOREM 3.3 *If $\mathcal{B}$ is any coreflective subconstruct of PRAP then exponential objects of $\mathcal{B}$ are ∞-pqs-metric spaces, i.e.,*

$$\mathrm{Exp}\,\mathcal{B} \subset \mathcal{B} \cap \mathrm{Exp}\,\mathit{PRAP}.$$

Proof. We use the notations explained in the previous propositions. The product in *PRAP* is denoted by $\times$ and the product in $\mathcal{B}$ by $\Box$. Let $X \in \mathcal{B}$. Then every Y_x for $x \in X$ and also $(X, \mathcal{A}_s)$ belongs to $\mathcal{B}$. If X is exponential in $\mathcal{B}$ then $(1_X \times s_x\colon X\Box Y_x \to X\Box(X,\mathcal{A}_s))_{x\in X}$ is final. It follows that id: $X\Box(X,\mathcal{A}_s) \to (X \times X, \mathcal{A}_{1\times s})$ is a contraction. So is the restriction to Δ,

$$\mathrm{id}_\Delta\colon \Delta_{X\Box(X,\mathcal{A}_s)} \longrightarrow \Delta_{(X\times X,\mathcal{A}_{1\times s})}.$$

It follows that $X \to \Delta_{(X\times X,\mathcal{A}_{1\times s})}$ is contractive too. Finally we can conclude that X and $\Delta_{(X\times X,\mathcal{A}_{1\times s})}$ are isomorphic. By Proposition 3.2 we can conclude that X is an ∞-*pqs*-metric space. □

In combination with the results obtained in [10] we can conclude the following.

COROLLARY 3.4 *pqs-MET^∞ is the largest coreflective cartesian closed subconstruct of PRAP.*

COROLLARY 3.5 *If $\mathcal{B}$ is coreflective and finitely productive in PRAP then we have*

$$\begin{aligned}\mathrm{Exp}\,\mathcal{B} &= \mathcal{B} \cap \mathrm{Exp}\,\mathit{PRAP}\\ &= \mathcal{B} \cap \mathit{pqs}\text{-}\mathit{MET}^\infty.\end{aligned}$$

3.6. EXAMPLES

1. Application of Corollary 3.5 to $\mathcal{B} = \mathit{PRTOP}$ gives us the result
   $\mathrm{Exp}\,\mathit{PRTOP} = \mathit{FING}$
   which was proved earlier in [7] and [12].

2. Application of Corollary 3.5 to *FING* gives the well known result that *FING* is cartesian closed.
3. Application of Theorem 3.3 to a coreflective subconstruct $\mathcal{B}$ of *PRTOP* implies $\mathrm{Exp}\,\mathcal{B} \subset FING$. It means that *FING* is the largest coreflective cartesian closed subconstruct in *PRTOP*. This result was established in [8] and [12].
4. Application of Corollary 3.5 to coreflective finitely productive subconstructs of *PRTOP* implies the earlier result $\mathrm{Exp}\,\mathcal{B} = \mathcal{B} \cap \mathrm{Exp}\,PRTOP$ obtained in [8] and [12].
5. Application of Corollary 3.5 to $\mathcal{B} = ps\text{-}MET^{\infty}$ yields the result that $ps\text{-}MET^{\infty}$ is cartesian closed.

3.7. COUNTEREXAMPLES IN *TOP*

Exponential objects in *TOP* have been internally characterized in [3] as the core compact spaces. Činčura [2] has shown that there is no largest coreflective cartesian closed subconstruct of *TOP*. In particular, $\mathcal{B}$ coreflective in *TOP* does not imply that the class of exponential objects in $\mathcal{B}$ is contained in the coreflective hull in *TOP* of the class of core compact spaces. Moreover the example constructed in [2] provides a coreflective and finitely productive subconstruct $\mathcal{B}$ of *TOP* for which $\mathrm{Exp}\,\mathcal{B} \not\subset \mathrm{Exp}\,TOP$.

4. Exponential Objects in Quotient Reflective Subconstructs

In [12] Richter established the result $\mathrm{Exp}\,\mathcal{B} = \mathcal{B} \cap \mathrm{Exp}\,PRTOP$ for quotient reflective subconstructs $\mathcal{B}$ of *PRTOP* containing an object with more than one point. In [9] one of the authors proved that for *TOP* too this equality holds. In this paragraph we adapt Richter's technique to the *PRAP* situation. Suppose $\mathcal{B} \subset PRAP$ is quotient reflective in *PRAP* and does contain a space X with more than one point. As in [12] it follows that $\mathcal{B}$ contains all discrete spaces and that $\mathcal{B}$ is closed under coproducts.

THEOREM 4.1 *If $\mathcal{B}$ is quotient reflective in PRAP and $\mathcal{B}$ contains a space with more than one point then* $\mathrm{Exp}\,\mathcal{B} = \mathcal{B} \cap \mathrm{Exp}\,PRAP$.

Proof. Since $\mathrm{Exp}\,\mathcal{B} \supset \mathcal{B} \cap \mathrm{Exp}\,PRAP$, as was pointed out in the introduction, it suffices to prove that $\mathrm{Exp}\,\mathcal{B} \subset pqs\text{-}MET^{\infty}$. We use the construction in Theorem 3.6 in [10], starting from $X = (X, \mathcal{A}) \in |\mathcal{B}|$, X exponential in $\mathcal{B}$. So as in [10], fix $x \in X$ and define

$$Z := \bigcup_{\varphi \in \mathcal{A}(x)} X \times \{\varphi\}$$

equipped with the coproduct structure. Y is the quotient of Z in *PRAP* by identification of the points (x, φ), $\varphi \in \mathcal{A}(x)$, i.e., the quotient with respect to the map

$$f\colon Z \to Y\colon (z, \mu) \to \begin{cases} (z, \mu) & \text{if } z \neq x, \\ a & \text{if } z = x. \end{cases}$$

The quotient Y belongs to $\mathcal{B}$. In order to prove this, remark that

$$\mathcal{S} = \left\{(x_\varphi)_{\varphi\in\mathcal{A}(x)} \mid x_\varphi = x \text{ except for at most one } \varphi\right\}$$

is a subspace of the product $\prod_{\mathcal{A}(x)} X$. Moreover by straightforward calculations we can verify that Y is finer than $\mathcal{S}$. It follows that $f\colon Z \to Y$ is a quotient in $\mathcal{B}$ and since X is exponential in $\mathcal{B}$, $1_X \times f\colon X \times Z \to X \times Y$ is a quotient in $\mathcal{B}$ and hence in *PRAP*. Finally, from the proof of Theorem 3.6 in [10] we deduce that X belongs to $pqs\text{-}MET^\infty$. □

4.2. EXAMPLES

We define a pre-approach space to be T_0 (T_1, Hausdorff) if its pretopological coreflection is T_0 (T_1, Hausdorff). Since the constructs $PRAP_0$, $PRAP_1$, and $PRAP_2$ of T_0, T_1, and Hausdorff pre-approach spaces are quotient reflective in *PRAP*, application of Theorem 4.1 implies

$$\begin{aligned}\operatorname{Exp} PRAP_0 &= PRAP_0 \cap \operatorname{Exp} PRAP\\ &= \text{the construct of all } pqs\text{-}MET^\infty\text{-spaces satisfying}\\ &\qquad d(x,y) = 0 \text{ and } d(y,x) = 0 \Rightarrow x = y,\end{aligned}$$

$$\operatorname{Exp} PRAP_1 = PRAP_1 \cap \operatorname{Exp} PRAP = qs\text{-}MET^\infty, \quad \text{and}$$

$$\operatorname{Exp} PRAP_2 = PRAP_2 \cap \operatorname{Exp} PRAP.$$

References

1. Adámek, J., Herrlich, H. and Strecker, G. E.: *Abstract and Concrete Categories*, Wiley 1990.
2. Činčura, J.: Cartesian closed coreflective subcategories of the category of topological spaces, *Topology Appl.* **41** (1991), 205–212.
3. Day, B. J. and Kelly, G. M.: On topological quotient maps preserved by pullbacks or products, *Proc. Camb. Phil. Soc.* **67** (1970), 553–558.
4. Lowen, E. and Lowen, R.: A quasitopos containing *CONV* and *MET* as full subcategories, *Internat. J. Math. Math. Sci.* **11**(3) (1988), 417–438.
5. Lowen, E. and Lowen, R.: Topological quasitopos hulls of categories containing topological and metric objects, *Cahiers Topologie Géo. Différentielle Catégoriques* **30**(3) (1989), 213–228.
6. Lowen, R.: *Approach Apaces: The Missing Link in the Topology-Uniformity-Metric Triad*, Oxford Mathematical Monographs, Oxford University Press, 1997.
7. Lowen-Colebunders, E. and Sonck, G.: Exponential objects and cartesian closedness in the construct *PRTOP*, *Appl. Cat. Struct.* **1** (1993), 345–360.
8. Lowen-Colebunders, E. and Sonck, G.: On the largest coreflective cartesian closed subconstruct of *PRTOP*, *Appl. Cat. Struct.* **4** (1996), 69–79.
9. Lowen-Colebunders, E.: Exponential objects in quotient reflective subconstructs of the category of topological spaces, preprint, 1996.
10. Lowen, E., Lowen, R. and Verbeeck, C.: Exponential objects in the construct *PRAP*, UIA Internal Report 96-09, 1996.

11. Nel, L.D.: Initially structured categories and cartesian closedness, *Canad. J. Math.* **27** (1975), 1361–1377.
12. Richter, G.: More on exponential objects in categories of pretopological spaces, *Appl. Cat. Struct.* **5** (1997), 309–319.

Applied Categorical Structures **8:** 257–269, 2000.
G. Brümmer & C. Gilmour (eds), Papers in Honour of Bernhard Banaschewski.

Completeness and Cocompleteness of ${}_R\mathbf{Smod}_{1N}$

Dedicated to Professor B. Banaschewski on the occasion of his seventieth birthday

HELMUT RÖHRL
9322 La Jolla Farms Road, La Jolla, CA 92037, USA

(Received: 10 July 1996; accepted: 30 June 1998)

Abstract. In this paper we formalize the concept of absolutely convergent infinite series by introducing the notion of prenormed R-semimodule with N-summation, where N is an arbitrary class. The morphisms between such semimodules are the contractive homomorphisms of R-semimodules that 'preserve' N-summation. The category ${}_R\mathbf{Smod}_{1N}$ of all prenormed R-semimodules with N-summation and their morphisms (composition being the set-theoretical one) fails to be an algebraic category. However, under quite general conditions it is both complete and cocomplete.

Key words: prenormed semimodule, N-summation, complete, cocomplete, free objects.

Mathematics Subject Classifications (2000): Primary: 16Y60; Secondary: 18A99.

1. Introduction

The set theory used in this paper is the Zermelo–Fraenkel set theory with one universe U as described in [2]; the terminology presented there will be used without further comment.

In [3] and [4] the notion of N-summation was employed to define the concept of N-convexity theory and to prove various statements involving N-convexity theories. In the second of these two papers N was assumed to be a set, that is an element of U, while in the first one N was chosen to be equal to U. In the present paper we let N be an arbitrary class, that is an arbitrary subset of U, and study the category ${}_C\mathbf{Srg}_{1N}$ of prenormed semirings with left N-summation over a fixed cone semiring C with left N-summation and the category ${}_R\mathbf{Smod}_{1N}$ of prenormed left R-semimodules with left N-summation where R is a fixed prenormed semiring with left N-summation; the subindex '1' indicates in either case that the morphisms are assumed to be contractive homomorphisms. Since we are only interested in the left concepts we will drop the epithet 'left'.

The definition of N-summation for the R-semimodule M involves the conglomerate M^N of all maps $N \to M$ whose support is a set. Since this is a legitimate conglomerate we may, and will, treat it as a class. Otherwise we would have to use two universes U and U', with $U \subset U'$, and introduce an additional hierarchy of set-types, say totalities, that reside above conglomerates.

In the first section we deal with the category ${}_C\mathbf{Srg}_1$ of prenormed semirings and their contractive homomorphisms and the category ${}_R\mathbf{Smod}_1$ of prenormed

R-semimodules over the prenormed semiring R and their contractive homomorphisms. We obtain necessary respectively sufficient conditions for the existence of arbitrary free objects in these categories. Furthermore we show that, under quite general conditions, both ${}_C\mathbf{Srg}_1$ and ${}_R\mathbf{Smod}_1$ are complete and cocomplete.

In the second section we introduce the concept of N-summation, where N is an arbitrary class, generalizing the corresponding notions in [3] and [4]. Similar concepts have been studied in automatic theory and in semiring theory (see [6, §6]). Using the notion of N-summation we obtain the category ${}_C\mathbf{Srg}_{1N}$ of prenormed semirings with N-summation and their contractive homomorphisms that preserve N-summation and the category ${}_R\mathbf{Smod}_{1N}$ of prenormed R-semimodules with N-summation over the prenormed semiring R with N-summation and their contractive homomorphisms that preserve N-summation. Again the existence of arbitrary free objects in these categories is discussed. Finally we show that, under quite general conditions, ${}_C\mathbf{Srg}_{1N}$ is complete and ${}_R\mathbf{Smod}_{1N}$ is both complete and cocomplete.

2. The Category of Prenormed Semimodules with Contractions

DEFINITION 2.1. Let R be a semiring (in the sense of [1]) and let C be a cone semiring (in the sense of [5]; equivalently, positive semiring in the sense of [1] with $0 \notin 1$). A *prenorm* of R with values in C is a map $\| \ \|\colon R \to C$ such that

(0) $\|0\| = 0$ (and $\|1\| = 1$, if $R \neq \{0\}$),

(i) $\|r_1 + r_2\| \leq \|r_1\| + \|r_2\|$, $r_1, r_2 \in R$,

(ii) $\|r_1 \cdot r_2\| \leq \|r_1\| \cdot \|r_2\|$, $r_1, r_2 \in R$.

C is also called the *value cone* of R.

DEFINITION 2.2. Let R and R' be prenormed semirings with value cone C and prenorms $\| \ \|$ respectively $\| \ \|'$. A homomorphism $f\colon R \to R'$ of semirings with $f(0) = 0$ and $f(1) = 1$ is said to be *contractive* if $\|f(r)\|' \leq \|r\|$, $r \in R$.

The category of all prenormed semirings with fixed value cone C and contractive homomorphisms (composition being the set-theoretical one) is denoted by ${}_C\mathbf{Srg}_1$ and is called the *category of prenormed semirings* (over C). Obviously, $\{0\}$ together with the standard prenorm is a terminal object of ${}_C\mathbf{Srg}_1$. However, there is also an initial object $\mathbb{N}_C$ of ${}_C\mathbf{Srg}_1$. Its underlying semiring is the subsemiring $\mathbb{N}_0 := \{z \in \mathbb{Z} : z \geq 0\}$ of $\mathbb{Z}$ and its prenorm is given by $\|n\| := 1 + \cdots + 1$ (n-times), $n \in \mathbb{N}_0$, where $1 \in C$ is the unit element of C.

PROPOSITION 2.3. *The category ${}_C\mathbf{Srg}_1$ has arbitrary free objects relative to the underlying-set functor if and only if C has a greatest element.*

Proof. Assume that ${}_C\mathbf{Srg}_{1C}$ has a free object $F(\{\phi\})$ on the one-point set $\{\phi\}$ and denote the canonical map $\{\phi\} \to F(\{\phi\})$ by δ. Since $\| \ \| := \mathrm{id}_C$ makes C a prenormed semiring with value cone C there is, for any map $\varphi\colon \{\phi\} \to C$, a contractive homomorphism $f\colon F(\{\phi\}) \to C$ of semirings with $\varphi = f \circ \delta$. Hence we have $\varphi(\phi) \leq \|\delta(\phi)\|$. Since $\varphi(\phi) \in C$ can be chosen arbitrarily it follows that $\|\delta(\phi)\|$ is the greatest element of C. Conversely let ∞ be the greatest element of C. If R is any semiring then the map $\| \ \|: R \to C$, given by

$$\|0\| := 0 \quad \text{and} \quad \|r\| = 1, \quad \text{for all } 0 \neq r \in R,$$

is a prenorm. Hence, if $F(A)$ is the free $_C\mathbf{Srg}_1$ object on the set A and $\delta: A \to F(A)$ the canonical map then the semiring $F'(A)$ underlying $F(A)$ is the free semiring on A with δ as its canonical map. Denoting the one-point set $\{\phi\}$ by A^0, $F'(A)$ is equal to the set of all maps $\chi: \dot{\bigcup}\{A^n : n \in \mathbb{N}_0\} \to \mathbb{N}_0$ with finite support and has its compositions given by

$$(\chi_1 + \chi_2)(x) = \chi_1(x) + \chi_2(x), \quad x \in \dot{\bigcup}\{A^n : n \in \mathbb{N}_0\},$$

$$(\chi_1 \cdot \chi_2)(x) = \Sigma\{\chi_1(x_1) \cdot \chi_2(x_1) : x_1 x_2 = x\}, \quad x \in \dot{\bigcup}\{A^n : n \in \mathbb{N}_0\},$$

where, for $x_1 \in A^{n_1}$ and $x_2 \in A^{n_2}$, the canonical image of $(x_1, x_2) \in A^{n_1} \times A^{n_2}$ in $A^{n_1+n_2}$ is written as $x_1 x_2$. Let $\overline{\delta}: \dot{\bigcup}\{A^n : n \in \mathbb{N}_0\} \to F'(A)$ be the Dirac map, that is the map which assigns to each x the Dirac function $\overline{\delta}_x$ based at x. Then $\delta := \overline{\delta}|A$ is the canonical map of the free semiring $F'(A)$ on A. Moreover $\mathbb{N}_0$ becomes a subsemiring of $F(A)$ via the map $\mathbb{N}_0 \ni n \mapsto n\delta_{\{\phi\}} \in F(A)$. Since there is a homomorphism $g: F(A) \to \mathbb{N}_C$ with $g \circ \delta = 0$ we obtain

$$\|n\| = 1 + \cdots + 1 \ (n\text{-times}), \quad n \in \mathbb{N}_0 \subseteq F(A).$$

In addition we have $\|\delta_a\| = \infty$, for all $a \in A$. Define $\| \ \|: F'(A) \to C$ by

$$\|x\| := \begin{cases} 1 + \cdots + 1 \ (n\text{-times}), & \text{if } x = n \in \mathbb{N}_0 \subseteq F'(A), \\ \infty, & \text{if } x \in F'(A)\backslash\mathbb{N}_0. \end{cases}$$

Since $n \cdot \infty = \infty$ and $\infty^n = \infty$, $n \in \mathbb{N}$, the map $\| \ \|: F'(A) \to C$ is a prenorm. It is routine work to verify that $F'(A)$ together with this prenorm is the free $_C\mathbf{Srg}_1$ object on A with δ as its canonical map. □

REMARK 2.4. $_C\mathbf{Srg}_1$ is not an algebraic category. This can be seen as follows. Define two maps $\| \ \|_1$ and $\| \ \|_2$ from $\mathbb{N}_0$ to C by

$$\|n\|_1 := \begin{cases} 0, & n = 0, \\ 1, & n \neq 0, \end{cases} \quad \text{respectively } \|n\|_2 := \begin{cases} 0, & n = 1, \\ 0, & n \neq 1. \end{cases}$$

Then both $\| \ \|_1$ and $\| \ \|_2$ are prenorms for $\mathbb{N}_0$ with value cone C. The identity map $\mathrm{id}_{\mathbb{N}_0}$ is a contraction when viewed as a map $(\mathbb{N}_0, \| \ \|_1) \to (\mathbb{N}_0, \| \ \|_2)$, but fails to be a contraction when viewed as a map $(\mathbb{N}_0, \| \ \|_1) \to (\mathbb{N}_0, \| \ \|_2)$, violating a well-known reflection principle for (concrete) algebraic categories.

Still we have

THEOREM 2.5. *Let C be conditionally complete. Then ${}_C\mathbf{Srg}_1$ is a complete category. If, in addition, C satisfies the Infimum Addition Law*

(IA) *for any two nonempty subsets S_1 and S_2 of C*

$$\wedge\{S_1 + S_2\} = \wedge\{S_1\} + \wedge\{S_2\}$$

and the Infimum Multiplication Law

(IM) *for any $0 \neq c \in C$ and any nonempty subset S of C*

$$\wedge\{cS\} = c \cdot \wedge\{S\}$$

then ${}_C\mathbf{Srg}_1$ is a cocomplete category.

Proof. In order to prove that ${}_C\mathbf{Srg}_1$ is complete we have to show that ${}_C\mathbf{Srg}_1$ has products and equalizers.

Let $\{R_i : i \in I\}$ be a family of semirings with prenorms $\| \|_i$: $R_i \to C, i \in I$. Form the subset

$$\{\chi \in X\{R_i : i \in I\}\colon \{\|\chi(i)\|_i : i \in I\} \text{ is bounded in } C\}$$

of the Cartesian product $X\{R_i : i \in I\}$. A simple computation shows that this set, equipped with the point-wise addition and multiplication, is a semiring R. Define $\| \|$: $R \to C$ by

$$\|\chi\| := \vee\{\|\chi(i)\|_i : i \in I\}, \quad \chi \in R.$$

A straight forward argument shows that $\| \|$ is a prenorm of R. Furthermore, the canonical projections p_j: $R \to R_j, j \in I$, are contractive homomorphisms of semirings. Let q_i: $S \to R_i, i \in I$, be a family of contractive homomorphisms of semirings. For $s \in S$ define $q(s) \in X\{R_i : i \in I\}$ by $q(s)(i) := q_i(s), i \in I$. Since $\|q(s)(i)\|_i = \|q_i(s)\|_i \leq \|s\|$, for all $i \in I$, we have $q(s) \in R$, for all $s \in S$. Obviously, q: $S \to R$ is the unique homomorphism of semirings satisfying $p_i \circ q = q_i, i \in I$. In addition

$$\|q(s)\| = \vee\{\|q(s)(i)\|_i : i \in I\} \leq \|s\|, \quad s \in S,$$

showing that q is contractive. Hence R is the product $\Pi\{R_i : i \in I\}$ in ${}_C\mathbf{Srg}_1$ of the family $\{R_i : i \in I\}$.

Let f_1 and f_2 be two contractive homomorphisms of prenormed semirings from S to R. Then a simple computation shows that the set-theoretical equalizer of f_1 and f_2 equipped with the induced semiring structure and the induced prenorm, is the equalizer of f_1 and f_2 in ${}_C\mathbf{Srg}_1$.

Now to the second part of our statement, the cocompleteness of the category ${}_C\mathbf{Srg}_1$. Let $\{R_i : i \in I\}$ be a family of semirings with prenorms $\| \|_i$: $R_i \to C, i \in I$. If ${}_C\mathbf{Srg}_1(R_i, X) \neq \phi$, for all $i \in I$, implies $X = \{0\}$ then $\{0\}$ is the coproduct of the family $\{R_i : i \in I\}$ of prenormed semirings. So assume that there is an object

$X \neq \{0\}$ with ${}_C\mathbf{Srg}_1(R_i, X) \neq \phi$, for all $i \in I$. Denote by R the coproduct of the family $\{R_i : i \in I\}$ in the category of semirings and let $k_j\colon R_j \to R$, $j \in J$, be the canonical homomorphisms. Then R is generated as a semiring by the set $\bigcup\{k_j(R_j) : j \in I\}$. If 1 denotes the unit element of R then every element $r \in R$ can be written as a finite sum of elements taken from the set

$$\{1 \cdot k_{j_1}(r_{j_1}) \cdot \cdots \cdot k_{j_n}(r_{j_n}) : n \in \mathbb{N}_0 \text{ and } r_j \in R_j\}$$

and we write

$$r = \Sigma\{1 \cdot k_{j_1}(r_{j_1}) \cdot \cdots \cdot k_{j_n}(r_{j_n})\}.$$

Define $\|\ \|\colon R \to C$ by

$$\|r\| := \wedge\{1 \cdot \|r_{j_1}\|_{j_1} \cdot \cdots \cdot \|r_{j_n}\|_{j_n} : r = \Sigma\{1 \cdot k_{j_1}(r_{j_1}) \cdot \cdots \cdot k_{j_n}(r_{j_n})\}\},$$
$$r \in R.$$

Obviously we have $\|0\| = 0$, and (IA) and (IM) imply

$$\|r_1 + r_2\| \leq \|r_1\| + \|r_2\| \quad \text{and} \quad \|r_1 \cdot r_2\| \leq \|r_1\| \cdot \|r_2\|, \quad r_1, r_2 \in R.$$

Let $\ell_j\colon R_j \to S$, $j \in I$, be a family of contractive homomorphisms of semirings. Then the resulting homomorphism $\ell\colon R \to S$ of semirings has the property that, whenever $r \in R$ is written as $r = \Sigma(1 \cdot k_{j_1}(r_{j_1}) \cdot \cdots \cdot k_{j_n}(r_{j_n})\}$, then

$$\ell(r) = \Sigma(1 \cdot \ell_{j_1}(r_{j_1}) \cdot \cdots \cdot \ell_{j_n}(r_{j_n})\}.$$

Hence, with $\|\ \|_S$ the prenorm on S

$$\|\ell(r)\|_S \leq \Sigma\{1 \cdot \|\ell_{j_1}(r_{j_1})\|_S \cdot \cdots \cdot \|\ell_{j_n}(r_{j_n})\|_S\} \leq \Sigma\{1 \cdot \|r_{j_1}\|_{j_1} \cdot \cdots \cdot \|r_{j_n}\|_{j_n}\}$$

and thus

$$\|\ell(r)\|_S \leq \|r\|,$$

showing that ℓ is a contraction. Since ℓ preserves unit elements we obtain

$$1 = \|1\|_S = \|\ell(1)\|_S \leq \|1\|.$$

On the other hand the definition of $\|\ \|\colon R \to C$ implies $\|1\| \leq 1$ and thus $1 \in R$ satisfies $\|1\| = 1$. Therefore $\|\ \|\colon R \to C$ is a prenorm and consequently the prenormed semiring R is the desired coproduct.

Let f_1 and f_2 be two contractive homomorphisms of prenormed semirings from S to R. Denote the coequalizer of f_1 and f_2 in the category **Srg** of semirings by $q\colon R \to R''$. Then q is a surjection. Define $\|\ \|''\colon R'' \to C$ by

$$\|r''\|'' := \wedge\{\|r\| : q(r) = r''\}.$$

As before, (IA) and (IM) imply that $\|\ \|''$ is a prenorm. Since $\|q(r)\|'' \leq \|r\|$, $r \in R$, q is a contractive homomorphism of prenormed semirings and therefore $R \xrightarrow{q} R''$ the coequalizer of f_1 and f_2 in ${}_C\mathbf{Srg}_1$. □

DEFINITION 2.6. Let R be a prenormed semiring with prenorm $\| \,\|: R \to C$. Let furthermore M be a (left) R-semimodule (in the sense of [1]). A *prenorm* of M is a map $\| \,\|: M \to C$ such that

(0) $\|0\| = 0$,
(i) $\|m_1 + m_2\| \leq \|m_1\| + \|m_2\|$, $m_1, m_2 \in M$,
(ii) $\|rm\| \leq \|r\| \cdot \|m\|$, $r \in R, m \in M$.

DEFINITION 2.7. Let M and M' prenormed R-semimodules with prenorms $\| \,\|$ respectively $\| \,\|'$. A homomorphism $f: M \to M'$ of R-semimodules with $f(0) = 0$ is said to be *contractive* if $\|f(m)\|' \leq \|m\|$, for all $m \in M$.

The category of all prenormed R-semimodules with fixed semiring R and contractive homomorphisms (composition being the set-theoretical one) is denoted by ${}_R\mathbf{Smod}_1$ and is called the *category of prenormed R-semimodules with contractions.*

PROPOSITION 2.8. *If the category ${}_R\mathbf{Smod}_1$ has a free object relative to the underlying-set functor on some nonempty set then $\{\|r\| : r \in R\}$ is bounded. Conversely, if the value cone C of R has a greatest element ∞ such that $c\infty = \infty$ for all $0 \neq c \in C$ then ${}_R\mathbf{Smod}_1$ has arbitrary free objects.*

Proof. The first claim follows as in the proof of 2.3. To prove our second assertion, let A be any set. If $A = \phi$ then $\{0\}$ is the free ${}_R\mathbf{Smod}_1$-object on A. So assume $A \neq \phi$ and let $F_S(A)$ be the free R-semimodule on A. Define $\| \,\|: F_S(A) \to C$ by

$$\|m\| := \begin{cases} 0, & \text{if } m = 0, \\ \infty, & \text{if } m \neq 0, \end{cases} \qquad m \in F_S(A).$$

Then $\| \,\|$ is a prenorm for $F_S(A)$ and a simple argument shows that the prenormed R-semimodule $F_S(A)$ is the free ${}_R\mathbf{Smod}_1$-object on A. □

However, ${}_R\mathbf{Smod}_1$ possesses always free objects relative to the closed unit ball functor.

PROPOSITION 2.9. *Let R be a prenormed semiring with prenorm $\| \,\|: R \to C$. For any set A let $R^{(A)}$ be the set of maps $\chi: A \to R$ with finite support. Equip $R^{(A)}$ with the point-wise addition and the point-wise scalar multiplication and define $\| \,\|: R^{(A)} \to C$ by*

$$\|\chi\| := \Sigma\{\|\chi(a)\| : a \in \operatorname{supp} \chi\}, \qquad \chi \in R^{(A)}.$$

Let furthermore $\delta: A \to R^{(A)}$ be the Dirac map. Then, for any prenormed R-semimodule with prenorm $\| \,\|: M \to C$ and any map

$$\varphi: A \to B(M) := \{m \in M : \|m\| \leq 1\}$$

there is a unique contractive homomorphism $f: R^{(A)} \to M$ of R-semimodules with $\varphi = f \circ \delta$.

Proof. Obviously, $f\colon R^{(A)} \to M$ is given by

$$f(\chi) := \Sigma\{\chi(a) \cdot \varphi(a) : a \in \operatorname{supp}\chi\}, \quad \chi \in R^{(A)}.$$

Hence

$$\begin{aligned} \|f(\chi)\| &\leq \Sigma\{\|\chi(a)\| \cdot \|\varphi(a)\| : a \in \operatorname{supp}\chi\} \\ &\leq \Sigma\{\|\chi(a)\| : a \in \operatorname{supp}\chi\} = \|\chi\|. \end{aligned}$$

□

REMARK 2.10. If $R \neq \{0\}$ then $_R\mathbf{Smod}_1$ is not an algebraic category. In this case the R-semimodule R carries at least two prenorms, namely the canonical one and the null prenorm (which maps all of R to 0). Again, this is a violation of a well-known reflection principle for (concrete) algebraic categories.

THEOREM 2.11. *Let R be a prenormed semiring with prenorm $\|\ \|\colon R \to C$. Suppose that the value cone C is conditionally complete. Then $_R\mathbf{Smod}_1$ is a complete category. If, in addition, C satisfies* (IA) *and* (IM) *then $_R\mathbf{Smod}_1$ is a cocomplete category.*

Proof. The completeness of $_R\mathbf{Smod}_1$ is proved similar to the corresponding part of the proof of 2.5.

In order to prove the existence of coproducts let $\{M_i : i \in I\}$ be a family of prenormed R-semimodules with prenorms $\|\ \|_i\colon M_i \to C, i \in I$, put $M := \oplus\{M_i : i \in I\}$, and denote the canonical homomorphism $M_j \to M$ by k_j, $j \in I$. Define $\|\ \|\colon M \to C$ by

$$\|\chi\| := \Sigma\{\|\chi(i)\|_i : i \in \operatorname{supp}\chi\}, \quad \chi \in M.$$

A simple computation shows that $\|\ \|\colon M \to C$ is a prenorm and that k_j, $j \in I$, is a contractive homomorphism (indeed an isometry). A straight forward argument shows that M together with the prenorm $\|\ \|$ and the homomorphisms k_j, $j \in I$, is the coproduct in $_R\mathbf{Smod}_1$ of the given family.

Finally we need the existence of coequalizers. So let f_1 and f_2 be two contractive homomorphisms of prenormed R-semimodules from M' to M. Denote by M'' the coequalizer of f_1 and f_2 in the category of R-semimodules and let $q\colon M \to M''$ be the canonical homomorphism. q is a surjection. Hence we obtain a map $\|\ \|''\colon M'' \to C$ by putting

$$\|m''\|'' := \wedge\{\|m\| : q(m) = m''\}, \quad m'' \in M''.$$

Then (IA) and (IM) imply that $\|\ \|''$ is a prenorm and, by definition, q is contractive. It is easy to verify that the prenormed R-semimodule M'' together with q is the coequalizer in $_R\mathbf{Smod}_1$ of f_1 and f_2. □

3. The Category of Prenormed Semimodules with N-Summation

Given any class N and any set A we denote by (N, A) the conglomerate of all maps $\varphi\colon N \to A$. If A has a distinguished element, denoted by 0, and φ is in (N, A) then

$\operatorname{supp}\varphi := N\backslash\varphi^{-1}(0)$ is called the *support* of φ. For any class N and any set A with distinguished element 0 the conglomerate $A^N := \{\varphi \in (N, A) : \operatorname{supp}\varphi \text{ is a set}\}$ is a legitimate conglomerate and hence is treated as a class. In this situation, if T is a subclass of N and φ is in A^N, we denote by φ^T the element of (N, A) given by

$$\varphi^T(a) := \begin{cases} \varphi(a), & \text{if } a \in T, \\ 0, & \text{if } a \notin T, \end{cases} \quad a \in A.$$

Clearly, φ^T is again in A^N.

Let C be a cone semiring with partial order relation $\leq$. Then the induced point-wise partial order relation $\leq^N$ on C^N makes C^N a partially ordered positive C-semimodule; for reason of convenience $\leq^N$ will be written as $\leq$. Next let R be a prenormed semiring with prenorm $\| \;\|$: $R \to C$. Then R^N, equipped with the point-wise addition and the point-wise scalar multiplication, is an R-semimodule. Moreover, $\| \;\|$ induces a map $\| \;\|^N$: $R^N \to C^N$ that satisfies the relations 1.6 (0)–(ii); again we will write $\| \;\|$ instead of $\| \;\|^N$. Similar considerations apply to prenormed R-semimodules.

Now we extend Definitions (1.1)–(1.3) of [3] to our present situation.

DEFINITION 3.1. Let C be a cone semiring. By a (left) N-*summation* for C is meant a pair (S_C, Σ_C) consisting of a (left) C-subsemimodule S_C of C^N and a homomorphism Σ_C: $S_C \to C$ of (left) C-semimodules such that

(0) $C^{(N)} := \{\alpha_* \in C^N : \operatorname{supp}\alpha_* \text{ is finite}\}$ is a (left) C-subsemimodule of S_C and, for all $\alpha_* \in C^{(N)}$, $\Sigma_C(\alpha_*) = \Sigma'\{\alpha_n : n \in \operatorname{supp}\alpha_*\}$, where Σ' stands for the sum in C of the finitely many elements $\{\alpha_n : n \in \operatorname{supp}\alpha_*\}$,

(ii) if $\alpha_* \in S_C$ and $\beta_* \in C^N$ satisfy $\beta_* \leq \alpha_*$ then β_* is in S_C and $\Sigma_C(\beta_*) \leq \Sigma_C(\alpha_*)$,

(iii) if $\alpha_* \in S_C$ and if φ: $N \to N$ is any map then $\alpha_*^{\varphi^{-1}(n)}$ is in S_C, for all $n \in N$, and the map $\Sigma_C(\alpha_*^{\varphi^{-1}})$, given by $N \ni n \mapsto \Sigma_C(\alpha_*^{\varphi^{-1}(n)}) \in C$, is in S_C and $\Sigma_C(\Sigma_C(\alpha_*^{\varphi^{-1}})) = \Sigma_C(\alpha_*)$,

(iv) if $\alpha_* \in C^N$ and if there is a map φ: $N \to N$ such that $\alpha_*^{\varphi^{-1}(n)}$ is in S_C, for all $n \in N$, and that $\Sigma_C(\alpha_*^{\varphi^{-1}})$ is in S_C, then α_* is in S_C.

Note that $(C^{(N)}, \Sigma')$ is an N-summation for C; in an obvious sense it is the smallest N-summation for C.

DEFINITION 3.2. Let R be a prenormed semiring with prenorm $\| \;\|$: $R \to C$. By a (left) N-*summation for* R is meant a (left) N-summation (S_C, Σ_C) for C together with a pair (S_R, Σ_R) consisting of a (left) R-subsemimodule S_R of R^N and a homomorphism Σ_R: $S_R \to R$ of (left) R-semimodules such that

(0) $\alpha_* \in S_R$ implies $\|\alpha_*\| \in S_C$,

(i) $R^{(N)} := \{\alpha_* \in R^N: \operatorname{supp}\alpha_* \text{ is finite}\}$ is a (left) R-subsemimodule of S_R and, for all $\alpha_* \in R^{(N)}$, $\Sigma_R(\alpha_*) = \Sigma'\{\alpha_n : n \in \operatorname{supp}\alpha_*\}$, where Σ' stands for the sum in R of the finitely many elements $\{\alpha_n : n \in \operatorname{supp}\alpha_*\}$,

(ii) for all $\alpha_* \in S_R$, $\|\Sigma_R(\alpha_*)\| \leq \Sigma_C(\|\alpha_*\|)$,
(iii) if $\alpha_* \in S_R$ and if $\varphi\colon N \to N$ is any map then $\alpha_*^{\varphi^{-1}(n)}$ is in S_R, for all $n \in N$, and the map $\Sigma_R(\alpha_*^{\varphi^{-1}})$, given by $N \ni n \mapsto \Sigma_R(\alpha_*^{\varphi^{-1}(n)}) \in R$, is in S_R and $\Sigma_R(\Sigma_R(\alpha_*^{\varphi^{-1}})) = \Sigma_R(\alpha_*)$.

Note that the map $|||\ |||\colon S_R \to C$, given by $S_R \ni \alpha_* \mapsto \Sigma_C(\|\alpha_*\|) \in C$, makes S_R a prenormed R-semimodule.

DEFINITION 3.3. Let R and R' be prenormed semirings with prenorms $\|\ \|\colon R \to C$ respectively $\|\ \|'\colon R' \to C$ and N-summations (S_R, Σ_R) respectively (S'_R, Σ'_R) over the same N-summation (S_C, Σ_C) for C. Then a contractive homomorphism $f\colon R \to R'$ of prenormed semirings is said to be a *contractive homomorphism of prenormed semirings with N-summation* if

(0) $f^N(S_R) \subseteq S_{R'}$,
(i) $f \circ \Sigma_R = \Sigma_{R'} \circ f^N$.

Note that if $f\colon R \to R'$ is a contractive homomorphism of prenormed semirings with N-summation then $f^N\colon S_R \to S_{R'}$ is a contractive homomorphism of prenormed R-semimodules, provided S_R is given the obvious R-semimodule structure and both S_R and $S_{R'}$ are equipped with the prenorm described after 3.2. In addition, 3.2(ii), and 3.3(i), imply

$$\|\Sigma_{R'}(f^N(\alpha_*))\|' \leq |||\alpha_*|||, \quad \alpha_* \in S_R.$$

The category of all prenormed semirings with N-summation over the fixed N-summation $NC := (S_C, \Sigma_C)$ and contractive homomorphisms of prenormed semirings with N-summation (composition being the set-theoretical one) is denoted by ${}_C\mathbf{Srg}_{1N}$ and is called the *category of prenormed semirings with N-summation* (over NC).

For any N-summation NC there is a full embedding $E_{NC}\colon {}_C\mathbf{Srg}_1 \to {}_C\mathbf{Srg}_{1N}$. It assigns to each prenormed semiring R the semiring R with N-summation $(R^{(N)}, \Sigma')$. This embedding sends $\{0\}$ to a terminal object and $\mathbb{N}_C$ to an initial object of ${}_C\mathbf{Srg}_{1N}$.

PROPOSITION 3.4. *Let NC be any N-summation for the cone semiring C. Then $F(A)$ is the free ${}_C\mathbf{Srg}_1$-object on A with canonical map $\delta\colon A \to F(A)$ if and only if $E_{NC}(F(A))$ is the free ${}_C\mathbf{Srg}_{1N}$-object on A with canonical map $\delta\colon A \to E_{NC}(F(A))$. In particular ${}_C\mathbf{Srg}_{1N}$ has arbitrary free objects if and only if C has a greatest element.*

Proof. Suppose that $F(A)$ is the free ${}_C\mathbf{Srg}_1$-object on A with canonical map $\delta\colon A \to F(A)$. Let furthermore R be in ${}_C\mathbf{Srg}_{1N}$ with N-summation (S_R, Σ_R) and let $\varphi\colon A \to R$ be any map. Then there is a contractive homomorphism $f\colon F(A) \to R$ of prenormed semirings satisfying $\varphi = f \circ \delta$. Since $E_{NC}(F(A))$ has N-summation

$(F(A)^{(N)}, \Sigma')$, 3.2(i), implies 3.3(0) and (i). The converse follows from the fact that E_{NC} is a full embedding. The balance is now implied by 2.3. □

Proposition 3.4 together with 2.4 shows that ${}_C\mathbf{Srg}_{1N}$ is not an algebraic category.

THEOREM 3.5. *Let C be conditionally complete. Then* ${}_C\mathbf{Srg}_{1N}$ *is a complete category.*

Proof. Let $\{R_i : i \in I\}$ be a family of prenormed semirings with prenorms $\| \ \|_i\colon R_i \to C$ and N-summations (S_i, Σ_i), $i \in I$. In the proof of 2.5 we constructed the product $R := \Pi\{R_i : i \in I\}$ in the category ${}_C\mathbf{Srg}_1$. Now we have to define the appropriate N-summation (S_R, Σ_R) for R. Given any map $\alpha_* \in (N, R)$, the associated map $\|\alpha_*\| \in (N, C)$ is well defined. Moreover, for $\alpha_* \in (N, R)$ and $i \in I$ we denote by $\alpha_*(i)$ the map $N \ni n \mapsto \alpha_n(i) \in R_i$. The definition of the prenorm for R, given in the proof of 2.5, leads to $\|\alpha_*(i)\|_i \leq \|\alpha_*\|$. Now put

$$S_R := \{\alpha_* \in (N, R) : \alpha_*(i) \in S_i, \text{ for all } i \in I, \text{ and } \|\alpha_*\| \in S_C\}.$$

Since I is a set and since each supp $\alpha_*(i), i \in I$, is a set it follows that supp $\alpha_* \subseteq \bigcup\{\text{supp } \alpha_*(i)\colon i \in I\}$ is also a set. Thus S_R is a subclass of R^N. Obviously S_R is an R-subsemimodule of R^N. Next let $\alpha_* \in S_R$. Then we can form the element $\Sigma_R(\alpha_*) \in X\{R_i : i \in I\}$ given by

$$(\Sigma_R(\alpha_*))(i) := \Sigma_i(\alpha_*(i)), \quad i \in I.$$

We have by 3.1(ii), and 3.2(ii),

$$\|(\Sigma_R(\alpha_*))(ii)\|_i = \|\Sigma_i(\alpha_*(i))\|_i \leq \Sigma_C \|(\alpha_*(i)\|_i \leq \Sigma_C \|\alpha_*\|, \tag{1}$$

showing that $\Sigma_R(\alpha_*)$ is an element of R. Hence we obtain a map $\Sigma_R\colon S_R \to R$. Obviously, Σ_R is a homomorphism of R-semimodules. We claim that (S_R, Σ_R) is an N-summation for R. 3.2(0) and (i) are trivially satisfied. The definition of the prenorm for R and (1) imply 3.2(ii). 3.2(iii), is valid because it is true for every $i \in I$. Hence (S_R, Σ_R) is an N-summation for R. Moreover the canonical projections $p_j\colon R \to R_j$, $j \in I$, are easily seen to be homomorphisms of prenormed semirings with N-summation. Let $q_i\colon S \to R_i$, $i \in I$, be a family of homomorphisms of prenormed semirings with N-summation. The resulting contractive homomorphism of prenormed semirings $q\colon S \to R$ satisfies, for every $\beta_* \in S_S$,

$$q^N(\beta_*)(i) = p_i^N \circ q^N(\beta_*) = q_i^N(\beta_*) \in S_i, \quad i \in I,$$

and

$$\|q^N(\beta_*)(i)\|_i = \|q_i^N(\beta_*)\| \leq \|(\beta_*)\|, \quad i \in I,$$

whence $\|q^N(\beta_*)\| \leq \|\beta_*\|$ holds. Thus $q^N(\beta_*)$ is in S_R and 3.3(0), is satisfied. Moreover,

$$\begin{aligned} p_i(\Sigma_R(q^N(\beta_*))) &= (\Sigma_R(q^N(\beta_*)))(i) = \Sigma_i(q^N(\beta_*))(i) \\ &= \Sigma_i q_i^N(\beta_*) = q_i(\Sigma_g(\beta_*)) = p_i(q(\Sigma_g(\beta_*))), \quad i \in I, \end{aligned}$$

and thus $\Sigma_R(q^N(\beta_*)) = q(\Sigma_S(\beta_*))$, which is 3.3(i). This means that ${}_C\mathbf{Srg}_{1N}$ has products.

In order to prove the existence of equalizers in ${}_C\mathbf{Srg}_{1N}$, let f_1 and f_2 be two homomorphisms of prenormed semirings with N-summation from S to R. Then the set-theoretical equalizer $S' := \{s \in S : f_1(s) = f_2(s)\}$, with the induced prenormed semiring structure and the inclusion $S' \subseteq S$, is the equalizer of f_1 and f_2 in ${}_C\mathbf{Srg}_1$. Let (S_S, Σ_S) respectively (S_R, Σ_R) be the N-summation for S respectively R and put $S_{S'} := S_S \cap S'^N$. If β_* is in $S_{S'}$, then $\beta_n, n \in N$, is in S' and thus $f_1(\beta_n) = f_2(\beta_n)$, implying $f_1^N(\beta_*) = f_2^N(\beta_*)$. Therefore by 3.3(i),

$$f_1(\Sigma_S(\beta_*)) = \Sigma_R(f_1^N(\beta_*)) = \Sigma_R(f_2^N(\beta_*)) = f_2(\Sigma_g(\beta_*))$$

and thus $\Sigma_g(\beta_*) \in S'$. Consequently the twosided restriction $\Sigma_{S'} := S'|\Sigma_S|S_{S'}$ is defined. Obviously, $S_{S'}$ is an S'-subsemimodule of S'^N and $\Sigma_{S'}$ is a homomorphism of S'-semimodules and it is easy to see that $(S_{S'}, \Sigma_{S'})$ is an N-summation for S'. A simple computation shows that the inclusion map $S' \subseteq S$ is a homomorphism of prenormed semirings with N-summation and, as such, is the equalizer of f_1 and f_2 in ${}_C\mathbf{Srg}_{1N}$. □

DEFINITION 3.6. Let M be a prenormed R-semimodule with prenorm $\| \,\|: M \to C$ over the prenormed semiring R with prenorm $\| \,\|: R \to C$ and N-summation (S_R, Σ_R) over the N-summation (S_C, Σ_C) for C. By a (left) *N-summation for M over the N-summation* (S_R, Σ_R) is meant a pair (S_M, Σ_M) consisting of an R-subsemimodule S_M of M^N and a homomorphism $\Sigma_M: S_M \to M$ of R-semimodules such that

(0) $\mu_* \in S_M$ implies $\|\mu_*\| \in S_C$,
(i) $M^{(N)} := \{\alpha_* \in M^N$: $\operatorname{supp} \mu_*$ is finite$\}$ is a (left) R-subsemimodule of S_M and, for all $\mu_* \in M^{(N)}$, $\Sigma_M(\mu_*) = \Sigma'\{\mu_n := n \in \operatorname{supp} \mu_*\}$, where Σ' stands for the sum in M of the finitely many elements $\{\mu_n : n \in \operatorname{supp} \mu_*\}$,
(ii) for all $\mu_* \in S_M$, $\|\Sigma_M(\mu_*)\| \leq \Sigma_C(\|\mu_*\|)$,
(iii) if $\mu_* \in S_M$ and if $\varphi: N \to N$ is any map then $\mu_*^{\varphi^{-1}(n)}$ is S_M, for all $n \in N$, and the map $\Sigma_M(\mu_*^{\varphi^{-1}})$, given by $N \ni n \mapsto \Sigma_M(\mu_*^{\varphi^{-1}(n)}) \in M$, is in S_M and $\Sigma_M(\Sigma_M(\mu_*^{\varphi^{-1}})) = \Sigma_M(\mu_*)$.

Again the map $|||\,|||: S_M \to C$, given by $S_M \ni \mu_* \mapsto \Sigma_C(\|\mu_*\|) \in C$, makes S_M a prenormed R-semimodule.

DEFINITION 3.7. Let M and M' be prenormed R-semimodules with prenorms $\| \,\|: M \to C$ respectively $\| \,\|': M' \to C$ and N-summation (S_M, Σ_M) respectively $(S_{M'}, \Sigma_{M'})$ over the same N-summation (S_R, Σ_R) over (S_C, Σ_C). Then a contractive homomorphism $f: M \to M'$ of prenormed R-semimodules is said to be a *contractive homomorphism of prenormed R-semimodules with N-summation* if

(0) $f^N(S_M) \subseteq S_{M'}$,
(i) $f \circ \Sigma_M = \Sigma_{M'} \circ f^N$.

The remarks following 3.3 remain valid in the present situation. The category of all prenormed R-semimodules with N-summation over the fixed N-summation (S_R, Σ_R) over (S_C, Σ_C) and contractive homomorphisms of prenormed R-semimodules with N-summation (composition being the set-theoretical one) is denoted by ${}_R\mathbf{Smod}_{1N}$ and is called the *category of prenormed R-semimodules with N-summation.*

As in the case of prenormed semirings there is, for any N-summation NR of R, a full imbedding H_{NR}: ${}_R\mathbf{Smod}_1 \to {}_R\mathbf{Smod}_{1N}$. Again, it assigns to each prenormed R-semimodule M the R-semimodule M with N-summation $(M^{(N)}, \Sigma')$.

LEMMA 3.8. *Let f: $M \to M''$ be a homomorphism of prenormed R-semimodules with N-summations (S_M, Σ_M) respectively $(S_{M''}, \Sigma_{M''})$. Put $\overline{M}'' := \operatorname{im} f$, equip $\overline{M}''$ with the prenorm induced from M'', and define $S := f^N(S_M)$ and $\Sigma := \Sigma_{M''}|S$. Then (S, Σ) is an N-summation for the prenormed R-semimodule $\overline{M}''$ and both the codomain restriction $\overline{f} := \overline{M}''|f$ and the inclusion $\overline{M}'' \subseteq M''$ are homomorphisms of prenormed R-semimodules with N-summation.*

Proof. Obviously, S is an R-subsemimodule of $\overline{M}''^N$ and Σ is a homomorphism of R-semimodules. The conditions 3.6(0)–(ii), are trivially satisfied. 3.6(iii), follows from

$$(f^N(\alpha_*))^{\varphi^{-1}(n)} = f^N(\alpha_*^{\varphi^{-1}(n)}) \in S$$

and

$$\Sigma_{M''}((f^N(\alpha_*)^{\varphi^{-1}}) = f^N(\Sigma_{M'}(\alpha_*^{\varphi^{-1}}))$$

by simple computation. The balance of the statement is now obvious. □

THEOREM 3.9. *Let R be a prenormed semiring with N-summation whose value cone is conditionally complete. Then ${}_R\mathbf{Smod}_{1N}$ is a complete category that has coproducts. If, in addition, $H_{NR}^{-1}(M)$ is a set for any prenormed R-semimodule M, then ${}_R\mathbf{Smod}_{1N}$ is a cocomplete category.*

Proof. The completeness of ${}_R\mathbf{Smod}_{1N}$ is proved similar to the corresponding part of the proof of 3.5.

Let $\{M_i : i \in I\}$ be a family of prenormed R-semimodules with prenorms $\| \|_i$: $M_i \to C$ and N-summations $(S_i, \Sigma_i), i \in I$. As in the proof of 2.11 we put $M := \bigoplus\{M_i : i \in I\}$, denote the canonical homomorphism $M_j \to M$ by $k_j, j \in I$, and equip M with the prenorm defined there. Furthermore define

$$S := \bigoplus\{k_i^N(S_i) : i \in I\}$$

and

$$\Sigma(\alpha_*) := \Sigma'\{k_i(\Sigma_i(\alpha_*(i)) : i \in \operatorname{supp} \alpha_*(-)\}, \quad \alpha_* \in S,$$

where Σ' stands for the sum in M of the finitely elements under Σ'. It is easy to verify that (S, Σ) is an N-summation for M, that the k_j, $j \in I$, are homomorphisms of prenormed R-semimodules with N-summation, and that M together with the k_j, $j \in I$, is the coproduct in ${}_R\mathbf{Smod}_{1N}$ of the given family.

Finally we have to prove the existence of coequalizers under the additional hypothesis. So let f_1 and f_2 be two homomorphisms of prenormed R-semimodules with N-summation from M' to M. Let X be the conglomerate of all surjective homomorphisms of prenormed R-semimodules with N-summation that have M as their domain and coequalize f_1 and f_2. The additional hypothesis implies that X is a set. If x is an element of X we write x also as $M \xrightarrow{g_x} M_x$ and make X a poset by putting $x_1 \leq x_2$ if there is a homomorphism $h_{x_1,x_2}\colon M_{x_1} \to M_{x_2}$ of prenormed R-semimodules with N-summation such that $g_{x_1} = h_{x_1,x_2} \circ g_{x_2}$. Let Δ be the diagram $\{M_x, h_{x_1,x_2} : x, x_1, x_2 \in X\}$ in ${}_R\mathbf{Smod}_{1N}$ and denote its limit by M'', with p_x the canonical image $M'' \to M_x, x \in X$. Then there is a unique homomorphism $g''\colon M \to M''$ of prenormed R-semimodules with N-summation such that $g_x = p_x \circ g'', x \in X$. Obviously, g'' coequalizes f_1 and f_2. Now replace $g''\colon M \to M''$ by $\overline{g}''\colon M \to \overline{M}''$ as described in 3.8 and denote $\overline{M}'' \subseteq M'' \xrightarrow{p_x} M_x$ by $\overline{p}_x$. Then the surjective homomorphism $\overline{g}''$ still coequalizes f_1 and f_2 and satisfies $g_x = \overline{p}_x \circ \overline{g}''$. Let $f\colon M \to N$ be any homomorphism of prenormed R-semimodules with N-summation that coequalizes f_1 and f_2. By 3.8 we have $f = \mathrm{inc} \circ \overline{f}$ where $\overline{f}$ is surjective and coequalizes f_1 and f_2. Hence $\overline{f}$ is an element of X and we obtain

$$f = \mathrm{inc} \circ \overline{f} = \mathrm{inc} \circ \overline{p}_{\overline{f}} \circ \overline{g}'' = f' \circ \overline{g}''.$$

Since $\overline{g}''$ is surjective, f' is uniquely determined by f. Thus $\overline{M}''$ together with $\overline{g}''$ is the coequalizer in ${}_R\mathbf{Smod}_{1N}$ of f_1 and f_2. □

References

1. Golan, J. S.: *The Theory of Semirings with Applications in Mathematics and Theoretical Computer Science*, Longman Sci. & Tech., Harlow, U.K., 1992.
2. Herrlich, H. and Strecker, G. E.: *Category Theory*, Heldermann, Berlin, 1979.
3. Kleisli, H. and Röhrl, H.: Convexity Theories 0 fin. Foundations, to be published.
4. Pareigis, B., Pumplün, D., and Röhrl, H.: Convexity theories 0 cont. Foundations, in *Proc. 2nd Gauss Symposium*, W. de Gruyter, Berlin, 1995, pp. 301–321.
5. Röhrl, H.: Convexity theories 0. Foundations, *Appl. Cat. Struct.* **2** (1994), 13–43.
6. Weinert, H. J.: Generalized semialgebras over semirings, Lecture Notes in Math. 1320, Springer, Berlin, 1986, pp. 380–416.

Applied Categorical Structures **8:** 271–277, 2000.
G. Brümmer & C. Gilmour (eds), Papers in Honour of Bernhard Banaschewski.

Cauchy Complete Nearness Spaces

Dedicated to Professor Bernhard Banaschewski on his seventieth birthday

SUNG SA HONG and YOUNG KYOUNG KIM
Department of Mathematics, Sogang University, Seoul, 121-742, Korea
e-mail: sshong@ccs.sogang.ac.kr

(Received: 3 October 1996; accepted: 11 April 1998)

Abstract. A nearness space is Cauchy complete if every regular Cauchy filter on the space is convergent. We show that the category **CCNear₂** of Cauchy complete N_2 spaces is reflective in the category **Near₂C** of N_2-spaces and Cauchy maps and that the reflection of an N_2-space is given by the strict extension associated with regular Cauchy filters on the space.

Mathematics Subject Classifications (2000): 18A40, 54D35, 54E17.

Key words: nearness spaces, regular Cauchy filters, Cauchy complete nearness spaces, Cauchy completions of nearness spaces, Cauchy maps.

1. Introduction

Using clusters or coclusters, Herrlich defined the concept of complete nearness spaces and then constructed the completions of nearness spaces, which give rise to reflections in the category **RNear** of regular nearness spaces and nearness preserving maps. Furthermore, he characterized the strict extensions of a topological space via the completion (see [5 – 8]).

Banaschewski and Pultr introduced the concepts of complete and Cauchy complete nearness frames and constructed the completions of nearness frames which furnish coreflections in the category of strong nearness frames and uniform homomorphisms (see [2 – 4]).

For the Cauchy complete nearness frames, they introduced the concept of regular Cauchy filters which are exactly minimal Cauchy filters in a *strong* nearness frame. We note that regular Cauchy filters are defined internally.

The present authors constructed the Cauchy completions of nearness frames and showed that the category of strong Cauchy complete nearness frames is coreflective in the category of strong nearness frames and uniform homomorphisms ([10]). To get this result, we used Cauchy homomorphisms between nearness frames introduced by Banaschewski and Pultr ([4]).

The purpose of this paper is to introduce Cauchy complete nearness spaces and construct Cauchy completions of nearness spaces.

A Cauchy filter $\mathcal{F}$ on a nearness space (X, μ) is called a regular Cauchy filter if for any $F \in \mathcal{F}$, there is $G \in \mathcal{F}$ with $G <_\mu F$ and (X, μ) is said to be Cauchy complete if every regular Cauchy filter on (X, μ) is convergent. We recall that there is a Cauchy complete but not complete uniform frame. But a regular nearness space is complete if and only if it is Cauchy complete. We give a Cauchy complete but not complete N_2-space. Using the fact that a regular Cauchy filter is a cocluster, we construct the Cauchy completion $(cX, c\mu)$ of a nearness space (X, μ) as a subspace of the completion (X^*, μ^*) of [6]. A map $f\colon (X, \mu) \to (Y, \mu')$ between nearness spaces is called a Cauchy map if for any regular Cauchy filter $\mathcal{F}$ on X, $f(\mathcal{F})$ contains a regular Cauchy filter. Since the inclusion map $c\colon (X, \mu) \to (cX, c\mu)$ is a strict extension, c is a Cauchy map. We show that the category of Cauchy complete N_2-spaces is reflective in the category of N_2-spaces and Cauchy maps via the Cauchy completions.

For the terminology, we refer to [1] for the category theory and [8] for nearness spaces. Since we use mostly uniform covers in a nearness space, the nearness structure μ on a nearness space (X, μ) is given by the family of uniform covers and its associated merotopic structure will be denoted by γ.

2. Cauchy Complete Nearness Spaces

First, we introduce a concept of regular Cauchy filters on a nearness space and using this, we define Cauchy complete nearness spaces.

DEFINITION 2.1. A filter $\mathcal{F}$ on a nearness space (X, μ) is called:

(1) a *Cauchy filter* if for any $\mathcal{A} \in \mu$, $\mathcal{F} \cap \mathcal{A} \neq \phi$,
(2) a *regular filter* if for any $F \in \mathcal{F}$, there is $G \in \mathcal{F}$ with $G <_\mu F$,
(3) a *regular Cauchy filter* if it is a regular and Cauchy filter.

REMARK 2.2. (1) Noting that if $\mathcal{A} = \text{stack}\,\mathcal{A} \in \gamma$ on a nearness space (X, μ), then for any $A <_\mu B$, $B \in \mathcal{A}$ or $X - A \in \mathcal{A}$, one can easily show that every regular Cauchy filter on (X, μ) is a cocluster and hence it is an open minimal Cauchy filter.

(2) The converse of (1) need not be true. Indeed, for any nonregular Hausdorff space X, there is $x \in X$ whose neighborhood filter is a cocluster but not a regular Cauchy filter (see also Example 2.5).

(3) Suppose that (X, μ) is a regular nearness space. Then for any Cauchy filter $\mathcal{F}$ on X, $\mathcal{F}(<_\mu) = \{B : A <_\mu B \text{ for some } A \in \mathcal{F}\}$ is the unique minimal Cauchy filter contained in $\mathcal{F}$ ([8]). Since a Cauchy filter $\mathcal{F}$ on X is a regular Cauchy filter if and only if $\mathcal{F} = \mathcal{F}(<_\mu)$, regular Cauchy filters on a *regular* nearness space are precisely minimal Cauchy filters or equivalently coclusters.

DEFINITION 2.3. A nearness space X is said to be *Cauchy complete* if every regular Cauchy filter on X is convergent.

Using Remark 2.2, one has the following immediately.

PROPOSITION 2.4. (1) *Every complete nearness space is Cauchy complete.*
(2) *A regular nearness space is complete if and only if it is Cauchy complete.*

EXAMPLE 2.5. Let N be the set of natural numbers and $X = N \times \{0, 1\}$. For any $(k, l) \in N \times N$, let $\mathcal{U}_{k,l} = \{\{(m, 0)\} : m < k\} \cup \{\{(m, 1)\} : m < k\} \cup \{\{(n, 0), (n, 1)\} : n \geq k\} \cup \{\{(m, 0) : m \geq l\}\}$. Then $\{\mathcal{U}_{k,l} : (k, l) \in N \times N\}$ generates a nearness structure $\mu = \{\mathcal{A} \subseteq P(X) : \mathcal{U}_{k,l} \prec \mathcal{A}$ for some $(k, l) \in N \times N\}$ and (X, μ) is an N_2-space whose underlying topology is the discrete one. Since the filter generated by $\{\{(m, 0) : m \geq l\} : l \in N\}$ is a nonconvergent cocluster, (X, μ) is not a complete nearness space (see [9] for the details).

Take any regular Cauchy filter $\mathcal{F}$ on (X, μ), then for any $(k, l) \in N \times N$, $\mathcal{F} \cap \mathcal{U}_{k,l} \neq \phi$ and therefore either there is $(m, n) \in X$ such that $\{(m, n)\} \in \mathcal{F}$ or $\{(m, 0) : m \geq l\} \in \mathcal{F}$ for all $l \in N$. In the first case, $\mathcal{F}$ clearly converges to (m, n). For the second case, there is $F \in \mathcal{F}$ with $F <_\mu \{(m, 0) : m \geq 1\}$; therefore there is $(p, q) \in N \times N$ such that $\mathcal{U}_{p,q} \prec \{X - F, \{(m, 0) : m \geq 1\}\}$. Since F can not be a finite set, $\{n : (n, 0) \in F\}$ is an infinite set and hence there is $k \in N$ such that $p \leq k$ and $(k, 0) \in F$. Since $\{(k, 0), (k, 1)\} \in \mathcal{U}_{p,q}$, we have a contradiction. Thus every regular Cauchy filter on (X, μ) is convergent, so that (X, μ) is a Cauchy complete but not complete N_2-space.

For a nearness preserving map $f\colon (X, \mu) \to (Y, \mu')$ and $A \subseteq X$, let $\operatorname{op} A = Y - \operatorname{cl}_{\mu'} f(X - A)$ and for $\mathcal{A} \subseteq P(X)$, $\operatorname{op} \mathcal{A} = \{\operatorname{op} A : A \in \mathcal{A}\}$.

DEFINITION 2.6. A nearness preserving map $f\colon (X, \mu) \to (Y, \mu')$ is said to be:

(1) *strict* if μ' is generated by $\{\operatorname{op} \mathcal{A} : \mathcal{A} \in \mu\}$,
(2) a *strict extension* if it is a strict and dense embedding.

In the following, we assume that an extension is always given by the inclusion map and for an extension $e\colon (X, \mu) \to (Y, \mu')$ and $A \subseteq X$, $\operatorname{op} A = Y - \operatorname{cl}_{\mu'}(X - A)$.

REMARK 2.7 ([5, 6, 8]). For a strict extension $e\colon (X, \mu) \to (Y, \mu')$, one has:

(1) For any $A \subseteq X$, $\operatorname{op} A$ is the largest open set W in (Y, μ') with $W \cap X = \operatorname{int} A$ and $\operatorname{op} A = \operatorname{op}(\operatorname{int} A)$. Moreover, $\operatorname{op}(A \cap B) = \operatorname{op} A \cap \operatorname{op} B$.
(2) $\{\operatorname{op} A : A \subseteq X\}$ forms a base for the underlying topology on (Y, μ').
(3) For any $A, B \subseteq X$, $A <_\mu B$ in (X, μ) if and only if $\operatorname{op} A <_{\mu'} \operatorname{op} B$ in (Y, μ').

PROPOSITION 2.8. *Let $e\colon (X, \mu) \to (Y, \mu')$ be a strict extension. Then (Y, μ') is Cauchy complete if and only if every regular Cauchy filter on (X, μ) is convergent in (Y, μ').*

Proof. Assume that (Y, μ') is Cauchy complete and $\mathcal{F}$ is a regular Cauchy filter on (X, μ). Since e is a strict extension, op $\mathcal{F}$ is again a regular Cauchy filter base on (Y, μ') by Remark 2.7 and hence it is convergent. Since op $\mathcal{F}$ generates a coarser filter than $\mathcal{F}$, $\mathcal{F}$ is also convergent. Conversely, take any regular Cauchy filter $\mathcal{G}$ on (Y, μ'). Since $\mathcal{G}$ is an open filter on Y and X is dense in Y, it has the trace filter $\mathcal{G}_X$ on X. Since e is an embedding, $\mathcal{G}_X$ is a regular Cauchy filter on (X, μ) and therefore convergent, say to y in (Y, μ'). Take any basic neighborhood op A of y, then there is an open $G \in \mathcal{G}$ with $G \cap X \subseteq \operatorname{int} A$. Since op A is the largest open set W in Y with $W \cap X = \operatorname{int} A$, $G \subseteq \operatorname{op} A$. Thus $\mathcal{G}$ converges to y. □

We now recall the completion of a nearness space (X, μ) (see [6–8] for the details).

Let $X^* = X \cup \{\mathcal{A} : \mathcal{A}$ is a nonconvergent cocluster on $(X, \mu)\}$ and for any $A \subseteq X$, let $\widetilde{\operatorname{op}}\, A = (\operatorname{int} A) \cup \{\mathcal{A} \in X^* - X : A \in \mathcal{A}\}$. Then $\{\widetilde{\operatorname{op}}\, \mathcal{A} = \{\widetilde{\operatorname{op}}\, A : A \in \mathcal{A}\} : \mathcal{A} \in \mu\}$ generates a nearness μ^* on X^*, the inclusion map j: $(X, \mu) \to (X^*, \mu^*)$ is a strict extension and (X^*, μ^*) is a complete nearness space. Further, for any $A \subseteq X$, $\widetilde{\operatorname{op}}\, A = \operatorname{op} A$ for the strict extension j.

For a nearness space (X, μ), let $cX = X \cup \{\mathcal{F} : \mathcal{F}$ is nonconvergent regular Cauchy filter on $(X, \mu)\}$, then by Remark 2.2, cX is a subset of X^*. Let $(cX, c\mu)$ denote the subspace of (X^*, μ^*) with the underlying set cX and c: $X \to cX$ the inclusion map. Since j: $(X, \mu) \to (X^*, \mu^*)$ is a strict extension and $(cX, c\mu)$ is a subspace of (X^*, μ^*), c: $(X, \mu) \to (cX, c\mu)$ is also a strict extension and for $A \subseteq X$, op A with respect to c is given by $(\operatorname{int} A) \cup \{\mathcal{F} \in cX - X : A \in \mathcal{F}\}$. We note that for a *regular* nearness space (X, μ), $(cX, c\mu) = (X^*, \mu^*)$.

Using the above notation, one has the following.

THEOREM 2.9. *For any nearness space* (X, μ), $(cX, c\mu)$ *is Cauchy complete.*

Proof. Since c: $(X, \mu) \to (cX, c\mu)$ is a strict extension, it is enough to show that every regular Cauchy filter on (X, μ) is convergent in $(cX, c\mu)$. Take any regular Cauchy filter $\mathcal{F}$ on (X, μ). If $\mathcal{F}$ is convergent in (X, μ), then it is convergent in $(cX, c\mu)$. In the case of $\mathcal{F}$ being nonconvergent in (X, μ), then $\mathcal{F} \in cX$. For any basic neighborhood op A of $\mathcal{F}$ in $(cX, c\mu)$ where A is an open set in X, $A \in \mathcal{F}$. Since $A \subseteq \operatorname{op} A$, op A belongs to the filter in cX generated by $\mathcal{F}$. Thus $\mathcal{F}$ converges to $\mathcal{F}$ in $(cX, c\mu)$. This completes the proof. □

DEFINITION 2.10. For a nearness space (X, μ), $(cX, c\mu)$ or the strict extension c: $(X, \mu) \to (cX, c\mu)$ is called the *Cauchy completion* of (X, μ).

Since the Cauchy completion $(cX, c\mu)$ of a nearness space (X, μ) is a subspace of (X^*, μ^*), the following is immediate from the result in [8].

PROPOSITION 2.11. *A nearness space* (X, μ) *is separated* (N_1, *regular, uniform, contigual, proximal, subtopological, respectively*) *if and only if* $(cX, c\mu)$ *is separated* (N_1, *regular, uniform, contigual, proximal, subtopological, respectively*).

3. Cauchy Completions and Cauchy Maps

In this section, we introduce a concept of Cauchy maps and then show that the Cauchy completions give rise to reflections in the category of N_2-spaces and Cauchy maps.

DEFINITION 3.1. Let (X, μ) and (Y, μ') be nearness spaces. A map $f\colon (X, \mu) \to (Y, \mu')$ is called a *Cauchy map* if for any regular Cauchy filter $\mathcal{F}$ on X, there is a regular Cauchy filter $\mathcal{G}$ on Y with $\mathcal{G} \subseteq f(\mathcal{F})$.

Clearly, the class of nearness spaces and Cauchy maps forms a category.

EXAMPLE 3.2. Let (X, μ) be an infinite nearness space endowed with the nearness structure μ generated by the cofinite topology. Then $\{X\}$ is the only regular filter on (X, μ) and hence there is no regular Cauchy filter on (X, μ). Let (X, μ') denote the discrete nearness space. Clearly the identity map $1_X\colon (X, \mu') \to (X, \mu)$ is a nearness preserving map but not a Cauchy map and $1_X\colon (X, \mu) \to (X, \mu')$ is a Cauchy map but not a nearness preserving map.

PROPOSITION 3.3. *A nearness preserving map $f\colon (X, \mu) \to (Y, \mu')$ between nearness spaces is a Cauchy map whenever*

(1) *(Y, μ') is regular, or*
(2) *f is onto initial.*

Proof. (1) follows from (3) of Remark 2.2.

For (2), we note that for any $A \subseteq X$, $Y - f(A) \subseteq f(X - A)$, because f is onto. Since f is onto initial, $A <_\mu B$ implies $f(A) <_{\mu'} f(B)$. Thus for any regular Cauchy filter $\mathcal{F}$ on (X, μ), $f(\mathcal{F})$ is also a regular Cauchy filter on (Y, μ'). □

PROPOSITION 3.4. *Suppose that (X, μ) is a regular subtopological nearness space. If $f\colon (X, \mu) \to (Y, \mu')$ is a Cauchy map, then it is a nearness preserving map.*

Proof. Take any $\mathcal{A} \in \gamma$. Since (X, μ) is subtopological, there is a Cauchy filter $\mathcal{F}$ on X which corefines $\mathcal{A}$. Since (X, μ) is regular, $\mathcal{F}$ contains a regular Cauchy filter $\mathcal{G}$. Thus $f(\mathcal{G})$ contains a regular Cauchy filter and hence $f(\mathcal{A}) \in \gamma'$. Thus f is a nearness preserving map. □

The following is immediate from the above two propositions.

COROLLARY 3.5. *Suppose that (X, μ) is a regular subtopological nearness space and that (Y, μ') is a regular nearness space, then a map $f\colon (X, \mu) \to (Y, \mu')$ is a Cauchy map if and only if it is a nearness preserving map.*

By the proof of Proposition 2.8, one has:

LEMMA 3.6. *Every strict extension is a Cauchy map and hence for any nearness space (X, μ), the Cauchy completion c: $(X, \mu) \to (cX, c\mu)$ is a Cauchy map.*

We now prove our main result.

THEOREM 3.7. *The subcategory* **CCNear₂** *of Cauchy complete N_2-spaces is reflective in the category of* **Near₂C** *of N_2-spaces and Cauchy maps.*

Proof. Let $(X, \mu) \in$ **Near₂C**, then by Proposition 2.11, $(cX, c\mu) \in$ **CCNear₂** and c: $(X, \mu) \to (cX, c\mu)$ is a Cauchy map by the above lemma. Take any Cauchy map f: $(X, \mu) \to (Y, \mu')$ with $(Y, \mu') \in$ **CCNear₂**. Since f is a Cauchy map and (Y, μ') is Cauchy complete, for any $\mathcal{F} \in cX - X$ $f(\mathcal{F})$ is convergent, say to $\bar{f}(\mathcal{F})$. We define $\bar{f}$: $cX \to Y$ as follows:

$$\bar{f}(z) = \begin{cases} f(z) & \text{if } z \in X, \\ \bar{f}(\mathcal{F}) & \text{if } z = \mathcal{F} \in cX - X. \end{cases}$$

Then $\bar{f}$ is a map and $\bar{f} \circ c = f$. We show that $\bar{f}$ is a Cauchy map. Take any regular Cauchy filter Ψ on cX. Then $\Psi_X = \{\Gamma \cap X : \Gamma \in \Psi\}$ is a regular Cauchy filter on X. Since f is a Cauchy map, there is a regular Cauchy filter $\mathcal{G}$ on Y with $\mathcal{G} \subseteq f(\Psi_X)$. We claim that $\mathcal{G} \subseteq \bar{f}(\Psi)$. Take any $G \in \mathcal{G}$. Since $\mathcal{G}$ is regular, there is $H \in \mathcal{G}$ with $H <_{\mu'} G$. So there is $\Gamma \in \Psi$ with $f(\Gamma \cap X) \subseteq H$. Take any $\mathcal{F} \in \operatorname{int}\Gamma - X$, then there is an open U in X with $\mathcal{F} \in \operatorname{op} U \subseteq \Gamma$; hence $U \in \mathcal{F}$. Since $f(U) = f(\operatorname{op} U \cap X) \subseteq f(\Gamma \cap X) \subseteq H$, $H \in f(\mathcal{F})$. Since $H <_{\mu'} G$, $\{\operatorname{int}(Y - H), \operatorname{int} G\} \in \mu'$. Suppose that $\bar{f}(\mathcal{F}) \in \operatorname{int}(Y - H)$, then $Y - H \in f(\mathcal{F})$, for $f(\mathcal{F})$ converges to $\bar{f}(\mathcal{F})$. Hence $H \cap (Y - H) = \phi \in f(\mathcal{F})$, which is a contradiction. Thus $\bar{f}(\mathcal{F}) \in \operatorname{int} G \subseteq G$. Since $\bar{f}(\Gamma \cap X) = f(\Gamma \cap X) \subseteq G$, $\bar{f}(\operatorname{int}\Gamma) \subseteq G$ and hence $G \in \bar{f}(\Psi)$, for $\operatorname{int}\Gamma \in \Psi$. So $\mathcal{G} \subseteq \bar{f}(\Psi)$ and therefore $\bar{f}$ is a Cauchy map. It remains to show the uniqueness of $\bar{f}$. Suppose that g: $cX \to Y$ is a Cauchy map with $g \circ c = f$. Take any $\mathcal{F} \in cX - X$, then $\operatorname{op}\mathcal{F}$ is a regular Cauchy filter on cX. For any $F \in \mathcal{F}$, $\mathcal{F} \in \operatorname{op} F$; hence $g(\mathcal{F})$ is a cluster point of $g(\operatorname{op}\mathcal{F})$. Since g is a Cauchy map and Y is Cauchy complete, $g(\operatorname{op}\mathcal{F})$ is convergent, say to y. Since $Y \in$ **Near₂**, $y = g(\mathcal{F})$. Thus $g(\operatorname{op}\mathcal{F})$ converges to $g(\mathcal{F})$; hence $g(c(\mathcal{F}))$ converges to $g(\mathcal{F})$ because $\operatorname{op}\mathcal{F} \subseteq c(\mathcal{F})$. Since $g \circ c = f$, $f(\mathcal{F})$ converges to $g(\mathcal{F})$, which implies $g(\mathcal{F}) = \bar{f}(\mathcal{F})$, for Y is an N_2-space. For any $x \in X$, $g(x) = g(c(x)) = f(x) = \bar{f}(x)$. In all, $g = \bar{f}$. This completes the proof. □

Acknowledgements

The authors are indebted to Prof. H. Herrlich for his valuable suggestions and examples. We also acknowledge the financial support by the Basic Research Institute Program, MOE, Korea, Project Number BSRI-94-1422 and TGRC-Kosef (TGRC Post Doc. Program) in 1996 for the second author.

References

1. Adámek, J., Herrlich, H., and Strecker, G. E.: *Abstract and Concrete Categories*, Wiley, New York, 1990.
2. Banaschewski, B.: *Completion in Point-Free Topology*, Lecture Notes in Math. and Appl. Math. 2/96, University of Cape Town, 1996.
3. Banaschewski, B., Hong, S. S., and Pultr, A.: The completions of nearness frame, *Quaestiones Math.* **21** (1998), 19–37.
4. Banaschewski, B., and Pultr, A.: Cauchy points of uniform and nearness frames, *Quaestiones Math.* **19** (1996), 101–127.
5. Bentley, H. L.: Nearness spaces and extensions of topological spaces, in N. M. Stavrakas and K. R. Allen (eds.), *Studies in Topology*, Academic Press, New York, 1975, pp. 47–66.
6. Bentley, H. L. and Herrlich, H.: Extensions of topological spaces, in *Topology*, Lecture Notes in Pure and Appl. Math. 24, Marcel Dekker, New York, 1976, pp. 129–184.
7. Herrlich, H.: A concept of nearness, *Gen. Topology Appl.* **4** (1974), 191–212.
8. Herrlich, H.: Topological structures I, *Math. Centre Tracts* **52** (1974), 59–122.
9. Herrlich, H.: *Topologie II: Uniforme Räume*, Heldermann Verlag, Berlin, 1988.
10. Hong, S. S. and Kim, Y. K.: Cauchy completions of nearness frames, *Appl. Categ. Struct.* **3** (1995), 371–377.

Applied Categorical Structures **8:** 279–305, 2000.
G. Brümmer & C. Gilmour (eds), Papers in Honour of Bernhard Banaschewski.

Categories of Relations and Functional Relations

Dedicated to Bernhard Banaschewski on the occasion of his 70th birthday

ROMAINE JAYEWARDENE
University of Colombo, Colombo, Sri Lanka

OSWALD WYLER
Carnegie Mellon University, Pittsburgh, PA, USA

(Received: 11 October 1996; accepted: 28 May 1998)

Abstract. We define relations and their composition in a category with $(\mathcal{E}, \mathcal{M})$-factorization structure, with $\mathcal{M}$ consisting of monomorphisms, but $\mathcal{E}$ not restricted to epimorphisms. We obtain an associativity criterion for composition of relations, and we study functional and induced relations. We show that under our assumptions, the categories of relations on functional and induced relations are isomorphic to the category of relations for the given category.

Mathematics Subject Classifications (2000): 18B10, 18A32.

Key words: subobject, relation, functional relation, map, induced relation.

Introduction

This paper presents the main results of the first author's thesis [10], with some related new results.

A relation $\rho : A \to B$ in a category is usually defined as a subobject of a product $A \times B$. Relations in this sense were introduced by S. Mac Lane [14] and D. Puppe [20] for Abelian categories, and by M. Barr [2] and P. A. Grillet [3] for regular categories. A. Klein discussed in [13] the general case of a category with finite limits, with an $(\mathcal{E}, \mathcal{M})$-factorization structure in the sense of [6] and [1], with $\mathcal{E}$ consisting of epimorphisms and $\mathcal{M}$ consisting of monomorphisms. Klein showed that in this generality, composition of relations is associative iff $\mathcal{E}$ is stable under pullbacks. G. M. Kelly [11] and others also discussed relations and functional relations in this generality. J. Meisen [15] and D. Pavlović [17, 18] discussed categories of relations for arbitrary $(\mathcal{E}, \mathcal{M})$-factorizations, retaining Klein's associativity condition.

The present paper considers primarily relations and functional relations for a category **C** with finite limits and an $(\mathcal{E}, \mathcal{M})$-factorization with $\mathcal{M}$ consisting of monomorphisms but $\mathcal{E}$ not restricted to epimorphisms, with a condition for associativity of composition which is weaker, and in an important example strictly weaker, than Klein's condition. We study the categories of functional and induced relations, and we show that these categories satisfy our assumptions and have essentially the same relations as **C**.

Subobjects and relations in a category can be used to obtain an internal logic. For topological categories over sets, with subspaces of a space X as subobjects, given by subsets of the underlying set of X, topological structures are completely lost in this process, and the resulting internal logic is classical two-valued logic. This is particularly embarrassing for the categories of fuzzy sets and 'crisp' maps considered by various authors. These categories are topological over sets, and with subspaces as subobjects, their internal logic is classical two-valued logic, with no fuzzyness left.

L. Stout [21] showed a way out of this dilemma, and it was shown in [23] that Stout's theory postulates an $(\mathcal{E}, \mathcal{M})$-factorization structure with $\mathcal{E}$ not consisting of epimorphisms. The results of [23] are generalized in [10] and this paper, with strictly categorical proofs.

We define subobjects in 1.2 as special morphisms in $\mathcal{M}$; this avoids using large equivalence classes of morphisms of **C** as relations, and several related complications. We review relations in **C** and their composition in Sections 1 and 2, with no conditions on the $(\mathcal{E}, \mathcal{M})$-factorization structure. From Section 3 on, we assume that $\mathcal{M}$ consists of monomorphisms, and that $\mathcal{E}$ is stable for pullbacks by legs r and s of relations (r, s). Composition of relations is associative in this setting. We define functional and induced relations and obtain their basic properties in Sections 3 through 6. We prove that with our assumptions, the categories of functional and induced relations have finite limits, and a (surjective, injective) factorization structure, and that relations for this structure are essentially the same as relations for **C**. Section 7 discusses some useful examples.

The most important examples in Section 7 are categories of fuzzy sets, with an appropriate lattice of truth values, and "crisp" maps. Thus our paper provides a contribution to a satisfactory mathematical theory of fuzzy sets. In the theory of fuzzy sets, there is strong evidence that functional relations rather than 'crisp' maps are the appropriate morphisms; see e.g. [19] and [9], and from a categorical viewpoint [8] and [16]. These functional relations are obtained from factorization structures defined in Section 7 of this paper.

NOTATIONS. Superfluous parentheses are sometimes omitted; we may write Fx instead of $F(x)$, especially if F is a functor. For a product $A \xleftarrow{p} A \times B \xrightarrow{q} B$ and morphisms $f : X \to A$ and $g : X \to B$, we denote by $\langle f, g \rangle : X \to A \times B$ the unique morphism $h : X \to A \times B$ such that $ph = f$ and $qh = g$.

We thank the referee for valuable suggestions.

1. Subobjects, Spans and Relations

1.1. ASSUMPTIONS

Throughout this paper, we work with a category **C** with finite limits, and with an $(\mathcal{E}, \mathcal{M})$-factorization structure, in the sense of [6] and [1]. This means that $\mathcal{E}$ and $\mathcal{M}$ are classes of morphisms of **C** with the following properties.

(i) Every morphism f of **C** factors $f = me$ with $e \in \mathcal{E}$ and $m \in \mathcal{M}$.
(ii) Every commutative square $mf = ge$ in **C** with $e \in \mathcal{E}$ and $m \in \mathcal{M}$ has a unique diagonal d with $f = de$ and $g = md$ in **C**.
(iii) Compositions ue in **C** with $e \in \mathcal{E}$ and u isomorphic are in $\mathcal{E}$, and compositions mv with $m \in \mathcal{M}$ and v isomorphic are in $\mathcal{M}$.

It follows (see [1, Section 14]) that $\mathcal{E}$ and $\mathcal{M}$ are closed under compositions, $\mathcal{E} \cap \mathcal{M}$ consists of all isomorphisms of **C**, and $\mathcal{M}$ is closed under pullbacks and products.

We usually assume that $\mathcal{M}$ consists of monomorphisms of **C**, and that $\mathcal{E}$ is leg-stable in the sense of 1.8 below, but we do not assume that $\mathcal{E}$ consists of epimorphisms. We note without proof the following variation on [1, 14.11].

PROPOSITION 1.1.1. *The following are equivalent.*

(i) *All strong monomorphisms of* **C** *are in* $\mathcal{M}$.
(ii) *All equalizers in* **C** *are in* $\mathcal{M}$.
(iii) *For every pullback square* $fu = gv$ *in* **C**, *the morphism* $\langle u, v\rangle$ *of* **C** *is in* $\mathcal{M}$.
(iv) *All morphisms* $\langle \mathrm{id}, f\rangle$ *of* **C** *are in* $\mathcal{M}$.
(v) *All morphisms* $\delta_A = \langle \mathrm{id}_A, \mathrm{id}_A\rangle$ *of* **C** *are in* $\mathcal{M}$.
(vi) $\mathcal{E}$ *consists of epimorphisms of* **C**.

1.2. SUBOBJECTS

Subobjects of an object A of **C** are usually defined as equivalence classes of morphisms in $\mathcal{M}$ with codomain A. Subobjects in this sense are often proper classes. We avoid this and related nuisances by postulating the existence of a subclass $\mathcal{M}_0$ of $\mathcal{M}$ with the following property.

(i) Every morphism f of **C** has a unique factorization $f = me$ with e in $\mathcal{E}$ and m in $\mathcal{M}_0$.

Morphisms in $\mathcal{M}_0$ with codomain A in **C** will then be called *subobjects* of A. We note that a morphism f of **C** is in $\mathcal{M}$ iff e in the $(\mathcal{E}, \mathcal{M}_0)$-factorization $f = me$ is an isomorphism.

Examples of classes $\mathcal{M}_0$ are subset insertions for sets and subspace inclusions for topological constructs. In these examples, $\mathcal{M}_0$ is closed under composition and contains all identity morphisms of **C**, but we shall not need these properties. Other examples are given in [23] and [10]; some of these are discussed in Section 7 of this paper.

1.3. SPANS

A *span* $(u, v) : A \to B$ in $\mathbf{C}$ is a pair of morphisms u, v of $\mathbf{C}$ with codomains A for u and B for v, and with the same domain. Spans $(u, v) : A \to B$ are the objects of a category $\mathrm{Span}(A, B)$, with $f : (u, v) \to (u', v')$ in $\mathrm{Span}(A, B)$ if $u = u'f$ and $v = v'f$ in $\mathbf{C}$. Composition in $\mathrm{Span}(A, B)$ is composition of the underlying morphisms of $\mathbf{C}$. We say that spans α and α' are *span-equivalent*, and write $\alpha \simeq \alpha'$, if α and α' are isomorphic in a category $\mathrm{Span}(A, B)$.

Categories $\mathrm{Span}(A, B)$ have finite limits. Pullbacks in $\mathrm{Span}(A, B)$ are lifted from $\mathbf{C}$, and projections $A \xleftarrow{p} A \times B \xrightarrow{q} B$ of a product $A \times B$ define a terminal span (p, q), with

$$\langle r, s\rangle : (r, s) \to (p, q)$$

for a span $(r, s) : A \to B$. Products in $\mathrm{Span}(A, B)$ are called *intersections*, with

$$(r, s) \cap (u, v) \simeq (rh, sh) = (uk, vk)$$

for a pullback square

$$\begin{array}{ccc} \cdot & \xrightarrow{k} & \cdot \\ \downarrow h & & \downarrow \langle u, v\rangle \\ \cdot & \xrightarrow{\langle r, s\rangle} & A \times B \end{array} .$$

1.4. RELATIONS

For objects A, B of $\mathbf{C}$, we define a *relation* $\rho : A \to B$ as a span $\rho = (r, s) : A \to B$ with $\langle r, s\rangle$ in $\mathcal{M}$, and we denote by $\mathrm{Rel}(A, B)$ the full subcategory of $\mathrm{Span}(A, B)$ with relations as its objects. It is often convenient to restrict relations to subobjects in the sense of 1.2, and we denote by $\mathrm{Rel}_0(A, B)$ the full subcategory of $\mathrm{Rel}(A, B)$ with objects (r, s) such that $\langle r, s\rangle \in \mathcal{M}_0$. This is a skeleton of $\mathrm{Rel}(A, B)$. If $\mathcal{M}$ consists of monomorphism, then $\mathrm{Rel}_0(A, B)$ is a meet semilattice, possibly large.

THEOREM AND DEFINITION. *For objects A and B of* $\mathbf{C}$, $\mathrm{Rel}(A, B)$ *and* $\mathrm{Rel}_0(A, B)$ *are $\mathcal{E}$-reflective full subcategories of* $\mathrm{Span}(A, B)$, *with reflections*

$$e : (a, b) \to \mathbf{r}(a, b) \tag{1}$$

given by $\mathbf{r}(a, b) = (r, s)$ *for the $(\mathcal{E}, \mathcal{M}_0)$-factorization $\langle a, b\rangle = \langle r, s\rangle \cdot e$. Thus* $\mathrm{Rel}(A, B)$ *is closed under limits in* $\mathrm{Span}(A, B)$.

This follows immediately from the definitions. We shall always use the notation $\mathbf{r}(a, b)$ for this reflection, and we note that e in (1) is an isomorphism iff (a, b) is a relation, and an identity morphism for $\langle a, b\rangle$ in $\mathcal{M}_0$.

1.5. DUAL SPANS AND RELATIONS

For a span $(r, s) : A \to B$, we put

$$(r, s)^{\mathrm{op}} = (s, r) : B \to A$$

and call $(r, s)^{\mathrm{op}}$ the *dual span* of (r, s). Then clearly

$$((r, s)^{\mathrm{op}})^{\mathrm{op}} = (r, s),$$

and the dual span of a relation is a relation.

Duality defines isomorphisms between the categories $\mathrm{Span}(A, B)$ and $\mathrm{Span}(B, A)$, and also between $\mathrm{Rel}(A, B)$ and $\mathrm{Rel}(B, A)$. We also note without proof that

$$(\mathbf{r}(a, b))^{\mathrm{op}} \simeq \mathbf{r}((a, b)^{\mathrm{op}})$$

for every span (a, b). If $(r, s) \in \mathrm{Rel}_0(A, B)$, then (s, r) need not be in $\mathrm{Rel}_0(B, A)$, but $(r, s) \mapsto \mathbf{r}(s, r)$ provides an isomorphism $\mathrm{Rel}_0(A, B) \to \mathrm{Rel}_0(B, A)$ with all properties of duality.

1.6. IMAGES AND INVERSE IMAGES OF RELATIONS AND SUBOBJECTS

For $f : A \to B$ and a relation $(r, s) : A \to C$, we put

$$f^{\to}(r, s) = \mathbf{r}(fr, s).$$

This clearly defines an *image functor*

$$f^{\to} := \mathrm{Rel}(A, C) \to \mathrm{Rel}(B, C)$$

for relations, with all values in $\mathrm{Rel}_0(B, C)$.

Now consider commutative squares

$$\begin{array}{ccc} \cdot & \xrightarrow{f'} & \cdot \\ \downarrow{\scriptstyle u'} & & \downarrow{\scriptstyle u} \\ A & \xrightarrow{f} & B \end{array} \quad \text{and} \quad \begin{array}{ccc} \cdot & \xrightarrow{f'} & \cdot \\ \downarrow{\scriptstyle \langle u', vf'\rangle} & & \downarrow{\scriptstyle \langle u, v\rangle} \\ A \times C & \xrightarrow{f \times \mathrm{id}_C} & B \times C \end{array} \qquad (1)$$

in $\mathbf{C}$. Then the lefthand square is a pullback iff the righthand square is one. As $\mathcal{M}$ is pullback-stable, putting

$$f^{\leftarrow}(u, v) = (u', vf')$$

defines an *inverse image relation* $f^{\leftarrow}(u, v)$ up to equivalence. We make the inverse image unique by choosing, as we may, pullbacks (1) with $\langle u', vf'\rangle$ in $\mathcal{M}_0$. Now we have a functor

$$f^{\leftarrow} : \mathrm{Rel}(B, C) \to \mathrm{Rel}(A, C)$$

for relations, with all values in $\mathrm{Rel}_0(A, C)$.

For a terminal object $\mathbf{1}$ of $\mathbf{C}$, a span $(f, t) : A \to \mathbf{1}$ is a relation iff $f \in \mathcal{M}$, since the projection $A \times \mathbf{1} \to A$ is an isomorphism. This defines isomorphisms between $\mathrm{Rel}(A, \mathbf{1})$ and the category $\mathcal{M}/A$ of morphisms in $\mathcal{M}$ with codomain A, and hence between $\mathrm{Rel}_0(A, \mathbf{1})$ and the category $\mathcal{M}_0/A$ of subobjects of A; we can use this to define direct and inverse images of subobjects.

If gf is defined in $\mathbf{C}$, then it is easily seen that

$$(gf)^{\rightarrow} = g^{\rightarrow} \circ f^{\rightarrow} \quad \text{and} \quad (gf)^{\leftarrow} = f^{\leftarrow} \circ g^{\leftarrow}. \tag{2}$$

We also note the following result.

PROPOSITION. *$f^{\rightarrow} \dashv f^{\leftarrow}$ for f in* $\mathbf{C}$.

Proof. From (1), we have a natural bijection of span morphisms

$$h : (fr, s) \to (u, v) \quad \text{and} \quad k : (r, s) \to f^{\leftarrow}(u, v),$$

given by $h = f'k$, and from Section 1.4 a natural bijection of morphisms

$$h : (fr, s) \to (u, v) \quad \text{and} \quad j : f^{\rightarrow}(r, s) \to (u, v).$$ □

1.7. LEGS

For a relation (r, s), we call r and s the *legs* of (r, s). Legs are compositions pm of a projection p of a product and a morphism m in $\mathcal{M}$. We note some basic properties of legs.

PROPOSITION. *Projections of products, and morphisms in $\mathcal{M}$, are legs. Legs form a pullback-stable subcategory of* $\mathbf{C}$, *and a product $r \times u$ of legs r and u is again a leg. If $\mathcal{E}$ consists of epimorphisms, then every morphism of* $\mathbf{C}$ *is a leg.*

Proof. The projections of $A \times B$ are the legs of a relation $(p, q) : A \to B$. A span $(m, t) : A \to \mathbf{1}$, for a terminal object $\mathbf{1}$ of $\mathbf{C}$, is a relation iff $m \in \mathcal{M}$, since the projection $A \times \mathbf{1} \to A$ is an isomorphism.

The construction of $f^{\leftarrow}(u, v)$ in 1.6(1) shows that every pullback u' of a leg u is again a leg.

If $(r, s) : A \to B$ and $(u, v) : X \to C$ are relations with ru defined in $\mathbf{C}$, then we have a morphism

$$(\langle r, s\rangle \times \mathrm{id}_C) \cdot \langle u, v\rangle = \langle ru, su, v\rangle : \blacksquare \to A \times B \times C$$

in $\mathcal{M}$, and hence a relation $(ru, \langle su, v\rangle) : A \to B \times C$. Thus ru is a leg.

If (r, s) and (u, v) are relations, then $(r \times u, s \times v)$ is a relation, since

$$\langle r \times u, s \times v\rangle = \pi \cdot (\langle r, s\rangle \times \langle u, v\rangle)$$

for a middle interchange isomorphism π in $\mathbf{C}$, with both sides in $\mathcal{M}$.

Finally, if $\mathcal{E}$ consists of epimorphisms, then every span (id, f) is a relation by Proposition 1.1, and thus every morphism of $\mathbf{C}$ is a leg. □

1.8. LEG STABILITY

We say that $\mathcal{E}$ is *stable* if $\mathcal{E}$ is closed under pullbacks, and *leg-stable* (*leg-regular* in [10]) if $\mathcal{E}$ is stable under pullbacks by legs of relations. Since legs are compositions pm with $m \in \mathcal{M}$ and p a projection of a product, $\mathcal{E}$ is leg-stable if and only if the following two conditions are satisfied.

(i) Every product $e_1 \times e_2$ of morphisms in $\mathcal{E}$ is in $\mathcal{E}$.
(ii) $\mathcal{E}$ is closed under pullbacks by morphisms in $\mathcal{M}$.

EXAMPLE 1.8.1. For the category of H-sets and crisp maps discussed in 7.3, the class $\mathcal{E}$ of covers is leg-stable, but not stable if H is not a singleton. Let A, B, C be the sets $\{0\}$, $\{1\}$, $\{0, 1\}$ with the coarse structure, and let $f : A \to C$ and $g : B \to C$ be the subset insertions. Then f and g are covers, and the pullback of f and g is empty, but the maps $\emptyset \to A$ and $\emptyset \to B$ are not covers.

2. Composition of Spans and Relations

2.1. SPAN COMPOSITION

We compose spans $(r, s) : A \to B$ and $(u, v) : B \to C$ in $\mathbf{C}$ by putting

$$(u, v)\Diamond(r, s) = (ru', vs') : A \to C$$

for a pullback square

$$\begin{array}{ccc} \bullet & \xrightarrow{s'} & \bullet \\ \big\downarrow{\scriptstyle u'} & & \big\downarrow{\scriptstyle u} \\ \bullet & \xrightarrow{s} & B \end{array}$$

in $\mathbf{C}$. This composition is defined up to span-equivalence.

Composition of spans is clearly functorial and associative, up to equivalence of spans. For an object A of $\mathbf{C}$, we put

$$\delta_A = (\mathrm{id}_A, \mathrm{id}_A).$$

Spans δ_A are identity spans, again up to span-equivalence. Thus the categories Span(A, B) form a bicategory in the sense of [4].

It is easily seen that

$$\alpha^{\mathrm{op}} \Diamond \beta^{\mathrm{op}} \simeq (\beta \Diamond \alpha)^{\mathrm{op}},$$

for spans α, β, if either composition is defined.

2.2. COMPOSITION OF RELATIONS

The *composition* $\sigma \circ \rho : A \to C$ of relations $\rho : A \to B$ and $\sigma : B \to C$ in $\mathbf{C}$ is defined by

$$\sigma \circ \rho = \mathbf{r}(\sigma \Diamond \rho).$$

This clearly defines a functor from Rel$(B, C) \times$ Rel(A, B) to Rel(A, C), with all values in Rel$_0(A, C)$. Thus composition of relations is order-preserving if $\mathcal{M}$ consists of monomorphisms. We note that

$$\sigma \circ (r, s) = r^{\to} s^{\leftarrow} \sigma \tag{1}$$

for relations $(r, s) : A \to B$ and $\sigma : B \to C$ in $\mathbf{C}$, and that always

$$(\sigma \circ \rho)^{\mathrm{op}} \simeq \rho^{\mathrm{op}} \circ \sigma^{\mathrm{op}},$$

by 1.5, 2.1 and the definitions, if either composition is defined. We also note that $\sigma' \circ \rho' = \sigma \circ \rho$ if $\rho' \simeq \rho$ and $\sigma' \simeq \sigma$, and the compositions are defined.

THEOREM 2.3. *For a category* $\mathbf{C}$ *with* $(\mathcal{E}, \mathcal{M})$*-factorization structure and finite limits, the following are equivalent.*

(i) $\sigma \circ \mathbf{r}\alpha = \mathbf{r}(\sigma \Diamond \alpha)$ *for all spans* α *and relations* σ *for which the compositions are defined.*

(ii) $\mathbf{r}\beta \circ \rho = \mathbf{r}(\beta \Diamond \rho)$ *for all relations* ρ *and spans* β *for which the compositions are defined.*

(iii) $\sigma \circ \mathbf{r}(a, b) = a^{\to} b^{\leftarrow} \sigma$ *for all spans* (a, b) *and relations* σ *such that the composition is defined.*

(iv) $e^{\to} e^{\leftarrow} \rho \simeq \rho$ *for all relations* ρ *in* $\mathbf{C}$ *and morphisms* e *in* $\mathcal{E}$ *with* $e^{\leftarrow}\rho$ *defined.*

If these conditions are satisfied, then composition of relations in $\mathbf{C}$ *is associative, with identity relations* $\mathbf{r}\delta_A$.

Proof. (ii) is clearly the same as (i) for the dual relations; thus (i) and (ii) are equivalent,

If $b^{\leftarrow}\sigma = (u, v)$, then

$$\sigma \Diamond (a, b) \simeq (au, v) \quad \text{and} \quad \mathbf{r}(\sigma \Diamond (a, b)) = a^{\to} b^{\leftarrow} \sigma;$$

thus (i) and (iii) are equivalent.

It follows from (i) and (ii) that

$$\rho \circ \mathbf{r}\delta_A \simeq \rho \simeq \mathbf{r}\delta_B \circ \rho \tag{2}$$

for a relation $\rho : A \to B$, with equality iff $\rho \in \mathrm{Rel}_0(A, B)$. Since $\mathbf{r}(e, e) = \mathbf{r}\delta_A$ for e in $\mathcal{E}$ with codomain A, (iv) follows from this and (iii). Conversely, if $\langle a, b\rangle = \langle r, s\rangle \cdot e$ is an $(\mathcal{E}, \mathcal{M})$-factorization of $\langle a, b\rangle$, then

$$a^{\to} b^{\leftarrow}\sigma = r^{\to} e^{\to} e^{\leftarrow} s^{\leftarrow} \sigma = r^{\to} s^{\leftarrow} \sigma = \sigma \circ (r, s)$$

if $\sigma \circ (r, s)$ is defined and (iv) is valid; thus (iv) $\Rightarrow$ (iii).

If (i) and (ii) are valid, then

$$\begin{aligned} \tau \circ (\sigma \circ \rho) &= \tau \circ \mathbf{r}(\sigma \Diamond \rho) = \mathbf{r}(\tau \Diamond (\sigma \Diamond \rho)) \\ &= \mathbf{r}((\tau \Diamond \sigma) \Diamond \rho) = \mathbf{r}(\tau \Diamond \sigma) \circ \rho = (\tau \circ \sigma) \circ \rho \end{aligned}$$

for relations ρ, σ, τ with $\sigma \circ \rho$ and $\tau \circ \sigma$ defined, and composition of relations is associative. □

2.4.

Leg-stability has been defined in 1.8; we note the following result.

PROPOSITION. *If $\mathcal{E}$ is leg-stable, then*

$$\mathbf{r}((b', c) \Diamond (a, b)) = \mathbf{r}(b', c) \circ \mathbf{r}(a, b) \tag{1}$$

for spans (a, b) and (b', c) with the compositions defined and with b a leg or b' a leg, and (i)–(iv) *of Theorem* 2.3(i)–(iv) *are valid for* $\mathbf{C}$. *Conversely, if* (i)–(iv) *in* 2.3 *are valid for* $\mathbf{C}$ *and* $\mathcal{M}$ *consists of monomorphisms, then* $\mathcal{E}$ *is leg-stable.*

Proof. (i) and (ii) of Theorem 2.3 clearly are special cases of (1).

For (1), let

$$\langle a, b\rangle = \langle r, s\rangle \cdot e_1 \quad \text{and} \quad \langle b', c\rangle = \langle u, v\rangle \cdot e_2$$

be $(\mathcal{E}, \mathcal{M})$-factorizations, and construct a diagram

$$\begin{array}{ccccc} \cdot & \xrightarrow{e_1''} & \cdot & \xrightarrow{s''} & \cdot \\ \big\downarrow{\scriptstyle e_2''} & & \big\downarrow{\scriptstyle e_2'} & & \big\downarrow{\scriptstyle e_2} \\ \cdot & \xrightarrow{e_1'} & \cdot & \xrightarrow{s'} & \cdot \\ \big\downarrow{\scriptstyle u''} & & \big\downarrow{\scriptstyle u'} & & \big\downarrow{\scriptstyle u} \\ \cdot & \xrightarrow{e_1} & \cdot & \xrightarrow{s} & \cdot \end{array} \tag{2}$$

of pullback squares. Then

$$(b', c)\Diamond(a, b) \simeq (re_1u''e_2'', ve_2s''e_1'') = (ru'e_1'e_2'', vs'e_1'e_2'').$$

Now s' and u' are legs, and e_1' and e_2' are in $\mathcal{E}$ if $\mathcal{E}$ is leg-stable. If $b = se_1$ is a leg, then so is $s'e_1'$, and thus e_2'' is in $\mathcal{E}$ if $\mathcal{E}$ is leg-stable. But then

$$\mathbf{r}((b', c)\Diamond(a, b)) = \mathbf{r}(ru', vs') = (u, v) \circ (r, s),$$

so that (1) is valid. The proof of (1) with b' a leg is exactly analogous.

For a relation (r, s) and a pullback square

$$\begin{array}{ccc} \cdot & \xrightarrow{e'} & \cdot \\ \downarrow{\scriptstyle r'} & & \downarrow{\scriptstyle r} \\ \cdot & \xrightarrow{e} & \cdot \end{array}$$

in $\mathbf{C}$, we have

$$e^{\rightarrow}e^{\leftarrow}(r, s) = \mathbf{r}(er', se') = \mathbf{r}(re', se').$$

If 2.3(iv) is valid, and the pullback shown above exists with e in $\mathcal{E}$, then we have an $(\mathcal{E}, \mathcal{M})$-factorization

$$\langle re', se' \rangle = \langle r, s \rangle \cdot e_1.$$

If $\mathcal{M}$ consists of monomorphisms, then $e' = e_1$; thus $\mathcal{E}$ is leg-stable. □

2.5. DISCUSSION AND PROBLEMS

If $\mathcal{E}$ is leg-stable, then the system with the objects of $\mathbf{C}$ as objects, $\mathrm{Rel}(A, B)$ as the class of morphisms $\rho : A \twoheadrightarrow B$ for objects A and B, and with compositions defined by 2.2 is almost, but not quite, a category, because we have span-equivalence instead of equality in 2.2(1). We obtain a *bona fide* category, which we denote by $\mathrm{Rel}\,\mathbf{C}$, if we restrict morphisms $\rho : A \twoheadrightarrow B$ to $\mathrm{Rel}_0(A, B)$. Every relation ρ in $\mathbf{C}$ is then span-equivalent to a unique morphism $\mathbf{r}\rho$ of $\mathrm{Rel}\,\mathbf{C}$.

The classes $\mathrm{Rel}_0(A, B)$ are categories with finite limits, and meet semi-lattices if $\mathcal{M}$ consists of monomorphisms, and composition of relations is functorial. Thus $\mathrm{Rel}\,\mathbf{C}$ is a bicategory, and in fact a 2-category in the sense of [4], if $\mathcal{E}$ is leg-stable. We note that the inclusion $\mathrm{Rel}\,\mathbf{C} \to \mathrm{Span}\,\mathbf{C}$ is a morphism of bicategories if $\mathcal{E}$ is leg-stable, and that the reflections 1.4(1) define a homomorphism of bicategories if $\mathcal{E}$ is stable.

The results of this Section are due to A. Klein [13] for the case that $\mathcal{E}$ consists of epimorphisms of $\mathbf{C}$ and $\mathcal{M}$ of monomorphisms. In this situation, all morphisms of $\mathbf{C}$ are legs, and composition of relations in $\mathbf{C}$ is associative, up to span-equivalence, iff $\mathcal{E}$ is stable for all pullbacks in $\mathbf{C}$.

If $\mathcal{E}$ does not consist of epimorphisms, then $\mathcal{E}$ can be stable for pullbacks by legs, but not for all pullbacks in **C**; see 1.8.1 for an important example. This leaves us with two problems.

Is it possible for composition of relations to be associative, with identity relations $\mathbf{r}\delta_A$, but with 2.3(i)–(iv) not valid? Can 2.3(i)–(iv) be valid if $\mathcal{E}$ is not leg-stable?

See 2.3 and Proposition 3.1 for related results.

3. Induced and Functional Relations

3.1. INDUCED RELATIONS

We assume from 3.3 on that $\mathcal{M}$ consists of monomorphisms, and that $\mathcal{E}$ is leg-stable.

We assign to every morphism $f : A \to B$ of **C** an *induced relation*

$$[f] = \mathbf{r}(\mathrm{id}_A, f) : A \rightarrow B.$$

Note that $[f]$ is always in $\mathrm{Rel}_0(A, B)$ for $f : A \to B$ in **C**, and that $[\mathrm{id}_A]$ is an identity relation in Rel **C** for every object A of **C**.

Relations $[f]$ were denoted by $\langle f \rangle$ in [23] and [10]; we have changed the notation for better distinction from morphisms $\langle f, g \rangle$.

PROPOSITION 3.2. *If* 2.3(i) *is valid, then*

(1) $f^{\to}\rho = \rho \circ [f]^{\mathrm{op}}$ *and* $g^{\leftarrow}\sigma = \sigma \circ [g]$,
(2) $[f] \circ \rho \simeq (f^{\to}\rho^{\mathrm{op}})^{\mathrm{op}}$ *and* $[f]^{\mathrm{op}} \circ \sigma \simeq (f^{\leftarrow}\sigma^{\mathrm{op}})^{\mathrm{op}}$,
(3) $[gf] = [g] \circ [f]$,
(4) $\mathbf{r}(f, g) = [g] \circ [f]^{\mathrm{op}}$,

for morphisms f, g and relations ρ, σ whenever the operations are defined.

Conversely, 2.3(i) *is valid if* (1) *holds and composition of relations is associative.*

Proof. (1), (2) and (4) are special cases of 2.3(i) and 2.3(ii), and (3) follows easily from (1) and 1.6(2). For the converse, we have

$$\begin{aligned} \sigma \circ \mathbf{r}(f, g) &= \sigma \circ f^{\to}[g] = \sigma \circ ([g] \circ [f]^{\mathrm{op}}) \\ &= (\sigma \circ [g]) \circ [f]^{\mathrm{op}} = f^{\to}g^{\leftarrow}\sigma = \mathbf{r}(\sigma \lozenge (f, g)), \end{aligned}$$

with the first and last equalities following easily from the definitions. □

3.3. SPECIAL RELATIONS

With terms borrowed from set theory, we say that a relation $\rho : A \rightarrow B$ is

single-valued if $\rho \circ \rho^{\mathrm{op}} \leqslant [\mathrm{id}_B]$,

total if $\rho^{\mathrm{op}} \circ \rho \geqslant [\mathrm{id}_A]$,
injective if $\rho^{\mathrm{op}} \circ \rho \leqslant [\mathrm{id}_A]$,
surjective if $\rho \circ \rho^{\mathrm{op}} \geqslant [\mathrm{id}_B]$.

We also say that ρ is *functional* if ρ is single-valued and total, and *bijective* if ρ is injective and surjective. We shall see in Proposition 3.7 that functional relations are the same as the maps of F. W. Lawvere [12] and G. M. Kelly [11].

Clearly ρ is single-valued if and only if ρ^{op} is injective, and ρ is total if and only if ρ^{op} is surjective.

An identity relation $[\mathrm{id}_A]$, span-equivalent to its dual, satisfies all four conditions. It is also easy to verify that each of the four classes of relations defined above is closed under span-equivalence, and under composition of relations.

THEOREM 3.4. *Every relation* $[f]$ *induced by a morphism* f *of* $\mathbf{C}$ *is functional.*

Proof. By 3.2(1) and 1.6, we have

$$\sigma \circ [f] \circ [f]^{\mathrm{op}} = f^{\rightarrow} f^{\leftarrow} \sigma \leqslant \sigma$$

for every relation σ with $f^{\leftarrow}\sigma$ defined; thus $[f]$ is single-valued. Totality of $[f]$ is obtained similarly. □

3.5.

Our next result shows that the pre-order of relations becomes span-equivalence when restricted to functional relations.

PROPOSITION. *If* $\rho \leqslant \sigma$ *for a total relation* ρ *and a single-valued relation* σ, *then* $\rho \simeq \sigma$.

Proof. We have

$$\sigma \leqslant \sigma \circ \rho^{\mathrm{op}} \circ \rho \leqslant \sigma \circ \sigma^{\mathrm{op}} \circ \rho \leqslant \rho$$

since $[\mathrm{id}] \leqslant \rho^{\mathrm{op}} \circ \rho$, $\rho^{\mathrm{op}} \leqslant \sigma^{\mathrm{op}}$, and $\sigma \circ \sigma^{\mathrm{op}} \leqslant [\mathrm{id}]$. □

PROPOSITION 3.6. *For a relation* $\rho = (r, s) : A \to B$, *the following are equivalent.*

(i) ρ *is total.*
(ii) $\sigma \circ \rho \geqslant [\mathrm{id}_A]$ *for some relation* $\sigma : B \to A$.
(iii) $r \in \mathcal{E}$.
(iv) $[r]$ *is surjective.*

Proof. (i) $\Rightarrow$ (ii) trivially; use $\sigma = \rho^{\mathrm{op}}$.

If $(u, v) \circ (r, s) = (a, a')$ and $(a, a') \geqslant [\mathrm{id}_A]$, let $su' = us'$ be a pullback square. Then ru' factors $ru' = ae$ with $e \in \mathcal{E}$, and there is t in $\mathbf{C}$ with $at = \mathrm{id}_A$. If we $(\mathcal{E}, \mathcal{M})$-factor $r = me'$, then $md = a$ and $de = e'u'$ for a morphism d, and

$mdt = \mathrm{id}_A$ follows. Thus m is a monomorphic retraction, hence an isomorphism, and $r \in \mathcal{E}$ follows.

If $r \in \mathcal{E}$, then $[r] \circ [r]^{\mathrm{op}} = [\mathrm{id}_A]$ by 3.2(4) and the definitions; thus $[r]$ is surjective.

If $[r]$ is surjective and $ru = ru'$ a pullback square, then

$$\rho^{\mathrm{op}} \circ \rho = \mathbf{r}(ru, ru') \geqslant \mathbf{r}(r, r) = [r] \circ [r]^{\mathrm{op}} = [\mathrm{id}_A],$$

by 3.2(4) and the definitions; thus ρ is total. □

COROLLARY 3.6.1. *The class of legs r with $[r]$ surjective is stable.*

Proof. For a relation $\rho = (r, s)$ and a pullback $rg = fu$, we have $\rho \circ [f] \simeq (u, sg)$. If $[r]$ is surjective, then ρ and $\rho \circ [f]$ are total; thus $[u]$ is surjective. □

PROPOSITION 3.7. *A relation $\rho : A \twoheadrightarrow B$ is functional if and only if there is a relation $\sigma : B \twoheadrightarrow A$ such that $\sigma \circ \rho \geqslant [\mathrm{id}_A]$ and $\rho \circ \sigma \leqslant [\mathrm{id}_B]$, and then $\sigma \simeq \rho^{\mathrm{op}}$.*

Proof. If ρ is functional, then $\sigma = \rho^{\mathrm{op}}$ satisfies the inequalities. Conversely, it follows from the first inequality, by 3.6 for ρ and σ^{op}, that ρ is total and σ surjective. But then

$$\sigma \leqslant \rho^{\mathrm{op}} \circ \rho \circ \sigma \leqslant \rho^{\mathrm{op}}$$

with the second inequality, and also

$$\rho \leqslant \rho \circ \sigma \circ \sigma^{\mathrm{op}} \leqslant \sigma^{\mathrm{op}}.$$

It follows that $\sigma \simeq \rho^{\mathrm{op}}$, and ρ is functional. □

3.8. ISOMORPHISMS

We say as usual that relations $\rho : A \twoheadrightarrow B$ and $\sigma : B \twoheadrightarrow A$ are *inverse isomorphisms of relations* if

$$\sigma \circ \rho = [\mathrm{id}_A] \quad \text{and} \quad \rho \circ \sigma = [\mathrm{id}_B].$$

It follows that inverses of relations are only determined up to equivalence if they exist. Inverses in Rel **C** are unique.

THEOREM. *A relation ρ is an isomorphism of relations if and only if ρ is functional and bijective, and then the inverses of ρ are the relations span-equivalent to ρ^{op}.*

Proof. If ρ is functional and bijective, then ρ^{op} is an inverse of ρ by the definitions. Conversely, if ρ and σ are inverse relations, then ρ and σ are functional, and $\sigma \simeq \rho^{\mathrm{op}}$, by 3.7. But then ρ and σ are also bijective. □

3.9. DISCUSSION

Since the relations $[\mathrm{id}_A] = \mathbf{r}\delta_A$ are identity relations, 3.2(3) says that the assignments $f \mapsto [f]$ define a functor, from $\mathbf{C}$ to Rel $\mathbf{C}$.

We note that only 2.3(i)–(iv) have been used in 3.2 and 3.4. The proofs of 3.6 and its corollary require only 3.2(4) if we replace 3.6(ii) by the condition that m is a split epimorphisms in an $(\mathcal{E}, \mathcal{M})$-factorization $r = me$. This has been stated in [17, 5.2].

Pavlović [17] has also shown that maps in the sense of [12] and [11] are functional relations for any $(\mathcal{E}, \mathcal{M})$-factorization structure with $\mathcal{E}$ stable (and not just leg-stable), but functional relations need not be maps if $\mathcal{M}$ does not consist of monomorphisms.

4. Surjective-Injective Factorizations

4.1.

We begin with some useful lemmas.

4.1.1. *For m in $\mathcal{M}$, the relation $[m]$ is injective.*
4.1.2. *An induced relation $[f]$ is surjective iff $f \in \mathcal{E}$.*
4.1.3. *If $[\mathrm{id}_A] = (a, b)$, then $[a] = [b]$ is an isomorphism of relations.*

Proof. For $m : A \to B$ in $\mathcal{M}$, we have

$$[m]^{\mathrm{op}} \circ [m] = \mathbf{r}((m, \mathrm{id}_A)\Diamond(\mathrm{id}_A, m)) = \mathbf{r}\delta_A = [\mathrm{id}_A],$$

by 1.7, 2.4, and the assumption that m is monomorphic.

For 4.1.2 $(\mathcal{E}, \mathcal{M}_0)$-factor $\langle \mathrm{id}, f\rangle = \langle r, s\rangle \cdot e$. Then $f = se$ is in $\mathcal{E}$ iff $s \in \mathcal{E}$, and hence iff $[f] = (r, s)$ is surjective, by 3.6 for $[f]^{\mathrm{op}}$.

For 4.1.3, we have $[a] \circ [e] = [b] \circ [e] = [\mathrm{id}_A]$ for a morphism e in $\mathcal{E}$. Since also $[e] \circ [e]^{\mathrm{op}} = [\mathrm{id}]$, $[e]$ and $[a] = [b]$ are inverse isomorphisms. □

PROPOSITION 4.2. *Every relation ρ factors $\rho \simeq [m] \circ \varepsilon$, with ε surjective and m in $\mathcal{M}_0$, hence $[m]$ injective, with ε total if and only if ρ is total, and ε single-valued if and only if ρ is single-valued.*

Proof. Put $\rho = (r, s)$ and factor $s = me$ with $m \in \mathcal{M}_0$ and $e \in \mathcal{E}$. Since $\langle r, s\rangle = (\mathrm{id} \times m)\langle r, e\rangle$ with $\langle r, s\rangle$ and $\mathrm{id} \times m$ in $\mathcal{M}$, this defines a relation $\varepsilon = (r, e)$, with $\rho \simeq [m] \circ \varepsilon$ by 3.2(2), and ε is surjective by 3.6 for $\varepsilon^{\mathrm{op}}$. By 3.6, ε is total iff ρ is total. We have $\varepsilon \simeq [m]^{\mathrm{op}} \circ \rho$, and $[m]$ and $[m]^{\mathrm{op}}$ are single-valued. Thus ε is single-valued iff ρ is single-valued. □

THEOREM 4.3. *For a commutative square $\mu \circ \rho = \sigma \circ \varepsilon$ of relations, with ε single-valued and surjective, and μ injective and total, the following four statements* *are logically equivalent for a relation τ, and each of them determines τ up to equivalence:*

$$\rho \simeq \tau \circ \varepsilon, \qquad \tau \simeq \rho \circ \varepsilon^{\mathrm{op}}, \qquad \sigma \simeq \mu \circ \tau, \qquad \tau \simeq \mu^{\mathrm{op}} \circ \sigma.$$

In this situation, ρ and τ are single-valued if σ is single-valued, and σ and τ are total if ρ is total.

Proof. If $\rho \simeq \tau \circ \varepsilon$, then $\tau \simeq \rho \circ \varepsilon^{\mathrm{op}}$ since $\varepsilon \circ \varepsilon^{\mathrm{op}} = [\mathrm{id}]$, and then

$$\mu \circ \tau = \mu \circ \rho \circ \varepsilon^{\mathrm{op}} = \sigma \circ \varepsilon \circ \varepsilon^{\mathrm{op}} \simeq \sigma.$$

Similarly, $\sigma \simeq \mu \circ \tau \Rightarrow \tau \simeq \mu^{\mathrm{op}} \circ \sigma \Rightarrow \rho \simeq \tau \circ \varepsilon$, using $\mu^{\mathrm{op}} \circ \mu = [\mathrm{id}]$, and the last part follows immediately from the assumptions and the displayed equivalences. □

THEOREM 4.4. *A relation $(r, s) : A \to B$ in $\mathbf{C}$ is total if and only if $[r]$ is surjective, single-valued if and only if $[r]$ is injective, and functional if and only if $[r]$ is bijective. Dually, (r, s) is surjective, injective or bijective if and only if $[s]$ has the same property.*

Proof. By 3.6 and the definitions, we must only prove the 'single-valued' part. We have

$$\rho \circ \rho^{\mathrm{op}} = [s] \circ [r]^{\mathrm{op}} \circ [r] \circ [s]^{\mathrm{op}} = [s] \circ [s]^{\mathrm{op}} \leqslant [\mathrm{id}_B]$$

if $[r]$ is injective. Conversely, if $ru = rv$ is a pullback square, and $[\mathrm{id}_B] = (a, b)$, then ρ is singlevalued iff there is a morphism t with $at = su$ and $bt = sv$. Since $[a] = [b]$ by 4.1.3, it follows that $[s][u] = [s][v]$, as well as $[r][u] = [r][v]$. Using 5.3 (whose proof does not use 4.4) and 3.2(4), we get

$$[\langle r, s\rangle] \circ [u] = \langle [ru], [su]\rangle = \langle [rv], [sv]\rangle = [\langle r, s\rangle] \circ [v],$$

and hence $[u] = [v]$ since $[\langle r, s\rangle]$ is injective. With this, we have

$$[r]^{\mathrm{op}} \circ [r] = \mathbf{r}(u, v) = [v] \circ [u]^{\mathrm{op}} = [u] \circ [u]^{\mathrm{op}} \leqslant [\mathrm{id}],$$

by 2.4 and 3.2, as claimed. □

4.5.

For injective functional relations μ_1 and μ_2, we put $\mu_1 \leqslant \mu_2$ if $\mu_1 \simeq \mu_2 \circ \tau$ for a relation τ. This determines $\tau \simeq \mu_2{}^{\mathrm{op}} \circ \mu_1$ up to span-equivalence, and τ is single-valued. From

$$\tau^{\mathrm{op}} \circ \tau = \tau^{\mathrm{op}} \circ \mu_2{}^{\mathrm{op}} \circ \mu_2 \circ \tau = \mu_1{}^{\mathrm{op}} \circ \mu_1 = [\mathrm{id}],$$

we see that τ is also total and injective. We call μ_1 and μ_2 *equivalent* if $\mu_1 \leqslant \mu_2$ and $\mu_2 \leqslant \mu_1$.

PROPOSITION. *For m_1 and m_2 in $\mathcal{M}$ with the same codomain, we have $m_1 \leqslant m_2$ if and only if $[m_1] \leqslant [m_2]$.*

Proof. We have $[m_1] \simeq [m_2] \circ (a, b)$ for a relation (a, b) iff

$$\langle \mathrm{id}, m_1 \rangle = \langle a, m_2 b \rangle \cdot e$$

for a morphism e in $\mathcal{E}$, and then $m_1 = m_2 be$. Conversely, if $m_1 = m_2 h$, then $[m_1] = [m_2] \circ [h]$. □

PROPOSITION 4.6. *If $\mu \simeq [m]$ and $\sigma \simeq [g]$ in 4.3, with $m \in \mathcal{M}$, then $\tau \simeq [t]$ for a morphism t of $\mathbf{C}$ with $mt = g$.*

Proof. If we $(\mathcal{E}, \mathcal{M})$-factor $g = m_1 e_1$, then $[m] \circ \tau \simeq [m_1] \circ [e_1]$, and $[m_1] \leqslant [m]$ follows by 4.3. But then $m_1 = ms$ for a morphism s by 4.5, and $g = mt$ for $t = se_1$. Now $\tau \simeq [t]$ since $[m]$ is monomorphic in $\mathrm{Rel}\,\mathbf{C}$. □

4.7.

We shall need the following result in Section 5.

PROPOSITION. *For a functional relation $\rho : A \to B$, compositions*

$$\tau = \rho \circ [m] \circ [m]^{\mathrm{op}},$$

with m in $\mathcal{M}_0$ with codomain A, define an order preserving bijection between morphisms m in $\mathcal{M}_0$ with codomain A, and relations $\tau \leqslant \rho$ in $\mathrm{Rel}_0(A, B)$.

Proof. For $\rho = (r, s)$, a composition $\rho \circ [m] \simeq (e, sh)$ is obtained from a pullback square

$$\begin{array}{ccc} \bullet & \xrightarrow{h} & \bullet \\ \Big\downarrow e & & \Big\downarrow r \\ \bullet & \xrightarrow{m} & A \end{array},$$

with $h \in \mathcal{M}$, and $e \in \mathcal{E}$ since $r \in \mathcal{E}$ by 3.6 and m is a leg. From these data, we get a relation

$$\tau = \rho \circ [m] \circ [m]^{\mathrm{op}} = m^{\to}(e, sh) \simeq (me, sh) = (rh, sh) \leqslant \rho.$$

We get m back from τ by $(\mathcal{E}, \mathcal{M}_0)$-factoring $rh = me$.

If $\tau \simeq (rh, sh)$ is given, with h in $\mathcal{M}$, then we obtain m from an $(\mathcal{E}, \mathcal{M}_0)$-factorization $rh = me$. In this situation,

$$\langle rh, sh \rangle = (m \times \mathrm{id}) \langle e, sh \rangle$$

in **C**. It follows that (e, sh) is a relation, total since $e \in \mathcal{E}$. From the construction of $\rho \circ [m]$ by a pullback, it follows that

$$(e, sh) \leqslant \rho \circ [m].$$

Since (e, sh) is total and $\rho \circ [m]$ single-valued, the two relations are equivalent by 3.5. But then τ is a composition $\rho \circ [m] \circ [m]^{\mathrm{op}}$ for this m.

If $m' \leqslant m$ in $\mathcal{M}_0$, with common codomain A, then $h' \leqslant h$ for the pullbacks h and h' of m and m' by r. Conversely, if $(rh', sh') \leqslant (rh, sh)$ in $\mathrm{Rel}(A, B)$, then $h' = hu$ for a morphism u, and $(\mathcal{E}, \mathcal{M})$-factoring $rh = me$ and $rh' = m'e'$ produces a commutative square $meu = m'e'$. With 1.1(ii), $m' \leqslant m$ follows. □

4.8.

The following result complements Proposition 1.1; we omit the easy proof.

PROPOSITION. *The following are equivalent.*

(i) *$\mathcal{E}$ consists of epimorphisms.*
(ii) *The functor $f \mapsto [f] : \mathbf{C} \to \mathrm{Rel}\,\mathbf{C}$ is faithful.*
(iii) *An induced relation $[f]$ is injective if and only if f is monomorphic in* **C**.
(iv) *A relation (r, s) is single-valued if and only if r is monomorphic in* **C**.
(v) *A relation (r, s) is injective if and only if s is monomorphic in* **C**.
(vi) *An induced relation $[f]$ is bijective if and only if f is monomorphic and in $\mathcal{E}$.*
(vii) *A relation (r, s) is functional if and only if r is monomorphic and in $\mathcal{E}$.*
(viii) *A relation (r, s) is bijective if and only if s is monomorphic and in $\mathcal{E}$.*

5. Finite Limits for Functional Relations

5.1. DEFINITIONS

We denote by $\mathrm{Fun}\,\mathbf{C}$ the subcategory of $\mathrm{Rel}\,\mathbf{C}$ with functional relations as morphisms, and by $\mathrm{Ind}\,\mathbf{C}$ the subcategory of $\mathrm{Fun}\,\mathbf{C}$ with induced relations $[f]$ as morphisms. The aim of this section is to construct finite limits in these categories.

5.2. TERMINAL OBJECTS

A terminal object **1** of **C** is also a terminal object of $\mathrm{Fun}\,\mathbf{C}$ and of $\mathrm{Ind}\,\mathbf{C}$.

Proof. Since the projection $A \times \mathbf{1} \to A$ is an isomorphism, a span $(m, t) : A \to \mathbf{1}$ is a relation iff $m \in \mathcal{M}$, and by 4.4 a functional relation iff also $m \in \mathcal{E}$, hence iff m is an isomorphism. Thus there is exactly one morphism $[t_A] = \mathbf{r}(\mathrm{id}_A, t_A) : A \to \mathbf{1}$ in $\mathrm{Fun}\,\mathbf{C}$, for the morphism $t_A : A \to \mathbf{1}$. □

5.3. BINARY PRODUCTS

The following result is independent of Section 4.

THEOREM. *If* $A \xleftarrow{p} A \times B \xrightarrow{q} B$ *are the projections of a product in* **C**, *then* $A \xleftarrow{[p]} A \times B \xrightarrow{[q]} B$ *are the projections of a product in* Fun **C** *and in* Ind **C**, *with*

$$\langle \rho, \sigma \rangle = [p]^{\mathrm{op}} \circ \rho \cap [q]^{\mathrm{op}} \circ \sigma : C \to A \times B \tag{1}$$

in Fun **C**, *for morphisms* $\rho : C \to A$ *and* $\sigma : C \to B$ *of* Fun **C**, *and*

$$\langle [f], [g] \rangle = [p]^{\mathrm{op}} \circ [f] \cap [q]^{\mathrm{op}} \circ [g] = [\langle f, g \rangle] \tag{2}$$

for $f : C \to A$ *and* $g : C \to B$ *in* **C**.

Proof. For a relation $\varphi : C \to A \times B$, we have

$$[p] \circ \varphi \leqslant \rho \quad \iff \quad \varphi \leqslant [p]^{\mathrm{op}} \circ \rho,$$

by 3.2(2) and 1.6 for the dual relations. In the same way,

$$[q] \circ \varphi \leqslant \sigma \quad \iff \quad \varphi \leqslant [q]^{\mathrm{op}} \circ \sigma.$$

It follows that

$$[p] \circ \varphi \leqslant \rho \quad \text{and} \quad [q] \circ \varphi \leqslant \sigma \quad \text{iff} \quad \varphi \leqslant \langle \rho, \sigma \rangle,$$

with $\langle \rho, \sigma \rangle$ defined by (1). By 3.5, these inequalities become equations if φ and $\langle \rho, \sigma \rangle$ are morphism of Fun **C**. Thus we must prove that (1) defines a functional relation $\langle \rho, \sigma \rangle$.

Let $\rho = (r, s) : C \to A$ and $\sigma = (u, v) : C \to B$. Then

$$[p]^{\mathrm{op}} \circ \rho \simeq (rp', s \times \mathrm{id}_B) \quad \text{and} \quad [q]^{\mathrm{op}} \circ \sigma \simeq (uq', \mathrm{id}_A \times v)$$

for pullback squares

$$\begin{array}{ccc} X \times B & \xrightarrow{p'} & X \\ \big\downarrow{\scriptstyle s \times \mathrm{id}_B} & & \big\downarrow{\scriptstyle s} \\ A \times B & \xrightarrow{p} & A \end{array} \quad \text{and} \quad \begin{array}{ccc} A \times Y & \xrightarrow{q'} & Y \\ \big\downarrow{\scriptstyle \mathrm{id}_A \times v} & & \big\downarrow{\scriptstyle v} \\ A \times B & \xrightarrow{q} & B \end{array},$$

with p' and q' projections of products. Thus a further pullback

$$\begin{array}{ccc} \bullet & \xrightarrow{\langle sx, y \rangle} & A \times Y \\ \big\downarrow{\scriptstyle \langle x, vy \rangle} & & \big\downarrow{\scriptstyle \langle uq', \mathrm{id}_A \times v \rangle} \\ X \times B & \xrightarrow{\langle rp', s \times \mathrm{id}_B \rangle} & C \times A \times B \end{array},$$

with x and y given by a pullback

$$\begin{array}{ccc} \bullet & \xrightarrow{y} & Y \\ \downarrow{\scriptstyle x} & & \downarrow{\scriptstyle u} \\ X & \xrightarrow{r} & C \end{array},$$

gives the desired intersection

$$\langle \rho, \sigma \rangle \simeq (rx, \langle sx, vy \rangle).$$

If ρ and σ are total, then $[r]$ and $[u]$ are surjective by 3.6, and so are $[x]$ and $[y]$ in the pullback diagram $rx = uy$, by 3.6.1. But then $[rx]$ is surjective, and $\langle \rho, \sigma \rangle$ is total, again by 3.6.

If ρ and σ are single-valued, then

$$\langle \rho, \sigma \rangle \circ \langle \rho, \sigma \rangle^{\mathrm{op}} \leqslant [p]^{\mathrm{op}} \circ \rho \circ \rho^{\mathrm{op}} \circ [p] \leqslant [p]^{\mathrm{op}} \circ [p],$$

and similarly $\langle \rho, \sigma \rangle \circ \langle \rho, \sigma \rangle^{\mathrm{op}} \leqslant [q]^{\mathrm{op}} \circ [q]$. Thus

$$\langle \rho, \sigma \rangle \circ \langle \rho, \sigma \rangle^{\mathrm{op}} \leqslant \langle [p], [q] \rangle.$$

If (2) is valid, the righthand side is $[\langle p, q \rangle] = [\mathrm{id}_{A \times B}]$, and $\langle \rho, \sigma \rangle$ is single-valued.

It remains to prove (2). For this, $(\mathcal{E}, \mathcal{M})$-factor

$$\langle \mathrm{id}_C, f \rangle = \langle r, s \rangle \cdot e_1 \quad \text{and} \quad \langle \mathrm{id}_C, g \rangle = \langle u, v \rangle \cdot e_2,$$

and construct a diagram of pullback squares

$$\begin{array}{ccccc} C & \xrightarrow{e_1} & \bullet & \xrightarrow{r} & C \\ \downarrow{\scriptstyle e_2} & & \downarrow{\scriptstyle e_2'} & & \downarrow{\scriptstyle e_2} \\ \bullet & \xrightarrow{e_1'} & \bullet & \xrightarrow{y} & \bullet \\ \downarrow{\scriptstyle u} & & \downarrow{\scriptstyle x} & & \downarrow{\scriptstyle u} \\ C & \xrightarrow{e_1} & \bullet & \xrightarrow{r} & C \end{array}$$

with morphisms id_C as sides of the outer square. Then e_1' and e_2' are in $\mathcal{E}$ since x and y are legs by 1.7, and we have

$$\langle rx, \langle sx, vy \rangle \rangle \cdot e_1' e_2 = \langle \mathrm{id}_C, \langle f, g \rangle \rangle.$$

This is an $(\mathcal{E}, \mathcal{M})$-factorization, and (2) follows. □

5.4. EQUALIZERS

For morphisms $\rho = (r, s) : A \to B$ and $\sigma = (u, v) : A \to B$ of Fun **C**, *and for a pullback square*

$$\begin{array}{ccc} \blacksquare & \xrightarrow{k} & \blacksquare \\ \downarrow h & & \downarrow \langle u, v\rangle \\ \blacksquare & \xrightarrow{\langle r, s\rangle} & A \times B \end{array} \tag{1}$$

and an $(\mathcal{E}, \mathcal{M})$-factorization $rh = m_1 e = uk$, the morphism $[m_1]$ is an equalizer of ρ and σ in Fun **C**. *If ρ and σ are morphisms of* Ind **C**, *this is also an equalizer in* Ind **C**,

Proof. For a functional relation $\xi = [m] \circ \varepsilon$ with codomain A, factored by 4.2, we have $\rho \circ \xi = \sigma \circ \xi$ iff $\rho \circ [m] = \sigma \circ [m]$, since ε is epimorphic in Rel **C**. This is the case iff $\rho \circ [m] \circ [m]^{\text{op}} = \sigma \circ [m] \circ [m]^{\text{op}}$, and hence iff the relations (rh, sh) and (uk, vk) corresponding to $[m]$ by 4.7 are equivalent. Thus we obtain the desired equalizer $[m_1]$, with $m_1 \in \mathcal{M}$, by forming a pullback diagram (1) and factoring $rh = uk$ as shown.

If we factor $[g] = [m_1] \circ \tau$ in Fun **C**, then τ is a morphism of Ind **C** by 4.6. Thus $[m_1]$ is also an equalizer in Ind **C** if ρ and σ are induced relations. □

5.5. PULLBACKS

For morphisms $\rho = (r, s) : A \to C$ and $\sigma = (u, v) : B \to C$ of Fun **C**, *we obtain a pullback square*

$$\begin{array}{ccc} P & \xrightarrow{[b]} & B \\ \downarrow [a] & & \downarrow \sigma \\ A & \xrightarrow{\rho} & C \end{array}, \tag{1}$$

in Fun **C** *by*

$$\sigma^{\text{op}} \circ \rho = (a, b)$$

in Rel **C**. *For morphisms ρ and σ of* Ind **C**, *this is a pullback square in* Ind **C**.

COROLLARY 5.5.1. *Every pullback square*

$$\begin{array}{ccc} P & \xrightarrow{\beta} & B \\ \downarrow \alpha & & \downarrow \sigma \\ A & \xrightarrow{\rho} & C \end{array}$$

in Fun **C** *or in* Ind **C** *satisfies the Beck condition*

$$\sigma^{\mathrm{op}} \circ \rho = \beta \circ \alpha^{\mathrm{op}}.$$

Proof. By the general theory, the components of the desired pullback are the components of an equalizer $[\langle a, b\rangle]$ of $\rho \circ [p]$, and $\sigma \circ [q]$, for the projections p and q of $A \times B$. For this, we need the intersection

$$\rho \circ [p] \cap \sigma \circ [q] : A \times B \to C.$$

This is dual to the intersection $\langle \rho^{\mathrm{op}}, \sigma^{\mathrm{op}}\rangle$ of 5.3, and thus $(\langle rh, uk\rangle, sh)$ for a pullback square $sh = vk$ in **C**. Now $(a, b) = \mathbf{r}(rh, uk)$; this is $\sigma^{\mathrm{op}} \circ \rho$.

For ρ and σ in Ind **C**, we have constructed an equalizer $[\langle a, b\rangle]$ in Ind **C**, and thus a pullback in Ind **C**.

The corollary is valid for the pullback (1) since $(a, b) = [b] \circ [a]^{\mathrm{op}}$, and hence for every pullback square in Fun **C**. □

THEOREM 5.6. *The following properties of* **C** *are equivalent.*

(i) $\mathcal{E}$ *is stable for all pullbacks in* **C**.

(ii) *All span compositions* $\beta\Diamond\alpha$ *in* **C** *satisfy*

$$\mathbf{r}(\beta\Diamond\alpha) = \mathbf{r}\beta \circ \mathbf{r}\alpha \qquad (1)$$

(iii) *The functor* $f \mapsto [f] : \mathbf{C} \to$ Fun **C** *preserves pullbacks.*

(iv) *The functor* $f \mapsto [f] : \mathbf{C} \to$ Fun **C** *preserves equalizers.*

(v) *The functor* $f \mapsto [f] : \mathbf{C} \to$ Fun **C** *preserves all finite limits.*

If these conditions are satisfied, then $[m]$ *is injective for every monomorphism* m *in* **C**.

Proof. If $\mathcal{E}$ is preserved by all pullbacks, then the proof of 2.4 works without any restrictions on b or b', and thus (1) is valid for all span compositions. If $m : A \to B$ is monomorphic, then

$$(\mathrm{id}_A, m)^{\mathrm{op}}\Diamond(\mathrm{id}_A, m) \simeq \delta_A,$$

and $[m]^{\mathrm{op}} \circ [m] = [\mathrm{id}_A]$ follows if (1) is valid.

Now assume (ii) and consider a pullback square

$$\begin{array}{ccc} \cdot & \xrightarrow{k} & B \\ \big\downarrow h & & \big\downarrow g \\ A & \xrightarrow{f} & C \end{array}$$

in **C**, with $(h, k) \simeq (\mathrm{id}_B, g)^{\mathrm{op}}\Diamond(\mathrm{id}_A, f)$. Thus

$$[g]^{\mathrm{op}} \circ [f] = \mathbf{r}(h, k)$$

by (1). If we $(\mathcal{E}, \mathcal{M}_0)$-factor $\langle h, k\rangle = \langle a, b\rangle \cdot e$, then $[f]\circ[a] = [g]\circ[b]$ is a pullback square in Fun **C** by 5.5. Now e in $\mathcal{E}$ is monomorphic since $\langle h, k\rangle$ is, and thus $[e]$ is injective as well as surjective. But then $[f] \circ [h] = [g] \circ [k]$ is also a pullback square, and so (iii) is valid.

Since $f \mapsto [f]$ preserves finite products, (iii), (iv) and (v) are equivalent. If they are valid and $fh = gk$ is a pullback square in **C** with f in $\mathcal{E}$, then $[f]\circ[h] = [g]\circ[k]$ is a pullback square, with $[f]$ surjective by 4.1.2. But then $[k]$ is surjective by 6.2 below, and k is in $\mathcal{E}$. □

REMARKS 5.7. If $\mathcal{E}$ consists of epimorphisms, then all morphisms of **C** are legs, and (i)–(v) of 5.6 are equivalent to leg-stability of **C**.

It does not follow from 5.6(iii) that the functor $f \mapsto [f] : \mathbf{C} \to \operatorname{Ind} \mathbf{C}$ preserves pullbacks and equalizers. An isomorphism $[f]$ in Fun **C** need not be an isomorphism in Ind **C**; there may be no morphism of **C** which induces the isomorphism $[f]^{-1}$ of Fun **C**.

6. Relations in Fun C and in Ind C

6.1. DISCUSSION

We have seen in Section 5 that Ind **C** and Fun **C** have finite limits, and that the embedding Ind **C** → Fun **C** preserves them. The functor $f \mapsto [f] : \mathbf{C} \to \operatorname{Fun} \mathbf{C}$ preserves products, but in general not equalizers or pullbacks. 1.8.1 is an example for this.

By 4.2, 4.3 and 4.6, the categories Ind **C** and Fun **C** have a (surjective,injective) factorization structure, preserved by the embedding Ind **C** → Fun **C**. We assume from now on that Fun **C** and Ind **C** are provided with this structure.

Every injective functional relation ρ satisfies $\rho = [m] \circ \tau$ for an isomorphism τ of relations and an induced $[m]$ with m in $\mathcal{M}$. By 4.5, this is the case for exactly one m in $\mathcal{M}_0$. Thus we can, and shall, use the relations $[m]$ with m in $\mathcal{M}_0$ as subobjects for Fun **C** and Ind **C**. With this convention, the functor $f \mapsto [f]$ also preserves and reflects subobjects.

Now only one thing is missing.

PROPOSITION 6.2. *Surjective morphisms are stable in* Fun **C** *and in* Ind **C**.

Proof. It is sufficient to prove this for the pullback square

$$\begin{array}{ccc} P & \xrightarrow{[b]} & B \\ \Big\downarrow{\scriptstyle [a]} & & \Big\downarrow{\scriptstyle \sigma} \\ A & \xrightarrow{\rho} & C \end{array},$$

constructed in 5.5, with $(a, b) = \sigma^{\mathrm{op}} \circ \rho$, and σ^{op} surjective since σ is total. If ρ is surjective, then (a, b) is surjective. But then $b \in \mathcal{E}$, and $[b]$ is surjective, by 3.6 and 4.1.2. □

6.3.

We can now construct categories of relations and of functional relations in Fun **C** and in Ind **C**, with all the properties obtained in this paper. However, we get nothing new. Every injective morphism $\langle \rho, \sigma \rangle : \blacksquare \to A \times B$ in Fun **C** is equivalent to exactly one morphism $[\langle r, s \rangle] = \langle [r], [s] \rangle$ with $\langle r, s \rangle : \blacksquare \to A \times B$ in $\mathcal{M}_0$. Thus the morphisms of Rel Fun **C** are just the pairs $([r], [s])$ with (r, s) a morphism of Rel **C**.

This leads up to the following result.

THEOREM. *The bijection* $(r, s) \mapsto ([r], [s])$ *from* Rel **C** *to* Rel Fun **C** *is an isomorphism of categories which preserves and reflects order and duality, and maps* Fun **C** *to* Fun Fun **C** = Ind Fun **C**.

Proof. For $(r, s) : A \to B$ and $(u, v) : B \to C$, consider pullbacks

$$\begin{array}{ccc} \blacksquare & \xrightarrow{k} & \blacksquare \\ \big\downarrow{\scriptstyle h} & & \big\downarrow{\scriptstyle u} \\ \blacksquare & \xrightarrow{s} & B \end{array} \qquad \text{and} \qquad \begin{array}{ccc} \blacksquare & \xrightarrow{[b]} & \blacksquare \\ \big\downarrow{\scriptstyle [a]} & & \big\downarrow{\scriptstyle [u]} \\ \blacksquare & \xrightarrow{[s]} & B \end{array}$$

in **C** and Fun **C**, with $(a, b) = [u]^{\mathrm{op}} \circ [s]$ by 5.5. Then $(u, v) \circ (r, s) = \mathbf{r}(rh, vk)$, and $([u], [v]) \circ ([r], [s])$ is obtained from an $(\mathcal{E}, \mathcal{M}_0)$-factorization of $\langle [ra], [vb] \rangle$, or equivalently of $\langle ra, vb \rangle$. Now

$$(h, k) \simeq (\mathrm{id}, u)^{\mathrm{op}} \Diamond (\mathrm{id}, s).$$

Since s and u are legs, we can apply 2.4 to this composition, getting

$$\mathbf{r}(h, k) = [u]^{\mathrm{op}} \circ [s] = (a, b).$$

Thus $\langle rh, vk \rangle = \langle ra, vb \rangle \cdot e$ with $e \in \mathcal{E}$, and $\mathbf{r}(rh, vk) = \mathbf{r}(ra, vb)$ follows. This shows that composition of relations is preserved, and we have an isomorphism of categories.

Since the isomorphism preserves order and duality, it also preserves the four properties defined in 3.3, and thus maps functional relations to functional relations. If (r, s) is functional, then $[r]$ is an isomorphism of Fun **C** by 4.4, and

$$([r], [s]) \simeq ([\mathrm{id}], [s] \circ [r]^{-1}) \quad \text{and} \quad ([r], [s]) = [[s] \circ [r]^{-1}]$$

follow. Thus every functional relation in Fun **C** is induced by a morphism of Fun **C**. □

REMARKS 6.4. The functor $f \mapsto [f] : \mathbf{C} \to \operatorname{Ind}\mathbf{C}$ is full, and by 4.8 an isomorphism of categories if $\mathcal{E}$ consists of epimorphism. Subobjects in Fun **C** are the same as subobjects in Ind **C**, and thus Rel Ind **C** and Fun Ind **C** are the same as Rel Fun **C** and Fun Fun **C**.

We have seen that every functional relation $([r], [s])$ in Fun **C** is an induced relation $[[s] \circ [r]^{-1}]$. However, $([r], [s])$ need not be induced by a morphism of Ind **C**. The relation $[r]^{-1} \simeq [r]^{\mathrm{op}}$ need not be induced by a morphism of **C**, even if $\mathcal{E}$ consists of epimorphisms.

All monomorphisms in Ind **C** and in Fun **C** are injective by 4.8; thus the surjective morphisms in these categories are the strong epimorphisms. There may be epimorphisms in Ind **C** or in Fun **C** which are not strong.

PROBLEM 6.5. Our results are only valid if $\mathcal{E}$ is leg-stable and $\mathcal{M}$ consists of monomorphisms. Do they remain valid in a more general situation if Fun **C** is replaced by the category of maps in the sense of [12] and [11]? Pavlović [17] has shown that this is indeed the case, with the category of maps isomorphic to **C**, if $\mathcal{E}$ consists of regular epimorphisms, i.e. coequalizers, but what happens in more general situations?

7. Examples

7.1. FINE MAPS

If **C** is topological over a category **A** with finite limits, with forgetful functor $U : \mathbf{C} \to \mathbf{A}$, then a morphism f of **C** is called a *fine map* if f by itself is a final sink. With fine maps as the class $\mathcal{E}$, and $\mathcal{M}_0$ consisting of all maps of the form $\mathrm{id}_S : A \to B$, for objects of **C** with $UA = S = UB$, we get an $(\mathcal{E}, \mathcal{M}_0)$-factorization structure of **C**.

A trivial example is given by **C** topological over itself. In this situation, every object A has exactly one subobject id_A, and there is always exactly one morphism $A \to B$ in Rel **C**. Fuzzy sets, discussed below, provide a nontrivial example.

We have not investigated leg-stability for topological categories with this factorization structure.

7.2. FUZZY SETS

Fuzzy sets were introduced by L. Zadeh [24], essentially as pairs $A = (|A|, \varepsilon_A)$, consisting of a set $|A|$ and a mapping $\varepsilon_A : |A| \to [0, 1]$, which assigns to every $x \in |A|$ a membership degree $\varepsilon_A(x)$, a number in the real unit interval.

Zadeh noted that membership degrees $\varepsilon_A(x)$ can be viewed as truth-values, with 1 meaning 'true' and 0 'false'. J. A. Goguen [7] pointed out that truth-values could be taken in any lattice H, preferably a complete Heyting algebra, with mappings $\varepsilon_A : |A| \to H$ defining truth-values of statements $x \in A$. Goguen also pointed out that H-valued fuzzy sets, with morphisms $f : A \to B$ given by mappings

$f : |A| \to |B|$ such that $\varepsilon_A(x) \leqslant \varepsilon_B(fx)$ for $x \in |A|$, define a topological construct. These morphisms are called *crisp maps*.

If we define a *fuzzy subset* of an H-valued fuzzy set A as an H-valued fuzzy set B with $|B| = |A|$, and $\varepsilon_B(x) \leqslant \varepsilon_A(x)$ for all $x \in |A|$, then subset insertions $\mathrm{id}_{|A|} : B \to A$ provide the class $\mathcal{M}_0$ for an $(\mathcal{E}, \mathcal{M}_0)$-factorization structure of H-valued fuzzy sets, with $\mathcal{E}$ consisting of all fine maps.

7.3. H-SETS

For a complete Heyting algebra H, we define an H-set as a pair $A = (|A|, \delta_A)$ consisting of an ordinary set $|A|$ and a mapping $\delta_A : |A| \times |A| \to H$ with the following two properties.

(1) $\delta_A(x, y) = \delta_A(y, x)$ for all x, y in $|A|$.
(2) $\delta_A(x, y) \wedge \delta_A(y, z) \leqslant \delta_A(x, z)$ for all x, y, z in $|A|$.

A *crisp map* $f : A \to B$ is defined as a mapping $f : |A| \to |B|$ of the underlying sets such that $\delta_A(x, y) \leqslant \delta_B(fx, fy)$ for all x, y in $|A|$.

With composition of the underlying mappings as composition of crisp maps, H-sets and crisp maps form a topological construct. For a set S, putting $\delta_{VS}(x, x') = \bot$ and $\delta_{CS}(x, x') = \top$, for the bottom element $\bot$ and the top element $\top$ of H and all x, x' in S, defines the discrete or *void* H-set VS and the indiscrete or *coarse* H-set CS with underlying set S. The empty set has only the void H-set structure.

For an H-set A, we interpret δ_A as H-valued equality. Putting $\varepsilon_A(x) = \delta_A(x, x)$ for $x \in |A|$ defines H-valued membership in an H-set, and thus an underlying H-valued fuzzy set.

We define an *H-subset structure* of an H-set A as a mapping $\alpha : |A| \to H$ satisfying the following two conditions.

(i) $\alpha(x) \leqslant \varepsilon_A(x)$ for all $x \in |A|$.
(ii) $\alpha(x) \wedge \delta_A(x, x') \leqslant \alpha(x')$ for all x, x' in $|A|$.

An H-subset structure α of an H-set A defines an *H-subset* of A, denoted by $A \restriction \alpha$, with $|A \restriction \alpha| = |A|$, $\varepsilon_{A\restriction\alpha}(x) = \alpha(x)$ for $x \in |A|$, and

$$\delta_{A\restriction\alpha}(x, x') = \alpha(x) \wedge \delta_A(x, x')$$

for x, x' in $|A|$, and also a crisp map $\mathrm{id}_{|A|} : A \restriction \alpha \to A$, called an *$H$-subset insertion*. We note that identity maps of H-sets are H-subset insertions, and that an H-subset of an H-subset of an H-set A is an H-subset of A.

Subset insertions provide the class $\mathcal{M}_0$ for an $(\mathcal{E}, \mathcal{M}_0)$-factorization structure of H-sets. Morphisms in $\mathcal{E}$ for this structure are called *covers*.

It is easily seen that a crisp map $f : A \to B$ of H-sets lifts to $f : A \to B \restriction \beta$ for an H-subset structure β of B iff

$$\varepsilon_A(x) \wedge \delta_B(fx, y) \leqslant \beta(y)$$

for all x in $|A|$ and y in $|B|$. It follows that

$$\beta(y) = \bigvee_{x \in |A|} (\varepsilon_A(x) \wedge \delta_B(fx, y)),$$

for all $y \in |B|$, provides the (cover, insertion) factorization

$$A \xrightarrow{f} B \restriction \beta \xrightarrow{\mathrm{id}_{|B|}} B$$

of $f : A \to B$.

7.4. DISCUSSION

Underlying fuzzy sets define a concrete functor from H-sets with crisp maps to fuzzy sets with crisp maps. This functor turns out to be topological. Its concrete left adjoint D assigns to a fuzzy set A a *discrete* H-set DA, with $|DA| = |A|$, $\varepsilon_{DA} = \varepsilon_A$, and $\delta_{DA}(x, y) = \bot$, the bottom element of H, for $x \neq y$ in A. Its concrete right adjoint assigns to a fuzzy set A an H-set CA with $|CA| = |A|$, and $\delta_{CA}(x, y) = \varepsilon_A(x) \wedge \varepsilon_A(y)$ for all x, y in A.

Covers for H-sets and crisp maps are stable for pullbacks by legs, by [10] 2.17, and covers for H-valued fuzzy sets and crisp maps are stable for all pullbacks. Thus categories Fun **C** and Ind **C** are defined for both categories. Example 1.8.1 shows that covers for H-sets are not stable for all pullbacks.

For H-sets and crisp maps, Fun **C** turns out to be the topos of H-sets and functional relations introduced by D. Higgs in [8], and also studied in detail by M. P. Fourman and D. S. Scott [5]. For this example, Ind **C** is a quasitopos, introduced independently by G. P. Monro [16] and D. Ponasse (see [19]). H-sets were called *totally fuzzy sets* by Ponasse. We note that $[f] = [g]$, for crisp maps $A \underset{g}{\overset{f}{\rightrightarrows}} B$ of H-sets, iff $\varepsilon_A(x) \leqslant \delta_B(fx, gx)$, for all x in $|A|$.

The categories Fun **C** and Ind **C** for H-valued fuzzy sets have also been studied, but they turn out to be much less interesting and well behaved than the corresponding categories for H-sets. This is not astonishing. For fuzzy sets, membership is fuzzy but equality crisp, with 'true' and 'false' as the only truth-values. For H-sets, membership and equality both have the same spectrum of truth-values. For the same reason, categories of functional relations, with fuzzy function values, are more interesting than categories of crisp maps, with crisp function values.

7.5. MORE EXAMPLES

Many examples of relations for $(\mathcal{E}, \mathcal{M})$-factorizations with $\mathcal{M}$ consisting of monomorphisms and $\mathcal{E}$ consisting of epimorphisms can be found in [2, 11, 13]. Spans provide an example with $\mathcal{E}$, the class of all isomorphisms of **C**, stable and consisting of epimorphisms, but $\mathcal{M}$, the class of all morphisms, not consisting of monomorphisms. Several other examples with $\mathcal{E}$ stable but not consisting of epimorphisms, and $\mathcal{M}$ not consisting of monomorphisms, are given in [17]. A more general concept of relations, outside the scope of this paper, is discussed in [18], with examples.

References

1. Adámek, J., Herrlich, H. and Strecker, G. E.: *Abstract and Concrete Categories*, Wiley, New York, 1990.
2. Barr, M.: Relational algebras, in *Reports of the Midwest Category Seminar IV*, Lecture Notes in Math. 137, 1970, pp. 39–55.
3. Barr, M., Grillet, P. A. and van Osdol, D. H.: Exact Categories and Categories of Sheaves, Lecture Notes in Math. 236, 1971.
4. Bénabou, J.: Introduction to bicategories, in *Reports of the Midwestern Category Seminar*, Lecture Notes in Math. 47, 1967, pp. 1–77.
5. Fourman, M. P. and Scott, D. S.: Sheaves and logic, in *Applications of Sheaves. Proceedings of the Durham Conference*, Lecture Notes in Math. 753, 1979, pp. 302–401.
6. Freyd, P. J. and Kelly, G. M.: Categories of continuous functors I, *J. Pure Appl. Algebra* **2** (1972), 169–191.
7. Goguen, J. A.: *L*-fuzzy sets, *J. Math. Anal. Appl.* **18** (1967), 145–174.
8. Higgs, D.: A category approach to Boolean-valued set theory, Preprint, 1973.
9. Höhle, U. and Stout, L. N.: Foundations of fuzzy sets, *Fuzzy Sets and Systems* **40** (1991), 257–296.
10. Jayewardene, R.: Relations and functional relations in categories, with examples from fuzzy set theory, Ph.D. Thesis, Carnegie Mellon University, 1995.
11. Kelly, G. M.: A note on relations relative to a factorization system, in *Category Theory, Proceedings, Como 1990*, Lecture Notes in Math. 1488, 1991, pp. 249–261.
12. Lawvere, F. W.: Metric spaces, generalized logic, and closed categories, *Rend. Sem. Mat. Fis. Milano* **43** (1973), 135–166.
13. Klein, A.: Relations in categories, *Illinois J. Math.* **14** (1970), 536–550.
14. Mac Lane, S.: An algebra of additive relations, *Proc. Nat. Acad. Sci. U.S.A.* **47** (1963), 1043–1051.
15. Meisen, J.: On bicategories of relations and pullback spans, *Comm. Algebra* **1** (1974), 377–401.
16. Monro, G. P.: Quasitopoi, logic and Heyting-valued models, *J. Pure Appl. Algebra* **42** (1986), 141–164.
17. Pavlović, D.: Maps I: Relative to a factorisation system, *J. Pure Appl. Algebra* **99** (1995), 9–34.
18. Pavlović, D.: Maps II: Chasing diagrams in categorical proof theory, *J. of the IGPL* **4** (1996), 159–194.
19. Ponasse, D.: Categorical studies of fuzzy sets, *Fuzzy Sets and Systems* **28** (1988), 235–244.
20. Puppe, D.: Korrespondenzen in abelschen Kategorien, *Math. Ann.* **148** (1962), 1–30.
21. Stout, L. N.: The logic of unbalanced subobjects in a category with two closed structures, in *Applications of Category Theory to Fuzzy Subsets*, Kluwer Academic Publishers, Dordrecht, 1992, pp. 73–105.
22. Wyler, O.: *Lecture Notes on Topoi and Quasitopoi*, World Scientific Publishing Co., Singapore, 1991.
23. Wyler, O.: Fuzzy logic and categories of fuzzy sets, in *Non-Classical Logics and Their Applications to Fuzzy Subsets*, Kluwer Academic Publishers, Dordrecht, 1995, pp. 235–268.
24. Zadeh, L. A.: Fuzzy sets, *Inform. and Control* **8** (1965), 338–353.

Applied Categorical Structures **8:** 307–316, 2000.
G. Brümmer & C. Gilmour (eds), Papers in Honour of Bernhard Banaschewski.

Weak Equivalence of Internal Categories

Dedicated to Bernhard Banaschewski on the occasion of his 70th birthday

RENATO BETTI
Politecnico di Milano, Italy

(Received: 20 September 1996; accepted: 14 December 1999)

Abstract. Weak equivalence is defined as equivalence in the bicategory of modules between internal categories. It is known that two categories are weakly equivalent if and only if their Cauchy completions are equivalent. We prove that this condition can be generalized to a suitable notion of *intermediate category*, stable under composition with weak equivalences. Applications to categorical Morita theory are given.

Mathematics Subject Classifications (2000): 18D35, 18A25, 18D05.

Key words: internal category, module, Morita equivalence.

Introduction

The equivalence of two categories in the bicategory of modules is weaker than their equivalence in *Cat*. This is our notion of *weak equivalence*. The categories we consider are internal to a given topos $\mathcal{E}$ and are regarded as one-object categories enriched in the bicategory *Span* $\mathcal{E}$ (as in Betti and Walters [1, 2]). We believe that the language of *Span* $\mathcal{E}$ enriched categories provides a precise and flexible tool for analysing internal structures. Here we concentrate on aspects of cocompleteness: it is known that two categories **C** and **D** are weakly equivalent exactly when their Cauchy-completions are equivalent and this in turn means exactly that they are Morita equivalent (i.e. their small cocompletions $\mathcal{E}^{\mathbf{C}^{op}}$ and $\mathcal{E}^{\mathbf{D}^{op}}$ are equivalent). We prove that this condition can be generalized to a suitable notion of *intermediate category* over **C**, i.e. a full subcategory of $\mathcal{E}^{\mathbf{C}^{op}}$ containing the representables. Applications to Morita theory follow.

1. Let $\mathcal{E}$ be an elementary topos. In this section we recall some basic facts and fix terminology. First, recall that an internal category **C** is a monad $\gamma\colon C_0 \longrightarrow\!\!\!\!\!\!\!+\;\; C_0$ in *Span* $\mathcal{E}$. Here, C_0 is the "object of objects", the category structure, given by "domain" and "codomain":

$$C_1 \rightrightarrows C_0$$

provides the span γ (here C_1 is the "object of morphisms" of $\mathbf{C}$), and the operations of $\mathbf{C}$ give rise to the analogous operations:

$$\gamma^2 \Rightarrow \gamma,$$
$$1_{C_0} \Rightarrow \gamma$$

of the monad. We write simply $\mathbf{C} = (C_0, \gamma)$ to denote internal categories.

An internal functor $\mathbf{C} \to \mathbf{D}$ is given by an arrow $f\colon C_0 \to D_0$ (the "correspondence on objects") endowed with a 2-cell $\phi\colon f\gamma \Rightarrow \delta f$ relating the homs (here δ is the monad providing the category structure of $\mathbf{D}$), which makes commute the usual diagrams of 2-cells in *Span* $\mathcal{E}$.

We need also *modules* between internal categories: a module $\mathbf{C} \nrightarrow \mathbf{D}$ is given by a span $\alpha\colon C_0 \nrightarrow D_0$ endowed with actions of $\mathbf{C}$ on the left and of $\mathbf{D}$ on the right, which are associative, unitary, and satisfy a "mixed associativity" law with respect to the simultaneous actions (also this laws can be expressed by commutative diagrams of 2-cells in *Span* $\mathcal{E}$). Moreover, because $\mathcal{E}$ has pullback stable coequalizers, modules compose in an associative and unitary way, giving rise to a bicategory *Mod* $\mathcal{E}$ whose objects are internal categories and whose arrows are modules between them: denote by $\psi \otimes_{\mathbf{B}} \varphi$ the composite $\mathbf{A} \overset{\varphi}{\nrightarrow} \mathbf{B} \overset{\psi}{\nrightarrow} \mathbf{C}$. In particular, the identity of module composition is provided by the category structure itself.

Finally, recall that an internal functor $f\colon \mathbf{C} \to \mathbf{D}$ gives rise to an adjoint pair of modules:

$$\mathbf{C} \underset{f^*}{\overset{f_*}{\rightleftarrows}} \mathbf{D}$$

in such a way that the assignment $f \mapsto f_*$ extends to a homomorphism (which is the identity on objects) $\mathit{Cat}\,\mathcal{E} \to (\mathit{Mod}\,\mathcal{E})^{\mathrm{co}}$ between the bicategory of internal categories and functors and that of internal categories and modules.

The category $\mathcal{E}^{\mathbf{C}^{\mathrm{op}}}$ is regarded as a locally internal category over $\mathcal{E}$ in the sense of Betti and Walters [1] (or as an indexed category in the sense of Paré and Schumacher [13]). This means that it is a category enriched in the bicategory *Span* $\mathcal{E}$ whose "objects over I" are the modules $\mathbf{C} \nrightarrow I$ (these are the I-indexed families of objects), substitution along a map $f\colon J \to I$ ("change of parametrization") is provided by composition with f^* (here I and J are regarded as discrete internal categories). If $\alpha\colon \mathbf{C} \nrightarrow I$ and $\beta\colon \mathbf{C} \nrightarrow J$ are two objects, their hom is the right Kan extension $\mathcal{E}^{\mathbf{C}^{\mathrm{op}}}(\alpha, \beta) = [\alpha, \beta]\colon I \nrightarrow J$ of β through α:

$$\begin{array}{ccc} \mathbf{C} & \overset{\alpha}{\nrightarrow} & \mathbf{D} \\ {\scriptstyle\beta}\big\downarrow & \Downarrow & \big\downarrow{\scriptstyle[\alpha,\beta]} \\ J & \overset{1}{\nrightarrow} & J \end{array}$$

For an internal category $\mathbf{C} = (C_0, \gamma)$, the *Yoneda embedding* $Y\colon \mathbf{C} \to \mathcal{E}^{\mathbf{C}^{op}}$ takes the unique object of $\mathbf{C}$ to the "representable module" $\gamma^\circ\colon \mathbf{C} \not\rightarrow C_0$, i.e. to the hom $\gamma\colon \mathbf{C} \not\rightarrow \mathbf{C}$, regarded as an object of $\mathcal{E}^{\mathbf{C}^{op}}$ over the "object of objects" C_0 of $\mathbf{C}$. This embedding is fully faithful in the sense that $\gamma \cong \mathcal{E}^{\mathbf{C}^{op}}(\gamma^\circ, \gamma^\circ)$.

It is easy to see that there is an equivalence of (ordinary) categories:

$$\frac{\mathbf{D} \xrightarrow{\ f\ } \mathcal{E}^{\mathbf{C}^{op}}}{\mathbf{C} \overset{\widehat{f}}{\not\rightarrow} \mathbf{D}} \tag{1}$$

which identifies composition of functors with composition of the corresponding modules. Namely: $\widehat{fg} \cong g^* \otimes_{\mathbf{D}} \widehat{f}$:

$$\frac{\mathbf{B} \xrightarrow{\ g\ } \mathbf{D} \xrightarrow{\ f\ } \mathcal{E}^{\mathbf{C}^{op}}}{\mathbf{C} \overset{\widehat{f}}{\not\rightarrow} \mathbf{D} \overset{g^*}{\not\rightarrow} \mathbf{B}}$$

Recall moreover that, by definition, when it exists, the *colimit* of a functor $g\colon \mathbf{A} \to \mathcal{X}$ (where $\mathbf{A}$ is internal) *indexed* by the module $\alpha\colon \mathbf{A} \not\rightarrow I$ is an object $\alpha * g$ of $\mathcal{X}$ over I, uniquely determined up to isomorphisms, which represents the right Kan extension $[\alpha, g]$ of α through g. By definition thus, the colimit $\alpha * g$ is characterized by a family of isomorphisms:

$$\mathcal{X}(\alpha * g, x') \cong \mathcal{E}^{\mathbf{A}^{op}}(\alpha, \mathcal{X}(g, x'))$$

natural with respect to x'.

The category $\mathcal{E}^{\mathbf{C}^{op}}$ is the free (small) cocompletion over $\mathbf{C}$ in the following sense:

- the colimit $\alpha * g$ is provided by the composite module $\alpha\widehat{g}$, where $\widehat{g}$ corresponds to g as in the equivalence (1),
- for any cocomplete $\mathcal{X}$, composition with Yoneda embedding provides an equivalence of categories:

$$\mathit{Cocont}(\mathcal{E}^{\mathbf{C}^{op}}, \mathcal{X}) \xrightarrow{\ \simeq\ } \mathit{Fun}(\mathbf{C}, \mathcal{X}). \tag{2}$$

The equivalence (2) associates to a functor $f\colon \mathbf{C} \to \mathcal{X}$ the cocontinuous functor $L_f\colon \mathcal{E}^{\mathbf{C}^{op}} \to \mathcal{X}$ described on objects by $\alpha \mapsto \alpha * f$. In particular L_f admits as right (enriched) adjoint the functor $R_f\colon \mathcal{X} \to \mathcal{E}^{\mathbf{C}^{op}}$ described on objects by $x' \mapsto \mathcal{X}(f, x')$.

It is easy to check that the Yoneda embedding Y is "dense", in the sense that any object α in $\mathcal{E}^{\mathbf{C}^{op}}$ can be written as the colimit of Y indexed by α itself.

Slightly extending definitions of Bunge [4] and Bourn [3], one has the following notion:

DEFINITION. A functor $f\colon \mathbf{C} \to \mathcal{X}$, with $\mathbf{C}$ internal and $\mathcal{X}$ cocomplete, is said to be *essential* if the cocontinuous functor $L_f\colon \mathcal{E}^{\mathbf{C}^{op}} \to \mathcal{X}$ associated to it admits a left adjoint. In this situation, we say also that the cocontinuous functor L_f is essential.

2. We now give the notion of weak equivalence.

DEFINITION. A *weak equivalence* of internal categories $\mathbf{C}$ and $\mathbf{D}$ is an equivalence in the bicategory *Mod* $\mathcal{E}$ of internal categories and modules between them.

In other words, to assign a weak equivalence means to assign a pair of modules:

$$\mathbf{C} \underset{\psi}{\overset{\varphi}{\rightleftarrows}} \mathbf{D}$$

endowed with isomorphisms $\varphi \otimes_{\mathbf{C}} \psi \cong 1_{\mathbf{D}}$ and $\psi \otimes_{\mathbf{D}} \varphi \cong 1_{\mathbf{C}}$.

Equivalences of internal categories (i.e. pairs of internal functors f and g such that $fg \cong 1_{\mathbf{D}}$ and $gf \cong 1_{\mathbf{C}}$) are weak equivalences: $f_* \dashv f^* \cong g_*$.

Other examples of weak equivalence are given by internal functors $f\colon \mathbf{C} \to \mathbf{D}$ which are fully-faithful (i.e. the correspondence on arrows is an invertible 2-cell) and surjective on objects (i.e. the correspondence on objects is an epimorphism in $\mathcal{E}$). In this case, one checks that the adjoint pair $f_* \dashv f^*$ is a weak equivalence.

A particular case of weak equivalence is obtained by taking an epimorphism $f\colon C_0 \to D_0$ in $\mathcal{E}$: consider the equivalence relation $f^\circ f$ induced by f (here f° denotes the right adjoint of f regarded in *Span* $\mathcal{E}$, i.e. the span $(f, 1_C)$). The monad $f^\circ f$ provides a category structure $\mathbf{C}$ on C_0, and $f\colon C_0 \to D_0$ is the correspondence on objects of a functor $\mathbf{C} \to D_0$ which is fully-faithful and surjective on objects (here D_0 is regarded as a discrete internal category).

The following result contains the so-called Eilenberg–Watts theorem and was stated by Gouzou and Grunig [7]:

THEOREM 1. *The correspondence L which associates to each $\varphi\colon \mathbf{C} \relbar\joinrel\mapstochar\joinrel\rightarrow \mathbf{D}$ the functor "composition with φ", i.e. the functor $L_\varphi\colon \mathcal{E}^{\mathbf{D}^{\mathrm{op}}} \to \mathcal{E}^{\mathbf{C}^{\mathrm{op}}}$ described on objects by $\alpha \mapsto \alpha \otimes_{\mathbf{D}} \varphi$ for any $\alpha\colon \mathbf{D} \relbar\joinrel\mapstochar\joinrel\rightarrow I$, is an equivalence of (ordinary) categories:*

$$Mod(\mathbf{C}, \mathbf{D}) \simeq Cocts(\mathcal{E}^{\mathbf{D}^{\mathrm{op}}}, \mathcal{E}^{\mathbf{C}^{\mathrm{op}}}). \tag{3}$$

Under the correspondence (3), *modules $\mathbf{C} \relbar\joinrel\mapstochar\joinrel\rightarrow \mathbf{D}$ with a right adjoint correspond to essential functors $\mathcal{E}^{\mathbf{D}^{\mathrm{op}}} \to \mathcal{E}^{\mathbf{C}^{\mathrm{op}}}$. Moreover, φ is a weak equivalence if and only if L_φ is an equivalence.*

Proof. For the equivalence (3), it is enough to consider (1) and to recall that $\mathcal{E}^{\mathbf{D}^{\mathrm{op}}}$ is the free cocompletion of $\mathbf{D}$.

Now, one can check that $\widehat{f}$ has a right adjoint ψ if and only if $- \cdot \widehat{f}$ admits $- \cdot \psi$ as a left adjoint. This proves that the equivalence (3) restricts to modules with a right adjoint from one side and to essential functors from the other one.

Moreover, when L_φ is an equivalence, the functor $L_\varphi g\colon \mathbf{A} \to \mathcal{E}^{\mathbf{C}^{op}}$ and the module $\widehat{g} \otimes_{\mathbf{D}} \varphi\colon \mathbf{C} \nrightarrow \mathbf{A}$ correspond each other in the equivalence (1):

$$\frac{\mathbf{C} \xrightarrow{\ \varphi\ } \mathbf{D} \xrightarrow{\ \widehat{g}\ } \mathbf{A}}{\mathbf{A} \xrightarrow{\ g\ } \mathcal{E}^{\mathbf{D}^{op}} \xrightarrow{\ L_\varphi\ } \mathcal{E}^{\mathbf{C}^{op}}}$$

for any $g\colon \mathbf{A} \longrightarrow \mathcal{E}^{\mathbf{D}^{op}}$. □

Observe that the "if" part of the proof relative to weak equivalences depends only on the fact that $\mathcal{E}^{\mathbf{C}^{op}}$ and $\mathcal{E}^{\mathbf{D}^{op}}$ contain the representable modules. Hence the theorem holds in a more general situation.

DEFINITION. An *intermediate category* over $\mathbf{C}$ is a full subcategory of $\mathcal{E}^{\mathbf{C}^{op}}$ which contains the representable module $\mathbf{C} \nrightarrow C_0$.

THEOREM 2. *The internal categories* $\mathbf{C}$ *and* $\mathbf{D}$ *are weakly equivalent if and only if there exist intermediate categories,* $\widetilde{\mathbf{C}}$ *over* $\mathbf{C}$ *and* $\widetilde{\mathbf{D}}$ *over* $\mathbf{D}$*, which are equivalent.*

Proof. If $\mathbf{C}$ and $\mathbf{D}$ are weakly equivalent, then their intermediate categories $\mathcal{E}^{\mathbf{C}^{op}}$ and $\mathcal{E}^{\mathbf{D}^{op}}$ are equivalent, by the previous theorem.

Conversely, suppose an equivalence of intermediate categories:

$$\sigma\colon \widetilde{\mathbf{C}} \xrightarrow{\ \simeq\ } \widetilde{\mathbf{D}}$$

is given. Then, by embedding $\widetilde{\mathbf{C}}$ and $\widetilde{\mathbf{D}}$ into their respective functor categories, one recovers two modules $\varphi\colon \mathbf{C} \nrightarrow \mathbf{D}$ and $\psi\colon \mathbf{D} \nrightarrow \mathbf{C}$, such that the composites with φ and ψ provide the opposite functors of the given equivalence σ.

Because $\widetilde{\mathbf{C}}$ and $\widetilde{\mathbf{D}}$ contain the representable modules and Yoneda embedding is dense, one has $\varphi \otimes_{\mathbf{C}} \psi \cong 1_{\mathbf{D}}$ and $\psi \otimes_{\mathbf{D}} \varphi \cong 1_{\mathbf{C}}$. □

3. We apply the previous result to Morita equivalence: recall that two internal categories are *Morita equivalent* when their functor categories $\mathcal{E}^{\mathbf{C}^{op}}$ and $\mathcal{E}^{\mathbf{D}^{op}}$ are equivalent. By Theorem 1, this is the same as weak equivalence.

Suppose now that a "notion" of intermediate category is given, i.e. suppose that to any internal category $\mathbf{C}$ one can associate an intermediate category $\widetilde{\mathbf{C}}$ over $\mathbf{C}$. Suppose moreover that this notion is "stable under composition with weak equivalences", i.e. if $\alpha\colon \mathbf{C} \nrightarrow I$ is in $\widetilde{\mathbf{C}}$ and $\psi\colon \mathbf{D} \nrightarrow \mathbf{C}$ is a weak equivalence, then $\alpha \otimes_{\mathbf{C}} \psi\colon \mathbf{D} \nrightarrow I$ is in $\widetilde{\mathbf{D}}$. Then:

COROLLARY. $\mathbf{C}$ *and* $\mathbf{D}$ *are Morita equivalent if and only if* $\widetilde{\mathbf{C}} \simeq \widetilde{\mathbf{D}}$ *for a notion of intermediate category which is stable under composition with weak equivalences.*

For example, a stable notion of intermediate category is provided by the assignment, to any internal category $\mathbf{C}$, of the free cocompletion $\mathcal{J}\mathbf{C}$ with respect to any family $\mathcal{J}$ of indexing modules which contains weak equivalences.

In particular, one can consider the family of adjoint pairs of modules which, after the work of Lawvere [10], is connected with the notion of "Cauchy completion" of a category. Namely, in [10], Lawvere shows that, for a monoidal base category suitably obtained by the preorder $\mathbf{R}^+$ of nonnegative real numbers, enriched categories are all the (nonsymmetric) metric spaces and moreover their adjoint pairs of modules correspond exactly to Cauchy sequences. In these terms, the Cauchy completion $\widetilde{\mathbf{C}}$ of a category $\mathbf{C}$ is a category enriched in *Span* $\mathcal{E}$ whose objects over I are pairs of adjoint modules:

$$I \underset{\beta}{\overset{\alpha}{\rightleftarrows}} \mathbf{C} \quad (\alpha \dashv \beta)$$

and the hom is provided by:

$$\widetilde{\mathbf{C}}((\alpha, \beta), (\alpha', \beta')) = \beta'\alpha.$$

It is known that two categories are Morita equivalent if and only if their Cauchy completions are equivalent (see, for instance, Lindner [11]).

Cauchy completion is a form of cocompleteness: in Street [14] it is shown that a category is Cauchy complete if and only if it admits all colimits indexed by right adjoints. Moreover, these are the "absolute colimits", i.e. those preserved by any functor (Street [15]). In accordance with the fact (Theorem 1) that essential functors $I \to \mathcal{E}^{\mathbf{C}^{\mathrm{op}}}$ correspond to adjoint pairs of modules, in Bunge [4] and in Bourn [3] it is proved that the Cauchy completion of an internal category $\mathbf{C}$ is the full subcategory of "essential points" of $\mathcal{E}^{\mathbf{C}^{\mathrm{op}}}$, i.e. its objects over I are essential functors $I \to \mathcal{E}^{\mathbf{C}^{\mathrm{op}}}$.

The question arises of a "smaller" notion of intermediate category which still could ensure Morita equivalence. In particular, one can ask what happens with the stack completion. Recall (Bunge [4], Bunge and Paré [5], Bourn [3]) that the *stack completion* of an internal category $\mathbf{C}$ is the full subcategory of $\mathcal{E}^{\mathbf{C}^{\mathrm{op}}}$ whose objects are "locally representable" modules, (or *torsors*), i.e. objects over I are modules $\alpha\colon \mathbf{C} \nrightarrow I$ for which there exist an epimorphism $p\colon J \to I$ in $\mathcal{E}$, a functor $f\colon J \to \mathbf{C}$ and a natural isomorphism $\alpha \cong p_* f^*$, where p_* denotes the weak equivalence induced by the epimorphism p. This is also a form of cocompleteness: a category $\mathcal{X}$ is stack complete (or simply a *stack*) exactly when it admits all colimits indexed by torsors.

Torsors are not stable under composition with weak equivalences. However this example suggests to consider those particular modules which are representable up to a general weak equivalence. Namely:

DEFINITION. Say that a module $\varphi\colon \mathbf{C} \nrightarrow I$ is *weakly representable* when there exist a functor $f\colon I \to \mathbf{A}$, a weak equivalence $\alpha\colon \mathbf{C} \nrightarrow \mathbf{A}$ and an isomorphism $\varphi \cong f^* \otimes_{\mathbf{A}} \alpha$.

Now, by considering the full subcategory $\mathcal{W}\mathbf{C}$ of $\mathcal{E}^{\mathbf{C}^{op}}$ given by the weakly representable modules of $\mathbf{C}$, one has a notion of intermediate category which is stable under composition with weak equivalences. Thus:

COROLLARY. *$\mathbf{C}$ and $\mathbf{D}$ are Morita equivalent if and only if their intermediate categories of weakly representable modules $\mathcal{W}\mathbf{C}$ and $\mathcal{W}\mathbf{D}$ are equivalent.*

Clearly, if $\mathbf{C}$ is Cauchy complete, then $\mathcal{W}\mathbf{C}$ is equivalent to the Cauchy completion $\widetilde{\mathbf{C}}$ of $\mathbf{C}$.

In order to find an intermediate category of $\mathbf{C}$ which is strictly contained in $\widetilde{\mathbf{C}}$, we limit ourselves to "Cauchy generators" in the following sense.

When x is an object of the locally internal category $\mathcal{X}$, the *internal full subcategory* it defines is the internal category $Int_{\mathcal{E}}(x)$ determined by the endomorphisms $\mathcal{X}(x, x)$. This category is endowed with an embedding into $\mathcal{X}$. In particular, if $\alpha\colon \mathbf{C} \dashrightarrow I$ is an object of $\mathcal{E}^{\mathbf{C}^{op}}$, the internal full subcategory generated by α in $\mathcal{E}^{\mathbf{C}^{op}}$ is $\mathbf{A} = [\alpha, \alpha]$. The embedding $\mathbf{A} \to \mathcal{E}^{\mathbf{C}^{op}}$ thus provides a canonical module: $\bar{\alpha}\colon \mathbf{C} \dashrightarrow \mathbf{A}$.

DEFINITION. An object x of the locally small category $\mathcal{X}$ over $\mathcal{E}$ is said to be a *generator* if the functor

$$\mathcal{X}(x, -)\colon \mathcal{X} \to \mathcal{E}^{\mathbf{A}^{op}}$$

is fully-faithful.

With the above terminology, one has immediately:

The object α of $\mathcal{E}^{\mathbf{C}^{op}}$ is a generator if and only if

$$\mathcal{E}^{\mathbf{C}^{op}}(\bar{\alpha}, -)\colon \mathcal{E}^{\mathbf{C}^{op}} \to \mathcal{E}^{\mathbf{A}^{op}} \tag{4}$$

is fully-faithful.

DEFINITION. A *Cauchy module* is a module with a left adjoint. A *Cauchy generator* is a Cauchy module which is also a generator of $\mathcal{E}^{\mathbf{C}^{op}}$.

THEOREM 3. *The Cauchy module $\alpha\colon \mathbf{C} \dashrightarrow I$ is a Cauchy generator if and only if the induced module $\bar{\alpha}\colon \mathbf{C} \dashrightarrow \mathbf{A}$ is invertible in $Mod\,\mathcal{E}$, i.e. $\mathbf{C}$ and $\mathbf{A}$ are weakly equivalent:*

$$\mathcal{E}^{\mathbf{C}^{op}} \simeq \mathcal{E}^{\mathbf{A}^{op}}.$$

Proof. Suppose that $\alpha\colon \mathbf{C} \dashrightarrow I$ is a Cauchy module, and denote by γ its left adjoint. If the functor (4) is an equivalence then it is fully-faithful and α is a generator.

Conversely, suppose that the functor (4) is fully-faithful, i.e. that the counit of the adjunction $L_{\bar{\alpha}} \dashv \mathcal{E}^{\mathbf{C}^{op}}(\bar{\alpha}, -)$ is an isomorphism, and consider the right Kan extension $[\bar{\alpha}, 1_{\mathbf{C}}]$ of $1_{\mathbf{C}}$ through $\bar{\alpha}$. One has:

$$[\bar{\alpha}, 1_{\mathbf{C}}] \otimes_{\mathbf{A}} \bar{\alpha} \cong 1_{\mathbf{C}}.$$

Moreover, a calculation shows that $\bar{\alpha} \otimes_{\mathbf{C}} [\bar{\alpha}, 1_{\mathbf{C}}] \cong 1_{\mathbf{A}}$ because $[\bar{\alpha}, 1_{\mathbf{C}}] \cong \gamma$ and $\alpha \otimes_I \gamma \cong [\alpha, \alpha]$. □

It is easy to check that, when ψ is a weak equivalence, $[\alpha, \alpha] \cong [\alpha\psi, \alpha\psi]$. Hence, Cauchy generators provide a notion of intermediate category which is stable under composition with weak equivalences and strictly contained in the Cauchy-completion.

4. We apply the previous result to classical Morita theory. In this case, internal categories are simply one object categories enriched in the monoidal category Ab of Abelian groups (i.e. rings) and a module $R \;+\!\!\!\!\longrightarrow S$ is a left-R, right-S module. Composition of modules is the usual tensor.

The applications we have in mind exploit the close relation sometimes occurring between a ring and the endomorphism ring of one of its modules. If $\alpha\colon R \;+\!\!\!\!\longrightarrow 1$ is a left-R module, the right Kan extension $[\alpha, \alpha]$ provides the endomorphism ring of α and the module α itself can be regarded as a right-$[\alpha, \alpha]$ module. This is $\bar{\alpha}\colon R \;+\!\!\!\!\longrightarrow Int_{\mathcal{E}}(\alpha)$, canonically associated to α as in Theorem 3.

(i) Theorem 3 specializes to a result of Hirata [8]. Namely, in Theorem 1.5 of [8], the assumptions of a module $\alpha\colon R \;+\!\!\!\!\longrightarrow 1$ to be a finitely generated projective generator gives rise to an equivalence between the categories of left-R modules and left $Int_{\mathcal{E}}(\alpha)$ modules, i.e. to the weak equivalence $\bar{\alpha}$ as in our Theorem 3.

(ii) Consider now the extension of a problem originally given by Morita [12].

Suppose that R and S are rings. If there exists an equivalence between their categories of modules, then the endomorphism rings of two corresponding modules are isomorphic. In [12] Morita studies the following question: when does the converse of this property hold?

In categorical terms, we have the following formulation of the problem: when $\mathbf{C}$ and $\mathbf{D}$ are Morita equivalent categories, consider two corresponding objects $\alpha\colon \mathbf{C} \;+\!\!\!\!\longrightarrow I$ and $\beta\colon \mathbf{D} \;+\!\!\!\!\longrightarrow I$, respectively, in $\mathcal{E}^{\mathbf{C}^{op}}$ and $\mathcal{E}^{\mathbf{D}^{op}}$. Then $\alpha \cong \beta \otimes_{\mathbf{D}} \varphi$ for a suitable $\varphi\colon \mathbf{C} \;+\!\!\!\!\longrightarrow \mathbf{D}$. Hence it is easy to verify that the endomorphism categories induced by the corresponding modules are isomorphic:

$$[\beta, \beta] \simeq [\alpha, \alpha]$$

because φ is invertible in $Mod\,\mathcal{E}$. The problem now regards the converse of this property. The answer is provided by the notion of Cauchy-generators. Indeed, as a direct consequence of Theorem 3, one has:

COROLLARY. *Let* $\mathbf{C}$ *and* $\mathbf{D}$ *be internal categories in* $\mathcal{E}$, *and suppose that* α: $\mathbf{C} \;+\!\!\!\!\longrightarrow I$ *and* β: $\mathbf{D} \;+\!\!\!\!\longrightarrow I$ *are Cauchy generators in* $\mathcal{E}^{\mathbf{C}^{op}}$ *and* $\mathcal{E}^{\mathbf{D}^{op}}$ *respectively.*

If the internal full subcategories $\mathbf{A} = [\alpha, \alpha]$ *and* $\mathbf{B} = [\beta, \beta]$ *are isomorphic, then* $\mathbf{C}$ *and* $\mathbf{D}$ *are weakly (i.e. Morita) equivalent.*

Proof. The Cauchy generator α gives rise to a module $\bar{\alpha}$: $\mathbf{C} \twoheadrightarrow \mathbf{A}$ which is invertible in *Mod* $\mathcal{E}$ and similarly for β. The composite module $[\bar{\beta}, 1_{\mathbf{D}}] \otimes_{\mathbf{A}} \bar{\alpha}$ is an invertible module $\mathbf{C} \twoheadrightarrow \mathbf{D}$, hence these categories are Morita equivalent. □

5. Now, let $\mathbf{C} = (C_0, \gamma)$ be a *promonoidal* internal category in the sense of Day [6]. This means that $\mathbf{C}$ is endowed with a co-monoidal structure in the bicategory of modules, i.e if $\mathbf{C} \times \mathbf{C} = (C_0 \times C_0, \gamma \times \gamma)$ is the product monad in *Span* $\mathcal{E}$, there are given modules P: $\mathbf{C} \twoheadrightarrow \mathbf{C} \times \mathbf{C}$ and J: $\mathbf{C} \twoheadrightarrow 1$ (here 1 is the terminal object of $\mathcal{E}$ regarded as a trivial internal category) which behave as co-multiplication and co-identity respectively (co-associativity and co-identity laws are assigned up to isomorphisms).

In particular, $\mathbf{C}$ could be *monoidal*, with tensor product T: $\mathbf{C} \times \mathbf{C} \to \mathbf{C}$ and identity e: $1 \to \mathbf{C}$. In this case the promonoidal structure is given by the modules $P = T^*$ and $J = e^*$.

From [6], there is a monoidal structure in $\mathcal{E}^{\mathbf{C}^{op}}$, with tensor product $\otimes$ defined on objects as follows: if α: $\mathbf{C} \twoheadrightarrow I$ and β: $\mathbf{C} \twoheadrightarrow J$ are in $\mathcal{E}^{\mathbf{C}^{op}}$ over I and J respectively, then $\alpha \otimes \beta$ is the object, over $I \times J$, defined by the composition:

$$\mathbf{C} \xrightarrow{P} \mathbf{C} \times \mathbf{C} \xrightarrow{\alpha \times \beta} I \times J$$

while the identity is provided directly by J: $\mathbf{C} \twoheadrightarrow 1$, regarded as an object of $\mathcal{E}^{\mathbf{C}^{op}}$.

Observe that, if P is represented as a functor:

$$\widehat{P}: \mathbf{C} \times \mathbf{C} \to \mathcal{E}^{\mathbf{C}^{op}}$$

by the equivalence (1) the computation of colimits in $\mathcal{E}^{\mathbf{C}^{op}}$ provides the "convolution" formula:

$$\alpha \otimes \beta \cong (\alpha \times \beta) * \widehat{P},$$

i.e. the colimit of $\widehat{P}$ indexed by $\alpha \times \beta$.

For monoidal categories, Johnson in [9] defines the notion of *monoidal Morita equivalence.* This notion is easily extended to promonoidal categories.

DEFINITION. The promonoidal categories $\mathbf{C}$ and $\mathbf{D}$ are said to be *monoidally Morita equivalent* if there is a monoidal equivalence $\mathcal{E}^{\mathbf{C}^{op}} \cong \mathcal{E}^{\mathbf{D}^{op}}$.

Applying Theorem 1 (Section 2) one has:

THEOREM 4. *The promonoidal categories* $\mathbf{C}$ *and* $\mathbf{D}$ *are monoidally Morita equivalent if and only if there is a weak equivalence* α: $\mathbf{C} \twoheadrightarrow \mathbf{D}$ *that preserves the promonoidal structures* (P, J) *and* (Q, K) *of* **C** *and* **D** *respectively, up to isomorphisms*:

$$\begin{array}{ccc} \mathbf{C} & \overset{\alpha}{\longrightarrow\!\!\!\!\!\!\!+\;\;\;} & \mathbf{D} \\ {\scriptstyle P}\big\downarrow & \cong & \big\downarrow{\scriptstyle Q} \\ \mathbf{C}\times\mathbf{C} & \overset{\alpha\times\alpha}{\longrightarrow\!\!\!\!\!\!\!+\;\;\;} & \mathbf{D}\times\mathbf{D} \end{array} \qquad \text{and} \qquad \begin{array}{ccc} \mathbf{C} & \overset{J}{\longrightarrow\!\!\!\!\!\!\!+\;\;\;} & 1 \\ {\scriptstyle \alpha}\big\downarrow & \cong & \big\downarrow{\scriptstyle 1} \\ \mathbf{D} & \overset{K}{\longrightarrow\!\!\!\!\!\!\!+\;\;\;} & 1 \end{array}$$

References

1. Betti, R. and Walters, R. F. C.: Closed bicategories and variable category theory, *Dip. Mat. Univ. Milano* **5** (1985).
2. Betti, R. and Walters, R. F. C.: On completeness of locally-internal categories, *J. Pure Appl. Algebra* **47** (1987), 105–117.
3. Bourn, D.: Une autre proprieté universelle pour le champ associé, *Cahiers Topologie Géom. Différentielle Catégoriques* **21** (1980), 403–410.
4. Bunge, M.: Stack completions and Morita equivalence for categories in a topos, *Cahiers Topologie Géom. Différentielle Catégoriques* **20** (1979), 401–436.
5. Bunge, M. and Paré, R.: Stacks and equivalence of indexed categories, *Cahiers Topologie Géom. Différentielle Catégoriques* **20** (1979), 373–400.
6. Day, B.: On closed categories of functors, Lecture Notes in Math. 137, 1970, pp. 1–38.
7. Gouzou, M.-F. and Grunig, R.: Caractérisation de Dist, *C.R. Acad. Sci. Paris* **276** (1973), 519–521.
8. Hirata, K.: Some types of separable extensions of rings, *Nagoya Math. J.* **33** (1968), 107–115.
9. Johnson, S. R.: Monoidal Morita equivalence, *J. Pure Appl. Algebra* **59** (1989), 169–177.
10. Lawvere, F. W.: Metric spaces, generalized logic, and closed categories, *Rend. Sem. Mat. Fis. Milano* **43** (1973), 135–166.
11. Lindner, H.: Morita equivalences of enriched categories, *Cahiers Topologie Géom. Différentielle Catégoriques* **15** (1974), 377–397.
12. Morita, K.: Category-isomorphisms and endomorphism ring of modules, *Trans. Amer. Math. Soc.* **103** (1962), 451–469.
13. Paré, R. and Schumacher, D.: Abstract families and the adjoint functor theorem, Lecture Notes in Math. 661, 1978, pp. 1–125.
14. Street, R. H.: Enriched categories and cohomology, *Quaestiones Math.* **6** (1983), 265–283.
15. Street, R. H.: Absolute colimits in enriched categories, *Cahiers Topologie Géom. Différentielle Catégoriques* **24** (1983), 377–379.

Applied Categorical Structures **8:** 317–326, 2000.
G. Brümmer & C. Gilmour (eds), Papers in Honour of Bernhard Banaschewski.

Isolated Submodules and Skew Fields

TEMPLE H. FAY
Department of Mathematics, University of Southern Mississippi, Hattiesburg, MS 39406-5045, U.S.A., e-mail: thfay@mozart.st.usm.edu

STEPHAN V. JOUBERT
Department of Mathematical Technology, Technikon Pretoria, Pretoria West 0002, South Africa e-mail: 100075.2440@compuserve.com

(Received: 11 October 1996; accepted: 17 October 1997)

Abstract. We study the generally distinct concepts of isolated submodule, honest submodule, and relatively divisible submodule for unital right R-modules, where R is an associative ring with identity. This is accomplished by studying a certain subset called the Q-torsion subset relative to a subset Q (sometimes a right ideal but not always) of R. The Q-isolator turns out to always to be a categorical closure operator and the notion of Q-honest is an 'operator' but need not be a closure operator. It is shown that the notions of Q-isolated and Q-honest coincide precisely when the Q-honest operator is a closure operator and this happens precisely when all submodules are Q-honest. As a corollary, we obtain when $Q = R$, every submodule is honest if and only if every submodule is isolated if and only if R is a skew field. We also determine a new characterization of a right Ore domain.

Mathematics Subject Classifications (2000): 16A90, 16A36.

Key words: isolated, honest, relatively divisible, skew field.

1. Introduction

Closure operators have been used with effectiveness in a variety of settings, particularly with respect to categorical compactness [5, 6, 11]. Not surprisingly, specific closure operators have a long history and have played a fundamental role in a variety of areas of mathematics, most notably in topology. However, in the category of all groups, the isolator of a subgroup (the intersection of all isolated subgroups containing the subgroup) is particularly useful in studying torsion-free groups and for torsion-free Abelian groups, purity of a subgroup is equivalent to that subgroup being isolated. Other types of closure have been used in categories of modules, for example, a submodule B of a module A is called T-*closed*, for a torsion-theory T, provided A/B is T-torsion-free [16, 9, 6].

In this note, we study the concept of isolated submodule and the related notions of honest [1, 15] and relatively divisible [13, 14] submodules for unital right R-modules, where R is an associative ring with identity. In doing so, we obtain characterizations of when R is a skew field and when R is a right Ore domain that appear to be new. In general, isolatedness implies honesty which in turn implies relatively divisible, and these are distinct notions. However, we will show that R is a skew field if and only if the three notions coincide on all R-modules. This is accomplished by studying a certain *subset* of an R-module called the *Q-torsion subset* relative to a *subset* Q (frequently a right ideal) of R. We end by showing that R is a right Ore domain iff for each right R-module the isolator of the zero submodule is precisely the *R-torsion subset.*

2. Preliminaries

We work in the category of all unital right modules and module homomorphisms over an associative ring R with identity. We let Q denote a *subset* of R which contains 0 and is not the 0 ideal. We say that a submodule B of module A is *Q-isolated* (or *Q-superhonest*) in A, if $0 \neq q \in Q$, $x \in A$ and $xq \in B$ implies $x \in B$. The submodule B is called *Q-honest* in A, if $q \in Q$, $x \in A$ and $0 \neq xq \in B$ implies $x \in B$ [14]. Finally, the submodule B is called *Q-RD* (*Q-relatively divisible*) in A, if $Bq = B \cap Aq$ for every $q \in Q$. If $Q = R$, we talk of the submodule B being, respectively, *isolated, honest* or *RD* in A.

It is clear that Q-isolation implies Q-honesty which in turn implies Q-relative divisibility. It follows that these are all distinct notions [14]

Isolated subgroups of an arbitrary group [18] and isolated subgroups of an Abelian group are well known subobjects (see [14, 11] and references therein). Honest subgroups of Abelian groups were introduced by [1].

Naturally the strongest results are obtained when we choose Q to be an ideal of R, but some interesting, classical examples in Abelian groups (cf. Example 2.1) and general modules (cf. Section 6) lead us to include the case when Q is not necessarily an ideal. So, in order to develop the most general theory and to cover some traditional concepts in Abelian group and module theory, we start out with Q being a subset and then see what minimum properties Q needs in order to develop the theory. For instance, if Q is any set of integers with the property that each positive divisor of each nonzero element of Q is in Q, then most of the theorems on pure subgroups of Abelian groups as given in [8] and the purities of Honda [12], Maranda [17], and Yahya [20] carry over *mutatis mutandis* to Q-RD (i.e. Q-pure) [19]. If Q is also multiplicatively closed, then it is possible to classify all Q-honest subgroups of an Abelian group [15].

EXAMPLE 2.1 For Abelian groups, consider the prime numbers $\mathbb{P}$ and let $Q = \mathbb{P} \cup \{1, 0\}$. Then Q-RD is exactly the concept of **neatness.** On the other hand, if we take $Q = \{p^k : k = 0, 1, 2, \ldots\} \cup \{0\}$ for some $p \in \mathbb{P}$, then Q-RD is exactly p-**purity** and because Q is multiplicatively closed, all Q-honest subgroups of a given group can be classified.

LEMMA 2.2 [13, p. 89] *A cyclic module has only the trivial submodules as honest submodules and at most the trivial submodules as isolated submodules.*

For any module, it is easy to see that the intersection of Q-isolated submodules is Q-isolated and that the intersection of Q-honest submodules is Q-honest. The intersection of Q-RD submodules need not be Q-RD as pure subgroups of Abelian groups are well known not to be closed under intersection.

Since Q-isolatedness, Q-honesty, and Q-relative divisibility are relative properties, it is of interest to note that each of these properties is transitive in the sense that, for example, if A is a module having submodules B and C with $C \leq B$, then C being Q-isolated in B and B being Q-isolated in A implies C is Q-isolated in A. The following proposition is easy to verify.

PROPOSITION 2.3 *Let A be a module with submodules B and C and assume $C \leq B$. Then the following hold:*

(i) *if B is Q-honest (respectively, Q-RD) in A, then B/C is Q-honest (Q-RD) in A/C.*
(ii) *if C is Q-honest (respectively, Q-RD) in A, B/C is Q-honest (Q-RD) in A/C, then B is Q-honest (Q-RD) in A.*
(iii) *B is Q-isolated in A if and only if B/C is Q-isolated in A/C.*

If $f\colon A \to B$ is a module homomorphism and C is a Q-isolated submodule of B, then the inverse image of C in A is Q-isolated in A. However, Q-honesty and Q-relative divisibility are not preserved under inverse images as the following example shows.

EXAMPLE 2.4 Let $R = Q = \mathbb{Z}$ (the ring of integers). If $0 \neq n \in R$, consider the map $f = \Delta\eta\colon \mathbb{Z} \to \mathbb{Z}(n) \oplus \mathbb{Z}(n)$ defined in the obvious way in the following diagram of $\mathbb{Z}$-modules and $\mathbb{Z}$-homomorphisms:

$$\mathbb{Z} \xrightarrow{\eta} \mathbb{Z}/n\mathbb{Z} \xrightarrow{\Delta} \mathbb{Z}(n) \oplus \mathbb{Z}(n).$$

Here $\mathbb{Z}(n) \oplus 0$ is Q-honest and Q-RD in $\mathbb{Z}(n) \oplus \mathbb{Z}(n)$, but $f^{\leftarrow}(\mathbb{Z}(n) \oplus 0) = n\mathbb{Z}$ is a proper submodule of a cyclic module and so cannot be Q-honest in $\mathbb{Z}$ by Lemma 2.2. It is clearly not Q-RD in $\mathbb{Z}$.

Following [6], we say that an *operator K on submodules* assigns for each module A and each submodule B of A, a submodule $K^A(B)$ of A such that the following hold:
(i) $B \leq K^A(B)$ (*expansive*);
(ii) $C \leq B \leq A$ implies $K^A(C) \leq K^A(B)$ (*monotone*).

An operator K is called *idempotent* if for every module A and submodule B, $K^A(B) = K^A(K^A(B))$; it is called *weakly hereditary* provided $K^{K^A(B)}(B) = K^A(B)$ for every module A and submodule B.

For an operator K, a submodule B of A is called *K-closed* if $K^A(B) = B$ and is called *K-dense* if $K^A(B) = A$. Similarly, for homomorphism $f\colon A \to B$ is called *K-closed* if $K^B(f(A)) = f(A)$ and *K-dense* if $K^B(f(A)) = B$. Being motivated by the classical notion of closure operator arising from topology, we have the analogue of Hausdorff space for operators: for an operator K, a module C is called *K-separated* provided for each K-dense homomorphism α: $A \to B$ and each pair of homomorphisms f, g: $B \to C$, $f\alpha = g\alpha$ implies that $f = g$. Finally, again in analogy with topological concepts, an operator K is called a *closure operator* provided it is expansive, monotone and for each homomorphism f: $A \to B$ and submodule C of A, $f(K^A(C)) \leq K^B(f(C))$; this latter condition is called the *continuity condition.*

3. Honesty and Isolatedness as Closure Operators

Now let us examine two examples of operators. For a module A with submodule B, define the *Q-honest-operator,* $h_Q(-)$, as follows:

$$h_Q^A(B) = \bigcap\{C \leq A \mid B \leq C, C \text{ is } Q\text{-honest in } A\}.$$

Similarly we define the *Q-isolator,* $I_Q(-)$, as follows:

$$I_Q^A(B) = \bigcap\{C \leq A \mid B \leq C, C \text{ is } Q\text{-isolated in } A\}.$$

Because the intersection of Q-isolated submodules are Q-isolated and similarly, the intersection of Q-honest submodules is Q-honest, we have a submodule B of the module A is Q-isolated in A if and only if $I_Q^A(B) = B$ and is Q-honest in A if and only if $h_Q^A(B) = B$. The next proposition is straight forward and its proof is omitted.

PROPOSITION 3.1 *The Q-honest-operator, $h_Q(-)$, is a weakly hereditary, idempotent (but not necessarily closure) operator. The Q-isolator is a weakly hereditary, idempotent closure operator.*

Example 2.4 provides us with an example to show that $h_Q(-)$ is not necessarily a closure operator. Indeed,

$$f\big(h_Q^{\mathbb{Z}}(n\mathbb{Z})\big) = f(\mathbb{Z}) = \Delta\mathbb{Z}(n) \supset h_Q^{\mathbb{Z}(n)\oplus\mathbb{Z}(n)}\big(f(n\mathbb{Z})\big) = h_Q^{\mathbb{Z}(n)\oplus\mathbb{Z}(n)}(0) = 0$$

and so $h_Q(-)$ does not necessarily satisfy the continuity condition.

Thus it is of interest to ask when the Q-honest-operator is a closure operator; that is, when does it satisfy the continuity condition?

THEOREM 3.2 *The following statements are equivalent:*

(i) *The Q-honest operator is a closure operator.*
(ii) *All submodules are Q-honest in any module A.*

Proof. (i) $\Rightarrow$ (ii) Let A be a module and B be a submodule. Let η: $A \to A/B$ denote the canonical surjection. Then it is clear that

$$\eta\big(h_Q^A(B)\big) \leq h_Q^{A/B}\big(\eta(B)\big) = h_Q^{A/B}(0) = 0,$$

because 0 is always a Q-honest submodule. Hence $h_Q^A(B) \leq B$, that is $B = h_Q^A(B)$.

(ii) $\Rightarrow$ (i) For any homomorphism $f\colon A \to C$, and submodule B of A,

$$f\left(h_Q^A(B)\right) = f(B) = h_Q^C\left(f(B)\right). \qquad \square$$

COROLLARY 3.3 *The following statements are equivalent:*

(i) *The honest-operator* ($Q = R$) *is a closure operator on the category of all right R-modules.*
(ii) *The honest-operator is a closure operator on the category of all left R-modules.*
(iii) *R is a skew field.*

Proof. We demonstrate that (i) is equivalent to (iii). For the necessity we see that, according to Lemma 2.2, R has only the trivial right ideals. For the sufficiency, assume that A is a right R-module with submodule B, and let $x \in A$, $q \in R$ with $0 \neq xq \in B$. The right annihilator $\rho(x)$ of x is a right ideal in R and so $\rho(x) = 0$. We have $xq \in xR$, thus

$$0 \neq xR \cap B \leq xR \cong R/\rho(x) \cong R.$$

Consequently $xR \cap B = xR$ and so $x \in B$. $\square$

4. The Q-isolator and Separation

A subset of the module A which has received much attention (see for example [14]), is

$$t_Q(A) = \{x \in A \mid \exists\, 0 \neq q \in Q \text{ such that } xq = 0\}$$

called the *Q-torsion subset* of A. It is in general not a submodule of A, even if we choose Q to be an ideal.

Let us call a module A a *Q-torsionfree* module provided $t_Q(A) = 0$. These considerations give rise to a class of modules of some interest, namely

$$\underline{t_Q} = \{A \mid A = t_Q(A)\}.$$

Along the same lines, we define an idempotent preradical τ_Q ([6, p. 53]) and a class $\mathcal{T}_Q$ as follows. For any module A, we define $\tau_Q(A) = I_Q^A(0)$ where, of course, $I_Q(-)$ denotes the Q-isolator. It is easy to see that $\tau_Q(-)$ is a preradical; the class $\mathcal{T}_Q$ is the so-called stabilizer class of $\tau_Q(-)$ and is defined by

$$\mathcal{T}_Q = \{A \mid A = \tau_Q(A)\}.$$

The class of all τ_Q-torsionfree modules:

$$\mathcal{F}_Q = \{A \mid \tau_Q(A) = 0\}$$

is a torsionfree class for some torsion theory because it is hereditary (i.e. closed under submodules), closed under formation of direct products and extensions.

PROPOSITION 4.1 *For any R-module A, the following hold:*

(i) *The map* $0 \to \tau_Q(A)$ *is* I_Q*-dense.*
(ii) $t_Q(A)$ *is contained in any Q-isolated submodule of B. In particular* $t_Q(A) \subseteq \tau_Q(A)$.
(iii) *If* $\underline{T}$ *is a non-trivial hereditary pretorsion class, then* $\underline{T} \subseteq \underline{t_R}$.

Proof. (i) follows from the idempotency of τ_Q, whereas (ii) and (iii) are easy to see. □

The equivalence of (vii) and (viii) in the next theorem has been proved in a more general setting in [4] and [7] and need not be proved here.

THEOREM 4.2 *The following statements are equivalent for any right R-module A:*

(i) *The notions of Q-isolatedness, Q-honesty and Q-relative divisibility coincide for submodules of A.*
(ii) *The notions of Q-isolation and Q-RD coincide for submodules of A.*
(iii) *The notions of Q-isolation and Q-honesty coincide for submodules of A.*
(iv) 0 *is Q-isolated in A.*
(v) $\tau_Q(A) = 0$.
(vi) $t_Q(A) = 0$.
(vii) *A is* I_Q*-separated.*
(viii) *The diagonal map* $\Delta\colon A \to A \times A$ *is* I_Q*-closed.*

Proof. Using Proposition 4.1 and the fact that 0 is always Q-honest in A, the equivalence of (i) through (vi) is easy to see.

(vi) $\Rightarrow$ (vii) Assume that

$$C \xrightarrow{e} B; \qquad I_Q^B(eC) = B; \qquad B \xrightarrow{f} A; \qquad B \xrightarrow{g} A; \quad \text{and} \quad fe = ge.$$

Now let

$$D = \{x \in B \mid f(x) = g(x)\}.$$

Clearly $eC \leq D$ and so $I_Q^B(D) = B$. Hence we need only show that D is Q-isolated in B whence $B = D$ and $f = g$. Indeed, if $0 \neq q \in Q$; $x \in B$ and $qx \in D$, then $f(qx) = g(qx)$. Thus $q(f(x) - g(x)) = 0$. The hypothesis now yields $f(x) = g(x)$, that is $x \in D$.

(vii) $\Rightarrow$ (v) Note that we have the following maps:

$$0 \xrightarrow{e} \tau_Q(A) \xrightarrow{0} A \quad \text{and} \quad 0 \xrightarrow{e} \tau_Q(A) \overset{\mu}{\hookrightarrow} A \quad \text{with } \mu e = 0e.$$

Thus Proposition 4.1 and the hypothesis yield $\mu = 0$. □

COROLLARY 4.3 *The class of all Q-torsionfree modules is exactly the class of τ_Q-torsionfree modules.*

5. The Honest Closure Operator

In this section we determine necessary and sufficient conditions for the honest-operator to be a closure operator.

PROPOSITION 5.1 *Consider the following statements:*

(i) *Every right R-module A satisfies any one of the equivalent conditions listed in Theorem* 4.2.
(ii) $h_Q(-) = I_Q(-)$ *on the category of all right R-modules.*
(iii) $h_Q(-)$ *is a closure operator in the category of all right R-modules.*
(iv) *For each right R-module A, all submodules are Q-honest in A.*

Then (i) $\Leftrightarrow$ (ii) $\Rightarrow$ (iii) $\Leftrightarrow$ (iv).

Proof. We need only demonstrate (iii) $\Leftrightarrow$ (iv) as the other implications are obvious:

(iii) $\Rightarrow$ (iv) Assume that A is a right R-module and that $B \leq A$. Then the natural epimorhism α: $A \to A/B$ is such that

$$\alpha\big(h_Q^A(B)\big) \leq h_Q^{A/B}\big(\alpha(B)\big) = h_Q^{A/B}(0) = 0$$

because 0 is always a Q-honest submodule. Hence $h_Q^A(B) \leq B$, that is $B = h_Q^A(B)$.

(iv) $\Rightarrow$ (iii) For any f: $A \to N$, and $B \leq A$,

$$f\big(h_Q^A(B)\big) = f(B) = h_Q^N\big(f(B)\big). \qquad \square$$

We need a motivating example for the next result.

EXAMPLE 5.2 Let $R = R_* \oplus R_*$ (ring direct sum) where R_* is a ring with identity e say. The ideal $K = R_* \oplus 0$ is a subring (with identity $y = (e, 0)$) of R.

LEMMA 5.3 *If $0 \neq Q$ is a right ideal in R so that $Q = yQ$ for some $y \in Q$, then as an R-module, Q has only the trivial submodules as Q-honest submodules and at most the trivial submodules as Q-isolated submodules.*

Proof. Assume that B is a Q-honest (Q-isolated) submodule of Q and $x \in B$. Then $x = yq$ for some $q \in Q$. If $x \neq 0$ (then $q \neq 0$), and so $y \in B$. $\square$

THEOREM 5.4 *Assume that Q is a right ideal in R and that, as subring, it has an identity. Then the following statements are equivalent:*

(i) *All R-submodules which are unital right Q-modules are Q-honest in each right R-module A.*

(ii) *Q is simple as a right R-module.*
(iii) *For each $0 \neq q \in Q$, $qR = Q$.*
(iv) *All R-submodules which are unital right Q-modules are Q-isolated in each right R-module A.*

Proof. (i) $\Rightarrow$ (ii) Assume that B is a right ideal of R contained in Q. Then B is a unital right Q-module, and so B is Q-honest in Q. The result follows from the previous lemma.

(iii) $\Rightarrow$ (iv) Assume that the identity of Q is e, $x \in A$ and that $B \leq_R A$ with B a right Q-module. If $0 \neq q \in Q$ with $xq \in B$, then according to the hypothesis there is an $r \in R$ such that $x = xe = xqr \in B$.

The rest of the proof is trivial. □

COROLLARY 5.5 *The following statements are equivalent:*

(i) $h_R(-) = I_R(-)$.
(ii) *0 is isolated in each right R-module A.*
(iii) $\tau_R(A) = 0$ *for each right R-module A.*
(iv) $t_R(A) = 0$ *for each right R-module A.*
(v) *The notions of isolatedness and honesty coincide for each right R-module A.*
(vi) *The notions of isolatedness and RD coincide for each right R-module A.*
(vii) *The three notions of isolatedness, honesty and RD coincide for each right R-module A.*
(viii) *The honest-operator $h(-)$ is a closure operator on the category of all right R-modules.*
(ix) *All submodules are honest in each right R-module A.*
(x) *All submodules are isolated in each right R-module A.*
(xi) *A is I_R-separated for each right R-module A.*
(xii) *The diagonal map Δ: $A \to A \times A$ is I_R-closed for each right R-module A.*
(xiii) *R is a skew field.*

Remark 1. All 'right' conditions can be replaced with 'left' conditions in Corollary 5.5 since (xiii) is left-right symmetric.

6. When is $\tau_Q(A) = t_Q(A)$ not Necessarily Trivially?

Recall that a subset S of R is called a *right permutable set* provided it is closed under multiplication and for each $s \in S$ and $r \in R$, there exists an $s' \in S$ and an $r' \in R$ such that $rs' = sr'$. It is folk-lore that if $Q\backslash 0$ is a right permutable set, then $t_Q(A)$ is a submodule of each module A and $t_Q(A/t_Q(A)) = 0$. In this case, according to Theorem 3.8 of [14], $t_Q(A)$ is a Q-isolated submodule of A, hence, according to 4.1, $\tau_Q(A) = t_Q(A)$. By choosing Q to be an ideal we obtain the converse.

THEOREM 6.1 *Assume that Q is an ideal of R. Then the following statements are equivalent:*

(i) $\tau_Q(A) = t_Q(A)$ *for each right R-module A.*
(ii) $Q\backslash 0$ *is a right permutable set.*
(iii) $Q\backslash 0$ *is multiplicatively closed and the intersection of two non-zero right ideals of R contained in Q is non-zero.*

Proof. If $\tau_Q(A) = t_Q(A)$, then $t_Q(A)$ is a Q-isolated (i.e. Q-superhonest) submodule of A and so the result follows from [14] Theorem 5.6. □

COROLLARY 6.2 $\tau_R(A) = t_R(A)$ *for each right R-module A if and only if R is a right Ore domain.*

Acknowledgement

The first author thanks the Technikon Pretoria and the Department of Mathematical Technology for financial support.

References

1. Abian, A. and Rinehart, D.: Honest subgroups of Abelian groups, *Rend. Circ. Math. Palermo* **12** (1963), 353–356.
2. Cassidy, C., Hebert, M. and Kelly, G. M.: Reflective subcategories, localizations and factorization systems, *J. Austral. Math. Soc. Ser. A* **38** (1985), 287–329.
3. Castellini, G.: Compact objects, surjectivity of epimorphisms and compactifications, *Cahiers Topologie Géom. Différentielle Catégoriques* **XXXI**(1) (1990), 53–65.
4. Clementino, M. M., Giuli, E. and Tholen, W.: Topology in a category: Compactness, *Port. Math.* **53** (1996), 397–433.
5. Dikranjan, D. and Giuli, E.: Closure operators I, *Topology Appl.* **27** (1987), 129–143.
6. Dikranjan, D. and Giuli, E.: Factorizations, injectivity and compactness in categories of modules, *Comm. Algebra* **19** (1991), 45–83.
7. Dikranjan, D. and Tholen, W.: *Categorical Structure of Closure Operators*, Kluwer, Dordrecht, 1995.
8. Fuchs, L.: *Infinite Abelian Groups I, II*, Academic Press, New York, 1971, 1973.
9. Fay, T. H.: Compact modules, *Comm. Algebra* **16** (1988), 1209–1219.
10. Fay, T. H.: Remarks on the Mal'cev completion of torsion-free locally nilpotent groups, *Cahiers Topologie Géom. Différentielle Catégoriques* **XXXV** (1994), 75–84.
11. Fay, T. H. and Walls, G. L.: Regular and normal closure operators and categorical compactness for groups, *Applied Categorical Structures* **3** (1995), 261–278.
12. Honda, K.: Realism in the theory of Abelian groups, I, *Comment. Math. Univ. St. Paul.* **5** (1956), 145–147.
13. Joubert, S. V. and Schoeman, M. J.: Superhonesty for modules and Abelian groups, *Chinese J. Math.* **12** (1984), 87–95.

14. Joubert, S. V. and Schoeman, M. J.: Generalized superhonesty for modules and Abelian groups, *Chinese J. Math.* **13** (1985), 1–13.
15. Joubert, S. V. and Schoeman, M. J.: A note on generalized honest subgroups of Abelian groups, *Comment. Math. Univ. St. Paul.* **36** (1987), 145–147.
16. Lambek, J.: *Torsion Theories, Additive Semantics, and Rings of Quotients*, Lecture Notes in Math. 117, Springer-Verlag, Berlin, 1971.
17. Maranda, J. M.: On pure subgroups of Abelian groups, *Arch. Math.* **II** (1960), 1–13.
18. Plotkin, B. I.: Generalized solvable and generalized nilpotent groups, *Uspekhi Mat. Nauk (N. S.)* **13** (1958), 89–172 (in Russian); *AMS Transl. (2)* **17** (1961), 29–115.
19. van Dyk, T. J.: Suiwerheid van ondergroepe in die Abelse groepteorie, Thesis, University of Pretoria, 1979.
20. Yahya, S. M.: P-pure exact sequences and the group of P-pure extensions, *Ann. Univ. Sci. Budapest* **5** (1962), 179–191.

Applied Categorical Structures **8:** 327–337, 2000.
G. Brümmer & C. Gilmour (eds), Papers in Honour of Bernhard Banaschewski.

Clone Segments of the Tychonoff Modification of Space

Dedicated to Bernhard Banaschewski on his 70th Birthday

A. TOZZI
Dipartimento di Matematica Pura a Applicata, Università Dell'Aquila, via Vetoio, 67100 l'Aquila, Italy

V. TRNKOVÁ
Mathematics Institute of Charles University, Sokolovská 83, 18600 Prague 8, Czech Republik

(Received: 14 October 1996; accepted: 7 July 1998)

Abstract. For every natural number $n > 1$, we construct a topological space X such that the clone segments of its Tychonoff modification TMX verify $Clo_k(X) = Clo_k(TMX)$ if $k < n$ and $Clo_n(X)$ is not isomorphic to $Clo_n(TMX)$.

Key words: Tychonoff modification, products, continuous maps, clones.

AMS Subject Classifications (2000): 08A40, 54C05.

1. Introduction

In [9] the second author studied the problem of [8] to find two topological spaces, X and Y, whose monoids of continuous selfmaps are isomorphic while their clones are not. We recall that the clone $Clo(X)$ of a topological space X is the full subcategory of *Top*, the category of topological spaces and continuous maps, generated by all finite powers X^n of the space X.

She introduced the notion of semirigid spaces and provided two metrizable (semirigid) spaces with the same monoid of continuous selfmaps and nonisomorphic clones.

A modification of this construction was given in [10] where it was proved that

– *For every natural number $n > 1$, there exists a metric space X such that $Clo_{n-1}(X) = Clo^{(u)}_{n-1}(X)$ but neither $Clo_n(X)$ admits a full embedding into $Clo^{(u)}_n(X)$ nor $Clo^{(u)}_n(X)$ into $Clo_n(X)$*

where $Clo(X)$ ($Clo^{(u)}(X)$) is the category of all finite topological (uniform) powers of X and all their continuous (uniformly continuous) maps and $Clo_n(X)$ ($Clo^{(u)}_n(X)$) their full subcategories generated by all X^k, $k < n$.

We present here another modification of these constructions. The aim of this paper is to prove the theorem below.

THEOREM. *For every* $a \in \{2, \ldots, \infty\}$ *there exists a space* X *such that* $Clo_n(X) = Clo_n(TMX)$ *iff* $n < a$ *and* $Clo_a(X)$ *is not isomorphic to* $Clo_a(TMX)$, *where* TMX *is the Tychonoff modification of* X.

The proof of the above theorem requires a finer manipulation of the constructions in [9 – 11] and [6] enriched by some reasoning concerning Tychonoff modifications.

Recall that the Tychonoff modification of a topological space X, TMX, consists of the same underlying set as that of X equipped with the topology generated by all cozero subsets of X. This is the largest Tychonoff topology contained in the topology of X.

PROBLEM. Let us mention explicitly that our space X satisfying the above theorem is Hausdorff but not regular. We do not know whether the theorem is true if we require X to be regular as well.

NOTATION. (a) If $f\colon X \to Y$ is a map, $\operatorname{Im} f$ will denote $f[X]$.

(b) If t, t' are topologies on a set P, we write $t \leq t'$ (and say t is finer than t' or t' is coarser than t, as usual) if the identity map $i\colon (P, t) \to (P, t')$ is continuous.

(c) If $\{t_\alpha\}$ is a family of topologies on a set P, then $\sup_\alpha t_\alpha$ is the finest topology on P coarser than all the t_α's.

2. The Basic Construction

2.1. STANDING HYPOTHESIS

Let P be a set, $B \subseteq P$ be a subset with $card\, B = card(P \setminus B) = 2^{\aleph_0}$, let $A \subseteq B$, $card\, A = \aleph_0$, and, for a natural number $n > 1$, $b\colon P^n \to P$ be a one-to-one map of P^n onto $B \setminus A$.

We set $G_0 = P \setminus \operatorname{Im} b$ and assume that b satisfies

$$\begin{aligned} B_1 &= b(G_0^n), \qquad G_k = G_{k-1} \cup B_k, \\ B_{k+1} &= b(G_k^n \setminus G_{k-1}^n), \\ B &= A \cup \bigcup_{k=1}^{\infty} B_k, \end{aligned}$$

where all the subsets are disjoint.

2.2. Let ϱ be a metric on P, $diam(P, \varrho) = 1$ and $\varrho(x, y) = 1$ for all $x, y \in B$, $x \neq y$ (we call it 1-*discrete on* B). Let w be a topology on B. We denote by $\varrho \star w$ the topology on P defined as follows:

(a) if $x \in P \setminus B$ the nbd-system of x is given by the ϱ-nbds of x;
(b) if $x \in B$ the nbd-system of x is given by $\{U_{\varrho,\varepsilon} \mid U \text{ is } w\text{-nbd of } x,\ \varepsilon > 0\}$, where $U_{\varrho,\varepsilon} = \{y \in P \mid \varrho(y, U) < \varepsilon\}$.

OBSERVATION. $TM(\varrho \star w) = \varrho \star TMw$.

2.3. Let (B, w) be a topological space. We say that it has the *R-property* if the closure of every neighborhood of any point contains its cozero-neighborhood.

LEMMA. *Let $B \subseteq P$, and ϱ be a metric on P which is 1-discrete on B and let w be a topology on B such that (B, w) has the* R*-property. Then $(P, \varrho \star w)$ also has the* R*-property.*

Proof. Consider $x \in P$, and let U be a neighborhood of x in $(P, \varrho \star w)$. We have to prove that the closure $\overline{U}$ contains a cozero-neighborhood of x. This is clear whenever $x \in P \setminus B$. Now suppose $x \in B$. We may assume that $U = V_{\varrho,\varepsilon}$ for some $\varepsilon > 0$ where V is a neighborhood of x in (B, w). Then the closure $\overline{V}$ contains a cozero-neighborhood of x in (B, w), say O. If we choose $\delta > 0$ such that $\delta < \min(\varepsilon, \frac{1}{2})$, then $O_{\varrho,\delta}$ is a cozero-neigborhood of x in $(P, \varrho \star w)$ and $O_{\varrho,\delta} \subseteq \overline{U}$, evidently. □

LEMMA 2.4. *Let $k \geq 1$ be a natural number, and (B, w) be a space with the* R*-property. Then $(B, w)^k$ also has the* R*-property. Hence $(TMw)^k = TMw^k$.*

Proof. (a) Let $x = (x_0, \ldots, x_{k-1})$ be in B^k, let U be a neighborhood of x in $(B, w)^k$. We may assume that $U = U_0 \times \cdots \times U_{k-1}$. Then the closure $\overline{U}$ is equal to $\overline{U}_0 \times \cdots \times \overline{U}_{k-1}$. Hence, for every $i \in k$, $\overline{U}_i$ contains a cozero-neighborhood of x_i, say O_i. Then $O = O_0 \times \cdots \times O_{k-1}$ is a cozero-neighborhood of x in $(B, w)^k$.

(b) We prove $(TMw)^k = TMw^k$. Let $x = (x_0, \ldots, x_{k-1})$ be in B^k, and O be a cozero-neighborhood of x in $(B, w)^k$. Then there exists a neighborhood U of x such that $\overline{U} \subseteq O$. We may assume that $U = U_0 \times \cdots \times U_{k-1}$ where U_i is a neighborhood of x_i in (B, w). Then, there exists a cozero-neighborhood O_i of x_i such that $O_i \subseteq \overline{U}_i$. Consequently, $O_0 \times \cdots \times O_{k-1} \subseteq O$. □

2.5. BASIC CONSTRUCTION

Let ϱ be a metric on P which is 1-discrete on B. Let u be a Hausdorff topology on A. Depending on these data and 2.1 we construct a space $X = (P, t)$ (which we also denote by $\langle (P, \varrho), b, (A, u) \rangle$), by induction, defining an increasing chain of topologies, as follows.

First we specify u_0 by requiring that u_0 equal u on A, u_0 be discrete on $B \setminus A$ and that A be clopen in (B, u_0). Then we set $t_0 = \varrho \star u_0$. For $\alpha = \beta + 1$, u_α is the topology for which $u_\alpha|_A = u$, A and $B \setminus A$ are clopen in (B, u_α), $u_\alpha|_{B \setminus A}$ is the topology making b a homeomorphism of $(P, t_\beta)^n$ onto $(B \setminus A, u_\alpha|_{B \setminus A})$, and $t_\alpha = \varrho \star u_\alpha$. For α limit ordinal, given u_β on B for all $\beta < \alpha$, we define $u_\alpha = \sup_{\beta < \alpha} u_\beta$ and $t_\alpha = \varrho \star u_\alpha$.

Clearly $u_0 \leq u_1 \leq \cdots$ and $t_0 \leq t_1 \leq \cdots$, so there exists α_0 such that $u_\gamma = u_{\alpha_0}$ and $t_\gamma = t_{\alpha_0}$ for all ordinals $\gamma \geq \alpha_0$. Let $t = t_{\alpha_0}$.

By construction, B is closed in (P, t), (A, u) is a clopen subspace in $(B, t|_B)$ and b is a homeomorphism of (P^n, t^n) onto $(B \setminus A, t|_{B \setminus A})$.

2.6. We recall the sets G_k introduced in 2.1.

LEMMA. *Every G_k is open in $X = (P, t)$ and $t|_{G_k} = t_k|_{G_k}$. Moreover, every $x \in G_k$ has a neighborhood in G_k such that its closure in (P, t) is still contained in G_k.*

Proof. (a) G_0 is open in $(P, \varrho \star s)$ for any topology s on B that renders A clopen on (B, s). Therefore G_0 is open in (P, t). Clearly, $t|_{G_0} = t_0|_{G_0}$. Next, we proceed by induction on k. We suppose that G_k is open in X and prove that G_{k+1} is also open in X. Since G_k is open in X, G_k^n is open in X^n. Since b: $(P^n, t^n) \to (B \setminus A, t|_{B \setminus A})$ is a homeomorphism, $b[G_k^n]$ is an open subset of $(B \setminus A, t|_{B \setminus A})$, which implies that $A \cup b[G_k^n]$ is an open subset of $(B, t|_B)$. Consequently, $G_{k+1} = (P \setminus B) \cup A \cup b(G_k^n)$ is an open subset of (P, t). The proof that $t|_{G_{k+1}} = t_{k+1}|_{G_{k+1}}$ is analogous.

(b) Let $x \in G_k$. If $k = 0$, i.e., $x \in G_0$, then $\varrho(x, B \setminus A) = \varepsilon > 0$. Evidently, there exists a neighborhood U of x that satisfies $\overline{U} \subseteq G_0$. Next, we proceed by induction on k again. Let $k > 0$ and let $x \in G_k$. Put $(x_0, \ldots, x_{n-1}) = b^{-1}(x)$. Then $x_i \in G_{k_i}$ with $k_i < k$, so that, by induction hypothesis, there exists a neighborhood U_i of x_i such that $\overline{U}_i \subseteq G_{k_i}$. Hence $U = U_0 \times \cdots \times U_{n-1}$ is a neighborhood of $b^{-1}(x)$ in (P^n, t^n) and $\overline{U} \subseteq G_{k-1}^n$. Therefore $b[\overline{U}]$ is a closed neighborhood of x in $(B \setminus A, t|_{B \setminus A})$, contained in $b[G_{k-1}^n] \subseteq G_k$. Hence $V = (b[U])_{\varrho,\varepsilon}$ is a neighborhood of x in (P, t) such that $\overline{V} \subseteq G_k$ whenever $0 < \varepsilon < \frac{1}{2}$. □

LEMMA 2.7. *If (A, u) is a Hausdorff space, then X is also Hausdorff.*

Proof. Let $x, y \in P$, $x \neq y$. We have to separate them by disjoint neighborhoods. This is clearly possible whenever $x, y \in G_0$. Now we proceed by induction. It suffices to prove the claim for $x, y \in Im\, b$, the other case is trivial. Let $x, y \in G_k$, $k > 0$ and $(x_0, \ldots, x_{n-1}) = b^{-1}(x)$, $(y_0, \ldots, y_{n-1}) = b^{-1}(y)$, $x_i, y_i \in G_{k-1}$ for all $i \in n$. We find $j \in n$ such that $x_j \neq y_j$ and neighborhoods $\tilde{U}$ of x_j, $\tilde{V}$ of y_j such that $\tilde{U} \cap \tilde{V} = \emptyset$. We set $U = b[\pi_j^{-1}(\tilde{U})]$, $V = b[\pi_j^{-1}(\tilde{V})]$. Then U, V are disjoint neighborhoods of x and y in $(B \setminus A, t|_{B \setminus A})$. Consequently, $U_{\varrho,\frac{1}{3}}$ and $V_{\varrho,\frac{1}{3}}$ are disjoint neighborhoods of x and y in (P, t). □

LEMMA 2.8. *If (A, u) has the* R*-property and $(P, t) = \langle (P, \varrho), b, (A, u) \rangle$, then $(P, TMt) = \langle (P, \varrho), b, (A, TMu) \rangle$.*

Proof. (a) First, we prove that for every natural number k, (B, u_k) and (P, t_k) also have the R-property. The space (B, u_0) has it, evidently; the space (P, t_0) has the R-property, by 2.3. Next, we proceed by induction. If (P, t_k) has the R-property, then (P^n, t_k^n) has it as well, by 2.4, so that (B, u_{k+1}) has the R-property. Hence (P, t_{k+1}) has the R-property, by 2.3 again.

(b) If (P, t_k) has the R-property for every natural number k, we use Lemma 2.6 to deduce that (P, t) also has the R-property.

(c) Let us denote by $\tilde{u}_\alpha$, $\tilde{t}_\alpha$ the topologies on B and P obtained by the basic construction 2.5 applied to (A, TMu). We prove that for every natural number k,

$$(B, \tilde{u}_k) = (B, TMu_k) \text{ and } (P, \tilde{t}_k) = (P, TMt_k).$$

The equality $(B, \tilde{u}_0) = (B, TMu_0)$ is evident; since B is C*-embedded in any $(P, \varrho \star w)$ for arbitrary topology w on B, $(P, t_0) = (P, TMt_0)$ is also evident. Let us suppose that $(P, \tilde{t}_k) = (P, TMt_k)$ so that $(P^n, \tilde{t}_k^n) = (P^n, (TMt_k)^n)$. By 2.4 and (a) above, $(P^n, (TMt_k)^n) = (P^n, TMt_k^n)$. Hence, $(B, \tilde{u}_{k+1}) = (B, TMu_{k+1})$ so that $(P, \tilde{t}_{k+1}) = (P, TMt_{k+1})$, because B is C*-embedded. Finally, we use 2.6 again. □

3. Extremally Semirigid Metrics

3.1. B-SEMIRIGID SPACES

DEFINITION. A point x of a topological space X is called *rigid* if every continuous selfmap $f\colon X \to X$ with $x \in \operatorname{Im} f$ is either the identity or constant. A space X is called *semirigid* if it has a rigid point.

Clearly, every semirigid space is connected. If we denote by B the complement of the subset of all rigid points of X, then every $f\colon X \to X$ is either the identity or constant or $f[X] \subseteq B$. Then we say that X is B-*semirigid.*

B-semirigid spaces satisfy the properties analogous to a result by Herrlich on rigid spaces [2].

PROPOSITION. *Let X be a B-semirigid space with $card(X \setminus B) \geq 3$ and let $g\colon X^n \to X$ be a continuous map. If n is finite or X is Hausdorff, then either g is constant or g is a projection or $g[X^n] \subseteq B$.*

For details and proofs, see [9].

DEFINITION 3.2. A metric ϱ on P is called *extremally B-semirigid* if
(a) $diam(P, \varrho) = 1$, $\varrho(x, y) = 1$ if $x \neq y$, $x, y \in B$;
(b) if w is a Hausdorff topology on B and $\tilde{w} = \varrho \star w$, then $(P, \tilde{w})$ is B-semirigid.

By [9], if $card\, B = card(P \setminus B) \geq 2^{\aleph_0}$, such a metric ϱ on P does exist (in the proof in [9], Cook continua [1] are essential).

3.3. Let P, B, b, n, A, G_0 be as in 2.1 and $\pi_i^{(k)}\colon P^k \to P$, $\text{const}_g^{(k)}\colon P^k \to P$ be the ith projection and the constant map onto $g \in G_0$, respectively. For every $k = 0, 1, \ldots$ we define $M(P^k, P) = \bigcup_{m=0}^{\infty} L_m^{(k)}$, where

$$L_0^{(k)} = \{\pi_i^{(k)} \mid i \in k\} \cup \{\text{const}_g^{(k)} \mid g \in G_0\},$$
$$L_{m+1}^{(k)} = L_m^{(k)} \cup \{b \circ (f_0 \dot{\times} \cdots \dot{\times} f_{n-1}) \mid f_j \in L_m^{(k)}\}$$

and $f_0 \dot{\times} \cdots \dot{\times} f_{n-1}\colon P^k \to P^n$ is the map sending every $p \in P^k$ to$(f_0(p), \ldots, f_{n-1}(p))$.

PROPOSITION 3.4. *If ϱ is an extremally B-semirigid metric on P, (A, u) is Hausdorff and totally disconnected, and $X = \langle (P, \varrho), b, (A, u) \rangle$, then, for every $k = 0, 1, \ldots,$ the set $\mathcal{C}(X^k, X)$ of all continuous maps from the kth power of X into X equals $M(P^k, P)$.*

Proof. Obviously, $M(P^k, P) \subseteq \mathcal{C}(X^k, X)$ since every $f \in M(P^k, P)$ belongs to some $L_m^{(k)}$, so it is continuous.

The proof of the converse is analogous to [6]. We recall (see 2.1) that $P = \bigcup_{i=0}^{\infty} G_i$. For any $f \in \mathcal{C}(X^k, X)$, let us denote by $r(f)$ the smallest i such that $f[X^k] \cap G_i \neq \emptyset$. We prove that f is in $M(P^k, P)$ by induction on $r(f)$.

If $r(f) = 0$, i.e., $f[X^k] \cap G_0 \neq \emptyset$, then, by Proposition 3.1, f is either a projection or a constant with the value in $P \setminus B$ or $f[X^k] \subseteq B$; if $f[X^k] \cap A \neq \emptyset$, then f is constant with a value in A because A is clopen in $(B, t|_B)$ and totally disconnected (and $f[X^k]$ is connected because X, being semirigid, is connected); we conclude that if $r(f) = 0$, then f is either a projection or constant with the value in G_0, i.e., $f \in L_0^{(k)}$.

The induction step: if $r(f) > 0$, then necessarily $\operatorname{Im} f \subseteq B \setminus A$; then, for every $j \in n$, the map $f_j = \pi_j^{(n)} \circ b^{-1} \circ f$ satisfies $r(f_j) < r(f)$ (see 2.1 and the requirements there), so that $f_j \in L_{m_j}^{(k)}$ for some m_j, so that $f = b \circ (f_0 \dot{\times} \cdots \dot{\times} f_{n-1})$ is in $\bigcup_{m=0}^{\infty} L_m^{(k)}$. □

COROLLARY. *If ϱ_1, ϱ_2 are extremally B-semirigid metrics on P and (A, u_1), (A, u_2) are totally disconnected Hausdorff spaces, then the clones of the spaces $\langle (P, \varrho_1), b, (A, u_1) \rangle$ and $\langle (P, \varrho_2), b, (A, u_2) \rangle$ consist of the same maps between finite powers of their common underlying set P.*

4. The Choice of (A, u)

4.1. Let $A = [0, 1] \cap Q$, where $[0, 1]$ is the interval of all real numbers $0 \leq x \leq 1$ and Q is the set of all rational numbers. Let $S = \{\frac{1}{n}\}_{n \in \mathbb{N}}$ be a sequence of the reciprocals of the natural numbers. We define on A a topology by means of the basis neighborhoods, $\mathcal{B}(x)$, of every point $x \in A$. Let $U_n(x) = (x - \frac{1}{n}, x + \frac{1}{n}) \cap [0, 1] \cap Q$

$$\mathcal{B}(x) = \begin{cases} \{U_n(x) \mid n \in \mathbb{N}\}, & \text{for } x \neq 0, \\ \{U_n(x) \setminus S \mid n \in \mathbb{N}\}, & \text{for } x = 0. \end{cases}$$

Let u be the resulting topology on A. (A, u) is Hausdorff but not regular. However, it has the R-property and the Tychonoff modification of u is the usual Euclidean topology. Furthermore, for every $k \in \mathbb{N}$, $TMu^k = (TMu)^k$.

Spaces for which the Tychonoff modification is preserved by products have been studied by many authors, in particular, Ishii [3], Oka [4] and Simon [7]. Maybe some constructions of these papers can be used to solve our Problem in Introduction, possibly by choosing a different space (A, u), but the authors did not succeed in this.

4.2. In P^n, the nth power of the set P, we consider the following subsets:

$$\begin{aligned} P[i,c] &= \{(x_0,\ldots,x_{n-1}) \in P^n \mid x_i = c\}, \quad \text{for } c \in P, i \in n, \\ P[i,j] &= \{(x_0,\ldots,x_{n-1}) \in P^n \mid x_i = x_j\}, \quad i,j \in n, i \neq j, \\ P[i, B\setminus A] &= \{(x_0,\ldots,x_{n-1}) \in P^n \mid x_i \in B\setminus A\}, \quad i \in n. \end{aligned}$$

We will refer to these sets and all their subsets as *small sets*.

5. Modification of the Basic Construction

5.1. Let $P, b, G_0, B, A, n > 1$ be as in 2.1, ϱ be an extremally B-semirigid metric on P, let (A, u) be the space as in 4.1. Let us denote by u_α and t_α the topologies on B and on P obtained by these data in the basic construction 2.5 and let us denote, as in 2.8, by $\tilde{u}_\alpha$ and $\tilde{t}_\alpha$ the topologies on B and on P applying the basic construction to the same $P, b, G_0, B, A, n > 1$ but taking the space (A, TMu). Now, we modify the basic construction and we get topologies u^*_α and t^*_α on B and on P as follows:

$$u^*_0 = \tilde{u}_0 \quad \text{and} \quad t^*_0 = \tilde{t}_0.$$

For α limit ordinal, $u^*_\alpha = \sup_{\beta<\alpha} u^*_\beta$ and $t^*_\alpha = \varrho \star u^*_\alpha$. For $\alpha = \beta + 1$, $u^*_\alpha|_A = TMu$, A and $B \setminus A$ are clopen in (B, u^*_α) and $u^*_\alpha|_{B\setminus A}$ is the topology for which $b\colon (P^n, z_\beta) \to (B \setminus A, u^*_\alpha|_{B\setminus A})$ is a homeomorphism where z_β is the following topology:

if $x \in P^n \setminus \{(0,\ldots,0)\}$, then the nbd-basis of x in z_β is just its nbd-basis in $(t^*_\beta)^n$;
the nbd-basis of $(0,\ldots,0)$ in z_β is $\{U \setminus S^*\}$ where U ranges through the nbd-basis of $(0,\ldots,0)$ in $(t^*_\beta)^n$ and $S^* = \{(1/m, 1/(m+1), \ldots, 1/(m+n-1)) \mid m = 1, 2, \ldots\}$.

Clearly, $u^*_0 \le u^*_1 \le \cdots$ and $t^*_0 \le t^*_1 \le \cdots$. Moreover, $u_\alpha \le u^*_\alpha \le \tilde{u}_\alpha$ and $t_\alpha \le t^*_\alpha \le \tilde{t}_\alpha$. Let γ be an ordinal sufficiently large such that $u_\gamma = u_{\gamma+1} = \cdots$, $\tilde{u}_\gamma = \tilde{u}_{\gamma+1} = \cdots$, $\tilde{t}_\gamma = \tilde{t}_{\gamma+1} = \cdots$, $z_\gamma = z_{\gamma+1} = \cdots$, $u^*_\gamma = u^*_{\gamma+1} = \cdots$, $t^*_\gamma = t^*_{\gamma+1} = \cdots$. Let us denote by $u, t, \tilde{u}, \tilde{t}, z, u^*, t^*$ the corresponding topologies. Then $t \le t^* \le \tilde{t}$ and, by 2.8, $\tilde{t} = TMt$. Hence $TMt^* = \tilde{t}$ as well. Therefore $CloTMt^* = CloTMt$. By 3.4 this implies

$$\mathcal{C}\big((P^k, (TMt^*)^k), (P, TMt^*)\big) = M(P^k, P) \quad \text{for all } k.$$

The construction also implies that b determines a homeomorphism of (P^n, z) onto $(B\setminus A, t^*|_{B\setminus A})$. Moreover, we have $z \le (t^*)^n$ and $z \neq (t^*)^n$ but z and $(t^*)^n$ coincide on all small subsets of P^n (see 4.2). This implies that the bijection $b\colon P^n \to B\setminus A$

(a) is **not** continuous from $(P^n, (t^*)^n)$ to $(B \setminus A, t^*|_{B\setminus A})$ but b^{-1} **is** continuous in the other direction;

(b) restricts to a continuous map on any small subset of P^n.

In fact, b is the composition of

$$\left(P^n, (t^*)^n\right) \xrightarrow{f} \left(P^n, z\right) \xrightarrow{g} \left(B \setminus A, t^*|_{B\setminus A}\right),$$

where f is carried by the identity of P^n and g is the above homeomorphism (with $g(x) = b(x)$ for all $x \in P^n$). And the map f is not continuous while f^{-1} and f restricted to small sets are continuous.

Compare the following result with Proposition 3.4.

PROPOSITION 5.2. *For* $k < n$, $\mathcal{C}((P^k, (t^*)^k), (P, t^*)) = M(P^k, P)$.

Proof. (a) If $f \in \mathcal{C}((P^k, (t^*)^k), (P, t^*))$, we proceed by induction on $r(f)$ as in 3.4, to conclude that $\mathcal{C}((P^k, (t^*)^k), (P, t^*)) \subseteq M(P^k, P)$; and this is true for all k.

(b) We have $M(P^k, P) = \bigcup_{m=0}^{\infty} L_m^{(k)}$. We must prove that any $f \in M(P^k, P)$ is continuous. It is evident if $f \in L_0^{(k)}$. Then we proceed by induction. If $f \in L_{m+1}^{(k)} \setminus L_0^{(k)}$, then f is of the form $f = b \circ (f_0 \dot{\times} \cdots \dot{\times} f_{n-1})$ with $f_0, \ldots, f_{n-1} \in L_m^{(k)}$; then $f_0, \ldots, f_{n-1}$ are continuous, by the induction hypothesis. Since $k < n$, necessarily some f_j is of the form $b \circ (f_0' \dot{\times} \cdots \dot{\times} f_{n-1}')$ or some f_j is a constant or $f_i = f_j$ for some $i \neq j$. This implies that either $\mathrm{Im}(f_0 \dot{\times} \cdots \dot{\times} f_{n-1})$ is a subset of $P^n[j, B \setminus A]$ or a subset of $P[j, c]$ or a subset of $P[i, j]$, i.e., it is a small set in any case. Since b restricted to any small set is continuous, f must be continuous. □

COROLLARY. $Clo_n(P, t^*) = Clo_n(P, TMt^*)$ $[= Clo_n(P, TMt)]$.

5.3. Now we prove that $Clo_{n+1}(P, t^*) \neq Clo_{n+1}(P, TMt^*)$ $[= Clo_{n+1}(P, TMt)]$.

However, b is continuous as a map b: $(P^n, TM(t^*)^n) \to (P, t^*)$ but it is not continuous as a map b: $(P^n, (t^*)^n) \to (P, t^*)$.

6. Nonisomorphism (and Nonembeddability)

In this section, we prove that (P, t^*) and (P, TMt^*) not only have distinct clones, but even have nonisomorphic clones.

6.1. First we note that, if $\mathbf{K}_1, \mathbf{K}_2 \subseteq Top$ are full subcategories of *Top* and Φ: $\mathbf{K}_1 \to \mathbf{K}_2$ is a full embedding (i.e., a full faithful functor), then any constant morphism in $\mathbf{K}_1(A, A)$, $A \in Obj\,\mathbf{K}_1$, is sent by Φ to a constant morphism in $\mathbf{K}_2(\Phi(A), \Phi(A))$. This follows by the well known fact that constant endomorphisms are precisely the left zeros of the endomorphism monoids in question. This implies that there exists a natural equivalence $\tau = \{\tau_A \mid A \in Obj\,\mathbf{K}_1\}$ of U_1 to $U_2 \circ \Phi$, where U_i: $\mathbf{K}_i \to Set$ are the forgetful functors (the domain-restriction of the natural forgetful functor $Top \to Set$). In fact, for every $A \in Obj\,\mathbf{K}_1$, $x \in U_1(A)$, we define $\tau_A(x) = y$, where Φ sends the constant map $A \to A$ with value x to the constant map $\Phi(A) \to \Phi(A)$ with value y. It is well known (and easy to see) that τ is really a natural equivalence.

6.2. If U_1, U_2 are the forgetful functors on $\mathbf{K}_1$ and $\mathbf{K}_2$, respectively, and $\tau = \{\tau_A | A \in Ob\,\mathbf{K}_1\}$ denotes the natural equivalence of U_1 to $U_2 \circ \Phi$, we have the following corollaries:

(1) $\tau_B(\operatorname{Im} U_1(\alpha)) = \operatorname{Im} U_2(\Phi(\alpha))$ since τ is bijective.
(2) $U_1(\alpha)$ is one-to-one or surjective iff $U_2(\Phi(\alpha))$ is one-to-one or surjective.
(3) $x \in U_1(B)$ is a rigid point of B in $\mathbf{K}_1$ iff $\tau_B(x)$ is a rigid point of $\Phi(B)$ in $\mathbf{K}_2$.
(4) $U_1(\alpha)$ is a surjection onto the set of all nonrigid points of B iff $U_2(\Phi(\alpha))$ is a surjection onto the set of all nonrigid points of $\Phi(B)$.

6.3. In all our constructions, we have a set P, a subset $B \subseteq P$ and, for $n > 1$, a one-to-one map $b\colon P^n \to P$ onto $B \setminus A$ that satisfy the hypothesis in 2.1. Under these hypotheses there exists a metric ϱ on P which is extremally B-semirigid.

Then we endow (P, ϱ) with the topologies t, TMt, $t^* = \tau$ which satisfy $t \leq \tau \leq TMt$ and turn

$$b\colon (P^n, t^n) \to (B \setminus A, t|_{B\setminus A}),$$
$$b\colon (P^n, TMt^n) = (P^n, (TMt)^n) \to (B \setminus A, TMt|_{B\setminus A})$$

into homeomorphisms, at the same time ensuring that

$$b\colon (P^n, \tau^n) \to (B \setminus A, \tau|_{B\setminus A})$$

is not continuous.

Let us write X and Y for the spaces (P, τ) and $(P, TMt) = (P, TM\tau)$.

We want to prove that G_0 is the set of rigid points both in X and in Y. Elements of G_0 are rigid in X and Y since these spaces are B-semirigid and A is totally disconnected. For an element p of the complement $P \setminus G_0$ we need to show that it is neither rigid in X nor in Y.

To prove this we must find a nonconstant, nonidentical continuous map $f\colon X \to X$ ($f\colon Y \to Y$) with $p \in \operatorname{Im} f$. Consider $q = b^{-1}(p)$, $q = (q_0, q_1, \ldots, q_{n-1})$ and define $h\colon X \to X^n$ ($g\colon Y \to Y^n$) by means of $h(x) = (x, q_1, \ldots, q_{n-1})$ ($g(y) = (y, q_1, \ldots, q_{n-1})$), where $q_1, \ldots, q_{n-1}$ are fixed while x varies in X (y varies in Y). We have that h (g) is continuous and $\operatorname{Im} h$ ($\operatorname{Im} g$) is a small set. The composition with b, $b \circ h\colon X \to X$ ($b \circ g\colon Y \to Y$) is continuous since b, restricted to the small sets, is continuous (b is a homeomorphism). In both cases, the composition is one-to-one, not the identity, and its image contains p.

PROPOSITION 6.4. *There exists neither a full embedding of $Clo_{n+1}(X)$ into $Clo(Y)$, nor a full embedding of $Clo_{n+1}(Y)$ into $Clo(X)$.*

Proof. (A) Let us suppose that there exists a full embedding $\Phi\colon \mathbf{K}_1 \to \mathbf{K}_2$ where $\mathbf{K}_1 = Clo_{n+1}(X)$, $\mathbf{K}_2 = Clo(Y)$. Then we have the natural equivalence $\tau\colon U_1 \to U_2 \circ \Phi$, where U_1, U_2 are the forgetful functors of $\mathbf{K}_1$, $\mathbf{K}_2$. Then $\tau_X\colon X \to \Phi(X)$ sends the set G_0 precisely onto the set of rigid points of $\Phi(X)$, by 3 in 6.2. Since Y^k has no rigid points whenever $k > 1$ (see, e.g., [11]), necessarily $\Phi(X) = Y$

and τ_X sends the subset G_0 of X bijectively onto the subset G_0 of Y. Moreover, Φ sends each X^k, $k \leq n$, to Y^k. In fact, both the spaces X, Y are semirigid so that $\mathcal{C}(X^k, X)$ and $\mathcal{C}(Y^k, Y)$ both precisely contain k surjective maps, namely, the projections. Then, by (2) in 6.2, necessarily $\Phi(X^k) = Y^k$ for all $k \leq n$; particularly, $\Phi(X^n) = Y^n$.

We show that this is impossible because $\mathcal{C}(X^n, X)$ contains a map which sends P^n onto the set $B \setminus A$ of all nonrigid points of X (namely, the map b) but $\mathcal{C}(Y^n, Y)$ does not contain any such map. This contradicts (4) in 6.2.

Thus, let f: $(P^n, \tau^n) \to (B \setminus A, \tau|_{B\setminus A})$ be continuous and surjective. Since b^{-1}: $(B \setminus A, \tau|_{B\setminus A}) \to (P^n, z)$ is continuous and $z \leq \tau^n$, we can compose it with the projections and get continuous maps

$$(P^n, \tau^n) \xrightarrow{f} (B \setminus A, \tau|_{B\setminus A}) \xrightarrow{b^{-1}} (P^n, z) \xrightarrow{\pi_i^{(n)}} (P, \tau).$$

Since (P, τ) is B-semirigid, $\pi_i^{(n)} \circ b^{-1} \circ f$ is either constant or a projection or its image is contained in B. But f is surjective, so the composition cannot be a projection. This implies that $b^{-1} \circ f$ can only permute the coordinates of the points. Let ψ: $n \to n$ be a bijection such that $\pi_i^{(n)} \circ b^{-1} \circ f = \pi_{\psi(i)}^{(n)}$. This implies that $b^{-1} \circ f$ is a homeomorphism. But this is impossible because $z \leq \tau^n$ and their nbd-systems of $(0, \dots, 0) = (\pi_{\psi(0)}^{(n)}(0), \dots, \pi_{\psi(n-1)}^{(n)}(0))$ are not equivalent. Hence, no surjective continuous function f: $(P^n, \tau^n) \to (B \setminus A, \tau|_{B\setminus A})$ exists.

(B) Let us suppose that there exists a full embedding Φ: $\mathbf{K}_1 \to \mathbf{K}_2$ where $\mathbf{K}_1 = Clo_{n+1}(Y)$, $\mathbf{K}_2 = Clo(X)$. Then we have the natural equivalence τ: $U_1 \to U_2 \circ \Phi$ again and the whole proof is quite analogous: necessarily $\Phi(Y) = X$, $\Phi(Y^n) = Y^n$ and $\mathcal{C}(Y^n, Y)$ does not contain a map which sends Y^n onto the set of all nonrigid points of Y while $\mathcal{C}(X^n, X)$ contains such a map. □

Concluding Remarks

In the paper, we investigated only the Tychonoff modification of space. However, the construction works, with some minor changes, also for other modifications, e.g., the sequential modification or the compactly generated modification. In both cases, instead of the space described in 4, one chooses for (A, u) a countable space with a unique nonisolated point $q \in A$, the neighborhood basis of which is $\{\{q\} \cup U | U \in \mathcal{U}\}$, where $\mathcal{U}$ is a free ultrafilter on $A \setminus \{q\}$. Then (A, mu) is discrete, if mu is the sequential or the compactly generated modification. If $X = \langle (P, \varrho), b, (A, u) \rangle$ and $Y = \langle (P, \varrho), b, (A, mu) \rangle$ are as in 2, then, $Y = mX$; the proof is analogous to 2 (and, in a sense easier). Since mu is finer than u, the construction of Section 5 has to be modified. The transfinite construction starts with (A, u) rather than (A, mu), and at each stage the set S^* of all n-tuples of pairwise distinct elements of $A \setminus \{q\}$ has to be removed from the neighborhoods of $(q, \dots, q)$. Then the resulting space and the space Y satisfy the claim of the theorem. Hence, in our theorem, the Tychonoff modification TM can be replaced by the sequential modification or by the compactly generated modification.

Acknowledgement

Financial supports from the Italian M.U.R.S.T. and from the Grant Agency of the Czech Republic under the grant 201/96/0119 are gratefully acknowledged.

References

1. Cook, H.: Continua which admit only the identity mapping onto non-degenerate subcontinua, *Fund. Math.* **60** (1967), 241–249.
2. Herrlich, H.: On the concept of reflection in general topology, in: *Contributions to Extension Theory of Topological Structures*, Proc. Conf. Berlin, 1967, VEB Deutscher Verlag d. Wiss., Berlin, 1969, pp. 105–114.
3. Ishii, T.: On the Tychonoff functor and w-compactness, *Topology Appl.* **11** (1980), 173–187.
4. Oka, S.: Tychonoff functor and product spaces, *Proc. Japan Acad.* **54** (1978), 97–100.
5. Pupier, R.: La complétion universelle d'un produit d'espaces complètement réguliers, *Publ. Dept. Math. Lyon* **6** (1969), 75–84.
6. Sichler, J. and Trnková, V.: Isomorphism and elementary equivalence of clone segments, *Period. Math. Hungar.* **32** (1996), 113–128.
7. Simon, P.: Completely regular modification and products, *Comment. Math. Univ. Carolin.* **25**(1) (1984), 121–128.
8. Taylor, W.: The clone of a topological space, in: *Research and Exposition in Math.* 13, Heldermann-Verlag, Berlin, 1986.
9. Trnková, V.: Semirigid spaces, *Trans. Amer. Math. Soc.* **343** (1994), 305–325.
10. Trnková, V.: Continuous and uniformly continuous maps of metric spaces, *Topology Appl.* **63** (1995), 189–200.
11. Trnková, V.: Clone segments in *Top* and *Unif*, in: *Categorical Topology, Proceedings of the L'Aquila Conference 1994*, Kluwer Academic Publishers, Dordrecht, 1996, pp. 249–268.

Applied Categorical Structures **8:** 339–350, 2000.
G. Brümmer & C. Gilmour (eds), Papers in Honour of Bernhard Banaschewski.

Representation of Relations by Partial Maps

Dedicated to Bernhard Banaschewski

XIAOMIN DONG and WALTER THOLEN
Department of Mathematics and Statistics, York University, North York, Canada M3J 1P3
e-mail: {xdong,tholen}@pascal.math.yorku.ca

(Received: 18 October 1996; accepted: 28 May 1998)

Abstract. With the notions of partial morphism and relation to be understood with respect to a class $\mathcal{M}$ of monomorphisms in a finitely complete category $\mathcal{C}$, we give sufficient conditions for the graph functor $Par(\mathcal{C}) \to Rel(\mathcal{C})$ to admit a right adjoint. Only under an additional condition is this right adjoint given by the naturally constructed 'pierced power objects'.

Mathematics Subject Classifications (2000): 18B10, 18B25, 18A30.

Key words: partial map, relation, pierced power object, element relation.

1. Introduction

Mostly, but not exclusively, motivated by the interest theoretical computer scientists have in them, in recent years many authors have considered relations from a categorical point of view (see, for example, [3, 4, 6, 8, 9, 11 – 13]). The global problem of presenting the assignment $\mathcal{C} \mapsto Rel(\mathcal{C})$ as a 'good functor' has been a prominent theme in many of these works. In this paper we revert back to the local problem of examining the embedding $\mathcal{C} \to Rel(\mathcal{C})$, more precisely its factorization through $Par(\mathcal{C})$, the category of objects of $\mathcal{C}$ with partial maps as morphisms.

In their monograph [10] Klein and Thompson distinguish carefully between a *relation* from a set A to a set B (that is: a subset $r \subseteq A \times B$) and a *correspondence* from a A to B (that is: a relation r for which $\xi_r(x) = \{y \in B \mid (x, y) \in r\}$ takes only non-empty values), and they provide a number of applications, both in mathematics and in economics. A correspondence is therefore equivalently described by a map $A \to P_0B$, with $P_0B = PB \backslash \{\emptyset\}$ the *pierced power set*, while a general relation r from A to B may be described by a *partial map* from A to P_0B, being defined only on the subset

$$D_r = \{x \in A \mid \xi_r(x) \neq \emptyset\}.$$

While the categorical rôle of PB in the representation of relations is well understood, as it may be used as the defining property of a topos (cf. [2]), that of P_0B

seems to be less clear. Let us first clarify our setting. In a *finitely complete* category $\mathcal{C}$, we take the notion of *subobject* to be given by a class $\mathcal{M}$ of monomorphisms in $\mathcal{C}$ satisfying the following condition:

(I) *$\mathcal{M}$ is closed under composition and stable under pullback, and $\mathcal{M}$ contains all split monomorphisms.*

(It actually suffices to ask that $\mathcal{M}$ contain every 'diagonal morphism'

$$\delta_B = \langle 1_B, 1_B\rangle\colon B \to B \times B \ (B \in \mathcal{C});$$

since every equalizer is a pullback of such a morphism, $\mathcal{M}$ must then contain all regular monomorphisms; in particular all split monomorphisms.) Two $\mathcal{M}$-morphisms with the same codomain A are equivalent if they are isomorphic in the comma category $\mathcal{C}/A$, and $\mathcal{M}$-subobjects are equivalence classes of $\mathcal{M}$-morphisms. We shall not distinguish between $\mathcal{M}$-subobjects and their representing morphisms in $\mathcal{M}$, and the prefix $\mathcal{M}$ is omitted in what follows. A *relation* $r\colon A \nrightarrow B$ is a subobject $r\colon R \to A \times B$; equivalently, a pair $(r_1\colon R \to A, r_2\colon R \to B)$ of morphisms with $\langle r_1, r_2\rangle\colon R \to A \times B$ in $\mathcal{M}$. A *partial morphism* $(m, f)\colon A \nrightarrow B$ is a pair with $m\colon D \to A$ and $f\colon D \to B$ arbitrary. Such a pair is in fact a relation since $\langle m, f\rangle$ is the composite

$$D \stackrel{\langle 1,f\rangle}{\longrightarrow} D \times B \stackrel{m\times 1_B}{\longrightarrow} A \times B$$

where the graph $\langle 1, f\rangle$ of $f\colon D \to B$ is a split monomorphism, and where $m \times 1_B$ is a pullback of m along the projection $A \times B \to B$; hence (I) gives $\langle m, f\rangle \in \mathcal{M}$. Moreover, there is a *composition* of a partial morphism $(m, f)\colon A \nrightarrow B$ followed by a relation $s\colon B \nrightarrow C$, given by pullback:

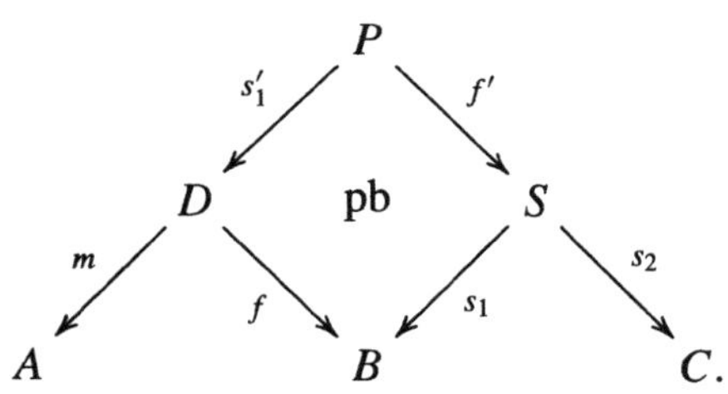

(1)

Putting $s \circ (m, f) := \langle m \cdot s_1', s_2 \cdot f'\rangle$ one indeed obtains a relation $A \nrightarrow C$ since $s \circ (m, f)$ is the composite

$$P \stackrel{\langle s_1', s_2\cdot f'\rangle}{\longrightarrow} D \times C \stackrel{m\times 1_C}{\longrightarrow} A \times C$$

where the first factor is a pullback of $\langle s_1, s_2\rangle \in \mathcal{M}$ along $f \times 1_C$. If (m, f) is *total*, i.e. if $m \cong 1_A$, we simply write $s \circ f$ for $s \circ (m, f)$.

One says that *relations are represented* (cf. [13]) if for any object B there is an object PB and a relation $\epsilon_B\colon PB \nrightarrow B$ such that every relation $r\colon A \nrightarrow B$

factors as $\epsilon_B \circ \xi_r = r$ with a unique morphism $\xi_r\colon A \to PB$ in $\mathcal{C}$. This is equivalent to saying that for every object B there is a contravariant representable functor $Rel(-, B)\colon \mathcal{C}^{op} \to Set$ which assigns to every object A the set of relations from A to B, and which on morphisms is given by pullback. This puts considerable constraints on $\mathcal{M}$ and $\mathcal{C}$: with 1 denoting a terminal object in $\mathcal{C}$, $\epsilon_1 = (t, 1_1)$ gives a *subobject classifier* $t\colon 1 \to P1 = \Omega$, and every object B is *exponentiable* w.r.t. Ω, with $\Omega^B = PB$; conversely, the existence of a subobject classifier with this property guarantees representation of relations. Furthermore, the existence of a subobject classifier makes every morphism in $\mathcal{M}$ a regular monomorphism (cf. [13]); hence, under condition (I), $\mathcal{M}$ then *is* the class of all regular monomorphisms in $\mathcal{C}$. For $\mathcal{M}$ the class of all monomorphisms, representation of relations already forces $\mathcal{C}$ to be a topos (cf. [2]), but for $\mathcal{M}$ the class of regular monomorphisms, also quasitopoi and other categories admit relations to be represented (cf. [13]).

In this paper, our concern is whether *relations are non-trivially represented*, that is: whether for every object B there is an object P_0B and a relation $\varepsilon_B\colon P_0B \nrightarrow B$ such that every relation $r\colon A \nrightarrow B$ factors as $\varepsilon \circ (m, f) = r$ with a unique partial morphism $(m, f)\colon A \nrightarrow P_0B$. Roughly we prove that in the presence of a strict initial object and of pseudo-complements of points, and if $\mathcal{M}$ is part of a stable factorization system, representation of relations already induces non-trivial representation. However, the situation is in fact more subtle than is suggested by this summary statement. One of our main points is that the naturally constructed *pierced power object* P_0B and the restriction $\varepsilon_B\colon P_0B \nrightarrow B$ of ϵ_B do not in general serve as a co-universal arrow for the non-trivial representation of relations $A \nrightarrow B$, unless the first projection $(\varepsilon_B)_1\colon E_B \to P_0B$ is epic. In the absence of this provision, the construction of P_0B and ε_B has to be modified, so as to enforce membership of $(\varepsilon_B)_1$ in the class of epimorphisms (see Theorem 3.3).

Representation of relations means the existence of a right adjoint to $\mathcal{C} \to Rel(\mathcal{C})$, whereas non-trivial representation of relations means the existence of a right adjoint to $Par(\mathcal{C}) \to Rel(\mathcal{C})$. However, we do not use this functorial interpretation until the end of the paper, since the definition of the category $Rel(\mathcal{C})$ would require us to assume the existence of an $(\mathcal{E}, \mathcal{M})$-factorization system *a priori*, which is irrelevant for large parts of the representation problem.

In the first author's Ph. D. Thesis [4], non-trivial representation of relations was established only under the additional condition on ε_B mentioned above (more precisely, under conditions (I)*, (II)–(V) found in Sections 2, 3 below). However, the proof of the crucial Proposition 2.3 below is essentially taken from [4].

2. Factorizing Relations through Pierced Power-Objects

In addition to the basic condition (I) of the introduction, we now assume that our finitely complete category

(II) $\mathcal{C}$ *has* a strict initial object 0 (*so that, every morphism with codomain* 0 *is an isomorphism*);
(III) *every point* (*that is: every morphism* $a\colon 1 \to A$, *with* 1 *denoting a terminal object*) *has* a pseudo-complement *in* $\mathcal{M}$ (*so that there is a morphism* $\bar{a}\colon A_a \to A$ *in* $\mathcal{M}$ *which is terminal among all morphisms* $B \to A$ *with* $B \times_A 1 \cong 0$);
(IV) *relations are represented; equivalently,* $\mathcal{C}$ *has a subobject classifier* Ω *such that all exponentials* Ω^B *exist.*

REMARK 2.1. (1) As Max Kelly observed, in a finitely complete category, any object A with the strictness property (making every morphism with codomain A iso) is necessarily initial.

(2) As a split monomorphism, every point $a\colon 1 \to A$ lies in $\mathcal{M}$. Furthermore, as a pseudo-complement of $1 \to 1$, condition (III) forces $0 \to 1$ to lie in $\mathcal{M}$; in fact as a pullback of $0 \to 1$, every morphism with domain 0 must lie in $\mathcal{M}$.

(3) As a special instance of a *pullback complement* (see [5]), the pseudo-complement of a exists in particular if a is exponentiable, i.e. if the pullback functor

$$a^*\colon \mathcal{C}/A \to \mathcal{C}/1 \cong \mathcal{C}$$

has a right adjoint a_*. In this case, the pseudo-complement is constructed as

$$\bar{a} = a_*(0 \to 1).$$

Furthermore, if $\mathcal{M}$ is part of an $(\mathcal{E}, \mathcal{M})$-factorization system and if $(0 \to 1) \in \mathcal{M}$, then $\bar{a}$ lies in $\mathcal{M}$ if a^* preserves morphisms in $\mathcal{E}$ (i.e. if $\mathcal{E}$ is stable under pullback along a), as was proved explicitly in [5], Corollary 5.4.

These observations show the essential part of:

PROPOSITION 2.2. *Conditions* (I)–(III) *are satisfied in every locally cartesian closed category with a strict object* 0 *and a proper, stable* $(\mathcal{E}, \mathcal{M})$*-factorization system such that the morphism* $0 \to 1$ *lies in* $\mathcal{M}$. *If, moreover,* $\mathcal{C}$ *has an* ($\mathcal{M}$-)*subobject classifier, then also condition* (IV) *is satisfied. In particular, any quasitopos with* $0 \to 1$ *regularly monic satisfies* (I)–(IV).

We note that there are quasitopoi for which $0 \to 1$ fails to be regularly monic (see 4.2(2) below).

The construction of the *pierced power object* P_0B and of the relation $\varepsilon_B\colon P_0B \nrightarrow B$ is quite simple. First we note that every subobject $m\colon M \to B$ induces a relation $\hat{m} = (!_M, m)\colon 1 \nrightarrow B$ and therefore a point $m^\bullet = \xi_{\hat{m}}\colon 1 \to PB$. This is true particularly for the subobject $0_B\colon 0 \to B$. Let now

$$i_B\colon P_0B \to PB$$

be a pseudo-complement of the morphism $0_B^\bullet$: $1 \to PB$, and let

$$\begin{array}{ccc} E_B & \xrightarrow{w} & F_B \\ {\scriptstyle \varepsilon_B}\downarrow & & \downarrow{\scriptstyle \epsilon_B} \\ P_0B \times B & \xrightarrow[i_B \times 1_B]{} & PB \times B \end{array} \tag{2}$$

be a pullback diagram. For any relation r: $A \nrightarrow B$, we construct a partial morphism $(d_r, r^\sharp)$: $A \to P_0B$ by the pullback diagram

$$\begin{array}{ccc} D_r & \xrightarrow{r^\sharp} & P_0B \\ {\scriptstyle d_r}\downarrow & & \downarrow{\scriptstyle i_B} \\ A & \xrightarrow[\xi_r]{} & PB \end{array} \tag{3}$$

PROPOSITION 2.3. *Every relation* r: $A \nrightarrow B$ *factors as* $r = \varepsilon_B \circ (d_r, r^\sharp)$.

Proof. We proceed in several steps.

Step 1. With $r = (r_1\colon R \to A, r_2\colon R \to B)$, one forms the composite $\tilde{r} = (r \circ r_1\colon R \nrightarrow B)$. Hence $\tilde{r} = \langle q_1, r_2 \cdot q_2 \rangle$, with q_1, q_2 as in the pullback diagram

$$\begin{array}{ccc} R \times_A R & \xrightarrow{q_2} & R \\ {\scriptstyle q_1}\downarrow & & \downarrow{\scriptstyle r_1} \\ R & \xrightarrow[r_1]{} & A \end{array} \tag{4}$$

We show that $\xi_{\tilde{r}}$: $R \to PB$ is disjoint from $0_B^\bullet$. In fact, given any commutative diagram

$$\begin{array}{ccc} X & \xrightarrow{!_X} & 1 \\ {\scriptstyle x}\downarrow & & \downarrow{\scriptstyle 0_B^\bullet} \\ R & \xrightarrow[\xi_{\tilde{r}}]{} & P_B \end{array} \tag{5}$$

we must necessarily have $X \cong 0$: examining the identity

$$\tilde{r} \circ x = \varepsilon_B \circ (\xi_{\tilde{r}} \cdot x) = \varepsilon_B \circ (0_B^\bullet \cdot !_X) = (!_0, 0_B) \circ !_X$$

one sees that there is a pullback diagram

$$\begin{array}{ccc} 0 & \longrightarrow & R \times_A R \\ \downarrow & & \downarrow{\scriptstyle q_1} \\ X & \xrightarrow[x]{} & R \end{array} \tag{6}$$

which gives a morphism $X \to 0$ since $q_1 \cdot \langle x, x \rangle = x = x \cdot 1_X$, and this implies $X \cong 0$.

Step 2. By definition of pseudo-complement, there is a morphism $c\colon R \to P_0 B$ with $i_B \cdot c = \xi_{\tilde{r}}$. Moreover, the diagram

$$\begin{array}{ccc} R & \xrightarrow{c} & P_0 B \\ {\scriptstyle r_1}\downarrow & & \downarrow{\scriptstyle i_B} \\ A & \xrightarrow[\xi_r]{} & PB \end{array} \tag{7}$$

commutes. In fact, this follows from

$$\epsilon_B \circ (\xi_r \cdot r_1) = r \cdot r_1 = \tilde{r} = \epsilon_B \circ \xi_{\tilde{r}} = \epsilon_B \circ (i_B \cdot c).$$

The pullback diagram (3) therefore gives a morphism $u\colon R \to D_r$ with $d_r \cdot u = r_1$ and $r^\sharp \cdot u = c$.

Step 3. For a subobject $m\colon M \to B$, let $\chi_m\colon B \to \Omega$ be the characteristic morphism. We claim that with

$$\mathrm{ev}_B = \chi_{\epsilon_B}\colon PB \times B \to \Omega \quad \text{and} \quad j_B = \mathrm{ev}_B \cdot (i_B \times 1_B),$$

there is a commutative diagram:

$$\begin{array}{ccc} R & \longrightarrow & 1 \\ {\scriptstyle \langle c, r_2\rangle}\downarrow & & \downarrow{\scriptstyle t} \\ P_0 B \times B & \xrightarrow[j_B]{} & \Omega \end{array} \tag{8}$$

In fact, since $i_B \cdot c = \xi_{\tilde{r}}$, one readily verifies $j_B \cdot \langle c, r_2 \rangle = \chi_{\tilde{r}} \cdot \langle 1_R, r_2 \rangle$; hence the assertion follows from the commutativity of (9):

$$\begin{array}{ccccc} R & \xrightarrow{\langle 1_R, 1_R \rangle} & R \times_A R & \longrightarrow & 1 \\ {\scriptstyle 1_R}\downarrow & & \downarrow{\scriptstyle \tilde{r}} & & \downarrow{\scriptstyle t} \\ R & \xrightarrow[\langle 1_R, r_2 \rangle]{} & R \times B & \xrightarrow[\chi_{\tilde{r}}]{} & \Omega \end{array} \tag{9}$$

Now the pullback property of

$$\begin{array}{ccccc} E_B & \xrightarrow{w} & F_B & \longrightarrow & 1 \\ {\scriptstyle \varepsilon_B}\downarrow & & \downarrow{\scriptstyle \epsilon_B} & & \downarrow{\scriptstyle t} \\ P_0 B \times B & \xrightarrow{i_B \times 1_B} & PB \times B & \xrightarrow{\mathrm{ev}_B} & \Omega \end{array} \tag{10}$$

gives a morphism $v\colon R \to E_B$ with $\varepsilon_B \cdot v = \langle c, r_2 \rangle$.

Step 4. The square

$$\begin{array}{ccc} R & \xrightarrow{v} & E_B \\ {\scriptstyle u}\downarrow & & \downarrow{\scriptstyle (\varepsilon_B)_1} \\ D_r & \xrightarrow{r^\sharp} & P_0 B \end{array} \tag{11}$$

is a pullback. Indeed, u, v are constructed in such a way that (11) commutes. With the commutativity of (8) one sees that the left part of

$$\begin{array}{ccccccc} R & \xrightarrow{v} & E_B & \xrightarrow{w} & F_B & \longrightarrow & 1 \\ {\scriptstyle r}\downarrow & & & & \downarrow{\scriptstyle \varepsilon_B} & & \downarrow{\scriptstyle t} \\ R \times B & & \xrightarrow{\xi_r \times 1_B} & & PB \times B & \xrightarrow{\mathrm{ev}_B} & \Omega \end{array} \tag{12}$$

commutes and is actually a pullback. Hence the back face of

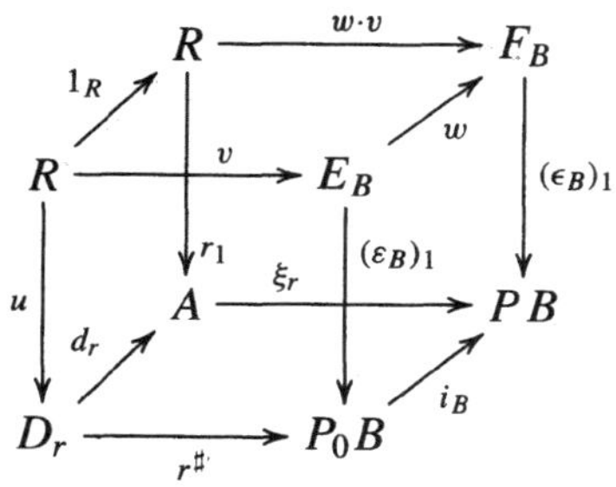

(13)

is a pullback. Since also the left and right faces are pullbacks, the front face is a pullback, as claimed.

This completes the proof of 2.3 since the composite $\varepsilon_B \circ (d_r, r^\sharp)$ is formed as in

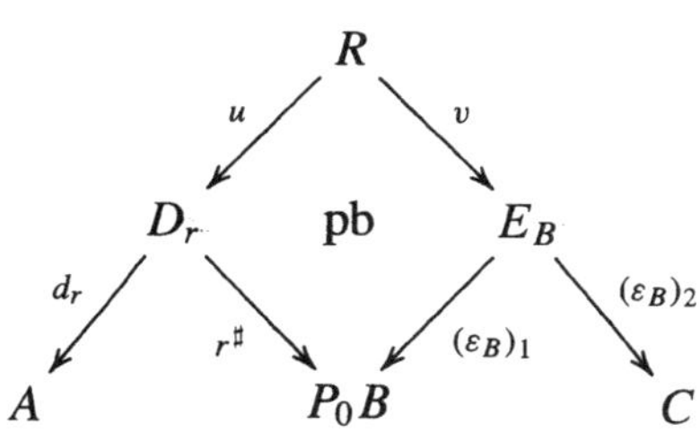

(14)

with $d_r \cdot u = r_1$ and $(\varepsilon_B)_2 \cdot v = r_2$. □

3. Non-Trivial Representation of Relations

Partial morphisms between two objects A and B are partially ordered: one writes $(m, f) \leq (m', f')$ if there is a morphism h with $m' \cdot h = f$ and $f' \cdot h = f$.

With respect to this order, the construction of Proposition 2.3 enjoys an additional property:

PROPOSITION 3.1. *Under conditions* (I)–(IV), *for every object B there is a relation ε_B: $P_0B \nrightarrow B$ such that for every relation r: $A \nrightarrow B$ there is a largest partial morphism (m, f): $A \nrightarrow P_0B$ with $\varepsilon \circ (m, f) = r$.*

Proof. In the notation of 2.3, it suffices to show that $(d_r, r^\sharp)$ is largest among all partial morphisms (m, f): $A \nrightarrow P_0B$ with $\varepsilon \circ (m, f) = r$. The latter equation means that when forming the pullback

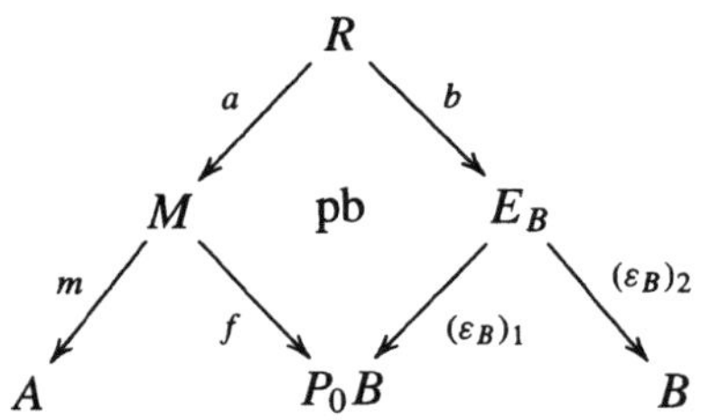

(15)

we have $m \cdot a = r_1$ and $(\varepsilon_B)_2 \cdot b = r_2$. In diagrams (16), (17) below, the right squares are pullbacks, and so are the left squares, as one easily checks when presenting their bottom arrows as pullbacks of f and m, respectively.

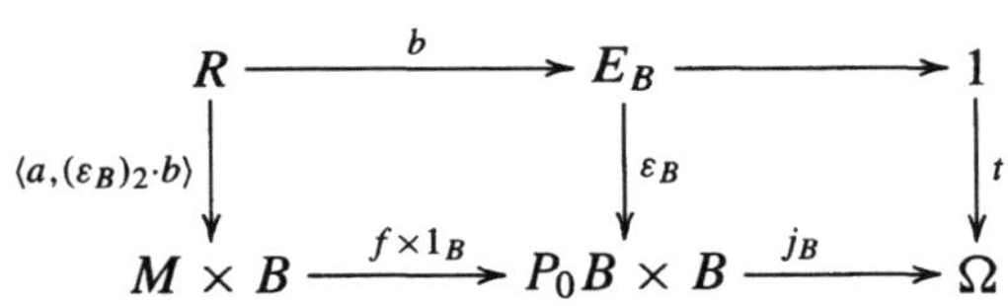

(16)

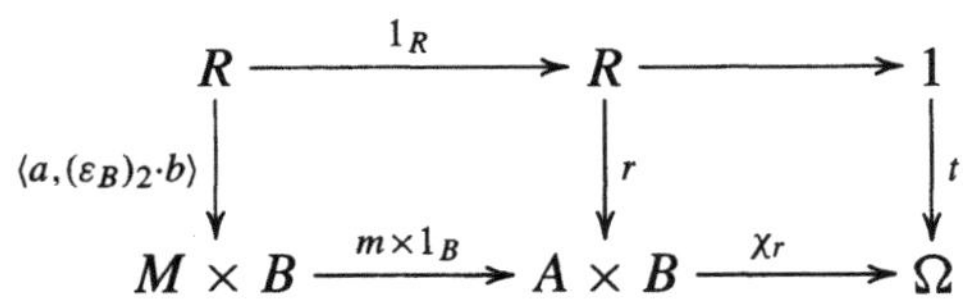

(17)

Hence, $j_B \cdot (f \times 1_B) = \chi_r \cdot (m \times 1_B)$ must be the characteristic morphism of $\langle a, (\varepsilon_B)_2 \cdot b \rangle$, and this implies $\xi_r \cdot m = i_B \cdot f$. We can now exploit the pullback property of (3) to obtain a morphism h: $M \to D_r$ with $d_r \cdot h = m$ and $r^\sharp \cdot h = f$, and this means exactly $(m, f) \leq (d_r, r^\sharp)$, as desired. □

When can we expect to obtain non-trivial relational representation by the construction given in 2.3? Let us strengthen condition (I) to:

(I*) *$\mathcal{M}$ belongs to a proper and pullback-stable $(\mathcal{E}, \mathcal{M})$-factorization system of $\mathcal{C}$.*

When $\mathcal{M}$ is the class of regular monomorphisms, $\mathcal{E}$ must then be the class of epimorphisms, which is required to be stable under pullback. We then have:

PROPOSITION 3.2. *Under conditions* (I*), (II)–(IV), *let the relations* $\varepsilon_B\colon P_0B \nrightarrow B$ *be constructed as in* 2.3, *and assume that*

(V) *the morphisms* $(\varepsilon_B)_1\colon E_B \to P_0B$ *are epic for all* B *in* $\mathcal{C}$.

Then relations are non-trivially represented.

Proof. Under conditions (I*) and (V), pullback stability of $\mathcal{E}$ forces $u\colon R \to D_r$ and $a\colon R \to M$ to belong to $\mathcal{E}$. Hence $d_r \cdot u = r_1 = m \cdot a$ are $(\mathcal{E}, \mathcal{M})$-factorizations of the same morphism, and h as constructed in the proof 3.1 must be an isomorphism. In other words: the factorization $r = \varepsilon_B \circ (d_r, r^\sharp)$ is unique. □

Given (I*) and (II)–(IV), we finally examine the question whether (V) is a necessary condition for relations to be non-trivially represented. The plain answer to this question is 'No': see 3.4 below. The reason for this is that $\varepsilon_B\colon P_0B \nrightarrow B$ is in fact *not* the natural candidate for providing non-trivial representation of relations with codomain B.

In fact, in the absence of (V), it seems natural to consider an $(\mathcal{E}, \mathcal{M})$-factorization of $(\varepsilon_B)_1$, for every B in $\mathcal{C}$:

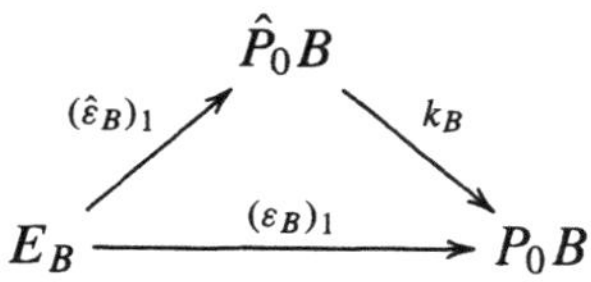

(18)

Trading ε_B for $\hat{\varepsilon}_B\colon \hat{P}_0B \nrightarrow B$ with $\hat{\varepsilon}_B = \langle(\hat{\varepsilon}_B)_1, (\varepsilon_B)_2\rangle$, we can indeed prove:

THEOREM 3.3. *Under conditions* (I*) *and* (II)–(IV), *relations are non-trivially represented. Moreover, this representation is provided by* ε_B *as constructed in* 2.3 *if and only if condition* (V) *of* 3.2 *holds.*

Proof. Given $r\colon A \nrightarrow B$, one finds $(\hat{d}_r, \hat{r}^\sharp)\colon A \nrightarrow \hat{P}_0B$ with $\hat{\varepsilon}_B \cdot (\hat{d}_r, \hat{r}^\sharp) = r$, as follows. Using the stability of $\mathcal{E}$, in the notation of 2.3 and with an $(\mathcal{E}, \mathcal{M})$-factorization $u = n \cdot \hat{u}$, one obtains two consecutive pullback diagrams as in (19):

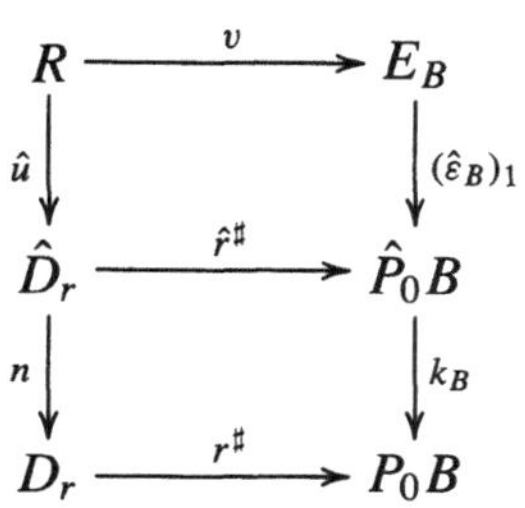

(19)

It is now obvious that with $\hat{d}_r := d_r \cdot n$ one obtains as in (14):

$$\hat{\varepsilon}_B \circ (\hat{d}_r, \hat{r}^\sharp) = r.$$

Given any partial morphism $(m, f)\colon A \nrightarrow \hat{P}_0 B$ with $\hat{\varepsilon}_B \circ (m, f) = r$, we form (15) (with $(\varepsilon_B)_1$ replaced by $(\hat{\varepsilon}_B)_1$) and modify (16), as follows:

$$\begin{array}{ccccccc} R & \xrightarrow{\ v\ } & E_B & \xrightarrow{\ 1_{E_B}\ } & E_B & \longrightarrow & 1 \\ {\scriptstyle \langle a,(\varepsilon_B)_2 \cdot b\rangle}\downarrow & & \downarrow{\scriptstyle \hat{\varepsilon}_B} & & \downarrow{\scriptstyle \varepsilon_B} & & \downarrow{\scriptstyle t} \\ M \times B & \xrightarrow{\ f \times 1_B\ } & \hat{P}_0 B \times B & \xrightarrow{\ k_B \times 1_B\ } & P_0 B \times B & \xrightarrow{\ j_B\ } & \Omega \end{array} \tag{20}$$

Since its middle square is trivially a pullback, comparison with (17) yields $\xi_r \cdot m = i_B \cdot k_B \cdot f$ as in the proof of 3.1. Hence there is a morphism $h\colon M \to D_r$ with $d_r \cdot h = m$ and $r^\sharp \cdot h = k_B \cdot f$, and as in the proof of 3.2 there is an isomorphism $h'\colon M \to \hat{D}_r$ with $\hat{d}_r \cdot h' = m$ and $h' \cdot a = \hat{u}$. Now $d_r \cdot n \cdot h' = m = d_r \cdot h$ gives $n \cdot h' = h$ and therefore

$$k_B \cdot \hat{r}^\sharp \cdot h' = r^\sharp \cdot n \cdot h' = r^\sharp \cdot h = k_B \cdot f.$$

Consequently, $\hat{r}^\sharp \cdot h' = f$, as desired.

The additional statement is now an immediate consequence of the uniqueness of co-universal arrows up to isomorphism: since $\varepsilon_B \circ k_B = \hat{\varepsilon}_B$, insisting on ε_B to represent relations with codomain B is insisting on k_B to be an isomorphism, which is the same as saying that $(\varepsilon_B)_1$ must lie in $\mathcal{E}$. □

We provide an easy example showing that condition (V) does not hold, in general.

EXAMPLE 3.4. Let $\mathcal{C}$ be the topos of $\mathbb{N}$-sets; we describe it as the category of universal algebras (X, α), with one unary operation $\alpha\colon X \to X$ (and no equations). Its subobject classifier may be described as $\Omega = (\mathbb{N} \cup \{\infty\}, \pi)$, with $\pi(0) = 0$, $\pi(\infty) = \infty$, $\pi(n) = n - 1$ for $n > 0$, with 0 being 'true'. The power object $PX = \Omega^X$ may then be realized as the set $\mathcal{C}(\mathbb{N} \times X, \Omega)$ (where $\mathbb{N}$ is a $\mathcal{C}$-object with the successor function), provided with the unary operation δ with

$$\delta(\varphi)(n, x) = \varphi(n + 1, x)$$

for all $n \in \mathbb{N}$, $x \in X$. The characteristic function of $\emptyset \to X$ is $X \to \Omega$ with $x \mapsto \infty$ for all $x \in X$, and its transpose $0^\bullet_X\colon 1 \to \Omega^X$ picks the function $\varphi_\infty\colon \mathbb{N} \times X \to \Omega$ with constant value ∞. Hence, the pseudo-complement of $0^\bullet_X$ gives

$$P_0 X = \{\varphi \in PX \mid \forall n \in \mathbb{N} : \delta^n(\varphi) \neq \varphi_\infty\}.$$

F_X as in (2) is given by $\{(\varphi, x) \mid \varphi(0, x) = 0\}$, and one easily verifies $F_X \subseteq P_0 X \times X$; hence, $E_X = F_X$. However, the projection $E_X \to P_0 X$ is in general not surjective. For example, for $X = 1 = \{0\}$ and $\varphi\colon \mathbb{N} \times X \to \Omega$ given by $\varphi(0, 0) = 1$ and $\varphi(n, 0) = 0$ for $n \neq 0$, one has $\varphi \in P_0 X$, but φ has no preimage in E_X.

4. Functorial Treatment of Relational Representation

In the presence of condition (I*), one defines the category

$$Rel(\mathcal{C})$$

as having the same objects as $\mathcal{C}$, but morphisms are relations where the composite $s \circ r$ of $r\colon A \nrightarrow B$ and $s\colon B \nrightarrow C$ is the $\mathcal{M}$-part of an $(\mathcal{E}, \mathcal{M})$-factorization of $\langle r_1 \cdot s_1', s_2 \cdot r_2' \rangle\colon R \times_B S \to A \times C$,

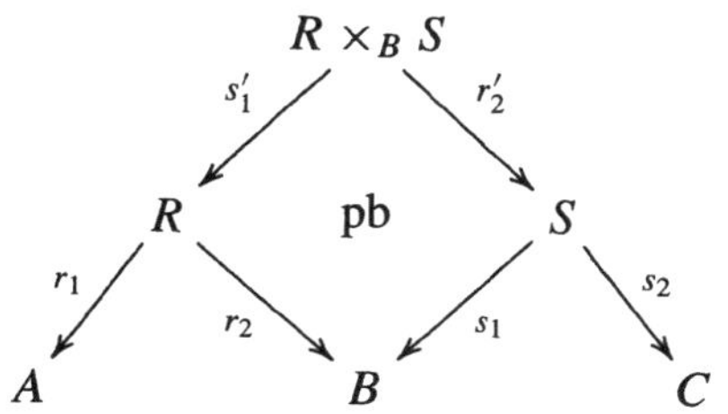

(21)

(In case (r_1, r_2) is a partial morphism, this definition coincides with the one given in the Introduction.) With $Par(\mathcal{C})$ denoting the non-full subcategory whose morphisms are partial morphisms, one has inclusion functors

$$\mathcal{C} \xrightarrow{J} Par(\mathcal{C}) \xrightarrow{I} Rel(\mathcal{C}).$$

The existence of a right adjoint for J is known to be equivalent to the existence of *$\mathcal{M}$-partial map classifiers* $t_B\colon B \to \widetilde{B}$, while the existence of a right adjoint for IJ is equivalent to asking relations to be represented. Finally, existence of a right adjoint to I means equivalently non-trivial representability of relations.

Recall that in the presence of (IV), the *singleton morphism* $\sigma_B\colon B \to PB$ is the morphism representing the trivial relation $(1_B, 1_B)$, which is known to be a monomorphism (cf. [13, 15.4]). From [13, 16.8] and Theorem 3.3 we obtain:

COROLLARY 4.1. *Assume that* (I*) *holds and that IJ has a right adjoint. Then*

(a) *J has a right adjoint if the singleton morphisms belong to $\mathcal{M}$, and*
(b) *I has a right adjoint if conditions* (II), (III) *are satisfied.*

REMARK 4.2. (1) Conditions (I*), (II), (III) do not guarantee the existence of a right adjoint for J. Indeed, consider any poset with finite infima and a least element 0 as a category $\mathcal{C}$ and let $\mathcal{M}$ be the class of all (mono)morphisms. Then I is the identity functor and has trivially a right adjoint, but the existence of a right adjoint for J would entail the existence of a subobject classifier, which is possible only for $0 = 1$.

(2) If one provides $\mathcal{C}$ as in (1) with $\mathcal{M}$ the class of all regular monomorphisms (= isomorphisms), then J is an isomorphism and I has a right adjoint (which takes

the constant value 1). All conditions but (III) are satisfied; failure of (III) is due to the fact that $0 \to 1$ fails to be in $\mathcal{M}$, unless $0 = 1$. This shows on the one hand that condition (III) is not a necessary condition for having non-trivial representation of relations. On the other hand, one does not need a quasitopos to have this kind of representation: any poset with finite infima and a bottom element 0 which is not a Heyting algebra yields an example.

Acknowledgement

This work has been supported by the Natural Sciences and Engineering Research Council of Canada.

References

1. Borceux, F.: *Handbook of Categorical Algebra*, Vols. 1–3, Cambridge University Press, Cambridge, 1994.
2. Barr, M. and Wells, C.: *Toposes, Triples and Theories*, Springer-Verlag, New York, 1985.
3. Carboni, A., Kelly, G. M., and Wood, R. J.: A 2-categorical approach to change of base and geometric morphisms I, *Cahiers Topologie Géom. Différentielle Catégoriques* **32** (1991), 47–95.
4. Dong, X.: Representation of Partial Maps and Relations, and the Span-Map Adjunction, PhD Thesis, York University, September 1996.
5. Dyckhoff, R. and Tholen, W.: Exponentiable morphisms, partial products and pullback complements, *J. Pure Appl. Algebra* **49** (1987), 103–116.
6. Fiore, M. P.: Order-enrichment for categories of partial maps, *Math. Struct. in Comp. Science*, **5** (1995), 533–562.
7. Freyd, P. J. and Kelly, G. M.: Categories of continuous functors I, *J. Pure Appl. Algebra* **2** (1972), 169–191.
8. Jay, C. B.: Extending Properties to Categories of Partial Maps, Technical Report, LFCS, Department of Computer Science, University of Edinburgh, February 1990.
9. Jay, C. B.: Partial functions, ordered categories, limits and cartesian closure, in *IV Higher Order Workshop*, Banff, 1991, pp. 151–161.
10. Klein, E. and Thompson, A. C.: *Theory of Correspondences. Including Applications to Mathematical Economics*, Wiley, New York, 1984.
11. Pavlović, D.: Maps I: Relative to a factorization system, *J. Pure Appl. Algebra* **99** (1995), 9–34.
12. Robinson, E. and Rosolini, G.: Categories of partial maps, *Inform. Computation* **79** (1988), 95–130.
13. Wyler, O.: *Lecture Notes on Topoi and Quasitopoi*, World Scientific, Singapore, 1991.

Applied Categorical Structures **8:** 351–366, 2000.
G. Brümmer & C. Gilmour (eds), Papers in Honour of Bernhard Banaschewski.

Structured Frames by Weil Entourages

In honour of Professor Bernhard Banaschewski on the occasion of his 70th birthday

JORGE PICADO
Departamento de Matemática, Universidade de Coimbra, Apartado 3008, 3000 Coimbra, Portugal
e-mail: picado@mat.uc.pt

(Received: 18 October 1996; accepted: 11 April 1998)

Abstract. In the framework of pointfree topology, we discuss the rôle of *Weil entourages* in the study of structures such as uniformities, quasi-uniformities, nearnesses, quasi-nearnesses, proximities and infinitesimal relations.

Mathematics Subject Classifications (2000): 06D20, 54E05, 54E15, 54E17.

Key words: frame, Weil entourage, uniform frame, localic group, quasi-uniform frame, nearness, quasi-nearness, proximity, infinitesimal relation.

In the same way as the notion of Boolean algebra appears as an abstraction of the power set $\mathcal{P}(X)$ of a set X, the notion of frame arises as an abstraction from the topology $\mathcal{T}$ of a topological space $(X, \mathcal{T})$: a *frame* is a complete lattice L satisfying the distributive law $x \wedge \bigvee S = \bigvee\{x \wedge s \mid s \in S\}$ for all $x \in L$ and $S \subseteq L$. A *frame homomorphism* is a map between frames which preserves finitary meets (including the unit 1) and arbitrary joins (including the zero 0). A standard reference for frames is Johnstone [18].

Pointfree topology deals with the category Frm of frames and frame homomorphisms (or, depending on the point of view, with its dual category of *locales*; in this 'localic' point of view one thinks of locales as generalized spaces; technically, however, we will work in Frm. Therefore, for example, statements on products of generalized spaces will appear as statements on coproducts of frames).

In topology, metrics, uniformities, nearnesses, proximities and such like structures are defined on a set and determine a topology on that set. In pointfree topology, "the specified object is the 'topology', that is, a frame; and consequently, metrics, uniformities, and nearnesses must appear as additional structures on a frame" [2].

The study of structured frames started in the early 1970's with Isbell [15], where uniformities on frames were introduced as the exact translation into frame terms of Tukey's approach [25] to uniform spaces via covers, and continued, in more detail, with Pultr [23, 24] and Frith [11]. Quasi-uniformities, which are uniformities without symmetry, were introduced by Frith [11] in terms of conjugate cover pairs of biframes (the translation into frame terms of the approach of Gantner and Steinlage [13] to quasi-uniform spaces via covers). Nearness, which is uniformity without the star-refinement condition, was introduced more recently by Banaschewski and Pultr [4], and in [5] and [6] Dube studied several properties of nearness frames.

As in the case of spaces, there are other ways of describing (quasi-)uniformities, such as the entourage (quasi-)uniformities of Fletcher and Hunsaker [9] and the Weil (quasi-)uniformities of Picado [20, 21]. The various descriptions are known to be equivalent.

The purpose of this article is to outline some results of my Ph.D. thesis, written under the supervision of Professors B. Banaschewski and M. Sobral, which I believe establish the naturality of an entourage approach to several frame structures (uniformities, quasi-uniformities, nearnesses, proximities etc.). The point that I wish to make here is that the notion of Weil entourage introduced in [20] is the exact translation into frame terms of Weil's classical notion of entourage and that it provides us with the appropriate environment to get frame extensions of spatial results envolving entourages, and to formulate them in a very similar way to the corresponding classical results. After a brief summary, in a preliminary section, on the needed tools, I describe some results of [21] which emphasize this point.

I would like to express my gratitude to Professor B. Banaschewski for his invaluable suggestions and comments.

Preliminaries

For subsets $A, B, \ldots$ and elements $x, y, \ldots$ of a frame L, let us recall the following notation and terminology:

$A \leq B$ if each member of A is below some member of B;

$$st(x, A) := \bigvee\{a \in A \mid a \wedge x \neq 0\}, \quad A^* := \{st(x, A) \mid x \in A\},$$
$$A \wedge B := \{a \wedge b \mid a \in A, b \in B\}.$$

Further, A is called a cover of L if $\bigvee A = 1$, and $Cov(L)$ denotes the set of all covers of L, taken as quasi-ordered by the above relation $\leq$. Next, for any $\mathcal{A} \subseteq Cov(L)$, $y \overset{\mathcal{A}}{\triangleleft} x$ means that $st(y, A) \leq x$ for some $A \in \mathcal{A}$, and $\mathcal{A}$ is called *admissible* whenever $x = \bigvee\{y \in L \mid y \overset{\mathcal{A}}{\triangleleft} x\}$ for all $x \in L$. Then, a *nearness* on L is an admissible filter $\mathcal{U}$ in $Cov(L)$. In particular, $\mathcal{U}$ is called a *uniformity* if it satisfies the refinement condition, that is, for each $U \in \mathcal{U}$, there exists $V \in \mathcal{U}$ such that $V^* \leq U$. A *nearness frame* (resp., *uniform frame*) is a frame together with a specified nearness (resp., uniformity). For nearness frames $(L, \mathcal{U})$ and $(L', \mathcal{U}')$, a frame map $f\colon L \to L'$ is *uniform* if $f[U] \in \mathcal{U}'$ for any $U \in \mathcal{U}$. Further, Nfrm (resp., Ufrm) will be the category of nearness frames and uniform homomorphisms (resp., uniform frames and uniform homomorphisms).

For the following, recall that a subset A of a poset $(X, \leq)$ is said to be a *down-set* if $A = \downarrow A$ where $\downarrow A$ denotes the set $\{x \in X \mid x \leq a \text{ for some } a \in A\}$.

Let L be a frame. Recall (cf., e.g., [18]) that the coproduct of the frame L by itself

$$L \xrightarrow{u_1^L} L \oplus L \xleftarrow{u_2^L} L$$

can be constructed as follows:

Take the Cartesian product $L \times L$ with the usual order. A down-set A of $L \times L$ is a *C-ideal* if $(\{x\} \times S \subseteq A \Rightarrow (x, \bigvee S) \in A)$ and $(S \times \{y\} \subseteq A \Rightarrow (\bigvee S, y) \in A)$. Put $L \oplus L$ as the frame of all C-ideals of $L \times L$. Observe that the case $S = \emptyset$ implies that every C-ideal contains the set $\mathbb{O} := \downarrow\{(1, 0)\} \cup \downarrow\{(0, 1)\}$. Obviously, each $\downarrow\{(x, y)\} \cup \mathbb{O}$ is a C-ideal. It is denoted by $x \oplus y$. Finally put $u_1^L(x) = x \oplus 1$ and $u_2^L(y) = 1 \oplus y$.

The following clear facts are useful:

- For every $A \in L \oplus L$, $A = \bigvee\{x \oplus y \mid (x, y) \in A\}$ and so every element of $L \oplus L$ is join-generated by some family of elements $x \oplus y$;
- $\mathbb{O} \neq x \oplus y \subseteq z \oplus w$ implies $x \leq z$ and $y \leq w$.

For any frame homomorphism $f\colon L \to M$, we write $f \oplus f\colon L \oplus L \to M \oplus M$ for the frame homomorphism given by $(f \oplus f) \cdot u_i^L = u_i^M \cdot f$ $(i = 1, 2)$. Obviously, $(f \oplus f)(\bigvee_\gamma (x_\gamma \oplus y_\gamma)) = \bigvee_\gamma (f(x_\gamma) \oplus f(y_\gamma))$.

Given A, B in the lattice $\mathcal{D}(L \times L)$ of all down-sets of $L \times L$ we denote by $k(A)$ the C-ideal generated by A and by $A \circ B$ the C-ideal generated by $\{(x, y) \in L \times L \mid \exists z \in L \setminus \{0\} : (x, z) \in A, (z, y) \in B\}$, that is, $\bigvee\{x \oplus y \mid \exists z \in L \setminus \{0\} : (x, z) \in A, (z, y) \in B\}$. The following technical lemma, proved in [20] (Lemma 3.1), will play a crucial rôle in our approach:

LEMMA A. *For any $A, B \in \mathcal{D}(L \times L)$, $k(A) \circ k(B) = A \circ B$.*

The importance of this result relies in the fact that it allows us to work with composition of frame entourages in a way very similar to the spatial one. This can be observed, for example, in the proof of Proposition 1.4 below.

Weil Structures on Frames

1. UNIFORMITIES

Uniform spaces were introduced by Weil in terms of the basic notion of entourage. An *entourage* of a set X is a subset E of $X \times X$ for which the diagonal $\Delta_X \colon X \to X \times X$ factorizes through the inclusion $E \hookrightarrow X \times X$:

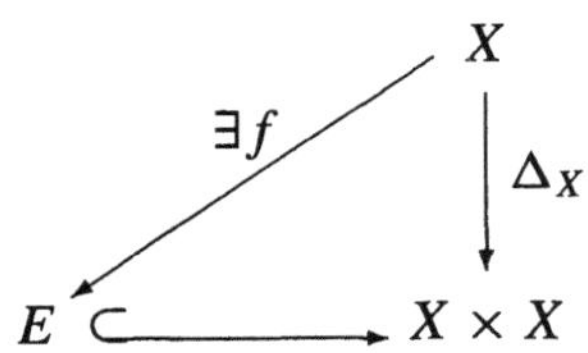

This motivates the study, on a frame L, of the elements E of the frame coproduct $L \oplus L$ for which the codiagonal $\nabla_L \colon L \oplus L \to L$ factorizes through the open sublocale $L \oplus L \to \downarrow\{E\}$ of $L \oplus L$:

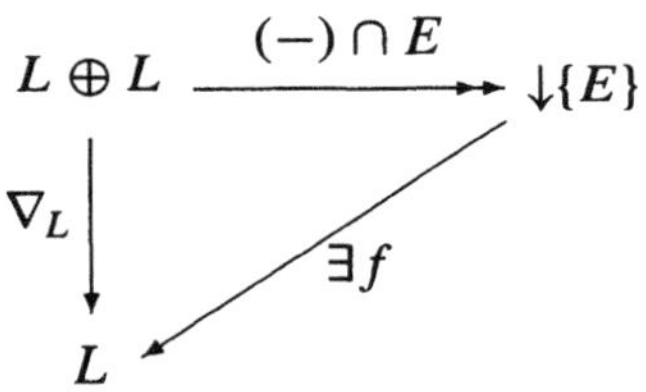

(here, $(-) \cap E(F) = E \cap F$, for every $F \in L \oplus L$ and ∇_L is the unique morphism such that $\nabla_L \cdot u_1^L = 1_L = \nabla_L \cdot u_2^L$). We call them *Weil entourages* of L. Evidently, $E \in L \oplus L$ is a Weil entourage if and only if $\nabla_L(E) = 1$. Since ∇_L is defined by

$$E = \bigvee_{(x,y)\in E} (x \oplus y) \longmapsto \bigvee_{(x,y)\in E} (x \wedge y),$$

an element E of $L \oplus L$ is a Weil entourage if and only if $\bigvee_{(x,x)\in E} x = 1$ or, in other words, if and only if there exists a cover U of L such that $\bigvee_{x\in U}(x \oplus x) \subseteq E$. This shows a perfect analogy between our notion of frame entourage (introduced in [20]) and the notion of entourage for sets and justifies the designation of Weil entourage for these frame entourages.

The collection $WEnt(L)$ of all Weil entourages of L may be partially ordered by inclusion. This is a partially ordered set with finitary meets (including a unit $1 = L \oplus L$).

We define the composition of Weil entourages as follows:

$$E \circ F := \bigvee\{x \oplus y \mid \exists z \in L \setminus \{0\} : (x, z) \in E, (z, y) \in F\}.$$

For the basic properties of the operation $\circ$ see [21].

The inverse of a Weil entourage E has the natural definition $E^{-1} = \{(y, x) \mid (x, y) \in E\}$. We also consider a new partial order in L, induced by a family $\mathcal{E}$ of Weil entourages:

$$y \overset{\mathcal{E}}{\lhd} x \text{ (read `}y\text{ is }\mathcal{E}\text{-strongly below }x\text{') if there is } E \in \mathcal{E} \text{ such that } E \circ (y \oplus y) \subseteq x \oplus x. \tag{1.1}$$

Of course, when $\mathcal{E}$ is symmetric (that is, $E \in \mathcal{E}$ implies $E^{-1} \in \mathcal{E}$) this is equivalent to saying that

$$\text{there is } E \in \mathcal{E} \text{ such that } (y \oplus y) \circ E \subseteq x \oplus x. \tag{1.2}$$

DEFINITION 1.3. Let L be a frame and let $\mathcal{E}$ be a nonempty filter of $(WEnt(L), \subseteq)$. Consider the following axioms:

(W1) [Admissibility Axiom] For any $x \in L$, $x = \bigvee\{y \in L \mid y \overset{\overline{\mathcal{E}}}{\lhd} x\}$ (where $\overline{\mathcal{E}}$ denotes the filter of $(WEnt(L), \subseteq)$ generated by $\mathcal{E} \cup \{E^{-1} \mid E \in \mathcal{E}\}$);

(W2) [Symmetry Axiom] For any $E \in \mathcal{E}$, $E^{-1} \in \mathcal{E}$;

(W3) [Refinement Axiom] For each $E \in \mathcal{E}$ there exists $F \in \mathcal{E}$ such that $F \circ F \subseteq E$.

The family $\mathcal{E}$ is called a *Weil nearness* on L if it satisfies (W1) and (W2) and it is called a *Weil uniformity* on L if it fulfils (W1), (W2) and (W3). The pair $(L, \mathcal{E})$ is called a *Weil nearness frame* (resp., *Weil uniform frame*) if $\mathcal{E}$ is a Weil nearness (resp., Weil uniformity) on L. A *Weil nearness base* (resp., *Weil uniformity base*) is just a filter base of some Weil nearness (resp., Weil uniformity).

If $(L, \mathcal{E})$ and $(L', \mathcal{E}')$ are Weil nearness frames, a frame homomorphism $f\colon L \to L'$ is called a *Weil homomorphism* from $(L, \mathcal{E})$ to $(L', \mathcal{E}')$ if $(f \oplus f)(E) \in \mathcal{E}'$ whenever $E \in \mathcal{E}$.

We shall denote by WNFrm (resp., WUFrm) the category of Weil nearness frames and Weil homomorphisms (resp., Weil uniform frames and Weil homomorphisms).

We proceed to give examples of Weil uniformities which illustrate the similarities with the corresponding spatial cases.

EXAMPLES. (a) Let (L, d) be a metric frame [24]. For any real $\epsilon > 0$ let $E_\epsilon = \bigvee\{x \oplus x \mid d(x) < \epsilon\}$. The family $(E_\epsilon)_{\epsilon > 0}$ is a base for a Weil uniformity on L.

(b) Consider a localic group [16] (a cogroup in the category of frames), that is, a frame L endowed with a multiplication $\mu\colon L \to L \oplus L$, an inverse $\iota\colon L \to L$ and a unit point $\varepsilon\colon L \to \mathbf{2}$ ($\mathbf{2}$ denotes the two-element lattice) satisfying the identities

$$(\mu \oplus 1_L) \cdot \mu = (1_L \oplus \mu) \cdot \mu,$$

$$(\varepsilon \oplus 1_L) \cdot \mu = 1_L = (1_L \oplus \varepsilon) \cdot \mu$$

and

$$\nabla_L \cdot (\iota \oplus 1_L) \cdot \mu = \nabla_L \cdot (1_L \oplus \iota) \cdot \mu = \sigma \cdot \varepsilon,$$

where σ is the morphism $\mathbf{2} \to L$. Note that the usual properties for groups

$$\varepsilon \cdot \iota = \varepsilon,$$
$$\iota \cdot \iota = 1_L$$

and

$$\mu \cdot \iota = \tau \cdot (\iota \oplus \iota) \cdot \mu,$$

where τ is the unique map from $L \oplus L$ to $L \oplus L$ satisfying $\tau \cdot u_1^L = u_2^L$ and $\tau \cdot u_L^2 = u_L^1$, are also valid here.

These groups have Weil uniformities that arise in a similar way as in the spatial setting of topological groups. For any $x \in L$ such that $\varepsilon(x) = 1$ put

$$E_x^l := (1_L \oplus \iota)(\mu(x)) \quad \text{and} \quad E_x^r := (\iota \oplus 1_L)(\mu(x)).$$

PROPOSITION 1.4. *$\mathcal{E}^l := \{E_x^l \mid x \in L, \varepsilon(x) = 1\}$ and $\mathcal{E}^r := \{E_x^r \mid x \in L, \varepsilon(x) = 1\}$ are bases for Weil uniformities.*

Proof. We only show that $\mathcal{E}^r$ is a Weil uniformity base. The proof for $\mathcal{E}^l$ is similar.

Each E_x^r is a Weil entourage because $\nabla_L(E_x^r) = \nabla_L \cdot (\iota \oplus 1_L) \cdot \mu(x) = \varepsilon(x) = 1$.

Obviously, $E_x^r \cap E_y^r = E_{x \wedge y}^r$. Since $\varepsilon(x \wedge y) = 1$ whenever $\varepsilon(x) = \varepsilon(y) = 1$, $\mathcal{E}^r$ is a filter base of $(WEnt(L), \subseteq)$.

The symmetry is a consequence of the fact that, for every x, $(E_x^r)^{-1} = E_{\iota(x)}^r$ which we prove next. Put $\mu(x) = \bigvee_{\gamma \in \Gamma}(x_\gamma \oplus y_\gamma)$. Then $E_x^r = \bigvee_{\gamma \in \Gamma}(\iota(x_\gamma) \oplus y_\gamma)$. On the other hand, since $\mu \cdot \iota = \tau \cdot (\iota \oplus \iota) \cdot \mu$, we have $\mu(\iota(x)) = \bigvee_{\gamma \in \Gamma}(\iota(y_\gamma) \oplus \iota(x_\gamma))$ and, therefore, $E_{\iota(x)}^r = \bigvee_{\gamma \in \Gamma}(y_\gamma \oplus \iota(x_\gamma))$.

Now, consider E_x^r with $\varepsilon(x) = 1$. We have $\varepsilon = 1_2 \oplus \varepsilon = (1_2 \oplus \varepsilon) \cdot (\varepsilon \oplus 1_L) \cdot \mu = (\varepsilon \oplus \varepsilon) \cdot \mu$. Thus $(\varepsilon \oplus \varepsilon) \cdot \mu(x) = 1$, that is, $\bigvee\{\varepsilon(a) \oplus \varepsilon(b) \mid (a, b) \in \mu(x)\} = 1$. Therefore there is some $(a, b) \in \mu(x)$ with $\varepsilon(a) = \varepsilon(b) = 1$. Also $(a \wedge b, a \wedge b) \in \mu(x)$ and $\varepsilon(a \wedge b) = 1$. Denote $a \wedge b$ by y. We claim that $E_y^r \circ E_y^r \subseteq E_x^r$. In fact, $E_y^r \circ E_y^r$ is the Weil entourage

$$\left(\bigvee_{(a,b) \in \mu(y)} (\iota(a) \oplus b) \right) \circ \left(\bigvee_{(a,b) \in \mu(y)} (\iota(a) \oplus b) \right)$$
$$= \left(\bigcup_{(a,b) \in \mu(y)} (\iota(a) \oplus b) \right) \circ \left(\bigcup_{(a,b) \in \mu(y)} (\iota(a) \oplus b) \right),$$

by Lemma A. Take $(\iota(a), b)$ with $(a, b) \in \mu(y)$ and $(\iota(c), d)$ with $(c, d) \in \mu(y)$ such that $b \wedge \iota(c) \neq 0$. From the inclusion $y \oplus y \subseteq \mu(x)$ it follows that

$$\mu(y) \oplus \mu(y) \subseteq (\mu \oplus \mu)(\mu(x)) = (1_L \oplus \mu \oplus 1_L) \cdot (\mu \oplus 1_L) \cdot (\mu(x)).$$

Therefore $a \oplus b \oplus c \oplus d \subseteq (1_L \oplus \mu \oplus 1_L) \cdot (\mu \oplus 1_L) \cdot (\mu(x))$. Applying $1_L \oplus \nabla_L \cdot (1_L \oplus \iota) \oplus 1_L$ to both sides we get

$$a \oplus (b \wedge \iota(c)) \oplus d \subseteq (1_L \oplus \sigma \oplus 1_L) \cdot (\mu(x)) = \bigvee_{(a,b)\in\mu(x)} (a \oplus 1_L \oplus b).$$

Since $b \wedge \iota(c) \neq 0$, then $(a, d) \in \mu(x)$ and $(\iota(a), d) \in E^r_x$. Hence $E^r_y \circ E^r_y \subseteq E^r_x$.

Finally, let us check the admissibility condition (W1). From the identity $(1_L \oplus \varepsilon) \cdot \mu = 1_L$, it follows that, for every $x \in L$,

$$\begin{aligned} x &= \bigvee\{a \oplus \varepsilon(b) \mid a \oplus b \subseteq \mu(x)\} \\ &= \bigvee\{a \in L \mid \exists b \in L : \varepsilon(b) = 1 \text{ and } a \oplus b \subseteq \mu(x)\}, \end{aligned}$$

whence it remains to show that a is $\mathcal{E}^r$-strongly below x whenever there is some $b \in L$ satisfying $\varepsilon(b) = 1$ and $a \oplus b \subseteq \mu(x)$. In order to conclude this, it suffices to show that, for any symmetric $F \in \mathcal{E}^r$ such that $F^2 \subseteq E^r_b$, $F \circ (a \oplus a) \subseteq x \oplus x$. So, assume $(\alpha, \beta) \in F$, $(\beta, \gamma) \leq (a, a)$ with $\alpha, \beta, \gamma \neq 0$. We need to show that both α and γ are below x. Clearly $\gamma \leq a \leq x$. Let us show that also $\alpha \leq x$. Of course, $(\alpha, \beta) \in F^2$ and, by the symmetry of F, $(\beta, \alpha) \in F^2$, which forces $(\alpha \vee \beta, \alpha \vee \beta) \in F^2 \subseteq E^r_b$, as (α, α) and (β, β) also belong to F^2. Moreover $(\alpha \vee \beta) \wedge a \neq 0$. Thus $\iota(\alpha \vee \beta) \oplus (\alpha \vee \beta) \subseteq \mu(b)$. On the other hand, $a \oplus b \subseteq \mu(x)$. Consequently, $a \oplus b \oplus a \subseteq (\mu \oplus 1_L)(x \oplus a)$ which, in turn, implies that $a \oplus \mu(b) \oplus a \subseteq (1_L \oplus \mu \oplus 1_L) \cdot (\mu \oplus 1_L)(x \oplus a)$. Thus, we have $a \oplus \iota(\alpha \vee \beta) \oplus a \subseteq (1_L \oplus \mu \oplus 1_L) \cdot (\mu \oplus 1_L)(x \oplus a)$. By the associativity of μ, $(1_L \oplus \mu \oplus 1_L) \cdot (\mu \oplus 1_L) = ((1_L \oplus \mu) \cdot \mu) \oplus 1_L = ((\mu \oplus 1_L) \cdot \mu) \oplus 1_L = (\mu \oplus 1_L \oplus 1_L) \cdot (\mu \oplus 1_L)$. Thus, $a \oplus \iota(\alpha \vee \beta) \oplus (\alpha \vee \beta) \oplus a \subseteq (\mu \oplus 1_L \oplus 1_L) \cdot (\mu \oplus 1_L)(x \oplus a)$. Applying $(\nabla_L \cdot (1_L \oplus \iota)) \oplus 1_L \oplus 1_L$ to both sides we get $(a \wedge (\alpha \vee \beta)) \oplus (\alpha \vee \beta) \oplus a \subseteq (\sigma \cdot \varepsilon \oplus 1_L \oplus 1_L) \cdot (\mu \oplus 1_L)(x \oplus a) = ((\sigma \cdot \varepsilon \oplus 1_L) \cdot \mu \oplus 1_L)(x \oplus a) = (((\sigma \oplus 1_L) \cdot (\varepsilon \oplus 1_L) \cdot \mu) \oplus 1_L)(x \oplus a) = (\sigma \oplus 1_L \oplus 1_L)(x \oplus a)$. But $(\sigma \oplus 1_L)(x) = 1 \oplus x$ so $(a \wedge (\alpha \vee \beta)) \oplus (\alpha \vee \beta) \oplus a \subseteq 1 \oplus x \oplus a$. Since $a \wedge (\alpha \vee \beta) \neq 0$, we finally obtain $(\alpha \vee \beta) \oplus a \subseteq x \oplus a$. In conclusion, $\alpha \leq \alpha \vee \beta \leq x$. □

We point out that, as for topological groups, the map ι is a Weil uniform frame isomorphism between the two structures: indeed, ι is a frame isomorphism and, for any $x \in L$, $(\iota \oplus \iota)(E^r_x) = (\iota \oplus \iota)(\iota \oplus 1_L)(\mu(x)) = (1_L \oplus \iota)(\mu(x)) = E^l_x$.

(c) Our approach via Weil entourages may also be useful to prove (or disprove) frame counterparts of spatial results envolving entourages. For an example in this direction see [21], where it is proved that whenever two uniformities with countable bases on a frame determine the same Samuel compactification they are equal. This is the localic version of a theorem of Efremovič of 1952 [8] (cf. Theorem 12.18 and Corollary 12.19 of [19]).

Weil entourages and covers interact in a very easy way: for any Weil entourage E the set $\{x \in L \mid x \oplus x \subseteq E\}$ is a cover of L and, conversely, for any cover U

of L, $\bigvee_{x \in U}(x \oplus x)$ is a Weil entourage of L. Moreover, for any nearness frame $(L, \mathcal{U})$, $\{\bigvee_{x \in U}(x \oplus x) \mid U \in \mathcal{U}\}$ forms a base for a Weil nearness $\Psi(\mathcal{U})$ on L and, conversely, for any Weil nearness frame $(L, \mathcal{E})$, $\{\{x \in L \mid x \oplus x \subseteq E\} \mid E \in \mathcal{E}\}$ forms a base for a nearness $\Phi(\mathcal{E})$ on L.

The correspondences $(L, \mathcal{U}) \overset{\Psi}{\longmapsto} (L, \Psi(\mathcal{U}))$ and $(L, \mathcal{E}) \overset{\Phi}{\longmapsto} (L, \Phi(\mathcal{E}))$ constitute a Galois connection between the partially ordered sets (by inclusion) of, respectively, nearnesses and Weil nearnesses on L: $1 \leq \Psi\Phi$ and $\Phi\Psi \leq 1$. This Galois connection induces a 'maximal' isomorphism between the partially ordered set of Weil nearnesses $\mathcal{E}$ on L satisfying the condition

$$\forall E \in \mathcal{E} \bigvee_{(x,x) \in E} (x \oplus x) \in \mathcal{E} \tag{1.3}$$

and the partially ordered set of nearnesses $\mathcal{U}$ on L satisfying the condition

$$\forall U \in \mathcal{U} \exists V \in \mathcal{U} : \left(x \oplus x \subseteq \bigvee_{x \in V} (x \oplus x) \Rightarrow \exists u \in U : x \leq u\right). \tag{1.4}$$

For the proofs and further details, the reader is referred to [21]. This extends the result of [20] that asserts the isomorphism between WUFrm and UFrm; in fact, in particular, this gives a bijection for uniformities since it can be readily seen that the refinement conditions in the definitions of Weil uniform frame and uniform frame are stronger than the corresponding conditions 1.3 and 1.4 above. However, I am not able to conclude whether WNFrm and NFrm are isomorphic. This is still an open question.

2. QUASI-UNIFORMITIES AND QUASI-NEARNESSES

One of the advantages of entourage-like theories is that the symmetry is explicit and so the corresponding nonsymmetric versions are evident and pleasantly manageable. By dropping the symmetry in Definitions 1.3 we have the notions of *Weil quasi-nearness frame* and *Weil quasi-uniform frame*.

With the lack of symmetry the equivalence between conditions 1.1 and 1.2 is no longer valid; whence, in the place of $\overset{\mathcal{E}}{\lhd}$ we have two order relations

$$y \overset{\mathcal{E}}{\lhd}_1 x \equiv E \circ (y \oplus y) \subseteq x \oplus x, \quad \text{for some } E \in \mathcal{E},$$

$$y \overset{\mathcal{E}}{\lhd}_2 x \equiv (y \oplus y) \circ E \subseteq x \oplus x, \quad \text{for some } E \in \mathcal{E},$$

which in turn, lead to two subframes of L,

$$L_1 := \left\{x \in L \mid x = \bigvee\{y \in L \mid y \overset{\mathcal{E}}{\lhd}_1 x\}\right\} \quad \text{and}$$
$$L_2 := \left\{x \in L \mid x = \bigvee\{y \in L \mid y \overset{\mathcal{E}}{\lhd}_2 x\}\right\}$$

(for spaces this corresponds to the two topologies defined by the quasi-uniformity).

Notice that the condition (W1) of admissibility says that the triple (L, L_1, L_2) is a biframe. This allows us to establish a Galois connection between Weil entourages and conjugate cover pairs of biframes:

Recall that a *biframe* [3] is a triple $B = (B_0, B_1, B_2)$ in which B_0 is a frame and B_1, B_2 are subframes of B_0 such that each $x \in B_0$ is the join of finite meets from $B_1 \cup B_2$.

Let $B = (B_0, B_1, B_2)$ be a biframe. The following definitions and notations are transcribed from [11]. A subset U of $B_1 \times B_2$ is a *conjugate cover pair* (or just *conjugate cover*) of B provided that $\{u_1 \wedge u_2 \mid (u_1, u_2) \in U\}$ is a cover of B_0. A conjugate cover U is *strong* if, for every $(u_1, u_2) \in U$, $u_1 \wedge u_2 \neq 0$ whenever $u_1 \vee u_2 \neq 0$. Let U and V be conjugate covers. Then $U \leq V$ provided that, for each $(u_1, u_2) \in U$, there is $(v_1, v_2) \in V$ with $u_1 \leq v_1$ and $u_2 \leq v_2$. Furthermore, $U \wedge V$ denotes the conjugate cover $\{(u_1 \wedge v_1, u_2 \wedge v_2) \mid (u_1, u_2) \in U, (v_1, v_2) \in V\}$ and U^* stands for the conjugate cover $\{(st_1(u_1, U), st_2(u_2, U)) \mid (u_1, u_2) \in U\}$, where $st_1(x, U) := \bigvee\{u_1 \mid (u_1, u_2) \in U, u_2 \wedge x \neq 0\}$ and $st_2(x, U) := \bigvee\{u_2 \mid (u_1, u_2) \in U, u_1 \wedge x \neq 0\}$ for any $x \in B_0$.

Let $B = (B_0, B_1, B_2)$ be a biframe and let $\mathcal{U}$ be a collection of conjugate covers. Then $\mathcal{U}$ is a *quasi-uniformity* for the biframe B and $(B, \mathcal{U})$ is a *quasi-uniform frame* provided that:

(i) The family of strong members of $\mathcal{U}$ is a filter base for $\mathcal{U}$ with respect to $\leq$;
(ii) For each $U \in \mathcal{U}$ there exists $V \in \mathcal{U}$ such that $V^* \leq U$;
(iii) For each $x \in B_i$ $(i \in \{1, 2\})$, $x = \bigvee\{y \in B_i \mid y \overset{\mathcal{U}}{\lhd}_i x\}$, where $y \overset{\mathcal{U}}{\lhd}_i x$ means that there is $U \in \mathcal{U}$ such that $st_i(y, U) \leq x$.

Let $(B, \mathcal{U})$ and $(B', \mathcal{U}')$ be quasi-uniform frames. A *uniform homomorphism* $f\colon (B, \mathcal{U}) \to (B', \mathcal{U}')$ is a biframe map $f\colon B \to B'$ such that, for every $U \in \mathcal{U}$, $f[U] := \{(f(u_1), f(u_2)) \mid (u_1, u_2) \in U\} \in \mathcal{U}'$. The category of quasi-uniform frames and uniform homomorphisms will be denoted by QUFrm.

The relationship between Weil entourages and conjugate covers is not so evident as the one between Weil entourages and covers. For any conjugate cover U of a biframe (B_0, B_1, B_2), $\bigvee\{u_1 \oplus u_2 \mid (u_1, u_2) \in U\}$ (meant in $B_0 \oplus B_0$) is a Weil entourage of B_0. Conversely, starting with a Weil entourage E of a quasi-nearness frame L, we may consider the biframe (L, L_1, L_2) and the pairs $(st_1(x, E), st_2(x, E))$ for $(x, x) \in E$, where $st_1(x, E) := \bigvee\{y \in L \mid (y, z) \in E, z \wedge x \neq 0\}$ and $st_2(x, E) := \bigvee\{y \in L \mid (z, y) \in E, z \wedge x \neq 0\}$. However there is a problem: these pairs may not belong to $L_1 \times L_2$. It may be solved by introducing two interior operators in $L \oplus L$ in the following manner: For every $I, J \in L \oplus L$, we define $I \overset{\mathcal{E}}{\sqsubseteq}_1 J$ by $E \circ I \subseteq J$ for some $E \in \mathcal{E}$, and similarly we define $I \overset{\mathcal{E}}{\sqsubseteq}_2 J$ by $I \circ E \subseteq J$ for some $E \in \mathcal{E}$. Further, for each $i \in \{1, 2\}$, let $int_i(E) := \bigvee\{I \in L \oplus L \mid I \overset{\mathcal{E}}{\sqsubseteq}_i E\}$. Then $\{(st_1(x, int_1(E)), st_2(x, int_2(E))) \mid x \oplus x \subseteq E\}$ is already a conjugate cover of (L, L_1, L_2). These correspondences define a Galois connection which gives an isomorphism for quasi-uniformities (see [22] for the details). For quasi-nearnesses the question of knowing whether the entourage approach and the cover approach are equivalent is still open.

Notice that Weil quasi-uniformities are defined on a frame; do not require – as the (biframe) quasi-uniformities – prior knowledge of the underlying biframe. The analogy with the spatial case is now evident. This clarifies the discussion in D. Doitchinov, "Some reflections on quasi-uniform frames", Topology with Applications (Szekszárd, Hungary, 1993), Bolyai Math. Soc., Math. Studies 4, 1995, pp. 151–158.

A natural question concerning Weil (quasi-)nearness frames is: which is the right spatial concept in analogy with the chosen frame concept, that is, which spaces correspond to them by an 'open'/'spectrum' adjunction?

The answer (see [21]) gives us the categories **WNear** (of '*Weil nearness spaces*') and **WQNear** (of '*Weil quasi-nearness spaces*'). We concluded in Chapter IV of [21] that **WNear** is an interesting category containing the categories of symmetric topological spaces and continuous maps, uniform spaces and uniformly continuous maps, and proximal spaces and proximal maps, in a nice way: they are either bireflective or bicoreflective full subcategories of **WNear**. However the category **Top** of topological spaces is not a subcategory of **WNear**. Other useful topological structures, namely the nonsymmetric ones of quasi-uniform spaces and quasi-proximal spaces are also not embeddable in **WNear**.

Nevertheless, in the realm of those 'Weil quasi-nearness spaces' mentioned above, it is possible to consider all those spaces of topological and uniform nature. It turns out that the category of Weil quasi-nearness spaces contains all the nonsymmetric categories referred to above as well as the category **Top** as nicely embedded full subcategories and so it is a unified theory of (nonsymmetric) topology and uniformity. The corresponding covering results were explained in the thesis [11].

3. PROXIMITIES

Let us now see how it goes with proximities. We recall that a binary relation $\ll$ on a frame L is a *strong inclusion* [1] if the following axioms are satisfied for x, y, z, w in L:

(P1) $x \leq y \ll z \leq w$ implies $x \ll w$;

(P2) $\ll$ is a sublattice of $L \times L$;

(P3) $x \ll y$ implies $x \prec y$ (where $x \prec y$ means that $x \wedge z = 0$ and $y \vee z = 1$ for some $z \in L$, alternatively expressed as $y \vee x^* = 1$ with the pseudocomplement $x^* = \bigvee\{z \in L \mid x \wedge z = 0\}$);

(P4) If $x \ll y$ then $x \ll z \ll y$ for some $z \in L$;

(P5) $x \ll y$ implies $y^* \ll x^*$;

(P6) For each $x \in L$, $x = \bigvee\{y \in L \mid y \ll x\}$.

Note that, in the presence of (P1), (P2) means: $0 \ll 0$, $1 \ll 1$, $x \ll y$ and $z \ll y$ imply $x \vee z \ll y$, and $x \ll y$ and $x \ll z$ imply $x \ll y \wedge z$.

The pair $(L, \ll)$ is called a *proximal frame* if $\ll$ is a strong inclusion on L. A frame map between proximal frames is said to be proximal if it preserves the strong inclusion. We denote by PFrm the category of proximal frames and proximal frame homomorphisms.

REMARKS 3.1. (a) The basic properties of Weil uniformities imply that, for any Weil uniform frame $(L, \mathcal{E})$, $(L, \overset{\mathcal{E}}{\lhd})$ is a proximal frame.

(b) When $x \vee y = 1$, the C-ideal $(x \oplus x) \vee (y \oplus y)$ is a Weil entourage of L (the converse is also true; for the proof the reader is referred to [21]). Thus, in any proximal frame $(L, \ll)$, since $\ll$ is stronger than $\prec$, $E_{x,y} := (x^* \oplus x^*) \vee (y \oplus y)$ is a Weil entourage whenever $x \ll y$.

The entourages $E_{x,y}$ are very important. Indeed, they enable us to formulate the concept of proximity in terms of Weil entourages:

PROPOSITION 3.2. *Let $(L, \ll)$ be a proximal frame. Then $\{E_{x,y} \mid x, y \in L, x \ll y\}$ is a subbase for a Weil uniformity $\mathcal{E}(\ll)$ on L. Moreover, the proximity $\overset{\mathcal{E}(\ll)}{\lhd}$ induced by $\mathcal{E}(\ll)$ coincides with $\ll$.*

Proof. It is an immediate consequence of Lemma IV.6.7 of [21] (which has a rather technical and long proof that we omit here). Indeed, this lemma affirms that, in any proximal frame $(L, \ll)$, $x \ll y$ whenever $\bigcap_{i=1}^{n} E_{x_i,y_i} \subseteq E_{x,y}$ and $x_i \ll y_i$ for every $i \in \{1, \dots, n\}$. □

The following proposition is also obvious:

PROPOSITION 3.3. *If $(L, \mathcal{E})$ is a Weil uniform frame satisfying*

$$\text{(W4)}\ \forall E \in \mathcal{E}\ \exists x_1, y_1, \dots, x_n, y_n \in L : \left(\bigcap_{i=1}^{n} E_{x_i,y_i} \subseteq E \text{ and } \bigcap_{i=1}^{n} E_{x_i,y_i} \in \mathcal{E}\right),$$

the Weil uniformity $\mathcal{E}(\overset{\mathcal{E}}{\lhd})$ induced by $\overset{\mathcal{E}}{\lhd}$ coincides with $\mathcal{E}$.

Let us denote by $\mathsf{WUFrm}_{(W4)}$ the full subcategory of WUFrm of all Weil uniform frames satisfying (W4). This is a bicoreflective subcategory of WUFrm.

The preceding propositions give us a characterization of frame proximities in terms of Weil entourages:

THEOREM 3.4. *The categories PFrm and $\mathsf{WUFrm}_{(W4)}$ are isomorphic.*

Proof. According to Remark 3.1(a) and Propositions 3.2 and 3.3 it remains to show that a frame map $f\colon L_1 \to L_2$ between two proximal frames $(L_1, \ll_1)$ and $(L_2, \ll_2)$ is a proximal frame homomorphism if and only if it is a Weil uniform homomorphism from $(L_1, \mathcal{E}(\ll_1))$ to $(L_2, \mathcal{E}(\ll_2))$. So, let f be a proximal frame homomorphism. For any $E \in \mathcal{E}(\ll_1)$ we may write $\bigcap_{i=1}^{n} E_{x_i,y_i} \subseteq E$, where $x_i \ll_1 y_i$ for every $i \in \{1, \ldots, n\}$. Then

$$\begin{aligned}(f \oplus f)(E) &\supseteq \bigcap_{i=1}^{n}(f \oplus f)(E_{x_i,y_i})\\ &= \bigcap_{i=1}^{n}\big((f(x_i^*) \oplus f(x_i^*)) \vee (f(y_i) \oplus f(y_i))\big).\end{aligned}$$

Now consider, for each $i \in \{1, \ldots, n\}$, $z_i, w_i \in L$ such that $x_i \ll_1 z_i \ll_1 w_i \ll_1 y_i$. There is a simple and well-known lemma (cf., for example, Lemma 2.8 of [11]), which affirms that, for any frame homomorphism f, $f(y)^* \leq f(x^*)$ whenever $x \prec y$. Applying this lemma, we obtain

$$\begin{aligned}&(f(x_i^*) \oplus f(x_i^*)) \vee (f(y_i) \oplus f(y_i))\\ &\quad\supseteq (f(z_i)^* \oplus f(z_i)^*) \vee (f(w_i) \oplus f(w_i)) = E_{f(z_i),f(w_i)}.\end{aligned}$$

So $(f \oplus f)(E) \supseteq \bigcap_{i=1}^{n} E_{f(z_i),f(w_i)}$. Since, by hypothesis, $f(z_i) \ll_2 f(w_i)$, then $(f \oplus f)(E) \in \mathcal{E}(\ll_2)$.

Conversely, if $x \ll_1 y$ then $(f \oplus f)(E_{x,y}) \in \mathcal{E}(\ll_2)$. As $(f \oplus f)(E_{x,y}) \subseteq E_{f(x),f(y)}$, the entourage $E_{f(x),f(y)}$ also belongs to $\mathcal{E}(\ll_2)$ and we may write $\bigcap_{i=1}^{n} E_{x_i,y_i} \subseteq E_{f(x),f(y)}$ for $x_1, y_1, \ldots, x_n, y_n \in L_2$ with $x_i \ll_2 y_i$ for every $i \in \{1, \ldots, n\}$. It suffices now to recall again Lemma IV.6.7 of [21] to conclude that $f(x) \ll_2 f(y)$. □

As it is well-known, the concepts of proximity and totally bounded uniformity are still equivalent for frames (the proof is due to Frith [12]). This motivates the following definition:

A Weil entourage E is *finite* provided there exist elements $x_1, y_1, \ldots, x_n, y_n$ in L such that $x_i \prec y_i$, for every $i \in \{1, \ldots, n\}$, and $\bigcap_{i=1}^{n} E_{x_i,y_i} = E$. By Remark 3.1(b), the conditions $x_i \prec y_i$ are redundant since each E_{x_i,y_i} by containing E is also a Weil entourage.

Now Theorem 3.4 affirms that

> "*The category* PFrm *is isomorphic to the full subcategory of* WUFrm *of Weil uniform frames with a base of finite entourages*",

which is the entourage version of Frith's result.

When the frame L is normal (i.e., $x \vee y = 1$ implies there exist $u, v \in L$ such that $x \vee u = 1 = y \vee v$ and $u \wedge v = 0$) our notion of finiteness may be stated in a way very likely to the spatial one:

PROPOSITION 3.5. *If L is normal, the filter $FWEnt(L)$ of $(WEnt(L), \subseteq)$ generated by all finite Weil entourages coincides with the filter generated by*

$$\left\{\bigvee_{i=1}^{n}(x_i \oplus x_i) \mid n \in \mathbb{N}, x_1, \ldots, x_n \in L, \bigvee_{i=1}^{n} x_i = 1\right\}.$$

Proof. Consider $E \in FWEnt(L)$. Then there are $x_1, y_1, \ldots, x_n, y_n \in L$ such that $x_i \prec y_i$ for every $i \in \{1, \ldots, n\}$ and $\bigcap_{i=1}^{n} E_{x_i,y_i} \subseteq E$. Since each E_{x_i,y_i} belongs to the set

$$\left\{\bigvee_{i=1}^{n}(x_i \oplus x_i) \mid n \in \mathbb{N}, x_1, \ldots, x_n \in L, \bigvee_{i=1}^{n} x_i = 1\right\},$$

E belongs to the filter generated by it.

Conversely, consider $\bigvee_{i=1}^{n}(x_i \oplus x_i)$ with $\bigvee_{i=1}^{n} x_i = 1$. From the normality of L we may ensure the existence of $y_1, \ldots, y_n \in L$ such that $\bigvee_{i=1}^{n} y_i = 1$ and $y_i \prec x_i$ for any $i \in \{1, \ldots, n\}$. Since $(L \oplus L, \cap, \bigvee)$ is a frame, one can easily prove, by induction on $n \geq 1$, that

$$\bigcap_{i=1}^{n} E_{y_i,x_i} = \bigvee_{z_1 \in \{y_1^*, x_1\}} \cdots \bigvee_{z_n \in \{y_n^*, x_n\}} (z_1 \wedge \cdots \wedge z_n \oplus z_1 \wedge \cdots \wedge z_n).$$

But $(y_1^* \wedge \cdots \wedge y_n^*) \oplus (y_1^* \wedge \cdots \wedge y_n^*) = \mathbb{O}$. Hence $\bigcap_{i=1}^{n} E_{y_i,x_i} \subseteq \bigvee_{i=1}^{n}(x_i \oplus x_i)$ and therefore $\bigvee_{i=1}^{n}(x_i \oplus x_i) \in FWEnt(L)$. □

REMARKS 3.6. (a) By dropping the symmetry of $\mathcal{E}$ in the definition of the objects of the category $\mathsf{WUFrm}_{(W4)}$ we obtain the theory of quasi-proximities in terms of Weil entourages. The isomorphism between WQUFrm and QUFrm yields an isomorphism between this category of '*quasi-proximal frames*' and the covering one of Frith ([11], p. 68).

(b) The isomorphism between PFrm and $\mathsf{WUFrm}_{(W4)}$ suggested to us the notion of finite Weil entourage. It also justifies the fact that we name the Weil nearness frames satisfying (W4) as '*Weil contigual frames*'. This is the analogous notion to the contigual frames of Dube [6] in the setting of nearness frames. One natural question is this: are these two categories equivalent? Another interesting problem is whether the corresponding category of '*Weil contigual spaces*' (as a full subcategory of WNear) is equivalent to the category of contigual nearness spaces of Herrlich [14] and, consequently, equivalent to the classical category of contigual spaces and contigual maps in the sense of Ivanova and Ivanov [17].

In this paper we have tried to emphasize the point that our language of Weil entourages allows us to get frame versions of spatial results concerning entourages, and to formulate them in a very similar way to the corresponding classical results. We end it with another illustration of this. Although not entirely surprising, it is still noteworthy that Weil entourages also serve the same purpose for uniform-like structures in frames as entourages for spaces and that, after all, covers much wider ground, since there are structured frames which are not spatial.

4. INFINITESIMAL RELATIONS

The notion of proximity in classical topology was originaly introduced by Efremovič [7] in terms of the infinitesimal relation 'A is near B' (usually denoted by $A\delta B$ [19]) for subsets A and B of any set. We conclude this article with a brief discussion of the corresponding axiomatization for frames.

PROPOSITION 4.1. *If $(L, \ll)$ is a proximal frame, the binary relation on L given by $x\delta y \equiv x \not\ll y^*$ satisfies the following properties:*

(I1) *$x\delta y$ implies $y\delta x$;*

(I2) *$x\delta(y \vee z)$ if and only if $x\delta y$ or $x\delta z$;*

(I3) *$x\delta y$ implies $x \neq 0$ and $y \neq 0$;*

(I4) *If $x\not\delta y$ then there is $z \in L$ such that $x\not\delta z$ and $y\not\delta z^*$;*

(I5) *$x^* \vee y^* \neq 1$ implies $x\delta y$;*

(I6) *For every $x \in L$, $x = \bigvee\{y \in L \mid y \leq x$ and $y\not\delta x^*\}$.*

Proof. Straightforward. □

Note that, since $x \wedge y \neq 0$ implies $x^* \vee y^* \neq 1$, (I5) says, in particular, that $x\delta y$ whenever $x \wedge y \neq 0$. When L is Boolean the converse is also true and condition (I5) is equivalent to the condition $(x \wedge y \neq 0 \Rightarrow x\delta y)$.

We say that a binary relation δ satisfying all properties (I1)–(I6) of 4.1 is an *infinitesimal relation* and that, in this case, (L, δ) is an *infinitesimal frame*.

The correspondence of 4.1 is invertible for Boolean frames:

PROPOSITION 4.2. *If (L, δ) is an infinitesimal frame, the binary relation $\ll$ given by $x \ll y \equiv x\not\delta y^*$ is a proximity on L if and only if L is Boolean.*

Proof. If $\ll$ is a proximity then, for any $x \in L$, $x^{**} = \bigvee\{y \in L \mid y \ll x^{**}\}$. But $y \ll x^{**} \Leftrightarrow y\not\delta x^{***} \Leftrightarrow y \ll x$. Consequently, $x^{**} = \bigvee\{y \in L \mid y \ll x\} = x$ and L is Boolean.

Since in any Boolean frame the DeMorgan law $(x_1 \wedge x_2)^* = x_1^* \vee x_2^*$ also holds, the proof that $\ll$ is a proximity when L is Boolean follows immediately from the properties of δ. □

These two results show that the description of proximities in terms of infinitesimal relations is also valid for Boolean frames.

It is a well-known result of proximal spaces that, given a uniform space $(X, \mathcal{E})$, there is a way of defining an infinitesimal relation on X by $A\delta B$ if and only if one of the following three equivalent conditions holds:

(i) For every $E \in \mathcal{E}$, $(A \times B) \cap E \neq \emptyset$;
(ii) For every $E \in \mathcal{E}$ there is $a \in A$ such that $E[a] \cap B \neq \emptyset$;
(iii) For every $E \in \mathcal{E}$ there is $b \in B$ such that $A \cap E[b] \neq \emptyset$.

In frames, we have a similar situation. In fact, the next proposition, together with Remark 3.1(a) and Proposition 4.1, says that any Weil uniformity $\mathcal{E}$ on L induces an infinitesimal relation δ on L by $x\delta y$ if and only if one of the following three equivalent conditions is satisfied:

(i) For every $E \in \mathcal{E}$, $(x \oplus y) \cap E \neq \mathbb{O}$;
(ii) For every $E \in \mathcal{E}$ there exists $x' \in \downarrow\{x\} \setminus \{0\}$ such that $E[x'] \wedge y \neq 0$;
(iii) For every $E \in \mathcal{E}$ there exists $y' \in \downarrow\{y\} \setminus \{0\}$ such that $x \wedge E[y'] \neq 0$.

PROPOSITION 4.3. *Let $\mathcal{E}$ be a Weil nearness on L and let $x, y \in L$. Denoting by $E[x]$ $(E \in \mathcal{E})$ the element $\bigvee\{y \in L \mid (x, y) \in E\}$, the following assertions are equivalent:*

(i) $x \overset{\mathcal{E}}{\lhd} y^*$;
(ii) $(x \oplus y) \cap E = \mathbb{O}$ *for some* $E \in \mathcal{E}$;
(iii) *There is $E \in \mathcal{E}$ such that, for every nonzero $x' \in \downarrow\{x\}$, $E[x'] \wedge y = 0$;*
(iv) *There is $E \in \mathcal{E}$ such that, for every nonzero $y' \in \downarrow\{y\}$, $x \wedge E[y'] = 0$.*

Proof. The equivalence between (i) and (ii) is a consequence of the equivalence between $(x \oplus y) \cap E^{-1} = \mathbb{O}$ and $E \circ (x \oplus x) \subseteq y^* \oplus y^*$, that we prove next. So, let $(x \oplus y) \cap E^{-1} = \mathbb{O}$ and consider $(a, b) \in E$ and $(b, c) \in x \oplus x$ with $a, b, c \neq 0$. Then $(b, a \wedge y) \in E^{-1} \cap (x \oplus y) = \mathbb{O}$ and $a \wedge y = 0$, that is, $a \leq y^*$. On the other hand $x \leq y^*$, i.e., $x \wedge y = 0$. In fact, $x \wedge y = \bigvee\{x \wedge y \wedge a \mid (a, a) \in E\}$ and, for any $(a, a) \in E$, $(x \wedge y \wedge a, x \wedge y \wedge a) \in (x \oplus y) \cap E = \mathbb{O}$. Thus $c \leq x \leq y^*$ and $(a, c) \in y^* \oplus y^*$. The reverse implication is also clear.

The equivalence (ii) $\Leftrightarrow$ (iii) is obvious. By symmetry, (ii) is then also equivalent to (iv). $\square$

Acknowledgement

Financial assistance from the Centro de Matemática da Universidade de Coimbra, the JNICT project PraxisXXI/ESCoLa and the Organising Committee of BBFest96 is gratefully acknowledged.

References

1. Banaschewski, B.: Compactifications of frames, *Math. Nachr.* **149** (1990), 105–115.
2. Banaschewski, B.: *Completion in Pointfree Topology*, Lecture Notes in Math. and Appl. Math. 2/96, University of Cape Town, 1996.
3. Banaschewski, B., Brümmer, G. C. L. and Hardie, K. A.: Biframes and bispaces, *Quaestiones Math.* **6** (1983), 13–25.
4. Banaschewski, B. and Pultr, A.: Cauchy points of uniform and nearness frames, *Quaestiones Math.* **19** (1996), 101–127.
5. Dube, T. A.: Structures in frames, Ph.D. Thesis, University of Durban, Westville, 1993.
6. Dube, T. A.: The Tamano–Dowker type theorems for nearness frames, *J. Pure Appl. Algebra* **99** (1995), 1–7.
7. Efremovič, V. A.: Infinitesimal spaces, *Dokl. Akad. Nauk* **76** (1951), 341–343.
8. Efremovič, V. A.: The geometry of proximity I, *Mat. Sb. N.S.* **31** (1952), 189–200.
9. Fletcher, P. and Hunsaker, W.: Entourage uniformities for frames, *Monatsh. Math.* **112** (1991), 271–279.
10. Fletcher, P., Hunsaker, W. and Lindgren, W.: Frame quasi-uniformities, *Monatsh. Math.* **117** (1994), 223–236.
11. Frith, J. L.: Structured frames, Ph.D. Thesis, University of Cape Town, 1987.
12. Frith, J. L.: The category of uniform frames, *Cahiers Topologie Géom. Différentielle Catég.* **31** (1990), 305–313.
13. Gantner, T. E. and Steinlage, R. C.: Characterizations of quasi-uniformities, *J. London Math. Soc. (2)* **5** (1972), 48–52.
14. Herrlich, H.: Topological structures, in *Topological Structures I*, Math. Centre Tracts 52, 1974, pp. 59–122.
15. Isbell, J.: Atomless parts of spaces, *Math. Scand.* **31** (1972), 5–32.
16. Isbell, J., Kříž, I., Pultr, A. and Rosícki, J.: Remarks on localic groups, in F. Borceux (ed.), *Categorical Algebra and its Applications* (Proc. Int. Conf. Louvain-La-Neuve 1987), Lecture Notes in Math. 1348, pp. 154–172.
17. Ivanova, V. M. and Ivanov, A. A.: Contiguity spaces and bicompact extensions, *Izv. Akad. Nauk. SSSR* **23** (1959), 613–634.
18. Johnstone, P. T.: *Stone Spaces*, Cambridge Stud. Adv. Math. 3, Cambridge University Press, Cambridge, 1982.
19. Naimpally, S. A. and Warrack, B. D.: *Proximity Spaces*, Cambridge University Press, Cambridge, 1970.
20. Picado, J.: Weil uniformities for frames, *Comment. Math. Univ. Carolin.* **36** (1995), 357–370.
21. Picado, J.: Weil entourages in pointfree topology, Ph.D. Thesis, University of Coimbra, 1995.
22. Picado, J.: Frame quasi-uniformities by entourages, in *Festschrift for GCL Brümmer on his 60th birthday*, Department of Mathematics, University of Cape Town, 1998.
23. Pultr, A.: Pointless uniformities I. Complete regularity, *Comment. Math. Univ. Carolin.* **25** (1984), 91–104.
24. Pultr, A.: Pointless uniformities II. (Dia)metrization, *Comment. Math. Univ. Carolin.* **25** (1984), 105–120.
25. Tukey, J. W.: *Convergence and Uniformity in Topology*, Ann. Math. Stud. 2, Princeton University Press, 1940.
26. Weil, A.: *Sur les espaces à structure uniforme et sur la topologie générale*, Publications de l'Institute Mathématique de l'Université de Strasbourg, Hermann, Paris, 1938.

Applied Categorical Structures **8:** 367–376, 2000.
G. Brümmer & C. Gilmour (eds), Papers in Honour of Bernhard Banaschewski.

Finite Intervals in the Lattice of Topologies

Dedicated to Bernhard Banaschewski on the occasion of his 70th birthday

JÜRGEN REINHOLD
Institut für Mathematik, Universität Hannover, Welfengarten 1, 30167 Hannover, Germany
e-mail: reinhold@math.uni-hannover.de

(Received: 18 October 1996; accepted: 28 May 1998)

Abstract. We discuss the question whether every finite interval in the lattice of all topologies on some set is isomorphic to an interval in the lattice of all topologies on a finite set – or, equivalently, whether the finite intervals in lattices of topologies are, up to isomorphism, exactly the duals of finite intervals in lattices of quasiorders. The answer to this question is in the affirmative at least for finite atomistic lattices. Applying recent results about intervals in lattices of quasiorders, we see that, for example, the five-element modular but non-distributive lattice cannot be an interval in the lattice of topologies. We show that a finite lattice whose greatest element is the join of two atoms is an interval of T_0-topologies iff it is the four-element Boolean lattice or the five-element non-modular lattice. But only the first of these two selfdual lattices is an interval of orders because order intervals are known to be dually locally distributive.

Mathematics Subject Classifications (2000): 06B15, 54D35, 54F05.

Key words: atomistic, interval, lattice of topologies, T_0-topology.

1. Introduction

The structure of the lattice $\mathcal{T}(X)$ of all topologies on a set X has been investigated by various authors (see, for example, [5, 7, 8] or [9]). It has been asked in [12] whether every finite lattice is isomorphic to some *interval* $\mathcal{T}(X)[\tau_0, \tau_1] = \{\tau \in \mathcal{T}(X) \mid \tau_0 \subseteq \tau \subseteq \tau_1\}$ in $\mathcal{T}(X)$. The complete sublattice $\mathcal{A}(X)$ of $\mathcal{T}(X)$ consisting of all *Alexandroff discrete* topologies on X, i.e. of all topologies on X which are not only closed under the formation of finite but of arbitrary intersections, is well known to be dually isomorphic to the lattice $\mathcal{Q}(X)$ of all *quasiorders*, i.e. reflexive, transitive relations on X (cf. [1] or [2]). If X is finite then, clearly, $\mathcal{A}(X)$ coincides with $\mathcal{T}(X)$. For elements x, y of a lattice L, we write $x \prec y$ if y is a *cover* of x, i.e. if the interval $L[x, y]$ has exactly two elements. An *atom* of a lattice L with smallest element $\bot$ is a cover of $\bot$. The lattice L is called *atomistic* if every element is a join of atoms. A *principal ideal* in a lattice L is a subset of the form $L(x] = \{y \in L \mid y \leq x\}$ and a *principal filter* in L is a subset of the form $L[x) = \{y \in L \mid x \leq y\}$. Using a result of F. Plastria [7] about covers in the lattice of topologies, we show that every finite atomistic interval I in $\mathcal{T}(X)$ is isomorphic to an interval in

$\mathcal{T}(Y)$ for some finite set Y and, therefore, dually isomorphic to an interval in $\mathcal{Q}(Y)$. In [4], intervals in lattices of the form $\mathcal{Q}(X)$ have been investigated extensively and many restrictive properties shared by all such intervals were found. We are thus able to exhibit a large class of finite atomistic lattices that are not isomorphic to any interval in some $\mathcal{T}(X)$. This class includes, for example, the five-element modular but non-distributive lattice M_3.

It is not surprising at all that structural investigations of the lattices $\mathcal{Q}(X)$ are much easier than those of the lattices $\mathcal{T}(X)$, because quasiorders are subsets of X^2 while topologies are subsets of 2^X. In [4], it was shown that any finite interval in a lattice of the form $\mathcal{Q}(X)$ is isomorphic to an interval in a lattice $\mathcal{Q}(Y)$ for some finite set Y. If we could show an analogous result not only for finite atomistic but for all finite intervals in lattices of the form $\mathcal{T}(X)$ then the finite intervals in lattices of topologies were, up to isomorphism, exactly the duals of finite intervals in lattices of quasiorders. We give some evidence that this is indeed the case, but the problem remains unsolved.

It is known that a finite lattice is isomorphic to an interval in $\mathcal{T}(X)$ consisting of T_1-topologies iff it is distributive (cf. [8, 11]). Moreover, if X is finite then every interval I in $\mathcal{T}(X)$ consisting of T_0-topologies is *locally distributive*, i.e. if $A \subseteq I$ is a finite set of covers of some element $\tau \in I$ then the subinterval $[\tau, \bigvee A]$ is distributive (cf. [4]). But we shall see that the finiteness assumption on X is essential here. Indeed, we shall prove that the four-element Boolean lattice 2^2 and the five-element non-modular lattice N_5 (which is not locally distributive) are exactly those lattices occuring as intervals of T_0-topologies whose greatest elements are joins of two atoms.

2. Atomistic Intervals

Let (X, τ) be a topological space. For $a \in X$, we denote by $\mathcal{U}_\tau(a)$ the neighborhood filter of a in (X, τ). The equivalence classes of the equivalence relation $\{(a, b) \in X^2 \mid \mathcal{U}_\tau(a) = \mathcal{U}_\tau(b)\}$ are called *T_0-classes* of (X, τ). The system of all T_0-classes of (X, τ) will be denoted by $\mathcal{C}_\tau$. The following simple lemma will be very useful.

LEMMA 2.1. (i) *Every interval (principal ideal) I in $\mathcal{T}(X)$ is isomorphic to an interval (principal ideal) I' in some $\mathcal{T}(Y)$ such that the greatest element of I' is T_0.*

(ii) *Let $I = \mathcal{T}(X)[\tau, \sigma]$ be an interval in $\mathcal{T}(X)$ such that σ is a finite topology (i.e. it consists of finitely many open sets). Then I is isomorphic to an interval in $\mathcal{T}(Y)$ for some finite set Y.*

Proof. (i) Let $I = \mathcal{T}(X)[\tau, \sigma]$ and $\varepsilon = \{(a, b) \in X^2 \mid \mathcal{U}_\sigma(a) = \mathcal{U}_\sigma(b)\}$. Then σ/ε is T_0, and it is easy to see that $\varphi : I \to I' = \mathcal{T}(X/\varepsilon)[\tau/\varepsilon, \sigma/\varepsilon]$, $\rho \mapsto \rho/\varepsilon$ is a well-defined isomorphism.

(ii) Let φ be the isomorphism from (i). Since σ is finite, so is $\varphi(\sigma) = \sigma/_{\varepsilon}$. Hence $\sigma/_{\varepsilon}$ is a finite T_0-topology on $X/_{\varepsilon}$ and, consequently, $X/_{\varepsilon}$ must be finite. □

For any set X, let $\mathcal{T}^*(X)$ be the set of all systems of subsets of X closed under arbitrary unions and finite non-void intersections. Notice that $\mathcal{T}^*(\{0\})$ is a two-element chain and that $\mathcal{T}^*(\{0,1\})$ is the seven-element lower but not upper semimodular lattice S_7^{op}, because $\mathcal{T}^*(\{0,1\})$ contains all subsets τ of the power set $\mathcal{P}(\{0,1\})$ with $\emptyset \in \tau$ except for $\{\emptyset, \{0\}, \{1\}\}$. Obviously, $\mathcal{T}(X)$ is a principal filter in $\mathcal{T}^*(X)$. On the other hand, if c is a new point then $\mathcal{T}^*(X)$ is isomorphic to a principal ideal in $\mathcal{T}(X \cup \{c\})$. Indeed, the mappings

$$\varphi: \mathcal{T}^*(X) \to \mathcal{T}(X \cup \{c\})(\mathcal{P}(X) \cup \{X \cup \{c\}\}], \sigma \mapsto \sigma \cup \{X \cup \{c\}\},$$

$$\psi: \mathcal{T}(X \cup \{c\})(\mathcal{P}(X) \cup \{X \cup \{c\}\}] \to \mathcal{T}^*(X), \tau \mapsto \tau \setminus \{X \cup \{c\}\}$$

are mutually inverse isomorphisms.

The following result is due to F. Plastria (see [7, Theorem 5.1]).

FACT 2.2. *For any set X and any topology $\tau \in \mathcal{T}(X)$, the set of all joins of covers of τ in $\mathcal{T}(X)$ is a sublattice of $\mathcal{T}(X)$ isomorphic to the direct product of lattices of the form $\mathcal{T}(Y)$ or $\mathcal{T}^*(Y)$.*

With the help of Fact 2.2, we can now prove

THEOREM 2.3. *Every finite atomistic interval $I = \mathcal{T}(X)[\tau, \sigma]$ in $\mathcal{T}(X)$ is isomorphic to an interval in $\mathcal{T}(Y)$ for some finite set Y.*

Proof. Let Γ be the lattice of all joins of covers of τ in $\mathcal{T}(X)$. Then I is a principal ideal in Γ and, by Fact 2.2, Γ is a sublattice of $\mathcal{T}(X)$ isomorphic to a direct product of lattices of the form $\mathcal{T}(Z)$ or $\mathcal{T}^*(Z)$. Since I is finite, it is isomorphic to a principal ideal in a direct product of finitely many lattices of the form $\mathcal{T}(Z)$ or $\mathcal{T}^*(Z)$. As mentioned above, any $\mathcal{T}^*(Z)$ is isomorphic to a principal ideal in some $\mathcal{T}(Z')$. Therefore, we can find a family $(X_j)_{j\in J}$ of finitely many pairwise disjoint sets X_j and topologies σ_j on X_j such that I is isomorphic to the direct product $\prod_{j\in J} T_j$ of principal ideals $T_j = \mathcal{T}(X_j)(\sigma_j]$ in $\mathcal{T}(X_j)$. Let $X' = \bigcup_{j\in J} X_j$ and let σ' be the sum topology $\sum_{j\in J} \sigma_j$. Since J is finite, so is the topology $\tau' = \{\bigcup_{j\in K} X_j \mid K \subseteq J\}$ on X'. The mapping

$$\varphi: \prod_{j\in J} T_j \to I' = \mathcal{T}(X')[\tau', \sigma'], (\tau_j)_{j\in J} \mapsto \sum_{j\in J} \tau_j$$

is easily checked to be an isomorphism. For every atom α in I', choose a set $A \in \alpha \setminus \tau'$, and let $\mathcal{A}$ be the collection of all these finitely many sets A. Then $\tau' \cup \mathcal{A}$ is a finite subbase of σ'. Thus σ' is a finite topology on X' and, by Lemma 2.1, I' and I are isomorphic to an interval in $\mathcal{T}(Y)$ for some finite set Y. □

Now, many results from [4] about finite intervals in lattices of quasiorders may be applied to finite atomistic intervals in lattices of topologies. We just sketch two of these applications (cf. [4, 5.5] and [4, 4.6]). The *height* of a finite lattice L is the maximal length of a chain contained in L. The *horizontal sum* of two lattices is obtained by forming their disjoint sum and identifying their greatest elements and their least elements, respectively. This sum is called *trivial* whenever one of the lattices has at most two elements.

COROLLARY 2.4. (i) *Let h be the height of a finite atomistic interval I in $\mathcal{T}(X)$. Then $|I| < 2(4h)^h$.*

(ii) *An atomistic non-trivial horizontal sum of two finite lattices is isomorphic to an interval in some $\mathcal{T}(X)$ iff it is a four-element Boolean lattice.*

Thus compared with their height finite atomistic intervals in lattices of topologies cannot be arbitrarily broad. For instance, no lattice of height 2 with more than 4 elements can be an interval in some $\mathcal{T}(X)$.

3. Finite Intervals of Topologies and Quasiorders

In [4], it was shown that for every finite interval $I = \mathcal{Q}(X)[\rho, \sigma]$ in the lattice of all quasiorders on a set X, there is a finite subset Y of X such that I is isomorphic to $\mathcal{Q}(Y)[\rho \cap Y^2, \sigma \cap Y^2]$. In the case of finite intervals in lattices of the form $\mathcal{T}(X)$, we cannot always get isomorphic intervals by forming finite subspaces. Indeed, it is known that every finite distributive lattice is isomorphic to an interval I in some $\mathcal{T}(X)$ such that all topologies in I are T_1 (cf. [11]), but all finite subspaces of a T_1-space are discrete. Nevertheless, every finite distributive lattice is isomorphic to an interval in some finite $\mathcal{T}(Y)$ (see [4]). In this section we present some first ideas that may help to decide whether the desired size reduction is always possible.

Let τ be a topology on the set X and let A be a subset of X. We denote the τ-closure of A by $\text{cl}_\tau(A)$, the τ-interior of A by $i_\tau(A)$ and the *τ-relative boundary* $\text{cl}_\tau(X \backslash A) \cap A = A \backslash i_\tau(A)$ by $\delta^*_\tau(A)$. Notice that $\tau(A) = \{(O \cap A) \cup O' \mid O, O' \in \tau\}$ is the smallest topology on X containing $\tau \cup \{A\}$. If σ is a cover of τ in $\mathcal{T}(X)$ then, clearly, $\sigma = \tau(A)$ for every $A \in \sigma \setminus \tau$.

Let us adopt a helpful fact from [7].

FACT 3.1. *Let $\sigma = \tau(A)$ be a cover of τ in $\mathcal{T}(X)$. Then $\delta^*_\tau(A)$ is contained in some T_0-class of (X, τ), and $\delta^*_\tau(B) = \delta^*_\tau(A)$ for every $B \in \sigma \setminus \tau$.*

Together with Lemma 2.1, the following result shows that for our structural investigations of finite intervals of the form $\mathcal{T}(X)[\tau, \sigma]$ we may always assume that there are only finitely many points whose T_0-classes or neighborhood filters in (X, τ) differ from those in (X, σ).

PROPOSITION 3.2. *Let h be the height of a finite interval $\mathcal{T}(X)[\tau, \sigma]$ in $\mathcal{T}(X)$ such that σ is T_0. Then $|\mathcal{C}_\sigma \setminus \mathcal{C}_\tau| \leq 2h$ and $|\{a \in X \mid \mathcal{U}_\tau(a) \neq \mathcal{U}_\sigma(a)\}| \leq h$.*

Proof. We proceed by induction on h. Clearly, the claim is true for $h = 0$. If $h > 0$ then we may choose a cover $\tau(A)$ of τ. We will apply the induction hypothesis to the interval $\mathcal{T}(X)[\tau(A), \sigma]$ which has height $h - 1$. By Fact 3.1, $\delta^*_\tau(A)$ is contained in some T_0-class C of (X, τ).

For all points a, b in X, the identity $\mathcal{U}_{\tau(A)}(a) = \mathcal{U}_{\tau(A)}(b)$ is equivalent to $\mathcal{U}_\tau(a) = \mathcal{U}_\tau(b)$ and $a \in A \Leftrightarrow b \in A$ which in turn is equivalent to $\mathcal{U}_\tau(a) = \mathcal{U}_\tau(b)$ and $a \in \delta^*_\tau(A) \Leftrightarrow b \in \delta^*_\tau(A)$. Thus $\mathcal{C}_{\tau(A)} = (\mathcal{C}_\tau \setminus \{C\}) \cup \{\delta^*_\tau(A), C \setminus \delta^*_\tau(A)\}$ and it follows that $|\mathcal{C}_\sigma \setminus \mathcal{C}_\tau| \leq |\mathcal{C}_\sigma \setminus \mathcal{C}_{\tau(A)}| + |\mathcal{C}_{\tau(A)} \setminus \mathcal{C}_\tau| \leq 2h$.

If $a \notin \delta^*_\tau(A)$ then every neighborhood $U = (A \cap O) \cup O' \in \mathcal{U}_{\tau(A)}(a)$ contains $(i(A) \cap O) \cup O' \in \tau$ and is therefore a member of $\mathcal{U}_\tau(a)$, whence $\mathcal{U}_\tau(a) = \mathcal{U}_{\tau(A)}(a)$. On the other hand, there is at most one point $a \in \delta^*_\tau(A)$ with $\mathcal{U}_{\tau(A)}(a) = \mathcal{U}_\sigma(a)$ since σ is T_0 and $\delta^*_\tau(A)$ is a T_0-class of $\tau(A)$. Thus we conclude that $|\{a \in X \mid \mathcal{U}_\tau(a) \neq \mathcal{U}_\sigma(a)\}| \leq |\{a \in X \mid \mathcal{U}_{\tau(A)}(a) \neq \mathcal{U}_\sigma(a)\} \cup \delta^*_\tau(A)| \leq |\{a \in X \mid \mathcal{U}_{\tau(A)}(a) \neq \mathcal{U}_\sigma(a)\}| + 1 \leq h$. □

Let S be any subset of the power set $\mathcal{P}(X)$. Then we denote by $\operatorname{top}_X(S)$ the smallest topology on X containing S. For any interval $I = \mathcal{T}(X)[\tau, \sigma]$, let

$$I_S = \mathcal{T}(X)[\tau \cap \operatorname{top}_X(S), \sigma \cap \operatorname{top}_X(S)] \quad \text{and}$$

$$\varphi^I_S \colon I \to I_S,\ \delta \mapsto \delta \cap \operatorname{top}_X(S).$$

Whenever S is finite then, clearly, $\operatorname{top}_X(S)$ is finite, too, and in this case, by Lemma 2.1, every interval of the form I_S is isomorphic to an interval in $\mathcal{T}(Y)$ for some finite set Y.

PROPOSITION 3.3. *Let $I = \mathcal{T}(X)[\tau, \sigma]$ be an interval in $\mathcal{T}(X)$ such that $\sigma \setminus \tau$ is finite. Then I is isomorphic to an interval in $\mathcal{T}(Y)$ for some finite set Y.*

Proof. For every subset $R \subseteq \sigma \setminus \tau$ and every element $O \in \operatorname{top}_X(\tau \cup R) \setminus \tau$, choose a finite subset $S_{R,O} \subseteq \tau$ such that $O \in \operatorname{top}_X(S_{R,O} \cup R)$. Then

$$S = \sigma \setminus \tau \cup \bigcup\{S_{R,O} \mid R \subseteq \sigma \setminus \tau,\ O \in \operatorname{top}_X(\tau \cup R) \setminus \tau\}$$

is a finite set. It suffices to show that φ^I_S is an isomorphism.

Obviously, φ^I_S is order preserving. For $\rho, \rho' \in I$ with $\rho \not\subseteq \rho'$, every set $O \in \rho \setminus \rho'$ is contained in $\sigma \setminus \tau \subseteq S \subseteq \operatorname{top}_X(S)$ whence $\varphi^I_S(\rho) \not\subseteq \varphi^I_S(\rho')$. Thus φ^I_S is an order embedding. Now, let δ be any element of I_S and let $D = \delta \setminus (\tau \cap \operatorname{top}_X(S))$. Then $D \subseteq \sigma \setminus \tau$. Suppose that there is a set $O \in \operatorname{top}_X(\tau \cup D) \cap \operatorname{top}_X(S) \setminus \delta$. Then $O \in \operatorname{top}_X(\tau \cup D) \setminus \tau$ implies $O \in \operatorname{top}_X(S_{D,O} \cup D) \subseteq \delta$, a contradiction. Thus $\operatorname{top}_X(\tau \cup D) \cap \operatorname{top}_X(S) = \delta$, and we conclude that φ^I_S is surjective. □

Let $I = \mathcal{T}(X)[\tau, \sigma]$ be a finite interval in $\mathcal{T}(X)$. For every pair $(\rho, \rho') \in I^2$ with $\rho \not\subseteq \rho'$ choose a set $O \in \rho \setminus \rho'$. Let S be the finite collection of all these sets O. One might guess that φ^I_S must always be an isomorphism. If this were true then we could drop the annoying finiteness assumption on $\sigma \setminus \tau$ in Proposition 3.3

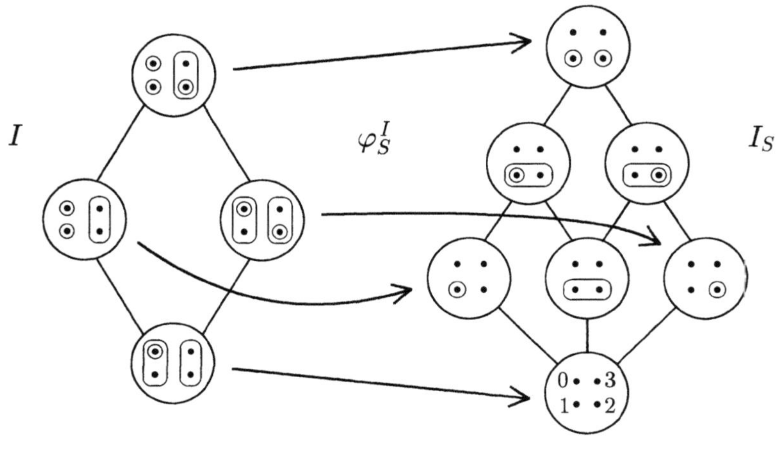

and thus answer the posed problem in the affirmative. But the following example shows that φ_S^I need not be surjective.

Let $X = \{0, 1, 2, 3\}$, let $\tau = \mathrm{top}_X(\{\{0\}, \{0, 1\}, \{2, 3\}\})$, $\sigma = \mathrm{top}_X(\{\{0\}, \{1\}, \{2\}, \{2, 3\}\})$ and let $I = \mathcal{T}(X)[\tau, \sigma]$. Then for every pair $(\rho, \rho') \in I^2$ with $\rho \not\subseteq \rho'$, we have $\{1\} \in \rho \setminus \rho'$ or $\{2\} \in \rho \setminus \rho'$. Let $S = \{\{1\}, \{2\}\}$. The mapping φ_S^I is shown in the picture, where all topologies in I and I_S are given by their smallest bases.

4. Finite Intervals of T_0-Topologies and Orders

One of the most intriguing results about finite intervals in lattices of quasiorders is that every finite interval of $\mathcal{Q}(X)$ whose greatest element (and thus every element) is an order must be dually locally distributive, i.e. if A is a finite set of elements covered by one and the same order ρ then the interval $[\bigwedge A, \rho]$ must be distributive (see [4]). Clearly, for a finite set Y, the intervals of $\mathcal{T}(Y)$ whose smallest elements (and thus all elements) are T_0-topologies are, up to isomorphism, exactly the duals of intervals in $\mathcal{Q}(Y)$ whose greatest elements are orders. But the following example shows that there are finite intervals of T_0-topologies on an infinite set which are not locally distributive and are therefore not dually isomorphic to intervals of orders. Recall that a subset D of an ordered set $(X, \leq)$ is *decreasing* if $x \leq d$ and $d \in D$ imply $x \in D$.

Let $X = \{(0, 0), (1, 0)\} \cup \{(0, \frac{1}{n}) \mid n = 1, 2, 3, \ldots\} \subseteq \mathbb{R}^2$ be ordered by $(x, y) \leq (x', y') \Leftrightarrow (x \leq x'$ and $y \leq y')$. Let τ be the set of all infinite or empty decreasing subsets of $(X, \leq)$. It is easily checked that τ as well as $\sigma = \tau \cup \{\{(0, 0)\}, \{(1, 0)\}, \{(0, 0), (1, 0)\}\}$ are T_0-topologies on X and that the interval $I = \mathcal{T}(X)[\tau, \sigma]$ is the pentagon shown in the picture below where all topologies $\rho \in I$ are given by the difference set $\rho \setminus \tau$.

Although finite intervals $\mathcal{T}(X)[\tau, \sigma]$ of T_0-topologies need not be locally distributive if the underlying set X is infinite, we shall see that they have a property

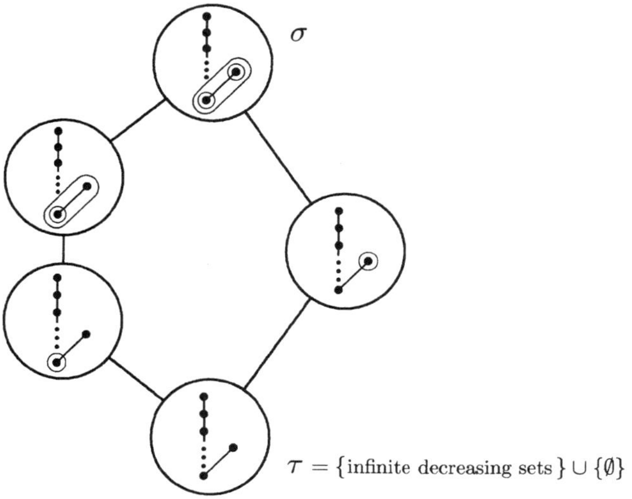

that may be viewed as a weak form of local distributivity. A short but subtle induction argument shows that a finite lattice L is locally distributive iff for every element $x \in L$ and every two different covers y, z of x the interval $L[x, y \vee z]$ is isomorphic to the Boolean lattice 2^2 (this result is implicitly contained in [3]). Now, we shall prove that for any two different covers τ_1, τ_2 of a T_0-topology τ on an arbitrary set X the interval $\mathcal{T}(X)[\tau, \tau_1 \vee \tau_2]$ is isomorphic to 2^2 or the pentagon N_5.

THEOREM 4.1. *Let L be a finite lattice such that the greatest element is the join of two distinct atoms. Then there is a T_0-space (X, τ) and an interval $\mathcal{T}(X)[\tau, \sigma]$ isomorphic to L iff L is isomorphic to 2^2 or N_5.*

Proof. As we have seen above, the pentagon is an interval of T_0-topologies. The same is true for the distributive lattice 2^2 (for example, take all topologies on $X = \{0, 1, 2, 3\}$ containing $\{\{0\}, \{1\}, \{0, 2\}, \{1, 3\}\}$).

Now, let $\tau_1 = \tau(A_1)$, $\tau_2 = \tau(A_2)$ be distinct covers of a T_0-topology τ in $\mathcal{T}(X)$. By Fact 3.1, there are points $x_1, x_2 \in X$ and sets $B_1, B_2 \in \tau$ such that $A_1 = B_1 \cup \{x_1\}$ and $A_2 = B_2 \cup \{x_2\}$. Let $\sigma = \tau_1 \vee \tau_2$ and $I = \mathcal{T}(X)[\tau, \sigma]$. If $\sigma = \tau_1 \cup \tau_2$ then for every topology $\rho \in I \setminus \{\tau, \tau_1, \tau_2\}$, there are sets $R_1 \in \rho \cap (\tau_1 \setminus \tau)$ and $R_2 \in \rho \cap (\tau_2 \setminus \tau)$. Then $\sigma \subseteq \tau(R_1)(R_2) \subseteq \rho \subseteq \sigma$, hence $\rho = \sigma$ and $I \cong 2^2$. Assume that $I \not\cong 2^2$ and choose any set $S \in \sigma \setminus (\tau_1 \cup \tau_2)$. Since $\sigma = \tau_1 \vee \tau_2 = \tau_1(A_2)$, there are sets $U, V \in \tau_1$ such that $S = (A_2 \cap U) \cup V$.

Claim 1. $S = T \cup \{x_1, x_2\}$ for some $T \in \tau$.

If $x_2 \notin S$ then $S = (B_2 \cap U) \cup V \in \tau_1$, a contradiction. Thus $x_2 \in S$ and, similarly, $x_1 \in S$. Since $\delta_\tau^*(U) \subseteq \{x_1\}$ and $\delta_\tau^*(V) \subseteq \{x_1\}$, we have $S = T \cup \{x_1, x_2\}$ where $T = (B_2 \cap i_\tau(U)) \cup i_\tau(V) \in \tau$.

Claim 2. $x_1 \neq x_2$.

Assume that $x_1 = x_2$. Then we have $x_1 \in A_2 \cap U$, for otherwise $S = (B_2 \cap U) \cup V \in \tau_1$, and $A_2 \cup T \in \tau_2 \setminus \tau_1$, for otherwise $S = T \cup ((A_2 \cup T) \cap U) \in \tau_1$. Thus there are $T_1, T_2 \in \tau$ such that $A_2 = ((A_2 \cup T) \cap T_1) \cup T_2 = ((B_2 \cup S) \cap T_1) \cup T_2$. Hence $A_2 \in \tau(S)$ and, similarly, $A_1 \in \tau(S)$. Thus $\tau(S) = \sigma$ and, since S was chosen arbitrarily in $\sigma \setminus (\tau_1 \cup \tau_2)$, it follows that $I = \{\tau, \tau_1, \tau_2, \sigma\}$, i.e. $I \cong 2^2$, a contradiction.

Claim 3. $S \setminus \{x_1\} = T \cup \{x_2\} \notin \tau$ and $S \setminus \{x_2\} = T \cup \{x_1\} \notin \tau$.

Assume that $S \setminus \{x_1\} \in \tau$. Since $x_1 \in S = (A_2 \cap U) \cup V$, we have $x_1 \in V$ or $x_1 \in A_2 \cap U$. If $x_1 \in V$ then $S = (S \setminus \{x_1\}) \cup V \in \tau_1$, and if $x_1 \in A_2 \cap U$ then $x_1 \neq x_2$ implies $x_1 \in B_2 = A_2 \setminus \{x_2\}$ thus $S = (B_2 \cap U) \cup (S \setminus \{x_1\}) \in \tau_1$, anyway, a contradiction. Thus $S \setminus \{x_1\} \notin \tau$ and, similarly, $S \setminus \{x_2\} \notin \tau$. It follows that $x_1 \notin T$, $x_2 \notin T$ whence $S \setminus \{x_1\} = T \cup \{x_2\}$, $S \setminus \{x_2\} = T \cup \{x_1\}$.

Claim 4. $\tau_1 = \tau(T \cup \{x_1\})$ and $\tau_2 = \tau(T \cup \{x_2\})$.

$S = (A_2 \cap U) \cup V = (A_2 \cap i_\tau(U)) \cup i_\tau(V) \cup \{x_1\} \notin \tau_2$ and $S \setminus \{x_1\} \notin \tau$ imply $T \cup \{x_2\} = S \setminus \{x_1\} = (A_2 \cap i_\tau(U)) \cup i_\tau(V) \in \tau_2 \setminus \tau$. Thus $\tau_2 = \tau(T \cup \{x_2\})$ and, similarly, $\tau_1 = \tau(T \cup \{x_1\})$.

Claim 5. $\tau_1 \subseteq \tau(S)$ or $\tau_2 \subseteq \tau(S)$.

Since τ is T_0, there is a set $O \in \tau$ such that $x_1 \in O, x_2 \notin O$ or $x_2 \in O, x_1 \notin O$. We only consider the first case. Then $T \cup \{x_1\} = (T \cup O) \cap S \in \tau(S)$ and, therefore, $\tau_1 = \tau(T \cup \{x_1\}) \subseteq \tau(S)$.

Claim 6. $S' \in \sigma \setminus (\tau_1 \cup \tau_2)$ implies $\tau(S') = \tau(S)$.

Since $\tau_2 = \tau(T \cup \{x_2\})$ and $\tau_1 = \tau(T \cup \{x_1\})$ there are sets $U', V' \in \tau_1$ and $U_1, U_2, V_1, V_2 \in \tau$ such that $S' = ((T \cup \{x_2\}) \cap U') \cup V'$, $U' = ((T \cup \{x_1\}) \cap U_1) \cup U_2$ and $V' = ((T \cup \{x_1\}) \cap V_1) \cup V_2$. As in Claim 1 we see that $x_1, x_2 \in S'$ and, therefore,

$$\begin{aligned} S' &= ((T \cup \{x_2\}) \cap (((T \cup \{x_1\}) \cap U_1) \cup U_2)) \cup ((T \cup \{x_1\}) \cap V_1) \cup V_2 \\ &= (S \cap ((S \cap U_1) \cup U_2)) \cup (S \cap V_1) \cup V_2 \in \tau(S). \end{aligned}$$

Interchanging the role of S and S' we also get $S \in \tau(S')$. Hence $\tau(S') = \tau(S)$. Now, let $\rho \in I \setminus \{\tau, \tau_1, \tau_2, \sigma\}$. Since $\rho \cap (\tau_1 \setminus \tau) \neq \emptyset \neq \rho \cap (\tau_2 \setminus \tau)$ would imply $\rho = \sigma$, w.l.o.g. we may assume that $\rho \cap (\tau_2 \setminus \tau) = \emptyset$. For every set $R \in \rho \setminus \tau_1$, it follows that $R \in \sigma \setminus (\tau_1 \cup \tau_2)$, hence $\tau(R) = \tau(S)$. Thus $\rho \subseteq \tau_1 \cup \tau(S)$. Since $\tau_2 \not\subseteq \rho \supseteq \tau(R) = \tau(S)$, it follows that $\tau_1 \subseteq \tau(S)$ and, therefore, $\rho = \tau(S)$. We conclude that $I = \{\tau, \tau_1, \tau_2, \tau(S), \sigma\}$ and $I \cong N_5$. □

Notice that the lattice S_7^{op} alias $\mathcal{T}^*(\{0,1\})$ shown in our first figure is an interval of topologies, but by the preceding theorem, it cannot be an interval of T_0-topologies.

5. Concluding Remarks

It will propably be very hard to find an 'abstract' or 'intrinsic' characterization of (finite) intervals in lattices of topologies. But it should not be too difficult to find further restrictive properties shared by these intervals. It would be very helpful to know whether all finite intervals in lattices of topologies are dually isomorphic to finite intervals in lattices of quasiorders. As we have seen, the class of all lattices isomorphic to finite intervals of topologies is greater than that of all lattices isomorphic to finite intervals of T_0-topologies which in turn is greater than that of all lattices dually isomorphic to finite intervals of orders. An abstract characterization for the lattices in the latter class has recently been announced by McIntyre and Watson but we are not aware of their results.

It is known that every finite interval I in the lattice $\mathcal{Q}(X)$ of all quasiorders on a set X is isomorphic to an interval in $\mathcal{Q}(Y)$ for some set Y with $|Y| \leq 2|J(I)|$, where $J(I)$ is the set of join-irreducible elements in I (see [4]). With the help of Proposition 3.2 it is possible to improve this result by showing that Y can be chosen to have size at most twice the height of I. If all finite intervals in $T(X)$ were representable on finite sets then the same size reduction would be possible, here.

Acknowledgement

The author wishes to thank the referee for pointing out the abovementioned improvement of the size reduction theorem for intervals in lattices of quasiorders and for the information that Good and McIntyre have announced a positive solution to the problem posed in this article, namely that, indeed, every finite interval in a lattice of topologies is representable on a finite set.

References

1. Alexandroff, P.: Sur les espaces discrets, *CR Acad. Sci. Paris* **200** (1935–36), 1649–1651.
2. Birkhoff, G.: Sur les espaces discrets, *CR Acad. Sci. Paris* **201** (1935–36), 19–20.
3. Dilworth, R. P. and Crawley, P.: Decomposition theory for lattices without chain conditions, *Trans. Amer. Math. Soc.* **96** (1960), 1–22.
4. Erné, M. and Reinhold, J.: Intervals in lattices of quasiorders, *Order* **12** (1995), 375–403.
5. Larson, R. E. and Andima, S. J.: The lattice of topologies: A survey, *Rocky Mountain J. Math.* **5** (1975), 177–198.
6. Larson, R. E. and Thron, W. J.: Covering relations in the lattice of T_1-topologies, *Trans. Amer. Math. Soc.* **168** (1972), 101–111.
7. Plastria, F.: Covers in the lattice of topologies, *Bull. Soc. Math. Belg.* **27** (1975), 295–307.
8. Rosický, J.: Modular, distributive and simple intervals of the lattice of topologies, *Arch. Math.* **2** (1975), 105–114.
9. Steiner, A. K.: The lattice of topologies: structure and complementation, *Trans. Amer. Math. Soc.* **122** (1966), 379–397.
10. Valent, R.: Basic intervals and related sublattices of the lattice of topologies, *Duke Math. J.* **40** (1973), 487–492.

11. Valent, R. and Larson, R. E.: Basic intervals in the lattice of topologies, *Duke Math. J.* **39** (1972), 401–411.
12. Watson, S.: Basic problems in general topology, Report no. 94-15, Dep. Math., York University, Ontario, Canada.

Applied Categorical Structures **8:** 377–390, 2000.
G. Brümmer & C. Gilmour (eds), Papers in Honour of Bernhard Banaschewski.

Local Connectedness Made Uniform

For Bernhard Banaschewski on his 70th Birthday

D. BABOOLAL
Department of Mathematics, University of Durban-Westville, Private Bag X54001, Durban 4000, South Africa, e-mail: dbab@pixie.udw.ac.za

(Received: 22 November 1996; accepted: 27 November 1997)

Abstract. The uniformly locally connected reflection for a locally connected uniform frame is constructed. Applications of this construction to the theory of locally connected completely regular frames are given. One such application is that if a completely regular frame is locally connected and pseudocompact then every compactification of it is locally connected.

Mathematics Subject Classifications (2000): 54A05, 54E15, 54D05.

Key words: local connectedness, property S, uniform local connectedness, connected congruences.

Introduction

In [11] Gleason constructs the uniformly locally connected coreflection of an arbitrary uniform space. The main purpose of this paper is to obtain an analogue of this result for the special case of those uniform frames for which the underlying frame is locally connected and to give some applications of our construction to the general theory of local connectedness for completely regular frames. One of our applications, in fact the main one, is that this construction allows a new proof of the sufficiency part of the local connectedness theorem of Baboolal and Banaschewski [1]: The Stone–Čech compactification of a completely regular frame L is locally connected iff L is locally connected and pseudocompact. At the same time we also obtain new results which may be of independent interest.

This approach turns around two uniform connection properties which we introduce into the theory of uniform frames, the concepts of property S and uniform local connectedness both of which have fruitful applications in the context of spaces (see e.g. [2]). This work is inspired by the well-known Banaschewski [4], Henriksen and Isbell [12] theorem on local connectedness of the Stone–Čech compactification for spaces, and in terms of approach via uniformities, we are indebted to the last authors who used uniform space techniques in their study of this particular theorem. A secondary goal of this paper is to demonstrate the usefulness of the uniform approach in studying questions about local connectedness of frames.

1. Preliminaries

Recall that a *frame* L is a complete lattice satisfying the infinite distributive law $x \wedge \bigvee S = \bigvee x \wedge s$ $(s \in S)$ for all $x \in L, S \subseteq L$. A *frame homomorphism* is a map $h\colon M \to L$ between frames preserving finitary meets, including the unit e and arbitrary joins, including the zero 0. A frame homomorphism $h\colon M \to L$ is called *dense* if $h(x) = 0$ implies $x = 0$ for all $x \in M$. A frame is called *regular* if for each $a \in L$ we have $a = \bigvee\{x \in L \mid x \prec a\}$, where $x \prec a$ means that $x \wedge y = 0$ and $a \vee y = e$ for some $y \in L$. An element $a \in L$ is called *compact* if $a \leq \bigvee S$ implies $a \leq \bigvee E$ for some finite $E \subseteq S$, for all $S \subseteq L$. L is called compact whenever the unit $e \in L$ is compact. A *compactification* of a frame L is a compact regular frame M together with a dense onto homomorphism $h\colon M \to L$.

A frame L is called *completely regular* if for each $a \in L$, $a = \bigvee\{x \in L \mid x \prec\prec a\}$ where $x \prec\prec a$ means that there exists a doubly indexed sequence $(x_{nk})_{n=0,1,\ldots;k=0,1,\ldots,2^n}$ such that

$$x = x_{n0}, \quad x_{nk} \prec x_{n\,k+1}, \quad x_{n2^n} = a, \quad x_{nk} = x_{n+1\,2k}$$

for all $n = 0, 1, \ldots$ and $k = 0, 1, \ldots, 2^n$.

For any frame L its *compact completely regular coreflection* is given by $\mathfrak{CRJ}L$ the frame of completely regular ideals $J \subseteq L$ together with the join map $\bigvee\colon \mathfrak{CRJ}L \to L$. By a *completely regular* ideal $J \subseteq L$ is meant an ideal J such that for each $a \in J$ there exists $b \in J$ such that $a \prec\prec b$. The dense homomorphism $\bigvee\colon \mathfrak{CRJ}L \to L$ is onto iff L is completely regular. Following, what by now is standard terminology, we shall denote $\mathfrak{CRJ}L$ in the case where L is completely regular by βL, the Stone–Čech compactification of L.

A cover of a frame L is a subset $A \subseteq L$ such that $\bigvee A = e$. We denote by $\mathrm{Cov}(L)$ the set of all covers of L. For $A, B \in \mathrm{Cov}(L)$ we use the following notation and terminology:

$A \leq B$ (A refines B) if for each $x \in A$, there exists $y \in B$ such that $x \leq y$

$$Ax = \bigvee\{y \in A \mid y \wedge x \neq 0\}$$

$$A \wedge B = \{x \wedge y \mid x \in A, y \in B\}$$

A uniformity on L is a set $\mathfrak{U}$ of covers of L which is a filter relative to $\leq$, satisfies the star-refinement property and which is admissible. We recall that $\mathfrak{U} \subseteq \mathrm{Cov}(L)$ is said to have the star-refinement property if for each $A \in \mathfrak{U}$ there exists a $B \in \mathfrak{U}$ such that $B^* \leq A$, i.e. $\{Bb \mid b \in B\} \leq A$. It is said to be admissible if for each $a \in L$, $a = \bigvee\{x \in L \mid x \lhd_{\mathfrak{U}} a\}$. Here $x \lhd_{\mathfrak{U}} a$ is read as 'x is $\mathfrak{U}$-strongly below a' means that $Cx \leq a$ for some $C \in \mathfrak{U}$. A uniform frame is a frame L together with a specified uniformity on it which we will denote by $\mathfrak{U}L$. The strongly below relation $\lhd_{\mathfrak{U}L}$ arising from a uniformity $\mathfrak{U}L$ will be simply denoted by $\lhd$ unless the context requires us to be more explicit. The members of $\mathfrak{U}L$ are referred to as the uniform covers of L.

The reader is referred to [6] for the basic material on uniform frames and to [13] for general background on frames.

2. Property S and Uniform Local Connectedness

Recall that an element $a \in L$ is said to be *connected* if whenever $a = b \vee c$ and $b \wedge c = 0$ then either $b = 0$ or $c = 0$. L is said to be *locally connected* if each $a \in L$ can be written as a join of connected elements (see e.g. [1]).

The concepts of property S and uniform local connectedness have proved quite useful in dealing with questions on local connectedness in the context of spaces and were introduced in the metric space setting many years ago (see e.g. Whyburn [16], Wilder [17]). Part of the usefulness of these concepts (both of which imply local connectedness for spaces) is that they are preserved by 'taking closures' which is not the case for local connectedness as a standard example in the plane shows. We now introduce these concepts in the setting of uniform frames.

DEFINITION 2.1 A uniform frame L is said to have property S if given any uniform cover A of L there exists a finite cover B of L consisting of connected elements such that $B \leq A$. It is said to be uniformly locally connected if for each uniform cover A there exists a uniform cover B consisting of connected elements such that $B \leq A$.

REMARK 2.2 The concept of property S (originally due to Sierpinski; see e.g. Whyburn [11]) and uniform local connectedness were first defined for metric spaces as mentioned above. Later, Gleason [11] introduced the concept of uniform local connectedness for uniform spaces and Collins [8] did likewise for the concept of property S.

PROPOSITION 2.3 *Every totally bounded uniformly locally connected uniform frame has property S.*

Proof. Let L be a uniform frame and let A be any uniform cover. By uniform local connectedness and total boundedness there exists a finite uniform cover C consisting of connected elements such that $C \leq A$. Hence L has property S. □

PROPOSITION 2.4 *If L has property S or if L is uniformly locally connected then L is locally connected.*

Proof. Let $a \in L$. Then $a = \bigvee\{b \in L \mid b \lhd a\}$. For each $b \in L$, $b \lhd a$, there exists a uniform cover C_b such that $C_b b \leq a$. If L has property S there exists a finite cover S_b of L consisting of connected elements such that $S_b \leq C_b$, or if L is uniformly locally connected there exists a cover $S_b \in \mathfrak{U}L$ consisting of connected elements such that $S_b \leq C_b$. In either case let $u_b = \bigvee\{s \in S_b \mid s \leq a\}$ and $v_b = \bigvee\{s \in S_b \mid s \nleq a\}$. Now for any $s \in S_b$, $s \nleq a$, we have

$s \wedge b = 0$ since $S_b \leq C_b$ and $C_b b \leq a$. Thus $b \wedge v_b = 0$, from which $b \leq u_b$ since $u_b \vee v_b = e$. Thus $b \leq u_b \leq a$ whenever $b \lhd a$, from which we get $a = \bigvee\{u_b \mid b \lhd a\}$. This means that a is a join of connected elements, proving that L is locally connected. □

3. The Basic Construction

We now obtain the analogue of Gleason's [11] construction of the uniformly locally connected coreflection of an arbitrary uniform space in the setting of those uniform frames whose underlying frame is locally connected.

PROPOSITION 3.1 *Let L be any locally connected uniform frame. For each $C \in \mathfrak{U}L$, let $\tilde{C} = \{d \in L \mid d$ is a component of some $c \in C\}$. Then $\{\tilde{C} \mid C \in \mathfrak{U}L\}$ is a basis for a uniformity $\mathfrak{U}\tilde{L}$ on L finer than $\mathfrak{U}L$ with respect to which L is uniformly locally connected.*

Proof. Firstly $\tilde{C} \in \mathrm{Cov}(L)$ for each $C \in \mathfrak{U}L$ since, L being locally connected, each element of L is a join of its connected components. Now let $C, D \in \mathfrak{U}L$. Then $C \wedge D \in \mathfrak{U}L$. It is easily checked that if $E = C \wedge D$ then $\tilde{E} \leq \tilde{C}$ and $\tilde{E} \leq \tilde{D}$. Now let $C \in \mathfrak{U}L$ and find $E \in \mathfrak{U}L$ such that $E^* \leq C$. We claim that $\tilde{E}^* \leq \tilde{C}$: Let $d \in \tilde{E}$. Then d is a component of some $s \in E$. Since $E^* \leq C$, there exists $c \in C$ such that $Es \leq c$. Now $d \leq s \leq Es \leq c$. Let v be the component of c above d. Then $\tilde{E}d \leq v$: For let $z \in \tilde{E}$ such that $z \wedge d \neq 0$. Now z is a component of some $s' \in E$. From $z \wedge d \neq 0$ and $d \leq s$ we can conclude that $s' \wedge s \neq 0$, whence $s' \leq c$ since $Es \leq c$. Thus $z \leq c$. Now $z \wedge d \neq 0$ implies $z \wedge v \neq 0$ and hence $z \leq v$ since v is a component of c and z is connected and below c. This proves the claim.

Now take any $a \in L$. Then $a = \bigvee\{b \mid b \lhd_{\mathfrak{U}L} a\}$. Now $b \lhd_{\mathfrak{U}L} a$ implies there exists $C \in \mathfrak{U}L$ such that $Cb \leq a$, from which we have $\tilde{C}b \leq a$. Thus $b \lhd_{\mathfrak{U}\tilde{L}} a$. Hence $a = \bigvee\{x \mid x \lhd_{\mathfrak{U}\tilde{L}} a\}$.

The above then shows that $\{\tilde{C} \mid C \in \mathfrak{U}L\}$ is a basis for some uniformity $\mathfrak{U}\tilde{L}$ on L, which clearly is finer than $\mathfrak{U}L$.

Lastly, let $D \in \mathfrak{U}\tilde{L}$. Then there exists $\tilde{C}$, $C \in \mathfrak{U}L$, such that $\tilde{C} \leq D$. As $\tilde{C} \in \mathfrak{U}\tilde{L}$ and consists only of connected elements we have $(L, \mathfrak{U}\tilde{L})$ is uniformly locally connected. □

Let $\mathbb{U}ni\mathbb{F}rm$ denote the category of uniform frames and uniform frame maps, $\mathbb{LCU}ni\mathbb{F}rm$, the full subcategory of locally connected uniform frames, and $\mathbb{ULCU}ni\mathbb{F}rm$ the full subcategory of uniformly locally connected uniform frames. We then have

THEOREM 3.2 *$\mathbb{ULCU}ni\mathbb{F}rm$ is a reflective subcategory of $\mathbb{LCU}ni\mathbb{F}rm$.*

Proof. That $\mathbb{ULCU}ni\mathbb{F}rm$ is a subcategory of $\mathbb{LCU}ni\mathbb{F}rm$ follows from the fact that uniform local connectedness implies local connectedness (Proposition 2.4).

Suppose now that $(L, \mathfrak{U}L) \in \mathbb{LCU}ni\mathbb{F}rm$. Consider $\mathfrak{U}\tilde{L}$ as given in the above proposition. Then $(L, \mathfrak{U}\tilde{L}) \in \mathbb{ULCU}ni\mathbb{F}rm$. The identity map $\mathrm{id}_L\colon (L, \mathfrak{U}L) \to (L, \mathfrak{U}\tilde{L})$ is a uniform frame map since $\mathfrak{U}L \subseteq \mathfrak{U}\tilde{L}$. Consider

$$\begin{array}{ccc} (L, \mathcal{U}) & \xrightarrow{f} & (M, \mathcal{U}M) \\ {\scriptstyle \mathrm{id}_L}\downarrow & & \downarrow{\scriptstyle \mathrm{id}_M} \\ (L, \mathcal{U}\tilde{L}) & \xrightarrow{g} & (M, \mathcal{U}M) \end{array}$$

Let $f\colon (L, \mathfrak{U}L) \to (M, \mathfrak{U}M)$ be any uniform frame map into a uniformly locally connected frame M. Let $g\colon (L, \mathfrak{U}\tilde{L}) \to (M, \mathfrak{U}M)$ be as given by f. It suffices to show that g is a uniform frame map.

Let then $\tilde{C} \in \mathfrak{U}\tilde{L}$ be a basic element, where $C \in \mathfrak{U}L$. Now $f(C) \in \mathfrak{U}M$, and by uniform local connectedness of M, there exists $D \in \mathfrak{U}M$ consisting of connected elements such that $D \leq f(C)$.

We claim that $D \leq g(\tilde{C})$: Let $d \in D$, and find $c \in C$ such that $d \leq f(c)$. Since L is locally connected, $c = \bigvee c_\alpha$, the join of components of c. Now $d \leq f(c) = \bigvee f(c_\alpha)$ so that there exists α such that $d \wedge f(c_\alpha) \neq 0$. Let $u = f(c_\alpha)$ and $v = \bigvee\{f(c_\beta) \mid \beta \neq \alpha\}$. Then $d \leq u \vee v$, $u \wedge v = f(c_\alpha) \wedge \bigvee\{f(c_\beta) \mid \beta \neq \alpha\} = \bigvee\{f(c_\alpha) \wedge f(c_\beta) \mid \beta \neq \alpha\} = \bigvee\{f(c_\alpha \wedge c_\beta) \mid \beta \neq \alpha\} = 0$. Since d is connected,we have either $d \leq u$ or $d \leq v$. Since $d \wedge u \neq 0$ we must have $d \leq u = f(c_\alpha) = g(c_\alpha)$ so that $D \leq g(\tilde{C})$. This shows that $g(\tilde{C}) \in \mathfrak{U}M$ and hence that g is a uniform frame map. □

4. Applications

Recall that a uniform frame L is said to be *totally bounded* iff the finite uniform covers forms a basis for the uniformity on L. Using the star-refinement property it is easy to see that a uniform frame L is totally bounded iff each $A \in \mathfrak{U}L$ is refined by a finite cover of L (not necessarily uniform).

One then has the following:

PROPOSITION 4.1 (a) *A uniform frame having property S must be totally bounded.*

(b) *If a uniform frame L has property S, then L has property S with respect to every coarser uniformity on L.*

If L has property S, L need not necesssarily have property S with respect to every uniformity on L finer than $\mathfrak{U}L$. It is of interest therefore to note that the following holds.

PROPOSITION 4.2 *Let $(L, \mathfrak{U}L)$ be a locally connected uniform frame and let $(L, \mathfrak{U}\tilde{L})$ be the uniformly locally connected reflection of $(L, \mathfrak{U}L)$. Then $(L, \mathfrak{U}L)$ has property S iff $(L, \mathfrak{U}\tilde{L})$ has property S.*

Proof. The sufficiency follows from Proposition 4.1. For the necessity suppose $(L, \mathfrak{U}L)$ has property S and let $\tilde{B} \in \mathfrak{U}\tilde{L}$ be a basic member. Then $B \in \mathfrak{U}L$. Since $(L, \mathfrak{U}L)$ has property S, there exists a finite connected refinement C of B. We claim that $C \leq \tilde{B}$: Take any $c \in C$. Then $c \leq b$ for some $b \in B$. Since c is connected, c lies below some component d of b. Thus $C \leq \tilde{B}$ showing that $(L, \mathfrak{U}\tilde{L})$ has property S. □

Recall that if $(L, \mathfrak{U}L)$ is a uniform frame then there is a finest uniformity $\mathfrak{U}_F L$ on L obtained in the analogous way as the fine uniformity for spaces. The fine uniformity on L is the uniformity on L consisting of all the normal covers of L, i.e. those covers A of L such that $A = A_1$ in some sequence $(A_n)_n$ of covers of L such that $A^*_{n+1} \leq A_n$ for all $n = 1, 2, 3, \ldots$ (see [15], for example).

Recall also that every completely regular frame is uniformizable, i.e. admits a uniformity, and that under the axiom of CDC every uniformizable frame is completely regular (Pultr [14], Frith [9]). We then obtain

PROPOSITION 4.3 *For any uniformizable frame L, L is locally connected iff L is uniformly locally connected with respect to its fine uniformity.*

Proof. If $(L, \mathfrak{U}_F L)$ is uniformly locally connected then L is certainly locally connected by Proposition 2.4. If, on the other hand L is locally connected, then by Proposition 3.1, $(L, \mathfrak{U}_F \tilde{L})$ is uniformly locally connected with $\mathfrak{U}_F L \subseteq \mathfrak{U}_F \tilde{L}$. Thus $\mathfrak{U}_F L = \mathfrak{U}_F \tilde{L}$ and hence $(L, \mathfrak{U}_F L)$ is uniformly locally connected. □

Recall that the totally bounded coreflection of a uniform frame $(L, \mathfrak{U}L)$ is determined by the uniformity on L having as a basis the finite uniform covers in $\mathfrak{U}L$ (see e.g. [15]). Denote this uniformity by $\mathfrak{U}_T L$. Recall also that every compact regular frame has a unique uniformity consisting of all the finite covers of the frame [6]. We then have

PROPOSITION 4.4 *If L is a frame with a unique uniformity (for example, if L is compact regular), then the following are equivalent:*

(i) *L has property S*
(ii) *L is locally connected*
(iii) *L is uniformly locally connected*

Proof. (i) $\Rightarrow$ (ii) follows from Proposition 2.4.

(ii) $\Rightarrow$ (iii) follows from Proposition 4.3

(iii) $\Rightarrow$ (i): Since L has a unique uniformity, this uniformity must be totally bounded (by taking its totally bounded coreflection) and thus by Proposition 2.3, L must have property S. □

LEMMA 4.5 *Let h: $M \to L$ be a dense onto map between frames with k: $L \to M$ its right adjoint. Then $a \prec b$ iff $kh(a) \prec b$.*

Proof. If $kh(a) \prec b$, then $a \leq kh(a) \prec b$ implies $a \prec b$. Now suppose $a \prec b$. Then there exists $t \in L$ such that $a \wedge t = 0, t \vee b = e$. Hence

$$h(kh(a) \wedge t) = hkh(a) \wedge h(t) = h(a) \wedge h(t) = 0$$

which implies that $kh(a) \wedge t = 0$ by the denseness of h. Hence $kh(a) \prec b$. □

Recall that a uniform frame map $h\colon M \to L$ is called a *surjection* if h is onto and $\mathfrak{U}L$ is generated by the $h(A)$, $A \in \mathfrak{U}M$ ([6]). We then have

PROPOSITION 4.6 *Let $h\colon M \to L$ be a dense surjection between uniform frames. If L is uniformly locally connected then so is M.*

Proof. Take any $A \in \mathfrak{U}M$ and find $A_1 \in \mathfrak{U}M$ such that $A_1^* \leq A$. Now $h(A_1) \in \mathfrak{U}L$. Hence we can find $C \in \mathfrak{U}L$ consisting of connected elements such that $C \leq h(A_1)$. By surjectivity we can find $D \in \mathfrak{U}M$ such that $h(D) \leq C$.

Consider $k(C) = \{k(c) \mid c \in C\}$. Each $k(c)$ is connected since h is dense and onto [1]. Furthermore, we have $k(C) \in \mathfrak{U}M$: In fact $D \leq k(C)$. For let $d \in D$. Then $h(d) \leq c$ for some $c \in C$ and hence $d \leq k(c)$ since k is the right adjoint of h.

Also $k(C) \leq A$: Take $c \in C$. Now $C \leq h(A_1)$, and so there exists $a_1 \in A$ such that $c \leq h(a_1)$. Also $A_1 a_1 \leq a$ for some $a \in A$ from which $a_1 \prec a$ by a familiar property. From the above lemma, $kh(a_1) \prec a$ so that $k(c) \leq kh(a_1) \leq a$. Hence M is uniformly locally connected. □

Recall that every uniform frame L has a unique completion $\gamma\colon CL \to L$, the completion map $\gamma\colon CL \to L$ being a dense surjection [6]. We thus have:

COROLLARY 4.7 *If L is uniformly locally connected, then so is its uniform completion.*

Recall that a lattice-theoretic characterization of pseudocompactness for a frame L originally due to Gilmour [10] and used for example in [1] is that whenever $a_0 \prec\prec a_1 \prec\prec \cdots$ is a sequence of elements of L such that $\bigvee a_n = e$, then $a_n = e$ for some n. Also, in terms of uniformities, one has the following characterization for completely regular L [15]: L is pseudocompact iff L is totally bounded with respect to each of its uniformities. The latter can then be used to deduce:

PROPOSITION 4.8 *For completely regular L, L is locally connected and pseudocompact iff L has property S with respect to each of its uniformities.*

Proof. For the necessity, suppose L is locally connected and pseudocompact. Then L is totally bounded with respect to its fine uniformity from the above mentioned result. Also by Proposition 4.3, L is uniformly locally connected in its fine uniformity. Since every totally bounded uniformly locally connected uniform frame has property S (Proposition 2.3), we have that L has property S with respect to its fine uniformity and hence with respect to all of its uniformities.

For the sufficiency, if L has property S with respect to each of its uniformities, then it must be totally bounded in each of its uniformities (Proposition 4.1) and hence is pseudocompact. The local connectedness of L follows from Proposition 2.4. □

The remainder of this section has one purpose. This is to show that if $h\colon M \to L$ is a dense onto map between completely regular frames and L is locally connected and pseudocompact, then M is locally connected. The idea of the proof is straightforward: If L is locally connected and pseudocompact, then L has property S in each of its uniformities (from above). If one can show that property S is reflected by dense surjections, then M would have property S, hence would be locally connected (by Proposition 2.4). The difficulty here is showing that property S is reflected by dense surjections. In order to do this we introduce (the seemingly weaker concept of) property S^*. We shall show in this section that in fact this concept is equivalent to property S and all the proofs here will be geared towards showing this equivalence.

DEFINITION 4.9 A uniform frame L is said to have property S^* if given any $U \in \mathfrak{U}L$ there exists a finite number of connected congruences $\Theta_1, \Theta_2, \ldots, \Theta_n$ such that $\bigwedge \Theta_i = \Delta$ and for each i there exists $u_i \in U$ such that $\Delta_{u_i} \leq \Theta_i$.

In the above definition recall from Chen [7], that a congruence Θ is said to be *connected* if the quotient frame L/Θ is connected.

PROPOSITION 4.10 *Let $h\colon M \to L$ be a dense uniform map. If L has property S^* then so does M.*

Proof. Let $W \in \mathfrak{U}M$. Find $U \in \mathfrak{U}M$ such that $U^* \leq W$. Since $h(U) \in \mathfrak{U}L$ there exist connected congruences $\Theta_1, \Theta_2, \ldots, \Theta_n$ such that $\bigwedge \Theta_i = \Delta$ and for each i there exists $u_i \in U$ such that $\Delta_{h(u_i)} \leq \Theta_i$. Let $\Phi_i = (h \times h)^{-1}(\Theta_i)$. Then Φ_i is connected and hence so is $\bar{\Phi}_i$ (Chen [7]), where $\bar{\Phi}_i$, recall, equals ∇_{a_i} where a_i is the largest element such that $(0, a_i) \in \Phi_i$. Now $\bar{\Phi}_1 \wedge \cdots \wedge \bar{\Phi}_n \leq \Phi_1 \wedge \cdots \wedge \Phi_n = \Delta$ and hence $\bar{\Phi}_1 \wedge \cdots \wedge \bar{\Phi}_n = \Delta$.

Now find, for each $i = 1, 2, \ldots, n$, $w_i \in W$ such that $Uu_i \leq w_i$. Thus for each i, $u_i \prec w_i$ and hence $u_i^* \vee w_i = e$. We claim that $\Delta_{w_i} \leq \bar{\Phi}_i = \nabla_{a_i}$: For this, note that since Δ_{w_i} is the smallest congruence containing (e, w_i), it is sufficient to show that $(e, w_i) \in \nabla_{a_i}$ or $a_i \vee w_i = e$. It would suffice to show that $u_i^* \leq a_i$ since $u_i^* \vee w_i = e$.

Now $\Delta_{u_i} \leq \Phi_i$ since $(x, y) \in \Delta_{u_i}$ implies $x \wedge u_i = y \wedge u_i$ which implies $h(x) \wedge h(u_i) = h(y) \wedge h(u_i)$. Thus $(h(x), h(y)) \in \Delta_{h(u_i)} \leq \Theta_i$ and hence $(x, y) \in (h \times h)^{-1}(\Theta_i) = \Phi_i$.

Hence $(e, u_i) \in \Phi_i$ and thus $(u_i^* \vee w_i, u_i) \in \Phi_i$. Meeting with u_i^* gives $(u_i^*, 0) \in \Phi_i$. Thus $u_i^* \leq a_i$ since a_i is the largest element such that $(0, a_i) \in \Phi_i$. This completes the proof. □

In order to prove the next proposition we require the following:

LEMMA 4.11 *If $a \prec b$ and $\Delta_a \leq \Theta$, then $\Delta_b \leq \bar{\Theta}$.*

Proof. If $a \prec b$ then there exists x such that $a \wedge x = 0$, $x \vee b = e$. Assume $\bar{\Theta} = \nabla_w$, where w is the largest element such that $(0, w) \in \Theta$. To show that $\Delta_b \leq \bar{\Theta}$ or $e = b \vee w$. Now $\Delta_a \leq \Theta$ implies $(e, a) \in \Theta$ and hence $(x, a \wedge x) \in \Theta$. Thus $(0, x) \in \Theta$ and hence $x \leq w$. Thus $b \vee w = e$ as required. □

PROPOSITION 4.12 *If L has property S^* then for each $a \in L$, Δ_a is a meet of connected closed congruences.*

Proof. Take any $a \in L$. Now $a = \bigvee x(x \lhd a)$. Now $x \lhd a$ implies there exists $A \in \mathfrak{U}L$ such that $Ax \leq a$. Find $B \in \mathfrak{U}L$ such that $B^* \leq A$. Now there exists a finite number of connected congruences $\Theta_1, \Theta_2, \ldots, \Theta_n$ such that $\bigwedge \Theta_i = \Delta$ and $\Delta_{b_i} \leq \Theta_i$ for all $i = 1, 2, \ldots, n$, and for suitable $b_i \in B$. Now $Bb_i \leq a_i \in A$ and hence $b_i \prec a_i$ for some a_i, $i = 1, 2, \ldots, n$. Hence $\Delta_{a_i} \leq \bar{\Theta}_i$ for all i from the above lemma.

Now $\bar{\Theta}_i \subseteq \Theta_i$ and hence $\bigwedge\{\bar{\Theta}_i \mid i = 1, 2, \ldots, n\} = \Delta$. Let $\bar{\Theta}_1, \ldots, \bar{\Theta}_n$ be those for which $\Delta_x \vee \bar{\Theta}_i \neq \nabla$. This is possible since if for all i we have $\Delta_x \vee \bar{\Theta}_i = \nabla$, then $\Delta_x \vee \bigwedge\{\bar{\Theta}_i \mid i = 1, \ldots, n\} = \bigwedge\{\Delta_x \vee \bar{\Theta}_i \mid i = 1, \ldots, n\} = \nabla$, and so $\Delta_x \vee \bigwedge\{\bar{\Theta}_i \mid i = 1, \ldots, n\} = \Delta_x \vee \Delta = \Delta_x$ so that $\Delta_x = \nabla$, i.e. $(0, e) \in \Delta_x$ i.e. $0 = x \wedge e = x$ which is not possible since it is assumed that $x \neq 0$ in the expression $a = \bigvee x(x \lhd a)$.

Now we have $\Delta_a \leq \bigwedge\{\bar{\Theta}_i \mid \Delta_x \vee \bar{\Theta}_i \neq \nabla\} \leq \Delta_x$: To show this note that we have $b_i \prec a_i$, $\Delta_{a_i} \leq \bar{\Theta}_i$. If $a_i \wedge x = 0$, then $\Delta_{a_i} \vee \Delta_x = \Delta_0 = \nabla$, so that $\Delta_x \vee \bar{\Theta}_i = \nabla$ which is not possible. Hence $a_i \wedge x \neq 0$, and thus $a_i \leq a$. Hence $\Delta_a \leq \Delta_{a_i} \leq \bar{\Theta}_i$ and hence $\Delta_a \leq \bigwedge\{\bar{\Theta}_i \mid \Delta_x \vee \bar{\Theta}_i \neq \nabla\}$. To show that $\bigwedge\{\bar{\Theta}_i \mid \Delta_x \vee \bar{\Theta}_i \neq \nabla\} \leq \Delta_x$, note that firstly we have $\bigwedge\{\bar{\Theta}_i \mid i = 1, 2, \ldots, n\} = \Delta \leq \Delta_x$ so that $\bigwedge\{\bar{\Theta}_i \mid \Delta_x \vee \bar{\Theta}_i \neq \nabla\} \wedge \bigwedge\{\bar{\Theta}_i \mid \Delta_x \vee \bar{\Theta}_i = \nabla\} \leq \Delta_x$. Hence we have $\bigwedge\{\bar{\Theta}_i \mid \Delta_x \vee \bar{\Theta}_i \neq \nabla\} \wedge \bigwedge\{\bar{\Theta}_i \vee \Delta_x \mid \Delta_x \vee \bar{\Theta}_i = \nabla\} \leq \Delta_x$ and thus $\bigwedge\{\bar{\Theta}_i \mid \Delta_x \vee \bar{\Theta}_i \neq \nabla\} \wedge \nabla \leq \Delta_x$, i.e. $\bigwedge\{\bar{\Theta}_i \mid \Delta_x \vee \bar{\Theta}_i \neq \nabla\} \leq \Delta_x$.

Thus we have $\Delta_a \leq \bigwedge\{\bar{\Theta}_i \mid \Delta_x \vee \bar{\Theta}_i \neq \nabla\} \leq \Delta_x$. But $a = \bigvee x(x \lhd a)$ so that $\Delta_a = \bigwedge \Delta_x(x \lhd a)$, which coupled with the above means that each Δ_a is a meet of connected closed congruences. □

Recall the following definitions and results:

DEFINITION 4.13 (Banaschewski [5]). A non-empty subset $S \subseteq L$ is said to be cochained if for all $x, y \in S$ there exists $x_1, x_2, \ldots, x_n \in S$ such that $x = x_1$, $y = x_n$ and $x_i \vee x_{i+1} < e$ for all $i = 1, 2, \ldots, n-1$. An element $a \in L$ is called coconnected if $\uparrow a$ is connected.

LEMMA 4.14 (Banaschewski [5]). *If $\emptyset \neq S \subseteq L$ is cochained and consists of coconnected elements, then $\bigwedge S$ is coconnected.*

DEFINITION 4.15 (Chen [7]). For $a, b \in L$ define $b \blacktriangleleft a$ if there exists a connected congruence Θ such that $\Delta_a \leq \Theta \leq \Delta_b$

LEMMA 4.16 (Chen [7]). *In any frame L, if $a = \bigvee c(c \blacktriangleleft a)$ then a is a join of connected elements.*

Using the above we can now prove

PROPOSITION 4.17 *If L has property S^* then L is locally connected.*

Proof. Take any $a \in L$. We have $a = \bigvee x(x \lhd a)$. Thus $\Delta_a = \bigwedge \Delta_x(x \lhd a)$. Also from Proposition 4.12, each Δ_a is a meet of closed connected congruences, ie. $\Delta_x = \bigwedge\{\nabla_d \mid d \in D_x \subseteq L, \nabla_d \text{ connected}\}$. Hence $\Delta_a = \bigwedge\{\nabla_d \mid d \in D_x \subseteq L, \nabla_d \text{ connected}, x \lhd a\}$.

Now take any ∇_d, $d \in D_x$, $x \lhd a$ for which $\Delta_a \leq \nabla_d$. Now $x \lhd a$ implies $\{x^*, a\} \in \mathfrak{U}L$. We have $\Delta_x \leq \nabla_d$ so that $e = x \vee d$. Hence $x^* \leq d$ and hence $\{x^*, a\} \leq \{a, d\}$. Thus $\{a, d\} \in \mathfrak{U}L$. Since L has property S^* there exists a finite number of $\nabla_{b_1}, \ldots, \nabla_{b_r}$ of closed connected congruences such that for each i, $1 \leq i \leq r$, either $\Delta_a \leq \nabla_{b_i}$ or $\Delta_d \leq \nabla_{b_i}$ and also $\bigwedge \nabla_{b_i} = \Delta$. (Note that we can assume these congruences are closed from the proof of Proposition 4.12.) Now if for all $i, 1 \leq i \leq r$, we have $\Delta_d \leq \nabla_{b_i}$, then $\Delta_d \leq \bigwedge \nabla_{b_i} = \Delta$. Hence $d = e$ and hence $\nabla_d = \nabla$. However from the way in which the ∇_d are produced as in Proposition 4.12 this is not possible since the $\nabla_d \neq \nabla$. Thus there exists i, $1 \leq i \leq r$ such that $\Delta_d \not\leq \nabla_{b_i}$. For this i, $\Delta_a \leq \nabla_{b_i}$.

Let $b_1, \ldots, b_n$ be those for which $\Delta_d \not\leq \nabla_{b_i}$. Then we have $\Delta_a \leq \nabla_{b_1} \wedge \cdots \wedge \nabla_{b_n}$. Now we have $\nabla_{b_1} \wedge \cdots \wedge \nabla_{b_r} = \Delta$ implies $b_1 \wedge \cdots \wedge b_r = 0$.
We claim that

$$\Delta_a \leq \nabla_{b_1} \wedge \cdots \wedge \nabla_{b_n} \leq \Delta_{b_{n+1} \wedge \cdots \wedge b_r} \leq \nabla_d. \tag{0.1}$$

For this note that since $b_1 \wedge \cdots \wedge b_n \wedge b_{n+1} \wedge \cdots \wedge b_r = 0$ we must have $\nabla_{b_1} \wedge \cdots \wedge \nabla_{b_n} \leq \Delta_{b_{n+1} \wedge \cdots \wedge b_r}$. To show $\Delta_{b_{n+1} \wedge \cdots \wedge b_r} \leq \nabla_d$ we must show $e = d \vee (b_{n+1} \wedge \cdots \wedge b_r)$. Now we have $\Delta_d \leq \nabla_{b_{n+1}}, \ldots, \nabla_{b_r}$ and hence $e = d \vee (b_{n+1} \wedge \cdots \wedge b_r)$ as required, thus proving the claim.

We now claim that $\nabla_{b_1 \wedge \cdots \wedge b_n}$ is connected. For this let $S = \{b_1, \ldots, b_n\}$. Since ∇_{b_i}, $1 \leq i \leq n$, are all connected, we have each b_i is coconnected since $L/\nabla_{b_i} \cong \uparrow b_i$. If S is not cochained then we may assume without loss of generality and by re-ordering of the indices if necessary that there exists k, $1 \leq k < n$ such that b_1 is cochained in S to b_j for all j, $1 \leq j \leq k$ and that b_1 is not cochained in S to b_i whenever $k < i \leq n$. Now for $1 \leq j \leq k$, $k < i \leq n$, we have $b_j \vee b_i = e$. Hence

$$\bigwedge_{j=1}^{k} b_j \vee \bigwedge_{i=k+1}^{n} b_i = e$$

i.e. $(b_1 \wedge \cdots \wedge b_k) \vee (b_{k+1} \wedge \cdots \wedge b_n) = e$.

Hence

$$[(b_1 \vee d) \wedge \cdots \wedge (b_k \vee d)] \vee [(b_{k+1} \vee d) \wedge \cdots \wedge (b_n \vee d)] = e.$$

Also

$$\begin{aligned} b_1 \wedge \cdots \wedge b_n &= b_1 \wedge \cdots \wedge b_n \wedge e \\ &= b_1 \wedge \cdots \wedge b_n \wedge (d \vee (b_{n+1} \wedge \cdots \wedge b_r)) \\ &= (b_1 \wedge \cdots \wedge b_n \wedge d) \vee (b_1 \wedge \cdots \wedge b_n \wedge \cdots \wedge b_r) \\ &= b_1 \wedge \cdots \wedge b_n \wedge d \leq d \end{aligned}$$

so that $d \vee (b_1 \wedge \cdots \wedge b_n) = d$ and hence

$$[(b_1 \vee d) \wedge \cdots \wedge (b_k \vee d)] \wedge [(b_{k+1} \vee d) \wedge \cdots \wedge (b_n \vee d)] = d.$$

Now since d is coconnected, either

$$\begin{aligned} (b_1 \vee d) \wedge \cdots \wedge (b_k \vee d) &= e \quad \text{or} \quad (b_{k+1} \vee d) \wedge \cdots \wedge (b_n \vee d)] = d \\ \text{i.e. } (b_1 \wedge \cdots \wedge b_k) \vee d &= e \quad \text{or} \quad (b_{k+1} \wedge \cdots \wedge b_n) \vee d = e. \end{aligned}$$

In either case there exists $t,\ 1 \leq t \leq n$, such that $b_t \vee d = e$ or $\Delta_d \leq \nabla_{b_t}$ which contradicts the fact that for all $i,\ 1 \leq i \leq n$ we have $\Delta_d \not\leq \nabla_{b_i}$. Thus S must be cochained. Hence by Lemma 4.14, $\bigwedge S$ must be coconnected, i.e. $\nabla_{b_1 \wedge \cdots \wedge b_n}$ is connected. From Proposition 4.12 Δ_a is a meet of the ∇_d so that Δ_a is a meet of elements of the form $\Delta_{b_{n+1} \wedge \cdots \wedge b_r}$ and hence a is a join of elements of the form $b_{n+1} \wedge \cdots \wedge b_r$ where $b_{n+1} \wedge \cdots \wedge b_r \blacktriangleleft a$. Hence a is a join of connected elements completing the proof. □

LEMMA 4.18 *Let $(\Theta_i)_{i \in I}$ be connected congruences such that there exists $i_0 \in I$ with the property that for any $i \neq i_0$ Θ_i is not separated from Θ_{i_0}, i.e. $\bar{\Theta}_i \vee \Theta_{i_0} \neq \nabla$ or $\Theta_i \vee \bar{\Theta}_{i_0} \neq \nabla$. Then $\bigwedge \Theta_i$ is connected.*

Proof. Suppose that $\Delta_{x \vee y} \leq \bigwedge \Theta_i$, $\nabla_{x \vee y} \leq \bigwedge \Theta_i$. We need to show that either $\Delta_x \leq \bigwedge \Theta_i$ or $\nabla_x \leq \bigwedge \Theta_i$ (see Chen [7]). Now $\Delta_{x \vee y} \leq \Theta_{i_0}$ and $\nabla_{x \vee y} \leq \Theta_{i_0}$. Since Θ_{i_0} is connected we have either $\Delta_x \leq \Theta_{i_0}$ or $\nabla_x \leq \Theta_{i_0}$. Assume $\Delta_x \leq \Theta_{i_0}$. Take any $i \neq i_0$. If $\nabla_x \leq \Theta_i$ then $\nabla_x \leq \bar{\Theta}_i$: To show this let $\bar{\Theta}_i = \nabla_a$, where a is the largest element such that $(0, a) \in \Theta_i$. Then $(0, x) \in \nabla_x \leq \Theta_i$ implies $x \leq a$ which implies $\nabla_x \leq \nabla_a$. Thus we have $\nabla = \Delta_x \vee \nabla_x \leq \Theta_{i_0} \vee \bar{\Theta}_i$ a contradiction. Thus $\Delta_x \leq \Theta_i$ for all $i \neq i_0$ which implies $\Delta_x \leq \bigwedge \Theta_i$. Also if $\nabla_x \leq \Theta_{i_0}$, then for any $i \neq i_0$ $\nabla_x \leq \Theta_i$, otherwise $\Delta_x \leq \Theta_i$ and again $\Theta_i \vee \bar{\Theta}_{i_0} = \nabla$. Thus $\nabla_x \leq \bigwedge \Theta_i$ in this case and hence $\bigwedge \Theta_i$ is connected. □

LEMMA 4.19 *If Θ is any congruence and $\{\Delta_t \mid t \in \Gamma\}$ is any collection of open congruences then*

$$\bigcap_{t \in \Gamma} (\Theta \vee \Delta_t) = \Theta \vee \bigcap_{t \in \Gamma} \Delta_t.$$

Proof.

$$\begin{aligned}(x, y) \in \bigcap_{t\in\Gamma}(\Theta \vee \Delta_t) &\Rightarrow (x, y) \in \Theta \vee \Delta_t \ \forall t \in \Gamma\\ &\Rightarrow (x \wedge t, y \wedge t) \in \Theta \ \forall t \in \Gamma\\ &\Rightarrow \left(\bigvee_{t\in\Gamma} x \wedge t, \bigvee_{t\in\Gamma} y \wedge t\right) \in \Theta\\ &\Rightarrow \left(x \wedge \bigvee_{t\in\Gamma} t, y \wedge \bigvee_{t\in\Gamma} t\right) \in \Theta\\ &\Rightarrow (x, y) \in \Theta \vee \Delta_{\vee t}\\ &\Rightarrow (x, y) \in \Theta \vee \bigwedge_{t\in\Gamma} \Delta_t = \Theta \vee \bigcap_{t\in\Gamma} \Delta_t\end{aligned}$$

and thus $\bigcap_{t\in\Gamma}(\Theta \vee \Delta_t) \leq \Theta \vee \bigcap_{t\in\Gamma} \Delta_t$. For the reverse inclusion note that

$$\Theta \vee \bigcap_{t\in\Gamma} \Delta_t \leq \Theta \vee \Delta_t \ \forall t \in \Gamma$$

$$\text{i.e. } \Theta \vee \bigcap_{t\in\Gamma} \Delta_t \leq \bigcap_{t\in\Gamma}(\Theta \vee \Delta_t).$$

Thus

$$\bigcap_{t\in\Gamma}(\Theta \vee \Delta_t) = \Theta \vee \bigcap_{t\in\Gamma} \Delta_t.$$ □

LEMMA 4.20 *In a locally connected frame L, if $\Delta_a \leq \Theta$, where Θ is a connected congruence and $\Theta \neq \nabla$, then there exists connected b such that $\Delta_a \leq \Delta_b \leq \Theta$.*

Proof. Since L is locally connected, $a = \bigvee c$ (c connected). Put $b = \bigvee c$ (c connected, $c \leq a$, $\Delta_c \vee \Theta \neq \nabla$). Note that in this definition if $\Delta_c \vee \Theta = \nabla$ for all such c, then $\nabla_c = (\nabla_c \wedge \Delta_c) \vee (\nabla_c \wedge \Theta) \leq \Theta$. This would imply that $\nabla_a = \bigvee \nabla_c$ (c connected, $c \leq a$) $\leq \Theta$, so that $\Delta_a \vee \nabla_a \leq \Theta$ showing that $\Theta = \nabla$, a contradiction. Hence there exists c connected, $c \leq a$ such that $\Delta_c \vee \Theta \neq \nabla$.

Now $\Delta_b = \bigwedge \Delta_c$ (c connected, $c \leq a$, $\Delta_c \vee \Theta \neq \nabla$). We claim that $\Delta_b \leq \Theta$: To show this note that

$$\begin{aligned}\Delta_a \leq \Theta &\Rightarrow \Delta_a \vee \Theta = \Theta\\ &\Rightarrow \bigwedge \Delta_c(c \text{ connected, } c \leq a) \vee \Theta = \Theta\\ &\Rightarrow \bigcap(\Theta \vee \Delta_c)(c \text{ connected, } c \leq a) = \Theta \text{ (from Lemma 4.19)}\\ &\Rightarrow \bigcap(\Theta \vee \Delta_c)(c \text{ connected, } c \leq a, \ \Theta \vee \Delta_c \neq \nabla) = \Theta\\ &\Rightarrow \Theta \vee \bigcap \Delta_c(c \text{ connected, } c \leq a, \ \Theta \vee \Delta_c \neq \nabla) = \Theta\\ &\Rightarrow \Theta \vee \Delta_b = \Theta\\ &\Rightarrow \Delta_b \leq \Theta.\end{aligned}$$

We need to show that $\Delta_b = \bigwedge \Delta_c(c \text{ connected}, c \leq a, \Theta \vee \Delta_c \neq \nabla)$ is connected. Note that from Lemma 4.18, $\Delta_c \wedge \Theta$ is connected for all such c. Fix c_1. Then for any c_2 we have $(\Delta_{c_1} \wedge \Theta) \vee (\Delta_{c_2} \wedge \Theta) = \Theta \wedge (\Delta_{c_1} \vee \Delta_{c_2}) \leq \Theta \neq \nabla$ so that again by the same lemma, $\bigwedge(\Delta_c \wedge \Theta)$ $(c$ connected, $c \leq a,\ \Theta \vee \Delta_c \neq \nabla)$ is connected, i.e. $\Theta \wedge \bigwedge \Delta_c$ is connected, i.e. Δ_b is connected. □

PROPOSITION 4.21 *$(L, \mathfrak{U}L)$ has property S iff $(L, \mathfrak{U}L)$ has property S^*.*

Proof. If L has property S then it clearly has property S^*, this fact resting on the fact that if an element $b \in L$ is connected then Δ_b is connected as a congruence.

For the reverse, suppose L has property S^* and take any $A \in \mathfrak{U}L$. Then there exists connected congruences $\Theta_1, \Theta_2, \ldots, \Theta_n$ such that $\Delta = \bigwedge \Theta_i$ and there exists $a_i \in A$ such that $\Delta_{a_i} \leq \Theta_i$ for all i. We can assume without loss of generality that $\Theta_i \neq \nabla$ for all i otherwise we could delete such Θ_i. Since L has property S^* and hence is locally connected by Lemma 4.17, there exists connected b_i such that $\Delta_{a_i} \leq \Delta_{b_i} \leq \Theta$ for all i from Lemma 4.20. Thus $\Delta = \Delta_{b_1} \wedge \cdots \wedge \Delta_{b_n} = \Delta_{b_1 \vee \cdots \vee b_n}$ and hence $b_1 \vee b_2 \vee \cdots \vee b_n = e$ with $b_i \leq a_i$ as required. □

THEOREM 4.22 *If $h\colon M \to L$ is a dense and onto map between completely regular frames and L is locally connected and pseudocompact, then M is locally connected.*

Proof. Take any uniformity $\mathfrak{U}M$ on M. This induces a uniformity $\mathfrak{U}L$ on L making h a dense surjection. Since L is locally connected and pseudocompact, $(L, \mathfrak{U}L)$ must have property S by Proposition 4.8. Thus $(L, \mathfrak{U}L)$ must have property S^* and since this property is reflected by dense uniform maps (Proposition 4.10), $(L, \mathfrak{U}M)$ must also have this property. Thus L must be locally connected by Proposition 4.17. □

COROLLARY 4.23 ([1]). *For completely regular L, if L is locally connected and pseudocompact then its Stone–Čech compactification βL is locally connected.*

5. Concluding Remarks

According to Chen [7], one can in fact dispense with the requirement that the map h in Theorem 4.22 be onto, but his proof makes use of the result in [1] that the local connectedness and pseudocompactness of a frame implies that its Stone–Čech compactification is locally connected, a result we are not of course entitled to use here.

Also, it would be nice if the methods of this paper could allow a proof of the general local connectedness theorem for frames due to [1] viz. that for completely regular L, βL is locally connected iff L is locally connected and pseudocompact. The difficulty here seems to be that there is no obvious way in which one could use our methods to show that if βL is locally connected then L is pseudocompact.

References

1. Baboolal, D. and Banaschewski, B.: Compactification and local connectedness of frames, *J. Pure Appl. Algebra* **70** (1991), 3–16.
2. Baboolal, D.: Local connectedness of the Stone–Čech compactification, *Canad. Math. Bull.* **31**(2) (1988), 111–115.
3. Banaschewski, B.: Local connectedness of extension spaces, *Canad. J. Math.* **8** (1956), 395–398.
4. Banaschewski, B.: *Lectures on Frames,* Univ. of Cape Town, 1988.
5. Banaschewski, B.: On locally connected frames, Preprint, 1995.
6. Banaschewski, B. and Pultr, A.: Samuel compactification and completion of uniform frames, *Math. Proc. Camb. Phil. Soc.* **108** (1990), 63–78.
7. Chen, X.: On the local connectedness of frames, *J. Pure Appl. Algebra* **79** (1992), 35–43.
8. Collins, P. J.: On uniform connection properties, *Amer. Math. Monthly* **78**(4) (1971), 372–374.
9. Frith, J.: Structured Frames, PhD thesis, Univ. of Cape Town, 1987.
10. Gilmour, C. R. A.: Private communication, 1989.
11. Gleason, A. M.: Universal locally connected refinements, *Illinois. J. Math.* **7** (1963), 521–531.
12. Henriksen, M. and Isbell, J. R.: Local connectedness in the Stone–Čech compactification, *Illinois. J. Math.* **1** (1957), 574–582.
13. Johnstone, P. T.: *Stone Spaces*, Cambridge Univ. Press, Cambridge, 1982.
14. Pultr, A.: Pointless uniformities I. Complete regularity, *Comm. Math. Univ. Carolinae* **25** (1984), 91–104.
15. Walters, J. L.: Completeness and Nearly Fine Uniform Frames, PhD thesis, Univ. Catholique de Louvain, 1996.
16. Whyburn, G. T.: *Analytic Topology*, Amer. Math. Soc. Colloq. Publ. 28, 1942.
17. Wilder, R. L.: *Topology of Manifolds*, Amer. Math. Soc. Colloq. Publ. 32, 1949.

G. Brümmer & C. Gilmour (eds), Papers in Honour of Bernhard Banaschewski.

Frölicher versus Differential Spaces: A Prelude to Cosmology

Dedicated, with thanks, to Bernhard Banaschewski on his seventieth birthday.

PAUL CHERENACK
Department of Mathematics and Applied Mathematics, University of Cape Town, 7701 Rondebosch, South Africa

(Received: 30 September 1996; accepted: 19 May 1999)

Abstract. Differential spaces due to R. Sikorski are compared with Frölicher spaces (earlier called smooth spaces and studied by A. Frölicher). The category of Frölicher spaces is known to be topological over sets and Cartesian closed under the usual set adjunction. We see that the category of differential spaces is topological over sets but not Cartesian closed under the usual set adjunction. The tangent and cotangent bundles of an arbitrary Frölicher space X are defined and these notions are seen to coincide with the usual ones if X is a smooth manifold. We show that one can define, in a fairly natural way, the notions metric, connection, Ricci tensor, Riemannian curvature tensor and finally Einstein tensor for an arbitrary Frölicher space.

Mathematics Subject Classifications (2000): 18F15, 57R55, 58D15.

Key words: Frölicher space, differential space, topological functor, Cartesian closed, tangent bundle, connection, Einstein tensor.

Introduction

In physics an object is often known by various scalar fields, such as temperature and pressure defined on it. Local estimates of temperature can be put together to obtain a global temperature function. Various scalar functions can be combined via a smooth function of several variables to determine other scalar fields. Thus, if I denotes the current in an electric line and R is resistance, then the function $V = f(I, R) = IR$ determines the scalar field V denoting potential. Such ideas lead to the notion of differential space and its generalizations which have played some role in mathematical physics. See [8, 12] among the wide variety of papers here. Penrose and Rindler [10] use related concepts. To be precise we now define the notion of differential space and map between differential spaces. Let X be a set and $\mathcal{F}$ be a set of real valued functions on X called *scalar fields* or just *scalars* which define an initial topology $\mathcal{T}$ on X. Thus, as one can show, $U \in \mathcal{T}$ if and only if U is the union of sets of the form $f^{-1}(V)$ where f is a finite (pointwise defined) product of elements in $\mathcal{F}$ and V is an open subset of the reals $\mathbb{R}$.

DEFINITION 0.1. The pair $(X, \mathcal{F})$ is a *differential space* (assumed to possess the topology $\mathcal{T}$) if:

- For each open covering $\{U_i\}_{i \in K}$ of X and function $g : X \to \mathbb{R}$, if $g|_{U_i} = h_i|_{U_i}$ for some $h_i : X \to \mathbb{R} \in \mathcal{F}$ and each $i \in K$, then $g \in \mathcal{F}$. Thus, if locally a function comes from $\mathcal{F}$, it must be in $\mathcal{F}$. This is a sheaf theoretic property.
- If $f_1, \ldots, f_n$ is a collection of functions in $\mathcal{F}$ and $g(x_1, \ldots, x_n)$ is a smooth real-valued function (i.e., infinitely differentiable), then $g \circ (f_i, \ldots, f_n)$ is again in $\mathcal{F}$.

A map $f : (X, \mathcal{F}_X) \to (Y, \mathcal{F}_Y)$ of differential spaces is a map $f : X \to Y$ such that $f^*\mathcal{F}_Y = \{g \circ f | g \in \mathcal{F}_Y\} \subset \mathcal{F}_X$. The class of such maps makes differential spaces into a category $\mathcal{DSP}$. A map in $\mathcal{DSP}$ will be called a *differentiable map*.

In the definition of differential spaces one would like to think of the scalar fields in $\mathcal{F}$ as smooth maps. There is only partial control though, over this property. To motivate our subsequent definition of Frölicher space, we note heuristically that one often considers curves on a physical surface where scalar fields (e.g., temperature or pressure) vary in a smooth way as a function of position. If such curves, as maps from $\mathbb{R}$ to the surface, are smooth, one expects of course their compositions with such scalar fields to be smooth maps $\mathbb{R} \to \mathbb{R}$. These intuitive notions are refined, in the case of $\mathbb{R}^n$, by the following result of Boman [1] which shows the important fact that the converse implication holds as well.

THEOREM 0.1. *Let $f : \mathbb{R}^n \to \mathbb{R}$ be a function. Then, f is smooth, on $\mathbb{R}^n$ if and only if $f \circ c : \mathbb{R} \to \mathbb{R}$ is smooth for each smooth curve $c : \mathbb{R} \to \mathbb{R}^n$.*

These ideas motivate the notion of Frölicher space which has been used in papers [3] by the author in smooth homotopy. Traditionally, Frölicher spaces were called smooth spaces by Frölicher and Kriegl [4]. Because such smooth spaces

- are determined by contours (inputs) into them and scalar fields (outputs) on them,
- as point subsets of the plane, they need not be smooth,
- Frölicher has worked extensively with them,

we now call them Frölicher spaces but call the maps smooth. Their definition, where $\mathcal{M}$ is the set of smooth maps from $\mathbb{R}$ to $\mathbb{R}$ follows.

DEFINITION 0.2. The triple $(X, \mathcal{F}, \mathcal{C})$ is a *Frölicher space* (assumed to possess the initial topology $\mathcal{T}$ defined by $\mathcal{F}$) if

- $\mathcal{F} \circ \mathcal{C} = \{f \circ g \mid f \in \mathcal{F}, g \in \mathcal{C}\} \subset \mathcal{M}$.
- $\Phi\mathcal{C} = \{f : X \to \mathbb{R} \mid f \circ c \in \mathcal{M} \text{ for all } c \in \mathcal{C}\} = \mathcal{F}$.
- $\Gamma\mathcal{F} = \{c : \mathbb{R} \to X \mid f \circ c \in \mathcal{M} \text{ for all } f \in \mathcal{F}\} = \mathcal{C}$.

The elements of $\mathcal{C}$ are called *contours*.
Frölicher and Kriegl [4] is our main reference for Frölicher spaces.

EXAMPLE 1. The most important Frölicher spaces are finite-dimensional smooth manifolds where if X is such a manifold, then $\mathcal{C}_X = \{c : \mathbb{R} \to X \mid c \text{ is smooth}\}$ and $\mathcal{F}_X = \{f : X \to \mathbb{R} \mid f \text{ is smooth}\}$.

Although there may at first glance seem little connection between Frölicher and differential spaces, one can show:

THEOREM 0.2. *Let $(X, \mathcal{C}, \mathcal{F})$ be a Frölicher space. Then, $(X, \mathcal{F})$ is a differential space.*

Proof. We use the notation in the definition of differential space. Let $c : \mathbb{R} \to X$ be a contour. As $h_i \circ c$ is smooth, so is $g|_{U_i} \circ c|_{c^{-1}(U_i)}$. Since the sets $c^{-1}(U_i)$ for $i \in K$ cover $\mathbb{R}$, $g \circ c$ must be smooth. But, since $(X, \mathcal{C}, \mathcal{F})$ is a Frölicher space, $g \in \mathcal{F}$.

Suppose that c is again a contour. Since $f_i \circ c$, $i = 1, \ldots, n$, is smooth, the composite $(f_1, \ldots, f_n) \circ c$ is smooth. But then $g \circ (f_1, \ldots, f_n) \circ c$ is smooth. This implies that $g \circ (f_1, \ldots, f_n)$ belongs to $\mathcal{F}$.

DEFINITION 0.3. A map of Frölicher spaces or just smooth map is a map arising from the underlying map of differential spaces. The resulting category of Frölicher spaces is denoted $\mathcal{FRL}$.

It follows that on identifying a Frölicher space with its corresponding differential space that one obtains:

PROPOSITION 0.1. *The category $\mathcal{FRL}$ is a full subcategory of $\mathcal{DSP}$.*

Let $\mathbf{C}$ be a category and $U : \mathbf{C} \to \mathcal{SETS}$ be a faithful functor to the category of sets. Here, the category $\mathbf{C}$ will be either $\mathcal{DSP}$ or $\mathcal{FRL}$ and U will be the forgetful functor. Let $\mathbf{S} = \{h_i | B \to U(A_i)\}_{i \in I}$ be a collection of set maps. Suppose that $B = U(A)$. Then A will be called an initial object for $\mathbf{S}$ if

(1) there are (unique) maps $\{g_i : A \to A_i\}_{i \in I}$ lying in $\mathbf{C}$ such that $U(g_i) = h_i$ for $i \in I$ and
(2) given an object A' and maps $\{k_i \mid A' \to A_i\}_{i \in I}$ such that $U(k_i) = h_i$ for $i \in I$ there is a necessarily unique map $\alpha : A' \to A$ in $\mathbf{C}$ such that $U(\alpha)$ is the identity map on B.

Dually, one can define the notion of final object (for U and a set of maps $\mathbf{S}' = \{h_i \mid U(A_i) \to B\}_{i \in I}$). If any set $\mathbf{S}$ has an initial object, then every set $\mathbf{S}'$ has a final object, U is said to be a topological functor and $\mathbf{C}$ is called topological over $\mathcal{SETS}$. For more detail, see Brümmer [2]. Now, for the forgetful functor $U : \mathcal{FRL} \to \mathcal{SETS}$, Frölicher and Kriegl [4] have shown that $\mathcal{FRL}$ is topological. They also show that $\mathcal{FRL}$ is Cartesian closed. We will show to start with (Section 1) that, for the forgetful functor $U : \mathcal{DSP} \to \mathcal{SETS}$, the category $\mathcal{DSP}$ is topological. If $\mathbf{C}$ is topological, it has all limits and colimits. One merely lifts the limit in $\mathcal{SETS}$ to

C. We will examine these limits and colimits and also initial subobjects in $\mathcal{FRL}$ and $\mathcal{DSP}$. We will see (Section 2) that initial subobjects in $\mathcal{FRL}$ compared to $\mathcal{DSP}$ better reflect any singularities which the subobject possesses. Furthermore, in $\mathcal{FRL}$, tangent cones can be defined for Frölicher spaces and tangent cones can be more appropriate than tangent spaces for the study of singularities in a Frölicher space. For instance, in algebraic geometry, tangent cones are used in 'blowing up' singularities. See [5]. In Section 3 we compare the products in $\mathcal{FRL}$ with those of $\mathcal{DSP}$. This enables us to see in Section 5 that unlike $\mathcal{FRL}$, if $\mathcal{DSP}$ is Cartesian closed (which seems unlikely), the Cartesian closedness does not arise from that of sets. In Section 4 we show that colimits of objects in $\mathcal{FRL}$ are the same irrespective of whether taken in $\mathcal{DSP}$ or $\mathcal{FRL}$. The Cartesian closedness of $\mathcal{FRL}$ enables us to construct in Section 6 the tangent bundle, cotangent bundle and associated type (p, g)-tensor bundles over an object in $\mathcal{FRL}$ in a natural way. We can also construct corresponding cone bundles in $\mathcal{FRL}$. Finally, in our last section, we show how to write down the Einstein tensor for an arbitrary Frölicher differential space.

Although much of the theory is abstract, it leads to many classical type questions.

The author would like to thank the referee for useful advice.

1. The Category $\mathcal{DSP}$ is Topological over Sets

We will show that, for the forgetful functor $U : \mathcal{DSP} \to \mathcal{SETS}$, the category of differential spaces is topological over sets.

If $\mathcal{F}_X^*$ is a collection of maps $X \to \mathbb{R}$, then $\mathcal{F}_X^*$ *generates* a collection of functions denoted $\Sigma\mathcal{F}_X^*$ of maps $X \to \mathbb{R}$ such that $(X, \Sigma\mathcal{F}_X^*)$ is a differential space and if $(X, \mathcal{F}_X')$ is a differential space with $\mathcal{F}_X^* \subset \mathcal{F}_X'$, then $\Sigma\mathcal{F}_X^* \subset \mathcal{F}_X'$. More precisely, one can show, see [11], that $\Sigma\mathcal{F}_X^*$ is the set of all $g : X \to \mathbb{R}$ such that, for each $P \in X$ and on some open set U with $P \in U$ (for the initial topology given by $\mathcal{F}_X^*$),

$$g|_U = h \circ (f_1|_U, \ldots, f_n|_U)$$

for some smooth function $h : \mathbb{R}^n \to \mathbb{R}$, $f_1, \ldots, f_n \in \mathcal{F}_x^*$ and $n \in \mathbb{N}$.

Let now $\{h_i : Y \to X_i\}_{i\in I}$ be a collection of set maps where X_i is a differential space with structure functions $\mathcal{F}_{X_i}$. Set

$$\mathcal{F}_Y^* = \{g : Y \to \mathbb{R} | g = f \circ h_i \text{ with } f \in \mathcal{F}_{X_i} \text{ and } i \in I\}.$$

Then, we show:

LEMMA 1.1. *The differential space $(Y, \Sigma\mathcal{F}_Y^*)$ is the initial object for the collection of maps $\{h_i : Y \to X_i\}_{i\in I}$.*

Proof. Clearly, the maps $h_i : (Y, \Sigma\mathcal{F}_Y^*) \to (X_i, \mathcal{F}_{X_i})$, where $i \in I$, are maps of differential spaces.

Let $(Y, \mathcal{F}')$ be a differential space such that

$$h_i : (Y, \mathcal{F}') \to (X_i, \mathcal{F}_{X_i})$$

is a map of differential spaces for each $i \in I$. Then, for $f \in \mathcal{F}_{X_i}$, $f \circ h_i \in \mathcal{F}'$ for each $i \in I$. But then $\mathcal{F}_Y^* \subset \mathcal{F}'$ and, from the above discussion, $\Sigma \mathcal{F}_Y^* \subset \mathcal{F}'$.

As we mentioned in the introduction, from the lemma it follows that $\mathcal{DSP}$ must thus also have final objects and be topological for the forgetful functor $U : \mathcal{DSP} \to \mathcal{SETS}$.

2. Initial Subobjects and Tangents in $\mathcal{FRL}$ and $\mathcal{DSP}$

Initial subobjects of an object in $\mathcal{FRL}$ exist. Extracting from the more general construction in Frölicher and Kriegl [4], let $(X, \mathcal{C}_X, \mathcal{F}_X)$ be a Frölicher space and A a subset of X. Then, the inclusion $i_A : A \to X$ places an initial structure on A where the resulting Frölicher space is $(A, \mathcal{C}_A, \mathcal{F}_A)$ with

$$\mathcal{C}_A = \{c : \mathbb{R} \to A \mid i_A \circ c \in \mathcal{C}_X\},$$
$$\mathcal{F}_A = \Phi\Gamma\{f \circ i_A \mid f \in \mathcal{F}_X\}.$$

With this structure A is called a Frölicher subspace of X.

Here the description of $\mathcal{C}_A$ is most useful. Initial here means that if

$$i_A : (A, \mathcal{C}^1, \mathcal{F}^1) \to (X, \mathcal{C}_X, \mathcal{F}_X)$$

is a map of Frölicher spaces, then the identity map $I_A : (A, \mathcal{C}^1, \mathcal{F}^1) \to (A, \mathcal{C}_A, \mathcal{F}_A)$ defines a smooth map.

EXAMPLE 2. We use the above notation.

- Let $X = \mathbb{R}$ and $A = \mathbb{Q}$ denote the rationals. Then, $\mathcal{C}_A$ consists of the constant maps and $\mathcal{F}_A$ thus consists of all functions. Since $\mathbb{Q}$ then has the discrete topology, we then call it a discrete Frölicher space. It is reasonably clear that there are examples of fractal curves which are Frölicher subsets of $\mathbb{R}^2$ and having the discrete Frölicher structure.
- Let C be the curve and subset of $\mathbb{R}^2$ defined in polar coordinates $r = 1/\theta, \theta \in (0, \infty)$ but also containing $(0, 0)$. A straightforward computation of the arc length $l(\theta)$ of the curve $r = 1/\theta$ from θ_0 to θ $(\theta > \theta_0)$ shows that $l_{(\theta)} \geq \ln \theta/\theta_0$. Since any smooth curve $c : [0, 1] \to \mathbb{R}^2$ must have finite arc length, it follows that there is no smooth curve connecting $(1/\theta_0, \theta_0)$ to $(0, 0)$. Although C is a connected subspace of $\mathbb{R}^2$ in the usual topology, a contour into C thus maps either to $(0, 0)$ or to the complement of $(0, 0)$. Hence, a scalar on C can take any value at $(0, 0)$ irrespective of values at other point. The Frölicher subspace $(A, \mathcal{C}_A, \mathcal{F}_A)$ thus is disconnected topologically.

Before we present other examples we define the tangent space at a point of a differential space according to the usual definition and an alternate definition of tangent at a point of a Frölicher space. This last definition follows a common way for defining tangents in the theory of differentiable manifolds.

DEFINITION 2.1. (1) Let $(X, \mathcal{F})$ be a differential space. A *tangent vector* V at $P \in X$ is a derivation $V : \mathcal{F} \to \mathbb{R}$ at P, i.e. a linear map such that

$$V(fg) = f(P)V(g) + g(P)V(f).$$

The set of such vectors form the tangent space TX_P of X at P.

(2) Let $(X, \mathcal{F})$ be a differential space and $P \in X$. Let $c \in \Gamma\mathcal{F}$ and suppose that $c(a) = P$ and $f \in \mathcal{F}$. Suppose that V_c is the derivation defined by setting

$$V_c(f) = \lim_{t \to a} \frac{f \circ c(t) - f \circ c(a)}{t - a}.$$

Set $TCX_P = \{V_c \mid c \in \Gamma\mathcal{F}, c(a) = P\}$. It is easy to see that $TCX_P \subset TX_P$. We call TCX_P the *tangent cone* to X at P.

In fact, as examples below show, TCX_P need not be a vector space.

EXAMPLES 3.

1. The rationals as a Frölicher subspace of $\mathbb{R}$ have trivial tangent spaces equal to their tangent cones: since contours must have constant values, the tangent cones must be trivial. Let $q \in \mathbb{Q}$. The function $f : \mathbb{Q} \to \mathbb{R}$ such that $f(q) = 1$ and $f(r) = 0$ if $r \neq q$ belongs to $\mathcal{F}_{\mathbb{Q}}$. Since $f^2 = f$, one can show that, for any derivation D at q, $D(f) = 0$. Let $g \in \mathcal{F}_{\mathbb{Q}}$ and $g' = fg$. Then, $D(g') = f(g)D(g)$. Since $g(q)g' = (g')^2$, $D(g') = 0$ and thus $D(g) = 0$. Hence, the tangent space at q is trivial.
2. Except at $(0, 0)$ the Frölicher curve C above has a one-dimensional tangent space (= tangent cone). At $(0, 0)$ the tangent space (= tangent cone) is trivial (as in 1).
3. Let B be the Frölicher subspace of $\mathbb{R}^2$ defined by $xy = 0$. The tangent cone to B agrees with the tangent space and is one-dimensional except at $(0, 0)$ where the tangent cone is B and the tangent space is $\mathbb{R}^2$. A scalar on B is a function $f : B \to \mathbb{R}$ such that f is smooth on the x and y-axes, respectively.
4. Let V be an algebraic subvariety of $\mathbb{R}^n$ defined by polynomial equations $f_1, \ldots, f_m$ and $P = (a_1, \ldots, a_n) \in V$. Let p_i be the degree of the first non-vanishing homogeneous term in f_i regarded as a polynomial in the $x_j - a_j$ for $i = 1, \ldots, m$ and $j = 1, \ldots, n$. Set $p = \min\{p_i\}_{i=1,\ldots,m}$. If g is a polynomial in the polynomial ring R generated by the $x_j - a_j$ for $j = 1, \ldots, n$, let $(g)_k$ denote the term of degree k in the same variables. Consider the ideal I generated by $(f_1)_p, \ldots, (f_m)_p$ in $\mathbb{R}$. The tangent cone to V at P is then the zero set of I in $\mathbb{R}^n$. One can show that the tangent space to V at P is the zero set of

$(f_1)_1, \ldots, (f_m)_1$. Thus, if $m = 1$ and $f_1 = x^2 + y^2$, then $TCV_{(0,0)} = \{(0, 0)\}$ and $TV_{(0,0)} = \mathbb{R}^2$.

Let now X be a differential space, $A \subset X$ and $\mathcal{F}_A^* = \{f \circ i_A \mid f \in \mathcal{F}_X\}$ where $i_A : A \to X$ is the inclusion. Then A with scalars $\Sigma(\mathcal{F}_A^*)$ is called a differential subspace of X.

EXAMPLES 4. We consider examples given above in $\mathcal{DSP}$.

1. $\mathbb{Q}$ as a differential subspace of $\mathbb{R}$ has scalars $f : \mathbb{Q} \to \mathbb{R}$ which are locally in the usual topology the restrictions of locally smooth functions on $\mathbb{R}$. Thus, the tangent space to a point $q \in \mathbb{Q}$ is the same as the tangent space when q is regarded as a point of $\mathbb{R}$ and one-dimensional.
2. The curve C in Examples 2, as a differential subspace of $\mathbb{R}^2$, is connected. Scalars at (0, 0) are the restrictions locally of smooth functions on $\mathbb{R}^2$ and hence cannot take on arbitrary values at (0, 0). For $f \in \mathcal{F}_C$, let

$$\left.\frac{\partial}{\partial x}\right|_{C,(0,0)} f = \lim_{h\to 0,(h,0)\in C} \frac{f(0+h, 0) - f(0, 0)}{h}.$$

Since f is the restriction locally of an element in $\mathcal{F}_{\mathbb{R}^2}$, this definition makes sense. One readily shows that $\frac{\partial}{\partial x}|_{C,(0,0)}$ is a derivation on $\mathcal{F}_C$. The derivation $\frac{\partial}{\partial y}|_{C,(0,0)}$ is defined similarly. As $\frac{\partial}{\partial x}|_{C,(0,0)} x|_C = 1$, $\frac{\partial}{\partial x}|_{C,(0,0)} y|_C = 0$, $\frac{\partial}{\partial y}|_{C,(0,0)} y|_C = 1$ and $\frac{\partial}{\partial y}|_{C,(0,0)} x|_C = 0$, the tangent space of C at (0, 0) is at least two-dimensional. To each derivation $\underline{D}$ on $\mathcal{F}_C$, one can associate a derivation D on $\mathcal{F}_{\mathbb{R}^2}$ by setting $D(g) = \underline{D}(g|_C)$. As D must have the form $D = a\frac{\partial}{\partial x}|_{(0,0)} + b\frac{\partial}{\partial y}|_{(0,0)}$, it follows that $\underline{D} = a\frac{\partial}{\partial x}|_{C,(0,0)} + b\frac{\partial}{\partial y}|_{C,(0,0)}$. Hence, the tangent space of C at (0, 0), as a differential space, is two-dimensional.

3. Products in $\mathcal{FRL}$ and $\mathcal{DSP}$

Suppose that one is given a collection $\{(X_i, \mathcal{F}_i)\}_{i\in I}$ of differential spaces or a collection $\{(X_i, \mathcal{C}_i, \mathcal{F}_i)\}_{i\in I}$ of Frölicher spaces. Let $\prod_{i\in I} X_i$ be the set product of the sets $\{X_i\}_{i\in I}$ and $\pi_j : \prod_{i\in I} X_i \to X_j$ for $j \in I$ denote the projection map. The initial structure on $\mathbf{P} = \prod_{i\in I} X_i$ in both $\mathcal{DSP}$ and $\mathcal{FRL}$ is generated by the set

$$\mathcal{F}_{\mathbf{P}}^* = \bigcup_{i\in I} \{f_i \circ \pi_i \mid f_i \in \mathcal{F}_i,\ i \in I\}.$$

In $\mathcal{DSP}$, as observed above, one obtains $\Sigma\mathcal{F}_{\mathbf{P}}^*$ as the initial structure on $\mathbf{P}$, whereas in $\mathcal{FRL}$ one obtains

$$\mathcal{C}_{\mathbf{P}} = \{c : \mathbb{R} \to \mathbf{P} \mid \text{ if } c(t) = (c_i(t))_{i\in I}, \text{ then } c_i \in \mathcal{C}_i\},$$
$$\mathcal{F}_{\mathbf{P}} = \Phi\Gamma\mathcal{F}_{\mathbf{P}}^*.$$

Here, the requirement that each component of c is a smooth map is most useful.

We now provide an example that shows that finite products in $\mathcal{FRL}$ and $\mathcal{DSP}$ need not be the same.

EXAMPLE 5. Let $\mathbb{R}^{\mathbb{N}}_{\oplus}$ denote the Frölicher space and thus differential space whose underlying set is $\mathbb{R}^{\mathbb{N}}$ and whose Frölicher space structure is generated by the set $\mathfrak{C}^0$ of curves which is equal to

$$\{(x_i(t))_{i\in\mathbb{N}} \in \mathfrak{C}_{\mathbb{R}^{\mathbb{N}}} \mid \text{ except for finitely many } i, x_i(t) \text{ is identically } 0\}.$$

Let $\alpha : \mathbb{R} \to \mathbb{R}$ be a smooth function such that $\alpha(t) = t$ if $-\frac{3}{4} \leq t \leq \frac{3}{4}$ and $|\alpha(t)| \leq 1$ otherwise. It is clear that the function $l : \mathbb{R}^{\mathbb{N}}_{\oplus} \to \mathbb{R}$ defined by setting

$$l((x_i)_{i\in\mathbb{N}}) = \sum_{i=1}^{\infty} \frac{1}{2^i}\alpha(x_i)$$

is a scalar on $\mathbb{R}^{\mathbb{N}}_{\oplus}$. Let $\mathbb{R}^{\mathbb{N}}_{\oplus\oplus}$ denote a second copy of $\mathbb{R}^{\mathbb{N}}_{\oplus}$ with coordinates $(y_i)_{i\in\mathbb{N}}$. Then we show:

LEMMA 3.1. *The function*

$$k((x_i, y_i)_{(i,j)\in\mathbb{N}\times\mathbb{N}}) = \sum_{i=1}^{\infty} \frac{1}{2^i}\alpha(x_i)\frac{1}{2^i}\alpha(y_i) \quad \textit{is}$$

(1) *a scalar on* $\mathbb{R}^{\mathbb{N}}_{\oplus} \times \mathbb{R}^{\mathbb{N}}_{\oplus\oplus}$ *for the Frölicher space product structure but*
(2) *not for the differential space product structure where the scalars near the zero* $\underline{0}\times\underline{0}$ *of* $\mathbb{R}^{\mathbb{N}}_{\oplus}\times\mathbb{R}^{\mathbb{N}}_{\oplus\oplus}$ *are of the form* $f(g_1, \ldots, g_n, h_1, \ldots, h_m)$ *with* $f : \mathbb{R}^{n+m} \to \mathbb{R}$ *a smooth function,* $g_1, \ldots, g_n \in \mathcal{F}_{\mathbb{R}^{\mathbb{N}}_{\oplus}}$ *and* $h_1, \ldots, h_m \in \mathcal{F}_{\mathbb{R}^{\mathbb{N}}_{\oplus\oplus}}$.

Proof. Since every contour $c : \mathbb{R} \to \mathbb{R}^{\mathbb{N}}_{\oplus} \times \mathbb{R}^{\mathbb{N}}_{\oplus\oplus}$ has the form $c(t) = (x_i(t), y_j(t))_{i,j\in\mathbb{N}}$ where all but finitely many $x_i(t)$ and $y_j(t)$ are identically 0, the first statement is clear. To see the last assertion, suppose that

$$f(g_1, \ldots, g_n, h_1, \ldots, h_m) = \sum_{i=1}^{\infty} \frac{1}{2^i}\alpha(x_i)\frac{1}{2^i}\alpha(y_i). \tag{1}$$

on some open set $U = F^{-1}(0, 1)$ where $F = \underline{f}(\underline{g}_1, \ldots, \underline{g}_{n_1}, \underline{h}_1, \ldots, \underline{h}_{m_1})$ with $\underline{f} : \mathbb{R}^{n_1+m_1} \to \mathbb{R}$ a smooth function, $\underline{g}_1, \ldots, \underline{g}_{n_1} \in \mathcal{F}_{\mathbb{R}^{\mathbb{N}}_{\oplus}}$ and $\underline{h}_1, \ldots, \underline{h}_{m_1} \in \mathcal{F}_{\mathbb{R}^{\mathbb{N}}_{\oplus\oplus}}$. Let $i^r_{\oplus} : \mathbb{R}^r \to \mathbb{R}^{\mathbb{N}}_{\oplus}$ be the inclusion where, if $i^r_{\oplus}(x_1, \ldots, x_r) = (z_1, \ldots, z_p, \ldots)$, then $z_j = x_j$ for $j = 1, \ldots, r$ and $z_j = 0$ if $j > r$. Defining $i^r_{\oplus\oplus} : \mathbb{R}^r \to \mathbb{R}^{\mathbb{N}}_{\oplus\oplus}$ in the same way, we write i^r instead of $i^r_{\oplus}$ or $i^r_{\oplus\oplus}$ when there is no confusion.

Applying the map $I^r = i^r_{\oplus} \times i^r_{\oplus\oplus} : \mathbb{R}^r \times \mathbb{R}^r \to \mathbb{R}^{\mathbb{N}}_{\oplus} \times \mathbb{R}^{\mathbb{N}}_{\oplus\oplus}$ to Equation (1), one obtains $f(g_1, \ldots, g_n, h_1, \ldots, h_m) \circ I^r(x_1, \ldots, x_r, y_1, \ldots, y_r) = f(g_1 \circ$

$i^r(x_1, \ldots, x_r), \ldots, g_n \circ i^r(x_1, \ldots, x_r), h_1 \circ i^r(y_1, \ldots, y_r), \ldots, h_m \circ i^r(y_1, \ldots, y_r)) = \sum_{i=1}^r \frac{1}{2^i}\alpha(x_i)\frac{1}{2^i}\alpha(y_i) = \frac{1}{2^i}(x_i)\frac{1}{2^i}(y_i)$ for $|x_i|, |y_i| < \frac{3}{4}$, $i = 1, \ldots, r$ and $(x_1, \ldots, x_r, y_1, \ldots, y_r) \in V^r = (I^r)^{-1}(U)$.

Let $\underline{0}_r$ be the zero of $\mathbb{R}^r$. Using Boman's theorem (see [1]) one sees that, for the usual differential structure on $\mathbb{R}^r$, the inclusions $i^r_\oplus$ and $i^r_{\oplus\oplus}$ are smooth in the category of differential spaces. Hence,

(1) $V^r = (I^r)^{-1}(U)$ is open for the usual topology on $\mathbb{R}^r \times \mathbb{R}^r$,
(2) the zero $\underline{0}_r \times \underline{0}_r$ of $\mathbb{R}^r \times \mathbb{R}^r$ is in V^r and
(3) $g_1 \circ i^r_\oplus, \ldots, g_n \circ i^r_\oplus, h_1 \circ i^r_{\oplus\oplus}, \ldots, g_m \circ i^r_{\oplus\oplus}$ are smooth functions on $\mathbb{R}^r$.

Thus,

$$\begin{aligned} H_1 &= f(g_1 \circ i^r(x_i), \ldots, g_n \circ i^r(x_i), h_i \circ i^r(y_i), \ldots, h_m \circ i^r(y_i)) \\ &= \sum_{i=1}^r \frac{1}{2^i}x_i\frac{1}{2^i}y_i \end{aligned} \tag{2}$$

on some open subset of $\underline{0}_r \times \underline{0}_r \in \mathbb{R}^r \times \mathbb{R}^r$ (for the usual topology). Similarly,

$$\begin{aligned} H_2 &= f(g_1 \circ i^r(y_i), \ldots, g_n \circ i^r(y_i), h_i \circ i^r(x_i), \ldots, h_m \circ i^r(x_i)) \\ &= \sum_{i=1}^r \frac{1}{2^i}x_i\frac{1}{2^i}y_i \end{aligned} \tag{3}$$

on some open subset of $\mathbb{R}^r \times \mathbb{R}^r$ containing $\underline{0}_r \times \underline{0}_r$. Note that here $g_1, \ldots, g_n$ are viewed as smooth functions on $\mathbb{R}^{\mathbb{N}}_{\oplus\oplus}$ and $h_1, \ldots, h_m$ are viewed as smooth functions on $\mathbb{R}^{\mathbb{N}}_\oplus$. Hence,

$$H = \frac{H_1 + H_2}{2} = \sum_{i=1}^r \frac{1}{2^i}x_i\frac{1}{2^i}y_i \tag{4}$$

on some open subset of $\mathbb{R}^r \times \mathbb{R}^r$ containing $\underline{0}_r \times \underline{0}_r$. Furthermore, H is of the form $F(k_1 \circ i^r_\oplus, \ldots, k_{n+m} \circ i^r_\oplus, k_1 \circ i^r_{\oplus\oplus}, \ldots, k_{n+m} \circ i^r_{\oplus\oplus})$ where

(1) $k_i = g_i$ for $i = 1, \ldots, n$ and $k_{i+n} = h_i$ for $i = 1, \ldots, m$ for the first $n + m$ coordinates are viewed as smooth functions on $\mathbb{R}^{\mathbb{N}}_\oplus$ and
(2) $k_i = g_i$ for $i = 1, \ldots, n$ and $k_{i+n} = h_i$ for $i = 1, \ldots, m$ for the second $n + m$ coordinates are viewed as smooth functions on $\mathbb{R}^{\mathbb{N}}_{\oplus\oplus}$.

Thus in Equation (2), one can assume that $n = m$, $g_i = h_i$ for $i = 1, \ldots, n$ and f is a smooth function with domain $\mathbb{R}^n \times \mathbb{R}^n$. Again, $\underline{0}$ is the zero vector of $\mathbb{R}^{\mathbb{N}}$. In addition, we let $X_1, \ldots, X_n$ be the coordinates on the first copy of $\mathbb{R}^n$ and $Y_1, \ldots, Y_n$ be the coordinates on the second copy of $\mathbb{R}^n$. Note that, using the above construction, one has

$$f(X_1, \ldots, X_n, Y_1, \ldots, Y_n) = f(Y_1, \ldots, Y_n, X_1, \ldots, X_n). \tag{5}$$

Suppose that $g_l(\underline{0}) = c_l$ for $l = 1, \ldots, n$. Let

$$\begin{aligned}\underline{f}(X_1, \ldots, X_n, Y_1, \ldots, Y_n) \\ = f(X_1 + c_1, \ldots, X_n + c_n, Y_1 + c_1, \ldots, Y_n + c_n).\end{aligned}$$

Then,

$$\underline{f}(g_1(x_i) - c_1, \ldots, g_n(x_i) - c_n, g_1(y_i) - c_1, \ldots, g_n(y_i) - c_n) = \sum_{i=1}^{r} \frac{1}{2^i} x_i \frac{1}{2^i} y_i$$

on some open subset of $\mathbb{R}^r \times \mathbb{R}^r$ containing $\underline{0}_r \times \underline{0}_r$. Thus, without loss of generality, one can assume that $g_l \circ i^r_{\oplus}(\underline{0}_r) = g_l(\underline{0}) = 0$ and $g_l \circ i^r_{\oplus\oplus}(\underline{0}_r) = g_l(\underline{0}) = 0$ for $l = 1, \ldots, n$.

Near $\underline{0}_r \in \mathbb{R}^r$ and, for $l = 1, \ldots, n$, $g_l(x_1, \ldots, x_r)$ has a Taylor expansion

$$g_l(x_1, \ldots, x_r) = \sum_{i=1}^{r} a_{il} x_i + H^1(g_l)(x_1, \ldots, x_r),$$

where $H^1(g_l)(x_1, \ldots, x_r)$ contains the higher-order nonlinear terms of $g_l(x_1, \ldots, x_r)$. A similar result holds on replacing $x_1, \ldots x_r$ by $y_1, \ldots, y_r$. Also, near $\underline{0}_{2n} \in \mathbb{R}^{2n}$, as a consequence of Equation (5), f has a Taylor expansion

$$f(X_1, \ldots, X_n, Y_1, \ldots, Y_n) = C + \sum_{p=1}^{n} \beta_p X_p + \sum_{q=1}^{n} \beta_1 Y_q + H^1(f),$$

where $H^1(f)$ consists of higher-order nonlinear terms. Substituting into f, one obtains

$$\begin{aligned}C + \sum_{p=1}^{n} \sum_{i=1}^{r} \beta_p (a_{ip} x_l + H^1(g_p)(x_1, \ldots, x_r)) + \sum_{q=1}^{n} \sum_{i=1}^{r} \beta_q (b_{iq} y_l + \\ + H^1(g_q)(y_1, \ldots, y_r)) + H^1(f)(x_1, \ldots, x_r, y_1, \ldots, y_r) = \sum_{i=1}^{r} \frac{1}{2^i} x_i \frac{1}{2^i} y_i.\end{aligned}$$

Since $H^1(g_p)(x_1, \ldots, x_r)$ (resp., $H^1(g_p)(y_1, \ldots, y_r)$) is a function of $x_1, \ldots, x_r$ alone (resp. $y_1, \ldots, y_r$ alone), one has

$$H^1(f)(x_1, \ldots, x_r, y_1, \ldots, y_r) = \sum_{i=1}^{r} \frac{1}{2^i} x_i \frac{1}{2^i} y_i$$

and thus one can assume that f has a Taylor expansion

$$f = \sum_{u=1}^{n} \sum_{v=1}^{n} \gamma^{uv} X_u Y_v + H^2(f), \tag{6}$$

where $H^2(f)$ contains terms of order strictly larger than 2.

Substituting into Equation (6), since higher degree terms must vanish, one obtains as before

$$\sum_{u=1}^{n}\sum_{v=1}^{n}\gamma^{uv}\Big(\sum_{i=1}^{r}a_{iu}x_i\Big)\Big(\sum_{j=1}^{r}a_{jv}y_j\Big)=\sum_{i=1}^{r}\frac{1}{2^i}x_i\frac{1}{2^i}y_i \tag{7}$$

or for all r

$$\sum_{i=1}^{r}\sum_{j=1}^{r}\Big(\sum_{u=1}^{n}\sum_{v=1}^{n}\gamma^{uv}a_{iu}a_{jv}\Big)x_iy_j=\sum_{i=1}^{r}\frac{1}{2^i}x_i\frac{1}{2^i}y_i. \tag{8}$$

From Equation (8) one obtains a matrix equation

$$\begin{pmatrix} a_{11} & \cdots & a_{1n}\\ \vdots & \vdots & \vdots\\ a_{r1} & \cdots & a_{rn}\end{pmatrix}(\gamma^{uv})_{u,v=1,\dots,n}\begin{pmatrix} a_{11} & \cdots & a_{1r}\\ \vdots & \vdots & \vdots\\ a_{n1} & \cdots & a_{nr}\end{pmatrix}=I_r \tag{9}$$

where I_r is a matrix whose only non-zero elements are diagonal elements and the element in the ith row and ith column is $1/2^i$. By taking $r=n$, one sees that

$$A^r=\begin{pmatrix} a_{11} & \cdots & a_{1n}\\ \vdots & \vdots & \vdots\\ a_{n1} & \cdots & a_{nn}\end{pmatrix}$$

and $\Gamma^r=(\gamma^{uv})_{u,v=1,\dots,n}$ are invertible.

Let now $r>n$. Returning to (8), one sees that

$$\sum_{u,v=1}^{n}\gamma^{uv}a_{iu}a_{jv}=0$$

if $i\neq j$. It follows that $A^n\Gamma^n(a_{vj})_{v=1,\dots,n;\,j=n+1,\dots,r}$ is a $n\times(n-r)$ zero matrix. As $A^n\Gamma^n$ is invertible, $(a_{vj})_{v=1,\dots,n;\,j=n+1,\dots,r}$ is a zero matrix. Hence, for all $r>n$ and $k=1,\dots,n$, $g_k\circ i^r_\oplus(x_1,\dots,x_r)=\sum_{j=1}^{n}a_{jk}y_j+H^1(g_k)$ where $H^1(g_k)$ contains the nonlinear terms of $g_k\circ i^r_\oplus(x_1,\dots,x_r)$. But then clearly Equalion (7) is impossible for r sufficiently large and we have a contradiction.

Let $\mathbb{R}_1$ denote the reals but with the smooth structure where

$$\mathcal{F}_{\mathbb{R}_1}=\Phi\Gamma(\mathcal{F}_{\mathbb{R}}\cup\{|t|\}).$$

Clearly, $|x|\in\mathcal{F}_{\mathbb{R}_1}$. It seems likely that $\sqrt{x^2+y^2}$ is a scalar for the Frölicher product $\mathbb{R}_1\times\mathbb{R}_1$ but not the differential product $\mathbb{R}_1\times\mathbb{R}_1$. No proof of this fact has been found.

4. Coequalizers and Coproducts in $\mathcal{FRL}$ and $\mathcal{DSP}$

Let $\{X_i\}_{i\in I}$ be a set of differential or Frölicher spaces with $\mathcal{F}_i$ the set of scalars on X_i $(i \in I)$, $\hat{\bigcup}_{i\in I} X_i$ the disjoint union and $i_{X_i} : X_i \to \hat{\bigcup}_{i\in I} X_i$ the inclusion map. Place the differential or Frölicher final structure on $\hat{\bigcup}_{i\in I} X_i$ given by the set $\{i_{X_i}\}_{i\in I}$. The resulting differential or Frölicher space is the coproduct of $\{X_i\}_{i\in I}$ in the respective category, denoted by $\coprod_{i\in I} X_i$ and

$$\mathcal{F}_{\coprod_{i\in I} X_i} = \left\{ f : \coprod_{i\in I} X_i \to \mathbb{R} \mid \text{ for each } i \in I,\ f|_{X_i} \in \mathcal{F}_i \right\}$$

is the set of scalars for the coproduct. See [4, 11]. Thus, from the above description, it follows that the inclusion of $\mathcal{FRL}$ into $\mathcal{DSP}$ is coproduct preserving.

EXAMPLES 6. In Examples 2.1 (the setting is the category $\mathcal{FRL}$)

1. $\mathbb{Q}$ can be identified with a countable union of points.
2. C can be identified with a coproduct of $\mathbb{R}$ and a point.

A similar analysis holds for coequalizers. Let $f, g : X \to Y$ be two differentiable maps between differential or Frölicher spaces and $Q = Y/\simeq$ where $\simeq$ is the smallest equivalence relation with the property that $y \simeq y'$ if $y = f(x)$ and $y' = g(x)$ for some $x \in X$. Furthermore, suppose that $q : Y \to Q$ is the set quotient map and Q acquires a final structure in differential spaces or Frölicher spaces via this quotient map. In both differential spaces, as it is easy to see, and Frölicher spaces (see [4]) $\mathcal{F}_Q = \{f : Q \to \mathbb{R} \mid f \circ q \in \mathcal{F}_Y\}$. Thus, the inclusion functor of $\mathcal{FRL}$ into $\mathcal{DSP}$ preserves coequalizers and, because every colimit can be formed from coproducts and coequalizers, the inclusion functor preserves colimits.

Now, for a differential space $\hat{X} = (X, \mathcal{F}_X)$ let $\Upsilon(\hat{X}) = (X, \Gamma\mathcal{F}_X, \Phi\Gamma\mathcal{F}_X)$ be the associated Frölicher space. One can prove:

LEMMA 4.1. *The association of a Frölicher space $\Upsilon(\hat{X})$ to a differential space $\hat{X}$ induces a functor $\mathcal{DSP} \to \mathcal{FRL}$.*

Proof. Let $f : \hat{X} \to \hat{Y}$ be a map of differential spaces. We need to show that f is a smooth map of the associated Frölicher spaces. The rest of the proof is transparent. Let $c \in \Gamma\mathcal{F}_X$. We must then show that $f \circ c \in \Gamma\mathcal{F}_Y$. But, $f \circ c \in \Gamma\mathcal{F}_Y$ if and only if $h \circ f \circ c$ is smooth for all $h \in \mathcal{F}_Y$. But, $h \circ f \in \mathcal{F}_X$ and this is thus the case.

For a differential space $\hat{X}$, the identity set map $I_X : \Upsilon(\hat{X}) \to \hat{X}$ is a map of differential spaces which one can use to satisfy the solution set condition in Freyd's Adjoint Functor Theorem (see [6]). One thus obtains:

PROPOSITION 4.1. *The inclusion functor $\mathcal{FRL} \to \mathcal{DSP}$ has a right adjoint.*

It is, in fact, easy to show that the right adjoint is the functor Υ.

Note that, for differential spaces, tangent cones are defined to be the tangent cones of the associated Frölicher spaces.

5. The Non-Cartesian Closedness of $\mathcal{DSP}$

The category of topological spaces is not Cartesian closed; so, in algebraic topology, where Cartesian closedness is useful, the category of compactly generated spaces was considered as an alternative (see [6]). The rationals form a non-compactly generated space and in the discussion below we try to see why hom-sets such as $Hom_{\mathcal{DSP}}(\mathbb{Q}, \mathbb{R})$ should provide examples where the Cartesian closedness of $\mathcal{DSP}$ fails. We will show that two possible ways to begin constructing Cartesian closedness for $\mathcal{DSP}$ (where $\mathbb{Q}$ plays a role) fail and that the set mappings defining sets and $\mathcal{FRL}$ to be Cartesian closed do not work for the category of differential spaces.

• Let X be a differential space and $\hat{X}$ the associated smooth space. The identity map $I : \hat{X} \to X$ is a map of differential spaces which induces a map

$$I : Hom_{\mathcal{DSP}}(X, \mathbb{R}) \to Hom_{\mathcal{DSP}}(\hat{X}, \mathbb{R}) = Hom_{\mathcal{FRL}}(\hat{X}, \mathbb{R}).$$

The set $Hom_{\mathcal{FRL}}(\hat{X}, \mathbb{R})$, since $\mathcal{FRL}$ is Cartesian closed, can be given an $\mathcal{FRL}$ structure and, via I, one can put an initial structure on $\mathcal{G} = Hom_{\mathcal{DSP}}(X, \mathbb{R})$ in $\mathcal{DSP}$. Will $\mathcal{G}$ with this initial structure be under set adjunction a Cartesian closed object in $\mathcal{DSP}$? In other words, is there a natural bijection

$$\theta : Hom_{\mathcal{DSP}}(Z, \mathcal{G}) \to Hom_{\mathcal{DSP}}(Z \times X, \mathbb{R}),$$

where $\theta(f)(z, x) = f(z)(x)$? Let $X = \mathbb{Q}$ be viewed as a differential subspace of $\mathbb{R}$. Since $\hat{\mathbb{Q}}$ is discrete, $\hat{\mathcal{G}} = Hom_{\mathcal{DSP}}(\hat{\mathbb{Q}}, \mathbb{R})$ consists of all maps $\hat{\mathbb{Q}} \to \mathbb{R}$. Also, $\underline{I} : \mathcal{G} \to \hat{\mathcal{G}}$ is the inclusion map. The scalars $\mathcal{F}_{\mathcal{G}}$ on $\mathcal{G}$ are generated by functions of the form $f \circ \underline{I}$ where $f \in \mathcal{F}_{\hat{\mathcal{G}}}$. However, $f \in \mathcal{F}_{\hat{\mathcal{G}}}$ if and only if

$$f \in Hom_{\mathcal{DSP}}\left(\coprod_{g \in \mathbb{Q}} q, \mathbb{R}\right) \cong \prod_{q \in \mathbb{Q}} \mathbb{R} = \hat{\mathbb{R}},$$

where f is identified with $(f(q)) \in \hat{\mathbb{R}}$. Let $Z = \mathbb{R}$. A map $l : \mathbb{R} \to \mathcal{G}$ sending t to l_t is smooth if and only if $\underline{I} \circ l$ is smooth and using our earlier statement about the functions on $\mathcal{G}$ if and only if

$$\underline{I}(l_t) = (l_t(q))_{q \in \mathbb{Q}} \tag{10}$$

is smooth for each t in each coordinate. However, the corresponding map $\hat{l} : \mathbb{R} \times \mathbb{Q} \to \mathbb{R}$ in differential spaces is differentiable if and only if $\hat{l}(t, q) = l_t(q)$ is smooth in an $\mathbb{R}^2$-neighborhood of each point $(t, q) \in \mathbb{R} \times \mathbb{Q}$. Clearly, however, smoothness of the expression in (10) above does not imply the smoothness required for $\hat{l}$. Thus, Cartesian closedness in $\mathcal{DSP}$ cannot be obtained by putting an initial structure on $\mathcal{G}$.

• We now attempt to make $\mathcal{G}$, as above, into a differential space by identifying $\mathcal{G}$ with the set of maps $f : \mathbb{R} - I \to \mathbb{R}$ where I ranges over the discrete subsets of $\mathbb{R}$ consisting exclusively of irrationals. This identification is possible since every $f \in \mathcal{G}$ is smooth in a neighborhood of each rational. Let I be a discrete subset of $\mathbb{R}$ consisting exclusively of irrationals. Then, set

$$\mathcal{H}_I = \{f : \mathbb{R} \to \mathbb{R} \mid f \text{ is smooth except at points in } I\}.$$

Notice that every map in $\mathcal{H}_I$ restricts to a unique map in $\mathcal{G}$ and that every map in $\mathcal{G}$ that extends to a map in $\mathcal{H}_I$ does so uniquely. One can thus view $\mathcal{G}$ as a direct limit in sets of the $\mathcal{H}_I$. Let

$$\tilde{I} = \{(i, j) \mid i, j \in I,\ i < j \text{ and no } l \in I \text{ is strictly between } i \text{ and } j\}.$$

It is easy to see that $\mathcal{H}_I$, since $\mathcal{FRL}$ is Cartesian closed, can be endowed with the Frölicher structure of the Frölicher space

$$Hom_{\mathcal{FRL}}\left(\coprod_{j \in \tilde{I}} \mathbb{R}^j, \mathbb{R}\right),$$

where $\mathbb{R}^j$ is a copy of $\mathbb{R}$. Since $\mathcal{DSP}$ is topological over sets, $\mathcal{G}$ can be viewed as a differential space and the direct limit of the $\mathcal{H}_I$ in $\mathcal{DSP}$, where $I < I'$ if $I \subset I'$. We then ask if $\mathcal{G}$ with this structure is a Cartesian object in $\mathcal{DSP}$. Since coproducts and coequalizers are the same taken over Frölicher spaces in $\mathcal{DSP}$, as we have seen earlier, $\mathcal{G}$ will also be an Frölicher space. Now take a path $c : \mathbb{R} \to \mathcal{G}$ where $c(a) \in \mathcal{H}_I$ in $\mathcal{DSP}$. A scalar $g : \mathcal{G} \to \mathbb{R}$ is essentially a collection of scalars $g_I : \mathcal{H}_I \to \mathbb{R}$ such that if $I \subset I'$, then the restriction of $g_{I'}$ to $\mathcal{H}_I$, is g_I. This implies that if $c(t) \in \mathcal{H}_I$ for t in an open interval O containing a, then the associated map $\hat{c} : O \times Q \to \mathbb{R}$ extends to a map $O \times \mathbb{R} \to \mathbb{R}$ which is smooth except possible on lines $y = i, i \in I$. However, without writing down the detail, it is not difficult, for a rational number q, to find a sequence r_n of irrational numbers converging to q and a sequence of numbers a_n converging to a such that $c(a_n)$ is not extendable to a smooth function at r_n for each $n \in \mathbb{N}$. Then, the associated map $\hat{c}$ will not be smooth at (a, q) which is impossible if one wishes $\mathcal{G}$ to be a Cartesian object under the map making sets Cartesian closed.

• Consider now the Frölicher space $\mathbb{R}^{\mathbb{N}}_{\oplus}$ defined earlier. Suppose that $\mathcal{DSP}$ is Cartesian closed under the usual set adjunction and K denotes the differential space with underlying set $Hom_{\mathcal{DSP}}(\mathbb{R}^{\mathbb{N}}_{\oplus}, \mathbb{R})$ and arising from the Cartesian closedness of $\mathcal{DSP}$. Let $\overline{K}$ be the associated Frölicher space and $g : \mathbb{R} \to \overline{K}$ a smooth map. Since $g : \mathbb{R} \to K$ is also a differentiable map, the map $\hat{g} : \mathbb{R} \times \mathbb{R}^{\mathbb{N}}_{\oplus} \to \mathbb{R}$ defined by setting $\hat{g}(t, x) = g(t)(x)$ is also a differentiable map (using the assumed Cartesian closedness of $\mathcal{DSP}$). Suppose that the product $\mathbb{R} \times \mathbb{R}^{\mathbb{N}}_{\oplus}$ taken in $\mathcal{DSP}$ is also the product in $\mathcal{FRL}$. Then, $\hat{g}$ is a smooth map. Hence, using the Cartesian closedness of $\mathcal{FRL}$, g must also define a smooth map $g : \mathbb{R} \to L$ where L denotes the Frölicher space with underlying set $Hom_{\mathcal{FRL}}(\mathbb{R}^{\mathbb{N}}_{\oplus}, \mathbb{R})$ arising from the Cartesian closedness of $\mathcal{FRL}$. Thus, every contour of $\overline{K}$ is a contour of L.

Let $g : \mathbb{R} \to L$ be a contour of L. Then $\hat{g} : \mathbb{R} \times \mathbb{R}^{\mathbb{N}}_{\oplus} \to \mathbb{R}$ is a smooth map with the product in $\mathcal{FRL}$. Again, since $\mathbb{R} \times \mathbb{R}^{\mathbb{N}}_{\oplus}$ is also the product in differential spaces, $g : \mathbb{R} \to K$ must be a differential map and induce a smooth map $g : \mathbb{R} \to \overline{K}$. Thus, L and $\overline{K}$ must have the same contours and are thus equal.

Finally, if $\mathcal{DSP}$ is Cartesian closed we have, using the natural bijection from sets:

$$\begin{aligned} Hom_{\mathcal{DSP}}(\mathbb{R}^{\mathbb{N}}_{\oplus} \times \mathbb{R}^{\mathbb{N}}_{\oplus}, \mathbb{R}) &\simeq Hom_{\mathcal{DSP}}(\mathbb{R}^{\mathbb{N}}_{\oplus}, K) \\ &\simeq Hom_{\mathcal{FRL}}(\mathbb{R}^{\mathbb{N}}_{\oplus}, \overline{K}) \simeq Hom_{\mathcal{FRL}}(\mathbb{R}^{\mathbb{N}}_{\oplus}, L) \\ &\simeq Hom_{\mathcal{FRL}}(\mathbb{R}^{\mathbb{N}}_{\oplus} \times \mathbb{R}^{\mathbb{N}}_{\oplus}, \mathbb{R}), \end{aligned}$$

where the products are taken in the appropriate categories. Since the product of $\mathbb{R}^{\mathbb{N}}_{\oplus}$ with itself is different in $\mathcal{DSP}$ and $\mathcal{FRL}$, it is clearly impossible to have all these bijections. Thus $\mathcal{DSP}$ is not Cartesian closed under the canonical set adjunction.

If the product $\mathbb{R} \times \mathbb{R}^{\mathbb{N}}_{\oplus}$ taken in $\mathcal{DSP}$ is not the product in $\mathcal{FRL}$, following similar arguments, one sees that $\mathcal{DSP}$ is not Cartesian closed under the canonical set adjunction.

- In a Cartesian closed category, products commute with coequalizers and it would be useful to show that $\mathcal{DSP}$ is not Cartesian closed using this fact.

6. Tangent and Cotangent Bundles on a Frölicher Space

Let X and Y be Frölicher spaces. Then (X, Y) will denote the Frölicher space whose underlying set is $Hom_{\mathcal{FRL}}(X, Y)$ and with structure arising from the Cartesian closedness of $\mathcal{FRL}$. We then write $\mathcal{F}_X = (X, \mathbb{R})$ to denote $\mathcal{F}_X$ with its associated Frölicher structure. Using the notation of Definition 2.1, the sets

$$\mathcal{D}_X = \{D : \mathcal{F}_X \to \mathbb{R} \mid D \text{ is a derivation at some } P \in X\}$$

and

$$\mathcal{D}_{X,\mathcal{C}} = \{V_c \in \mathcal{D}_X \mid c \in \mathcal{C}_X, c(0) = P \text{ and } P \in X\}$$

can be viewed as Frölicher subspaces of $(\mathcal{F}_X, \mathbb{R})$ and the product $X \times \mathcal{D}_X$ is a Frölicher space. We are now able to make the following definition:

DEFINITION 6.1. The *tangent bundle* TX (resp., *tangent cone bundle* TCX) on X is the Frölicher subspace of $X \times \mathcal{D}_X$ (resp., $X \times \mathcal{D}_{X,\mathcal{C}}$ consisting of all (P, D) such that D is a derivation at P. The projection map $\pi : TX \to X$ (resp., $\Pi : TCX \to X$) is the smooth map sending (P, D) to P. A *vector field* on X is (most properly) a section of Π or (more generally) π.

REMARKS. The notion of tangent bundle can clearly be extended to differential spaces. In fact, for differential spaces, there are other, not quite so canonical ways to define tangent bundles. See [9]. However, looking at the rationals $\mathbb{Q}$, it is clear that the tangent bundle of a differential space need not coincide with the tangent bundle of the associated Frölicher space. The tangent cone bundle of a differential space can be defined as the tangent cone bundle of the associated Frölicher space. This may be useful in some circumstances, but, as in the case of $\mathbb{Q}$, information might be lost in certain cases.

EXAMPLE 7. Let $X = \mathbb{R}^n$. Then there is a bijection

$$\delta : TX \to \mathbb{R}^n \times \mathbb{R}^n,$$

where if $(P, D) \in TX$ and $D = \sum_{i=1}^n d_i e_i$, using the notation in Nakahara [7], then $\delta(P, D) = (P, (d_i))$. A map $\zeta : \mathbb{R} \to TX$ is a contour if and only if, in

$$\zeta(t) = \left(c(t), \sum_{i=1}^{n} d_i(t) e_{i,c(t)}\right),$$

$c(t)$ is a contour of X and $\sum_{i=1}^n d_i(t)e_{i,c(t)}$ is a contour of $\mathcal{D}_X$ where $e_{i,c(t)}$ denotes the partial $\frac{\partial}{\partial x_i}$ taken at $c(t)$. But, $\sum_{i=1}^n d_i(t)e_{i,c(t)}$ is smooth if and only if the map $\mu : \mathbb{R} \times \mathcal{F}_X \to \mathbb{R}$ sending (t, f) to $\sum_{i=1}^n d_i(t)e_i(f)(c(t))$ is smooth. Taking $f = x_i$, one sees that $d_i(t)$ must be a smooth function of t for $i = 1, \ldots, n$. The map μ is smooth if and only if, for any smooth map $\mathbb{R} \to \mathcal{F}_X$ sending $s \to f_s$, $d_i(t)$ and $e_i(f_s)(c(t))$, for $i = 1, \ldots, n$ are smooth in s and t. But, $e_i(f_s)(c(t)) = \frac{\partial \hat{f}(s,c(t))}{\partial x_i}$ and $\hat{f}(s, t) = f_s(t)$ is smooth in s and t. It follows that ζ is smooth if and only if $c(t)$ and $d_i(t)$, $i = 1, \ldots, n$, are smooth. In that event, the bijection δ is a smooth isomorphism.

Let X be an n-dimensional smooth manifold and U an open coordinate neighborhood of X isomorphic as a smooth manifold to $\mathbb{R}^n$. One can extend the above arguments to show that $\pi^{-1}(U)$ is isomorphic to $\mathbb{R}^n \times \mathbb{R}^n$ as a Frölicher space. It is then clear that the tangent bundle defined for X as a smooth manifold is smooth isomorphic to TX when X is viewed as a Frölicher space.

Let E be a Frölicher space and $\pi : E \to X$ be a smooth map such that $\pi^{-1}(x)$ is a given vector space or a vector space aside from the fact that the addition is only partially defined, for each $x \in X$. Then, by $\mathbb{LIN}(E, R)$ we mean the Frölicher subspace of $(E, \mathbb{R})$ consisting of all f such that f restricted to $\pi^{-1}(x)$ satisfies $f(ca) = cf(a) (c \in \mathbb{R})$ and $f(a + b) = f(a) + f(b)$, whenever $a + b$ is defined, for each $x \in X$.

Define a relation $\simeq$ on

$$\underline{\mathbf{LIN}} = X \times \mathbf{LIN}(TX, \mathbb{R})$$

by setting $(P, f) \simeq (Q, g)$ if and only if $P = Q$ and, for some neighborhood U of P, $f|_{\pi^{-1}(U)} = g|_{\pi^{-1}(U)}$.

DEFINITION 6.2. The *cotangent bundle* T^*X to a Frölicher space is $\underline{\mathbf{LIN}}/\simeq$. The projection map $\pi_1^0 : T^*X \to X$ is the smooth map induced from the projection of $\underline{\mathbf{LIN}}$ onto its first factor. Replacing TX by TCX, one obtains the notion of *cotangent cone bundle*, denoted T^*CX and where the projection is Π_1^0. A section s of π_1^0 or Π_1^0 will be called a (smooth) *1-form* provided that, for each point $P \in X$, P has a neighborhood U such that $s(x) = [(x, s')]$ for some fixed $s' \in \mathbf{LIN}(TX, \mathbb{R})$ and all $x \in U$. Thus, s is represented locally by a global section.

REMARKS. Notice that the fiber $(\pi_1^0)^{-1}(P)$ above P has an induced vector space structure. One commonly writes $T^*X_P = (\pi_1^0)^{-1}(P)$. We would prefer to omit the condition given on the section s but it is not clear whether this is possible and, in practice, one needs to work at $P \in X$ with an element of $\mathbf{LIN}(TX, \mathbb{R})$ restricted to a suitable neighborhood of P.

EXAMPLE 8. Let $X = \mathbb{R}^n$ and $q : \underline{\mathbf{LIN}} \to T^*X$ be the quotient map. A map $\tau : \mathbb{R} \to T^*X$ is smooth if and only if $f \circ \tau$ is smooth for each smooth map $f : T^*X \to \mathbb{R}$. But f is smooth if and only if $f \circ q$ is smooth on $\underline{\mathbf{LIN}}$. Let $[(P, L)]$ be an equivalence class in T^*X and $L|_{\pi^{-1}(P)} = \sum_{i=1}^n a_{i,P} dx^i|_P$. There is then a bijection $v : T^*X \to \mathbb{R}^n \times \mathbb{R}^n$ sending $[(P, L)]$ to $(P, (a_{i,P}))$. We wish to show that this bijection is a smooth isomorphism. Let $c : \mathbb{R} \to T^*X$ be a contour and write $v(t) = v \circ c(t) = (P_t, (a_{i,P_t}(t)))$.

Using the fact that q is a quotient, one sees that there is a smooth map $\psi : T^*X \to X$ such that $\psi \circ q = \lambda_1$ where λ_1 is projection onto the first factor of $\underline{\mathbf{LIN}}$. As $\lambda_1(P_t, L_t) = P_t$ is smooth in t and $\psi = P_1 \circ v = \pi_1^0$, where P_1 is the projection of $\mathbb{R}^n \times \mathbb{R}^n$ onto the first of two factors, v is smooth in its first coordinate.

Define

$$\Theta_i : \underline{\mathbf{LIN}} \to \mathbb{R}$$

by setting $\Theta_i(P, L) = a_{i,P}$ where $L|_{\pi^{-1}(P)} = \sum_{i=1}^n a_{i,P} dx^i|_P$. Again, as q is a set quotient, there is a map $\rho_i : T^*X \to \mathbb{R}$ such that $\rho_i \circ q = \Theta_i$. We wish first to show that Θ_i is smooth. Let $d : \mathbb{R} \to \underline{\mathbf{LIN}}$ be a smooth map and $d(t) = (P_t, L_t)$. Then L_t is smooth in t if and only if the associated map $\hat{L} : \mathbb{R} \times TX \to \mathbb{R}$ defined by $\hat{L}(t, v) = L_t(v)$ is smooth. Identifying TX with $\mathbb{R}^n \times \mathbb{R}^n$, as above, one can write

$$\hat{L}(t, P, w) = A(t, P) \cdot w,$$

where $A(t, P)$ is a matrix with entries which are smooth functions of t and P, w is written as a column vector and $\cdot$ is matrix multiplication. Let $w = \frac{\partial}{\partial x_i}$. Then, with

w viewed as a column vector, $a_{i,P_t}(t) = \Theta_i \circ d(t) = A(t, P_t) \cdot w$. Hence, Θ_i and thus also the ρ_i are smooth. Since $\rho_i = P_{n+i} \circ v$, where P_{n+i} is projection on the $n+i$-th coordinate of $\mathbb{R}^n \times \mathbb{R}^n$, v must be smooth.

Conversely, suppose that the $a_{i,P_t}(t)$ for $i = 1, \ldots, n$ and P_t are smooth functions of t. Let $\beta(t) = (P_t, \sum_{i=1}^n a_{i,P_t}(t) dx^i|_{P_t})$. Then the second component of β is smooth since the induced map sending $(t, (P_t, b_i))$ to $\sum a_{i P_t} b_i$, arising from the Cartesian closedness of $\mathcal{FRL}$ and using the identification of TX with $\mathbb{R}^n \times \mathbb{R}^n$, is clearly smooth. As β is then smooth, so is $q \circ \beta$. Hence, v^{-1} is smooth and the conclusion sought follows.

Let $E_1, \ldots, E_n$ be Frölicher spaces and $\pi_i : E_i \to X$ be a smooth map for $i = 1, \ldots, n$ such that $\pi_i^{-1}(P)$ is a given vector space or a vector space aside from the fact that the addition is only partially defined, for each $P \in X$. Let $\Pi^* : E_1 \oplus E_2 \oplus \cdots \oplus E_n \to X$ be the pullback of $\pi_1 \times \pi_2 \times \cdots \times \pi_n : E_1 \times E_2 \times \cdots \times E_n \to X^n$ along the diagonal map $\delta : X \to X^n$ which is clearly smooth. The set $E_1 \oplus E_2 \oplus \cdots \oplus E_n$ is usually called the *Whitney product* of $E_1, E_2, \ldots, E_n$. By **MULTILIN**$(E_1, \ldots, E_n; \mathbb{R})$ we mean the Frölicher subspace of

$$(E_1 \oplus E_2 \oplus \cdots \oplus E_n, \mathbb{R})$$

consisting of all f such that

1.
$$f(a_1, \ldots, ca_i, \ldots, a_n) = cf(a_1, \ldots, a_i, \ldots, a_n) \quad \text{for} \quad c \in \mathbb{R}, (a_1, \ldots, a_n) \in (\pi^*)^{-1}(P) \text{ and } P \in X.$$

2.
$$\begin{aligned} &f((a_1, \ldots, a_{i-1}, b_i, a_{i+1}, \ldots, a_n) + (a_1, \ldots, a_{i-1}, c_i, a_{i+1}, \ldots, a_n)) \\ &\quad = f(a_1, \ldots, a_{i-1}, b_i, a_{i+1}, \ldots, a_n) + f(a_1, \ldots, a_{i-1}, c_i, a_{i+1}, \ldots, a_n) \end{aligned}$$

for

$$(a_1, \ldots, a_{i-1}, b_i, a_{i+1}, \ldots, a_n), (a_1, \ldots, a_{i-1}, c_i, a_{i+1}, \ldots, a_n)$$

in $(\pi^*)^{-1}(P)$ and $P \in X$ but only if

$$(a_1, \ldots, a_{i-1}, b_i, a_{i+1}, \ldots, a_n) + (a_1, \ldots, a_{i-1}, c_i, a_{i+1}, \ldots, a_n)$$

is defined.

Define a relation $\approx$ on

$$\underline{\mathbf{MULTILIN}}(E_1, \ldots, E_n) = X \times \mathbf{MULTILIN}(E_1, \ldots, E_n; \mathbb{R})$$

by setting $(P, f) \approx (Q, g)$ if and only if $P = Q$ and there is a neighborhood U of P such that

$$f|_{(\Pi^*)^{-1}(U)} = g|_{(\Pi^*)^{-1}(U)}.$$

DEFINITION 6.3. The *higher level tensor bundle* $\underline{T}^p_q X$ of *type* (p, q) is equal to the Frölicher space

$$\underline{\mathbf{MULTILIN}}(\underbrace{T^*X, \ldots, T^*X}_{p}, \underbrace{TX, \ldots, TX}_{q})/\approx,$$

where T^*X appears in the first p factors and TX appears in the last q factors. Replacing TX by TCX and T^*X by T^*CX, one obtains the tensor cone bundle. The projections π^p_q and Π^p_q are the evindent ones. A section s of π^p_q or Π^p_q is called a (p, q)-*tensor field* if, for each $P \in X$, there is a neighborhood U of P such that $s(Q) = [(Q, K)]$ for some $K \in \mathbf{MULTILIN}(T^*X, \ldots, T^*X, TX, \ldots, TX; \mathbb{R})$ and all $Q \in U$. A *pointwise cone Riemannian metric* g is a (0, 2)-tensor field such that, if g_P is the restriction of g to $(\underline{\Pi}^0_2)^{-1}(P)$, then

(1) $g_P(U, V) = g_P(V, U)$.
(2) the induced fiber preserving smooth map

$$(g_*) : TCX \to \mathbf{LIN}(TCX, \mathbb{R}),$$

where $(g_*)_P(X)(Y) = g_P(X, Y)$, $X, Y \in \pi^{-1}(P)$ and $P \in X$, is a smooth isomorphism onto its image. A *pointwise Riemannian metric* is defined by taking $\underline{\pi}^0_2$ instead of $\underline{\Pi}^0_2$, TX instead of TCX and requiring, in addition, $(q_*)_P$ to be bijective.

REMARKS. In the case of finite-dimensional smooth manifolds, condition (2) of Definition 6.3 is usually replaced by the condition: $gp(U, V) = 0$ for all $U \in TX_P$ implies $V = 0$. However, a Frölicher space may have infinite-dimensional tangent spaces and this necessitates the revised requirement. Since the fibers of $\pi : TX \to X$ need not be finite-dimensional, $\underline{T}^0_1$ need not be the same as TX. Thus, we distinguish between ordinary and higher level tensor bundles, which involve taking duals. Application of g_* in tensor analysis is lowering of indices.

Let X be an Frölicher space and $c : I = [0, 1] \to X$ be a smooth map. Suppose that $c(a) = P$. Then P has a neighborhood U such that $g(x) = [(x, g^*)]$ for each $x \in U$. As $g^* \in \mathbf{LIN}(TX, TX; \mathbb{R})$ and

$$d : \mathbb{R} \to TX \times TX$$

defined by setting $d(t) = (c'(t), c'(t))$ is smooth, $g^*(c'(t), c'(t))$ is smooth for t in a neighborhood of a. Letting $g_P = g_P|_{\pi^{-1}(P)\times\pi^{-1}(P)}$ if $g = [(P, g_P)]$,

$$\underline{g}_{c(t)}(c'(t), c'(t)) = g^*(c'(t), c'(t))$$

is a real valued smooth function for t near a and hence for all $t \in I$. Thus, $\int_c ds$ can be defined as

$$\int_0^1 \sqrt{\underline{g}_{c(t)}(c'(t), c'(t))}dt.$$

EXAMPLE 9. Let $X \subset \mathbb{R}^2$ be the union of the x and y axes. Let $\alpha : \mathbb{R} \to \mathbb{R}$ be a smooth function such that $\alpha(t) = 0$ for $|t| \leq \frac{1}{8}$ and $\alpha(t) = t$ for $t \geq \frac{1}{2}$. Define $c : \mathbb{R} \to X$ by setting $c(t) = (\alpha(t), \alpha(-t))$. The curve $c(t)$ starts at $(0, 1)$ and passing through $(0, 0)$ proceeds to $(1, 0)$. Define the Riemannian metric in the usual way except at $(0, 0)$. Thus, $g = dx$ along the x axis and $g = dy$ along, the y axis. At $(0, 0)$, let $g = (dx + dy)$. Then,

$$\int_c ds = -\int_{-1}^{0} \alpha'(-y)dy + \int_0^1 \alpha'(x)dx.$$

It is now clear that the notion of geodesic can be extended to Frölicher spaces.

7. Derivation of Einstein Tensor

We will show how one can formally define the Einstein tensor by looking at the tensor bundles of a Frölicher space. On singular spaces it would be better to look at the tangent cone bundle but the definitions require more care. Our references are [7, 13].

DEFINITION 7.1. Let A be an Frölicher space, $f \in \mathcal{F}_A$ and $\mathcal{X}[A]$ be the set of smooth sections of the projection $\pi : TA \to A$, i.e., the smooth vector fields of A. The *covariant derivative* $\nabla_X(f)$ of f or the *directional derivative* $X[f]$ of f is the element in $\mathcal{F}_A$ defined by setting $\nabla_X(f)(a) = X[f]_a = q(X[a])(f)$ where, if $X(a) = (a, D_a)$, then $q(X(a)) = D_a$. A smooth map

$$\nabla : \mathcal{X}[A] \times \mathcal{X}[A] \to \mathcal{X}[A]$$

is a *covariant derivative* or *connection* if

(1) ∇ is bilinear.
(2) $\nabla_{fX} Y = f\nabla_X Y$ for $f \in \mathcal{F}_A$.
(3) $\nabla_X(fY) = X[f]Y + f\nabla_X Y$ (the Leibniz rule).
(4) ∇ is induced from a smooth map

$$\nabla' : \mathcal{X}[A] \times TA \to TA,$$

mapping fiber to fiber, on setting $(\nabla_X)Y(P) = \nabla'_X(Y(P))$.

REMARK. The last condition in our definition of connection is normally implicit in the theory of connections on a finite-dimensional smooth manifold.

Let $\mathcal{F}^{(0,1)}$ be the collection of smooth $(0, 1)$-tensor fields or, in other words, smooth 1-forms. Define $h : \mathcal{F}^{(0,1)} \to \mathbf{LIN}(TA; \mathbb{R})$ by letting $h(\omega)(X(Q)) = \underline{\omega}_P(X(Q))$, for Q in a neighborhood of P, where $\omega(P) = [(P, \underline{\omega}_P)]$ and $\underline{\omega}_P$ defines ω in a neighborhood of P. Here, we write $X(P)$ merely to indicate that $X(P) \in TX_P$. As $h \circ q(P, \underline{\omega}_P)(X(Q)) = \underline{\omega}_P(X(Q))$, for Q in a neighborhood of

$P, h \circ q$ and hence h is smooth. In order for ∇_Y, where Y is a smooth vector field, to satisfy the Leibniz rule

$$\nabla_Y(h(\omega)(X)) = \nabla_Y h(\omega)(X) + h(\omega)(\nabla'_Y X) \tag{11}$$

for $X \in TA$. Using condition (4) in the definition of connection (Definition 7.1), $\nabla_Y h(\omega) : TA \to \mathbb{R}$, defined by Equation (11) directly above, is smooth. Thus, $(P, \nabla_Y h(\omega)) \in A \times \mathbf{LIN}(TA; \mathbb{R})$ and hence the assignment

$$P \to [(P, \nabla_Y h(\omega))]$$

defines a smooth 1-form $\nabla_Y \omega$.

In a similar way, one defines

$$h : \mathcal{F}^{(0,2)} \to \mathbf{MULTILIN}(TA, TA; \mathbb{R}).$$

The relation

$$\nabla_Z(h(g)(X, Y)) = (\nabla_Z h(g))(X, Y) + h(g)(\nabla'_Z X, Y) + h(g)(X, \nabla'_Z(Y))$$

implies that the map $\nabla_Z h(g) : TA \times TA \to \mathbb{R}$ is smooth and finally one lets $\nabla_Z g$ be the map sending P to $[(P, \nabla_Z h(g))]$.

We note, for the following definition, that the set of vector fields $\mathcal{X}[A]$ on a Frölicher space A has a Lie product. Let $X, Y \in \mathcal{X}[A]$, $X(P) = (P, D_P^X)$ and $Y(P) = (P, D_P^Y)$. Since multiplication on $\mathbb{R}$ is smooth, the map $A \to TA$ defined by setting $[X, Y](P) = (P, D_P^X D_P^Y - D_P^Y D_P^X)$ is smooth and we have our Lie product.

DEFINITION 7.2.

(1) The covariant derivative ∇ is a *metric connection* if and only if $\nabla_X g = 0$ for each $X \in \mathcal{X}[A]$.
(2) The smooth map $T : \mathcal{X}[A] \times \mathcal{X}[A] \to \mathcal{X}[A]$ defined by setting

$$T(X, Y) = \nabla_X Y - \nabla_Y X - [X, Y]$$

is called the *torsion tensor.*
(3) The smooth map $R : \mathcal{X}[A] \times \mathcal{X}[A] \times \mathcal{X}[A] \to \mathcal{X}[A]$ defined by seltting

$$R(X, Y, Z) = \nabla_X \nabla_Y Z - \nabla_Y \nabla_X Z - \nabla_{[X,Y]} Z$$

is called the *Riemannian curvature tensor.*

Let A be a smooth finite-dimensional manifold. Let $\mathbf{LIN}(\mathcal{X}[A]; \mathcal{X}(A))$ be the set of smooth linear maps $\mathcal{X}[A] \to \mathcal{X}[A]$ induced from pointwise linear maps of the tangent bundle. One can show that there is a trace function $TR : \mathbf{LIN}(\mathcal{X}[A]; \mathcal{X}[A]) \to \mathcal{F}_A$, additive and invariant under coordinate change. The Riemannian curvature tensor $R : \mathcal{X}[A] \times \mathcal{X}[A] \times \mathcal{X}(A) \to \mathcal{X}[A]$ induces, because of the Cartesian closedness of $\mathcal{FRL}$, a map $\overline{R} : \mathcal{X}[A] \times \mathcal{X}[A] \to \mathbf{LIN}(\mathcal{X}[A]; \mathcal{X}[A])$. The map $TR \circ \overline{R} : \mathcal{X}[A] \times \mathcal{X}[A] \to \mathcal{F}_A$ is a *contraction* of R and one calls $\mathcal{RIC} = TR \circ R$ the Ricci tensor. Let g be a Riemannian metric on a smooth space.

DEFINITION 7.3.

(1) Suppose that $\overline{R} : \mathcal{X}[A] \times \mathcal{X}[A] \to \mathbf{LIN}(\mathcal{X}[A]; \mathcal{X}[A])$ is the map induced from R (using the Cartesian closedness of $\mathcal{FRL}$ and condition 4 of Definition 7.1) and one possesses a trace function $TR : \mathbf{LIN}(\mathcal{X}[A]; \mathcal{X}[A]) \to \mathcal{F}_A$. Then, the *Ricci tensor* $\mathcal{RIC} : \mathcal{X}[A] \times \mathcal{X}[A] \to \mathcal{F}_A$ is defined by setting $\mathcal{RIC} = TR \circ \overline{R}$.

(2) Because of the Cartesian closedness of $\mathcal{FRL}$, the Ricci tensor $\mathcal{RIC}(X, Y)$ induces a smooth linear map

$$\widehat{\mathcal{RIC}} : \mathcal{X}[A] \to \mathbf{LIN}(\mathcal{X}[A], \mathcal{F}_A).$$

(3) If the smooth map $g_{**} : \mathcal{X}[A] \to \mathbf{LIN}(\mathcal{X}[A], \mathcal{F}_A)$ (defined by setting $g_{**}(Q)(X)(Y) = h(X(Q), Y(Q))$ where $g = [(Q, h)]$ for Q in a neighborhood of P) is a smooth isomorphism, then g is called a global Riemannian metric.

(4) For a global Riemannian metric, the scalar curvature $\mathcal{R}$ on a smooth space A is defined by setting

$$\mathcal{R} = TR(g_{**}^{-1} \circ \widehat{\mathcal{RIC}}).$$

(5) Finally, for a global Riemannian metric, $\mathcal{G} = \mathcal{RIC} - \frac{1}{2}\mathcal{R}g$ is called the *Einstein tensor*.

REMARK. In the case of finite-dimensional smooth manifolds, a point-wise Riemannian metric is a global Riemannian metric. The same assertion is unclear for general smooth spaces.

In conclusion, one sees that categorical language surrounding the category $\mathcal{FRL}$ can fix many of the terms in mathematical physics in a natural way and provide some clues to the direction of future endeavors.

References

1. Boman, J.: Differentiability of a function and of its compositions with functions of one variable, *Math. Scand.* **20** (1967), 249–268.
2. Brümmer, G. C. L.: Topological categories, *Topology and its Applications* **18** (1984), 27–41.
3. Cherenack, P.: Smooth homotopy, *Bolyai Society Mathematical Studies: Topology with Applications, Szekszárd* (Hungary), (1993), 47–70,4.
4. Frölicher, A. and Kriegl, A.: *Linear Spaces and Differentiation Theory*, Wiley, New York, 1988.
5. Hartshorne, R.: *Algebraic Geometry*, Springer-Verlag, New York, 1977.
6. Mac Lane, S.: *Categories for the Working Mathematician*, Springer-Verlag, New York. 1971.

7. Nakahara, M.: *Geometry, Topology and Physics*, Adam Hilger, Bristol, 1990.
8. Heller, M. and Sasin, W.: Anatomy of the elementary quasi-regular singularity, *Acta Cosmologica* **21** (1995), 47–59.
9. Kowalczyk, A.: Tangent differential spaces and smooth forms, *Demonstratio Math.* **23** (1980), 893–905.
10. Penrose, R. and Rindler, W.: *Spinors and Space-Time*, Vol. 1: *Two-Spinor Calculus and Relativistic Fields*, Cambridge University Press, Cambridge, 1984.
11. Sasin, W.: Differential spaces and singularities in differential space time, *Demonstratio Math.* **24** (1991), 601–634.
12. Sikorski, R.: Differential modules, *Colloq. Math.* **24** (1971), 45–79.
13. Wald, R. M.: *General Relativity*, The University of Chicago Press, Chicago, 1984.

G. Brümmer & C. Gilmour (eds), Papers in Honour of Bernhard Banaschewski.

Weakly Hereditary Initial Closure Operators

TEMPLE H. FAY
Department of Mathematics, University of Southern Mississippi, MS 39406-9045, Hattiesburg, U.S.A. e-mail: thfay@mozart.st.usm.edu

(Received: 19 June 1996; accepted: 28 May 1998)

Abstract. We extend some recent of M. M. Clementino (*Topology and its Applications* **49** (1993)) and of the author (*Cahiers de Topologie et Géométrie Différentielle Catégoriques* **XXXVII**(4) (1996)) concerning when a regular closure operator is weakly hereditary. In this work, we study initial closure operators which include both regular and normal closure operators. Working in a quasi-additive regular category in which pullbacks of cokernels are cokernels, we characterize when they initial closure operators are weakly hereditary. For the category of all groups, we show that a non-trivial normal closure operator is never weakly hereditary.

Mathematics Subject Classifications (2000): 18A40, 18E40.

Key words: closure operator, regular closure operator, normal closure operator, weakly hereditary closure operator, generalized torsion theory.

1. Introduction

In the theory of categorical closure operators, weakly hereditary closure operators hold an important place. They arise from factorization structures for morphisms and are extremely well behaved with regards to categorical compactness (see [15]). However, there are other types of closure operators which are of interest and which lead to interesting classes of categorically compact objects. These include the regular and normal closure operators determined by an epireflective class [13].

Recently, the author considered when a regular closure operator is weakly hereditary [12]. This work illuminated and extended work of Clementino [7], who showed that for **TOP**, an extremal-epireflective subcategory $\mathcal{F}$ is a disconnectedness (see [1]) if and only if the regular closure operator induced by $\mathcal{F}$ is weakly hereditary. Clementino also showed that for an epireflective subcategory $\mathcal{F}$ of an Abelian category (with suitable completeness), $\mathcal{F}$ is a torsion-free subcategory if and only if the regular closure operator induced by $\mathcal{F}$ is weakly hereditary. In [12], this latter result was extended to the category of all groups **GRP**, as well as to pointed quasi-additive categories for which the pullback of a cokernel is a cokernel.

We further explore this connection between a closure operator $c(-)$ being weakly hereditary and properties of the class $\mathcal{F}_c = \{G \mid c^G(0) = 0\}$. This class turns out to be closed under products and subobjects and thus is quotient-reflective. This means that for each object G, there is a quotient map (= coequalizer) $\theta_G\colon G \to RG$ with $RG \in \mathcal{F}_c$ satisfying a universal mapping property. The closure operator $c(-)$ is called *initial* when, for each subobject A of G, we have

$$c^G(A) = \theta_G^{\leftarrow}(c^{RG}(\theta_G(A))).$$

Regular and normal closure operators turn out to be initial. We characterize when an initial closure operator is weakly hereditary. We also show that, in the category of all groups, a non-trivial normal closure operator is never weakly hereditary.

2. Preliminaries

A category $\mathcal{C}$ is called *regular* if it is finitely complete, finitely cocomplete, enjoys the (strong epi, mono) factorization structure for morphisms, and strong epimorphisms are stable under pullbacks. In a regular category, the strong epimorphisms and regular epimorphisms coincide. A pointed category is called *quasi-additive* when a morphism f is a monomorphism whenever $\ker(f) = 0$. Throughout this paper we shall assume $\mathcal{C}$ to be quasi-additive, finitely complete, finitely cocomplete, has the property that all pullbacks of cokernels are cokernels, and admits all intersections of subobjects. These categories enjoy many properties of additive and Abelian categories with cokernels in the place of coequalizers [2, Proposition 8.12]. In this situation, every morphism f factorizes as $f = me$ with m monomorphic and e a cokernel; all strong epimorphisms are cokernels; and composites of cokernels are cokernels. If $f\colon X \to Y$ is a morphism and $f = me$ is the (cokernel, mono)-factorization of f, then we denote the codomain of e (domain of m) by $f(X)$.

We will also want to have classes closed under both products and subobjects to be quotient-reflective and to have certain intersections and joins of subobjects to behave nicely. Thus we also require $\mathcal{C}$ to be well-powered.

As usual, if A is a subobject of an object G, then we write $A \leq G$. A *closure operator* on $\mathcal{C}$ assigns to each subobject A of G, a subobject $c^G(A)$ of G such that for each pair of subobjects A and B of G and each morphism $f\colon G \to H$, the following hold:

(i) $A \leq c^G(A)$ (expansive);
(ii) $A \leq B$ implies $c^G(A) \leq c^G(B)$ (monotone);
(iii) $f(c^G(A)) \leq c^H(f(A))$ (continuity condition).

A closure operator $c(-)$ is called *idempotent* provided $c^G(c^G(A)) = c^G(A)$ for each subobject A of G; a *weakly hereditary* closure operator satisfies $c^{c^G(A)}(A) = c^G(A)$.

The general theory of closure operators has been developed by several authors, for example see [3, 9]. It is of importance to note that the continuity condition implies that $c^G(0)$ is always fully invariant in G.

Let $\mathcal{E}$ and $\mathcal{M}$ be isomorphism closed classes of morphisms. The pair $(\mathcal{E}, \mathcal{M})$ is called a *factorization structure* provided:

(i) for every commutative square $mf = ge$ with $e \in \mathcal{E}$ and $m \in \mathcal{M}$, there exists a unique morphism d so that $de = f$ and $md = g$ (the $(\mathcal{E}, \mathcal{M})$-diagonalization property);

(ii) every morphism f can be factored as $f = me$ with $e \in \mathcal{E}$ and $m \in \mathcal{M}$ (the $(\mathcal{E}, \mathcal{M})$-factorization property). This idea encapsulates and generalizes the usual (surjective, injective) factorization for group homomorphisms.

For a closure operator $c(-)$, we let $\mathcal{E}_c$ be the class of all *c-dense* morphisms (the c-closure of the image equals the codomain) and we let $\mathcal{M}_c$ be the class of all *c-closed embeddings* (monomorphisms with c-closed image). It follows that the category $\mathcal{C}$ enjoys the $(\mathcal{E}_c, \mathcal{M}_c)$-diagonalization property [9, Proposition 3.1], but that the composition of c-closed embeddings need not be a c-closed embedding. The pair $(\mathcal{E}_c, \mathcal{M}_c)$ determines a factorization structure when $c(-)$ is weakly hereditary and idempotent, see [9, 10].

Let $\mathcal{F}$ be any class of objects containing the zero object 0. We call a subobject A of an object G an *$\mathcal{F}$-regular subobject* of G provided there exist an $F \in \mathcal{F}$ and a pair $f, g\colon G \to F$ so that A is the equalizer of f and g. For categories algebraic over the base category **SET** in the sense of [14], this simply means $A = \{x \in G \mid f(x) = g(x)\}$.

If $c(-)$ is an arbitrary closure operator, one has a descending chain of closure operators defined in the obvious way, if $A \leq G$, then

$$c_0^G(A) = G,\ c_{\alpha+1}^G(A) = c^{c_\alpha^G(A)}(A),\ c_\beta^G(A) = \bigcap_{\alpha\in\beta} c_\alpha^G(A),\ \ \beta \text{ a limit ordinal.}$$

From the well-poweredness assumption, for any object G, there is a least ordinal $\mu(G)$ so that $c_{\mu(G)+1}^G(A) = c_{\mu(G)}^G(A)$ for all subobjects A of G. Define $c_\infty(-)$ by $c_\infty^G(A) = c_{\mu(G)}^G(A)$ for each $A \leq G$. This closure operator satisfies the inequality $c_\infty^G(A) = c^G(A)$ and is weakly hereditary; $c_\infty(-)$ is called the *weakly hereditary core* of $c(-)$ [9]. If $c(-)$ is idempotent, then so is $c_\infty(-)$ [9, Theorem 4.2]. This is the only place we need the hypothesis of the existence of intersections of subobjects for the category $\mathcal{C}$, to guarantee the existence of the weakly hereditary core. This assumption may be weakened to only require intersections of certain families of subobjects.

3. Separation

In this section, we discuss several classes of objects that arise naturally from a closure operator which play the role of separated objects, the analogue of Hausdorff.

Closure operators have been used to generalize the well known result that a topological space X is Hausdorff if and only if the diagonal Δ_X is closed in $X \times X$. Let $c(-)$ be an arbitrary closure operator. We let $\Delta_c = \{G \mid \Delta_G \text{ is closed in } G \times G\}$.

We obtain a preradical $\tau_c(-)$ on the category $\mathcal{C}$ by defining, for each object G, $\tau_c G = c^G(0)$. It is the continuity condition that makes $\tau_c(-)$ behave properly. Clearly, $\tau_c G$ is the smallest c-closed subobject of G. Associated with this preradical is its annihilator class $\mathcal{F}_c = \{G \mid \tau_c G = 0\}$.

A morphism $e\colon A \to B$ is called *c-dense* provided $c^B(e(A)) = B$. An object G is called *c-separated* provided for every c-dense map $e\colon A \to B$ and pair of morphisms $f, g\colon B \to G$, $fe = ge$ implies $f = g$. We denote the class of all c-separated objects by *c-Sep*. Of course, this class also is a good candidate to play the role of 'Hausdorff' as it is well known that two continuous functions having a Hausdorff codomain and agreeing on a dense subspace of their common domain are equal.

THEOREM 3.1. *For an arbitrary closure operator $c(-)$, the following hold*:

(i) *each of the classes Δ_c, $\mathcal{F}_c$, and c-Sep are closed under products and subobjects, and thus are quotient-reflective*;

(ii) $\Delta_c \subseteq \mathcal{F}_c \subseteq \mathcal{F}_{c_\infty}$;

(iii) $\Delta_c \subseteq c\text{-}Sep = \Delta_{c_\infty} = c_\infty\text{-}Sep \subseteq \mathcal{F}_{c_\infty}$.

Proof. (ii) Let $\mu\colon \tau_c G \to G$ be the inclusion map and observe that there is a natural map $\phi\colon \tau_c G \to G \times G$ defined by $\pi_1\phi = \mu$ and $\pi_2\phi = 0$. Clearly, the image of this map is simply $\tau_c G \times 0$. By the continuity property,

$$\tau_c G \times 0 = \phi(\tau_c G) \le c^{G\times G}(\phi(0)) = \tau_c(G \times G).$$

If $G \in \Delta_c$, then $\tau_c(G \times G) \le \Delta_c$ and thus

$$\tau_c G = \pi_1(\tau_c G \times 0) = \pi_2(\tau_c G \times 0) = 0.$$

Hence $G \in \mathcal{F}_c$. Clearly $\tau_c G = 0$ implies $\tau_{c_\infty} G = 0$ and thus $\mathcal{F}_c \subseteq \mathcal{F}_{c_\infty}$.

(iii) If $G \in \Delta_c$, $e\colon A \to B$ is c-dense, and $f, g\colon B \to G$ with $fe = ge$, then there is a map $h\colon A \to G$ so that $\{f, g\}e = \Delta_G h$. Here $\{f, g\}\colon B \to G \times G$ is that unique map having $\pi_1\{f, g\} = f$ and $\pi_2\{f, g\} = g$. From the continuity property,

$$\begin{aligned}\{f, g\}(B) &= \{f, g\}(c^B(e(A))) \le c^{G\times G}(\{f, g\}e(A)) \\ &= c^{G\times G}(\Delta_G h(A)) \le c^{G\times G}(\Delta_G(G)) = \Delta_G(G).\end{aligned}$$

Thus $\{f, g\}$ factors through Δ_G from which it follows that $f = g$. This means that $G \in c\text{-}Sep$.

If $G \in c\text{-}Sep$, and $\beta\colon c_\infty^{G\times G}(\Delta_G(G)) \to G \times G$ and $\alpha\colon G \to c_\infty^{G\times G}(\Delta_G(G))$ are inclusions, then we have $\Delta_G = \beta\alpha$. But c_∞ is weakly hereditary, so α is c-dense. From $\pi_1\beta\alpha = \pi_2\beta\alpha$, we have $\pi_1\beta = \pi_2\beta$ since $G \in c\text{-}Sep$. This implies $c_\infty^{G\times G}(\Delta_G(G)) \le \Delta_G(G)$ and thus the diagonal in $G \times G$ is c_∞-closed. This means $c\text{-}Sep \subseteq \Delta_{c_\infty}$. We now have $\Delta_c \subseteq c\text{-}Sep \subseteq \Delta_{c_\infty} \subseteq c_\infty\text{-}Sep$. But if $G \in c_\infty\text{-}Sep$, then $G \in c\text{-}Sep$ since c-dense maps and c_∞-dense maps coincide. □

In general, the classes Δ_c, $\mathcal{F}_c$, and *c-Sep* are distinct. For example, in the category of all groups **GRP**, a subgroup A of a group G is called *isolated* provided

$x^n \in A$ implies $x \in A$. It follows that the intersection of isolated subgroups is isolated, and thus we define the *isolator of* A *in* G by $I^G(A) = \bigcap\{B \leq G \mid A \leq B, B \text{ isolated}\}$. This defines a very well behaved closure operator $I(-)$, it is idempotent and weakly hereditary. The isolator of the trivial subgroup, denoted $\tau_I G$, is called the *generalized torsion subgroup*. It follows that $I\text{-}Sep = \Delta_I$ consists of the class of all *R-groups*; i.e., those groups having unique roots. However, $\mathcal{F}_I$ is the class of all torsion-free groups.

Also in **GRP**, if we declare a subgroup A of a group G to be c-closed if A is normal and G/A is torsion-free, then Δ_c is the class of all torsion-free Abelian groups and $\mathcal{F}_c$ is the class of all torsion-free groups. This is an example of a *normal closure operator determined by a quotient reflective class*; we shall say a great deal more about these operators in the next section. If we let $\nu(-)$ denote the *normal closure*, so that the smallest normal subgroup of G containing A is $\nu^G(A)$, then the closure operator $c(-)$ is precisely the composition operator $I\nu(-)$. This follows from the fact that the isolator of a normal subgroup is normal. For this well behaved closure operator, $\mathcal{F}_c = \mathcal{F}_{c_\infty}$.

In the category of all Abelian groups **AB**, let p be a prime and for $A \leq G$, define $c_p^G(A) = A + pG$. It is easy to compute that $c_p^{c_p^G(A)}(A) = A + p^2G$. It follows that $G \in \Delta_{c_p}$ if and only if $pG = 0$ and $G \in \Delta_{c_p\infty}$ if and only if $p^\infty G = 0$ for some ordinal α. Thus $\mathcal{F}_{c_p} = \Delta_{c_p} \subset \Delta_{c_p\infty} = c_p\text{-}Sep = \mathcal{F}_{c_p\infty}$ and $\mathcal{F}_{c_p\infty} = \{G \mid p^\alpha G = 0 \text{ for some ordinal } \alpha\}$.

4. Initial Closure Operators

Let $c(-)$ be an arbitrary closure operator and $\mathcal{F}$ be a reflective subcategory. For each object G, let $\theta_G\colon G \to RG$ denote the reflection map. Following [8], we call $c(-)$ *initial with respect to the class* $\mathcal{F}$, or more simply *$\mathcal{F}$-initial*, provided for each object G and arbitrary subobject $A \leq G$,

$$c^G(A) = \theta_G^{\leftarrow}(c^{RG}(\theta_G(A))).$$

That is, for each $A \leq G$, we have a pullback square:

$$\begin{array}{ccc} c^G(A) & \longrightarrow & G \\ \downarrow & PB & \downarrow{\scriptstyle \theta_G} \\ c^{RG}(\theta_G(A)) & \longrightarrow & RG \end{array}$$

When $c(-)$ is initial with respect to the class $\mathcal{F}_c$ we simply say $c(-)$ is *initial*. This latter terminology is slightly different from the term 'initial' used in [8] in that Dikranjan calls $c(-)$ initial if it is initial with respect to the reflective class Δ_c. If $\mathcal{F}$ is a quotient-reflective subcategory, then the regular closure operator $reg_{\mathcal{F}}(-)$ induced by $\mathcal{F}$ enjoys $\mathcal{F} = \mathcal{F}_{reg_{\mathcal{F}}} = \Delta_{reg_{\mathcal{F}}}$ and the two notions coincide.

To see that this definition of $\mathcal{F}_c$-initial is different, recall the example of the isolator $I(-)$ from **GRP**. It follows that Δ_I is the class of R-groups and the class

$\mathcal{F}_I$ is the class of all torsion-free groups. It is easy to see that $I(-)$ is $\mathcal{F}_I$-initial, but as there are torsion-free groups which are not R-groups, $I(-)$ fails to be Δ_I-initial.

We let $\mathcal{F}$ denote an arbitrary class of objects. Recall that a subobject A of an object G is called *$\mathcal{F}$-regular* provided there exist an object $F \in \mathcal{F}$ and pair of maps f, g: $G \to F$ so that A is the equalizer of the pair $\{f, g\}$. In the category of groups, this simply means $A = \{x \in G \mid f(x) = g(x)\}$.

Now for an arbitrary subobject A of an object G, define $reg_{\mathcal{F}}^G(A)$ to be the intersection of all $\mathcal{F}$-regular subobjects of G containing A. Salbany [16] was the first to introduce such a closure operator for the category of topological spaces; further and more general developments can be found in [4] and [9].

Since $\mathcal{C}$ has coequalizers, if $\widehat{\mathcal{F}}$ denotes the quotient-reflective hull of the class $\mathcal{F}$ (we shall use the terminology 'quotient' from 'quotient map' and 'coequalizer' interchangeably), then $reg_{\widehat{\mathcal{F}}}(-) = reg_{\mathcal{F}}(-)$. The quotient-reflective hull is simply the smallest class containing $\mathcal{F}$ which is closed under formation of products and subobjects. Thus there is no loss of generality in assuming $\mathcal{F}$ to be closed under products and subobjects when considering a regular closure operator. In this case, $reg_{\mathcal{F}}^G(A)$ is $\mathcal{F}$-regular. Note $reg_{\mathcal{F}}(-)$ is an idempotent closure operator.

We call a subobject A of an object G an *$\mathcal{F}$-normal subobject* of G provided A is normal in G, and $G/A \in \mathcal{F}$. For an arbitrary subobject A of an object G, we define $norm_{\mathcal{F}}^G(A)$ to be the intersection of all $\mathcal{F}$-normal subobject of G which contain A. Naturally, we say that $norm_{\mathcal{F}}(-)$ is the *normal closure operator induced by the class $\mathcal{F}$*. In this case, $norm_{\widehat{\mathcal{F}}}(-) = norm_{\mathcal{F}}(-)$ and thus there is no loss of generality in assuming $\mathcal{F}$ to be closed under products and subobjects when considering a normal closure operator. Again, in this case, $norm_{\mathcal{F}}^G(A)$ is $\mathcal{F}$-normal. Note $norm_{\mathcal{F}}(-)$ is an idempotent closure operator. Also note that $norm_{\mathcal{F}}^G(A)$ is always an $\mathcal{F}$-regular subobject.

Simple examples in the category **GRP** exist to show that a normal $\mathcal{F}$-regular subgroup need not be $\mathcal{F}$-normal. For example, take $\mathcal{F}$ to be the class of all torsion-free groups. Then the group of integers $\mathbb{Z}$ is normal $\mathcal{F}$-regular in the group of rationals $\mathbb{Q}$ since the amalgamated coproduct $\mathbb{Q}\amalg_{\mathbb{Z}}\mathbb{Q}$ is torsion-free, but $\mathbb{Z}$ certainly is not $\mathcal{F}$-normal in $\mathbb{Q}$. This suggests considering the closure operator $nreg_{\mathcal{F}}(-)$ where, for an arbitrary subobject A of an object G, we define $nreg_{\mathcal{F}}^G(A)$ to be the intersection of all normal $\mathcal{F}$-regular subobjects of G which contain A. If $\mathcal{F}$ is closed under products and subobjects, then $nreg_{\mathcal{F}}^G(A)$ is a normal $\mathcal{F}$-regular subobject and $nreg_{\mathcal{F}}^G(-)$ is an idempotent closure operator. In general, we have

$$reg_{\mathcal{F}}(-) \leq nreg_{\mathcal{F}}(-) \leq norm_{\mathcal{F}}(-).$$

PROPOSITION 4.1. *For the quotient-reflective class $\mathcal{F}$,*

(i) $\Delta_{reg_{\mathcal{F}}} = \mathcal{F}_{reg_{\mathcal{F}}} = \mathcal{F}$;
(ii) $\Delta_{norm_{\mathcal{F}}} \subseteq \mathcal{F}_{norm_{\mathcal{F}}} = \mathcal{F}$;
(iii) $\Delta_{nreg_{\mathcal{F}}} \subseteq \mathcal{F}_{nreg_{\mathcal{F}}} = \mathcal{F}$;
(iv) *in* **GRP**, $\Delta_{norm_{\mathcal{F}}} = \Delta_{nreg_{\mathcal{F}}} = \mathcal{F} \cap$ **AB**.

Let $\mathcal{F}$ be a quotient reflective class and let θG be the normal subobject of G for which $RG = G/\theta G$ is the reflection of G into $\mathcal{F}$. Again, we let θ_G: $G \to G/\theta G = RG$ denote the canonical (cokernel) reflection map.

The following result is adapted from [11, Proposition 3.11].

LEMMA 4.2 The Fundamental Lemma for Regular Closure Operators. *If $\mathcal{F}$ is a quotient-reflective subcategory of $\mathcal{C}$ and $A \leq G$, then $\theta_G^{\leftarrow}(reg_F^{RG}(\theta_G(A))) = reg_{\mathcal{F}}^G(A)$.*

In view of this lemma, if A is an $\mathcal{F}$-regular subobject of G, then $\theta G \leq A$. Moreover, if $\theta G \leq A$, then A is $\mathcal{F}$-regular in G if and only if $A/\theta G$ is $\mathcal{F}$-regular in $G/\theta G$. More generally, for an arbitrary object G and $\mathcal{F}$-regular B in RG, the inverse image $A = \theta_G^{\leftarrow}(B)$ is $\mathcal{F}$-regular in G. And since the pullback of a normal subobject is normal, we have:

THEOREM 4.3. *If $\mathcal{F}$ is a quotient-reflective subcategory of $\mathcal{C}$, then each of $reg_{\mathcal{F}}(-)$, $nreg_{\mathcal{F}}(-)$, and $norm_{\mathcal{F}}(-)$ is an idempotent $\mathcal{F}$-initial closure operator.*

PROPOSITION 4.4. *If $c(-)$ is initial, then $\tau_c(-)$ is a radical.*

Proof. From $c(-)$ being initial we have $G/\tau_c G \cong RG$ and thus $\theta G = \tau_c G$. But now, since $\tau_c(G/\tau_c G)$ is the pullback of 0 in $R(G/\tau_c G) \cong RG$, we have $\tau_c(G/\tau_c G) = 0$. □

If $c(-)$ is an arbitrary closure operator, for each $A \leq G$, define $\hat{c}^G(A) = \theta_G^{\leftarrow}(c^{RG}(\theta_G(A)))$. Then clearly $\hat{c}(-)$ is expansive and monotone. If f: $G \to H$, then there is an induced map $\hat{f}$: $RG \to RH$ so that $\theta_H f = \hat{f}\theta_G$. Computing,

$$\begin{aligned}\theta_H f(\hat{c}^G(A)) &= \hat{f}\theta_G(\hat{c}^G(A)) = \hat{f}(c^{RG}(\theta_G(A)))\\ &\leq c^{RH}(\hat{f}\theta_G(A)) = c^{RH}(\theta_H f(A)).\end{aligned}$$

Thus we have

$$f(\hat{c}^G(A)) \leq \theta_H^{\leftarrow} c^{RH}(\theta_H f(A)) = \hat{c}^H(f(A)),$$

so $\hat{c}(-)$ *is* a closure operator. We also have immediately, $\mathcal{F}_{\hat{c}} = \{G \mid \tau_{\hat{c}} G = \theta G = 0\} = \mathcal{F}_c$ and $\tau_{\hat{c}}$ is the smallest radical containing τ_c. If $k(-)$ is an initial closure operator with $\mathcal{F}_c = \mathcal{F}_k$ and $c(-) \leq k(-)$ then for every $A \leq G$, $\hat{c}^G(A) \leq k^G(A)$. Thus $\hat{c}(-)$ is called the *initial hull* of $c(-)$.

PROPOSITION 4.5. *Let $\{c_i(-) \mid i \in I\}$ be a family of closure operators each initial with respect to the reflective class $\mathcal{F}$. Then $c(-) = \bigcap_{i \in I} c_i(-)$ is $\mathcal{F}$-initial.*

Proof. For any object G, and $A \leq G$,

$$\begin{aligned}c^G(A) &= \bigcap_{i\in I} c_i^G(A) = \bigcap_{i\in I} \theta_G^{\leftarrow}(c_i^{RG}(\theta_G(A)))\\ &= \theta_G^{\leftarrow}\left(\bigcap_{i\in I} c_i^{RG}(\theta_G(A))\right) = \theta_G^{\leftarrow}(c^{RG}(\theta_G(A))).\end{aligned}$$

Thus $c(-)$ is $\mathcal{F}$-initial. □

PROPOSITION 4.6. *If $c(-)$ and $k(-)$ are initial closure operators with respect to the quotient reflective class $\mathcal{F}$, then the composition $kc(-)$ is initial with respect to $\mathcal{F}$.*

Proof. For each $A \leq G$,

$$\begin{aligned} k^G(c^G(A)) &= \theta_G^{\leftarrow} k^{RG}(\theta_G(c^G(A))) = \theta_G^{\leftarrow} k^{RG}(\theta_G(\theta_G^{\leftarrow}(c^{RG}(A)))) \\ &= \theta_G^{\leftarrow} k^{RG}(c^{RG}(A)). \end{aligned}$$

The uniqueness of the (cokernel, mono)-factorization shows that for any $B \leq RG$, $\theta_G \theta_G^{\leftarrow}(B) = B$. □

For an arbitrary closure operator $c(-)$, there is an *idempotent hull* $c^\infty(-)$ defined in the following natural way [9]. For $A \leq G$, inductively define

$$c^0(A) = A, \qquad c^{\alpha+1}(A) = c^G(c^\alpha(A)),$$
$$c^\beta(A) = \bigvee_{\alpha \in \beta} c^\alpha(A), \quad \beta \text{ a limit ordinal.}$$

From the well poweredness assumption, for each object G, there is an least ordinal $\mu(G)$ so that $c^{\mu(G)+1}(A) = c^{\mu(G)}(A)$ for all subobjects A of G. Define $c^\infty(-)$ by $c^\infty(A) = c^{\mu(G)}(A)$ for each $A \leq G$.

Let G be an arbitrary object and $A \leq G$. We say that A is *saturated* (*with respect to the quotient-reflective subcategory* $\mathcal{F}$) provided $\theta_G^{\leftarrow}\theta_G(A) = A$. We say that $\mathcal{C}$ has *saturated joins of saturated chains* (*with respect to the quotient-reflective subcategory* $\mathcal{F}$) provided, whenever $\{A_i \mid i \in I\}$ is a well-ordered chain of saturated subobjects of an object G, the smallest subobject A of G containing all the A_i's is also saturated.

THEOREM 4.7. *If $\mathcal{C}$ has saturated joins of saturated chains (with respect to the quotient-reflective subcategory $\mathcal{F}$), then for any closure operator $c(-)$ initial with respect to $\mathcal{F}$, its idempotent hull $c^\infty(-)$ is also initial with respect to $\mathcal{F}$.*

Proof. If $c(-)$ is $\mathcal{F}$-initial, then for each natural number n, $c^n(-)$ is $\mathcal{F}$-initial by Proposition 6.4. The hypothesis implies $c^\omega(-)$ is $\mathcal{F}$-initial. The proof now proceeds by induction. □

In the category of all groups, **GRP**, the join of a well-ordered chain of subgroups is simply their union. Thus the next result is straight forward.

PROPOSITION 4.8. *The category* **GRP** *has saturated joins of saturated chains for any quotient-reflective subcategory*

THEOREM 4.9. *If $\mathcal{C}$ has saturated joins of saturated chains with respect to the quotient reflective subcategory $\mathcal{F}$, then for any closure operator $c(-)$, the idempotent hull of its initial hull is the initial hull of its idempotent hull; i.e., $\hat{c}^\infty(-) = \widehat{c^\infty}(-)$.*

Proof. By Theorem 4.7, $\hat{c}^\infty(-)$ is $\mathcal{F}$-initial and as $c^\infty(-) \le \hat{c}^\infty(-)$, we have $\widehat{c^\infty}(-) \le \hat{c}^\infty(-)$.

On the other hand, we will compute directly that $\widehat{c^\infty}^2(-) = \widehat{c^\infty}(-)$. For if $A \le G$, then

$$\begin{aligned}
\widehat{c^\infty}(\widehat{c^\infty}(A)) &= \theta_G^{\leftarrow}(c^{\infty RG}(\theta_G(\widehat{c^\infty}(A)))) = \theta_G^{\leftarrow}(c^{\infty RG}(\theta_G\theta_G^{\leftarrow}c^{\infty RG}(A))) \\
&= \theta_G^{\leftarrow}(c^{\infty RG}(c^{\infty RG}(A))) = \theta_G^{\leftarrow}(c^{\infty RG}(A)) \\
&= \widehat{c^\infty}(A).
\end{aligned}$$

Thus $\hat{c}^\infty(-) \le \widehat{c^\infty}(-)$, since $\hat{c}(-) \le \widehat{c^\infty}(-)$. □

5. Weakly Hereditary Initial Closure Operators

In this section we explore the relationship between weakly hereditary initial closure operators and properties of the quotient-reflective class $\mathcal{F}$. In doing so, we extend some recent results of the author in [12]. Let $c(-)$ be an initial closure operator with $\mathcal{F}_c = \mathcal{F}$.

The next definition is adapted from Clementino [7]. We say the *$\mathcal{F}$-reflections are hereditary* (*with respect to the class* $\{0 \to G \mid G \in \mathcal{C}\}$) provided for each $\mathcal{C}$-object G, the map $P \to 0$ in the pullback diagram:

$$\begin{array}{ccc}
P & \longrightarrow & 0 \\
\downarrow & PB & \downarrow \\
G & \longrightarrow & RG
\end{array}$$

is the $\mathcal{F}$-reflection of P. It is clear that in the above pullback diagram, $P = \theta G \times 0$. Thus P is canonically isomorphic to $\theta G = \tau_c G$. To have $P \to 0$ be the $\mathcal{F}$-reflection is the same as saying $\theta P = P$; i.e., $\tau_c P = P$. This is the same as saying $\tau_c^2 G = \tau_c G$.

Following the notation and concepts of [2], denote by $\mathcal{F}^{\leftarrow}$ the class $\{G \in \mathcal{C} \mid Hom_{\mathcal{C}}(G, F) = 0 \text{ for all } F \in \mathcal{F}\}$. Since this class is always a class of torsion objects for a generalized torsion theory, we denote $\mathcal{F}^{\leftarrow}$ by $\mathcal{T}$. We observe that the class $\mathcal{T}^{\rightarrow} = \{G \in \mathcal{C} \mid Hom_{\mathcal{C}}(T, G) = 0 \text{ for all } T \in \mathcal{T}\}$ contains the class $\mathcal{F}$, is closed under formation of products, subobjects, and extensions, and that the pair $(\mathcal{T}, \mathcal{T}^{\rightarrow})$ is a generalized torsion theory, see [2] for more details. Note that it is easy to identify the class $\mathcal{T}$, as it is clear that $\mathcal{T} = \{G \in \mathcal{C} \mid \tau_c G = G\}$. The class $\mathcal{T}$ is closed under formation of coproducts, images, and extensions.

Since $\tau_c^\infty(G) \in \mathcal{T}$, $G \in \mathcal{T}^{\rightarrow}$ implies $\tau_c^\infty(G) = 0$. On the other hand, if $T \in \mathcal{T}$ and $f\colon T \to G$, then $f(T) \subset \tau_c^\infty(G)$. Thus $\tau_c^\infty(G) = 0$ implies $G \in \mathcal{T}^{\rightarrow}$. This shows that $\mathcal{T}^{\rightarrow} = \{G \mid \tau_c^\infty(G) = 0\}$, and consequently we have $\mathcal{F} = \mathcal{T}^{\rightarrow}$ if and only if $\tau_c^2 = \tau_c$.

If $A \leq G$ is c-closed, then by initiality, we have a pullback diagram:

$$\begin{array}{ccccc} RA & \longleftarrow & A & \longrightarrow & G \\ \downarrow{\scriptstyle \rho_A} & & \downarrow{\scriptstyle \theta} & PB & \downarrow{\scriptstyle \theta_G} \\ B & = & c^{RG}(\theta_G(A)) & \longrightarrow & RG \end{array}$$

From $B \in \mathcal{F}$, there is a unique map ρ_A: $RA \to B$ with $\rho_A \theta_A = \theta$: $A \to c^{RG}(\theta_G(A))$. Note that this map ρ_A is necessarily a cokernel since θ is a cokernel. It follows that ρ_A is monic if and only if $\tau_c A = \tau_c G$. If $\tau_c^2 = \tau_c$ and A is c-closed in G, then $\tau_c G \leq A$. From τ_c being a preradical, we have $\tau_c^2 G \leq \tau_c A \leq \tau_c G$. Thus $\tau_c^2 G = \tau_c G$ and this implies that ρ_A is a monomorphism. On the other hand, if each ρ_A is monic, then $\rho_{\tau_c G}$ is monic, which forces $\tau_c^2 G = \tau_c G$.

Finally note that if $c(-)$ is weakly hereditary, then

$$\tau_c^2 G = c_c^{\tau_c G}(0) = c_c^{c_c^G(0)}(0) = c_c^G(0) = \tau_c G.$$

Thus we have proved the following theorem.

THEOREM 5.1. *The following are equivalent for an initial closure operator $c(-)$ and imply that $\mathcal{F} = \mathcal{F}_c$ is closed under extensions*:

(i) $\tau_c^2 = \tau_c$, *that is, $\tau_c(-)$ is an idempotent radical*;
(ii) $\mathcal{F}$*-reflections are hereditary*;
(iii) $\mathcal{F} = \mathcal{T}^{\rightarrow}$;
(iv) ρ_A *is monic for every c-closed $A \leq G$.*

Moreover, if $c(-)$ is weakly hereditary, then these conditions hold.

For an idempotent initial closure operator we can say a bit more. Let A be c-closed in B and let B be c-closed in G. Then we have the commutative diagram:

$$\begin{array}{ccccc} A & \longrightarrow & B & \longrightarrow & G \\ \downarrow & 2 & \downarrow & 1 & \downarrow{\scriptstyle \theta_G} \\ \theta_G(A) & \longrightarrow & \theta_G(B) & \longrightarrow & RG \end{array}$$

It follows that $\theta_G(A)$ is c-closed in RB and $\theta_G(B)$ is c-closed in RG by initiality. In particular, square 1 is a pullback square by initiality. Since B is c-closed in G, $\tau_c B \leq \tau_c G \leq B$. Thus $\tau_c G/\tau_c B \leq B/\tau_c B$. If $\mathcal{F} = \mathcal{T}^{\rightarrow}$, then $\tau_c G/\tau_c B \in \mathcal{T} \cap \mathcal{F} = 0$ and thus $\tau_c G = \tau_c B$. In this case then, $RB \cong \theta_G(B) = B/\tau_c G = B/\tau_c B$. Hence it follows that $RA = \theta_G(A) = A/\tau_c A = A/\tau_c B = A/\tau_c G$. This means that square 2 is a pullback square. Thus the outer rectangle is a pullback square. Hence, by initiality, A is c-closed in G if and only if RA is c-closed in RG. This argument yields a sharpening of [7, Proposition 4.2] and of [12, Theorem 4.4] as follows:

THEOREM 5.2. *The idempotent initial closure operator $c(-)$ is weakly hereditary on $\mathcal{C}$ if and only if $c(-)$ is weakly hereditary on $\mathcal{F}$ and $\mathcal{F} = \mathcal{T}^{\rightarrow}$.*

Proof. An idempotent closure operator is weakly hereditary if and only if the composition of c-closed embeddings is a c-closed embedding. If $c(-)$ is weakly hereditary, then certainly it is weakly hereditary on $\mathcal{F}$ and by the previous theorem $\mathcal{F} = \mathcal{T}^{\rightarrow}$. Conversely, it follows that the composition of two c-closed embeddings is a c-closed embedding for objects belonging to $\mathcal{F}$. Thus by the above arguments, this implies the composition of c-closed embeddings is a c-closed embedding in general. That is, $c(-)$ is weakly hereditary. □

It is easy to see that for an idempotent closure operator $c(-)$, we can replace the condition '$c(-)$ is weakly hereditary on $\mathcal{F}$' by the condition 'for each $G \in \mathcal{F}$, c-closed $B \leq G$, and for any $A \leq B$, $c^B(A) = c^G(A) \cap B$' (cf. [9, Proposition 3.5.1]); however, the author knows of no example where it is easier to verify this condition than it is to verify the composition of two c-closed embeddings is a c-closed embedding.

We wish to relate the requirement that $\mathcal{F} = \mathcal{T}^{\rightarrow}$ to $\mathcal{F}$ being closed under formation of extensions. Clementino's theorem states that when $\mathcal{C}$ is an Abelian category, $reg_{\mathcal{F}}(-)$ is weakly hereditary precisely when $\mathcal{F}$ is closed under extensions. A similar result [12, Theorem 4.9] shows this result to hold for the category of all groups, **GRP**, but the verification is somewhat more involved than for the Abelian situation.

We say that a monomorphism $A \rightarrow B$ and a cokernel $\alpha\colon A \rightarrow F$ have a *semiextension* provided there exist an H, with $F \leq H$, and a map $\beta\colon B \rightarrow H$ so that the following diagram commutes:

$$\begin{array}{ccc} A & \xrightarrow{\alpha} & F \\ \downarrow & & \downarrow \\ B & \xrightarrow{\beta} & H \end{array}$$

Note that $\alpha(A) = F \leq \beta(B)$, so that without loss of generality, we may assume β to be a cokernel. We call the above semiextension an $\mathcal{F}$*-semiextension* provided $F \in \mathcal{F}$ implies H can be chosen to belong to $\mathcal{F}$. We call a monomorphism $A \rightarrow B$ $\mathcal{F}$*-semiextendible* provided for every cokernel $\alpha\colon A \rightarrow F$, with $F \in \mathcal{F}$, has an $\mathcal{F}$*-semiextension.*

PROPOSITION 5.3. *If $\tau_c G \rightarrow G$ is $\mathcal{F}$-semiextendible for every G, then $\mathcal{F} = \mathcal{T}^{\rightarrow}$.*

Proof. The hypothesis implies $\tau_c G/\tau_c^2 G = 0$. □

We say that the category $\mathcal{C}$ enjoys the *(cokernel, kernel)-Pushout Property* or more simply the *Pushout Property*, if in the following pushout square

$$\begin{array}{ccc} A & \xrightarrow{m} & F \\ e \downarrow & PO & \downarrow f \\ B & \xrightarrow{n} & H \end{array}$$

e a cokernel and m a normal monomorphism imply that n is a monomorphism.

THEOREM 5.4. *If $\mathcal{C}$ enjoys the Pushout Property, then $\mathcal{F}$ closed under extensions implies for every G, $\tau_c G \to G$ is $\mathcal{F}$-semiextendible.*

Proof. Consider the following diagram with indicated pushout square:

$$\begin{array}{ccccc} \tau_c G & \longrightarrow & G & \to & G/\tau_c G \\ \downarrow & PO & \downarrow & & \| \\ \tau_c G/\tau_c^2 G & \longrightarrow & P & \longrightarrow & Q \end{array}$$

The image of a normal subobject is normal, and it is easy to see that the quotient Q is isomorphic to $G/\tau_c G$. If $\mathcal{F}$ is closed under extensions, then P belongs to $\mathcal{F}$. □

The next proposition also shows that closure under extensions implies $\mathcal{F} = \mathcal{T}^{\to}$ under common circumstances.

PROPOSITION 5.5. *If fully invariant implies normal, then $\mathcal{F}$ closed under extensions implies $\mathcal{F} = \mathcal{T}^{\to}$.*

Proof. Under this hypothesis, $\tau_c^2 G$ is normal in G, and thus we have the short exact sequence

$$0 \to \tau_c G/\tau_c^2 G \to G/\tau_c^2 G \to G/\tau_c G \to 0$$

and $\mathcal{F}$ closed under extensions implies $G/\tau_c^2 G$ belongs to $\mathcal{F}$. This in turn implies $\tau_c = \tau_c^2$. □

COROLLARY 5.6. *In the category of groups,* **GRP**, *the following are equivalent for an idempotent initial closure operator $c(-)$:*

(i) *$c(-)$ is weakly hereditary;*
(ii) *$c(-)$ is weakly hereditary on $\mathcal{F}$ and $\mathcal{F}$ is closed under extensions.*

Each of our main examples of initial closure operators, $reg_{\mathcal{F}}(-)$, $nreg_{\mathcal{F}}(-)$, $norm_{\mathcal{F}}(-)$, and the isolator $I(-)$ are idempotent. Thus it is of interest to classify when these operators are weakly hereditary. The isolator is weakly hereditary for $\mathcal{F}_I$ is the class of torsion-free groups, which is closed under extensions, and the composition of two isolated monos is an isolated mono. The characterization for $reg_{\mathcal{F}}(-)$ was the main result for **GRP** contained in [12] already mentioned.

COROLLARY 5.7. *In the category of groups,* **GRP**, *$reg_{\mathcal{F}}(-)$ is weakly hereditary if and only if $\mathcal{F}$ is closed under extensions.*

Proof. If $reg_{\mathcal{F}}(-)$ is weakly hereditary, then $\mathcal{F} = \mathcal{T}^{\to}$ and $\mathcal{F}$ is closed under extensions. Conversely, it was shown in [12] that $\mathcal{F}$ being closed under extensions implies that $reg_{\mathcal{F}}(-)$ acts discretely on $\mathcal{F}$ and thus is weakly hereditary on $\mathcal{F}$. □

Except for trivial cases, the idempotent initial closure operator $nreg_{\mathcal{F}}(-)$ and $norm_{\mathcal{F}}(-)$ are never weakly hereditary.

THEOREM 5.8. *Let $\mathcal{F}$ be a non-trivial quotient reflective class of groups closed under extensions. Then neither $nreg_{\mathcal{F}}(-)$ nor $norm_{\mathcal{F}}(-)$ is weakly hereditary.*

Proof. Since $\mathcal{F}$ is non-trivial and is closed under extensions, it must contain all free groups [12, Theorem 4.9]. Let $F = F\{x_1, x_2, \ldots\}$ be the free group on countably many generators. The group F has an automorphism ϕ defined by $\phi(x_1) = x_2$, $\phi(x_{2k}) = x_{2k+2}$, and $\phi(x_{2k+1}) - \phi(x_{2k-1})$ for $k \geq 1$. Let G be the semidirect product of F by $\langle\phi\rangle \cong \mathbb{Z}$. Then we have a short exact sequence

$$0 \to F \to G \to \mathbb{Z} \to 0,$$

and thus as $\mathcal{F}$ is closed under extensions, $G \in \mathcal{F}$. Furthermore, F is both $\mathcal{F}$-normal and normal $\mathcal{F}$-regular in G; i.e., $nreg^G_{\mathcal{F}}(F) = F$ and $norm^G_{\mathcal{F}}(F) = F$. Let N be the $\mathcal{F}$-normal closure of $x_1 \in F$ and observe that $F/N \cong F\{x_2, x_3, \ldots\}$ and thus is both $\mathcal{F}$-normal and normal $\mathcal{F}$-regular in F. But N is not normal in G since $\phi^{-1}x_1\phi = \phi(x_1) = x_2$ and $x_2 \notin N$. Thus the composition of c-closed embeddings is not a c-closed embedding for $c(-) = nreg_{\mathcal{F}}(-)$ and $c(-) = norm_{\mathcal{F}}(-)$. □

6. Categorical Compactness

Categorical compactness with respect to a closure operator is a topic of current interest. First developed for weakly hereditary closure operators in [15] (although this terminology was not used), Castellini [5] developed the notion for regular closure operators and building on their work on closure operators in general [9], Dikranjan and Giuli thoroughly investigated categorical compactness for categories of modules [10].

The definition of categorical compactness with respect to a closure operator is motivated by the Kuratowski–Mrówka theorem which characterizes compact topological spaces. Let $c(-)$ be a closure operator. An object G is called *c-compact* provided for every object H, the second projection map $\pi_2\colon G \times H \to H$ preserves c-closed subobjects under the (cokernel, mono) factorization.

In this section, we derive two results for categorical compactness for an initial closure operator $c(-)$, which reduce the testing for compactness to objects belonging to the quotient reflective class $\mathcal{F}_c$. To do so, we assume that the category $\mathfrak{C}$ enjoys the following property for commutative diagrams, having short exact rows, of the following form:

$$\begin{array}{ccccccccc} 0 & \longrightarrow & K & \longrightarrow & A & \longrightarrow & B & \longrightarrow & 0 \\ & & \| & & \downarrow{\scriptstyle\lambda} & & \| & & \\ 0 & \longrightarrow & K & \longrightarrow & P & \longrightarrow & B & \longrightarrow & 0. \end{array} \tag{6.1}$$

In such a diagram, the map λ is necessarily monic. We require that it also be a cokernel, and hence an isomorphism. Many familiar categories enjoys this property including the category of all groups GRP, any Abelian category, and the category of associative rings. This property has the useful application:

LEMMA 6.1. *Given a commutative diagram with short exact rows of the form:*

$$\begin{array}{ccccccccc} 0 & \longrightarrow & K & \longrightarrow & A & \xrightarrow{\phi} & B & \longrightarrow & 0 \\ & & \| & & \downarrow{\scriptstyle f} & & \downarrow{\scriptstyle g} & & \\ 0 & \longrightarrow & K & \longrightarrow & G & \xrightarrow{\psi} & H & \longrightarrow & 0 \end{array}$$

the right hand square is a pullback square.

The proof is obtained straight forwardly by taking the pullback of ψ and g and obtaining a diagram of the form (6.1) above.

Another useful consequence of this property is, given any short exact sequence

$$0 \to K \to G \xrightarrow{\phi} G/K \to 0 \tag{6.2}$$

and subobject $A \leq G$, the smallest subobject of G containing both A and K is $\phi^{\leftarrow}\phi(A)$. We formalize this by letting $A \bullet K$ denote the smallest subobject of G containing both A and K in the following.

PROPOSITION 6.2. *For any short exact sequence of the form* (6.2) *and subobject* $A \leq G$, $A \bullet K = \phi^{\leftarrow}\phi(A)$.

Proof. Since $\phi(A) \leq \phi(A \bullet K) \leq \phi\phi^{\leftarrow}\phi(A) = \phi(A)$, we have $\phi(A) = \phi(A \bullet K)$. This means we have a commutative square with short exact rows

$$\begin{array}{ccccccccc} 0 & \longrightarrow & K & \longrightarrow & A \bullet K & \longrightarrow & \phi(A) & \longrightarrow & 0 \\ & & \| & & \downarrow{\scriptstyle \lambda} & & \| & & \\ 0 & \longrightarrow & K & \longrightarrow & \phi^{\leftarrow}\phi(A) & \longrightarrow & \phi(A) & \longrightarrow & 0 \end{array}$$

The assumption concerning such squares implies λ is an isomorphism.

Now let $c(-)$ be an initial closure operator and for any object G, let $G \xrightarrow{\theta_G} RG = G/\theta G$ denote the reflection of G into $\mathcal{F}_c$. Let G and H be arbitrary and let $A \leq G \times H$ be c-closed. Since $c(-)$ is initial, we have a pullback square

$$\begin{array}{ccc} A & \longrightarrow & \theta_G \times \theta_H(A) = B \\ \downarrow & & \downarrow \\ G \times H & \longrightarrow & G/\theta G \times H/\theta H \end{array} \tag{6.3}$$

with the vertical maps inclusions. Applying the second projection map π_2 to this pullback square, we obtain a commutative square

$$\begin{array}{ccc} \pi_2(A) & \longrightarrow & \pi_2(B) \\ \downarrow & & \downarrow \\ H & \longrightarrow & H/\theta H. \end{array} \tag{6.4}$$

Since (6.3) is a pullback, $\theta(G \times H) \leq A$ which implies $\theta H \leq \pi_2(A)$. This implies we have a commutative square with short exact rows

$$\begin{array}{ccccccccc} 0 & \longrightarrow & \theta H & \longrightarrow & \pi_2(A) & \longrightarrow & \pi_2(B) & \longrightarrow & 0 \\ & & \| & & \downarrow & & \downarrow & & \\ 0 & \longrightarrow & \theta H & \longrightarrow & H & \longrightarrow & H/\theta H & \longrightarrow & 0 \end{array} \tag{6.5}$$

By the previous lemma, we have (6.4) is a pullback square.

This observation provides simple proofs to the next two theorems, which subsume Theorems 4.1, 4.2, 4.3 and 4.4 of [13] for $\mathcal{F}$-regular and $\mathcal{F}$-normal closure operators. □

THEOREM 6.3. *For an initial closure operator $c(-)$ and object G, the following are equivalent*:

(i) *G is c-compact*;
(ii) *$G/\theta G$ is c-compact*;
(iii) *every $\mathcal{F}_c$-image of G is c-compact.*

Proof. Since the class of categorically compact objects is closed under formation of images, it suffices to show that (ii) implies (i). If G and H are arbitrary and $A \leq G \times H$ is c-closed, then we have a commutative square

$$\begin{array}{ccc} \theta_G \times \theta_H(A) = B & \longrightarrow & \pi_2(B) \\ \downarrow & & \downarrow \\ G/\theta G \times H/\theta H & \xrightarrow{\pi_2} & H/\theta H \end{array}$$

where the vertical maps are inclusions. Since A is c-closed, B is c-closed in $G/\theta G \times H/\theta H$. The c-compactness of $G/\theta G$ implies that $\pi_2(B)$ is c-closed in $H/\theta H$. From the observation above and the fact that (6.4) is a pullback square, this implies $\pi_2(A)$ is c-closed in H. Thus G is c-compact. □

THEOREM 6.4. *For an initial closure operator $c(-)$ and object G belonging to $\mathcal{F}_c$, the following are equivalent*:

(i) *G is c-compact*;
(ii) *for every $H \in \mathcal{F}_c$, π_2: $G \times H \to H$ preserves c-closed subobjects.*

Proof. We need only observe that (ii) implies (i). To that end, let H be arbitrary and $A \leq G \times H$ be c-closed. Note that $\theta G = 0$, so $G \cong G/\theta G$. From the initiality of $c(-)$, $\theta_G \times \theta_H(A) = 1_G \times \theta_H(A) = B$ is c-closed in $G/\theta G \times H/\theta H \cong G \times H/\theta H$. The hypothesis of (ii) implies that $\pi_2(B)$ is c-closed in $H/\theta H$ and the above observation implies that $\pi_2(A)$ is c-closed in H. Thus G is c-compact. □

The next two results are direct consequences of Theorems 6.3 and 6.4.

COROLLARY 6.5 [13, Theorem 5.18]. *In the category of all groups* **GRP**, *let* $I(-)$ *be the isolator. Then the following are equivalent*:

(i) *G is I-compact*;
(ii) *$G/\tau_I G$ is I-compact*;
(iii) *every torsion-free image of G is I-compact.*

Proof. In this case, $\mathcal{F}_I$ is the class of all torsion-free groups and $\theta G = \tau_I G$ is the generalized torsion subgroup of G, namely $I^G(0)$. □

COROLLARY 6.6. *In the category of all groups* **GRP**, *the following are equivalent for a torsion-free group G*:

(i) *G is I-compact*;
(ii) *for each torsion-free group H, π_2: $G \times H \to H$ preserves isolated subgroups.*

The class of all I-compact groups contains the class of all groups G having $G/\tau_I G$ radicable (every element has an nth-root for each n). It is an open question to decide if these classes coincide. For other details, see [13, Section 5].

Acknowledgment

The author wishes to thank Dikran Dikranjan and Gary L. Walls for helpful conversations. Support from the Technikon Pretoria and the FRD is gratefully acknowledged.

References

1. Arhangel'skiĭ, A. and Wiegant, R.: Connectednesses and disconnectednesses in topology, *Gen. Topology Appl.* **5** (1975), 9–33.
2. Cassidy, C., Hébert, M. and Kelly, G. M.: Reflective subcategories, localizations and factorization systems, *J. Austral. Math. Soc. Ser. A* **38** (1985), 287–329.
3. Castellini, G.: Closure operators, monomorphisms and epimorphisms in categories of groups, *Cahiers Topologie Géom. Différentielle Catégoriques* **XXVII**-2 (1986), 151–167.
4. Castellini, G.: Compact objects, surjectivity of epimorphisms and compactifications, *Cahiers Topologie Géom. Différentielle Catégoriques* **XXXI**-1 (1990), 53–65.
5. Castellini, G.: Regular closure and compactness, *Cahiers Topologie Géom. Différentielle Catégoriques* **XXXIII**-1 (1992), 21–31.
6. Castellini, G. and Giuli, E.: Hereditarity of closure operators and injectivity, *Comment. Math. Univ. Carolin.* **33** (1992), 149–157.
7. Clementino, M. M.: Weakly hereditary closure operators, *Topology Appl.* **49** (1993), 129–139.
8. Dikranjan, D.: Semiregular closure operators and epimorphisms in topological categories, *Supplemento ai Renconti del Circolo Matematico di Palermo, Serie* II **29** (1992), 105–160.
9. Dikranjan, D. and Giuli, E.: Closure operators I, *Topology Appl.* **27** (1987), 129–143.
10. Dikranjan, D. and Giuli, E.: Factorizations, injectivity and compactness in categories of modules, *Comm. Algebra* **19** (1991), 45–83.

11. Dikranjan, D., Giuli, E. and Tholen, W.: Closure operators II, in *Categorical Topology and its Relation to Analysis, Algebra, and Combinatorics, Proceedings International Conference Prague, 1988*, World Scientific, Singapore, 1989, pp. 297–335.
12. Fay, T. H.: Weakly hereditary regular closure operators, *Cahiers Topologie Géom. Différentielle Catégoriques* **XXXVII**(4) (1996), 279–293.
13. Fay, T. H. and Walls, G. L.: Regular and normal closure operators and categorical compactness for groups, *Appl. Categ. Struct.* **3** (1995), 261–278.
14. Herrlich, H. and Strecker, G. E.: *Category Theory*, Allyn and Bacon Inc., Boston, 1973.
15. Herrlich, H., Salicrup, G. and Strecker, G. E.: Factorizations, denseness, separation, and relatively compact objects, *Topology Appl.* **27** (1987), 157–169.
16. Salbany, S.: Reflective subcategories and closure operators, in *Proceedings Conference Categorical Topology*, Mannheim, 1975, Lecture Notes in Mathematics 540, Springer-Verlag, Berlin, pp. 548–565.

G. Brümmer & C. Gilmour (eds), Papers in Honour of Bernhard Banaschewski.

Strongly ϑ-Continuous Functions and Topologies on Function Spaces

D. N. GEORGIOU
Department of Civil Engineering, University of Thessaly, Pedion Areos, 383 34 Volos, Greece

B. K. PAPADOPOULOS
Department of Mathematics, Democritus University of Thrace, 67100 Xanthi, Greece

(Received: 21 May 1996; accepted: 24 March 1999)

Abstract. In this paper we study the function space of strongly ϑ-continuous functions and generalize some basic results of R. Arens, D. Dugundji and A. Di Concilio.

Mathematics Subject Classifications (2000): 54C35, 54A20.

Key words: strongly ϑ-continuous maps and function spaces.

Introduction

Let Y, Z be topological spaces and let f be a map of Y into Z. Then f is *strongly ϑ-continuous* (respectively, *ϑ-continuous*) at $y \in Y$ if for every open neighbourhood V of $f(y)$ there exists an open neighbourhood U of y such that $f(\mathrm{Cl}(U)) \subseteq V$ (respectively, $f(\mathrm{Cl}(U)) \subseteq \mathrm{Cl}(V)$.) (Let Y be a space, then by $\mathrm{Cl}(A)$ we denote the closure of A in Y.) The map f is *strongly ϑ-continuous* (respectively, *ϑ-continuous*) on Y if it is strongly ϑ-continuous (respectively, ϑ-continuous) at each point of Y. (See, for example, [7, 8, 11] and [3].) Clearly, every strongly ϑ-continuous function is continuous and every continuous function is ϑ-continuous. In what follows by $S\Theta(Y, Z)$ (respectively, by $\Theta(Y, Z)$) we denote the set of all strongly ϑ-continuous (respectively, ϑ-continuous) maps of Y into Z. If τ is a topology on the set $S\Theta(Y, Z)$ (respectively, $\Theta(Y, Z)$), then the corresponding topological space is denoted by $S\Theta_\tau(Y, Z)$ (respectively, $\Theta_\tau(Y, Z)$). Also by $C(Y, Z)$ we denote the set of all continuous maps of a space Y into the space Z.

The pointwise topology τ_p on $C(Y, Z)$ is generated by the sets of the form

$$(\{y\}, U) = \{f \in C(Y, Z) : f(y) \in U\},$$

where $y \in Y$ and U is an open set of Z.

The compact open topology τ_c on $C(Y, Z)$ is generated by the sets of the form

$$(K, U) = \{f \in C(Y, Z) : f(K) \subseteq U\},$$

where K is a compact subset of Y and U is an open set of Z.

Let X be a set. A *net* in X is a map $S : \Lambda \to X$ of a directed set Λ into X. The net S is also denoted by $\{s_\lambda,\ \lambda \in \Lambda\}$, where $s_\lambda = S(\lambda)$.

A net $\{y_\lambda : \lambda \in \Lambda\}$ in a space Y *converges* (respectively, ϑ*-converges*) to y ($y_\lambda \to y$) (respectively, ($y_\lambda \stackrel{\vartheta}{\to} y$)) if for each neighbourhood U of y there is some $\lambda_0 \in \Lambda$ such that $\lambda \geq \lambda_0$ implies $y_\lambda \in U$ (respectively, $y_\lambda \in \mathrm{Cl}(U)$). Evidently any net which converges to y ϑ-converges to y. (See [4, 9] and [5].)

A filter $\mathfrak{R}$ on Y ϑ-converges to y ($\mathfrak{R} \stackrel{\vartheta}{\to} y$) if any closed neighbourhood of y belongs to $\mathfrak{R}$. (See [4].)

A function f is continuous (respectively, ϑ-continuous) at $y \in Y$ if and only if whenever $\mathfrak{R} \to y$ (respectively, $\mathfrak{R} \stackrel{\vartheta}{\to} y$) in Y, then $f(\mathfrak{R}) \to f(y)$ (respectively, $f(\mathfrak{R}) \stackrel{\vartheta}{\to} f(y)$) in Z. (See [9] and [4].)

A function f is continuous (respectively, ϑ-continuous) at $y \in Y$ if and only if whenever $y_\lambda \to y$ (respectively, $y_\lambda \stackrel{\vartheta}{\to} y$) in Y, then $f(y_\lambda) \to f(y)$ (respectively, $f(y_\lambda) \stackrel{\vartheta}{\to} f(y)$) in Z. (See [9] and [4].)

A net $\{f_\mu, \mu \in M\}$ in $C(Y, Z)$ (respectively, in $\Theta(Y, Z)$) *continuously converges* (respectively, ϑ*-continuously converges*) to $f \in \Theta(Y, Z)$ if for any net $\{y_\lambda : \lambda \in \Lambda\}$ in Y such that $y_\lambda \to y$ (respectively, $y_\lambda \stackrel{\vartheta}{\to} y$) the net $f_\mu(y_\lambda) \to f(y)$ (respectively, $f_\mu(y_\lambda) \stackrel{\vartheta}{\to} f(y)$) in Z. (See [1, 6, 10] and [4].)

A net $\{f_\mu, \mu \in M\}$ in $C(Y, Z)$ (respectively, in $\Theta(Y, Z)$) continuously converges (respectively, ϑ-continuously converges) to $f \in C(Y, Z)$ (respectively, $f \in \Theta(Y, Z)$) if for any $y \in Y$ and any neighbourhood V of $f(y)$ there is a $\mu_0 \in M$ and an open neighbourhood U of y such that $f_\mu(U) \subseteq V$ (respectively, $f_\mu(\mathrm{Cl}(U)) \subseteq \mathrm{Cl}(V)$), for every $\mu \in M$, $\mu \geq \mu_0$. (See [1] and [4].)

By $\mathcal{C}^*$ (respectively, $\mathcal{C}^*_\vartheta$) we denote the class of all pairs $(\{f_\lambda, \lambda \in \Lambda\}, f)$, where $\{f_\lambda, \lambda \in \Lambda\}$ is a net in $C(Y, Z)$ (respectively, in $\Theta(Y, Z)$) which continuously converges (respectively, ϑ-continuously converges) to $f \in C(Y, Z)$ (respectively, $f \in \Theta(Y, Z)$). If τ is a topology on $C(Y, Z)$ (respectively, on $\Theta(Y, Z)$) then by $\mathcal{C}(\tau)$ (respectively, by $(\mathcal{C}(\tau))_\vartheta$) we denote the class of all pairs $(\{f_\lambda, \lambda \in \Lambda\}, f)$, where $\{f_\lambda, \lambda \in \Lambda\}$ is a net in $C(Y, Z)$ (respectively, in $\Theta(Y, Z)$) which converges (respectively, ϑ-converges) to $f \in C(Y, Z)$ (respectively, to $f \in \Theta(Y, Z)$) in the topology τ.

The following criteria are given in [1] and [4]:

(1) A topology τ on $C(Y, Z)$ (respectively, on $\Theta(Y, Z)$) is splitting (respectively, ϑ-splitting) if and only if $\mathcal{C}^* \subseteq \mathcal{C}(\tau)$ (respectively, $\mathcal{C}^*_\vartheta \subseteq (\mathcal{C}(\tau))_\vartheta$).

(2) A topology τ on $C(Y, Z)$ (respectively, on $\Theta(Y, Z)$) is jointly continuous (respectively, ϑ-jointly continuous) if and only if $(\mathcal{C}(\tau)) \subseteq \mathcal{C}^*$ (respectively, $(\mathcal{C}(\tau))_\vartheta \subseteq \mathcal{C}^*_\vartheta$).

Throughout this paper the word “space” means “topological space”.

1. Strongly ϑ-Continuous Functions

THEOREM 1.1. *The following propositions are true*:

(1) *A map f of a space Y into a space Z is strongly ϑ-continuous at $y \in Y$ if and only if for every net $\{y_\lambda, \lambda \in \Lambda\}$ in Y for which $y_\lambda \overset{\vartheta}{\to} y$ we have that $f(y_\lambda) \to f(y)$ in Z.*

(2) *A map f of a space Y into a space Z is strongly ϑ-continuous at $y \in Y$ if and only if for every filter $\mathfrak{R}$ which ϑ-converges to $y \in Y$ we have that $f(\mathfrak{R}) \to f(y)$.*

The proof of this theorem is clear.

DEFINITION 1.2. A net $\{f_\mu, \mu \in M\}$ in $S\Theta(Y, Z)$ *strongly ϑ-continuously converges to $f \in S\Theta(Y, Z)$* if and only if for every net $\{y_\lambda, \lambda \in \Lambda\}$ in Y which ϑ-converges to $y \in Y$ we have that the net $\{f_\mu(y_\lambda), (\lambda, \mu) \in \Lambda \times M\}$ converges to $f(y)$ in Z.

THEOREM 1.3. *A net $\{f_\mu, \mu \in M\}$ in $S\Theta(Y, Z)$ strongly ϑ-continuously converges to $f \in S\Theta(Y, Z)$ if and only if for every $y \in Y$ and for every open neighbourhood V of $f(y)$ in Z there exists an element $\mu_0 \in M$ and an open neighbourhood U of y in Y such that*

$$f_\mu(\mathrm{Cl}(U)) \subseteq V,$$

for every $\mu \geq \mu_0$, $\mu \in M$.

Proof. Let $y \in Y$ and let V be an open neighbourhood of $f(y)$ in Z such that for any $\mu \in M$ and for every open neighbourhood U of $y \in Y$ there exists $\mu' \geq \mu$ such that

$$f_{\mu'}(\mathrm{Cl}(U)) \nsubseteq V.$$

Then for every open neighbourhood U of y in Y we can choose a point $y_U \in \mathrm{Cl}(U)$ such that $f_{\mu'}(y_U) \notin V$. Let $\mathcal{N}(y)$ be the set of all open neighbourhoods of y in Y. The set $\mathcal{N}(y)$ with the relation of inverse inclusion (that is, $U_1 \leq U_2$ if and only if $U_2 \subseteq U_1$) forms a directed set. Clearly, the net $\{y_U, U \in \mathcal{N}(y)\}$, ϑ-converges to y but the net $\{f_\mu(y_U), (U, \mu) \in \mathcal{N}(y) \times M\}$ does not converge to $f(y)$ in Z.

Conversely, let $\{y_\lambda, \lambda \in \Lambda\}$ be a net in $S\Theta(Y, Z)$ which ϑ-converges to y in Y and let V be an arbitrary open neighbourhood of $f(y)$ in Z. By assumption there exist an open neighbourhood U of y in Y and an element $\mu_0 \in M$ such that $f_\mu(\mathrm{Cl}(U)) \subseteq V$, for every $\mu \geq \mu_0$, $\mu \in M$. Since the net $\{y_\lambda, \lambda \in \Lambda\}$ ϑ-converges to y in Y. There exists $\lambda_0 \in \Lambda$ such that $y_\lambda \in \mathrm{Cl}(V)$, for every $\lambda \in \Lambda$, $\lambda \geq \lambda_0$. Let $(\lambda_0, \mu_0) \in \Lambda \times M$. Then for every $(\lambda, \mu) \in \Lambda \times M$, $(\lambda, \mu) \geq (\lambda_0, \mu_0)$ we have $f_\mu(y_\lambda) \in f(\mathrm{Cl}(U)) \subseteq V$. Thus the net $\{f_\mu(y_\lambda), (\lambda, \mu) \in \Lambda \times M\}$ converges to $f(y) \in Z$. □

DEFINITION 1.4. Let $\mathcal{P}(Y)$ be the set of all subsets of a space Y. If Λ is a directed set, then by $\vartheta\text{-}\overline{\lim}_{\Lambda}(A_\lambda)$, where $A_\lambda \subseteq Y$, we denote the *ϑ-upper limit* of the net $\{A_\lambda, \lambda \in \Lambda\}$ in $\mathcal{P}(Y)$, that is, the set of all points y of Y such that for every $\lambda_0 \in \Lambda$ and for every open neighbourhood U of y in Y there exists an element $\lambda \in \Lambda$ for which $\lambda \geq \lambda_0$ and $A_\lambda \cap \mathrm{Cl}(U) \neq \emptyset$.

THEOREM 1.5. *A net $\{f_\lambda, \lambda \in \Lambda\}$ in $S\Theta(Y, Z)$ strongly ϑ-continuously converges to $f \in S\Theta(Y, Z)$ if and only if*

$$\vartheta\text{-}\overline{\lim}_{\Lambda}(f_\lambda^{-1}(K)) \subseteq f^{-1}(K), \tag{1}$$

for every closed subset K of Z.

Proof. Let $\{f_\lambda, \lambda \in \Lambda\}$ be a net in $S\Theta(Y, Z)$, which strongly ϑ-continuously converges to f and let K be an arbitrary closed subset of Z. Let $y \in \vartheta\text{-}\overline{\lim}_{\Lambda}(f_\lambda^{-1}(K))$ and let W be an arbitrary neighbourhood of $f(y)$ in Z. Since the net $\{f_\lambda, \lambda \in \Lambda\}$ strongly ϑ-continuously converges to f, there exist an open neighbourhood V of y in Y and an element $\lambda_0 \in \Lambda$ such that $f_\lambda(\mathrm{Cl}(V)) \subseteq W$, for every $\lambda \in \Lambda$, $\lambda \geq \lambda_0$. (See Theorem 1.3.) On the other hand, there exists an element $\lambda \geq \lambda_0$ such that $\mathrm{Cl}(V) \cap f_\lambda^{-1}(K) \neq \emptyset$. Hence, $f_\lambda(\mathrm{Cl}(V)) \cap K \subseteq W \cap K \neq \emptyset$. This means that $f(y) \in \mathrm{Cl}(K) = K$. Thus, $y \in f^{-1}(K)$.

Conversely, let $\{f_\lambda, \lambda \in \Lambda\}$ be a net in $S\Theta(Y, Z)$ and $f \in S\Theta(Y, Z)$ such that the relation (1) holds for every closed subset K of Z. We prove that the net $\{f_\lambda, \lambda \in \Lambda\}$ strongly ϑ-continuously converges to f. Let $y \in Y$ and W be an open neighbourhood of $f(y)$ in Z. Since $y \notin f^{-1}(K)$, where $K = Z \setminus W$ we have $y \notin \vartheta\text{-}\overline{\lim}_{\Lambda}(f_\lambda^{-1}(K))$. This means that there exists an element $\lambda_0 \in \Lambda$ and an open neighbourhood V of y in Y such that $f_\lambda^{-1}(K) \cap \mathrm{Cl}(V) = \emptyset$, for every $\lambda \in \Lambda$, $\lambda \geq \lambda_0$. Then we have $\mathrm{Cl}(V) \subseteq Y \setminus f_\lambda^{-1}(K) = f_\lambda^{-1}(Z \setminus K) \subseteq f_\lambda^{-1}(W)$ and, therefore, $f_\lambda(\mathrm{Cl}(V)) \subseteq W$, for every $\lambda \in \Lambda$, $\lambda \geq \lambda_0$, that is, $\{f_\lambda, \lambda \in \Lambda\}$ strongly ϑ-continuously converges to f. □

THEOREM 1.6. *The following propositions are true*:

(1) *If $\{f_\lambda, \lambda \in \Lambda\}$ is a net in $S\Theta(Y, Z)$ such that $f_\lambda = f$, for every $\lambda \in \Lambda$, then the $\{f_\lambda, \lambda \in \Lambda\}$ strongly ϑ-continuously converges to $f \in S\Theta(Y, Z)$.*
(2) *If $\{f_\lambda, \lambda \in \Lambda\}$ is a net in $S\Theta(Y, Z)$ which strongly ϑ-continuously converges to $f \in S\Theta(Y, Z)$ and $\{g_\mu, \mu \in M\}$ is a subnet of $\{f_\lambda, \lambda \in \Lambda\}$, then the net $\{g_\mu, \mu \in M\}$ strongly ϑ-continuously converges to f.*

The proof of this theorem is clear.

2. Strongly ϑ-Notions on Function Spaces

NOTATIONS 2.1. Let X be a space and $F : X \times Y \to Z$ be a strongly ϑ-continuous map. Then by F_x, where $x \in X$, we denote the strongly ϑ-continuous map of Y into Z, for which $F_x(y) = F(x, y)$, for every $y \in Y$. By $\widehat{F}$ we denote the map of X into the set $S\Theta(Y, Z)$ for which $\widehat{F}(x) = F_x$, for every $x \in X$.

Let G be a map of the space X into the set $S\Theta(Y, Z)$. By $\widetilde{G}$ we denote the map of the space $X \times Y$ into the space Z, for which $\widetilde{G}(x, y) = G(x)(y)$, for every $(x, y) \in X \times Y$.

DEFINITIONS 2.2. A topology τ on $S\Theta(Y, Z)$ is called *strongly ϑ-splitting* if and only if for every space X, the strong ϑ-continuity of a map $F : X \times Y \to Z$ implies the strong ϑ-continuity of the map $\widehat{F} : X \to S\Theta_\tau(Y, Z)$.

A topology τ on $S\Theta(Y, Z)$ is called *strongly ϑ-jointly continuous* if and only if for every space X, the strong ϑ-continuity of a map $G : X \to S\Theta_\tau(Y, Z)$ implies the strong ϑ-continuity of the map $\widetilde{G} : X \times Y \to Z$.

Let $\mathcal{A}$ be a family of spaces. A topology τ on the set $S\Theta(Y, Z)$ is called *strongly $(\mathcal{A} - \vartheta)$-splitting* (respectively, *strongly $(\mathcal{A} - \vartheta)$-jointly continuous*) if and only if for every element X of $\mathcal{A}$, the strong ϑ-continuity of a map $F : X \times Y \to Z$ (respectively, a map $G : X \to S\Theta_\tau(Y, Z)$) implies the strong ϑ-continuity of the map $\widehat{F} : X \to S\Theta_\tau(Y, Z)$ (respectively, of the map $\widetilde{G} : X \times Y \to Z$).

Obviously, if $\mathcal{A}$ is the family of all spaces, then the notions strongly $(\mathcal{A} - \vartheta)$-splitting and strongly $(\mathcal{A} - \vartheta)$-jointly continuous concide with the notions strongly ϑ-splitting and strongly ϑ-jointly continuous, respectively.

If $\mathcal{A} = \{X\}$, then instead of "strongly $(\mathcal{A} - \vartheta)$-splitting" and "strongly $(\mathcal{A} - \vartheta)$-jointly continuous" we write "strongly $(X - \vartheta)$-splitting" and "strongly $(X - \vartheta)$-jointly continuous".

In what follows, unless indicated, $\mathcal{A}$ is an arbitrary family of spaces.

REMARK 2.3. Clearly, the above definition of strongly ϑ-jointly continuous topology does not coincide with the definition of jointly strongly ϑ-continuous given in [7].

EXAMPLES 2.4. (1) The trivial topology on $S\Theta(Y, Z)$ is strongly ϑ-splitting.

(2) Let $\mathcal{A}$ be an arbitrary family of spaces. Then, the pointwise topology τ_p on $S\Theta(Y, Z)$ is strongly $(\mathcal{A} - \vartheta)$-splitting.

Indeed, let $X \in \mathcal{A}$ and let $F : X \times Y \to Z$ be a strongly ϑ-continuous map. We prove that the map $\widehat{F} : X \to S\Theta_{\tau_p}(Y, Z)$ is strongly ϑ-continuous. Let $x \in X$ and let $\widehat{F}(x) \in (\{y\}, U)$, where $y \in Y$ and U be an open set of the space Z. Then we have $\widehat{F}(x)(y) = F(x, y) \in U$. Since the map F is strongly ϑ-continuous, there exist open neighbourhoods W_1 and W_2 of x and y respectively such that $F(\mathrm{Cl}(W_1) \times \mathrm{Cl}(W_2)) \subseteq U$. This means that $\widehat{F}(\mathrm{Cl}(W_1)) \in (\{y\}, U)$ and the map $\widehat{F}$ is strongly ϑ-continuous. Thus the pointwise topology τ_p is strongly $(\mathcal{A} - \vartheta)$-splitting.

(3) The compact open topology τ_c on $S\Theta(Y, Z)$ is strongly ϑ-splitting.

First we prove the following proposition:

Let $F : X \times Y \to Z$ be a strongly ϑ-continuous map, O be an open set of Z, K be a compact subset of Z and $x \in X$ such that $\{x\} \times K \subseteq F^{-1}(O)$. Then there exists an open neighbourhood V_x of x in X such that $\mathrm{Cl}(V_x) \times K \subseteq F^{-1}(O)$.

Indeed, for every $y \in K$ we have $(x, y) \in F^{-1}(O)$ and therefore $F(x, y) \in O$. Since the map F is strongly ϑ-continuous there exist open neighbourhoods V_x^y and V_y of x and y respectively such that $\mathrm{Cl}(V_x^y) \times \mathrm{Cl}(V_y) \subseteq F^{-1}(O)$.

Also, we have $K \subseteq \bigcup\{V_y : y \in K\}$. Since the subspace K is compact there exist open sets $V_{y_{i_1}}, \ldots, V_{y_{i_k}}$ such that $K \subseteq V_{y_{i_1}} \cup \cdots \cup V_{y_{i_k}}$.

Let $V_x = V_x^{y_{i_1}} \cap \cdots \cap V_x^{y_{i_k}}$ and $V_y' = V_{y_{i_1}} \cup \cdots \cup V_{y_{i_k}}$. We prove that $\mathrm{Cl}(V_x) \times K \subseteq F^{-1}(O)$.

Let $(x_1, y_1) \in \mathrm{Cl}(V_x) \times K \subseteq \mathrm{Cl}(V_x) \times \mathrm{Cl}(V_y')$. Then $x_1 \in \mathrm{Cl}(V_x^{y_{i_j}})$, for every $j = 1, \ldots, k$ and $y_1 \in \mathrm{Cl}(V_{y_{i_m}})$, for some $m = 1, \ldots, k$. Thus $(x_1, y_1) \in \mathrm{Cl}(V_x^{y_{i_m}}) \times \mathrm{Cl}(V_{y_{i_m}}) \subseteq F^{-1}(O)$ and therefore $\mathrm{Cl}(V_x) \times K \subseteq F^{-1}(O)$.

Now, we prove that the topology τ_c is strongly ϑ-splitting. Let X be an arbitrary space and let $F : X \times Y \to Z$ be a strongly ϑ-continuous map. We must prove that the map $\widehat{F} : X \to S\Theta_{\tau_c}(Y, Z)$ is strongly ϑ-continuous. Let $x \in X$ and let $\widehat{F}(x) \in (K, U)$, where K is a compact set of Y and U be an open set of the space Z. We prove that there exists an open subset W containing x in X such that $\widehat{F}(\mathrm{Cl}(W)) \subseteq (K, U)$. We have $\{x\} \times K \subseteq F^{-1}(U)$. By the above proposition there exists an open neighbourhood V of x such that $\mathrm{Cl}(V) \times K \subseteq F^{-1}(U)$. Thus $F(\mathrm{Cl}(W) \times K) \subseteq U$ and hence $\widehat{F}(\mathrm{Cl}(W)) \subseteq (K, U)$.

(4) Let $\mathcal{A}$ be the family of all discrete spaces. Then every topology on $S\Theta(Y, Z)$ is strongly $(\mathcal{A} - \vartheta)$-splitting and strongly $(\mathcal{A} - \vartheta)$-jointly continuous.

THEOREM 2.5. *Let $\mathcal{A}_1$ and $\mathcal{A}_2$ be families of spaces such that $\mathcal{A}_1 \subseteq \mathcal{A}_2$. If τ is strongly $(\mathcal{A}_2 - \vartheta)$-splitting (respectively, strongly $(\mathcal{A}_2 - \vartheta)$-jointly continuous) topology, then τ is also strongly $(\mathcal{A}_1 - \vartheta)$-splitting (respectively, strongly $(\mathcal{A}_1 - \vartheta)$-jointly continuous) topology.*

The proof of this theorem is clear.

THEOREM 2.6. *On the set $S\Theta(Y, Z)$ there exists the greatest strongly ϑ-splitting topology.*

Proof. Let T be the set of all strongly ϑ-splitting topologies on the set $S\Theta(Y, Z)$. Let $\tau = \bigvee T$. We prove that τ is the greatest strongly ϑ-splitting topology. It is sufficient to prove that τ is an strongly ϑ-splitting topology. Let $F : X \times Y \to Z$ be a strongly ϑ-continuous map. We prove that the map $\widehat{F} : X \to S\Theta_\tau(Y, Z)$ is strongly ϑ-continuous. Let $x \in X$ and let U be an open neighbourhood of $\widehat{F}(x)$ in $S\Theta_\tau(Y, Z)$. Since $\tau = \bigvee T$ we have that $U \in \tau'$, $\tau' \in T$. Also, since $\widehat{F} : X \to S\Theta_{\tau'}(Y, Z)$ is strongly ϑ-continuous there exists an open neighbourhood V of X such that $\widehat{F}(\mathrm{Cl}(V)) \subseteq U$. Thus the map $\widehat{F}$ is strongly ϑ-continuous and the topology τ is strongly ϑ-splitting. □

COROLLARY 2.6.1. *On the set $S\Theta(Y, Z)$ there exists the greatest strongly $(\mathcal{A} - \vartheta)$-splitting topology, which is denoted by $\tau_{s-\vartheta}(\mathcal{A})$.*

THEOREM 2.7. *The following propositions are true:*

(1) *Let* $\mathcal{A}_i$, $i \in I$, *be a family of spaces and let* $\mathcal{A} = \bigcup\{\mathcal{A}_i : i \in I\}$. *Then,* $\tau_{s-\vartheta}(\mathcal{A}) = \bigcap\{\tau_{s-\vartheta}(\mathcal{A}_i) : i \in I\}$.

(2) *Let* $\mathcal{A}_i$, $i \in I$, *be a family of spaces and let* $\mathcal{A} = \bigcap\{\mathcal{A}_i : i \in I\}$. *If* $\mathcal{A} \neq \emptyset$, *then* $\bigvee\{\tau_{s-\vartheta}(\mathcal{A}_i) : i \in I\} \subseteq \tau_{s-\vartheta}(\mathcal{A})$.

The proof of this theorem is clear.

NOTATIONS 2.8. By $\mathcal{C}^*_{s-\vartheta}$ we denote the class of all pairs $(\{f_\lambda, \lambda \in \Lambda\}, f)$, where $\{f_\lambda, \lambda \in \Lambda\}$ is a net in $S\Theta(Y, Z)$ which strongly ϑ-continuously converges to $f \in S\Theta(Y, Z)$. If τ is a topology on $S\Theta(Y, Z)$ then by $(\mathcal{C}(\tau))_{s-\vartheta}$ we denote the class of all pairs $(\{f_\lambda, \lambda \in \Lambda\}, f)$, where $\{f_\lambda, \lambda \in \Lambda\}$ is a net in $S\Theta(Y, Z)$ which converges to $f \in S\Theta(Y, Z)$ in the topology τ.

THEOREM 2.9. *A topology* τ *on* $S\Theta(Y, Z)$ *is strongly* ϑ*-splitting if and only if*

$$\mathcal{C}^*_{s-\vartheta} \subseteq (\mathcal{C}(\tau))_{s-\vartheta}.$$

Proof. Let τ be a strongly ϑ-splitting topology on $S\Theta(Y, Z)$ and let $(\{f_\lambda, \lambda \in \Lambda\}, f) \in \mathcal{C}^*_{s-\vartheta}$. We prove that the net $\{f_\lambda, \lambda \in \Lambda\}$ converges to f in the topology τ. Indeed, we consider the set $X = \Lambda \cup \{\infty\}$, where ∞ is a symbol such that $\infty \geq \lambda$, for every $\lambda \in \Lambda$. Then we topologize $X = \Lambda \cup \{\infty\}$ defining any singleton $\{\lambda\}$, $\lambda \in \Lambda$ to be open and neighbourhoods of ∞ the sets $\{\lambda \in X : \lambda \geq \lambda_0$ for some $\lambda_0 \in \Lambda\}$. (See [1] and [4].) Let $F : X \times Y \to Z$ be a map, for which $F(\lambda, y) = f_\lambda(y)$, $\lambda \neq \infty$ and $F(\infty, y) = f(y)$, for every $y \in Y$. The map F is strongly ϑ-continuous. Obviously $\widehat{F}(\lambda) = f_\lambda$ and $\widehat{F}(\infty) = f$. Since the topology τ is strongly ϑ-splitting, the map $\widehat{F} : X \to S\Theta_\tau(Y, Z)$ is strongly ϑ-continuous.

By the strongly ϑ-continuity of $\widehat{F}$ we have that for every open neighbourhood U of f in $S\Theta_\tau(Y, Z)$, there exists an open neighbourhood V of ∞ in X such that

$$\widehat{F}(\mathrm{Cl}(V)) \subseteq U.$$

By the definition of the topology on X, there exists an element $\lambda_0 \in \Lambda$ such that $\lambda \in V \subseteq \mathrm{Cl}(V)$, for every $\lambda \in \Lambda$, $\lambda \geq \lambda_0$. Hence $f_\lambda \in U$, for every $\lambda \in \Lambda$, $\lambda \geq \lambda_0$, that is the net $\{f_\lambda, \lambda \in \Lambda\}$ converges to f in the topology τ. Thus $\mathcal{C}^*_{s-\vartheta} \subseteq ((\mathcal{C}(\tau))_{s-\vartheta}$.

Conversely, let τ be a topology on $S\Theta(Y, Z)$ such that $\mathcal{C}^*_{s-\vartheta} \subseteq ((\mathcal{C}(\tau))_{s-\vartheta}$. We prove that the topology τ is strongly ϑ-splitting.

Let X be an arbitrary space and let $F : X \times Y \to Z$ be a strongly ϑ-continuous map. Consider the map $\widehat{F} : X \to S\Theta_\tau(Y, Z)$. We must prove that the map $\widehat{F}$ is strongly ϑ-continuous. Let $\{x_\lambda, \lambda \in \Lambda\}$ be a net in X which ϑ-converges to x. We prove that the net $\widehat{F}(x_\lambda) \to \widehat{F}(x)$. Let $\{y_\mu, \mu \in M\}$ be a net in Y which ϑ-converges to y in Y. Since the map F is strongly ϑ-continuous and the net $\{(x_\lambda, y_\mu), (\lambda, \mu) \in \Lambda \times M\}$ of $X \times Y$ ϑ-converges to (x, y) in $X \times Y$ we have $F(x_\lambda, y_\mu) \to F(x, y)$. This means that $F_{x_\lambda}(y_\mu) \to F_x(y)$. Thus the net $\{\widehat{F}(x_\lambda), \lambda \in \Lambda\}$ strongly ϑ-continuously converges to $\widehat{F}(x)$. By assumption the net $\{\widehat{F}(x_\lambda), \lambda \in \Lambda\}$ converges to $\widehat{F}(x)$. Thus the map $\widehat{F}$ is strongly ϑ-continuous and the topology is τ strongly ϑ-splitting. □

THEOREM 2.10. *A topology τ on $S\Theta(Y, Z)$ is strongly ϑ-jointly continuous if and only if*

$$(\mathcal{C}(\tau))_{s-\vartheta} \subseteq \mathcal{C}^*_{s-\vartheta}.$$

Proof. Let τ be a strongly ϑ-jointly continuous topology, X be the space which was defined in the proof of Theorem 2.9 and $(\{f_\lambda, \lambda \in \Lambda\}, f) \in (\mathcal{C}(\tau))_{s-\vartheta}$. Clearly, the map $G : X \to S\Theta_\tau(Y, Z)$, where $G(\lambda) = f_\lambda$ and $G(\infty) = f$ is strongly ϑ-continuous. Thus, the map $\widetilde{G} : X \times Y \to Z$ is strongly ϑ-continuous.

Now, we prove that $(\{f_\lambda, \lambda \in \Lambda\}, f) \in \mathcal{C}^*_{s-\vartheta}$. It is sufficient to prove that if $\{y_\mu, \mu \in M\}$ is a net in Y which ϑ-converges to $y \in Y$, then the net $\{f_\lambda(y_\mu), (\lambda, \mu) \in \Lambda \times M\}$ converges to $f(y)$. The net $\{\lambda, \lambda \in \Lambda\}$ in X strongly ϑ-converges to ∞. Hence, the net $\{(\lambda, y_\mu), (\lambda, \mu) \in \Lambda\times M\}$ strongly ϑ-converges to (∞, y). Since the map $\widetilde{G}$ is strongly ϑ-continuous the net $\{\widetilde{G}(\lambda, y_\mu) \equiv G(\lambda)(y_\mu) \equiv f_\lambda(y_\mu), (\lambda, \mu) \in \Lambda \times M\}$ converges to $\widetilde{G}(\infty, y) \equiv f(y)$.

Conversely, let τ be a topology on $S\Theta(Y, Z)$ such that

$$(\mathcal{C}(\tau))_{s-\vartheta} \subseteq \mathcal{C}^*_{s-\vartheta}.$$

We prove that the topology τ is strongly ϑ-jointly continuous. Let X be an arbitrary space and let $G : X \to S\Theta_\tau(Y, Z)$ be a strongly ϑ-continuous map. We prove that the map $\widetilde{G} : X \times Y \to Z$ is strongly ϑ-continuous. Let $\{(x_\lambda, y_\mu), (\lambda, \mu) \in \Lambda \times M\}$ be a net in $X \times Y$ which ϑ-converges to (x, y). We prove that the net $\{\widetilde{G}(x_\lambda, y_\mu), (\lambda, \mu) \in \Lambda \times M\}$ in Z converges to $\widetilde{G}(x, y)$.

Since the net $\{x_\lambda, \lambda \in \Lambda\}$ ϑ-converges to x in X and the map G is strongly ϑ-continuous, the net $\{G(x_\lambda), \lambda \in \Lambda\}$ converges to $G(x)$. Thus by assumption the net $\{G(x_\lambda), \lambda \in \Lambda\}$ strongly ϑ-continuously converges to $G(x)$. Now, since the net $\{y_\mu, \mu \in M\}$ ϑ-converges to y the net $\{G(x_\lambda)(y_\mu) = \widetilde{G}(x_\lambda, y_\mu), (\lambda, \mu) \in \Lambda \times M\}$ converges to $G(x)(y) = \widetilde{G}(x, y)$. Hence the topology τ is strongly ϑ-jointly continuous. □

THEOREM 2.11. *A topology τ on $S\Theta(Y, Z)$ is simultaneously, strongly ϑ-splitting and strongly ϑ-jointly continuous if and only if*

$$(\mathcal{C}(\tau))_{s-\vartheta} = \mathcal{C}^*_{s-\vartheta}.$$

The proof of this theorem follows by Theorems 2.9 and 2.10.

PROBLEM 2.12. Does there exist a characterization of the finest strongly ϑ-splitting topology on $S\Theta(Y, Z)$ as a set-open topology?

NOTATIONS AND DEFINITIONS 2.13. Let Ω be a family of directed sets. Then for every $\Lambda \in \Omega$ we consider the set $\mathrm{Sp}(\Lambda) \equiv \Lambda \cup \{\infty\}$, where ∞ is a symbol such that $\infty \geq \lambda$, for every $\lambda \in \Lambda$. Then we topologize $\mathrm{Sp}(\Lambda) = \Lambda \cup \{\infty\}$ defining any singleton $\{\lambda\}$, $\lambda \in \Lambda$ to be open and neighbourhoods of ∞ the sets $\{\lambda \in \mathrm{Sp}(\Lambda) : \lambda \geq \lambda_0$ for some $\lambda_0 \in \Lambda\}$. (See [1] and [4].) By $\mathrm{Sp}(\Omega)$ we denote the family of all spaces $\mathrm{Sp}(\Lambda)$, where $\Lambda \in \Omega$. Instead of "strongly $(\mathrm{Sp}(\Omega) - \vartheta)$-splitting" or "strongly $(\mathrm{Sp}(\Omega) - \vartheta)$-jointly continuous" we write "strongly $(\Omega - \vartheta)$-splitting" or "strongly $(\Omega - \vartheta)$-jointly continuous", respectively.

By $\mathcal{C}^*_{s-\vartheta-\Omega}$ (respectively, by $(\mathcal{C}(\tau))_{s-\vartheta-\Omega}$) we denote the subclass of all elements $(\{f_\lambda, \lambda \in \Lambda\}, f)$ of $\mathcal{C}^*_{s-\vartheta}$ (respectively, of $(\mathcal{C}(\tau))_{s-\vartheta}$) for which $\Lambda \in \Omega$.

THEOREM 2.14. *The following propositions are true*:

(1) *A topology τ on $S\Theta(Y, Z)$ is strongly $(\Omega - \vartheta)$-splitting if and only if*
$$\mathcal{C}^*_{s-\vartheta-\Omega} \subseteq (\mathcal{C}(\tau))_{s-\vartheta-\Omega}.$$
(2) *A topology τ on $S\Theta(Y, Z)$ is strongly $(\Omega - \vartheta)$-jointly continuous if and only if*
$$(\mathcal{C}(\tau))_{s-\vartheta-\Omega} \subseteq \mathcal{C}^*_{s-\vartheta-\Omega}.$$
(3) *A topology τ on $S\Theta(Y, Z)$ is simultaneously strongly $(\Omega - \vartheta)$-splitting and strongly $(\Omega - \vartheta)$-jointly continuous if and only if*
$$(\mathcal{C}(\tau))_{s-\vartheta-\Omega} = \mathcal{C}^*_{s-\vartheta-\Omega}.$$

The proof of this theorem is similar to that of the proof of Theorems 2.9, 2.10 and 2.11.

THEOREM 2.15. *The following propositions are true*:

(1) *A topology τ on $S\Theta(Y, Z)$ is strongly ϑ-splitting if and only if is strongly $(\mathcal{A} - \vartheta)$-splitting, where $\mathcal{A}$ is the family of all spaces having exactly one non-isolated point.*
(2) *A topology τ on $S\Theta(Y, Z)$ is strongly ϑ-jointly continuous if and only if is strongly $(\mathcal{A} - \vartheta)$-jointly continuous, where $\mathcal{A}$ is the family of all spaces having exactly one non-isolated point.*

Proof. It is sufficient to prove that if τ is strongly $(\mathcal{A} - \vartheta)$-splitting, where $\mathcal{A}$ is the family of all spaces having exactly one non-isolated point, then the topology τ is strongly ϑ-splitting.

Let $(\{f_\lambda, \lambda \in \Lambda\}, f) \in \mathcal{C}^*_{s-\vartheta}$. We must prove that the net $\{f_\lambda, \lambda \in \Lambda\}$ converges to the element f in the topology τ.

Let $X = \Lambda \cup \{\infty\}$, where ∞ is a symbol such that $\infty \geq \lambda$, for every $\lambda \in \Lambda$. Then we topologize $X = \Lambda \cup \{\infty\}$ defining any singleton $\{\lambda\}$, $\lambda \in \Lambda$ to be open and neighbourhoods of ∞ the sets $\{\lambda \in X : \lambda \geq \lambda_0$ for some $\lambda_0 \in \Lambda\}$. (See [1] and [4].) Clearly the element ∞ is the unique non-isolated point of the space X and $X \in \mathcal{A}$.

We consider the map $F : X \times Y \to Z$ setting $F(\lambda, y) = f_\lambda(y)$, $\lambda \neq \infty$ and $F(\infty, y) = f(y)$. Obviously the map F is strongly ϑ-continuous. Now we prove that $\{f_\lambda, \lambda \in \Lambda\}$ converges to f in the topology τ.

Indeed, let $U \in \tau$ be an open neighbourhood of f. By assumption the topology τ is strongly $(\mathcal{A} - \vartheta)$-splitting. Hence, the map $\widehat{F} : X \to S\Theta_\tau(Y, Z)$ is strongly ϑ-continuous. Also, we have $\widehat{F}(\infty) = f$ and $\widehat{F}(\lambda) = f_\lambda$, $\lambda \neq \infty$. Thus, there exists an open neighbourhood V of ∞ such that

$$\widehat{F}(\mathrm{Cl}(V)) \subseteq U.$$

Since the set V is an open neighbourhood of ∞ in X, there exists an element $\lambda_0 \in \Lambda$ such that $\lambda \in V \subseteq \mathrm{Cl}(V)$, for every $\lambda \geq \lambda_0$, $\lambda \in \Lambda$.

Hence, $\widehat{F}(\lambda) = f_\lambda \in U$, for every $\lambda \geq \lambda_0$, $\lambda \in \Lambda$. Thus, the net $\{\widehat{F}(\lambda) = f_\lambda, \lambda \in \Lambda\}$ converges to f in the topology τ and the topology τ is strongly ϑ-splitting.

Similarly, we can prove the proposition (2). □

2.1. THE STRONGLY $(\mathcal{A} - \vartheta)$-EXPONENTIAL FUNCTION

Let $\mathcal{A}$ be an arbitrary family of topological spaces, $X \in \mathcal{A}$ and Y, Z be fixed topological spaces, then the *strongly* $(\mathcal{A} - \vartheta)$*-exponential function*

$$e^{\vartheta-\mathcal{A}}_{XYZ} : S\Theta(X \times Y, Z) \to S\Theta(X, S\Theta_\tau(Y, Z))$$

is defined by $e^{\vartheta-\mathcal{A}}_{XYZ}(F) = \widehat{F}$, for every $F \in S\Theta(X \times Y, Z)$.

Clearly we have the following propositions:

(1) If τ is a strongly ϑ-splitting topology on $S\Theta(Y, Z)$ then the $(\mathcal{A} - \vartheta)$-exponential function $e^{\vartheta-\mathcal{A}}_{XYZ}$ is well defined.

(2) If $S\Theta(X, S\Theta_\tau(Y, Z)) \subseteq e^{\vartheta-\mathcal{A}}_{XYZ}(S\Theta(X \times Y, Z))$, then the topology τ on $S\Theta(Y, Z)$ is strongly $(\mathcal{A} - \vartheta)$-jointly continuous.

(3) If $e^{\vartheta-\mathcal{A}}_{XYZ}(S\Theta(X \times Y, Z)) \subseteq S\Theta(X, S\Theta_\tau(Y, Z))$, then the topology τ is strongly $(\mathcal{A} - \vartheta)$-splitting.

3. Strongly ϑ^*-Notions on Function Spaces

DEFINITIONS 3.1. A topology τ on $S\Theta(Y, Z)$ is *called strongly* ϑ^*-splitting if and only if for every space X, the strong ϑ-continuity of a map $F : X \times Y \to Z$ implies the ϑ-continuity of the map $\widehat{F} : X \to S\Theta_\tau(Y, Z)$.

A topology τ on $S\Theta(Y, Z)$ is called *strongly* ϑ^**-jointly continuous* if and only if for every space X, the ϑ-continuity of a map $G : X \to S\Theta_\tau(Y, Z)$ implies the strong ϑ-continuity of the map $\widetilde{G} : X \times Y \to Z$.

Let $\mathcal{A}$ be a family of spaces. A topology τ on the set $S\Theta(Y, Z)$ is called *strongly* $(\mathcal{A} - \vartheta^*)$*-splitting* (respectively, *strongly* $(\mathcal{A} - \vartheta^*)$-jointly continuous) if and only if for every element X of $\mathcal{A}$, the strong ϑ-continuity of a map $F : X \times Y \to Z$

(respectively, the ϑ-continuity of a map $G : X \to S\Theta_\tau(Y, Z)$) implies the ϑ-continuity of the map $\widehat{F} : X \to S\Theta_\tau(Y, Z)$ (respectively, the strong ϑ-continuity of the map $\widetilde{G} : X \times Y \to Z$).

Obviously, if $\mathcal{A}$ is the family of all spaces, then the notions strongly $(\mathcal{A} - \vartheta^*)$-splitting and strongly $(\mathcal{A} - \vartheta^*)$-jointly continuous concide with the notions strongly ϑ^*-splitting and strongly ϑ^*-jointly continuous, respectively.

In what follows, unless indicated, $\mathcal{A}$ is an arbitrary family of spaces.

THEOREM 3.2. *The following propositions are true*:

(1) *Let τ be a strongly ϑ-splitting topology on $S\Theta(Y, Z)$. Then the topology τ is also strongly ϑ^*-splitting.*
(2) *Let τ be a strongly ϑ^*-jointly continuous topology on $S\Theta(Y, Z)$. Then the topology τ is also strongly ϑ-jointly continuous.*

The proof of this theorem is clear.

THEOREM 3.3. *A topology τ on $S\Theta(Y, Z)$ is strongly ϑ^*-jointly continuous if and only if the evaluation map $e : S\Theta(Y, Z) \times Y \to Z$, defined by $e(f, y) = f(y)$, is strongly ϑ-continuous.*

Proof. Clearly, the identity map $G \equiv 1 : S\Theta_\tau(Y, Z) \to S\Theta_\tau(Y, Z)$ is ϑ-continuous. Since the topology τ is strongly ϑ^*-jointly continuous, the map $\widetilde{G} \equiv e : S\Theta_\tau(Y, Z) \times Y \to Z$ is strongly ϑ-continuous.

Conversely, let X be a space, $G : X \to S\Theta_\tau(Y, Z)$ be a ϑ-continuous map and $1 : Y \to Y$ be the identity map. Clearly, the map $G \times 1 : X \times Y \to S\Theta_\tau(Y, Z) \times Y$ is ϑ-continuous. Also the map $e : S\Theta_\tau(Y, Z) \times Y \to Z$ is strongly ϑ-continuous. It is not difficult to prove that the map $e \circ (G \times 1) : X \times Y \to Z$ is strongly ϑ-continuous and $\widetilde{G} = e \circ (G \times 1)$. Thus the topology τ is strongly ϑ^*-continuous. □

REMARKS 3.4. (1) In general Theorem 3.3 is not satisfied for a strongly ϑ-jointly continuous topology.

(2) By (1) of Theorem 3.2 we have that all the examples in Section 2 remain true for the case of strongly ϑ^*-notions.

(3) Theorem 5 in Section 2 remains true for the case of strongly ϑ^*-notions.

3.1. THE STRONGLY $(\mathcal{A} - \vartheta^*)$-EXPONENTIAL FUNCTION

Let $\mathcal{A}$ be an arbitrary family of topological spaces, $X \in \mathcal{A}$ and Y, Z be fixed topological spaces, then the *strongly $(\mathcal{A} - \vartheta^*)$-exponential function*

$$e_{XYZ}^{\vartheta^* - \mathcal{A}} : S\Theta(X \times Y, Z) \to \Theta(X, S\Theta_\tau(Y, Z))$$

is defined by $e_{XYZ}^{\vartheta^* - \mathcal{A}}(F) = \widehat{F}$, for every $F \in S\Theta(X \times Y, Z)$.

Clearly we have the following propositions:

(1) If τ is a strongly ϑ^*-splitting topology on $S\Theta(Y, Z)$ then the $(\mathcal{A} - \vartheta^*)$-exponential function $e_{XYZ}^{\vartheta^*-\mathcal{A}}$ is well defined.

(2) If $\Theta(X, S\Theta_\tau(Y, Z)) \subseteq e_{XYZ}^{\vartheta^*-\mathcal{A}}(S\Theta(X \times Y, Z))$, then the topology τ on $S\Theta(Y, Z)$ is strongly $(\mathcal{A} - \vartheta^*)$-jointly continuous.

(3) If $e_{XYZ}^{\vartheta^*-\mathcal{A}}(S\Theta(X \times Y, Z)) \subseteq \Theta(X, S\Theta_\tau(Y, Z))$, then the topology τ is strongly $(\mathcal{A} - \vartheta^*)$-splitting.

Acknowledgment

The authors are grateful to the referees for their fruitful comments for the improvement of this paper.

References

1. Arens, R. and Dugundji, J.: Topologies for function spaces, *Pacific J. Math.* **1** (1951), 5–31.
2. Dugundji, J.: *Topology*, Allyn and Bacon, Rockleigh, NJ, 1966.
3. Abd El-Monsef, M. E. and Hanafy, I. M.: Semistrongly ϑ-continuous functions, *Indian J. Pure Appl. Math.* **21**(8) (1990), 721–728.
4. Di Concilio, A.: On ϑ-continuous convergence in function spaces, *Rend. Mat. (Serie 7)* **4**(1) (1984), 85–94.
5. Di Concilio, A.: Exponential law and ϑ-continuous functions, *Quaestiones Mathematicae* **8** (1985), 131–142.
6. Frink, O.: Topology in lattices, *Trans. Amer. Math. Soc.* **51** (1942), 569–582.
7. Ganguly, S. and Bandyopadhyay, S.: A new topology on function spaces, *Bull. Calcutta Math. Soc.* **84** (1992), 344–350.
8. Iliadis, S. D. and Fomin, S.: The method of centred systems in the theory of topological spaces, *Uspekhi Mat. Nauk* **21** (1966), 47–76 (in Rusian). Translation: *Russian Math. Surveys* **21** (1966), 37–62.
9. Kelley, J. L.: *General Topology*, Springer, Berlin, 1955.
10. Kuratowski, C.: Sur la notion de limite topologique d' ensembles, *Ann. Soc. Polon. Math.* **21** (1948–49), 219–225.
11. Noiri, T.: On δ-continuous functions, *J. Korean Math. Soc.* **16** (1980), 161–166.

(Strongly) Zero-Dimensional Partially Ordered Spaces ⋆

K. R. NAILANA
Department of Mathematics, Applied Mathematics and Astronomy, University of South Africa, P.O. Box 392, Pretoria 0001, South Africa, e-mail: nailakr@alpha.unisa.ac.za

(Received: 8 July 1996; accepted: 29 December 1998)

Abstract. We give an internal characterization of the notion of zero-dimensionality in preordered topological spaces. It is shown that for compact partially ordered topological spaces this property is equivalent to total order-disconnectedness. We define the notion of strong zero-dimensionality in preordered topological spaces and give several internal characterizations.

Mathematics Subject Classifications (2000): 54A05, 54B30, 54F05, 54E15, 54E55.

Key words: preordered spaces, pospaces, quasi-uniform spaces, zero-dimensional preordered spaces, strongly zero-dimensional pospaces.

Introduction

We refer the reader to [24] for basic concepts concerning preordered topological spaces and partially ordered topological spaces that are not explained here. A *preordered topological space* (or *preordered space*) is a triple $(X, \tau, \leq)$, where X is a set, τ a topology on X and $\leq$ is a preorder (i.e., a reflexive and transitive relation) with a closed graph in the product topology on $X \times X$. If $\leq$ is antisymmetric then we call $(X, \tau, \leq)$ a *partially ordered topological space* (or *pospace*). We will denote the category of preordered spaces and continuous order preserving maps by **QoTop** and the category of pospaces and continuous order preserving maps by **PoTop**. If τ is a topology on X then the order $\leq_\tau$ defined by $x \leq_\tau y \Leftrightarrow x \in \mathrm{cl}_\tau\{y\}$ is called the *specialization order* with respect to τ. We will denote the real line by $\mathbf{R}$ and we will denote the pospace ($\mathbf{R}$, usual topology, usual order) by $\mathbf{R}_0$. For a subset $\mathbf{A}$ of $\mathbf{R}$, $\mathbf{A}_0$ denotes an ordered subspace of $\mathbf{R}_0$, e.g., $\mathbf{I}_0 =$ ([0, 1], usual topology, usual order), $\mathbf{D}_0 =$ ({0, 1}, discrete topology, usual order). A preordered space $(X, \tau, \leq)$ is a *completely regular preordered space* iff the source $\mathbf{QoTop}(X, \mathbf{I}_0)$ is initial, where the initial structure with respect to the family of maps is obtained by forming the initial structures in the categories of topological spaces (with continuous maps)

⋆ A draft of this paper was written at the University of Cape Town (UCT) in December 1993 using the results from the author's M.Sc. thesis (UCT 1993, produced under the supervision of Professor G. C. L. Brümmer). Grants from the Foundation for Research Development and from UCT via the Topology and Category Theory Research Group are acknowledged.

and preordered spaces (with order preserving maps) separately. In this case the initial topology is the weak topology induced by the continuous maps and the initial preorder is represented as the intersection of the preimages of the given, closed preorders, cf. [34]. The category of completely regular preordered spaces and continuous order preserving maps will be denoted by **CTopOrd**.

The reader is referred to [21] and [31] for basic concepts concerning bitopological spaces. The category of bitopological spaces and bicontinuous maps will be denoted by **BiTop**. Let i be the topology on $\mathbf{R}$ having as a base $\{(a, \infty) \mid a \in \mathbf{R}\}$ and d be the topology having as a base $\{(-\infty, a) \mid a \in \mathbf{R}\}$. Then we denote the bitopological space $(\mathbf{R}, i, d)$ by $\mathbf{R}_b$. For a subset $\mathbf{A}$ of $\mathbf{R}$, $\mathbf{A}_b$ denotes the corresponding subbispace of $\mathbf{R}_b$; we shall apply this in particular with $\mathbf{D} = \{0, 1\}$ and $\mathbf{I} = [0, 1]$ for the set $\mathbf{A}$. A bitopological space (X, τ_1, τ_2) is initial with respect to a family f_α of maps on X if τ_1 and τ_2 are initial with respect to f_α, i.e., the bitopology (τ_1, τ_2) is the coarsest bitopology on the set X for which the maps are continuous. A bitopological space (X, τ_1, τ_2) is called *completely regular* iff the source $\mathbf{BiTop}((X, \tau_1, \tau_2), \mathbf{I}_b)$ is initial. These bispaces together with the bicontinuous maps form the category **Cr2Top**.

We refer the reader to [13] for basic concepts concerning quasi-uniform spaces. If $\mathcal{Q}$ is the quasi-uniformity on X then $\tau(\mathcal{Q})$ is the topology generated by $\{Q[x] \mid x \in X, Q \in \mathcal{Q}\}$ where $Q[x] = \{y \mid (x, y) \in Q\}$. A quasi-uniformity $\mathcal{Q}$ on a set X is said to be *transitive* if it has a (sub-)base of transitive elements of $\mathcal{Q}$. We denote the category of quasi-uniform spaces and quasi-uniformly continuous maps by **Quu**. We now consider the following forgetful functors: V: **Quu** $\to$ **CTopOrd** denotes the concrete functor given by $V(X, \mathcal{Q}) = (X, \tau(\mathcal{Q} \vee \mathcal{Q}^{-1}), \bigcap \mathcal{Q})$ for all $(X, \mathcal{Q}) \in$ **Quu** where $\mathcal{Q} \vee \mathcal{Q}^{-1}$ is the coarsest uniformity finer than $\mathcal{Q}$ and $\mathcal{Q}^{-1}$ [33, 34]. The functor M: **Cr2Top** $\to$ **CTopOrd** given by $M(X, \tau_1, \tau_2) = (X, \tau_1 \vee \tau_2, \leq_{\tau_1})$ for $(X, \tau_1, \tau_2) \in$ **Cr2Top**, where $\tau_1 \vee \tau_2$ is the coarsest topology finer than τ_1 and τ_2 [32]. The functor $\overline{T}$: **Quu** $\to$ **Cr2Top** given by
$\overline{T}(X, \mathcal{Q}) = (X, \tau(\mathcal{Q}), \tau(\mathcal{Q}^{-1}))$ [6].

The functor M preserves initial sources and $M\overline{T} = V$ [33], i.e., the diagram

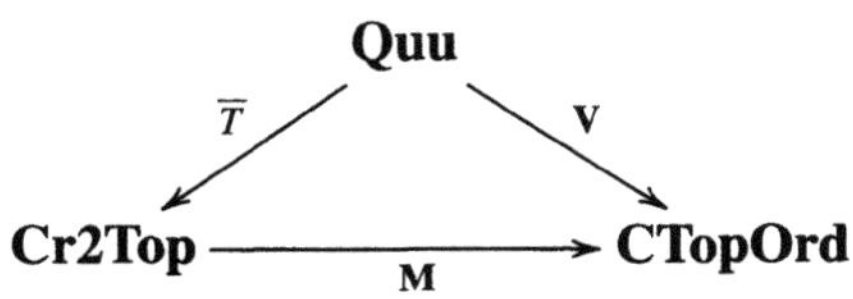

commutes. We recall from [4, 17] that a bitopological space (X, τ_1, τ_2) is zero-dimensional in the sense of Reilly [30] iff the source $\mathbf{BiTop}((X, \tau_1, \tau_2), \mathbf{D}_b)$ is initial. Equivalently (X, τ_1, τ_2) is a zero-dimensional bitopological space if and only if τ_1 is zero-dimensional with respect to τ_2 and τ_2 is zero-dimensional with respect to τ_1, where τ_i is zero-dimensional with respect to τ_k if τ_i has a base of τ_k-closed sets, i.e., for each $x \in X$ and each τ_i-open set A containing x there is a τ_k-closed τ_i-open set G such that $x \in G \subseteq A$, where $i, k = 1, 2$ and $i \neq k$. A bitopological space is zero-dimensional if and only if it admits a transitive quasi-uniformity [4, 17]. We have a similar situation in preordered spaces, see Corollary 1.5 below.

1. Zero-Dimensional Preordered Spaces

DEFINITION 1.1. A preordered space $(X, \tau, \leq)$ is a *zero-dimensional preordered space* iff the source $\mathbf{QoTop}((X, \tau, \leq), \mathbf{D}_0)$ is initial.

This definition is equivalent to the one given in [11] in which a 3-point chain was used.

LEMMA 1.2. *The functor M: $\mathbf{Cr2Top} \to \mathbf{CTopOrd}$ sends zero-dimensional bispaces to zero-dimensional preordered spaces.*

Proof. Let X be a zero-dimensional bitopological space. Then the source $\mathbf{Cr2Top}(X, \mathbf{D}_b)$ is initial. Since the functor M preserves initial sources and any source containing an initial source is again initial, the source $\mathbf{CTopOrd}(MX, M\mathbf{D}_b)$ is initial. Thus MX is zero-dimensional. □

THEOREM 1.3. *A preordered space $(X, \tau, \leq)$ is zero-dimensional if and only if τ has a subbase consisting of clopen increasing and clopen decreasing sets and $\leq = \leq_{\mathcal{U}}$ where $\mathcal{U}$ is the topology generated by the clopen increasing sets, i.e., $\leq$ is the specialization order with respect to $\mathcal{U}$.*

Proof. Suppose that $(X, \tau, \leq)$ is a zero-dimensional preordered space. Then the source $\mathcal{A} = \mathbf{CTopOrd}((X, \tau, \leq), \mathbf{D}_0)$ is initial. Thus τ has a subbase $\mathcal{S} = \{f^{-1}[V] \mid V \text{ open in } \mathbf{D}_0, f \in \mathcal{A}\}$. It is clear that $\mathcal{S}$ consists of clopen increasing and clopen decreasing sets. Let $\mathcal{U}$ be the topology which has the subbase (in fact a base since the finite intersection of clopen increasing sets is again clopen increasing) $\{f^{-1}(\{1\}) \mid f \in \mathcal{A}\}$. Clearly every clopen increasing set belongs to this family. We now show that $\leq = \leq_{\mathcal{U}}$, where the order $\leq$ is given by the initiality of the source, i.e., $\forall x, y \in X, x \leq y \Leftrightarrow \forall f \in \mathcal{A}, f(x) \leq^{\mu} f(y)$. Let $x, y \in X$. Now if $x \nleq y$ then there exists f in $\mathcal{A}$ such that $f(x) \nleq^{\mu} f(y)$, i.e., $f(x) = 1$ and $f(y) = 0$. Since $x \in f^{-1}(\{1\})$ and $y \in f^{-1}(\{0\})$, $x \notin \mathrm{cl}_{\mathcal{U}}\{y\}$. Therefore $x \nleq_{\mathcal{U}} y$. Conversely suppose that $x \nleq_{\mathcal{U}} y$. Then $x \notin \mathrm{cl}_{\mathcal{U}}\{y\}$. Thus there exists a $U \in \mathcal{U}$ such that $x \in U$ and $y \notin U$. Therefore $x \nleq y$ since U is $\leq$-increasing. Hence $\leq = \leq_{\mathcal{U}}$.

Suppose τ has a subbase consisting of clopen increasing and clopen decreasing sets and $\leq = \leq_{\mathcal{U}}$ where $\mathcal{U}$ is the topology generated by all clopen increasing sets. Let $\mathcal{L}$ be the topology generated by all clopen decreasing sets. Then $\tau = \mathcal{U} \vee \mathcal{L}$. Thus $(X, \tau, \leq) = (X, \mathcal{U} \vee \mathcal{L}, \leq_{\mathcal{U}})$, since $\leq = \leq_{\mathcal{U}}$. By Lemma 1.2 it suffices to show that $(X, \mathcal{U}, \mathcal{L})$ is a zero-dimensional bitopological space. As remarked in the first part of the proof, we have the bases $\mathcal{B}_1$ and $\mathcal{B}_2$ for $\mathcal{U}$ and $\mathcal{L}$ respectively, where $\mathcal{B}_1$ consists of all clopen increasing sets and $\mathcal{B}_2$ consists of all clopen decreasing sets. Let $B \in \mathcal{B}_1$. Then $X - B$ is τ-clopen and decreasing, and hence $X - B \in \mathcal{B}_2$, so that B is $\mathcal{L}$-closed. Therefore $\mathcal{U}$ has a base consisting of $\mathcal{L}$-closed sets. Similarly

$\mathcal{L}$ has a base consisting of $\mathcal{U}$-closed sets. Hence $(X, \mathcal{U}, \mathcal{L})$ is a zero-dimensional bitopological space. □

COROLLARY 1.4. *A preordered space* $(X, \tau, \leq)$ *is zero-dimensional iff there exists a zero-dimensional bitopological space* (X, τ_1, τ_2) *such that* $M(X, \tau_1, \tau_2) = (X, \tau, \leq)$.

Proof. This follows from the Lemma 1.2 and the proof of 1.3. □

By analogy with standard terminology we shall say that a preordered space $(X, \tau, \leq)$ *admits* a zero-dimensional bitopology if there exists a zero-dimensional bitopological space (X, τ_1, τ_2) such that $M((X, \tau_1, \tau_2)) = (X, \tau, \leq)$. Similarly $(X, \tau, \leq)$ *admits* a transitive quasi-uniformity if there is a transitive quasi-uniform space $(X, \mathcal{Q})$ such that $V(X, \mathcal{Q}) = (X, \tau, \leq)$. As in bitopological space we have the following corollary:

COROLLARY 1.5. *Let* $(X, \tau, \leq)$ *be a preordered space. Then the following are equivalent*:

(a) $(X, \tau, \leq)$ *is zero-dimensional*;
(b) $(X, \tau, \leq)$ *admits a zero-dimensional bitopology*;
(c) $(X, \tau, \leq)$ *admits a transitive quasi-uniformity.*

Proof. (a) $\Rightarrow$ (b) Follows from Corollary 1.4.

(b) $\Rightarrow$ (c) If there exists a zero-dimensional bitopological space (X, τ_1, τ_2) with $M(X, \tau_1, \tau_2) = (X, \tau, \leq)$, then there is a transitive quasi-uniformity $\mathcal{U}$ with $\overline{T}(X, \mathcal{U}) = (X, \tau_1, \tau_2)$. So $(X, \tau, \leq) = M\overline{T}(X, \mathcal{U}) = V(X, \mathcal{U})$. Therefore (c) holds.

(c) $\Rightarrow$ (a) Suppose $(X, \tau, \leq)$ admits a transitive quasi-uniformity $\mathcal{U}$, i.e., $V(X, \mathcal{U}) = (X, \tau, \leq)$. Thus $(X, \tau, \leq) = M(\overline{T}(X, \mathcal{U}))$. We know that $\overline{T}(X, \mathcal{U})$ is a zero-dimensional bitopological space. By 1.4 $(X, \tau, \leq)$ is a zero-dimensional preordered space. □

Let **ZTopOrd** be the full subcategory of **QoTop** which consists of zero-dimensional preordered spaces and order preserving maps. It is obvious that a zero-dimensional preordered space is a completely regular preordered space. So **ZTopOrd** is a full subcategory of **CTopOrd**.

PROPOSITION 1.6. **ZTopOrd** *is a topological category.*

Proof. By Definition 1.1 and the above argument, **ZTopOrd** is the initial hull of $\{D_0\}$ in **CTopOrd**. Since over **Set**, initial hulls are bireflective hulls, we have that **ZTopOrd** is a bireflective subcategory of **CTopOrd**. Since **CTopOrd** is topological (see Schauerte [33, p. 18]), **ZTopOrd** is topological (because any bireflective subcategory of a topological category over **Set** is topological). □

DEFINITION 1.7 [28]. A pospace $(X, \tau, \leq)$ is said to be *totally order-disconnected* if, given $x, y \in X$ with $x \not\leq y$, there exist disjoint τ-clopen sets U, L with U increasing and L decreasing such that $x \in U$, $y \in L$.

PROPOSITION 1.8. *For a compact partially ordered space $(X, \tau, \leq)$ the following are equivalent*:

(a) $(X, \tau, \leq)$ *is totally order-disconnected.*
(b) $(X, \tau, \leq)$ *is zero-dimensional.*
(c) *The unique quasi-uniformity of* $(X, \tau, \leq)$ *is transitive.*
(d) *The unique bitopology of* $(X, \tau, \leq)$ *is zero-dimensional.*

Proof. (b) $\Leftrightarrow$ (c) $\Leftrightarrow$ (d) Corollary 1.5.
(a) $\Leftrightarrow$ (c) This is proved in [20, Proposition 1]. □

2. Strongly Zero-Dimensional Partially Ordered Spaces

Nachbin–Stone–Čech partially ordered compactification. For the original construction of the Nachbin–Stone–Čech ordered compactification we refer the reader to [24]. We give an outline of a construction of the Nachbin–Stone–Čech ordered compactification due to Choe and Hong [11].

For the following, let $X = (X, \tau, \leq)$ be any completely regular pospace.

A filter $\mathcal{F}$ on X is said to be *o-completely regular* if $\mathcal{F}$ has an open filterbase $\mathcal{B}$ such that for each $U \in \mathcal{B}$ there exists a $V \in \mathcal{B}$ with $V \subseteq U$, and there exist $f_1, \ldots, f_n \in C_0(X, [-1, 1]_0)$ such that $f_i[V] = \{0\}$ for $i \in \{1, \ldots, n\}$ and $X - U \subseteq \bigcup_{i=1}^{n} f_i^{-1}[\{-1, 1\}]$, where $C_0(X, [-1, 1]_0)$ consists of continuous order preserving functions from X to $[-1, 1]_0$.

The following results were shown in [11]:

(a) Every neighbourhood filter on X is a maximal o-completely regular filter.
(b) A filter $\mathcal{M}$ on X contains a maximal o-completely regular filter iff $f(\mathcal{M})$ is convergent for each $f \in C_0(X, [-1, 1]_0)$.
(c) X is compact iff every maximal o-completely regular filter on X is convergent.

Given a maximal o-completely regular filter $\mathcal{M}$ on X and $f \in C_0(X, [-1, 1]_0)$ we will denote the limit of the filter $f(\mathcal{M})$ by $\lim f(\mathcal{M})$. Let $\beta^{\uparrow}X$ be the set of all maximal o-completely regular filters on X, endowed with the topology τ^* generated by the base $\{U^* \mid U \in \tau\}$, where $U^* = \{\mathcal{M} \in \beta^{\uparrow}X \mid U \in \mathcal{M}\}$, and with an order $\leq$ defined as follows: $\mathcal{M} \leq \mathcal{N}$ in $\beta^{\uparrow}X$ if and only if $\lim f(\mathcal{M}) \leq \lim f(\mathcal{N})$ for all $f \in C_0(X, [-1, 1]_0)$.

Let e: $X \to \beta^{\uparrow}X$ be the map defined by $e(x) = \mathcal{N}_x$, where $\mathcal{N}_x$ is the neighbourhood filter at x (see (a)). Then e is a dense **PoTop**-embedding and $\beta^{\uparrow}X$ is a compact pospace. Also for every continuous order preserving function f: $X \to [-1, 1]_0$ there is a unique extension $\overline{f}$: $\beta^{\uparrow}X \to [-1, 1]_0$ of f defined by $\overline{f}(\mathcal{M}) =$

$\lim f(\mathcal{M})$. Thus the pair $(e, \beta^{\uparrow} X)$ is equivalent to the Nachbin–Stone–Čech ordered compactification of X.

The easy proof of the following lemma is omitted.

LEMMA 2.1. *If U is open in $(X, \tau, \leq)$ then $e^{-1}[U^*] = U$.*

THEOREM 2.2. *Let $(X, \tau, \leq)$ be a completely regular pospace. If U is clopen and increasing in $(X, \tau, \leq)$, then $\mathrm{cl}_{\beta^{\uparrow} X} e[U]$ is clopen and increasing in $\beta^{\uparrow} X$.*

Proof. Let U be clopen and increasing in $(X, \tau, \leq)$. Define $\chi_U \colon X \to [-1, 1]$ by:

$$\chi_U(x) = \begin{cases} 1 & \text{if } x \in U, \\ 0 & \text{if } x \notin U, \end{cases}$$

χ_U is continuous and order preserving. Thus there is an extension, $\overline{\chi}_U$, of χ_U from $\beta^{\uparrow} X$ into $[-1, 1]$ given by $\overline{\chi}_U(\mathcal{M}) = \lim \chi_U(\mathcal{M})$ for $\mathcal{M} \in \beta^{\uparrow} X$. We now show that $\mathrm{cl}_{\beta^{\uparrow} X}[U] = \overline{\chi}_U{}^{-1}[\{1\}]$. We know that $(\overline{\chi}_U \circ e)[U] = \chi_U[U] = \{1\}$, therefore $e[U] \subseteq \overline{\chi}_U{}^{-1}[\{1\}]$. This implies that $\mathrm{cl}_{\beta^{\uparrow} X} e[U] \subseteq \overline{\chi}_U{}^{-1}[\{1\}]$ since $\overline{\chi}_U{}^{-1}[\{1\}]$ is closed in $\beta^{\uparrow} X$. We show that $\overline{\chi}_U{}^{-1}[\{1\}] \subseteq \mathrm{cl}_{\beta^{\uparrow} X} e[U]$.

$$\begin{aligned} \mathcal{M} \in \overline{\chi}_U{}^{-1}[\{1\}] &\Longrightarrow \lim \chi_U(\mathcal{M}) = 1 \\ &\Longrightarrow (\exists\, M \in \mathcal{M})\,(M \subseteq U) \text{ since } \chi_U \text{ is 2-valued} \\ &\Longrightarrow U \in \mathcal{M} \\ &\Longrightarrow \mathcal{M} \in U^*. \end{aligned}$$

Suppose $\mathcal{M} \notin \mathrm{cl}_{\beta^{\uparrow} X} e[U]$. Since $\{V^* \mid V \in \tau\}$ is a base for τ^*, there exists an open set V such that $\mathcal{M} \in V^*$ and $V^* \cap e[U] = \emptyset$. Since e is injective, $e^{-1}[V^*] \cap U = \emptyset$, and so $V \cap U = \emptyset$ by Lemma 2.1. Since $V \in \mathcal{M}$ and $\mathcal{M}$ is a filter we have a contradiction. Therefore $\mathcal{M} \in \mathrm{cl}_{\beta^{\uparrow} X} e[U]$ and we conclude that $\mathrm{cl}_{\beta^{\uparrow} X} e[U] = \overline{\chi}_U{}^{-1}[\{1\}]$.

Since $\overline{\chi}_U$ is continuous, $\mathrm{cl}_{\beta^{\uparrow} X} e[U]$ is clopen. We now show that $\mathrm{cl}_{\beta^{\uparrow} X} e[U]$ is increasing. Let $\mathcal{M} \leq \mathcal{N}$ with $\mathcal{M} \in \mathrm{cl}_{\beta^{\uparrow} X} e[U]$. Then $1 = \overline{\chi}_U(\mathcal{M}) \leq \overline{\chi}_U(\mathcal{N}) \leq 1$ by definition of $\leq$. Thus $\overline{\chi}_U(\mathcal{N}) = 1$ and hence $\mathcal{N} \in \mathrm{cl}_{\beta^{\uparrow} X} e[U]$. So $\mathrm{cl}_{\beta^{\uparrow} X} e[U]$ is clopen and increasing. □

REMARK 2.3. The implications in the proof of Theorem 2.2 can be reversed. Thus one has $\overline{\chi}_U{}^{-1}[\{1\}] = U^*$.

In analogy with the definition of strong zero-dimensionality in spaces [12] and bispaces [3], we give the following definition:

DEFINITION 2.4. We define a completely regular pospace $(X, \tau, \leq)$ to be *strongly zero-dimensional* if $\beta^{\uparrow} X$ is a zero-dimensional pospace.

The proof of the following lemma is omitted:

LEMMA 2.5. *If U is clopen and increasing in $(X, \tau, \leq)$ then $U^* = \mathrm{cl}_{\beta^\uparrow X} e[U]$.*

PROPOSITION 2.6. *Let $(X, \tau, \leq)$ be a compact zero-dimensional pospace. If A is closed and decreasing and B is closed and increasing, with $A \cap B = \emptyset$, then there exists a clopen increasing set U such that $B \subseteq U$ and $A \cap U = \emptyset$.*

Proof. Consider any arbitrary point $a \in A$. We have $y \nleq a$ for every $y \in B$. Since $(X, \tau, \leq)$ is totally order-disconnected by 1.8, corresponding to every point $y \in B$, there exist a clopen increasing set U_y containing y and a clopen decreasing set L_y containing a which are disjoint. Then $B \subseteq \bigcup_{y \in B} U_y$. Since B is closed and $(X, \tau, \leq)$ is compact, B is compact. Thus there exist finitely many points $y_1, \ldots, y_n \in B$ such that $B \subseteq \bigcup_{i=1}^n U_{y_i} = U$. Clearly U is clopen and increasing. Put $L = \bigcap_{i=1}^n L_{y_i}$. Then L is clopen and decreasing, contains a and is disjoint from U. Thus for every $x \in A$ we can find a clopen increasing set V_x containing B and a clopen decreasing set L_x containing x, with $V_x \cap L_x = \emptyset$. Since A is compact there exist finitely many $x_1, \ldots, x_m \in A$ such that $A \subseteq \bigcup_{i=1}^m L_{x_i}$. Now put $L = \bigcup_{i=1}^m L_{x_i}$ and $V = \bigcap_{i=1}^m V_{x_i}$. Then $B \subseteq V$ and $A \cap V = \emptyset$ since $V \cap L = \emptyset$. Since V is clopen and increasing we are finished. □

DEFINITION 2.7. Let $(X, \tau, \leq)$ be a pospace. A subset Z of X is called a *decreasing (increasing) zero set* in $(X, \tau, \leq)$ when there is an order preserving continuous function from X to $\mathbf{R}_0$ such that $Z = \{x \in X \mid f(x) \leq 0\}$ $(Z = \{x \in X \mid f(x) \geq 0\})$. The set of all decreasing (increasing) zero sets will be denoted by $\mathcal{A}_0(\mathcal{B}_0)$.

Let $(X, \tau, \leq)$ be a completely regular preordered space and let $a \in \mathbf{R}$. Set $A := \{x \in X \mid f(x) \leq a\}$ and set $B := \{x \in X \mid f(x) \geq a\}$, where $f\colon (X, \tau, \leq) \to \mathbf{R}_0$ is a continuous order preserving function. Then $A \in \mathcal{A}_0$ and $B \in \mathcal{B}_0$. This follows by observing that the function g from X into $\mathbf{R}_0$ defined by $g(x) = f(x) - a$ is continuous and order preserving.

LEMMA 2.8 [23]. *If $(X, \tau, \leq)$ is a pospace, $A \in \mathcal{A}_0$, $B \in \mathcal{B}_0$ with $A \cap B = \emptyset$, then there exists an order preserving continuous function $f\colon (X, \tau, \leq) \to \mathbf{R}_0$ such that $f[A] \subseteq \{0\}$ and $f[B] \subseteq \{1\}$.*

THEOREM 2.9. *Let $(X, \tau, \leq)$ be a completely regular pospace. Then $(X, \tau, \leq)$ is a strongly zero-dimensional pospace if and only if for any $A \in \mathcal{A}_0$ and for any $B \in \mathcal{B}_0$ such that $A \cap B = \emptyset$ there exists a clopen increasing set U in $(X, \tau, \leq)$ such that $B \subseteq U$ and $U \cap A = \emptyset$.*

Proof. Suppose that $A \in \mathcal{A}_0$, $B \in \mathcal{B}_0$ are such that $A \cap B = \emptyset$ and also that $\beta^\uparrow X$ is a zero-dimensional pospace. By Lemma 2.8 there exists a continuous order preserving function $f\colon (X, \tau, \leq) \to \mathbf{I}_0$ such that $A \subseteq f^{-1}[\{0\}]$ and $B \subseteq f^{-1}[\{1\}]$. Then $e[A] \subseteq \bar{f}^{-1}[\{0\}]$ and $e[B] \subseteq \bar{f}^{-1}[\{1\}]$ where $\bar{f}\colon \beta^\uparrow X \to \mathbf{R}_0$ is the extension of f. Since $\bar{f}^{-1}[\{0\}]$ is closed and decreasing and $\bar{f}^{-1}[\{1\}]$ is closed and increasing, and $\bar{f}^{-1}[\{0\}] \cap \bar{f}^{-1}[\{1\}] = \emptyset$, by Proposition 2.6 there exists a clopen and increasing set $\underline{U}$ in $\beta^{\uparrow} X$ such that $\bar{f}^{-1}[\{1\}] \subseteq \underline{U}$ and $\underline{U} \cap \bar{f}^{-1}[\{0\}] = \emptyset$. Thus $e[B] \subseteq \underline{U}$ and $\underline{U} \cap e[A] = \emptyset$. Then $B \subseteq e^{-1}[\underline{U}]$ and $e^{-1}[\underline{U}] \cap A = \emptyset$. Since $U := e^{-1}[\underline{U}]$ is clopen and increasing the result follows.

Conversely suppose that for any $A \in \mathcal{A}_0$ and for any $B \in \mathcal{B}_0$ such that $A \cap B = \emptyset$ there exists a clopen increasing set U in X such that $B \subseteq U$ and $U \cap A = \emptyset$. We shall show that $\beta^{\uparrow} X$ is totally order-disconnected. Let $\mathcal{M}, \mathcal{N} \in \beta^{\uparrow} X$ and let $\mathcal{M} \not\leq \mathcal{N}$. Thus $\lim f(\mathcal{M}) \not\leq \lim f(\mathcal{N})$ for some $f \in C_0(X, [-1, 1]_0)$. Then $\lim f(\mathcal{N}) < \lim f(\mathcal{M})$. Let $a, b \in [-1, 1]$ with $\lim f(\mathcal{N}) < a < b < \lim f(\mathcal{M})$. Since $f^{-1}[[-1, a]] \in \mathcal{A}_0$, $f^{-1}[[b, 1]] \in \mathcal{B}_0$ and $f^{-1}[[-1, a]] \cap f^{-1}[[b, 1]] = \emptyset$, by our assumption there exists a clopen increasing set U in X such that $f^{-1}[[b, 1]] \subseteq U$ and $U \cap f^{-1}[[-1, a]] = \emptyset$. Since $f^{-1}[(b, 1]] \subseteq U$ and $b \leq \lim f(\mathcal{M})$, $U \in \mathcal{M}$, i.e., $\mathcal{M} \in U^*$, $U \cap f^{-1}[[-1, a]] = \emptyset$ and $\lim f(\mathcal{N}) \leq a$ imply that $\mathcal{N} \notin U^*$. By 2.2 and 2.4, U^* is clopen and increasing in $\beta^{\uparrow} X$ with $\mathcal{M} \in U^*$ and $\mathcal{N} \notin U^*$. Then $\beta^{\uparrow} X - U^*$ is clopen and decreasing and contains $\mathcal{N}$. Therefore $\beta^{\uparrow} X$ is totally order-disconnected. By 1.8 $\beta^{\uparrow} X$ is a zero-dimensional pospace. Therefore X is strongly zero-dimensional. □

REMARK 2.10. We give further characterizations of strongly zero-dimensional pospaces similar to the ones given in [14] for strongly zero-dimensional bitopological spaces. We shall call the complements of the increasing zero sets and decreasing zero sets in a completely regular pospace, the *decreasing cozero sets* and the *increasing cozero sets* respectively.

LEMMA 2.11. *Let $(X, \tau, \leq)$ be a completely regular pospace. Then,*

(a) *A subset A of X is a decreasing zero set in $(X, \tau, \leq)$ iff there is a continuous order preserving function $f\colon (X, \tau, \leq) \to \mathbf{I}_0$ such that $A = f^{-1}[\{0\}]$.*
(b) *A subset B of X is an increasing zero set in $(X, \tau, \leq)$ iff there is a continuous order preserving function $f\colon (X, \tau, \leq) \to \mathbf{I}_0$ such that $B = f^{-1}[\{1\}]$.*

Proof. (a) Let A be a decreasing zero set. Then $A = \{x \in X \mid h(x) \leq 0\}$ for some continuous order preserving function $h\colon (X, \tau, \leq) \to \mathbf{R}_0$. Let $g\colon \mathbf{R}_0 \to \mathbf{I}_0$ be given by:

$$g(x) = \begin{cases} 0 & \text{if } x \leq 0, \\ x & \text{if } 0 < x < 1, \\ 1 & \text{if } x \geq 1. \end{cases}$$

It is obvious that $g\colon \mathbf{R}_0 \to \mathbf{I}_0$ is continuous and order preserving. The function $f\colon (X, \tau, \leq) \to \mathbf{I}_0$ defined by $f = g \circ h$ is continuous and order preserving, and $A = f^{-1}[\{0\}]$ as required.

Conversely, let $f\colon (X, \tau, \leq) \to \mathbf{I}_0$ be a continuous order preserving function such that $A = f^{-1}\{0\}$. Let $g\colon (X, \tau, \leq) \to \mathbf{R}_0$ be given by $g = i \circ f$, where

$i\colon \mathbf{I}_0 \hookrightarrow \mathbf{R}_0$ is the inclusion map. Then the function g is continuous and order preserving, and furthermore $A = \{x \in X \mid g(x) \leq 0\}$.

(b) This can be proved in a similar way with the function g above given by:

$$g(x) = \begin{cases} 0 & \text{if } x < -1, \\ x+1 & \text{if } -1 \leq x < 0, \\ 1 & \text{if } x \geq 0. \end{cases}$$

□

THEOREM 2.12. *For a completely regular pospace $(X, \tau, \leq)$ the following are equivalent*:

(a) *X is strongly zero-dimensional.*
(b) *For any $A \in \mathcal{A}_0$, $B \in \mathcal{B}_0$ such that $A \cap B = \emptyset$ there exists a clopen decreasing set V in $(X, \tau, \leq)$ and there exists a clopen increasing set U in $(X, \tau, \leq)$ such that $A \subseteq V$, $B \subseteq U$ and $U \cap V = \emptyset$.*
(c) *For any $A \in \mathcal{A}_0$, $B \in \mathcal{B}_0$ such that $A \cap B = \emptyset$, there exists a set V which is a countable intersection of clopen decreasing sets and there exists a set U which is a countable intersection of clopen increasing sets such that $A \subseteq V$, $B \subseteq U$, $V \cap B = \emptyset$ and $U \cap A = \emptyset$.*
(d) *For any $A \in \mathcal{A}_0$, $B \in \mathcal{B}_0$ such that $A \cap B = \emptyset$ there exists a set V which is a countable union of clopen increasing sets and there exists a set U which is a countable union of clopen decreasing sets such that $A \subseteq U$, $B \subseteq V$, $U \cap B = \emptyset$ and $V \cap A = \emptyset$.*
(e) *Every $A \in \mathcal{A}_0$ is a countable intersection of clopen decreasing sets and every $B \in \mathcal{B}_0$ is a countable intersection of clopen increasing sets.*
(f) *Every increasing cozero set of X is a countable union of clopen increasing sets and every decreasing cozero set of X is a countable union of clopen decreasing sets.*

Proof. (a) $\Leftrightarrow$ (b) Follows from Theorem 2.9.

(b) $\Rightarrow$ (c) Trivial.

(c) $\Rightarrow$ (d) This follows by taking complements.

(d) $\Rightarrow$ (e) Suppose (d) holds and let $B \in \mathcal{B}_0$. By Lemma 2.11 there exists a continuous order preserving function $f\colon (X, \tau, \leq) \to \mathbf{I}_0$ such that $B = f^{-1}[\{1\}]$. Fix $n \in \mathbf{N}$. Now $f^{-1}[[0, \frac{n-1}{n}]]$ is a decreasing zero set and by (d) there exist clopen decreasing sets V_{kn}, $k \in \mathbf{N}$ such that $f^{-1}[[0, \frac{n-1}{n}]] \subseteq \bigcup_{k=1}^{\infty} V_{kn}$ and $\bigcup_{k=1}^{\infty} V_{kn} \cap B = \emptyset$. We show that $B = \bigcap_{n=1}^{\infty} \bigcap_{k=1}^{\infty} (X - V_{kn})$. Clearly $B \subseteq \bigcap_{n=1}^{\infty} \bigcap_{k=1}^{\infty} (X - V_{kn})$. Now, for each $n \in \mathbf{N}$

$$\begin{aligned} x \in \bigcap_{k=1}^{\infty} (X - V_{kn}) &\Longrightarrow x \in (X - V_{kn}) \quad \text{for all } k \\ &\Longrightarrow x \notin V_{kn} \quad \text{for all } k \\ &\Longrightarrow x \notin f^{-1}[[0, 1 - 1/n]] \\ &\Longrightarrow f(x) > 1 - 1/n. \end{aligned}$$

This holds for every $n \in \mathbf{N}$, and so $f(x) \geq 1$, i.e., $f(x) = 1$, i.e., $x \in B$. Hence $B = \bigcap_{n=1}^{\infty} \bigcap_{k=1}^{\infty}(X - V_{kn})$. Since each V_{kn} is clopen and decreasing, each $X - V_{kn}$ is clopen and increasing. If $A \in \mathcal{A}_0$, then by Lemma 2.11 there exists a continuous order preserving $f\colon (X, \tau, \leq) \to \mathbf{I}_0$ such that $A = f^{-1}[\{0\}]$. Since $f^{-1}[[\frac{1}{n}, 1]]$ is an increasing zero set, by (d) there exist clopen increasing sets $U_{kn}, k \in \mathbf{N}$ such that $f^{-1}[\frac{1}{n}, 1] \subseteq \bigcup_{k=1}^{\infty} U_{kn}$ and $\bigcup_{k=1}^{\infty} U_{kn} \cap A = \emptyset$. As in the previous case $A = \bigcap_{n=1}^{\infty} \bigcap_{k=1}^{\infty}(X - U_{kn})$ and each $X - U_{kn}$ is clopen and decreasing.

(e) $\Rightarrow$ (f) Obvious.

(f) $\Rightarrow$ (b) Suppose that (f) holds. Let $A \in \mathcal{A}_0$, $B \in \mathcal{B}_0$ and $A \cap B = \emptyset$. By (f) there exist clopen increasing sets $U_n, n \in \mathbf{N}$ and there exist clopen decreasing sets $V_n, n \in N$ such that $X - A = \bigcup_{n=1}^{\infty} U_n$ and $X - B = \bigcup_{n=1}^{\infty} V_n$. Since $A \cap B = \emptyset$, we have $X = X - (A \cap B) = (X - A) \cup (X - B) = \bigcup_{n=1}^{\infty}(U_n \cup V_n)$. Now define $U_n' = U_n - (V_1 \cup V_2 \cup \cdots \cup V_{n-1})$, and define $V_n' = V_n - (U_1 \cup U_2 \cup \cdots \cup U_n)$. Furthermore let $U_1' = U_1$. Then U_n' and V_n' are clearly clopen for all n. We show that U_n' is increasing and V_n' is decreasing. The case $n = 1$ is obvious since U_1 and V_1 are increasing and decreasing respectively. Let $x \leq y$ and $x \in U_n'$. Then $x \in U_n$ and $x \notin (V_1 \cup V_2 \cup \cdots \cup V_{n-1}), n \neq 1$. If $y \notin U_n'$ then we have the following two cases:

Case 1: $y \in U_n$ and $y \in V_1 \cup V_2 \cup \cdots \cup V_{n-1}$. But this will imply that $x \in V_1 \cup V_2 \cup \cdots \cup V_{n-1}$ since $V_1 \cup V_2 \cup \cdots \cup V_{n-1}$ is decreasing, which contradicts our assumption.

Case 2: $y \notin U_n$. This cannot happen since U_n is increasing.

In both cases we have a contradiction and thus $y \in U_n'$. Hence U_n' is increasing. Similarly V_n' is decreasing. It is clear that $U_n' \cap V_n' = \emptyset$ for all $n \in \mathbf{N}$. Letting $U = \bigcup_{n=1}^{\infty} U_n'$ and $V = \bigcup_{n=1}^{\infty} V_n'$, we have $U \cap V = \emptyset$. We now show that $A \subseteq V$ and $B \subseteq U$. Let $x \in A$. Then $x \notin \bigcup_{n=1}^{\infty} U_n$, so that $x \in \bigcup_{n=1}^{\infty} V_n$ since $X = \bigcup_{n=1}^{\infty}(U_n \cup V_n)$. Now we have a smallest $n_0 \in \mathbf{N}$ for which $x \in V_{n_0}$. Therefore $x \in V_{n_0}'$, and thus $x \in V$. Thus $A \subseteq V$, and similarly $B \subseteq U$. Furthermore we have $X = U \cup V$. The sets U and V then satisfy (b). □

Acknowledgement

The author is grateful to the referee for suggested changes which improved the paper.

References

1. Adámek, J., Herrlich, H. and Strecker, G. E.: *Abstract and Concrete Categories*, Pure Appl. Math., Wiley, New York, 1990.
2. Banaschewski, B.: Spaces of dimension zero, *Canad. J. Math.* **9** (1957), 38–46.
3. Banaschewski, B. and Brümmer, G. C. L.: Strong zero-dimensionality of biframes and bispaces, *Quaestiones Math.* **13** (1990), 273–290.

4. Bîrsan, T.: Transitive quasi-uniformities and zero-dimensional bitopological spaces, *An. Ştiinţ. Univ. "Al. I. Cuza" Iaşi Sect. Ia Mat.* **20** (1974), 317–322.
5. Brümmer, G. C. L.: A categorial study of initiality in uniform topology, PhD Thesis, Univ. Cape Town, 1971.
6. Brümmer, G. C. L.: On the non-unique extension of topological to bitopological properties, in *Categorical Aspects of Topology and Analysis (Conference Proceedings, Ottawa, 1980)*, Lecture Notes in Math. 915, 1982, pp. 50–67.
7. Brümmer, G. C. L.: Functorial transitive quasi-uniformities, in *Categorical Topology (Conference Proceedings, Toledo, 1983)*, Sigma Series in Pure Math. 5, 1984, pp. 163–184.
8. Burgess, D. C. J. and McCartan, S. D.: Order-continuous functions and order-connected spaces, *Proc. Cambridge Philos. Soc.* **68** (1970), 27–31.
9. Charlton, J. R. H.: Topology, bitopology and order, M.Sc. Thesis, Univ. Cape Town, 1973.
10. Choe, T. H.: Partially ordered topological spaces, *An. Acad. Brasil. Ciênc.* **51** (1979), 53–63.
11. Choe, T. H. and Hong, Y. H.: Extensions of completely regular ordered spaces, *Pacific J. Math.* **66** (1976), 37–48.
12. Engelking, R.: *General Topology*, Polish Scientific Publishers, Warsaw, 1977.
13. Fletcher, P. and Lindgren, W. F.: *Quasi-Uniform Spaces*, Marcel Dekker, New York, 1982.
14. Fora, A. A.: Strongly zero-dimensional bitopological spaces, *J. Univ. Kuwait Sci.* **11** (1984), 180–189.
15. Gillman, L. and Jerison, M.: *Rings of Continuous Functions*, D. Van Nostrand, Princeton, 1960.
16. Halpin, M. N.: Transitive quasi-uniform spaces, M.Sc. Thesis, Univ. Cape Town, 1974.
17. Halpin, M. N.: Transitive quasi-uniformities for bitopological spaces, *Math. Colloq. Univ. Cape Town* **11** (1977), 47–64.
18. Künzi, H. P. A.: Completely regular ordered spaces, *Order* **7** (1990), 283–293.
19. Künzi, H. P. A.: Strongly zero-dimensional bispaces, *J. Austral. Math. Soc. Ser. A* **53** (1992), 327–337.
20. Künzi, H. P. A. and Brümmer, G. C. L.: Sobrification and bicompletion of totally bounded quasi-uniform spaces, *Math. Proc. Cambridge Philos. Soc.* **101** (1987), 237–247.
21. Lane, E. P.: Bitopological spaces and quasi-uniformities, *Proc. London Math. Soc. (3)* **17** (1967), 241–256.
22. McCartan, S. D.: Separation axioms for topological ordered spaces, *Proc. Cambridge Philos. Soc.* **64** (1968), 965–973.
23. McCallion, T.: Compactifications of ordered topological spaces, *Proc. Cambridge Philos. Soc.* **64** (1972), 463–474.
24. Nachbin, L.: *Topology and Order*, Van Nostrand, Princeton, 1965.
25. Nailana, K. R.: (Strongly) zero-dimensional ordered spaces, M.Sc. Thesis, Univ. Cape Town, 1993.
26. Nyikos, P.: A survey of zero-dimensional spaces, in *Topology (Proc., Memphis State Univ. Conf.)*, Marcel Dekker, New York, 1976, pp. 87–114.
27. Pervin, W.: Connectedness in bitopological spaces, *Nederl. Akad. Wetensch. Proc. Ser. A* **70** = *Indag Math.* **29** (1967), 369–372.
28. Priestley, H. A.: Representation of distributive lattices by means of ordered Stone spaces, *Bull. London Math. Soc.* **2** (1970), 186–190.
29. Priestley, H. A.: Ordered topological spaces and the representation of distributive lattices, *Proc. London Math. Soc. (3)* **24** (1972), 507–530.
30. Reilly, I. L.: Zero-dimensional bitopological spaces, *Nederl. Akad. Wetensch., Proc. Ser. A* **76** = *Indag Math.* (1973), 127–131.
31. Salbany, S.: Bitopological spaces, compactifications and completions, Ph.D. Thesis, Univ. Cape Town, 1970.
32. Salbany, S.: A bitopological view of topology and order, in *Categorical Topology (Proc. Conf. Toledo, 1983)*, Heldermann, Berlin, 1984, pp. 481–504.

33. Schauerte, A.: Functorial quasi-uniformities over partially ordered spaces, M.Sc. Thesis, Univ. Cape Town, 1988.
34. Schauerte, A.: On the MacNeille completion of the category of partially ordered topological spaces, *Math. Nachr.* **163** (1993), 281–288.
35. Schwartz, F. and Weck-Schwartz, S.: Is every partially ordered space with a completely regular topology already a completely regular partially ordered space? *Math. Nachr.* **161** (1993), 199–201.
36. Swart, J.: Total disconnectedness in bitopological spaces and product bitopological spaces, *Nederl. Akad. Wetensch., Proc. Ser. A* **74** = *Indag Math.* **33** (1971), 135–145.
37. Walker, R. C.: The Stone–Čech compactification, Springer-Verlag, New York, 1974.

G. Brümmer & C. Gilmour (eds), Papers in Honour of Bernhard Banaschewski.

Structure Sheaves for Additive Categories

Dedicated to Professor Bernhard Banaschewski on his 70th birthday

SHU-HAO SUN
School of Mathematics F07, University of Sydney, NSW 2006, Australia

(Received: 23 October 1996; accepted: 27 May 1998)

Abstract. In this paper, we establish a general theory of structure presheaves for any small additive category. We will show that this presheaf functor is exact and will give several characterizations for the presheaves *to be* sheaves.

Mathematics Subject Classifications (2000): 18E05, 18E40, 14F20, 18F20.

Key words: structure sheaf, additive category, Grothendieck topology.

Introduction

Some effort has been made in the literature to extend Grothendieck's scheme theory and sheaf cohomology theory to non-commutative rings (e.g., [8, 2]). To proceed, one may have the following two main difficulties:

(1) The first is suggested by Deligne's formula $\widetilde{M}(U) = \lim_n \hom(I^n, M)$, where $U = D(I)$ and $\widetilde{M}$ is the structure sheaf associated to a module over a commutative Noetherian ring R, and by using non-commutative localization theory (i.e., torsion theory), one obtains the so-called *structure presheaf* over Spec$R$ for non-commutative Noetherian rings R (see [8, 2]). If R is commutative, then the structure presheaf is just the classical structure sheaf. But the structure presheaf is not necessarily a sheaf in general even for (non-commutative) Noetherian rings. One question arising here is when is the structure presheaf a sheaf?

(2) In the classical theory, the exactness of the structure sheaf functor is extremely important. But for non-commutative rings, it was open question as to whether or not the structure presheaf functor is exact.

However, during the last few years, some real progress has been made towards solving the above two problems.

Regarding the first problem, paper [1] has presented several useful necessary and sufficient conditions for the structure presheaf to be a sheaf when R is Noetherian. In [5], the author extended the above results to not-necessarily Noetherian rings. Relating to the second problem, the author has shown in [6] that the structure presheaf functor is exact by choosing a suitable non-commutative version of classical sheaf category.

The main aim of this paper is to extend the new theory just mentioned above to rings with many objects (i.e., additive categories). More precisely, we will establish a general theory of structure presheaves for any small additive category (see Section 3). Then we will show that this presheaf functor is exact (see Section 4) and will give several characteriztions for the presheaves to be sheaves (see Section 5).

For the definition and basic discussions about ideals, prime ideals and prime spectra of additive categories, refer to the author's joint paper [7].

1. Preliminaries

First, let us briefly recall the classical theory. Let X denote the prime spectrum Spec R of R (i.e., the set of all prime ideals of R with the Zariski topology) and $D(I) = \{P \in \mathrm{Spec} R\colon I \not\subseteq P\}$ the typical open set determined by I, and $\mathrm{Psh}(X) = [X^{op}, R\text{-Mod}]$ the presheaf category and $\mathrm{Sh}(X)$ the sheaf category. Then the classical sheaf category $\widetilde{R}$-Mod, where $\widetilde{R}$ is the structure sheaf associated to R, is naturally isomorphic to a (not-necessarily full) subcategory of $\mathrm{Sh}(X)$ since each component at a quasi-compact open subset U of an object in $\widetilde{R}$-Mod is an $\widetilde{R}(U)$-module which clearly has an R-module structure along the ring morphism $R = \widetilde{R}(\mathrm{Spec}\, R) \to \widetilde{R}(U)$.

Let $\mathcal{B}$ be the basis consisting of all principal ideals of a commutative ring R. Then $\mathrm{Sh}(X) = \mathrm{Sh}(\mathcal{B})$, since $\mathcal{B}$ is closed under products.

Now let τ_a be the Gabriel topology generated by a (that is, an ideal I is in τ_a iff I contains a^n for some n) and recall that an R-module M is called τ_a-closed if for each ideal $I \in \tau_a$ and any R-module morphism $f\colon I \to M$, there is a unique extension to $R \to M$.

Write $Q_{\tau_a}(M)$ for the localizing module. That is, $Q_{\tau_a}(M) = E_{\tau_a}(E_{\tau_a}(M))$ where $E_{\tau_a}(M) = \lim_{I \in \tau_a} \hom(I, M)$. Then we see $Q_{\tau_a}(M) = \widetilde{M}(a)$ where $\widetilde{M}$ is the structure sheaf associated to the R-module M and $\widetilde{M}(a)$ is the set of all local sections over the open set $D(a)$.

LEMMA 1.1. *If R is commutative and $a, b \in R$, then we have the following*:

(1) *The category R_a-Mod of all R_a-modules is equivalent to the full subcategory (R, τ_a)-Mod consisting of all τ_a-closed modules*;
(2) *if $Ra \subseteq Rb$ and M is τ_a-closed and N is τ_b-closed, then each R-linear morphism $f\colon M \to N$ has the property that for each $\alpha \in Q_{\tau_a} R$ and $m \in M$, $f(\alpha m) = Q^{\tau_a}_{\tau_b}(\alpha) f(m)$, where $Q^{\tau_a}_{\tau_b}$ is the restriction morphism $\widetilde{R}(a) = Q^{\tau_a}(R) \to Q^{\tau_b}(R) = \widetilde{R}(b)$.*

Proof. (1) is known (e.g., see [2]).

To show (2), let α be represented by $h\colon I \to R/t_a R$, where $I \in \tau_a$. Then αm is represented by the composition $\hat{m}h\colon I \to R/t_a R \to M$ which can be extended to a morphism $R \to M$ and, hence, is determined by a unique element in M, say m'. That is, for each $i \in I$, $im' = r_i m$, where $h(i) = r_i + t_a R$. Thus, $\alpha m \in Q_{\tau_a} M$ corresponds bijectively to $m' \in M$. On the other hand, $Q^{\tau_a}_{\tau_b}(\alpha) f(m)$ is represented by the composition $\widehat{f(m)}\cdot q\cdot h\colon I \to R/\tau_a R \to R/\tau_b R \to N$ which is determined by a unique element $n \in N$. Now, for each $i \in I$, since $h(i) = r_i + \tau_a R$, we have $in = \hat{n}(i) = \widehat{f(m)}(r_i + \tau_b R) = r_i f(m) = f(r_i m) = f(im') = if(m')$ and, hence, $n = f(m')$ since N is τ_b-closed and $I \in \tau_b$. □

By Lemma 1.1, we have the following proposition:

PROPOSITION 1.2. *Let R be a commutative ring and $\mathcal{B}$ the basis consisting of all principal ideals. Then the classical sheaf category $\widetilde{R}$-Mod is isomorphic to the full subcategory* Sh($\mathcal{B}$*-closed*) *consisting of those objects $F \in$* Sh($\mathcal{B}$) *whose component at $D(a)$ is τ_a-closed for each $a \in R$.*

Let C be any small additive category and $\widetilde{C} = [C^{op}, \mathbf{Ab}]$ the category of all additive presheaves of Abelian groups.

Recall that an additive sieve K of c, with $c \in C$, is a set of arrows whose codomain is c with the properties that K is closed under addition and under composition on the left (i.e., $f \in K$ implies $hf \in K$ for any h). It is clear that the class of all sieves of c corresponds bijectively to the class of all additive subobject of $\hom(-, c)$.

A topology τ on an additive category C is defined by specifying, for each object c of C, a set $\tau(c)$ of additive sieves at c, called *covering subobjects* of the topology, such that

T1. For any $c \in C$, the maximal sieve $C(-, c)$ is in $\tau(c)$.

T2. If $R \in \tau(c)$ and $f\colon c' \to c$, then
$$f^*(R) = \{g \mid fg \in R\}$$
is in $\tau(c')$.

T3. An additive sieve S of $C(-, c)$ is in $\tau(c)$ iff there is an $R \in \tau(c)$ such that for each $f\colon v \to c$ in R, one has $f^*(S) \in \tau(v)$.

These axioms imply that

T4. For each $c \in C$, if $R \in \tau(c)$ and $T \supseteq R$, then $T \in \tau(c)$.

T5. For each $c \in C$, if $R_1, R_2 \in \tau(c)$, then $R_1 \cap R_2 \in \tau(c)$.

noindent τ is called a *pretopology* if it satisfies only T1 and T2.

Let τ be a topology and σ_1 and σ_2 are pretopologies on C. The smallest topology containing σ will be denoted by $\bar{\sigma}$. Then we have the following technical lemma:

LEMMA 1.3. *If $\tau \supseteq \sigma_1 \cap \sigma_2$, then $\tau \supseteq \bar{\sigma}_1 \cap \bar{\sigma}_2$.*

Proof. It suffices to show that $\tau \supseteq \bar{\sigma}_1 \cap \sigma_2$. Recall that the topology $\bar{\sigma}_1$ can be constructed as follows: for each small ordinal β, for each object c of C, we define a set $\mathcal{S}_\beta(c)$ of sieves at c, starting with $\mathcal{S}_0(c) = \sigma_1(c)$; and we put a sieve L at c in $\mathcal{S}_{\beta+1}(c)$ if there exists some $K \in \mathcal{S}_\beta(c)$ such that $k^*(L) \in \mathcal{S}_\beta(a)$ for each $k\colon a \to c$ in K; and we set $\mathcal{S}_\alpha(c) = \bigcup_{\beta \leqslant \alpha} \mathcal{S}_\beta(c)$ for a limit-ordinal α. Then the topology $\bar{\sigma}_1$ is $\mathcal{S}_\alpha$ for some large enough α such that $\mathcal{S}_\alpha = \mathcal{S}_{\alpha+1}$.

We will use transfinite induction to prove our lemma. Assume $\tau(c) \supseteq \mathcal{S}_\beta(c) \cap \sigma_2(c)$ for all c in C, and let L be in $\mathcal{S}_{\beta+1}(c) \cap \sigma_2(c)$. Then L is in $\sigma_2(c)$ and in $\mathcal{S}_{\beta+1}(c)$ and, hence, there exists a $K \in \mathcal{S}_\beta(c)$ such that $k^*(L) \in \mathcal{S}_\beta(a)$ for each $k\colon a \to c$ in K. Therefore, we see $K + L \in \mathcal{S}_\beta(c) \cap \sigma_2(c) \subseteq \tau(c)$. Moreover, for each $k+l\colon a \to c \in K+L$, where $k \in K$ and $l \in L$, we have $(k+l)^*(L) \supseteq k^*(L) \cap l^*(L) = k^*(L)$ and, hence, $(k+l)^*(L) \in \mathcal{S}_\beta(a)$. On the other hand, $(k+l)^*(L)$ is in $\sigma_2(a)$ since σ_2 satisfies T2. Thus, we see that $(k+l)^*(L)$ is in $\mathcal{S}_\beta(a) \cap \sigma_2(a)$ and, hence, is in $\tau(a)$ by induction assumption. Thus, we finally have that L itself is in $\tau(c)$ since τ satisfies T3. That is, we have shown that $\tau(c) \supseteq \mathcal{S}_{\beta+1}(c) \cap \sigma_2(c)$. For a limit ordinal α case, we also have $\tau(c) \supseteq \bigcup_{\beta<\alpha}(\mathcal{S}_\beta(c) \cap \sigma_2(c)) = (\bigcup_{\beta<\alpha}(\mathcal{S}_\beta(c)) \cap \sigma_2(c) = \mathcal{S}_\alpha(c) \cap \sigma_2(c)$. □

Recall that if C is a small additive category, then the set of all topologies on C, denoted by $L(C)$, is a locale.

In order to proceed, we introduce the following somewhat more general notion:

DEFINITION 1.4. A quasi-spectrum of an additive category C is a sub-lattice with a bottom element of $L(C)$, where join commutes with any meets if it exists and where $L(C)$ is the locale of all topologies on C.

Let S be a quasi-spectrum, and write $\mathrm{Psh}_{\widetilde{C}}(S)$ for the functor category $[S, \widetilde{C}]$ and $P(S\text{-closed})$ the full subcategory of $\mathrm{Psh}_{\widetilde{C}}(S)$ consisting of those objects $F \in \mathrm{Psh}_{\widetilde{C}}(S)$ whose component $F(\tau)$ is a τ-sheaf for each $\tau \in S$. Write $\mathrm{Sh}(S)$ for the full subcategory of $\mathrm{Psh}_{\widetilde{C}}(S)$ consisting of all sheaves, and $\mathrm{Sh}(S\text{-closed})$ for the full subcategory of $\mathrm{Sh}(S)$ consisting of those objects whose component at τ is a τ-sheaf for each $\tau \in S$. Then $\mathrm{Sh}(S\text{-closed})$ is the intersection of $P(S\text{-closed})$ and $\mathrm{Sh}(S)$.

Then we have the following easy but useful lemmas:

LEMMA 1.5. *The category $P(S\text{-closed})$ is a localization of* $\mathrm{Psh}(S)$.

LEMMA 1.6. $\mathrm{Sh}(S\text{-}closed)$ *is a localization of* $\mathrm{Psh}(S)$.

Now we are going to show that each basis of Spec C is isomorphic to a quasi-spectrum. The definition of Spec C is similar to the classical one, which consists of all prime ideals of C with the hull-kernel topology (the details see [7]). For an ideal J of C, $\sqrt{J}$ denotes the intersection of all prime ideals containing J.

Let I be an ideal of C and $T(I)$ be the set of ideals J such that $\sqrt{J} \supseteq I$. Then $T(I) = T(J)$ if $D(I) = D(J)$ (recall $D(I) = \{P \in \mathrm{Spec}\, C\colon I \not\subseteq P\}$). Let $\widetilde{T}(I)$ be the family of all the sieves S, say at c, which contains $J(c)$ for some $J \in T(I)$, where $J(c)$ denotes the family of arrows whose codomains are c. Thus, $\widetilde{T}(I)$ is the smallest pretopology containing $T(I)$. Let τ_I be the smallest topology containing $T(I)$.

Then we have the following useful lemma:

LEMMA 1.7. *For any pair I, J of ideals of C, we have $\tau_{I+J} = \tau_I \cap \tau_J$.*

Proof. Let $\widetilde{T}(K)$ be defined as above (recall that a sieve L at c is in $\widetilde{T}(K)(c)$ iff L contains some $J(c)$ for some $J \in T(K)$). Then we have the following equality $\widetilde{T}(I+J) = \widetilde{T}(I) \cap \widetilde{T}(J)$. In fact, let $L \in \widetilde{T}(I)(c) \cap \widetilde{T}(J)(c)$. Then there are ideals $K_1 \in T(I)$ and $K_2 \in T(J)$ such that $L \supseteq K_1(c)$ and $L \supseteq K_2(c)$. It is easy to see that $(K_1 + K_2)$ is in $T(I+J)$ and, hence, L is in $\widetilde{T}(I+J)$. The other inclusion is clear. Now the conclusion follows from Lemma 1.3. □

For each basis $\mathcal{B}$ of Spec C, write Sh($\mathcal{B}$-closed) for the full subcategory of Sh($\mathcal{B}$), consisting of those sheaves whose component at $D(I)$ is a τ_I-sheaf for each $D(I) \in \mathcal{B}$. Then by Propositions 1.7, 1.6 and Lemma 1.2, we have the following proposition:

PROPOSITION 1.8. (1) *Each basis of Spec C is isomorphic to a quasi-spectrum of C, which sends each $D(I)$ to the topology τ_I;*

(2) Sh(*$\mathcal{B}$-closed) is a localization of* Sh($\mathcal{B}$);

(3) *If C is a commutative ring, then* Sh(*$\mathcal{B}$-closed) is the classical sheaf category.*

LEMMA 1.9. *An open subset $U = D(J)$ is quasi-compact iff $U = D(I)$ for some finitely generated ideal I of C.*

2. Torsion Theory in $\widetilde{C}$

To proceed, we need some additive categorical version of torsion theory.

For each topology τ on C, we may define an endofunctor t_τ on $\widetilde{C} = [C^{op}, \mathbf{Ab}]$ as follows: for each $P \in \widetilde{C}$ and each $c \in C$, define $t_\tau P(c)$ to be the subgroup of $P(c)$ consisting of those elements $x \in P(c)$ for which there exists a sieve I_x at c such that $P(f)(x) = 0$ for each $f \in I_x$; or equivalently, $\tilde{x} \mid I = 0$, where $\tilde{x}$ is the unique morphism hom$(-, c) \to P$ determined by x, by the Yoneda Lemma.

It is easy to check that t_τ is left exact. We call $P \in \widetilde{C}$ τ-torsion if $t_\tau P = P$ and τ-torsion-free if $t_\tau P = 0$.

First we have the following lemma:

LEMMA 2.1. *For each $P \in \widetilde{C}$, $P/\tau P$ is τ-torsion-free.*

Proof. Suppose that $x \in P(c)$ has the property that there is a sieve I at c such that for each $h\colon b \to c$ in I, the induced morphism $P(c)/tP(c) \to P(b)/tP(b)$ sends $x + tP(c)$ to zero; i.e., $P(h)(x) \in tP(b)$. Then there is a sieve I_h at b such that $P(g)(P(h)(x)) = 0$ for each $g\colon a \to b$ in I_h, which implies $P(hg)(x) = 0$. Let $J = \sum_{h \in I, g \in I_h} hg$. Then J is in $\tau(c)$ by T3, and $\tilde{x} \mid J = 0$. Thus $x \in tP(c)$ and P/tP is τ-torsion-free. □

Let τ be a topology on C. Let $E_\tau P$ denote the presheaf specifying $E_\tau P(c) = \mathrm{colim}_{I \in \tau(c)} \hom(I, P)$ for each $c \in C$, and $\phi_\tau(P)$ is the canonical morphism $P \to$

$E_\tau P$. Then $Q_\tau P = E_\tau E_\tau P$ is the sheafification of P with respect to the topology τ. Let η_τ denote the canonical morphism from the identity to Q_τ. Then $\eta_\tau(P)$ is the composition of $\phi_\tau(P)$ and $\phi_\tau(E_\tau(P))$. Then we have the following lemma:

LEMMA 2.2. *Let τ be any given topology on C. Then we have the following for any presheaf P:*

(1) *For each $c \in C$ and each $s \in E_\tau(P)(c)$, there is an $K \in \tau(c)$ such that $\tilde{s} \mid K \subseteq \operatorname{im} \phi_\tau(P)$.*

(2) *$t_\tau P$ is the kernel of $\phi_\tau(P)$ and is also the kernel of $\eta_\tau(P)$.*

(3) *$E_\tau P$ is zero iff P is τ-torsion;*

(4) *$Q_\tau P \cong E_\tau(P/tP)$;*

(5) *any monic j: $M \to P$ induces an isomorphism $Q_\tau(j)$ iff P/M is τ-torsion.*

Proof. (1) It suffices to prove that the following diagram is commutative for each $c \in C$ and each $s \in E_\tau(P)(c)$ represented by f: $K \to P$, where $K \in \tau(c)$.

$$\begin{array}{ccc} K & \longrightarrow & \hom(-, c) \\ {\scriptstyle f}\downarrow & & \downarrow{\scriptstyle \tilde{s}} \\ P & \xrightarrow[\phi_\tau(P)]{} & E_\tau(P) \end{array} .$$

We want to show that the following is commutative for each $a \in C$:

$$\begin{array}{ccc} K(a) & \longrightarrow & \hom(a, c) \\ {\scriptstyle f_a}\downarrow & & \downarrow{\scriptstyle \tilde{s}_a} \\ P(a) & \xrightarrow[\phi_\tau(P_a)]{} & E_\tau(P)(a). \end{array}$$

In fact, for each h: $a \to c$ in K, $f_a(h)$ is in $P(a)$ and is sent to the element in $E_\tau(P)(a)$ represented by $\widetilde{f_a(h)}$. However, $\tilde{s}_a(h) = E_\tau(P)(h)(s)$ represented by h^*f: $h^*K \to P$, where $(h^*f)_a(w\colon b \to a) = f_b(hw) \in P(b)$. Now

$$\widetilde{(f_a(h))}_b(w\colon b \to a) = P(w)(f_a(h))$$

by Yoneda. However, the quality $P(w)(f_a(h)) = f_b(hw)$ follows from the following commutative diagram:

$$\begin{array}{ccc} K(a) & \xrightarrow{(-)w} & K_b \\ {\scriptstyle f_a}\downarrow & & \downarrow{\scriptstyle f_b} \\ P(a) & \xrightarrow[P(w)]{} & P(b) \end{array} .$$

(2) We have to prove that $t_\tau P(c)$ is the kernel of $\phi_\tau(c)$: $P(c) \to E_\tau(P)(c)$ for each $c \in C$: saying $x \in t_\tau P(c)$ is equivalent to saying that there is a sieve $I \in \tau(c)$ with $\tilde{x} \mid I = 0$, which is also equivalent to saying that $\phi_\tau(P)(c)(x)$ represented by

$\tilde{x}$: hom$(-, c) \to P$ is zero. The second part follows from the fact that $E_\tau P$ is a separated presheaf and $\phi_\tau(E_\tau P)$ is monic.

(3) It is clear that $E_\tau P = 0$ iff $Q_\tau P = 0$. Now if $Q_\tau P$ is zero, then the kernel of η is the whole P and, hence, is τ-torsion by (2). Now support $tP = P$. We want to show that $E_\tau(P)(c) = 0$ for any $c \in C$. Let f: $I \to P$ with $I \in \tau(c)$. We have to find a sieve $J \in \tau(c)$ with $f \mid J = 0$. Now for each component f_a: $I(a) \to P(a)$ and each h: $a \to c$ in $I(a)$, there is a sieve $S_h \in \tau_a$ such that $P(x)(f_a(h)) = 0$ for each x: $b \to a$ in S_h. By the naturality of f, we see that $f_b(hx) = 0$.

Now let $J = \sum_{h \in I} h \cdot S_h$. Then J is a subsieve of I. Note that any arrow z: $b \to c \in J$ is a finite sum of those elements which are of the form hy, where h: $a \to c$ is in I and y is in some sieve S at a with the property that $P(y)(f_a(h)) = 0$. It follows that $f \mid J = 0$.

The fact that J is in $\tau(c)$ follows from the fact that τ satisfies T3. Thus $E_\tau(P)(c) = 0$ and, hence, $E_\tau(P) = 0$ and $Q_\tau(P) = E_\tau(E_\tau(P)) = 0$.

(4) It follows from (1) that the cokernel of $\phi_\tau(P)$ (or, equivalently, of $P/tP \to E_\tau(P)$) is τ-torsion and from (2) that $E_\tau(\text{cok}\,\phi_\tau) = 0$. Now consider the following diagram:

$$0 \to P/tP \to E_\tau(P) \to \text{cok}(\phi_\tau(P)) \to 0.$$

Since E_τ is left exact, we have $E_\tau(P/tP) \cong E_\tau E_\tau(P)$.

(5) Consider the following commutative diagram:

$$\begin{array}{ccccccccc}
0 & \longrightarrow & M & \xrightarrow{j} & P & \longrightarrow & P/M & \longrightarrow & 0 \\
 & & \downarrow & & \downarrow{\scriptstyle \eta_\tau P} & & \downarrow{\scriptstyle \eta_\tau(P/M)} & & \\
0 & \longrightarrow & Q_\tau M & \xrightarrow{Q_\tau(j)} & Q_\tau P & \xrightarrow{\pi} & Q_\tau(P/M). & &
\end{array}$$

If P/M is τ-torsion, then $Q_\tau(P/M) = 0$ and, hence, $Q_\tau(j)$ is an isomorphism. Conversely, if $Q_\tau(j)$ is isomorphic, then $\pi = 0$ and, hence, $\eta_\tau(P/M)$ is zero since $P \to P/M$ is epi. Thus, P/M is τ-torsion. □

From Lemma 2.2(1), we have the following lemma:

LEMMA 2.3. *If P is a τ-torsion-free C-module and ξ: $Q_\tau M \to P$ is such a morphism that the composition $M \to Q_\tau M \to P$ (or the composition $M \to E_\tau M \to P$) is zero, then $\xi = 0$.*

Proof. First show that $\xi = 0$ if the composition $M \to E_\tau M \to P$) is zero. It needs to show that $\xi_c = 0$ for each $c \in C$. In fact, for each $\alpha \in Q_\tau(M)(c)$, there is an $I \in \tau(c)$ such that for each h: $a \to c$ in I, $E_\tau M(h)(\alpha)$ is in im$(\phi_\tau(M)(c))$ by Lemma 2.2(1). So we have $\xi_c(E_\tau M(h)(\alpha)) = 0$. But by the naturality of ξ, we see that $P(h)\xi_c(\alpha) = \xi_c(E_\tau M(h)(\alpha)) = 0$. It follows from the fact that P is torsion-free that $\xi_c(\alpha)) = 0$. Thus, $\phi_c = 0$.

Now suppose that the composition $M \to Q_\tau M \to P$ is zero. Then the composition $M/tM \to E_\tau(M/tM) \cong Q_\tau M \to P$ is zero. It follows from the first paragraph that $\xi = 0$. □

The following is another technical lemma which will be used in the sequel.

LEMMA 2.4. *Let τ and σ be two topologies on C with $\tau \subseteq \sigma$. Let $P \in \widetilde{C}$ and $\alpha \in Q_\tau P(c)$ represented by $f\colon I \to P/\tau P$ be in the kernel of the natural morphism $Q_\tau P(c) \to Q_\sigma P(c)$. Then there is a $J \in \sigma(c)$ with $J \subseteq I$ such that $f \mid J = 0$.*

Proof. Since the composition of f and the quotient morphism $q\colon P/\tau P \to P/\sigma P$ represents zero in $Q_\sigma P(c)$, there is a $K \in \sigma(c)$ with $K \subseteq I$ such that $qf \mid K$ is zero. For each $k\colon a \to c \in K$, let $f_a(k) = m_k + t_\tau P(a)$, then m_k is in $t_\sigma P(a)$. So there is a $J_k \in \sigma(a)$ such that $P(x)(m_k) = 0$ (and, hence, is in $\tau P(b)$) for each $x\colon b \to a$ in J_k. (Note that it is independent of the choice of m_k; in fact, if $f_a(k) = m'_k + t_\tau P(a)$, then for each $x\colon b \to a \in J_k$ and so we have $P(x)(m'_k) = P(x)(m'_k - m_k) \in t_\tau P(b)$, since $m'_k - m_k \in \tau P(a)$.) It is clear that $P(x)\colon P(a) \to P(b)$ induces a morphism denoted by $\bar{P}(x)$ from $P(a)/tP(a) \to P(b)/tP(b)$ Then $\bar{P}(x)(f_a(k)) = \bar{P}(x)(m_k + tP(a)) = P(x)(m_k) + tP(b) = 0$ in $P(b)/tp(b)$. By the naturality of f, we see that $P(x)(f_a(k)) = f_b(kx)$. So we have $f \mid kJ_k = 0$. Let $J = \sum_{k\in K} kJ_k$. Then J is in $\sigma(c)$, since σ satisfies T3, and $f \mid J = 0$. □

Recall that a subgroup D of an Abelian group G is called *essential* if $D \cap G_1 = 0$ implies $G_1 = 0$ for any subgroup G_1 of G.

DEFINITION 2.5. Let $P \in \widetilde{C}$. A subobject $E \to P$ is called *essential* if, for each subobject P_1 of P, the condition that the pullback

$$\begin{array}{ccc} E \cap P_1 & \longrightarrow & P_1 \\ \downarrow & & \downarrow \\ E & \longrightarrow & P \end{array}$$

is zero implies $P_1 = 0$.

LEMMA 2.6. *For each τ-torsion-free $P \in \widetilde{C}$, the subobject $P \to E_\tau P$ is essential.*

LEMMA 2.7. *If $E \to P$ is an essential subobject, then E is τ-torsion-free iff P is.*

Proof. One way is easy. Now assume that E is τ-torsion-free. Since the pullback

$$\begin{array}{ccc} E \cap tP & \longrightarrow & tP \\ \downarrow & & \downarrow \\ E & \longrightarrow & P \end{array}$$

is equal to tE, since for each $c \in C$, $(E \cap tP)(c) = E(c) \cap tP(c) = tE(c)$. But $tE = 0$, since E is τ-torsion-free and, hence, $tP = 0$ since E is essential. Thus P is τ-torsion-free. □

3. Structure Presheaves over Spec C

For each $M \in \widetilde{C}$, each open subset $D(I)$ of Spec C, write $Q_I M = Q_{\tau_I} M$ for the quotient C-module of M. For any basis $\mathcal{B}$ of Spec C, we can define a presheaf Q_M over $\mathcal{B}$ by setting $Q_M(D(I)) = Q_I M$. The Q_M will be called *the structure presheaf* associated to M.

Let $\widetilde{M}_P$ be the stalk of the presheaf Q_M at $P \in$ Spec R. That is,

$$\widetilde{M}_P = \operatorname{colim}_{P \in D(I)} Q_I M.$$

Then for each $c \in C$,

$$\begin{aligned}\widetilde{M}_P(c) &= \operatorname{colim}_{P \in D(I)} \operatorname{colim}_{K \in \tau_I(c)} \hom(K, M/t_I M)\\ &= \operatorname{colim}_{P \in D(I)} Q_I(M)(c).\end{aligned}$$

The following lemma is obvious.

LEMMA 3.1. *A presheaf $F\colon \mathcal{B}^{op} \to \widetilde{C}$ is a sheaf (resp. separated presheaf) iff for each $c \in C$, the presheaf $F(-)(c)$ of Abelian groups over $\mathcal{B}$ is a sheaf (resp. separated presheaf).*

For any additive category C and each $M \in \widetilde{C}$, for each $c \in C$, let $\mathcal{S}_M(c)$ be the disjoint union of $\widetilde{M}_P(c)$'s and η the natural projection.

For each open set $D(I) \in \mathcal{B}$ and each $a \in Q_I(M)(c)$ represented by $f_a\colon K \to M/t_I M$ with $K \in \tau_I(c)$, we may define $\hat{a}\colon D(I) \to \mathcal{S}_M(c)$ by setting $\hat{a}(P)$ to be the element in $\widetilde{M}_P(c)$ represented by f_a. It is clear that each $\hat{a}$ is a local section of η in **Set**.

We may have $\widehat{a_1 + a_2} = \widehat{a_1} + \widehat{a_2}$ for any $a_1, a_2 \in Q_I(M)(c)$, where the sums are defined pointwisely. Thus we have that

$$\widehat{Q_I}(M)(c) = \{\widehat{a} \mid a \in Q_I(M)(c)\}$$

is an Abelian group.

We will call the elements in $\widehat{Q}_I(M)(c)$ *quotients of elements* of M at c.

For any open subset $D(I)$, we write $\widetilde{M}_c(D(I))$ for the set of all functions $\sigma\colon D(I) \to \mathcal{S}_M(c)$ which is *locally a quotient of elements* of M at c; that is, those functions σ for each $P \in D(I)$ there is a finitely generated ideal I_P with the properties that $D(I_P) \subseteq D(I)$, $P \in D(I_P)$ and there is an $s_P \in Q_{I_P}(M)(c)$ such that $\widehat{s_P}$ and σ are equal on $D(I_P)$.

If $D(I) \supseteq D(J)$, let $(\widetilde{M}_c)^U_V$ denote the obvious restriction map $\widetilde{M}_c(U) \to \widetilde{M}_c(V)$. Then by the standard content of sheaf theory, we see that $\widetilde{M}_c$ is the sheafification of $Q_M(c)$(regarded as a presheaf of Abelian groups over $\mathcal{B}$) and that the set $\Gamma(D(I), \mathcal{S}_M(c))$ of all continuous sections over $D(I)$ is canonically isomorphic to $\widetilde{M}_c(D(I))$ (where $\mathcal{S}_M(c)$ is considered as a sheaf space in a usual way).

Note that each arrow $h\colon c \to d$ in C induces a morphism $Q_\tau(M)(h)\colon Q_\tau(M)(d) \to Q_\tau(M)(c)$ in such a way that each $f\colon I \to M/tM)$ with $I \in \tau(d)$ is sent to

$h^*f\colon h^*I \to M/tM$, where $h^*f(g) = f(hg)$. In turn, it induces a morphism $\widetilde{M}_P(d) = \operatorname{colim} Q_I(M)(d) \to \operatorname{colim} Q_I(c) = \widetilde{M}_P(c)$ for each $P \in D(I)$, which induces a morphism $\widetilde{M}_d \to \widetilde{M}_c$. Thus, $\widetilde{M}$ is in fact an object in $[S^{op}, \widetilde{C}]$. Now it follows from Lemma 3.1 that $\widetilde{M}$ is a sheaf, since $\widetilde{M}_c$ is for each $c \in C$.

We will call $\widetilde{M}$ *the structure sheaf associated to* M, which is the sheafification of Q_M.

PROPOSITION 3.2. *For any additive category C and any basis $\mathcal{B}$, Q_M is a separated presheaf for each $M \in \widetilde{C}$.*

Proof. It suffices to show that, for each $c \in C$, the natural morphism

$$\wedge\colon\ Q_I(M)(c) \to \widetilde{M}(D(I))(c),$$

which sends $a \in Q_I(M)(c)$ to $\hat{a}$, is injective.

Suppose $s \in Q_I(M)(c)$ represented by $f\colon K \to M/t_IM$ with $K \in \tau_I(c)$ and $\hat{s} = 0$. Then for each $P \not\supseteq I$, the element represented by $f\colon K \to M/t_IM$ in $\widetilde{M}_P(c)$ is zero. So there is an open neighborhood $D(I_P)$ of P with $D(I_P) \subseteq D(I)$, such that the image of s in $Q_{I_P}M(c)$ is zero. By Lemma 2.4, there is a $J_P \in \tau_{I_P}(c)$ such that $f \mid I_P = 0$. Let $J = \sum_{P\in D(I)} J_P$. Then we still have $f \mid J = 0$. Since $D(I)$ is quasi-compact, there is a finite cover of $D(I)$, say $\{I_{P_i}\}$. Then $J \in \bigcap_P \tau_{I_P} \subseteq \bigcap_i \tau_{I_{P_i}}(c) = \tau_I(c)$ by Lemma 1.7. Now we conclude that $s = 0$, since $f \mid J$ and f represent the same element s. □

PROPOSITION 3.3. *For any additive category C and any ideal I of C, if an $M \in \widetilde{C}$ is τ_I-torsion, then $\widetilde{M}_P = 0$ for each $P \in D(I)$. The converse is also true if $D(I)$ is quasi-compact or Spec C has a basis $\mathcal{U}$ with $\tau_J = \widetilde{T}(J)$ for each $D(J) \in \mathcal{U}$.*

Proof. Suppose that $t_IM = M$. For any open subset $D(J) \subseteq D(I)$, we see that $\tau_I \subseteq \tau_J$ gives $t_JM = M$. Thus, we have $Q_M(D(J)) = Q_J(M) = 0$ and, hence, each stalk $\widetilde{M}_P = \operatorname{colim}_{P\in D(J)} Q_JM = 0$.

Suppose that for each $P \in D(I)$, $\widetilde{M}_P = 0$. Then for each $c \in C$ and each $m \in M(c) \cong (Q_R(M))(c)$ and each $P \in D(I)$, there exists an open neighborhood $D(I_P)$ of P such that $Q^R_{I_P}M(m) = 0$. By Lemma 2.4, there is a $K_P \in \tau_{I_P}(c)$ such that $\tilde{m} \mid K_P = 0$. If $D(I)$ is quasi-compact, then there is a finite cover $\{D(I_{P_i}) \mid i < n\}$ of $D(I)$. By Lemma 1.7, the ideal $K = \sum_P K_P \supseteq \sum_i K_{P_i}$ is in $\cap\tau_{I_{P_i}} = \tau_I$ and satisfies $\tilde{m} \mid K = 0$. Thus, $m \in t_IM$ and, hence, $t_I(M) = M$. If Spec R has a basis $\{D(J_i)\}$ with $\tau_{J_i} = \widetilde{T}(J_i)$, then we can choose $D(I_P)$ to be in $\mathcal{U}$, so there are ideals $\bar{K}_P \in T(I_P)$ with $\bar{K}_P \subseteq K_P$. Now it is clear that $K \supseteq \sum_P \bar{K}_P$ and, hence, is in $T(I)$ and $\tilde{m} \mid K = 0$ as well. □

4. The Exactness of the Structure Presheaf Functor

Let C be an additive category and $\widetilde{C} = [C^{op}, \mathbf{Ab}]$ the category of all additive presheaves of Abelian groups. Given a quasi-spectrum S of C, recall that $\mathrm{Psh}_{\widetilde{C}}(S)$

is the functor category S^{op}, $\widetilde{C}$, and $P(S\text{-closed})$ the full subcategory consisting of those objects $F \in \mathrm{Psh}_{\widetilde{C}}(S)$ whose component at τ is a τ-sheaf for each $\tau \in S$.

We may define the so-called *structure presheaf functor* Q_S: $\widetilde{C} \to P(S\text{-closed})$:

> For each $P \in \widetilde{C}$, define Q_P to be the object in $P(S)$ whose component at $\tau \in S$ is $Q_\tau(P)$ – the sheafification of P with respect to the topology τ. It is clear that Q_S is left exact. We shall prove that, for any quasi-spectrum S of C, the structrue presheaf functor Q_S is left adjoint to the global section functor and, hence, is *exact*.

The *global section functor* from $P(S)$ to $\widetilde{C}$ is defined by sending objects F to $F(\perp)$, where $\perp$ is the bottom element of $L(C)$, and sending natural tranformations ϕ: $F \to G$ to $\phi_\perp$: $F_\perp \to G_\perp$.

First, for any $P \in \widetilde{C}$ and any topology $\tau \in L(C)$, let $\eta_\tau(P)$ denote the natural morphism $P \to Q_\tau P$. Then η is a natural transformation from identity functor to Q_τ, and so we have the following lemma:

LEMMA 4.1. *For any $P \in \widetilde{C}$ and any topology $\tau \in L(C)$, the following two morphisms are equal*

$$Q_\tau(P) \underset{\eta_\tau(Q_\tau(P))}{\overset{Q_\tau(\eta_\tau(P))}{\rightrightarrows}} Q_\tau(Q_\tau(P)),$$

and, hence, both are invertible.

Proof. By the naturality of η_τ, we have the following commutative diagram:

$$\begin{array}{ccc} P & \xrightarrow{\eta_\tau(P)} & Q_\tau(P) \\ {\scriptstyle \eta_\tau(P)}\downarrow & & \downarrow{\scriptstyle Q_\tau(\eta_\tau(P))} \\ Q_\tau(P) & \xrightarrow[\eta_\tau(Q_\tau(P))]{} & Q_\tau Q_\tau(P). \end{array}$$

That is, the composition $(Q_\tau(\eta_\tau(P)) - \eta_\tau(Q_\tau(P))) \cdot \eta_\tau(P) = 0$. Now the conclusion follows from Lemma 2.3. □

THEOREM 4.2. *For any quasi-spectrum S of C, the functor Q_S is a right adjoint of the global section functor.*

Proof. Note that $\eta_\tau = \{\eta_\tau(P) \mid P \in \widetilde{C}\}$ is a natural transformation from the identity functor to Q_τ.

For any $F \in P(S\text{-}closed)$, any $P \in \widetilde{C}$ and any morphism f: $P \to F(\perp)$. We want to show that there is a unique morphism $\tilde{f}$: $Q_P \to F$ such that $f = \tilde{f}_\perp \eta_\perp P$, where $\eta_\perp P$ is the canonical isomorphism $P \to Q_\perp(P)$.

Note that $\eta_\tau(F(\tau))$: $F(\tau) \to Q_\tau F(\tau)$ is invertible, since $F(\tau)$ is τ-closed.

Define $\tilde{f}$: $Q_P \to F$ as follows: for any $\tau \in S$, $\tilde{f}(\tau)$ is the composition of

$$Q_\tau(M) \xrightarrow{Q_\tau(f)} Q_\tau(F(\perp)) \xrightarrow{Q_\tau(F_\tau^\perp)} Q_\tau F(\tau) \xrightarrow{\eta_\tau^{-1}(F(\tau))} F(\tau).$$

To check that $\tilde{f}$ is a natural transformation. Let $\tau_1 \supseteq \tau_2$. Then we have the following commutative diagram:

$$\begin{CD}
Q_{\tau_2}(P) @>{Q_{\tau_2}(f)}>> Q_{\tau_2}(F(\bot)) @>{Q_{\tau_2}(F^{\bot}_{\tau_2})}>> Q_{\tau_2}(F(\tau_2)) @>{\eta^{-1}_{\tau_2}(F(\tau_2))}>> F(\tau_2)\\
@V{Q^{\tau_2}_{\tau_1}(P)}VV @V{Q^{\tau_2}_{\tau_1}(F(\bot))}VV @V{Q^{\tau_2}_{\tau_1}(F^{\tau_2}_{\tau_1})}VV @V{F^{\tau_2}_{\tau_1}}VV\\
Q_{\tau_1}(P) @>>{Q_{\tau_1}(f)}> Q_{\tau_1}(F(\bot)) @>>{Q_{\tau_1}(F^{\bot}_{\tau_1})}> Q_{\tau_1}(F(\tau_1)) @>>{\eta^{-1}_{\tau_1}(F(\tau_1))}> F(\tau_1)
\end{CD}$$

where $Q^{\tau_2}_{\tau_1}(F^{\tau_2}_{\tau_1})$ is the composition of $Q_{\tau_2}(F^{\tau_2}_{\tau_1})$ and $Q^{\tau_2}_{\tau_1}(F(\tau_1))$. The left square commutes by the naturality of $Q^{\tau_2}_{\tau_1}$. The middle one commutes because the following two diagrams commute by the naturality of Q_{τ_2} and of $Q^{\tau_2}_{\tau_1}$:

$$\begin{array}{ccc}
Q_{\tau_2}(F(\bot)) & \xrightarrow{Q_{\tau_2}(F^{\bot}_{\tau_2})} & Q_{\tau_2}(F(\tau_2))\\
& {\scriptstyle Q_{\tau_2}(F^{\bot}_{\tau_1})}\searrow \quad \swarrow{\scriptstyle Q_{\tau_2}(F^{\tau_2}_{\tau_1})} & \\
& Q_{\tau_2}(F(\tau_1)) &
\end{array}$$

and

$$\begin{CD}
Q_{\tau_2}(F(\bot)) @>{Q_{\tau_2}(F^{\bot}_{\tau_1})}>> Q_{\tau_2}(F\tau_1)\\
@V{Q^{\tau_2}_{\tau_1}(F(\bot))}VV @VV{Q^{\tau_2}_{\tau_1}(F(\tau_1))}V\\
Q_{\tau_1}(F(\bot)) @>{Q_{\tau_1}(F^{\bot}_{\tau_1})}>> Q_{\tau_1}(F(\tau_1)).
\end{CD}$$

The right square commutes because the following diagram does:

$$\begin{array}{ccc}
Q_{\tau_2}(F(\tau_2)) & \xleftarrow{\eta_{\tau_2}(F(\tau_2))} & F(\tau_2)\\
{\scriptstyle Q_{\tau_2}(F^{\tau_2}_{\tau_1})}\downarrow & & \downarrow{\scriptstyle F^{\tau_2}_{\tau_1}}\\
Q_{\tau_2}(F(\tau_1)) & \xleftarrow{\eta_{\tau_2}(F(\tau_1))} & F(\tau_1)\\
& {\scriptstyle Q^{\tau_2}_{\tau_1}(F(\tau_1))}\searrow \quad \swarrow{\scriptstyle \eta_{\tau_1}(F(\tau_1))} & \\
& Q_{\tau_1}(F(\tau_1)) &
\end{array}\quad .$$

Now we want to show that the following diagram commutes

$$\begin{array}{ccc}
P & \xrightarrow{\eta_{\bot}P} & Q_{\bot}(P)\\
& {\scriptstyle f}\searrow \quad \swarrow{\scriptstyle \tilde{f}_{\bot}} & \\
& F(\bot) &
\end{array}\quad .$$

In fact, we have the following commutative diagram

$$\begin{array}{ccc} P & \xrightarrow{\eta_\perp P} & Q_\perp(P) \\ {\scriptstyle f}\downarrow & & \downarrow{\scriptstyle Q_\perp(f)} \\ F(\perp) & \xrightarrow{\eta_\perp(F(\perp))} & Q_\perp(F(\perp)) \end{array}$$

that is, $\eta_\perp(F(\perp))f = Q_\perp(f)\eta_\perp(P)$ and, hence, $f = \eta_\perp^{-1}(F(\perp))Q_\perp(f)\eta_\perp(P) = \tilde{f}_\perp$.

To show the uniqueness, suppose that ϕ: $Q_P \to F$ is a morphism such that the following diagram commutes:

$$\begin{array}{ccc} P & \xrightarrow{\eta_\perp P} & Q_\perp(P) \\ & {\scriptstyle f}\searrow & \downarrow{\scriptstyle \phi_\perp} \\ & & F(\perp) \end{array} .$$

Then we have the following commutative diagram for each $\tau \in S$:

$$\begin{array}{ccccc} Q_\tau P & & \xrightarrow{Q_\tau(f)} & & Q_\tau(F(\perp)) \\ & \searrow & & \nearrow{\scriptstyle Q_\tau(\phi_\perp)} & \\ {\scriptstyle Q_\tau(\eta_\tau P)}\downarrow & & Q_\tau Q_\perp(P) & & \downarrow{\scriptstyle Q_\tau(F_\tau^\perp)} \\ & \swarrow{\scriptstyle Q_\tau(Q_\tau^\perp P)} & & & \\ Q_\tau Q_\tau(P) & & \xrightarrow[Q_\tau\phi_\tau]{} & & Q_\tau F(\tau) \\ {\scriptstyle \eta_\tau^{-1}(Q_\tau P)}\downarrow & & & & \downarrow{\scriptstyle \eta_\tau^{-1}F(\tau)} \\ Q_\tau(P) & & \xrightarrow[\phi_\tau]{} & & F(\tau) \end{array} .$$

On the other hand, we also have $\eta_\tau(Q_\tau(P)) = Q_\tau(\eta_\tau(P))$ by Lemma 2.2. Thus, we have shown that $\phi_\tau = \tilde{f}_\tau$ and, hence, completed the proof. □

COROLLARY. *Q_S is exact.*

Let $\mathrm{Sh}_{\tilde{C}}(S)$ denote the full subcategory of $\mathrm{Psh}_{\tilde{C}}(S)$ consisting of all sheaves over S with respect to the canonical topology, and $\mathrm{Sh}(S)$ the intersection of $\mathrm{Sh}_{\tilde{C}}(S)$ and $P(S)$.

Let $\widetilde{Q}_F$ be the sheafification of Q_F with respect to the canonical topology on S. Then we have

THEOREM 4.3. *For any quasi-spectrum S of C, if for each $F \in \widetilde{C}$ and for each $\tau \in S$, $\widetilde{Q}_F(\tau)$ is a τ-sheaf, then the sheaf functor $\widetilde{Q}$ has a right adjoint.*

5. Characterizations of Structure Sheaves

The main purpose of this section is to characterize when the presheaf Q_M is a sheaf, where $M \in \widetilde{C}$.

First consider a special case: Let R be commutative and B the set of all basic open subsets. Then for each R-module M, $\tau_{Ra}M = M_a$ and $M_{ab} = \tau_{Rb}\tau_{Ra}M$. Furthermore, one has the exact sequence

$$0 \to Q_{\tau_a \cap \tau_b} M \to Q_{\tau_a} M \oplus Q_{\tau_b} M \to Q_{\tau_{ab}} M.$$

Thus, for any τ_1 and τ_2 in $S = \theta(B)$, we have the exact sequence

$$0 \to Q_{\tau_1 \cap \tau_2} \to Q_{\tau_1} \oplus Q_{\tau_2} \to Q_{\tau_1 \vee \tau_2}. \qquad (*)$$

DEFINITION 5.1. An element τ in a quasi-spectrum S is called *compact* if $\tau = \bigcap_{i \in \Lambda} \tau_i$ implies that τ is an finite intersection of $\tau_i' s$. A quasi-spectrum is called *compact* if each element of it is compact.

EXAMPLE. The quasi-spectrum $S = \theta_R(B)$ above is compact. More general, if R is an S-ring and B is a basis of some quasi-compact open subset of Spec$R$ which is closed under binary intersections, then $S = \theta_R(B)$ is compact.

THEOREM 5.2. *Let S be a compact quasi-spectrum of an additive category C. Then each Q_F with $F \in \widetilde{C}$ is a sheaf over S iff $(*)$ holds for each pair τ_1 and τ_2 in S.*

LEMMA 5.3. *For any quasi-spectrum S of C and any pair τ and σ in S, $t_\tau Q_\sigma = Q_\sigma t_\tau$ iff Q_σ preserves τ-torsion and τ-torsion-free objects.*

Proof. Assume $t_\tau Q_\sigma = Q_\sigma t_\tau$. Then it is clear that Q_σ preserves both τ-torsion and τ-torsion-free presheaves over C.

To show the other inclusion, let P be any given presheaf in $\widetilde{C}$. Consider the exact commutative diagram:

$$\begin{array}{ccccccc}
0 & \longrightarrow & t_\tau P & \longrightarrow & P & \longrightarrow & P/t_\tau P \\
 & & \downarrow & & \downarrow & & \downarrow \\
0 & \longrightarrow & Q_\sigma(t_\tau P) & \longrightarrow & Q_\sigma P & \longrightarrow & Q_\sigma(P/t_\tau P).
\end{array}$$

By assumption, $Q_\sigma(t_\tau P)$ is τ-torsion, so we have that $Q_\sigma(t_\tau P) \subseteq t_\tau(Q_\sigma P)$. Moreover, since $Q_\sigma(P/t_\tau P)$ is τ-torsion-free, and

$$t_\tau Q_\sigma P/Q_\sigma t_\tau P \subseteq Q_\sigma P/Q_\sigma t_\tau P \subseteq Q_\sigma(P/t_\tau P),$$

we see that $t_\tau Q_\sigma p/Q_\sigma t_\tau P$ is τ-torsion-free and, hence, is zero. So we have the other inclusion as well. □

LEMMA 5.4. *For any quasi-spectrum S of C and any pair τ and σ in S, the following are equivalent*:

(1) *Q_τ preserves σ-torsion objects*;

(2) *If P is τ-torsion-free, then $P/\sigma P$ is τ-torsion-free;*
(3) *Q_σ preserves τ-torsion-free objects.*

Proof. To prove that (2) implies (1), let $P \in \widetilde{C}$ be σ-torsion. Then $P/\tau P$ is also σ-torsion and, hence, is a subobject of $\sigma Q_\tau P$. So there is a surjection

$$Q_\tau P/(P/\tau P) \to Q_\tau(P)/\sigma Q_\tau(P).$$

Since the first object is τ-torsion by Lemma 2.2(1), we see that so is the second one. But since $Q_\tau(P)$ is τ-torsion-free, by assumption (2), we see that $Q_\tau(P)/\sigma Q_\tau(P)$ is τ-torsion-free as well and, hence, is zero. That is, $Q_\tau(P) = \sigma Q_\tau(P)$ is σ-torsion.

To prove (2) $\Leftrightarrow$ (3), note that $P/\sigma P$ is an essential subobject of $Q_\sigma P = E_\sigma(P/\sigma P)$. By Lemma 2.9, $P/\sigma P$ is τ-torsion iff $Q_\sigma P = E_\sigma(P/\sigma P)$ is. Thus (2) is true iff (3) is true.

To prove that (1) implies (3), let P be a τ-torsion-free object. Then we have the inclusion $P/\sigma P \subseteq Q_\tau P/\sigma Q_\tau P$.

By assumption (1), we also have the inclusion $Q_\tau(\sigma P) \subseteq \sigma Q_\tau P$ and, hence, the (epi)-morphism $Q_\tau P/Q_\tau(\sigma P) \to Q_\tau P/\sigma Q_\tau P$. Note that the following diagram is commutative:

$$\begin{array}{ccc} P/\sigma P & \longrightarrow & Q_\tau(P)/\sigma Q_\tau(P) \\ & \searrow \quad \nearrow & \\ & Q_\tau P/Q_\tau(\sigma P) & \end{array}$$

Thus we have the inclusion $P/\sigma P \to Q_\tau P/Q_\tau(\sigma P) \to Q_\tau(P/\sigma P)$ and claim that $P/\sigma P$ is τ-torsion-free and, hence, so is $Q_\sigma(P) = E_\sigma(P/\sigma P)$ by Lemma 2.9. □

LEMMA 5.5. *The kernel of the canonical morphism $Q_\tau Q_\sigma P \to Q_{\tau\vee\sigma} P$ is $(\tau \vee \sigma)Q_\tau Q_\sigma P$. If P is $(\tau \vee \sigma)$-torsion-free, then the above morphism is a monomorphism.*

LEMMA 5.6. *For any quasi-spectrum S of C and any pair τ and σ in S, the following are equivalent:*
(1) *both $Q_\tau Q_\sigma$ and $Q_\sigma Q_\tau$ are subfunctors of $Q_{\tau\vee\sigma}$;*
(2) *$\tau Q_\sigma = Q_\sigma \tau$.*

Proof. (2) $\Rightarrow$ (1): First note that assumption implies that Q_σ preserves τ-torsion-free objects. For any $P \in \widetilde{C}$, $P/\tau P$ is τ-torsion-free and, hence, so is $Q_\sigma(P/\tau P)$, which also implies that $Q_\sigma(P/\tau P)$ is $(\tau\vee\sigma)$-torsion-free. By Lemma 5.5, we have

$$\begin{aligned} Q_\tau Q_\sigma(P/\tau P) &= Q_\tau Q_\sigma(Q_\sigma(P/\tau P)) \subseteq Q_{\tau\vee\sigma}(Q_\sigma(P/\tau P)) \\ &= Q_{\tau\vee\sigma}(P/\tau P) = Q_{\tau\vee\sigma}(Q_\tau(P/\tau P)) \\ &= Q_{\tau\vee\sigma}(Q_\tau P) = Q_{\tau\vee\sigma}(P). \end{aligned}$$

Now applying $Q_\tau Q_\sigma$ to the exact sequence

$$0 \to \tau P \to P \to P/\tau P \to 0,$$

we have the following exact sequence

$$0 \to Q_\tau Q_\sigma \tau P \to Q_\tau Q_\sigma P \to Q_\tau Q_\sigma P/\tau P \subseteq Q_{\tau\vee\sigma}(P).$$

But $Q_\tau Q_\sigma \tau P = Q_\tau \tau Q_\sigma P = 0$ implies that $Q_\tau Q_\sigma P \subseteq Q_{\tau\vee\sigma}(P)$. Thus we have proved $Q_\tau Q_\sigma$ is a subfunctor of $Q_{\tau\vee\sigma}$.

For any given $P \in \widetilde{C}$, $Q_\tau P$ is τ-torsion-free and, hence, so is $Q_\sigma Q_\tau P$ by assumption (2). Thus $Q_\sigma Q_\tau P$ is $(\tau \vee \sigma)$-torsion-free and, hence, is a subobject of $Q_{\sigma\vee\tau} P$ by Lemma 5.5. Thus $Q_\tau Q_\sigma$ is a subfunctor of $Q_{\tau\vee\sigma}$.

(1) $\Rightarrow$ (2): We want to show that Q_τ preserves σ-torsion objects and Q_σ preserves τ-torsion objects: let $P \in \widetilde{C}$ be σ-torsion. Then P is also $(\sigma \vee \tau)$-torsion and, hence, $Q_{\tau\vee\sigma} P = 0$. It follows that $Q_\sigma Q_\tau P = 0$ and, hence, $Q_\tau P$ is σ-torsion. Similarly, we can prove that Q_σ preserves τ-torsion objects. Now the conclusion follows from Lemmas 5.4 and 5.3. □

THEOREM 5.7. *Let C be an additive category and $\mathcal{B}$ a basis of Spec C consisting of some quasi-compact open subsets. Then, for each $M \in \widetilde{C}$, the following are equivalent*:

(1) *the associated presheaf Q_M is a sheaf over $\mathcal{B}$*;

(2) *for each pair of $D(I)$ and $D(J)$ in $\mathcal{B}$, the sequence*

$$0 \to Q_{I+J} \to Q_I \oplus Q_J \to Q_{IJ}$$

is exact;

(3) *for each pair of $D(I)$ and $D(J)$ in $\mathcal{B}$, the sequence*

$$0 \to Q_{\tau_I \cap \tau_J} \to Q_I \oplus Q_J \to Q_{IJ}$$

is exact;

(4) *both $Q_I Q_J$ and $Q_J Q_I$ are subfunctors of Q_{IJ}.*

Proof. The fact that (2) $\Leftrightarrow$ (3) follows from Proposition 1.7. It suffices to show (3) $\Leftrightarrow$ (4). To show (3) $\Rightarrow$ (4), we will first prove the following fact that (3) is true iff the following two sequences are exact:

(i) $0 \to Q_I \to Q_I \oplus Q_I Q_J \to Q_{IJ}$,

(ii) $0 \to Q_J \to Q_J Q_I \oplus Q_J \to Q_{IJ}$.

In fact, the fact that condition (3) implies the exactness of both (i) and (ii) follows from the left exactness of Q_I and Q_J.

To prove the converse, for each $P \in \widetilde{C}$, let $K(P)$ be the kernel of $Q_I(P) \oplus Q_J(P) \to Q_{IJ}(P)$. Then $K(P)$ is $\tau_I \cap \tau_J$-closed. Let j_P: $Q_{\tau_I \cap \tau_J}(P) \to K(P)$ be the canonical injection. Applying Q_I (resp. Q_J), we deduce that both $Q_I(j_P)$ and $Q_J(j_P)$ are isomorphisms. By Lemma 2.2(5), we see that that Cok j_P is both τ_I-torsion and τ_J-torsion and, hence, is $\tau_I \cap \tau_J$-torsion. Using Lemma 2.2(5) again, we have that $Q_{\tau_I\cap\tau_J}(j_P)$ is an isomorphism too and, hence, so is j_P since both $K(P)$ and $Q_{\tau_I\cap\tau_J}P$ are $\tau_I \cap \tau_J$-closed.

Now we show (3) ⇒ (4): The morphisms involved in (i) are

$$\psi_P\colon\ Q_I(P) \to Q_I(P) \oplus Q_I Q_J(P)$$

defined by $\psi_P(x) = (x, \alpha_P(x))$ and

$$\phi_P\colon\ Q_I(P) \oplus Q_I Q_J(P) \to Q_{IJ}(P)$$

defined by $\phi_P((y, z)) = \beta_P(y) - \gamma_P(z)$, where α_P, β_P and γ_P are obvious morphisms. We also note that the kernel of γ_P is $t_{IJ} Q_I Q_J(P)$. Thus, showing that $Q_I Q_J$ is a subfunctor of Q_{IJ} is equivalent to showing that γ_P is monic for each $P \in \underset{\sim}{C}$: Let $z \in \ker \gamma_P$. Then $(0, z)$ is in the kernel of ϕ_P. By the exactness of (ii), there is an $x \in Q_I(P)$ such that $\psi_P(x) = (0, z)$, but $\psi_P(x) = (x, \alpha_P(x))$ which implies $x = 0$ and so is z. By symmetry (using the exactness of (ii)) we see that $Q_J Q_I$ is a subfunctor of Q_{IJ}.

(4) ⇒ (3): It is clear that the composition of $Q_I(M) \to Q_I(M) \oplus Q_I Q_J(M) \to Q_{IJ}(M)$ is zero. Now let $(y, z) \in Q_I(M) \oplus Q_I Q_J(M)$ be such that $\beta(y) - \gamma(z) = 0$. Since $\beta = \gamma\alpha$, we have $\gamma(\alpha(y) - z) = 0$. Since saying that $Q_I Q_J$ is a subfunctor of Q_{IJ} is saying that γ is monic, it follows $\alpha(y) - z = 0$ and $(y, z) = (y, \alpha(y)) = \psi(y)$ and, hence, that (i) is exact. Similarly, we see that (ii) is also exact and, hence, (3) follows. □

References

1. Bueso, J. L., Jara, P. and Verschoren, A.: *Compatibility, Stability, and Sheaves*, in Monogr. Textbooks Pure Appl. Math. 185, Marcel Dekker, New York, 1995.
2. Bueso, J. L., Torrecillas, B. and Verschoren, A.: *Local cohomology and localizations*, in Pitman Res. Notes in Math. Series 226, Longman Scientific & Technical, Harlow, 1989.
3. Hartshorne, R.: *Algebraic Geometry*, Graduate Texts in Mathematics 52, Springer-Verlag, New York, 1977.
4. Stenstrom, B.: *Rings of Quotients*, Springer-Verlag, New York, 1975.
5. Sun, S. H.: On structure sheaves of non-commutative rings, to appear (*Comm. Algebra*).
6. Sun, S. H.: Adjunctions of structure sheaf functors for non-commutative rings, *J. Pure Appl. Algebra* **102** (1995), 195–206.
7. Sun, S. H. and Dhillon, A.: Prime spectra of additive categories, *J. Pure Appl. Algebra* **122** (1997), 135–157.
8. von Oystaeyen, F. and Verschoren, F.: *Non-commutative Algebraic Geometry*, Lecture Notes in Math. 887, Springer-Verlag, New York, 1981.

G. Brümmer & C. Gilmour (eds), Papers in Honour of Bernhard Banaschewski.

Lindelöf Degree and Function Spaces

Dedicated to Professor B. Banaschewski on the occasion of his 70th birthday

VESKO VALOV
University of Swaziland, Department of Mathematics, Private Bag 4, Kwaluseni, Swaziland
e-mail: valov@realnet.co.sz

DUMISANI VUMA
University of Zimbabwe, Department of Mathematics, P.O.Box 167, Mount Pleasant, Harare, Zimbabwe, e-mail: vuma@maths.uz.zw

(Received: 10 September 1996; accepted: 26 November 1997)

Abstract. We prove that if X and Y are completely regular perfect spaces and Φ is a continuous linear homeomorphism between $C_p(X)$ and $C_p(Y)$ (resp., $C_p^*(X)$ and $C_p^*(Y)$) then the Lindelöf degrees of X and Y are the same. When Φ is positive the above result remains true for any completely regular X and Y. It is also shown that pseudocompactness and compactness are preserved by continuous linear homeomorphisms between $C_p^*(X)$ and $C_p^*(Y)$.

Mathematics Subject Classifications (2000): 54C35, 57N17.

Key words: function spaces, Lindelöf degree, linear homeomorphism.

1. Introduction

For a completely regular space X, $C(X)$ (resp., $C^*(X)$) denotes the set of all real-valued continuous (resp., continuous and bounded) functions on X. When $C(X)$ and $C^*(X)$ are endowed with the pointwise convergence topology we write $C_p(X)$ and $C_p^*(X)$, respectively. Two spaces X and Y (we shall consider only completely regular spaces) are called l_p-equivalent if $C_p(X)$ and $C_p(Y)$ are linearly homeomorphic.

In [1], Arkhangel'skiî announced the following result proved by Veličko: if X and Y are l_p-equivalent then X is Lindelöf if and only if Y is. According to Arkhangel'skiî [1], Veličko's proof does not work for higher Lindelöf degrees (the Lindelöf degree $l(Z)$ of a space Z is the least cardinal $\tau \geq \omega$ such that each open cover of Z has a subcover of cardinality at most τ). Baars [4] proved that Veličko's theorem is true for higher Lindelöf degrees provided X and Y are first countable paracompact spaces. Recently the first author [15] generalized Baars' result in different directions: in the class of μ-complete wq-spaces the Lindelöf degree is preserved by linear homeomorphisms between $C_p(X, E)$ and $C_p(Y, F)$, as well as between $C_k(X, E)$ and $C_k(Y, F)$, where E and F are normed spaces.

Here $C_p(X, E)$ (resp., $C_k(X, E)$) denotes the space of all continuous E-valued maps on X with the pointwise convergence (resp., compact open) topology. The class of wq-spaces [15] is closely related to that of q-spaces [8], and it contains all locally compact and all first countable spaces. Let us also note that μ-complete spaces are substantially more general than paracompact ones. In the present paper we prove that if Φ is a linear homeomorphism between $C_p(X)$ and $C_p(Y)$ (resp., $C_p^*(X)$ and $C_p^*(Y)$), where X and Y are perfect spaces, then $l(X) = l(Y)$. It is shown that the requirement for X and Y to be perfect can be omitted if Φ is positive.

We also prove that compactness and pseudocompactness are preserved by continuous linear homeomorphisms between $C_p^*(X)$ and $C_p^*(Y)$.

2. Preliminaries

Let $L_p(X)$ (resp., $L_p^*(X)$) be the continuous dual of $C_p(X)$ (resp., $C_p(X)$) with the pointwise convergence topology. The common notation $D_p(X)$ will be adopted for either of the spaces $L_p(X)$ and $L_p^*(X)$. It is well known that, for every $\mu \in D_p(X) - \{0\}$ there are finitely many points $x_1, \ldots, x_n$ from X and $\lambda_{x_1}, \ldots, \lambda_{x_n} \in \mathbf{R} - \{0\}$ (see [2] or [5]) such that $\mu = \Sigma\{\lambda_{x_i} \cdot \delta_{x_i} : i = 1, \ldots, n\}$, where δ_{x_i} denotes the evaluation map at x_i defined by $\delta_{x_i}(f) = f(x_i)$. We define supp(μ) to be the set $\{x_i : i = 1, \ldots, n\}$ if $\mu \neq 0$, and supp$(\mu) = \emptyset$ if $\mu = 0$. In the sequel the following notations will be considered:

$$\text{supp}(\mu^+) = \{x \in \text{supp}(\mu) : \lambda_x > 0\}, \quad \|\mu^+\| = \Sigma\{\lambda_x : x \in \text{supp}(\mu^+)\},$$

$$\text{supp}(\mu^-) = \{x \in \text{supp}(\mu) : \lambda_x < 0\}, \quad \|\mu^-\| = \Sigma\{|\lambda_x| : x \in \text{supp}(\mu^-)\}.$$

For every $x \in \beta X$ let $\mathcal{A}(x) = \{f \in C(\beta X, I) : x \in \text{Int}_{\beta X}(f^{-1}(1))\}$, where βX is the Čech–Stone compactification of X and $I = [0, 1]$. When $x \in \beta X$ and $\mu \in D_p(X)$ we define $s_\mu(x) = \inf\{\mu(f) : f \in \mathcal{A}(x)\}$ and $\mathcal{S}_\mu(x) = \sup\{\mu(f) : f \in A(x)\}$, where $\mu(f)$ stands for $\mu(f|X)$ for all $f \in A(x)$.

LEMMA 2.1 *For every $\mu \in D_p(X)$ we have:*

(i) $s_\mu(x) = \lambda_x - \|\mu^-\|$ *and* $\mathcal{S}_\mu(x) = \|\mu^+\|$ *if* $x \in \text{supp}(\mu^+)$*;*
(ii) $s_\mu(x) = -\|\mu^-\|$ *and* $\mathcal{S}_\mu(x) = \lambda_x + \|\mu^+\|$ *if* $x \in \text{supp}(\mu^-)$*;*
(iii) $s_\mu(x) = -\|\mu^-\|$ *and* $\mathcal{S}_\mu(x) = \|\mu^+\|$ *if* $x \notin \text{supp}(\mu)$.

Proof. Follows from the representation $\mu = \Sigma\{\lambda_x \cdot \delta_x : x \in \text{supp}(\mu)\}$. □

For any $a \in \mathbf{R}$ and $\mu \in D_p(X)$ let $\varphi_a(\mu) = \{x \in \beta X : s_\mu(x) \geq a\}$ and $\psi_a(\mu) = \{x \in \beta X : \mathcal{S}_\mu(x) \leq a\}$. Observe that φ_a and ψ_a are set-valued maps from $D_p(X)$ into βX. Recall that a set-valued map $\theta\colon Y \to P(Z)$ is upper-semi continuous (written, *usc*) if the set $\theta^\#(U) = \{y \in Y : \theta(y) \subset U\}$, is open in Y whenever U is open in Z. Note that $P(Z)$ stands for the set of all subsets of Z, so the possibility $\theta(y) = \emptyset$, for some $y \in Y$, is not excluded. We say that θ is *usco* if θ is usc and $\theta(y)$ is compact for every $y \in Y$.

LEMMA 2.2 *The set-valued maps* φ_a *and* ψ_a, $a \in \mathbf{R}$, *are usc.*

Proof. We shall consider only φ_a, the other case is similar. Suppose $\varphi_a(\mu) \subset U$ for some open $U \subset \beta X$, $a \in \mathbf{R}$ and $\mu \in D_p(X)$. For every $x \in \beta X - U$ we have $s_\mu(x) < a$ so there is $f_x \in \mathcal{A}(x)$ with $\mu(f_x) < a$. Then $\{V_x\colon x \in \beta X - U\}$ is an open cover of $\beta X - U$, where $V_x = \mathrm{Int}_{\beta X}(f_x^{-1}(1))$. Choose finitely many points $x(i) \in \beta X - U$, $i = 1, \ldots, n$, such that $\{V_{x(i)} : i = 1, \ldots, n\}$ covers $\beta X - U$. Since $\mu(f_{x(i)}) < a$ for all i, there exists a neighborhood $W \subset D_p(X)$ of μ such that $\nu(f_{x(i)}) < a$ for every $\nu \in W$ and $i = 1, \ldots, n$. The proof will be completed if $\nu \in W$ implies $\varphi_a(\nu) \subset U$. To prove this implication it suffices to show that for a fixed $\nu \in W$ we have $\varphi_a(\nu) \cap V_{x(i)} = \emptyset$, $i = 1, \ldots, n$. Suppose not. Then there exist $j \in \{1, \ldots, n\}$ and $x_0 \in \varphi_a(\nu) \cap V_{x(j)}$. Since $s_\nu(x_0) \geq a$, and $x_0 \in V_{x(j)}$ yields $f_{x(j)} \in \mathcal{A}(x_0)$, we obtain $\nu(f_{x(j)}) \geq a$. The last inequality contradicts $\nu \in W$. □

Let us note the following observations:

LEMMA 2.3 *Let* $a \in \mathbf{R}$ *and* $\mu \in D_p(X)$. *Then* $\varphi_a(\mu)$ *is either* βX *or a finite subset of* $\mathrm{supp}(\mu^+)$ *and:*

(i) $\varphi_a(\mu) \cap \mathrm{supp}(\mu)$ *consists of all* $x \in \mathrm{supp}(\mu^+)$ *with* $-\|\mu^-\| < a \leq \lambda_x - \|\mu^-\|$;
(ii) $\varphi_a(\mu) = \emptyset$ *iff* $\lambda_x - \|\mu^-\| < a$ *for every* $x \in \mathrm{supp}(\mu^+)$;
(iii) $\varphi_a(\mu) = \beta X$ *iff* $-\|\mu^-\| \geq a$.

LEMMA 2.4 *For* $a \in \mathbf{R}$ *and* $\mu \in D_p(X)$ *either* $\psi_a(\mu)$ *is a finite subset of* $\mathrm{supp}(\mu^-)$ *or* $\psi_a(\mu) = \beta X$ *and:*

(i) $\psi_a(\mu) \cap \mathrm{supp}(\mu)$ *consists of all* $x \in \mathrm{supp}(\mu^-)$ *with* $\lambda_x + \|\mu^+\| \leq a < \|\mu^+\|$;
(ii) $\psi_a(\mu) = \emptyset$ *iff* $\lambda_x + \|\mu^+\| > a$ *for every* $x \in \mathrm{supp}(\mu^-)$;
(iii) $\psi_a(\mu) = \beta X$ *iff* $\|\mu^+\| \leq a$.

LEMMA 2.5 *For every* $\mu \in D_p(X)$ *and* $a, b \in \mathbf{R}$ *with* $a < b$ *we have* $\varphi_b(\mu) \subset \varphi_a(\mu)$ *and* $\psi_a(\mu) \subset \psi_b(\mu)$.

3. Lindelöf Degrees and Linear Homeomorphisms

Let $\Phi\colon C_p(X) \to C_p(Y)$ (resp., $\Phi\colon C_p^*(X) \to C_p^*(Y)$) be a linear continuous map. For every $y \in Y$ the function $\mu_y\colon C_p(X) \to \mathbf{R}$ (resp., $\mu_y\colon C_p^*(X) \to \mathbf{R}$), where $\mu_y(f) = \Phi(f)(y)$, is continuous and linear. Therefore we have a map $\Phi^*\colon Y \to D_p(X)$ with $\Phi^*(y) = \mu_y$. It is easily seen that Φ^* is continuous, and Φ^* is an embedding when Φ is surjective. We shall consider the set-valued maps $\varphi_a^*, \psi_a^*\colon Y \to P(\beta X)$, $a \in \mathbf{R}$, defined by $\varphi_a^*(y) = \varphi_a(\mu_y)$ and $\psi_a^*(y) = \psi_a(\mu_y)$.

According to Lemma 2.2, φ_a^* and ψ_a^* are usc. Let $U_a = \{y \in Y : \varphi_a^*(y) \subset X\}$ and $V_a = \{y \in Y : \psi_a^*(y) \subset X\}$.

LEMMA 3.1 *Let Φ: $C_p(X) \to C_p(Y)$ (resp., Φ: $C_p^*(X) \to C_p^*(Y)$) be continuous and linear. Then U_a and V_a, $a \in \mathbf{R}$, are open in Y.*

Proof. Let $a \in \mathbf{R}$. By Lemma 2.3, $Y - U_a$ consists of all $y \in Y$ with $\varphi_a^*(y) = \beta X$. We claim that $Y - U_a \subset Y$ is closed. Let $\{y_\alpha : \alpha \in \Lambda\}$ be a net in $Y - U_a$ converging to a point $y \in Y$. If $\varphi_a^*(y) \neq \beta X$, we can choose an open set $W \subset \beta X$ such that $\varphi_a^*(y) \subset W$ and $W \neq \beta X$. Then, since φ_a^* is usc and $\{y_\alpha : \alpha \in \Lambda\}$ converges to y, there is $\gamma \in \Lambda$ with $\varphi_a^*(y_\gamma) \subset W$. This is a contradiction because $\varphi_a^*(y_\gamma) = \beta X$. Hence $Y - U_a \subset Y$ is closed. By using Lemma 2.4 instead of Lemma 2.3 in this proof we can show that V_a is open. □

LEMMA 3.2 *Let Φ: $C_p(X) \to C_p(Y)$ (resp., Φ: $C_p^*(X) \to C_p^*(Y)$) be continuous, linear. Then, for every $y \in Y$, there exist rational numbers $p, q \in Q$ such that $\varphi_p^*(y) = \operatorname{supp}(\mu_y^+)$ and $\psi_q^*(y) = \operatorname{supp}(\mu_y^-)$.*

Proof. Suppose both $\operatorname{supp}(\mu_y^+)$ and $\operatorname{supp}(\mu_y^-)$ are non-empty. Then $\|\mu_y^-\| \neq 0$ and $\|\mu_y^+\| \neq 0$, so we can take $p, q \in Q$ such that $-\|\mu_y^-\| < p \leq \lambda_x - \|\mu_y^-\|$ for all $x \in \operatorname{supp}(\mu_y^+)$ and $\lambda_x + \|\mu_y^+\| \leq q < \|\mu_y^+\|$ for all $x \in \operatorname{supp}(\mu_y^-)$. By Lemma 2.3(i) and Lemma 2.4(i), we have $\varphi_p^*(y) = \operatorname{supp}(\mu_y^+)$ and $\psi_q^*(y) = \operatorname{supp}(\mu_y^-)$. If $\operatorname{supp}(\mu_y^+) = \emptyset$ (resp., $\operatorname{supp}(\mu_y^-) = \emptyset$) we choose $p \in Q$ (resp., $q \in Q$) with $-\|\mu_y^-\| < p$ (resp., $q < \|\mu_y^+\|$) and apply Lemma 2.3(ii) (resp., Lemma 2.4(ii)). □

Recall that a space X is perfect if every open set in X is a union of countably many closed subsets of X.

THEOREM 3.3 *Let $\mathcal{P}$ be a class of spaces with the following properties:*

(a) *if $Y \in \mathcal{P}$ and $F \subset Y$ is closed, then $F \in \mathcal{P}$;*
(b) *if $X = \bigcup\{X_n : n \in \mathbf{N}\}$ with each $X_n \in \mathcal{P}$, then $X \in \mathcal{P}$;*
(c) *if $F \in \mathcal{P}$ and there is usco θ: $F \to P(X)$ such that $X = \bigcup\{\theta(z) : z \in F\}$, then $X \in \mathcal{P}$.*

Let $Y \in \mathcal{P}$ be a perfect space. If $C_p(X)$ and $C_p(Y)$ (resp., $C_p^*(X)$ and $C_p^*(Y)$) are linearly homeomorphic, then $X \in \mathcal{P}$.

Proof. Let Φ: $C_p(X) \to C_p(Y)$ (resp., Φ: $C_p^*(X) \to C_p^*(Y)$) be a linear homeomorphism. Let $H_a = \bigcup\{\varphi_a^*(y) : y \in U_a\}$ and $F_a = \bigcup\{\psi_a^*(y) : y \in V_a\}$ for all $a \in \mathbf{R}$. By Lemma 3.1, U_a and V_a are open in Y, hence each of them is a union of countably many closed subsets of Y. Consequently, $U_a, V_a \in \mathcal{P}$ for any $a \in \mathbf{R}$. Since φ_a^*, considered as a set-valued map from U_a onto H_a, is usco, we have $H_a \in \mathcal{P}$. Similarly, $F_a \in \mathcal{P}$, $a \in \mathbf{R}$. To prove that $X \in \mathcal{P}$ it is enough to show that X is a union of the countable family $\gamma = \{H_q, F_q : q \in Q\}$. To this end, fix $x \in X$. Because Φ is a linear homeomorphism, there is $y \in Y$ with $x \in \operatorname{supp}(\mu_y)$ (see [5]).

Then, by Lemma 3.2, we have $\varphi_p^*(y) = \mathrm{supp}(\mu_y^+)$ and $\psi_q^*(y) = \mathrm{supp}(\mu_y^-)$ for some $p, q \in Q$. Thus $x \in \mathrm{supp}(\mu_y) = \mathrm{supp}(\mu_y^+) \cup \mathrm{supp}(\mu_y^-) \subset H_p \cup F_q$, so $X = \bigcup \gamma$. □

COROLLARY 3.4 *Let $C_p(X)$ and $C_p(Y)$ (resp., $C_p^*(X)$ and $C_p^*(Y)$) be linearly homeomorphic with X and Y perfect. Then $l(X) = l(Y)$.*

Proof. By Theorem 3.3, it suffices to show that $l(X) \leq \tau$, where $\tau = l(Y)$. Let $\mathcal{P}_\tau$ be the class of all spaces having Lindelöf degree not greater that τ. It is easily seen that $\mathcal{P}_\tau$ satisfies the hypotheses of Theorem 3.3, hence $X \in \mathcal{P}_\tau$. □

PROPOSITION 3.5 *Let $\mathcal{P}$ be a class of spaces such that:*

(a) *$\mathcal{P}$ is hereditary with respect to closed subspaces;*
(b) *if $X \in \mathcal{P}$ and Y is a continuous image of X, then $Y \in \mathcal{P}$;*
(c) *if $X, Y \in \mathcal{P}$, then $X \times Y \in \mathcal{P}$;*
(d) *$\mathcal{P}$ is closed with respect to taking countable unions;*
(e) $\mathbf{R} \in \mathcal{P}$.

Suppose Φ: $C_p^*(X) \to C_p^*(Y)$ is a quotient, continuous linear surjection. Then $Y \in \mathcal{P}$ provided $X \in \mathcal{P}$.

Proof. Arkhangel'skiĭ [2] proved that, for any class $\mathcal{B}$ satisfying (b)–(e), $X \in \mathcal{B}$ implies $L_p(X) \in \mathcal{B}$. So, $L_p(X) \in \mathcal{P}$. Since the inclusion map i: $C_p^*(X) \to C_p(X)$ induces a continuous surjection i^*: $L_p(X) \to L_p^*(X)$, we have $L_p^*(X) \in \mathcal{P}$.

We claim that Φ^*: $Y \to L_p^*(X)$ is a closed embedding. Indeed, since Φ is a continuous surjection, Φ^* is an embedding. To show that $\Phi^*(Y) \subset L_p^*(X)$ is closed choose a net $\{\mu_{y(\alpha)} : \alpha \in \Lambda\} \subset \Phi^*(Y)$ converging to $\mu \in L_p^*(X)$. Then for any $f, g \in C_p^*(X)$ with $\Phi(f) = \Phi(g)$ we have $\mu(f) = \mu(g)$ because $\mu_{y(\alpha)}(f) = \mu_{y(\alpha)}(g)$, $\mu(f) = \lim \mu_{y(\alpha)}(f)$ and $\mu(g) = \lim \mu_{y(\alpha)}(g)$. Since Φ is quotient, there is $\nu \in L_p^*(Y)$ with $\mu = \nu \circ \Phi$. It is easily seen that the evaluation map $y \to \delta_y$ is a closed embedding from Y into $L_p^*(Y)$. Then $\{\mu_{y(\alpha)}\}$ converges to μ yields $\{\delta_{y(\alpha)}\}$ converges to ν. Therefore $\nu = \delta_y$ for some $y \in Y$, so $\mu = \mu_y \in \Phi^*(Y)$.

Because Y is homeomorphic to the closed set $\Phi^*(Y) \subset L_p^*(X)$ and $L_p^*(X) \in \mathcal{P}$, we finally obtain $Y \in \mathcal{P}$. □

Note that the requirement in Proposition 3.5 for Φ to be quotient cannot be dropped. If, for example, X is not σ-compact, then the restriction map Φ: $C_p(\beta X) \to C_p^*(X)$ is a continuous, linear surjection, βX is σ-compact and the class of σ-compact spaces has all above properties (a)–(e).

COROLLARY 3.6 *Let $C_p^*(Y)$ be an image of $C_p^*(X)$ under a continuous quotient linear surjection and let X have one of the following properties: X is σ-compact, X is a Lindelöf Σ-space, X^n is Lindelöf for every $n \in \mathbf{N}$. Then Y has the same property.*

The analog of Corollary 3.6 for the spaces $C_p(X)$ and $C_p(Y)$ is also true, even in the more general case when $C_p(X)$ and $C_p(Y)$ are homeomorphic (see Okunev [9]).

Uspenskiî [14] proved that if $C_p(X)$ and $C_p(Y)$ are uniformly homeomorphic, then X is compact iff Y is. Let us note that every linear homeomorphism between function spaces is uniformly continuous. Concerning the bounded functions, the authors know only the following result of Baars, de Groot and Pelant [6]: if X and Y are metrizable and there is a continuous linear surjection from $C_p^*(X)$ onto $C_p^*(Y)$, then X compact yields Y compact. The next theorem shows that this result remains true for any X and Y in the case of linear homeomorphisms.

THEOREM 3.7 *Let $C_p^*(X)$ and $C_p^*(Y)$ be linearly homeomorphic. Then X is compact if and only if Y is compact.*

Proof. Suppose X is compact. By Corollary 3.6, Y is σ-compact. Since pseudocompactness is preserved by linear homeomorphisms between $C_p^*(X)$ and $C_p^*(Y)$ (see Theorem 4.3 below), Y is pseudocompact. Hence Y is compact. □

The remarks following Proposition 3.5 show that we cannot extend Theorem 3.7 to continuous linear surjections. The question is whether it is true for continuous quotient linear surjections.

4. Appendix

First, let us show that the requirement in Theorem 3.3 and Corollary 3.4 for X and Y to be perfect can be omitted if $C_p(X)$ and $C_p(Y)$ (resp., $C_p^*(X)$ and $C_p^*(Y)$) are homeomorphic under positive or negative linear homeomorphisms.

THEOREM 4.1 *Let $\mathcal{P}$ be a class of spaces satisfying the hypotheses of Theorem 3.3. If there is either a positive or negative linear homeomorphism between $C_p(X)$ and $C_p(Y)$ (resp., $C_p^*(X)$ and $C_p^*(Y)$), then $X \in \mathcal{P}$ if and only if $Y \in \mathcal{P}$.*

Proof. Suppose $\Phi\colon C_p(X) \to C_p(Y)$ (resp., $\Phi\colon C_p^*(X) \to C_p^*(Y)$) is a positive linear homeomorphism. Since all μ_y are positive, we have $\operatorname{supp}(\mu_y^-) = \emptyset$ and $\|\mu_y^-\| = 0$ for each $y \in Y$. Then $s_{\mu_y}(x) = \lambda_x$ if $x \in \operatorname{supp}(\mu_y^+)$ and $s_{\mu_y}(x) = 0$ otherwise. Consequently, $\varphi_a^*(y)$ is a finite subset of $\operatorname{supp}(\mu_y^+)$ if $a > 0$ and $\varphi_a^*(y) = \beta X$ if $a \le 0$. Let $F_a = \{y \in Y : \varphi_a^*(y) \neq \emptyset\}$ for $a > 0$. Since $F_a = Y - \{y \in Y : \varphi_a^*(y) = \emptyset\}$ and φ_a^* is usc, $F_a \subset Y$ is closed. Consider the family $\gamma = \{F_q : q \in Q^+\}$, where Q^+ is the set of all positive rational numbers. The arguments from Lemma 3.2 show that, for every $y \in Y$, there is $q(y) \in Q^+$ with $\varphi_{q(y)}^*(y) = \sup p(\mu_y^+)$. This yields $y \in F_{q(y)}$. We claim that $X = \bigcup\{\varphi_q^*(F_q) : q \in Q^+\}$. Indeed, for every $x \in X$ there exists $y \in Y$ such that $x \in \operatorname{supp}(\mu_y^+)$ (see the proof of Theorem 3.3). Choosing $q \in Q^+$ with $\varphi_q^*(y) = \operatorname{supp}(\mu_y^+)$ we obtain $x \in \varphi_q^*(F_q)$.

To finish the proof it suffices to show that $Y \in \mathcal{P}$ implies $X \in \mathcal{P}$. Since each $F_q \subset Y$ is closed and $\varphi_q^*\colon F_q \to P(\varphi_q^*(F_q))$ is usco, we have $\varphi_q^*(F_q) \in \mathcal{P}$. Therefore $X \in \mathcal{P}$.

If Φ is a negative linear homeomorphism, then $-\Phi$ is positive and we can apply the proof above. □

COROLLARY 4.2 *If there exists a positive continuous linear homeomorphism between $C_p(X)$ and $C_p(Y)$ (resp., $C_p^*(X)$ and $C_p^*(Y)$), then $l(X) = l(Y)$.*

Now, let us show that pseudocompactness is preserved by linear homeomorphisms between $C_p^*(X)$ and $C_p^*(Y)$. It is well known that if f is a map from a space X onto a set Y, then the quotient topology on Y (with respect to f) is not always completely regular. But the strongest among all completely regular topologies on Y with respect to which f is continuous exists and it is called the **R**-*quotient* [2]. The set Y with the **R**-quotient topology is denoted by Y^q and $q(f)$ is the corresponding continuous map from X onto Y^q. We also adopt the following notations. For a space X let $\mathcal{F}_X$ denote the family of all continuous maps from X onto spaces of countable weight, and direct this family by the relation: $g \succ f$ if there is a continuous surjection $h(g, f)$ from $g(X)$ onto $f(X)$ such that $h(g, f) \circ g = f$. For $f \in \mathcal{F}_X$ let $X_f = f(X)$, $C_p(f) = f_p^*(C_p^*(X))$ and $C_p^q(f) = q_p^*(f)(C_p^*(X_f^q))$, where $f_p^*\colon C_p^*(X_f) \to C_p^*(X)$ and $q_p^*(f)\colon C_p^*(X_f^q) \to C_p^*(X)$ are the continuous duals of f and $q(f)$, respectively. When the uniform convergence topology is considered similar notation is used, the only difference is that the subscript p is replaced by u. Note that f_p^* , $q_p^*(f)$, f_u^* and $q_u^*(f)$ are embeddings. We need the following observations:

FACT 1 $C_p^*(X) = \bigcup\{C_p(f)\colon f \in \mathcal{F}_X\}$ *and* $C_p(f) \subset C_p^q(f)$, $f \in \mathcal{F}_X$.

FACT 2 *For every countable $A \subset C^*(X)$ there exists $f \in \mathcal{F}_X$ such that $A \subset C_p(f)$.*

FACT 3 $C_p^q(f) \subset C_p^*(X)$ *is closed and separable for every $f \in \mathcal{F}_X$.*

Proof. $C_p^q(f) \subset C_p^*(X)$ is closed because $q(f)$ is **R**-quotient [2]. Since there is a one-to-one map from X_f^q onto X_f, $C_p(f) \subset C_p^q(f)$ is dense [2]. On the other hand $C_p(f)$ is homeomorphic to $C_p^*(X_f)$ and $C_p^*(X_f)$ is separable because X_f has a countable weight. □

An analog of the following theorem was proved in [16] for function spaces with the compact-open topology. The same arguments work in the case of pointwise convergence topology.

THEOREM 4.3 *Let $C_p^*(X)$ be linearly homeomorphic to $C_p^*(Y)$. Then X is pseudocompact if and only if Y is.*

Proof. Suppose X is pseudocompact and $\Phi\colon C_p^*(X) \to C_p^*(Y)$ is a linear homeomorphism. To show that Y is pseudocompact it is enough to prove that $g(Y)$ is compact for every $g \in \mathcal{F}_Y$. So, let $g_1 \in \mathcal{F}_Y$ be fixed. Since, by Fact 3, $C_p^q(g_1) \subset C_p^*(Y)$ is closed and separable, Facts 1 and 2 imply that $\Phi^{-1}(C_p^q(g_1)) \subset C_p^q(f_1)$ for some $f_1 \in \mathcal{F}_X$. Again using Facts 1, 2 and 3 we can find $g_2 \in \mathcal{F}_Y$ such that $\Phi(C_p^q(f_1)) \subset C_p^q(g_2)$. We can assume $g_2 \succ g_1$ (otherwise instead of g_2 we take the diagonal product of g_1 and g_2). Proceeding in this way two sequences $\{f_n\} \subset \mathcal{F}_X$, $\{g_n\} \subset \mathcal{F}_Y$ will be constructed such that:

(1) $f_{n+1} \succ f_n$ and $g_{n+1} \succ g_n$;

(2) $\Phi(C_p^q(f_n)) \subset C_p^q(g_{n+1})$ and $\Phi^{-1}(C_p^q(g_n)) \subset C_p^q(f_n)$.

Let f (resp., g) be the diagonal product of all f_n (resp., g_n). It is easily seen that $\bigcup_{n\in\mathbf{N}} h_p^*(f, f_n)(C_p^*(f_n(X)))$ is dense in $C_p^*(X_f)$, where $h_p^*(f, f_n)$: $C_p^*(f_n(X))) \to C_p^*(X_f)$ is the dual of the map $h(f, f_n)$: $X_f \to f_n(X)$. Because $C_p(f)$ is dense in $C_p^q(f)$, we have that $\bigcup_{n\in\mathbf{N}} C_p^q(f_n)$ is dense in $C_p^q(f)$ and, similarly, $\bigcup_{n\in\mathbf{N}} C_p^q(g_n)$ is dense in $C_p^q(g)$. Then, by (2), $\Phi(C_p^q(f)) = C_p^q(g)$. By the closed graph theorem, Φ considered as a map between $C_u^*(X)$ and $C_u^*(Y)$, is a homeomorphism, so $C_u^q(f)$ and $C_u^q(g)$ are homeomorphic. Consequently, $C_u^*(X_f^q)$ and $C_u^*(Y_g^q)$ are linearly homeomorphic (recall that $q_u^*(f)$ and $q_u^*(g)$ are also linear homeomorphisms). But X pseudocompact implies X_f^q is homeomorphic to X_f ([3], see also [12]). Hence $C_u^*(X_f)$ is homeomorphic to $C_u^*(Y_g^q)$. So, $C_u^*(Y_g^q)$ is separable (since X_f compact yields $C_u^*(X_f)$ separable). Therefore, since Y_g^q is compact [11], so is Y_g. Then $g_1(Y)$ is also compact because $g \succ g_1$. □

Note Added in Proof

Recently the first author proved that Theorem 3.3 remains true if $C_p(X, E)$ and $C_p(Y, F)$ (resp., $C_p^*(X, E)$ and $C_p^*(Y, F)$) are uniformly homeomorphic, where E and F are normed spaces. We also observed some new results treating invariance of Lindelöf degree under (linear) homeomorphisms between $C_p(X)$ and $C_p(Y)$. First of all, let mention that the C_p-part of Corollary 3.4 was obtained by J. Baars and H. Gladdines [7]. Unfortunately, we couldn't see this paper and to compare with our. V. Tkačuk [13] showed that hereditary Lindelöf degree is l_p-invariant which also implies Corollary 3.4 (more precisely, that part concerning $C_p(X)$ and $C_p(Y)$; but Tkačuk's proof seems to be valid in case of bounded function spaces). In this context, we may have to note that our proof of Corollary 3.4 works for arbitrary X and Y when hereditary Lindelöf degree is considered. The most general result in this direction was proved by O. Okunev [10]; particularly, it states that if $C_p(X)$ and $C_p(Y)$ (resp., $C_p^*(X)$ and $C_p^*(Y)$) are homeomorphic, then X^n and Y^n have the same hereditary Lindelöf degree for all $n \in \mathbf{N}$. Therefore Okunev's result also yields Corollary 3.4.

Acknowledgements

Authors' thanks are due to the referee and the Editor Prof. R. Lowen for their valuable suggestions.

References

1. Arkhangel'skiĭ, A. V.: C_p-theory, in M. Hušek and J. van Mill (eds.), *Recent Progress in General Topology*, Elsevier Science Publishers, 1992, pp. 1–56.

2. Arkhangel'skiĭ, A. V.: *Topological Function Spaces*, Kluwer Academic Publishers, Dordrecht, 1992.
3. Arkhangel'skiĭ, A. V.: On R-quotient mappings of spaces with a countable base, *Soviet Math. Dokl.* **33** (1986), 302–305.
4. Baars, J.: Function spaces on first countable paracompact spaces, *Bull. Pol. Acad. Sci.* **42**(1) (1994), 29–35.
5. Baars, J. and de Groot, J.: On Topological and Linear Equivalence of Certain Function Spaces, CWI-tract 86, Centre for Mathematics and Computer Science, Amsterdam, 1992.
6. Baars, J., de Groot, J. and Pelant, J.: Function spaces of completely metrizable spaces, *Trans. Amer. Math. Soc.* **340**(2) (1993), 871–879.
7. Baars, J. and Gladdiness, H.: On the linear invariance of Lindelöf degree, *Canad. Math. Bull.* **39**(2) (1996), 129–137.
8. Michael, E.: A quintuple quotient quest, *Gen. Topology Appl.* **2** (1972), 91–138.
9. Okunev, O.: Weak topology of an associated space and t-equivalence, *Math. Notes* **46**(1–2) (1989), 534–538.
10. Okunev, O.: Homeomorphisms of function spaces and hereditary cardinal invariants, Preprint.
11. Semadeni, Z.: *Banach Spaces of Continuous Functions*, PWN, Warszawa, 1971.
12. Tkačuk, V. V.: Spaces projective with respect to classes of mappings, *Trudy Moskov. Mat. Obshch.* **50** (1987), 138–155 (in Russian).
13. Tkačuk, V. V.: Some non-multiplicative properties are l_p-invariant, Preprint.
14. Uspenskiĭ, V.: Characterization of compactness in terms of the uniform structure in function spaces, *Uspekhi Mat. Nauk* **37**(4) (1982), 183–184 (in Russian).
15. Valov, V.: Function spaces, *Topology. Appl.*, to appear.
16. Valov, V.: Spaces of bounded functions with the compact-open topology, *Bull. Polish Acad. Sci.* **44**(2) (1996), 169–177.

G. Brümmer & C. Gilmour (eds), Papers in Honour of Bernhard Banaschewski.

Connectednesses and Disconnectednesses of Petri Nets

STEFAN VELDSMAN
Department of Mathematics, University of Port Elizabeth, P.O. Box 1600, 6000 Port Elizabeth, South Africa. e-mail: maassv@upe.ac.za

(Received: 31 July 1996; accepted: 26 November 1997)

Abstract. A theoretical framework is provided which ensures that any net can be decomposed into subnets which are maximal with respect to certain prescribed properties.

Mathematics Subject Classifications (2000): 68Q85, 05C99, 18B99, 16N99.

Key words: Petri net, connectedness, disconnectedness.

1. Introduction

It is our intention here to provide an analysis of the structural properties of a net by using the theory of connectednesses and disconnectednesses (i.e. a radical theory for Petri nets). We will deal only with the structural properties of nets and not with any dynamic properties. It is perceived that the theory presented will also have an influence on the dynamic properties of a net, but such investigations are not our aim here.

One of the most useful tools in the analysis of algebras is radical theory. Its roots go back to the famous Wedderburn structure theorem (1908), see for example, [2] or [10]. The underlying principle is to identify a certain 'pathological' subalgebra, factor it out and then describe the quotient algebra by some other well-behaved algebras. This process has subsequently been axiomatised and radical theory per se has been studied extensively. During the seventies these ideas were extended to non-algebraic structures. Arhangelskii and Wiegandt [1] showed that the connectednesses and disconnectednesses studied by Preuss [7 – 9] to clarify the relationship between connectedness and separation properties of topological spaces, coincide with the radical theory of topological spaces. Subsequently Fried and Wiegandt [3] developed a radical theory for the category of undirected graphs which admit loops (here also called connectednesses). The mathematical structures of algebras and those of graphs are quite different; consequently the radical theory and the practical applications thereof are quite different. In the case of graph theory, the radical theory identifies in each graph subgraphs which are maximal with respect to some prescribed property of graphs. Radical theory also provides a useful method of comparing classes of graphs.

Every net (as far as the structural properties are concerned), can be considered as a directed bipartite graph which admits multiple edges and as such we will develop here a connectedness and disconnectedness theory for Petri nets.

Our theory will show, amongst others, the well-known fact that any net can be decomposed into maximal strongly connected subnets – maximal in the sense that any connected subnet of the net must be contained in one of these maximal ones. Moreover, associated with this decomposition is a certain 'quotient net' which contains no strongly connected subnet which has more than one place and one transition.

2. Preliminaries

A Petri net N is a quintuple $N = (P, T, B, F, M)$, where P and T are finite, disjoint and nonempty sets, called the *places* and *transitions* respectively, B and F are the *backwards* and *forwards incidence functions* respectively, i.e.,

$$B, F\colon P \times T \to \mathbb{N}_0 := \{0, 1, 2, 3, \ldots\}$$

where $B(p, t)$ denotes the number of arcs from $p \in P$ to $t \in T$ and $F(p, t)$ denotes the number of arcs from $t \in T$ to $p \in P$. M is the *initial marking*, i.e. $M\colon P \to \mathbb{N}_0$ is a function where $M(p)$ denotes the number of tokens at place $p \in P$. We do not allow isolated places and transitions so, in addition, we require that for any $p \in P$ there exists a $t \in T$ such that $B(p, t) + F(p, t) \geqslant 1$ and for any $t \in T$ there exists an $p \in P$ such that $B(p, t) + F(p, t) \geqslant 1$. For more information on the basics and dynamics of Petri nets, consult [4, 5] or [6].

To concretize our notation, we give an example of a small net:

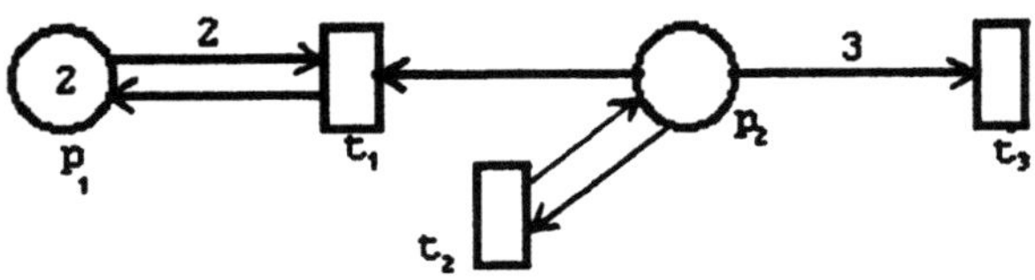

Here $P = \{p_1, p_2\}$, $T = \{t_1, t_2, t_3\}$, $M(p_1) = 2$, $M(p_2) = 0$, $B(p_1, t_1) = 2$, $B(p_2, t_3) = 3$, $B(p_2, t_2) = F(p_1, t_1) = F(p_2, t_1) = F(p_2, t_2) = 1$ and $B(p_i, t_j) = F(p_i, t_j) = 0$ otherwise.

We often use N to denote $P \cup T$ and $x \in N$ will be referred to as a *node* of N. Usually the context in which it is used will show the type of x, i.e. if x represents a place or a transition. For a net N, we sometimes use P_N, T_N, B_N, etc. to denote its set of places, transitions, backwards incidence function, etc. respectively. A net N is *trivial* if $|N| = 2$ (for a set X, $|X|$ denotes its cardinality), i.e. it is of the form $N = (\{p\}, \{t\}, B, F, M)$ with $B(p, t) = n$, $F(p, t) = m, n + m \geqslant 1$ and $M(p) = l \geqslant 0$, denoted by $\omega(n, m, l)$:

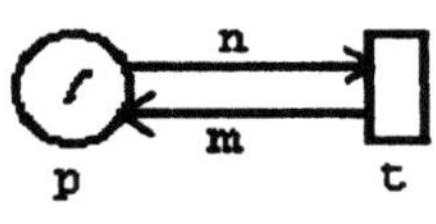

The (infinite) class of all trivial nets will be denoted by $\mathcal{T}$.

A *weak subnet* $N' = (P', T', B', F', M')$ of a net $N = (P, T, B, F, M)$ is a net with $P' \subseteq P, T' \subseteq T, B'(p', t') \leqslant B(p', t'), F'(p', t') \leqslant F(p', t')$ and $M'(p') \leqslant M(p')$ for all $p' \in P',\ t' \in T'$. A weak subnet N' of a net N is called a *subnet* if $B'(p', t') = B(p', t'), F'(p', t') = F(p', t')$ and $M'(p') = M(p')$ for all $p' \in P',\ t' \in T'$. Intuitively, a subnet N' of a net N is a subset of the set of all nodes of N which do not 'forget' arcs and tokens – care should be taken not to isolate nodes since a subnet itself must be a net. A *trivial subnet* N' of a net N is a subnet with $|N'| = 2$; otherwise it is called *nontrivial.*

The intersection of two (or more) subnets of a net (i.e. the intersection of the place sets and the intersection of the transition sets endowed with the arcs induced by the net) need not exist as a subnet – the intersections could be empty and even if it is nonempty, it may have isolated nodes and so it is not a net. The union $N_1 \cup N_2$ of two (or more) nets N_1 and N_2 is just the concatenation of the nets which is of course again a net. The union of subnets is more tricky. To wit, if N_1 and N_2 are subnets of N, then we may consider them as nets in their own right and form their union $N_1 \cup N_2$. In general $N_1 \cup N_2$ is only a weak subnet of N (in N there may be arcs between nodes from N_1 and N_2). If we endow $N_1 \cup N_2$ with all the arcs induced by N, we obtain a subnet of N, called the *subnet union* of N_1 and N_2 and we denote it by $N_1 \vee N_2$. In the natural way we can extend these ideas to unions or subnet unions of more than two nets. A net N is called *simple* if every subnet of N is either trivial or coincides with N. The simple nets N must have $|N| \in \{2, 3, 4\}$. Any net N with $|N| \in \{2, 3\}$ must be simple, but there are nets N with $|N| = 4$ which are not simple, e.g.

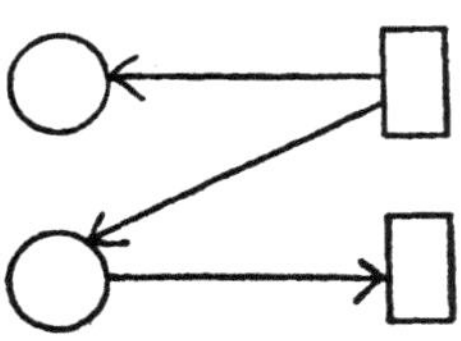
and
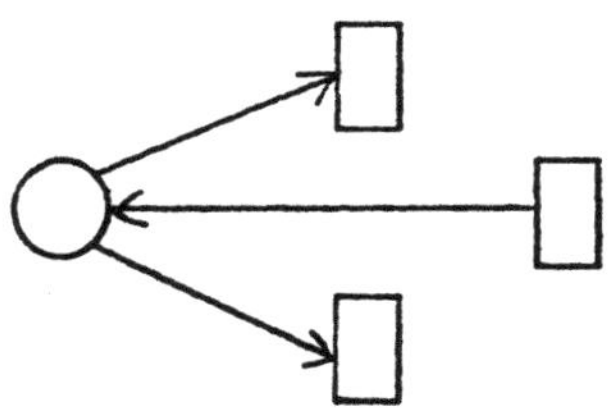
are not simple

while
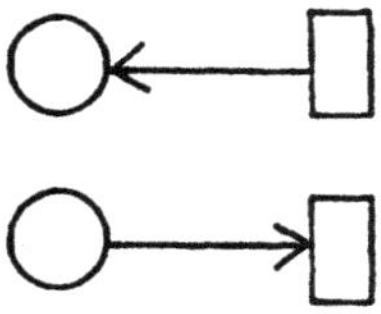
is simple.

Our standard notation for the simple nets N with $|N| = 3$ are:

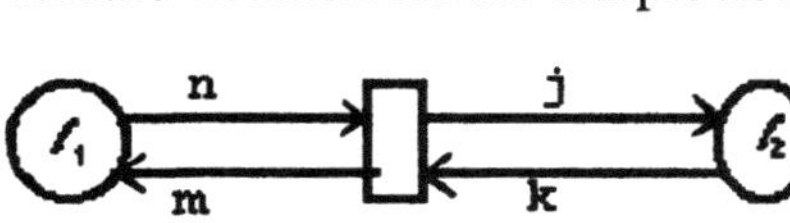

and
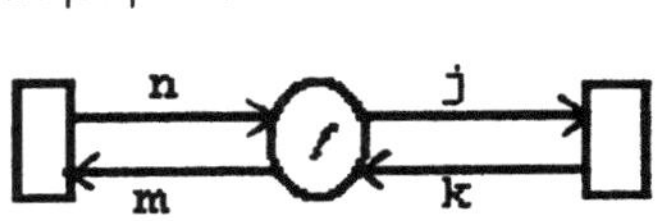

$\pi(n, m, j, k, l_1, l_2)$ $\tau(n, m, j, k, l)$

where $n, m, j, k, l_1, l_2, l \in \mathbb{N}_0$ subject to $n + m \neq 0,\ j + k \neq 0$.

Let N_1 and N_2 be nets. A *homomorphism* f from N_1 to N_2 is a function $f\colon N_1 \to N_2$ that sends places to places (i.e. $f(P_1) \subseteq P_2$), sends transitions to transitions (i.e. $f(T_1) \subseteq T_2$) and does not forget arcs and tokens (i.e. $B_1(p,t) \leqslant B_2(f(p), f(t))$, $F_1(p,t) \leqslant F_2(f(p), f(t))$ and $M_1(p) \leqslant M_2(f(p))$ for all $p \in P_1,\ t \in T_1$). A homomorphism $f\colon N_1 \to N_2$ is called a *strong homomorphism* if f is surjective (i.e. $f(P_1) = P_2$ and $f(T_1) = T_2$) and for any $q \in P_2,\ s \in T_2$ there exists $p_1, p_2, p_3 \in f^{-1}(q)$ and $t_1, t_2 \in f^{-1}(s)$ such that $B_2(q,s) = B_1(p_1, t_1)$, $F_2(q,s) = F_1(p_2, t_2)$ and $M_2(q) = M_1(p_3)$. This means such homomorhisms f have $B_2(q,s) = \max\{B_1(p,t) \mid p \in f^{-1}(q),\ t \in f^{-1}(s)\}$, $F_2(q,s) = \max\{F_1(p,t) \mid p \in f^{-1}(q),\ t \in f^{-1}(s)\}$ and $M_2(q) = \max\{M_1(p) \mid p \in f^{-1}(q)\}$. A homomorphism $f\colon N_1 \to N_2$ is called *trivial* if $|f(N_1)| = 2$; otherwise it is *nontrivial*. If f is surjective, then N_2 is called a *homomorphic image* of N_1.

Let $N_1 = (P_1, T_1, B_1, F_1, M_1)$ be a net and let P_2 and T_2 be any nonempty disjoint finite sets. Let $f\colon P_1 \cup T_1 \to P_2 \cup T_2$ be a surjective function with $f(P_1) = P_2$ and $f(T_1) = T_2$. Then P_2 and T_2 can be endowed with a net structure, say $N_2 = (P_2, T_2, B_2, F_2, M_2)$, such that f becomes a strong homomorphism: Define $B_2, F_2\colon P_2 \times T_2 \to \mathbb{N}_0$ and $M_2\colon P_2 \to \mathbb{N}_0$ by $B_2(q,s) = \max\{B_1(p,t) \mid p \in f^{-1}(q),\ t \in f^{-1}(s)\}$, $F_2(q,s) = \max\{F_1(p,t) \mid p \in f^{-1}(q),\ t \in f^{-1}(s)\}$ and $M_2(q) = \max\{M_1(p) \mid p \in f^{-1}(q)\}$ for all $q \in P_2,\ s \in T_2$. The net N_2 is called the *quotient net* determined by the function f. A strong homomorphism $f\colon N_1 \to N_2$ is called an *isomorphism* (and then N_1 and N_2 are said to be *isomorphic*) if it is injective. As usual, we do not distinguish between isomorphic nets. For a net N, the identity function $1_N\colon N \to N$ is an isomorphism. Note that if $f\colon N \to K$ and $g\colon K \to N$ are homomorphisms such that $g \circ f = 1_N$ and $f \circ g = 1_K$, then both f and g are isomorphisms.

We conclude this section with a last remark on homomorphisms $f\colon N_1 \to N_2$. Let U be a subnet of N_1. The set $W = f(U)$ can be given a net structure in two natural ways: Firstly, it can be given all the arcs and tokens induced by N_2 in which case it is a subnet of N_2 and secondly, it can be given the quotient net structure determined by the restriction of f to U in which case it is only a weak subnet of N_2. If, however, f is a strong homomorphism, then the two coincide.

3. Definitions, Constructions and Examples

Let $\mathcal{C}$ be a class of nets. $\mathcal{C}$ could be the class of all nets representing a certain property, e.g. the class of all strongly connected nets, or the class of all nets which have only one transition or the class of all nets which only have arcs from places to transitions, etc. When considering classes of nets, we shall always assume them to be abstract in the sense that any isomorphic copy of a net from the class is also in the class. A subnet N' of a net N is called a $\mathcal{C}$*-subnet* of N if $N' \in \mathcal{C}$.

A class of nets $\mathcal{C}$ is called a *connectedness* if it satisfies: A net N is in $\mathcal{C}$ if and only if every nontrivial homomorphic image of N has a nontrivial $\mathcal{C}$-subnet. A class of nets $\mathcal{D}$ is called a *disconnectedness* if it satisfies: A net N is in $\mathcal{D}$ if and only if every nontrivial subnet of N has a nontrivial homomorphic image which is in $\mathcal{D}$. We will now establish some simple properties of such classes and also give various methods for constructing them. The proofs of these results are standard in general radical theory; consequently only a small sample of proofs will be provided to demonstrate the techniques.

PROPOSITION 3.1. *Let $\mathcal{C}$ be a connectedness. Then $\mathcal{T} \subseteq \mathcal{C}$ and $\mathcal{C}$ is homomorphically closed (i.e. if $N_1 \in \mathcal{C}$ and $f\colon N_1 \to N_2$ is a surjective homomorphism, then $N_2 \in \mathcal{C}$).*

Proof. Let $N \in \mathcal{T}$. Then certainly every nontrivial homomorphic image of N contains a nontrivial $\mathcal{C}$-subnet; hence $N \in \mathcal{C}$. Let $f\colon N_1 \to N_2$ be a surjective homomorphism with $N_1 \in \mathcal{C}$. We show $N_2 \in \mathcal{C}$. If $N_2 \in \mathcal{T}$, then $N_2 \in \mathcal{C}$ since $\mathcal{T} \subseteq \mathcal{C}$. Suppose $N_2 \notin \mathcal{T}$. To show that $N_2 \in \mathcal{C}$, consider any nontrivial surjective homomorphism $g\colon N_2 \to N_3$. Then $g \circ f\colon N_1 \to N_3$ is a nontrivial surjective homomorphism and since $N_1 \in \mathcal{C}$, N_3 has a nontrivial $\mathcal{C}$-subnet. Thus $N_2 \in \mathcal{C}$ by the definition of a connectedness. □

Note that if $\mathcal{C}$ is a connectedness and $N \in \mathcal{C}$, then also $N' \in \mathcal{C}$ for any net N' obtained from N by adding more arcs and/or tokens. This follows since $\mathcal{C}$ is homomorphically closed. A class of nets $\mathcal{D}$ is *hereditary* (respectively *strongly hereditary*) if it satisfies: whenever $N \in \mathcal{D}$ and U is a subnet (respectively weak subnet) of N, then $U \in \mathcal{D}$.

PROPOSITION 3.2. *Let $\mathcal{D}$ be a disconnectedness. Then $\mathcal{T} \subseteq \mathcal{D}$ and $\mathcal{D}$ is strongly hereditary (and thus also hereditary).*

Proof. Let $N \in \mathcal{T}$. Then every nontrivial subnet of N has a nontrivial homomorphic image in $\mathcal{D}$; hence $N \in \mathcal{D}$ and so $\mathcal{T} \subseteq \mathcal{D}$. Let $N \in \mathcal{D}$ and let N_1 be a weak subnet of N. We show $N_1 \in \mathcal{D}$. If $N_1 \in \mathcal{T}$ we are done (by the first part); suppose $N_1 \notin \mathcal{T}$. Let N_2 be any nontrivial subnet of N_1. Let $\overline{N_2}$ be the subnet of N determined by N_2, i.e. $\overline{N_2}$ has the same set of nodes as N_2 but has all the arcs and tokens induced by N. Since $N \in \mathcal{D}$ and $\overline{N_2}$ is a nontrivial subnet of N, $\overline{N_2}$ has a nontrivial homomorphic image, say N_3, in $\mathcal{D}$. Then N_3 is also a nontrivial homomorphic image of N_2. By the definition of a disconnectedness, we get $N_1 \in \mathcal{D}$. □

Note that if $\mathcal{D}$ is a disconnectedness and $N \in \mathcal{D}$, then also $N' \in \mathcal{D}$ where N' is any net obtained from N by deleting nodes, arcs or tokens (just be careful not to isolate nodes).

CONSTRUCTIONS OF CONNECTEDNESSES AND DISCONNECTEDNESSES 3.3 *Let $\mathcal{A}$ be a hereditary class of nets. Then:*

(1) $\mathcal{UA}$:= {*all nets* N | N *has no nontrivial homomorphic image in* $\mathcal{A}$} *is a connectedness and it is the largest connectedness for which* $\mathcal{A} \cap \mathcal{UA} = \mathcal{T}$.
(2) $\mathcal{DA}$:= {*all nets* N | *every nontrivial subnet of* N *has a nontrivial homomorphic image in* $\mathcal{A}$} *is a disconnectedness and it is the smallest disconnectedness which contains* $\mathcal{A}$.

Let $\mathcal{B}$ *be a homomorphically closed class of nets. Then:*

(3) $\mathcal{LB}$ = {*all nets* N | *every nontrivial homomorphic image of* N *has a nontrivial* $\mathcal{B}$*-subnet*} *is a connectedness and it is the smallest connectedness which contains* $\mathcal{B}$.
(4) $\mathcal{SB}$ = {*all nets* N | N *has no nontrivial* $\mathcal{B}$*-subnets*} *is a disconnectedness and it is the largest disconnectedness for which* $\mathcal{B} \cap \mathcal{SB} = \mathcal{T}$.

We tacitly assumed in the above that both $\mathcal{A}$ and $\mathcal{B}$ contain the class $\mathcal{T}$ of all trivial nets. This rider will in fact hardly ever be mentioned. Finding hereditary classes of nets is easy – let $\mathcal{A}$ be any class of simple nets. Then $\mathcal{A}$ is hereditary. (Strictly speaking we should say: Let $\mathcal{A}$ be any class of simple nets which contains all the trivial nets.)

As could be expected, there is a close connection between connectednesses and disconnectednesses: If $\mathcal{C}$ is a connectedness, then $\mathcal{SC}$ is a disconnectedness. If $\mathcal{D}$ is a disconnectedness, then $\mathcal{UD}$ is a connectedness. Moreover, a class of nets $\mathcal{C}$ is a connectedness if and only if $\mathcal{C} = \mathcal{USC}$ and a class of nets $\mathcal{D}$ is a disconnectedness if and only if $\mathcal{D} = \mathcal{SUD}$. It can also be shown that a class of nets $\mathcal{D}$ is a disconnectedness if and only if it satisfies: $N \in \mathcal{D}$ if and only if every nontrivial weak subnet of N has a nontrivial homomorphic image in $\mathcal{D}$.

EXAMPLES 3.4. The verifications that the classes below are what they are claimed to be follow easily from 3.3 above or subsequent results (4.5 and 5.2).

(1) The class $\mathcal{T}$ is both a connectedness and a disconnectedness; thus $\mathcal{ST}$ = {all nets N} is a disconnectedness and $\mathcal{UT}$ = {all nets N} is a connectedness.

(2) A net N is *strongly connected* (or *path connected*) if for any $x, y \in N$, there is a path from x to y. By a path from x to y we understand the existence of a finite sequence of nodes $x = z_1, z_2, z_3, \dots, z_n = y$ in N, where $z_i \in \bullet z_{i+1}$, $i = 1, 2, \dots, n-1$. As usual for $p, t \in N$, $\bullet p = \{t' \in T \mid F(p, t') \geqslant 1\}$ and $\bullet t = \{p' \in P \mid B(p', t) \geqslant 1\}$. The class $\mathcal{C}$ = {all nets N | N is trivial or N is path connected} is a connectedness.

(3) Let $\mathcal{A} = \{\pi(n, m, j, k, l_1, l_2) \mid nmjk = 0,\ n+m \neq 0,\ j+k \neq 0$, where $n, m, j, k, l_1, l_2 \in \mathbb{N}_0\} \cup \{\tau(n, m, j, k, l) \mid nmjk = 0,\ n+m \neq 0,\ j+k \neq 0$, where $n, m, j, k, l \in \mathbb{N}_0\}$, i.e. $\mathcal{A}$ is the class of all the simple nets with three nodes which have at least one of n, m, j or k equal to 0. Then $\mathcal{A}$ is hereditary and $\mathcal{C} = \mathcal{UA}$ is a connectedness. It can be verified that $\mathcal{C}$ = {all nets N | if N is nontrivial and $B(p, t) \geqslant 1$ for some $p, t \in N$, then there is an $s \in T$ such that $F(p, s) \geqslant 1$ and there is a $q \in P$ such that $F(q, t) \geqslant 1$ and if $F(p, t) \geqslant 1$ for some $p, t \in N$, then there is an $s \in T$ such that $B(p, s) \geqslant 1$ and there is a $q \in P$ such that $B(q, t) \geqslant 1$}.

(4) C = {all nets $N \mid N \in \mathcal{T}$ or if $N \notin \mathcal{T}$, then for all $p \in P$ there exists a $t \in T$ such that $B(p,t) \geqslant 1$ and for all $t \in T$ there exists a $p \in P$ such that $F(p,t) \geqslant 1$} is a connectedness.

(5) The class $\mathcal{B}$ = {all nets $N \mid N$ has at most three nodes} is homomorphically closed; hence $\mathcal{S}\mathcal{B}$ = {all nets $N \mid N$ has no subnet with three nodes} is a disconnectedness. The corresponding connectedness $\mathcal{C} = \mathcal{U}\mathcal{S}\mathcal{B}$ is $\mathcal{C}$ = {all nets $N \mid N$ is line connected}. A net N is *line connected* if for any $x, y \in N$, there is a *line* between x and y, i.e. there is a finite sequence of nodes $x = z_1, z_2, z_3, \ldots, z_n = y$ in N where $z_i \in (\bullet z_{i+1} \cup z_{i+1}\bullet)$ for $i = 1, 2, 3, \ldots, n-1$. Recall, for $p, t \in N$, $p\bullet = \{s \in T \mid B(p,s) \geqslant 1\}$ and $t\bullet = \{q \in P \mid F(q,t) \geqslant 1\}$.

(6) A net N is *place connected* if for all $p, q \in P$, there is a path from p to q. Then C = {all nets $N \mid N$ is place connected} is a connectedness. Likewise, the class of all transition connected nets is a connectedness.

(7) Let $\mathcal{A}$ = {all nets $N \mid N$ has one place and two transitions}$\cup\{\pi(n, m, j, k, l_1, l_2) \mid n + m \neq 0,\ j + k \neq 0$ subject to: $k = 0$ (and then $j \neq 0$) or if $k \neq 0$, then $m = j = 0\}$. Since $\mathcal{A}$ consists of simple nets, it is hereditary and $\mathcal{C} = \mathcal{U}\mathcal{A}$ is a connectedness. It can be shown that $\mathcal{C}$ = {all nets $N \mid N$ has one transition and if N is nontrivial, then there is $p_0, t_0 \in N$ such that $F(p_0, t_0) \geqslant 1$ and for any $p, t \in N$, if $F(p,t) \geqslant 1$, then $B(p,t) \geqslant 1$}.

(8) Both the classes {all nets $N \mid |T_N| = 1$} and {all nets $N \mid |P_N| = 1$} are connectednesses.

(9) Let $\mathcal{B} = \{\pi(n, m, j, k, l_1, l_2) \mid n \neq 0,\ m \neq 0,\ j + k \neq 0$, where $n, m, j, k, l_1, l_2 \in \mathbb{N}_0\} \cup \{\tau(n, m, j, k, l) \mid n \neq 0,\ m \neq 0, j + k \neq 0$ where $n, m, j, k, l \in \mathbb{N}_0\}$. Then $\mathcal{B}$ is homomorphically closed, hence $\mathcal{S}\mathcal{B}$ = {all nets $N \mid$ if N is nontrivial, then $B(p_0, t_0) \geqslant 1$ and $F(p_0, t_0) \geqslant 1$ implies $B(p_0, t) = F(p_0, t) = F(p, t_0) = B(p, t_0) = 0$ for all $p, t \in N \setminus \{p_0, t_0\}$} is a disconnectedness.

(10) Let $B = \{(n, m, j, k, l_1, l_2) \mid n+m \neq 0,\ j+k \neq 0$, where $n, m, j, k, l_1, l_2 \in \mathbb{N}_0\} \cup \{\tau(n, m, j, k, l) \mid n + m \neq 0, j + k \neq 0$, where $n, m, j, k, l \in \mathbb{N}_0\}$. Then $\mathcal{B}$ is homomorphically closed, hence $\mathcal{S}\mathcal{B}$ = {all nets $N \mid$ if N is nontrivial, then $B(p_0, t_0) \geqslant 1$ or $F(p_0, t_0) \geqslant 1$ implies $B(p_0, t) = F(p_0, t) = F(p, t_0) = B(p, t_0) = 0$ for all $p, t \in N \setminus \{p_0, t_0\}$} is a disconnectedness.

(11) The classes {all nets $N \mid$ if N is nontrivial, then $|\bullet p| = |p \bullet| \leqslant 1$ for all $p \in P$} and {all nets $N \mid$ if N is nontrivial, then $|\bullet t| = |t \bullet| \leqslant 1$ for all $t \in T$} are both disconnectednesses.

The first six examples and the last three examples should give credence to the use of the terminology connectedness and disconnectedness. One more source of connectednesses comes from the next proposition which is, from a general radical theoretic viewpoint, of some importance:

PROPOSITION 3.5. *Let $\mathcal{W}$ be a nonempty class of nets. Let $\mathcal{C}(\mathcal{W})$ = {all nets $N \mid N \in \mathcal{T}$ or there exists a net $W \in \mathcal{W}$ and a homomorphism $W \to N$}. Then $\mathcal{C}(\mathcal{W})$ is a* complementary connectedness *(i.e. a connectedness for which*

$\mathcal{C}(\mathcal{W}) \cup \mathcal{SC}(\mathcal{W})$ is the class of all nets) and the corresponding disconnectedness is $\mathcal{SC}(\mathcal{W}) = \{$all nets $N \mid N \in \mathcal{T}$ or there is no homomorphism from W to N for all $W \in \mathcal{W}\}$.

Proof. $\mathcal{C}(\mathcal{W})$ is clearly homomorphically closed; hence if $N \in \mathcal{C}(\mathcal{W})$, then every nontrivial homomorphic image of N has a nontrivial $\mathcal{C}(\mathcal{W})$-subnet. Conversely, let N be a net such that every nontrivial homomorphic image of N has a nontrivial $\mathcal{C}(\mathcal{W})$-subnet. We must show $N \in \mathcal{C}(\mathcal{W})$. If $N \in \mathcal{T}$ we are done. Suppose $N \notin \mathcal{T}$. Since N is a nontrivial homomorphic image of itself, N has a nontrivial $\mathcal{C}(\mathcal{W})$-subnet, say N_1. By the definition of $\mathcal{C}(\mathcal{W})$, there is a $W \in \mathcal{W}$ and a homomorphism $f\colon W \to N_1$. Thus there is a homomorphism $W \to N$ and so $N \in \mathcal{C}(\mathcal{W})$. We conclude that $\mathcal{C}(\mathcal{W})$ is a connectedness. It is complementary: Let N be any net such that $N \notin \mathcal{SC}(\mathcal{W})$. Then N has a nontrivial $\mathcal{C}(\mathcal{W})$-subnet and as above, we then find a homomorphism $W \to N$ for some $W \in \mathcal{W}$. Hence $N \in \mathcal{C}(\mathcal{W})$. □

The interesting question is to find classes $\mathcal{W}$ for which $\mathcal{C}(\mathcal{W})$ is nontrivial, i.e. $\mathcal{T} \subset \mathcal{C}(\mathcal{W}) \subset \{$all nets$\}$. Now $\mathcal{C}(\mathcal{W}) = \mathcal{T}$ if and only if whenever there is a homomorphism $W \to N$, $W \in \mathcal{W}$, then $N \in \mathcal{T}$. Furthermore, $\mathcal{C}(\mathcal{W}) = \{$all nets$\}$ if and only if for all nets $N \notin \mathcal{T}$, there exists a $W \in \mathcal{W}$ and a homomorphism $W \to N$. So, for examples of nontrivial $\mathcal{C}(\mathcal{W})$'s with $|\mathcal{W}| = 1$ (in which case we write $\mathcal{C}(\mathcal{W})$ as $\mathcal{C}(W)$ where $\mathcal{W} = \{W\}$), we have to look for nets W for which there are two nontrivial nets N_1 and N_2 such that there is a homomorphism $W \to N_1$ but none from W to N_2. There are infinitely many such choices. Some are:

(1) $W = \omega(1, 0, 0)$, $N_1 = \pi(1, 0, 1, 0, 0, 0)$ and $N_2 = \pi(0, 1, 1, 0, 0, 0)$.

(2) $W = \pi(0, 1, 1, 0, 0, 0)$, $N_1 = \pi(0, 1, 1, 1, 0, 0)$ and $N_2 = \pi(1, 0, 0, 1, 0, 0)$.

(3) For each $n \in \mathbb{N} = \{1, 2, 3, \ldots\}$, let W_n be the net $W_n = (P_n, T_n, B_n, F_n, M_n)$, where

$$P_n = \{p_1, p_2, \ldots, p_{n+1}\}, \qquad T_n = \{t_1, t_2, \ldots, t_{n+1}\},$$

$$F_n(p_i, t_j) = 0 = M_n(p_i) \quad \text{and} \quad B_n(p_i, t_j) = \begin{cases} 1 & \text{if } i = j, \\ 0 & \text{otherwise.} \end{cases}$$

Then $\mathcal{C}(W_1) \subset \mathcal{C}(W_2) \subset \mathcal{C}(W_3) \subset \cdots$ is an infinite chain of complementary connectednesses.

4. Properties and Characterization of Connectednesses

Let Λ be some index set and for each $\alpha \in \Lambda$, let $N_\alpha = (P_\alpha, T_\alpha, B_\alpha, F_\alpha, M_\alpha)$ be a subnet of the net $N = (P_N, T_N, B_N, F_N, M_N)$. We refer to $\mathcal{A} = \{N_\alpha \mid \alpha \in \Lambda\}$ as a *sink of subnets* of N. Since a net is a finite structure, Λ must necessarily be finite.

A homomorphism $f\colon N \to K$ is called a *cokernel* of the sink of subnets $\{N_\alpha \mid \alpha \in \Lambda\}$ if $f(N_\alpha)$ is trivial for all $\alpha \in \Lambda$ and if $g\colon N \to L$ is any other homomorphism such that $g(N_\alpha)$ is trivial for all $\alpha \in \Lambda$, then there is a unique homomorphism $h\colon K \to L$ such that $h \circ f = g$. As the notion of a cokernel is important in the subsequent considerations, it will be beneficial to investigate its properties in some detail.

4.1. PROPERTIES OF COKERNELS

4.1.1. *A Cokernel is Always a Strong Homomorphism*

Let $f: N \to K = (P_K, T_K, B_K, F_K, M_K)$ be the cokernel of the sink of subnets $\{N_\alpha \mid \alpha \in \Lambda\}$. Suppose $f(N) \subset K$. Let $n = \max\{B_K(p,t) \mid p \in P_K,\ t \in T_K\}$, $m = \max\{F_K(p,t) \mid p \in P_K,\ t \in T_K\}$ and $\ell = \max\{M_K(p) \mid p \in P_K\}$. Let q and s be fixed symbols not in $P_K \cup T_K$. Let $L = (P_L, T_L, B_L, F_L, M_L)$ be the net with $P_L = \{q\} \cup \{p \in P_K \mid p \notin f(P_N)\}$, $T_L = \{s\} \cup \{t \in P_K \mid t \notin f(T_N)\}$, $B_L(p,t) = n$, $F_L(p,t) = m$ and $M_L(p) = \ell$ for all $p \in P_L$, $t \in T_L$.

Define functions g, h: $K \to L$ by $g(P_K) = \{q\} = h(f(P_N))$, $g(T_K) = \{s\} = h(f(T_N))$, $h(p) = p$ for all $p \in P_K \setminus f(P_N)$ and $h(t) = t$ for all $t \in T_K \setminus f(T_N)$.

Then g and h are homomorphisms and e: $N \to L$ defined by $e = h \circ f (= g \circ f)$ is a homomorphism with $e(N_\alpha)$ trivial for all $\alpha \in \Lambda$. By the definition of a cokernel, there is a unique homomorphism b: $K \to L$ such that $b \circ f = e$. But $h \circ f = e = g \circ f$; hence $h = b = g$. But this contradicts $f(N) \subset K$. Hence f must be surjective.

Next we show that f has the strong property. Define a function g: $N \to f(N)$ by $g(x) = f(x)$ for all $x \in N$. Then g is surjective and we may give $f(N)$ the quotient net structure determined by g, call this net L. Thus g: $N \to L$ is a strong homomorphism. The identity map $e: L \to K$ is a homomorphism and $e \circ g = f$. Since $g(N_\alpha)$ is trivial for all $\alpha \in \Lambda$, there is a unique homomorphism h: $K \to L$ such that $h \circ f = g$. But both f and g are surjective; hence $e \circ h = 1_K$ and $h \circ e = 1_L$. Thus e is an isomorphism and consequently f must be a strong homomorphism.

4.1.2. *Any Sink of Subnets of a Net Always Has a Cokernel, and it Is Unique up to Isomorphism*

We start with the construction of the product of certain nets (for which we will have a later use as well). For each $i \in I$, I some index set (not necessarily finite), let f_i: $N \to N_i$ be a homomorphism. Let $\Pi(P_i \mid i \in I)$ and $\Pi(T_i \mid i \in I)$ be the set theoretic product of the sets $\{P_i \mid i \in I\}$ and $\{T_i \mid i \in I\}$ respectively. This means, for example, that $\Pi(P_i \mid i \in I) = \{p \mid p\colon I \to \bigcup(P_i \mid i \in I)$ is a function with $p(i) \in P_i$ for all $i \in I\}$. We write p_i for $p(i)$. Clearly $(\Pi P_i) \cap (\Pi T_i) = \phi$. Define a function f: $P_N \cup T_N \to (\Pi P_i) \cup (\Pi T_i)$ by: For $p \in P_N$, let $f(p)$: $I \to \bigcup P_i$ be defined by $f(p)(i) = f_i(p)$. Likewise for $t \in T$, let $f(t)(i) = f_i(t)$ for all $t \in T_N$. Let $P_\pi = f(P_N)$ and $T_\pi = f(T_N)$. Then P_π and T_π are nonempty, finite and disjoint sets and f: $P_N \cup T_N \to P_\pi \cup T_\pi$ is a surjective function. We then give P_π and T_π the quotient net structure determined by f, say this net is $\Pi N_i = (P_\pi, T_\pi, B_\pi, F_\pi, M_\pi)$. Thus f: $N \to \Pi N_i$ is a strong homomorphism.

The pair $(f, \Pi N_i)$ is called the *product* of the class $\{(f_i, N_i) \mid i \in I\}$. For each $j \in I$, let $\pi_j \colon \Pi N_i \to N_j$ be the function defined by $\pi_j(f(x)) = f(x)(j) = f_j(x)$ for all $x \in N$. Thus $\pi_j \circ f = f_j$ for all $j \in I$ and π_j is called the jth *projection.* It is a homomorphism: We verify, for example, that $B_\pi(q, s) \leqslant B_j(\pi_j(q), \pi_j(s))$ for all $q \in P_{\pi'}$, $s \in T_\pi$. By definition of P_π and T_π, there are $p \in P_N$ and $t \in T_N$ such that $q = f(p)$ and $s = f(t)$. Furthermore, $B_\pi(q, s) = \max\{B_N(p', t') \mid p' \in f^{-1}(q),\ t' \in f^{-1}(s)\}$. Let $p' \in f^{-1}(q)$, $t' \in f^{-1}(s)$. Then $f(p') = q = f(p)$ and $f(t') = s = f(t)$; hence $f_j(p') = f_j(p)$ and $f_j(t') = f_j(t)$. Since f_j is a homomorphism, $B_N(p', t') \leqslant B_j(f(p'), f(t')) = B_j(f(p), f(t)) = B_j(\pi_j(f(p)), \pi_j(f(t))) = B_j(\pi_j(q), \pi_j(s))$. Since $p' \in f^{-1}(q)$ and $t' \in f^{-1}(s)$ were arbitrary, we conclude that $B_\pi(q, s) \leqslant B_j(\pi_j(q), \pi_j(s))$.

We return to the existence of the cokernel of a sink of subnets $\{N_\alpha \mid \alpha \in \Lambda\}$ of the net N. Let $\{(f_i, N_i) \mid i \in I\}$ be the class of all pairs with $f_i \colon N \to N_i$ a homomorphism such that $f_i(N_\alpha)$ is trivial for all $\alpha \in \Lambda$. Note that there is at least one such $f_i \colon N \to N_i$ since any net can be mapped onto a suitable trivial net. Let $(f, \Pi N_i)$ be the product of this class. We claim $f \colon N \to \Pi N_i$ is the cokernel of the given sink. Firstly note that $f(N_\alpha)$ is trivial for all α. Indeed, let $p, q \in P_\alpha$. For every $i \in I$, $f_i(N_\alpha)$ is trivial and so $f(p)(i) = f_i(p) = f_i(q) = f(q)(i)$. Thus $f(p) = f(q)$. Secondly, let $g \colon N \to L$ be any homomorphism such that $f(N_\alpha)$ is trivial for all $\alpha \in \Lambda$. Then (g, L) must be one of the (f_i, N_i)'s, say $g = f_j$ and $L = N_j$. Then $\pi_j \circ f = f_j = g$ and since f is surjective, π_j is unique with respect to this property. Thus f is the cokernel of the given sink. The uniqueness of f, as cokernel of the sink, follows trivially.

Although the above guarantees the existence of the cokernel, we now give an explicit construction of the cokernel of the sink. Let $\mathbf{P} = \{P_\alpha \mid \alpha \in \Lambda\}$ and $\mathbf{T} = \{T_\alpha \mid \alpha \in \Lambda\}$. On $\mathbf{P}$ define a relation $\sim$ by: $P_\alpha \sim P_\beta$ if and only if there is a sequence $P_\alpha = P_{\gamma_0}, P_{\gamma_1}, P_{\gamma_2}, \ldots, P_{\gamma_n} = P_\beta$ of elements in $\mathbf{P}$ with $P_{\gamma_{i-1}} \cap P_{\gamma_i} \neq \phi$ for all $i = 1, 2, \ldots, n$. This relation is an equivalence relation on $\mathbf{P}$; let $\{P_{\alpha_1}, P_{\alpha_2}, \ldots, P_{\alpha_n}\}$ be a class of representatives for the equivalence classes. For each α_i, let $Q_{\alpha_i} = \bigcup(P_\beta \in \mathbf{P} \mid P_\beta \sim P_{\alpha_i})$. Then $Q_{\alpha_i} \cap Q_{\alpha_j} = \phi$ for $\alpha_i \neq \alpha_j$ and every $P_\beta \in \mathbf{P}$ is contained in some Q_{α_i}. In exactly the same way define an equivalence relation $\sim$ on $\mathbf{T}$ with $\{T_{\beta_1}, T_{\beta_2}, \ldots, T_{\beta_k}\}$ as a class of representatives for the equivalence classes. For each β_j, let $S_{\beta_j} = \bigcup(T_\alpha \in \mathbf{T} \mid T_\alpha \sim T_\beta)$. For $\beta_i \neq \beta_j$, $S_{\beta_i} \cap S_{\beta_j} = \phi$ and for every $T_\alpha \in \mathbf{T}$ there is a β_j such that $T_\alpha \subseteq S_{\beta_j}$.

For each α_i, $i = 1, 2, \ldots, n$, fix $q_{\alpha_i} \in Q_{\alpha_i}$ and for each β_j, $j = 1, 2, \ldots, k$, fix $s_{\beta_j} \in S_{\beta_j}$. Let $P_c = \{q_{\alpha_1}, q_{\alpha_2}, \ldots, q_{\alpha_n}\} \cup \{p \in P_N \mid p \notin \bigcup(P_\alpha \mid \alpha \in \Lambda)\}$ and let $T_c = \{s_{\beta_1}, s_{\beta_2}, \ldots, s_{\beta_k}\} \cup \{t \in T_N \mid t \notin \bigcup(T_\alpha \mid \alpha \in \Lambda)\}$. Then P_c and T_c are nonempty, finite and disjoint. Define a function $f \colon P \cup T \to P_c \cup T_c$ by

$$f(p) = \begin{cases} q_{\alpha_i} & \text{if } p \in Q_{\alpha_i} \text{ for some } \alpha_i, \\ p & \text{if } p \in P_N \setminus \bigcup(P_\alpha \mid \alpha \in \Lambda) \end{cases}$$

and

$$f(t) = \begin{cases} s_{\beta_j} & \text{if } t \in S_{\beta_j} \text{ for some } \beta_j, \\ t & \text{if } t \in T_N \setminus \bigcup(T_\alpha \mid \alpha \in \Lambda). \end{cases}$$

Then f is a surjective function and we give P_c, T_c the quotient net structure determined by f to obtain a net $N_c = (P_c, T_c, B_c, F_c, M_c)$ and a strong homomorphism $f\colon N \to N_c$. We claim f is the cokernel of the sink $\{N_\alpha \mid \alpha \in \Lambda\}$. Clearly $f(N_\alpha)$ is trivial for all $\alpha \in \Lambda$. Let $g\colon N \to L$ be a homomorphism such that $g(N_\alpha)$ is trivial for all $\alpha \in \Lambda$. Define a function $h\colon N_c \to L$ as follows: Let $q \in P_c$ and let $h(q) = g(p)$, where $p \in f^{-1}(q)$. Note that h is well-defined, for if $q = q_{\alpha_i}$ for some α_i and $p, p' \in f^{-1}(q_{\alpha_i}) = Q_{\alpha_i}$, then $g(p) = g(p')$. Indeed, $p, p' \in Q_{\alpha_i}$ implies $p \in P_\gamma$, $p' \in P_\delta$ and $P_\gamma \sim P_{\alpha_i} \sim P_\delta$. Since $g(N_\alpha)$ is trivial for all α, this means $g(P_\gamma) = g(P_\delta)$ and so $g(p) = g(p')$. In the same way we define h on T_c so that $h\colon N_c \to L$ becomes a function with $h \circ f = g$. Since f is surjective, h is unique with respect to this property. Finally we show that h is a homomorphism. Let $q \in P_c$ and $s \in T_c$. Then

$$\begin{aligned} B_c(q, s) &= \max\{B(p, t) \mid p \in f^{-1}(q), t \in f^{-1}(s)\} \\ &= B(p_0, t_0) \quad \text{for some } p_0 \in f^{-1}(q),\ t_0 \in f^{-1}(s) \\ &\leqslant B_L(g(p_0), g(t_0)) \quad \text{(since } g \text{ is a homomorphism)} \\ &= B_L(h(q), h(s)) \quad \text{(by the definition of } h\text{)}. \end{aligned}$$

The remaining two requirements for h to be a homomorphism, follow in a similar way. For later reference, we record some of the above in

4.1.3. Let $f\colon N \to K$ be the cokernel of the sink of subnets $\mathcal{A} = \{N_\alpha \mid \alpha \in \Lambda\}$ of N

(i) $f(P_\alpha) = f(P_\beta)$ if and only if $P_\alpha \sim P_\beta$.
(ii) $f(T_\alpha) = f(T_\beta)$ if and only if $T_\alpha \sim T_\beta$.
(iii) The restriction of f to the set $N \setminus \bigcup(N_\alpha \mid \alpha \in \Lambda)$ is a bijection onto the set $N_c \setminus \bigcup(f(N_\alpha) \mid \alpha \in \Lambda)$.
(iv) Let $\mathcal{A}' = \{N_\alpha \in \mathcal{A} \mid N_\alpha \notin \mathcal{T}\}$. Then f is also the cokernel of the sink $\mathcal{A}'$.

A net N is called a *chain* of the sink of subnets $\{N_\alpha \mid \alpha \in \Lambda\}$ if the cokernel of this sink is trivial. Chains can be characterized by:

PROPOSITION 4.1. *Let $\{N_\alpha \mid \alpha \in \Lambda\}$ be a sink of subnets of the net $N = (P, T, B, F, M)$. Then N is a chain of the subnets $\{N_\alpha \mid \alpha \in \Lambda\}$ if and only if $N = \bigvee(N_\alpha \mid \alpha \in \Lambda)$, $P_\alpha \sim P_\beta$ in $\mathbf{P} = \{P_\alpha \mid \alpha \in \Lambda\}$ and $T_\alpha \sim T_\beta$ in $\mathbf{T} = \{T_\alpha \mid \alpha \in \Lambda\}$ for all $\alpha, \beta \in \Lambda$.*

Proof. Let $f\colon N \to K$ be the cokernel of the sink $\{N_\alpha \mid \alpha \in \Lambda\}$. Suppose firstly that N is a chain of this sink. Then f is trivial by definition. We now show $P = \bigcup(P_\alpha \mid \alpha \in \Lambda)$. Suppose this is not the case. Let $Q = \{q_1, q_2\}$ and let $S = \{s\}$. Define a function $g\colon P \cup T \to Q \cup S$ by:

$$g(p) = \begin{cases} q_1 & \text{if } p \in \bigcup(P_\alpha \mid \alpha \in \Lambda), \\ q_2 & \text{if } p \in P \setminus \bigcup(P_\alpha \mid \alpha \in \Lambda) \end{cases}$$

and $g(t) = s$ for all $t \in T$. Then g is a surjective function and we may turn Q, S into a net L by giving Q and S the quotient net structure determined by g. This makes g: $N \to L$ a homomorphism and $g(N_\alpha)$ is trivial for all $\alpha \in \Lambda$. By the definition of a cokernel we then know that there is a homomorphism h: $K \to L$ such that $h \circ f = g$. Since f is trivial, g must be trivial which is only possible if $P = \bigcup(P_\alpha \mid \alpha \in \Lambda)$. Likewise it follows that $T = \bigcup(T_\alpha \mid \alpha \in \Lambda)$.

Let $\alpha, \beta \in \Lambda$. Since f is trivial, $f(P_\alpha) = f(P_\beta)$, $f(T_\alpha) = f(T_\beta)$ and from 4.1.3, we have $P_\alpha \sim P_\beta$ and $T_\alpha \sim T_\beta$ for all $\alpha, \beta \in \Lambda$.

Conversely, suppose the conditions are satisfied. We now show that f is trivial. Let $x, y \in P$. Then $x \in P_\alpha$ and $y \in P_\beta$ for some $\alpha, \beta \in \Lambda$. Since $P_\alpha \sim P_\beta$ we have from the 4.1.3 that $f(P_\alpha) = f(P_\beta)$ and thus $f(x) = f(y)$ (as $f(N_\alpha)$ is trivial). Likewise, for $x \in T_\alpha$ and $y \in T_\beta$ we get $f(x) = f(y)$ and so f is trivial. □

A class of nets $\mathcal{C}$ is a *component class* if it satisfies: Whenever a net N is a chain of a sink of $\mathcal{C}$-subnets $\{N_\alpha \mid \alpha \in \Lambda\}$, then $N \in \mathcal{C}$.

PROPOSITION 4.2. *Any connectedness $\mathcal{C}$ is a component class.*

Proof. Let N be a chain of a sink of $\mathcal{C}$-subnets $\{N_\alpha \mid \alpha \in \Lambda\}$. To show that $N \in \mathcal{C}$, we will show that for any nontrivial surjective homomorphism g: $N \to L$, the net L has a nontrivial $\mathcal{C}$-subnet. If $g(N_\alpha)$ is trivial for each $\alpha \in \Lambda$, then g must be trivial (because the cokernel of the sink $\{N_\alpha \mid \alpha \in \Lambda\}$ is trivial). But this contradicts our choice of g; hence there is an N_β such that $g(N_\beta)$ is nontrivial. But $N_\beta \in \mathcal{C}$ and $\mathcal{C}$ is homomorphically closed; thus $g(N_\beta)$ is a nontrivial $\mathcal{C}$-subnet of L. □

A class of nets $\mathcal{C}$ is *closed under extensions* if it satisfies: Whenever $\{N_\alpha \mid \alpha \in \Lambda\}$ is a sink of $\mathcal{C}$-subnets of a net N and f: $N \to K$ is the cokernel of this sink with $K \in \mathcal{C}$, then $N \in \mathcal{C}$.

PROPOSITION 4.3. *Any connectedness $\mathcal{C}$ is closed under extensions.*

Proof. Let f: $N \to K$ be the cokernel of a sink of $\mathcal{C}$-subnets $\{N_\alpha \mid \alpha \in \Lambda\}$ of the net N and suppose $K \in \mathcal{C}$. Let g: $N \to L$ be any nontrivial surjective homomorphism. If $g(N_\beta)$ is nontrivial for some $\beta \in \Lambda$, then $g(N_\beta)$ is a nontrivial $\mathcal{C}$-subnet of L. If $g(N_\alpha)$ is trivial for all $\alpha \in \Lambda$, then there is a unique homomorphism h: $K \to L$ such that $h \circ f = g$. Since $K \in \mathcal{C}$, $h(K)$, regarded as a subnet of L, is a $\mathcal{C}$-subnet of L. It is a nontrivial subnet since g is nontrivial. We conclude that $N \in \mathcal{C}$. □

Since nets are finite structures, we have an easy characterization of connectednesses by closure conditions:

THEOREM 4.4. *A class of nets $\mathcal{C}$ is a connectedness if and only if it satisfies:*

(1) $\mathcal{T} \subseteq \mathcal{C}$,

(2) $\mathcal{C}$ *is homomorphically closed and*

(3) $\mathcal{C}$ *is closed under extensions.*

Proof. The necessity of the conditions follow from 3.1 and 4.3. Suppose thus $\mathcal{C}$ is a class of nets which satisfies conditions (1), (2) and (3). If $N \in \mathcal{C}$, then every nontrivial homomorphic image of N has a nontrivial $\mathcal{C}$-subnet (by (2)). Conversely, let N be a net such that every nontrivial homomorphic image of N has a nontrivial $\mathcal{C}$-subnet. We show $N \in \mathcal{C}$. If $N \in \mathcal{T}$ we are done (by (1)). Suppose $N \notin \mathcal{T}$. Then N itself must have a nontrivial $\mathcal{C}$-subnet, say N_1. If $N_1 = N$ we are done; hence we assume $N_1 \subset N$. Consider the cokernel of the one-element sink of subnets $\{N_1\}$ of N, say it is $f_1\colon N \to N_2$. If N_2 is trivial, then $N_1 = N$ which contradicts our assumption. Thus N_2 is nontrivial. Also, if $N_2 \in \mathcal{C}$, then $N \in \mathcal{C}$ (by (3)). Suppose $N_2 \notin \mathcal{C}$. Since N_2 is a homomorphic image of N and since N_1 is nontrivial, $|N_2| < |N|$. Now N_2 is a nontrivial homomorphic image of N and by our assumption on N, it has a nontrivial $\mathcal{C}$-subnet, say N_3. As above, $N_3 \subset N_2$ since $N_2 \notin \mathcal{C}$. Let $f_2\colon N_2 \to N_4$ be the cokernel of the one element sink $\{N_3\}$ of N_2. Then $|N_4| < |N_2|$ and $N_4 \notin \mathcal{C}$ (since $N_2 \notin \mathcal{C}$). Now N_4 is a nontrivial homomorphic image of N and consequently it must have a nontrivial $\mathcal{C}$-subnet, say N_5. Repeating the above arguments, we eventually have a chain of nets $N, N_2, N_4, N_6, \ldots, N_{2k}$ for some $k \geqslant 1$, such that $|N| > |N_2| > |N_4| > \cdots > |N_{2k}|$ but N_{2k}, although being a nontrivial homomorphic image of N, can have no nontrivial $\mathcal{C}$-subnet. Thus $N = N_1 \in \mathcal{C}$. □

For a homomorphically closed class $\mathcal{C}$ containing $\mathcal{T}$, we call a subnet U of a net N a *proper* $\mathcal{C}$*-subnet of* N if there is a nontrivial net $L \in \mathcal{C}$ and a surjective homomorphism $g\colon L \to U$. Any proper $\mathcal{C}$-subnet is a $\mathcal{C}$-subnet and it could be trivial. Any trivial subnet of a net N is a $\mathcal{C}$-subnet but it need not be a proper $\mathcal{C}$-subnet. Also, any nontrivial $\mathcal{C}$-subnet of N is a proper $\mathcal{C}$-subnet. We need to consider one more condition to facilitate decompositions by connectednesses. A class of nets $\mathcal{C}$ is *closed under forming of strings* if it satisfies: Whenever N_1 and N_2 are trivial proper $\mathcal{C}$-subnets of a net N with $N_1 \cap N_2 \neq \phi$, then $N_1 \vee N_2 \in \mathcal{C}$. Our choice of terminology is made clear in:

PROPOSITION 4.5. *Let* $\mathcal{C}$ *be a connectedness. Then* $\mathcal{C}$ *is closed under forming of strings if and only if it satisfies: Whenever* $\mathcal{A} = \{N_\alpha \mid \alpha \in \Lambda\}$ *is a sink of proper* $\mathcal{C}$*-subnets of* N *such that* $N = \bigvee(N_\alpha \mid \alpha \in \Lambda)$ *and for all* $\alpha, \beta \in \Lambda$*, there is a sequence* $N_\alpha = N_{\alpha_1}, N_{\alpha_2}, \ldots, N_{\alpha_k} = N_\beta$ *in the sink* $\mathcal{A}$ *such that* $N_{\alpha_{i-1}} \cap N_{\alpha_i} \neq \phi$ *for all* $i = 2, 3, \ldots, k$*, then* $N \in \mathcal{C}$.

Proof. Suppose $\mathcal{C}$ is closed under forming of strings. Let N_1 and N_2 be two proper $\mathcal{C}$-subnets of N with $N_1 \cap N_2 \neq \phi$. We show $N_1 \vee N_2 \in \mathcal{C}$. If $N_1 \subseteq N_2$ or vice versa, then clearly $N_1 \vee N_2 \in \mathcal{C}$. Suppose $N_1 \nsubseteq N_2$ and $N_2 \nsubseteq N_1$. Let $g\colon N_1 \vee N_2 \to N_3$ be the cokernel of the sink of subnets $\{N_1, N_2\}$ of $N_1 \vee N_2$. If $P_1 \cap P_2 \neq \phi$ and $T_1 \cap T_2 \neq \phi$, then N_3 is of the form $\omega(n, m, l)$; if $P_1 \cap P_2 \neq \phi$ but $T_1 \cap T_2 = \phi$, then N_3 is of the form $\tau(n, m, j, k, l)$ and if $P_1 \cap P_2 = \phi$ but

$T_1 \cap T_2 \neq \phi$ then N_3 is of the form $\pi(n, m, j, k, l_1, l_2)$ where $n, m, j, k, l, l_1, l_2 \in \mathbb{N}_0$ with $n+m \neq 0$ and $j+k \neq 0$. In the first case, $N_3 \in \mathcal{T} \subseteq \mathcal{C}$. In the second case (and similarly for the third case) we also have $N_3 \in \mathcal{C}$ since $N_3 = N_1' \vee N_2'$, where N_1' is the net $\omega(m, n, l)$ and N_2' is the net $\omega(j, k, l)$ with the place in common. They are both proper $\mathcal{C}$-subnets of N_3 with $N_1' \cap N_2' \neq \phi$ and, since $\mathcal{C}$ is closed under the forming of strings, $N_3 \in \mathcal{C}$ holds. Since $\mathcal{C}$ is closed under extensions we then have $N_1 \vee N_2 \in \mathcal{C}$. Any net can have only a finite number of subnets and an inductive argument (as above) will then establish the general case. The converse is of course trivial. □

Finding classes $\mathcal{C}$ which are closed under forming of strings is not hard. Let $\mathcal{C}$ be a homomorphically closed class which contains the nets $\pi(1, 1, 1, 1, 0, 0)$ and $\tau(1, 1, 1, 1, 0)$. In addition, suppose every nontrivial net N in $\mathcal{C}$ has $B(p_1, t_1) \geqslant 1$ and $F(p_2, t_2) \geqslant 1$ for some $p_1, p_2, t_1, t_2 \in N$. Then $\mathcal{C}$ is closed under forming of strings. Indeed, let N_1 and N_2 be two trivial proper $\mathcal{C}$-subnets of N such that $N_1 \cap N_2 \neq \phi$, $N_1 \neq N_2$. Then they are of the form $\omega(n, m, l_1)$ and $\omega(j, k, l_2)$ respectively, where $l_1, l_2, n, m, j, k \in \mathbb{N}_0$, $nmjk \neq 0$ (by our assumption on the nets in $\mathcal{C}$). If $P_1 \cap P_2 \neq \phi$, then $N_1 \vee N_2$ is of the form $\tau(m, n, j, k, l)$, where $l = l_1 = l_2$ and if $T_1 \cap T_2 \neq \phi$, then it is of the form $\pi(n, m, k, j, l_1, l_2)$. Both these nets are in $\mathcal{C}$ by our assumption that the two simple nets above are in $\mathcal{C}$ and $\mathcal{C}$ is homomorphically closed.

PROPOSITION 4.6. *Let $\mathcal{C}$ be a class of nets which is closed under the formation of strings. Then $\mathcal{C}$ is a component class.*

Proof. Let N be a chain of the sink of $\mathcal{C}$-subnets $\mathcal{A} = \{N_\alpha \mid \alpha \in \Lambda\}$. Let $f \colon N \to K$ be the cokernel of this sink, which we know must be trivial (by definition). Let $\mathcal{A}' = \{N_\alpha \in \mathcal{A} \mid N_\alpha \notin \mathcal{T}\}$. Then f is also the cokernel of the sink $\mathcal{A}'$. By 4.1 it follows that N is a chain of the sink of subnets $\mathcal{A}'$ and 4.5 concludes the proof. □

Let $\mathcal{C}$ be a component class and let $\mathcal{A} = \{N_\alpha \mid \alpha \in \Lambda\}$ be the sink of all the $\mathcal{C}$-subnets of the net N. On $\mathcal{A}$ we define a relation $\approx$ by: $N_\alpha \approx N_\beta$ if and only if there is a sequence $N_\alpha = N_{\alpha_0}, N_{\alpha_1}, N_{\alpha_2}, \ldots, N_{\alpha_n} = N_\beta$ in $\mathcal{A}$ such that for $i, j \in \{0, 1, 2, \ldots, n\}$, $P_{\alpha_i} \sim P_{\alpha_j}$ in $\{P_{\alpha_0}, P_{\alpha_1}, P_{\alpha_2}, \ldots, P_{\alpha_n}\}$ and $T_{\alpha_i} \sim T_{\alpha_j}$ in $\{T_{\alpha_0}, T_{\alpha_1}, T_{\alpha_2}, \ldots, T_{\alpha_n}\}$. Here, for example, $P_{\alpha_i} \sim P_{\alpha_j}$ in $\{P_{\alpha_0}, P_{\alpha_1}, \ldots, P_{\alpha_n}\}$ means there is a subset $\{P_{\beta_0}, P_{\beta_1}, \ldots, P_{\beta_k}\}$ of $\{P_{\alpha_0}, P_{\alpha_1}, \ldots, P_{\alpha_n}\}$ such that $P_{\alpha_i} = P_{\beta_0}$, $P_{\alpha_j} = P_{\beta_k}$ and $P_{\beta_{u-1}} \cap P_{\beta_u} \neq \phi$ for all $u = 1, 2, \ldots, k$. Likewise for the transitions. Then $\approx$ is an equivalence relation and if $N_\alpha \approx N_\beta$ with corresponding sequence $N_\alpha = N_{\alpha_0}, N_{\alpha_1}, N_{\alpha_2}, \ldots, N_{\alpha_n} = N_\beta$, then $N_\alpha \approx N_{\alpha_i}$ for any $i \in \{0, 1, 2, \ldots, n\}$. Let $N_{\tau_1}, N_{\tau_2}, \ldots, N_{\tau_k}$ be representatives for the equivalence classes and for each τ_i, let $\overline{N}_{\tau_i} = \bigvee(N_\alpha \in \mathcal{A} \mid N_\alpha \approx N_{\tau_i})$. Then $\overline{N}_{\tau_i}$ is a chain of the sink of subnets $\{N_\alpha \in \mathcal{A} \mid N_\alpha \approx N_{\tau_i}\}$ of $\overline{N}_{\tau_i}$ and since $\mathcal{C}$ is a component class, $\overline{N}_{\tau_i} \in \mathcal{C}$ (cf. 4.2). These $\overline{N}_{\tau_i}$'s are called the *maximal $\mathcal{C}$-components* of N. This terminology is justified by:

(1) Any $\mathcal{C}$-subnet of N is contained in one of the $\overline{N}_{\tau_i}$'s.
(2) If N_α is any $\mathcal{C}$-subnet of N with $\overline{P}_{\tau_i} \cap P_\alpha \neq \phi$ and $\overline{T}_{\tau_i} \cap T_\alpha \neq \phi$, then $N_\alpha \subseteq \overline{N}_{\tau_i}$.
(3) If $\overline{N}_{\tau_i} \subseteq N_\alpha$ for some $\mathcal{C}$-subnet N_α, then $\overline{N}_{\tau_i} = N_\alpha$.

Note that for $\tau_i \neq \tau_j$, it could be the case that either $\overline{P}_{\tau_i} \cap \overline{P}_{\tau_j} \neq \phi$ holds or $\overline{T}_{\tau_i} \cap \overline{T}_{\tau_j} \neq \phi$ holds but not both. If however, $\mathcal{C}$ is a connectedness which is closed under the forming of strings, then $\overline{N}_{\tau_i} \cap \overline{N}_{\tau_j} = \phi$ for any two different nontrivial maximal $\mathcal{C}$-components.

Let $f\colon N \to N_c$ be the cokernel of the sink $\mathcal{A} = \{N_\alpha \mid \alpha \in \Lambda\}$ of all $\mathcal{C}$-subnets of N. Then f is called the *decomposition of* N *by* $\mathcal{C}$. We remark that f is also the cokernel of the sink $\mathcal{A}' = \{\overline{N}_{\tau_1}, \overline{N}_{\tau_2}, \ldots, \overline{N}_{\tau_k}\}$ as well as the cokernel of the sink $\{\overline{N}_{\tau_i} \in \mathcal{A}' \mid \overline{N}_{\tau_i} \notin \mathcal{T}\}$, i.e. the sink of all the nontrivial maximal $\mathcal{C}$-components of N.

THEOREM 4.7. *Let* $\mathcal{C}$ *be a connectedness which is closed under forming of strings. For any net* N, *if* $f\colon N \to N_c$ *is the decomposition of* N *by* $\mathcal{C}$, *then* $N \in \mathcal{SC}$.

Proof. If all the $\mathcal{C}$-subnets of N are trivial, then $N \in \mathcal{SC}$ and f is an isomorphism. Thus $N \in \mathcal{SC}$ and we are done. Suppose N has nontrivial $\mathcal{C}$-subnets. Let $\overline{N}_{\alpha_1}, \overline{N}_{\alpha_2}, \ldots, \overline{N}_{\alpha_k}$ be the nontrivial maximal $\mathcal{C}$-components of N. Then they all are proper $\mathcal{C}$-subnets of N, $\overline{N}_{\alpha_i} \cap \overline{N}_{\alpha_j} = \phi$ for all $i \neq j$ and f is the cokernel of the sink $\{\overline{N}_{\alpha_1}, \overline{N}_{\alpha_2}, \ldots, \overline{N}_{\alpha_k}\}$. Suppose that $f(\overline{P}_{\alpha_i}) = \{q_i\}$ and $f(\overline{T}_{\alpha_i}) = \{s_i\}$ for all $i = 1, 2, \ldots, k$. Since $\overline{N}_{\alpha_i} \cap \overline{N}_{\alpha_j} = \phi$ for $i \neq j$, we have $q_i \neq q_j$ and $s_i \neq s_j$. Suppose $N_c \notin \mathcal{SC}$. Then N_c has a nontrivial $\mathcal{C}$-subnet and thus also a nontrivial maximal $\mathcal{C}$-component, say U. It is of course a proper $\mathcal{C}$-subnet of N_c. If $U \subseteq N_c \setminus \{q_1, q_2, \ldots, q_k, s_1, s_2, \ldots, s_k\}$, then U can be regarded as a subnet of N (4.1.3(iii)); in fact it is then a nontrivial $\mathcal{C}$-subnet of N. Then $U \subseteq \overline{N}_{\alpha_i}$ for some i which is not possible. Thus U must contain some of the q_i's and/or some of the s_i's.

We now show $q_i \in U$ implies $s_i \in U$. Let $q_i \in U$. Then $f(\overline{N}_{\alpha_i}) = \{q_i\} \cup \{s_i\}$, with the subnet structure induced by N_c, is a proper $\mathcal{C}$-subnet of N_c with $f(\overline{N}_{\alpha_i}) \cap U \neq \phi$. Thus $f(\overline{N}_{\alpha_i}) \vee U$ is a $\mathcal{C}$-subnet of N_c. But the maximality of U then implies $s_i \in f(\overline{N}_{\alpha_i}) \subseteq U$. The converse is also true, i.e. $s_i \in U$ implies $q_i \in U$.

This means U is of the form $U = \{q_1, q_2, \ldots, q_n, p_1, p_2, \ldots, p_m\} \cup \{s_1, s_2, \ldots, s_n, t_1, t_2, \ldots, t_a\}$ for $1 \leqslant n \leqslant k$, $m \geqslant 0$ and $a \geqslant 0$ where $p_i, t_j \in N_c \setminus \{q_1, q_2, \ldots, q_k, s_1, s_2, \ldots, s_k\}$. Let W be the subnet of N with set of nodes given by $W = \overline{N}_{\alpha_1} \cup \overline{N}_{\alpha_2} \cup \cdots \cup \overline{N}_{\alpha_n} \cup \{p_1, p_2, \ldots, p_m\} \cup \{t_1, t_2, \ldots, t_a\}$ (by 4.1.3(iii) we may identify p_i and t_j with their pre-images in N) and arcs and tokens those induced by N. W must be a subnet of N since each $\overline{N}_{\alpha_i}$ is a subnet of N, U is a subnet of N_c and f is a strong homomorphism. Then the restriction of f to W, say g, is of the form $g\colon W \to U$ and it is the cokernel of the sink $\{\overline{N}_{\alpha_1}, \overline{N}_{\alpha_2}, \ldots, \overline{N}_{\alpha_n}\}$ of $\mathcal{C}$-subnets of W. Since $\mathcal{C}$ is closed under extensions, we get $W \in \mathcal{C}$. Thus W is a nontrivial $\mathcal{C}$-subnet of N and so $W \subseteq \overline{N}_{\alpha_i}$ for some i. But then $U = g(W) = f(W) \subseteq f(\overline{N}_{\alpha_i})$ is trivial which contradicts our choice of U. Thus $N_c \subset \mathcal{SC}$. □

EXAMPLE 4.8. (A connectedness $\mathcal{C}$ for which the decomposition of a net N by $\mathcal{C}$ is not always in $\mathcal{S}\mathcal{C}$). Let $\mathcal{C}$ be the connectedness of Example 3.4(7), i.e. $\mathcal{C} = \{$all nets $N \mid |T_N| = 1$ and if $N \notin \mathcal{T}$, then there is a $p_0, t_0 \in N$ such that $F(p_0, t_0) \geqslant 1$ and for any $p, t \in N$, if $F(p,t) \geqslant 1$, then $B(p,t) \geqslant 1\}$. Consider the net N given by

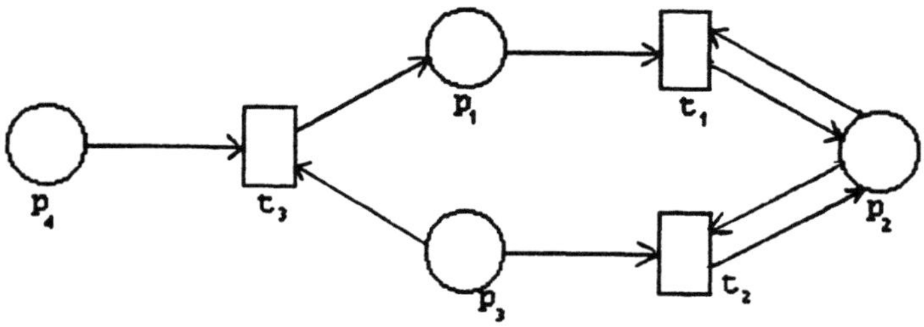

It has two nontrivial maximal $\mathcal{C}$-components namely $\overline{N}_1 = \{p_1, p_2\} \cup \{t_1\}$ and $\overline{N}_2 = \{p_2, p_3\} \cup \{t_2\}$ with $\overline{N}_1 \cap \overline{N}_2 \neq \phi$, but $\overline{N}_1 \vee \overline{N}_2 \notin \mathcal{C}$ (since it has more than one transition). Thus $\mathcal{C}$ is not closed under forming of strings. The decomposition of N by $\mathcal{C}$, say $f\colon N \to N_c$, is the cokernel of $\{\overline{N}_1, \overline{N}_2\}$, and is:

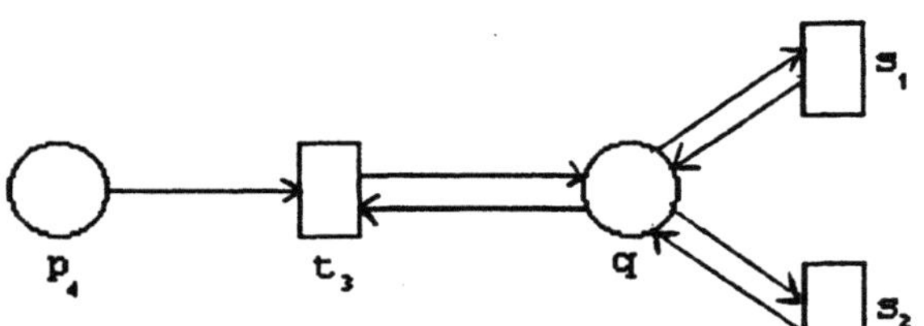

where $f(\overline{N}_i) = \{q\} \cup \{s_i\}$, $i = 1, 2$, $f(t_3) = t_3$, $f(p_4) = p_4$. But $N_c \notin \mathcal{S}\mathcal{C}$ since it has the nontrivial $\mathcal{C}$-subnet:

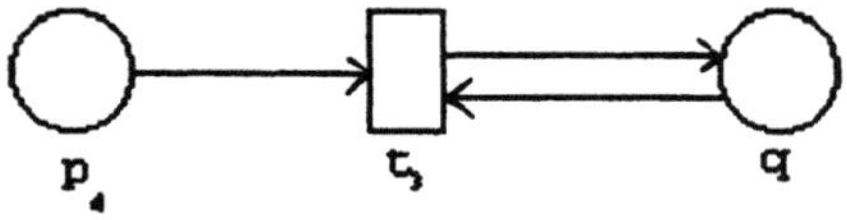

EXAMPLE 4.9. (A connectedness which is not closed under forming of strings but the decomposition of any net is in $\mathcal{S}\mathcal{C}$). Let $\mathcal{C}$ be the connectedness $\mathcal{C} = \{$all nets $N \mid |T_N| = 1\}$. Let N be the net:

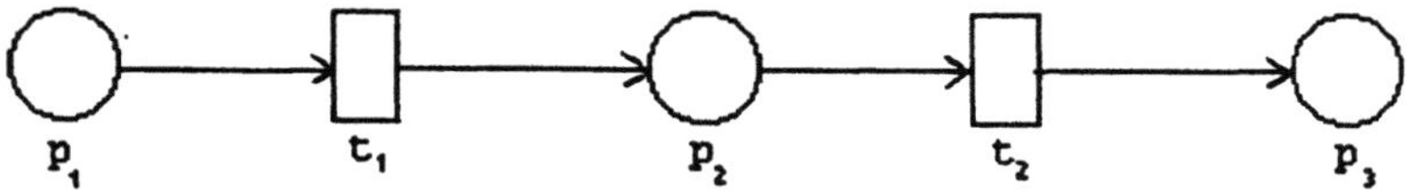

It has the two proper $\mathcal{C}$-subnets $N_1 = \{p_1, p_2\} \cup \{t_1\}$ and $N_2 = \{p_2, p_3\} \cup \{t_2\}$ with $N_1 \cap N_2 \neq \phi$, but $N = N_1 \vee N_2 \notin \mathcal{C}$; hence $\mathcal{C}$ is not closed under forming of strings. However, the decomposition of any net by $\mathcal{C}$ is always in $\mathcal{S}\mathcal{C}$:

For any net N, let $f: N \to N_c$ be the decomposition of N by $\mathcal{C}$. Suppose $N_c \notin \mathcal{SC}$ and let U be a nontrivial $\mathcal{C}$-subnet of N_c. Since $U \in \mathcal{C}$, $T_U = \{s\}$, say. As U is nontrivial, we may choose $q_1, q_2 \in P_U$ with $q_1 \neq q_2$. Since U is a subnet of N_c, there must be at least one arc between s and q_1 and at least one arc between s and q_2. But f is a strong homomorphism, so there are places p_1 and p_2 in N with $f(p_i) = q_i$, $i = 1, 2$ and a transition t in N with $f(t) = s$ such that there is at least one arc between t and p_1 and at least one arc between t and p_2. (Note that here the restriction of f to T must necessarily be injective.) Thus $W = \{p_1, p_2\} \cup \{t\}$, with the arcs and tokens induced by N, is a nontrivial $\mathcal{C}$-subnet of N. Then $f(W)$ must be trivial; a contradiction. Hence $N_c \in \mathcal{SC}$.

The connectednesses of Example 3.4(2), (3), (4), (5) and (6) are all closed under forming of strings. Consequently, if $\mathcal{C}$ is any one of them, then any net N can be decomposed into disjoint subnets, each of which is in $\mathcal{C}$ and they are maximal with respect to the properties describing membership of $\mathcal{C}$.

5. Disconnectednesses

A class of nets $\mathcal{D}$ is said to be *extensive* if it satisfies: Whenever $f: N \to L$ is a surjective homomorphism with $L \in \mathcal{D}$ and $U \in \mathcal{D}$ for all subnets U of N such that $f(U)$ is trivial, then $N \in \mathcal{D}$.

PROPOSITION 5.1. *Every disconnectedness $\mathcal{D}$ is extensive.*

Proof. Let $f: N \to L$ be a surjective homomorphism such that $L \in \mathcal{D}$ and $U \in \mathcal{D}$ for every subnet U of N for which $f(U)$ is trivial. We show $N \in \mathcal{D}$. Let U be any nontrivial subnet of N. If $f(U)$ is nontrivial, then the restriction of f to U is a nontrivial homomorphic image of U onto the nontrivial net $f(U)$ (with the subnet structure induced by L) and $f(U) \in \mathcal{D}$ (since $L \in \mathcal{D}$ and $\mathcal{D}$ is strongly hereditary (3.2)). If $f(U)$ is trivial, then $U \in \mathcal{D}$ (by assumption) and in this case also U has a nontrivial homomorphic image (namely U itself under the identity map) in $\mathcal{D}$. Thus $N \in \mathcal{D}$. □

Once again, since nets are finite, we have an easy characterization of disconnectednesses by means of closure conditions.

THEOREM 5.2. *A class of nets $\mathcal{D}$ is a disconnectedness if and only if it satisfies:*

(1) $\mathcal{T} \subseteq \mathcal{D}$,
(2) *$\mathcal{D}$ is strongly hereditary and*
(3) *$\mathcal{D}$ is extensive.*

Proof. The necessity of the conditions follows from 3.2 and 5.1. Suppose thus $\mathcal{D}$ satisfies conditions (1), (2) and (3). If $N \in \mathcal{D}$, then every nontrivial subnet of N has a nontrivial homomorphic image in $\mathcal{D}$ (by (2)). Conversely, let N be a net such that every nontrivial subnet of N has a nontrivial homomorphic image in $\mathcal{D}$. We show $N \in \mathcal{D}$. By (1) we may assume $N \notin \mathcal{T}$. Then N itself has a nontrivial homomorphic image, say $g: N \to D$ with $D \in \mathcal{D}$. We show $N \in \mathcal{D}$ by using the fact that $\mathcal{D}$ is extensive. Let U_1 be a subnet of N such that $g(U_1)$ is trivial. We show $U_1 \in \mathcal{D}$. If $U_1 \in \mathcal{T}$ we are done; so suppose $U_1 \notin \mathcal{T}$. Note that $U_1 \subset N$ (otherwise $D = g(N) = g(U_1)$ is trivial; a contradiction). By our assumption on N, U_1 has a nontrivial homomorphic image in $\mathcal{D}$, say $g_1: U_1 \to D_1$, $D_1 \in \mathcal{D}$, $D_1 \notin \mathcal{T}$. We proceed to show that $U_1 \in \mathcal{D}$ by using the fact that $\mathcal{D}$ is extensive. Let U_2 be a nontrivial subnet of U_1 such that $g_1(U_2)$ is trivial. Then $U_2 \subset U_1$ and since U_2 is a nontrivial subnet of N, it must have a nontrivial homomorphic image $g_2: U_2 \to D_2 \in \mathcal{D}$, $D_2 \notin \mathcal{T}$ (by our assumption on N). We show U_2 is in $\mathcal{D}$. Let U_3 be any nontrivial subnet of U_2 such that $g_2(U_3)$ is trivial. Then $U_3 \subset U_2$. Since U_3 is a nontrivial subnet of N, there is a nontrivial surjective homomorphism $g_3: U_3 \to D_3$ with $D_3 \in \mathcal{D}$, $D_3 \notin \mathcal{T}$. To show that U_3 is in $\mathcal{D}$, we once again will use the fact that $\mathcal{D}$ is extensive. Let U_4 be a nontrivial subnet of U_3 such that $g_3(U_4)$ is trivial. Then $U_4 \subset U_3$. Continuing in this way, we obtain a finite sequence of subnets $U_1, U_2, U_3, \ldots, U_k$ of N with $U_k \subset U_{k-1} \subset \cdots \subset U_2 \subset U_1$, and a nontrivial surjective homomorphism $g_k: U_k \to D_k$, $D_k \in \mathcal{D}$, $D_k \notin \mathcal{T}$ such that U_k has no nontrivial subnets. Then $U_k \in \mathcal{D}$ since $\mathcal{D}$ is extensive; hence $U_{k-1} \in \mathcal{D}$, hence $U_{k-2} \in \mathcal{D}, \ldots,$ hence $U_1 \in \mathcal{D}$. We conclude that $N \in \mathcal{D}$. □

To link disconnectednesses $\mathcal{D}$ with connectednesses $\mathcal{C}$ for which the decomposition is in $\mathcal{SC}$, we need to consider a further condition.

A class of nets $\mathcal{D}$ is *productive* if it satisfies: For every net N and class of pairs $\{(f_i, N_i) \mid i \in I\}$, where for each $i \in I$, I some index set, $f_i: N \to N_i$ is a homomorphism with $N_i \in \mathcal{D}$, then $\Pi N_i \in \mathcal{D}$ where $(f, \Pi N_i)$ is the product of the class $\{(f_i, N_i) \mid i \in I\}$ as defined in Section 4.

PROPOSITION 5.3. *Any disconnectedness $\mathcal{D}$ is productive.*

Proof. For the net N, consider a class of homomorphisms $f_i: N \to N_i$ with $N_i \in \mathcal{D}$ for all $i \in I$. Let W be any nontrivial subnet of ΠN_i. We show W has a nontrivial homomorphic image in $\mathcal{D}$. Now $\pi_j(W)$ is nontrivial for some $j \in I$. Indeed, if $\pi_i(W)$ is trivial for all $i \in I$, then W is trivial (for $p, q \in W$, $p_i = \pi_i(p) = \pi_i(q) = q_i$ for all $i \in I$ and so $p = q$). By assumption W is nontrivial; hence $\pi_j(W)$ is nontrivial for some $j \in I$. Since $\mathcal{D}$ is strongly hereditary $\pi_j(W)$, regarded as a subnet of $N_j \in \mathcal{D}$, becomes a nontrivial homomorphic image of W which is in $\mathcal{D}$. Hence $\Pi N_i \in \mathcal{D}$. □

Let $\mathcal{D}$ be a hereditary and productive class of nets with $\mathcal{T} \subseteq \mathcal{D}$. For any net N, let $\{(f_i, N_i) \mid i \in I\}$ be the class of all pairs with $f_i: N \to N_i$ a surjective homomorphism and $N_i \in \mathcal{D}$. This class is nonempty since any net can be mapped onto a suitable trivial net (which is in $\mathcal{D}$ by assumption). Let $(f, \Pi N_i)$ be the product of the class $\{(f_i, N_i) \mid i \in I\}$. We denote ΠN_i by N_d. Then $N_d \in \mathcal{D}$ and $f: N \to N_d$ is called the *maximal $\mathcal{D}$-image of N*. By definition, f is a strong homomorphism and if $g\colon N \to L$ is any homomorphism with $L \in \mathcal{D}$, then there is a homomorphism $h\colon N_d \to L$ such that $h \circ f = g$.

PROPOSITION 5.4. *Let $\mathcal{C}$ be a connectedness which is closed under the forming of strings. Let $\mathcal{D} = \mathcal{S}\mathcal{C}$ be the corresponding disconnectedness. Then the decomposition of any net N by $\mathcal{C}$ and the maximal $\mathcal{D}$-image of N coincide (up to isomorphism).*

Proof. Let $f\colon N \to N_c$ be the decomposition and let $g\colon N \to N_d$ be the maximal $\mathcal{D}$-image. Suppose $\{N_\alpha \mid \alpha \in \Lambda\}$ is the sink of all the $\mathcal{C}$-subnets of N. Then f is the cokernel of this sink. Since $g(N_\alpha)$ is a $\mathcal{C}$-subnet of $N_d \in \mathcal{D} = \mathcal{S}\mathcal{C}$, we must have $g(N_\alpha)$ trivial for all α. By the definition of a cokernel, there is a homomorphism $h\colon N_c \to N_d$ such that $h \circ f = g$. By 4.7, $N_c \in \mathcal{S}\mathcal{C} = \mathcal{D}$ and the fact that N_d is the maximal $\mathcal{D}$-image implies that there is a homomorphism $k\colon N_d \to N_c$ such that $k \circ g = f$. This makes h an isomorphism. □

A class of nets $\mathcal{D}$ is called a *separation class* if it satisfies: Let N_1 and N_2 be two nontrivial subnets of a net N such that $P_1 \cap P_2 \neq \phi$ or $T_1 \cap T_2 \neq \phi$ but not both. If there is a nontrivial surjective homomorphism $g\colon N_1 \vee N_2 \to D \in \mathcal{D}$, then there must be a surjective nontrivial homomorphism $g_1\colon N_1 \to D_1 \in \mathcal{D}$ or a surjective nontrivial homomorphism $g_2\colon N_2 \to D_2 \in \mathcal{D}$. It is straightforward to verify that if $\mathcal{C}$ is a connectedness with corresponding disconnectedness $\mathcal{D} = \mathcal{S}\mathcal{C}$, then $\mathcal{D}$ is a separation class if and only if $\mathcal{C}$ is closed under forming of strings. Combining the results of Sections 4 and this section, we conclude with:

THEOREM 5.5. *Let $\mathcal{C}$ be a connectedness which is closed under the forming of strings and let $\mathcal{D} = \mathcal{S}\mathcal{C}$ be the corresponding disconnectedness (or equivalently, let $\mathcal{D}$ be a disconnectedness which is also a separation class and let $\mathcal{C} := \mathcal{U}\mathcal{D}$ be the corresponding connectedness). For any net N there is a strong homomorphism g of N onto a net $N_d \in \mathcal{D}$ such that:*

(1) *N_d is the maximal $\mathcal{D}$-image of N.*
(2) *For every proper $\mathcal{C}$-subnet U of N, $g(U)$ is trivial and $g^{-1}(g(U))$ is a maximal $\mathcal{C}$-component of N.*

References

1. Arhangelskii, A. V. and Wiegandt, R.: Connectednesses and disconnectednesses in topology, *Topology Appl.* **5** (1975), 9–33.
2. Divinsky, N.: *Rings and Radicals*, Allen and Unwin, London, 1965.
3. Fried, E. and Wiegandt, R.: Connectednesses and disconnectednesses of graphs, *Algebra Universalis* **5** (1975), 411–428.
4. Murata, T.: Petri nets: Properties, analysis and applications, *Proc. of the IEEE* **77** (1989), 541–580.
5. Peterson, J. L.: Petri nets, *ICM Computing Surveys* **9** (1977), 223–252.

6. Peterson, J.L.: *Petri Nets and the Modeling of Systems*, MIT Press Series in Computer Science, 1981.
7. Preuss, G.: Trennung und Zusammenhang, *Monatsh. Math.* **74** (1970), 70–87.
8. Preuss, G.: E-zusammenhängende Räume, *Manuscripta Math.* **3** (1970), 331–342.
9. Preuss, G.: Eine Galois-Korrespondenz in der Topologie, *Monatsh. Math.* **75** (1971), 447–452.
10. Wiegandt, R.: *Radical and Semisimple Classes of Rings*, Queen's Papers in Pure and Appl. Math. No 37, Kingston, 1974.

MIX
Papier aus verantwortungsvollen Quellen
Paper from responsible sources
FSC® C105338

FSC
www.fsc.org

If you have any concerns about our products,
you can contact us on
ProductSafety@springernature.com

In case Publisher is established outside the EU,
the EU authorized representative is:
Springer Nature Customer Service Center GmbH
Europaplatz 3, 69115 Heidelberg, Germany

Printed by Libri Plureos GmbH
in Hamburg, Germany